AF476850

ADVANCES IN POLYAMINE RESEARCH

Volume 4

This volume is dedicated to the memory of Carl C. Levy whose untimely death has created an unfillable void in the "Polyamine Community." A scientist's scientist, Carl embodied the qualities to which we all aspire—curiosity, intelligence, creativity, and dedication. Combined with the ability to meld these qualities into achievement, Carl's scientific accomplishments stand as a lasting testament. However, Carl's humane characteristics are perhaps his more important legacy. He was one of those unique individuals who deeply touched those who worked with and knew him. He helped form their futures and was always willing to sacrifice on their behalf. Carl left us much richer for his being, a claim few can make.

Laurence J. Marton

Advances in Polyamine Research
Volume 4

Volume Editors

Uriel Bachrach
Department of Molecular Biology
Hebrew University-Hadassah Medical School
Jerusalem, Israel

Alvin Kaye
Department of Hormone Research
Weizmann Institute of Science
Rehovot, Israel

Ralph Chayen
Endocrine Laboratory
Tel Aviv Medical Center
Ichilov Hospital
Tel Aviv, Israel

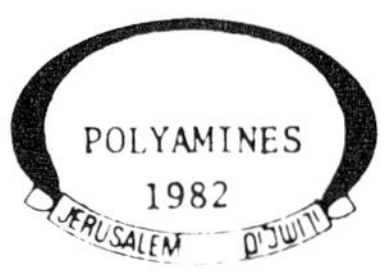

Raven Press ■ New York

Raven Press, 1140 Avenue of the Americas, New York, New York 10036

Made in the United States of America

International Standard Book Number 0-89004-890-8
Library of Congress Catalog Number 77-83687

Preface

The almost explosive growth in publications on polyamines, noted in the preface to Volume 3 of *Advances in Polyamine Research* is continuing with no sign of abatement as yet. The current tendencies within the field of polyamine research can be discerned from the range of topics covered in this volume. The upsurge of activity in the field of plant polyamines is demonstrated by the lengthy section devoted to this topic. Another feature of much of the research presented herein is the almost ubiquitous use of the inhibitor of ornithine decarboxylase, α-difluoromethyl ornithine.

Every new conference increases our wonder at the exquisite control of ornithine decarboxylase, and of its product putrescine. We have become accustomed to the antizyme and the ornithine decarboxylase inhibiting factor. Last year we learned how the enzyme can be modulated by tranaglutaminase, and this year we are introduced to polyamine-activated protein kinases and an antizyme inhibitor. Putrescine may interact with ornithine decarboxylase, may undergo acetylation, or be degraded by diamine oxidase, apart from the possibility of continuing along the biosynthetic pathway to spermidine and spermine.

One of the challenging results of increasing knowledge on amine oxidases is that we must rethink classical concepts. No longer is it sufficient to rely on classifications of the enzymes according to substrate affinity or coenzyme requirements. As a practical guide, we are now advised to think in terms of aminoguanidine-sensitive amine oxidases rather than those that are more susceptible to other inhibitors. However, in the long term, it is obvious that the individual enzymes will have to be characterized more rigorously.

The clinical field is well represented in this volume, with confirmatory indications that polyamine analyses can be helpful in diagnosis and treatment of disease. Even more encouraging is the demonstration that α-difluoromethyl ornithine in combination with other drugs may be useful in treatment of disease, both parasitic and neoplastic.

The Editors

Acknowledgment

Symptomatic of the increasing importance of polyamines in biochemistry is the formation of the polyamine section within the Italian Biochemical Society. This lead has now been followed in Israel. While the new wave of interest in polyamines gave rise to a number of symposia in the USA in the 1970s, culminating in the biennial Gordon Conferences, younger researchers in Europe have had difficulty in attending these meetings. Thus, the new trend for researchers in the field of polyamines to meet in European and Mediterranean localities has allowed for the greater participation of non-American scientists, and for the subsequent inclusion of important contributions to volumes such as this one.

This volume represents the proceedings of the 41st Bat-Sheva Seminar, on the subject of "Polyamines in growth and differentiation processes." The Seminar, held July 4–9, 1982 at Kiryat Anavim (near Jerusalem) in Israel was made possible by a generous grant from the Bat-Sheva de Rothschild Foundation for the Advancement of Science in Israel. We wish to thank the Bat-Sheva Foundation for their sponsorship of this meeting, and all the participants for their contributions.

Contents

1 Roles of Amines in the Action of Antibiotics—Bleomycin, Spergualin, and Aminoglycosides
Hamao Umezawa

Polyamines in Cancer (and in Other Pathological Conditions)

17 Experimental Approaches to the Systemic and Topical Use of Polyamine Antimetabolites as Antiproliferative Agents
J. Jänne, L. Alhonen-Hongisto, K. Käpyaho, and P. Seppänen

33 Effects of Polyamine Depletion on the Cytotoxicity of Cancer Chemotherapeutic Agents
Laurence J. Marton, Stina M. Oredsson, David T. Hung, and Dennis F. Deen

41 The Influence of Polyamine Depletion on Alkylation- and Carbamoylation-Induced Cytotoxicity
Stina M. Oredsson, Dennis F. Deen, and Laurence J. Marton

45 Polyamines as Biochemical Markers in Cancer
Yoav Horn, Stuart L. Beal, Natalio Walach, Warren P. Lubich, Lina Spigel, and Laurence J. Marton

49 Polyamines and Histamine in Serum from Patients with Hematological Diseases
H. Desser, R. Waldner, W. Kläring, and D. Lutz

59 Polyamine Excretion as Prognostic Marker in Radiologic Therapy of Breast Carcinoma in Metastatic Phase
M. Romano, L. Cecco, M. Cerra, and A. De Matteis

65 Urinary Excretion of Spermine in Breast Cancer
D. Chayen, R. Chayen, R. Dvir, S. Goldberg, A. Harell, D. Lichtenstein, and M. Stavorovsky

73 Determination of Free Polyamines in Blood and Urine During Choriocarcinoma Chemotherapy
Kazimierz Kamiński

79 Decreased Activity of Ornithine Decarboxylase Antizyme During Hyperplastic and Neoplastic Growth of Rat Liver
G. Scalabrino, M.E. Ferioli, and D. Modena

97 Enhancement of Thermal Killing by Spermine in EMT6 Multicellular Tumor Spheroids Versus Monolayer Cells
E. Ben-Hur, J.J. Shaw, and N.M. Bleehen

107 Myeloid Body Formation Under Conditions of Polyamine Stress
R.A. Campbell, T. LaBerge, M. Campbell-Boswell, R.E. Brooks, and Y.B. Talwalkar

127 Evidence That an Elevated Level of Ornithine Decarboxylase Activity is an Essential Component of Tumor Promotion
R.K. Boutwell

135 Diamine Oxidase and Polyamine Catabolism
N. Seiler, B. Knödgen, G. Bink, S. Sarhan, and F. Bolkenius

155 Polyamine Oxidation and Human Pregnancy
D.M.L. Morgan

169 Polyamine Oxidation and the Killing of Intracellular Parasites
D.M.L. Morgan and J. Christensen

175 Induction of Diamine Oxidase Activity in Some Processes of Growth
A. Perin, A. Sessa, and M.A. Desideric

183 Amine Oxidase Activity in Malignant Human Brain Tumors
B. Mondovì, P. Riccio, A. Riccio, and G.S. Marcozzi

193 Diamines and Polyamines in Mammalian Reproduction
A.-Ch. Henningsson, S. Henningsson, and O. Nilsson

209 Polyamines and Methionine Sulfoximine-Induced Seizures
R. Porta, M. Camardella, A. De Santis, V. Gentile, and O.Z. Sellinger

221 Antagonism of Polyamine Metabolism—A Critical Factor in Chemotherapy of African Trypanosomiasis
Cyrus J. Bacchi, Peter P. McCann, Henry C. Nathan, Seymour H. Hutner, and Albert Sjoerdsma

233 Effect of Putrescine Derivatives on Growth and Protein Synthesis
Sara H. Goldemberg, I.D. Algranati, J.J. Miret, D.O. Alonso Garrido, and B. Frydman

245 Mammalian Propylamine Transferases
A. Raina, T. Eloranta, T. Hyvönen, and R.-L. Pajula

255 Effect of Magnesium on the Interaction of Polyamines with Nucleic Acids
A.K. Abraham and T. Flatmark

267 Effect of 1,25-Dihydroxycholecalciferol on Intestinal Spermine-Binding Proteins in Chicks
G. Mezzetti, M.S. Moruzzi, M.G. Monti, G. Moruzzi, and B. Barbiroli

279 The Use of Carboxymethyl Cellulose Resin in the Competitive Radiospermine-Binding Assay
G. Mezzetti, M.G. Monti, M.S. Moruzzi, G. Piccinini, and B. Barbiroli

285 Polyamine Compartmentalization and DNA Synthesis in Polyamine Depleted and Restored Mammalian Cells
M. Mach, P. Ebert, and W. Kersten

297 Polyamines, Carnitine, and Fatty Acid Metabolism in Human Platelets
Victor R. Villanueva

307 The Polyamine Specificity of the Valyl-tRNA Synthesis Reaction Catalysed by Procaryotic *(Escherichia coli)* and Eucaryotic (plant, *Lupinus luteus*) Valyl-tRNA Synthetases
Hieronim Jakubowski

321 Acetylation of Spermidine in Chicken Tissues: Effect of Starvation, Insulin, and Glucagon
M.A. Grillo

331 Involvement of Polyamines in Phospholipid-Induced Changes of Cyclic Nucleotide Contents and Macromolecular Synthesis in Cultured Heart Cells
C. Pignatti, B. Tantini, C. Rossoni Caldarera, E. Turchetto, and C. Clô

Polyamines and Plants

347 The Cinnamic Acid Amides of the Di- and Polyamines
Terence A. Smith, Jonathan Negrel, and Colin R. Bird

371 Some Properties of the Spermidine Synthase of Chinese Cabbage
Ram K. Sindhu and Seymour S. Cohen

381 The Control of Arginine Decarboxylase Activity in Higher Plants
A.W. Galston, Y.-r. Dai, H.E. Flores, and N.D. Young

395 Alternative Metabolic Pathways for Polyamine Biosynthesis in Plant Development
Arie Altman, Ra'anan Friedman, and Nitsa Levin

409 *In Vitro* and *In Vivo* Effect of Ornithine and Arginine Decarboxylase Inhibitors in Plant Tissue Culture
N. Bagni, P. Torrigiani, and P. Barbieri

419 Polyamines and Morphogenesis in *Helianthus Tuberosus* Explants
Donatella Serafini Fracassini and Milena Alessandri

427 Macromolecular Effectors of Ornithine Decarboxylase Activity in Germinating Barley Seeds
Dimitrios A. Kyriakidis

437 Applied Polyamines Inhibit Macromolecular Synthesis in Plant Tissue
Akiva Apelbaum and Isaac Icekson

443 Involvement of Polyamines in Cell Division of Various Plant Tissues
E. Cohen, Y.M. Heimer, S. Malis-Arad, and Y. Mizrahi

445 Effect of ODC and ADC Inhibitors on Tomato Fruit Development
E. Cohen, S. Malis-Arad, Y.M. Heimer, and Y. Mizrahi

447 Inhibition of Tomato Fruit Growth *In Vitro*: Possible Involvement of Polyamines
D.C. Teitel, Shoshana Malis-Arad, E. Birnbaum, and Y. Mizrahi

449 The Involvement of Polyamines in the Cell Cycle of *Chlorella*
E. Cohen, Shoshana Malis-Arad, Y.M. Heimer, and Y. Mizrahi

451 Polyamines in Roots of *Zea Mays* Seedlings
M. Schwartz, T. Arzee, Y. Cohen, and A. Altman

Polyamines and Microorganisms

455 Polyamine Biosynthesis and Function in *Escherichia Coli*
Herbert Tabor and Celia White Tabor

467 Biochemical and Genetic Studies of Polyamines in *Saccharomyces Cerevisiae*
Celia White Tabor, Herbert Tabor, and Anil K. Tyagi

479 Unusual Polyamines in an Extreme Thermophile, *Thermus Thermophilus*
Tairo Oshima

489 Ornithine Decarboxylase Activity from an Extremely Thermophilic Bacteria, *Clostridium Thermohydrosulfuricum*
Hannu Pösö and Lars Paulin

495 Inhibition of Bacterial Polyamine Biosynthesis and Consequent Effects on Cell Proliferation
Alan J. Bitonti, Arja Kallio, Peter P. McCann, and Albert Sjoerdsma

507 The Antiviral Effect of Inhibitors of Polyamine Metabolism
A. Stanley Tyms, Bandhana K. Rawal, Helen M. Naim, and John D. Williamson

519 Level of Polyamines in *Escherichia Coli* Carrying the *MetA* Gene on a Multicopy Plasmid
Shulamit Michaeli, Sonia Rozenhak, and Eliora Z. Ron

521 A Mutant of *Escherichia Coli* Deficient in *S*-Adenosylmethionine
Bracha Kimchi and Eliora Z. Ron

Ornithine Decarboxylase and Protein Kinase

525 Evidence for Sulfhydrilic Involvement in the Inactivation of Rat Heart Ornithine Decarboxylase by Oxygen Radicals and Hyperoxia
C.M. Caldarera, F. Flamigni, C. Muscari, C. Clô, and C. Guarnieri

537 ODC Activity Evaluated in Perfused Rat Heart in Response to Catecholamine
C. Guarnieri, F. Flamigni, C. Muscari, and C.M. Caldarera

549 Catecholamine-Stimulated β_2-Receptors Coupled to Ornithine Decarboxylase Induction and to Cellular Hypertrophy and Proliferation
J.R. Womble and Diane Haddock Russell

563 Effect of Spermidine Depletion on the Response of Hepatic Pyruvate Kinase and Lipogenesis to Refeeding
Margaret E. Brosnan and Yu-Wan Hu

571 Immunocytochemical Demonstration of Ornithine Decarboxylase
Lo Persson, Elsa Rosengren, Frank Sundler, and Rolf Uddman

585 Visualization of Rhodamine-Labeled α-Difluoromethylornithine in Viable Neuroblasts: Towards *In Vivo* Localization of Ornithine Decarboxylase
Gad M. Gilad and Varda H. Gilad

591 Mammalian Cell Mutants with Altered Levels of Ornithine Decarboxylase Activity
Carolyn Steglich, Jung Choi, and Immo E. Scheffler

603 The Use of α-Difluoromethylornithine to Study Eukaryotic Ornithine Decarboxylases
Hannu Pösö, James E. Seely, Ian S. Zagon, and Anthony E. Pegg

615 New Perspectives on Polyamine-Dependent Protein Kinase and the Regulation of Ornithine Decarboxylase by Reversible Phosphorylation
Glenn D. Kuehn and Valerie J. Atmar

631 Effects of Spermidine and Its Monoacetylated Derivatives on Phosphorylation by Nuclear Protein Kinase NII
Samson T. Jacob, Kathleen M. Rose, and Zoe N. Canellakis

647 Multiple-Protein Complex with [Calmodulin]-Polyamine Responsive Protein Kinase Activity
W.E. Criss, Y. Morishita, Q. Watanabe, C. Akogyeram, A. Sahai, B. Deu, and T. Oka

655 Hormonal Regulation and Function of Polyamine Responsive Protein Kinase Activity in the Mouse Mammary Gland
Lisa J. Leiderman, Wayne E. Criss, Yorihiko Morishita, and Takami Oka

667 Regulation of Cyclic Nucleotide Metabolism in Heart Cell Cultures
C. Clô, B. Tantini, C. Pignatti, C. Guarnieri, and C.M. Caldarera

683 Regulation of Hepatic Ornithine Decarboxylase by Antizyme and Antizyme Inhibitor
K. Fujita, Y. Murakami, T. Kameji, S. Matsufuji, K. Utsunomiya, R. Kanamoto, and S. Hayashi

693 Ornithine Decarboxylase and Ornithine Decarboxylase Inactivating Factor in Rat Organs
A.M. Kaye and M. Yariv

705 Requirement for Polyamines in the Response of Sympathetic Neurons to Axonal Injury
Gad M. Gilad

707 Cell Cycle Dependent Induction of Ornithine Decarboxylase and Its Antizyme in Ehrlich Ascites Tumor Cells Separated by Centrifugal Elutriation
S. Anehus, M. Linden, Å. Borg, E. Långström, and O. Heby

713 Ornithine Decarboxylase Activity in Cultured Bone Cells is Activated by Bone-Seeking Hormones and Physical Stimulation
D. Sömjen, M. Yariv, A.M. Kaye, R. Korenstein, H. Fischler, and I. Binderman

719 Whole Body- and Microautoradiographic Localization of Ornithine Decarboxylase in the Kidneys of Nandrolone-Treated Mice Using Tritium-Labeled α-Difluoromethyl Ornithine
A.-Ch. Henningsson, S. Henningsson, T. Tjälve, L. Hammar, and G. Löwendahl

Polyamines in Differentiation

727 A Role for the Polyamines in Mouse Embryonal Carcinoma (F9 and PCC3) Cell Differentiation but Not in Human Promyelocytic Leukemia (HL-60) Cell Differentiation
O. Heby, S.M. Oredsson, I. Olsson, and L.J. Marton

743 The Role of Polyamines in the Differentiation of Mouse Neuroblastoma Cells
Kuang Yu Chen and Alice Y.-C. Liu

757 Turnover and Analysis of Polyamines and Their Acetyl Derivatives in Differentiating Friend Erythroleukemia Cells
Yair Gazitt and Uriel Bachrach

769 Diacetylputrescine Induces Differentiation and Is Metabolized in Friend Erythroleukemia Cells
Zoe Nakos-Canellakis and Philip K. Bondy

779 Effect of 5′-Methylthioadenosine and Its Analogs on Friend Erythroleukemic Cell Proliferation and Differentiation
M. Cartenì-Farina, M. Porcelli, G. Cacciapuoti, V. Zappia, M. Grieco, and P.P. Di Fiore

793 *Subject Index*

Contributors

Abraham K. Abraham
Department of Biochemistry
Arstadveien 19
5000 Bergen, Norway

C. Akogyeram
Department of Pharmacology
Howard University
Washington, D.C. 20059

Milena Alessandri
Istituto Botanico
Universita di Bologna
40126 Bologna, Italy

I.D. Algranati
Instituto de Investigaciones Bioquimicas "Fundacion Campomar" and Facultad de Ciencias Exactas y Naturales
1428 Buenos Aires, Argentina

Arie Altman
Department of Horticulture
Hebrew University of Jerusalem
P.O. Box 12
Rehovot, Israel

S. Anehus
Department of Zoophysiology
University of Lund
Helgonavagen 3
S-223 62 Lund, Sweden

Akiva Apelbaum
Division of Fruit and Vegetable Storage
ARO
The Volcani Center
P.O. Box 6
Bet Dagan 50250, Israel

Shoshana Malis-Arad
Applied Research Institute
Ben-Gurion University of the Negev
P.O. Box 1025
Beer-Sheva 84110, Israel

Tova Arzee
Department of Botany
The George S. Wise Faculty of Life Sciences
Tel-Aviv University
Tel-Aviv, Israel

Valerie J. Atmar
Department of Chemistry
New Mexico State University
Las Cruces, New Mexico 88003

Cyrus J. Bacchi
Haskins Laboratories and Biology Department
Pace University
New York, New York 10038

Uriel Bachrach
Department of Molecular Biology
The Hebrew University-Hadassah Medical School
Jerusalem, Israel

Nello Bagni
Istituto Botanico
Universita di Bologna
40126 Bologna, Italy

P. Barbieri
Istituto Botanico
Universita di Bologna
40126 Bologna, Italy

Bruno Barbiroli
Istituto di Chimica Biologica
Universita di Modena
41100 Modena, Italy

Stuart L. Beal
Department of Laboratory Medicine
School of Medicine
University of California
San Francisco, California 94143

Ehud Ben-Hur
Nuclear Research Center
Negev, P.O. Box 9001
Beer-Sheva 84190, Israel

I. Binderman
Hard Tissue Unit
Ichilov Hospital
64239 Tel-Aviv, Israel

G. Bink
Centre de Recherche Merrell International
16, rue d'Ankara
67084 Strasbourg Cedex, France

Colin R. Bird
Long Ashton Research Station
University of Bristol
Long Ashton
Bristol, BS18 9AF, England

Elliott Birnbaum
Applied Research Institute
Ben-Gurion University of the Negev
P.O. Box 1025
Beer-Sheva 84110, Israel

Alan J. Bitonti
Merrell Dow Research Center
2110 E. Galbraith Road
Cincinnati, Ohio 45215

N. M. Bleehen
MRC Unit of Clinical Oncology and Radiotherapeutics
The Medical School
Cambridge University
Cambridge CB2 2QH, England

F. Bolkenius
Centre de Recherche Merrell International
16, rue d'Ankara
67084 Strasbourg Cedex, France

Philip K. Bondy
West Haven VA Medical Center
West Haven, Connecticut and
Department of Internal Medicine
Yale University School of Medicine
New Haven, Connecticut 06510

M. Campbell-Boswell
Institute of Pathology
School of Medicine
Case Western Reserve University
Cleveland, Ohio 44100

Å. Borg
Department of Oncology
University of Lund
Lasarettet S-221 85 Lund, Sweden

R. K. Boutwell
McArdle Laboratory for Cancer Research
450 North Randall Avenue
University of Wisconsin
Madison, Wisconsin 53706, U.S.A.

R. E. Brooks
Departments of Pediatrics and Anatomical Pathology
Oregon Health Sciences University
Portland, Oregon 97201

Margaret E. Brosnan
Department of Biochemistry
Memorial University of Newfoundland
St. John's
Newfoundland, Canada A1B 3X9

G. Cacciapuoti
Department of Biochemistry
II Chair, First Medical School
University of Naples
80138 Naples, Italy

Claudio M. Caldarera
Istituto di Chimica Biologica
Facolta di Medicina e Chirurgia
Universita di Bologna
40126 Bologna, Italy

C. Rossoni Caldarera
Istituto di Chimica Biologica
Facolta di Medicina e Chirurgia
Universita di Bologna
40126 Bologna, Italy

M. Camardella
Department of Biochemistry
First Medical School
University of Naples
80138 Naples, Italy

Robert A. Campbell
Departments of Pediatrics and Anatomical Pathology
Oregon Health Sciences University
Portland, Oregon 97201

Zoe N. Canellakis
Departments of Pharmacology and Medicine
Yale University and West Haven VA Medical Center
New Haven, Connecticut 06510

L. Cecco
Department of Biochemistry
Istituto dei Tumori
80131 Naples, Italy

M. Cerra
Department of Biochemistry
Istituto dei Tumori
80131 Naples, Italy

David Chayen
Department of Surgery C
Medical Center and Sackler School of Medicine
Tel-Aviv 64239, Israel

Ralph Chayen
Endocrine Laboratory
Medical Center and Sackler School of Medicine
Tel-Aviv 64239, Israel

Kuang Yu Chen
Department of Chemistry
Rutgers University
Wright Chemistry Laboratory
Piscataway, New Jersey 08854

Jung Choi
Department of Biology, B-022
University of California, San Diego
La Jolla, California 92093

J. Christensen
Division of Clinical Cell Biology
MRC Clinical Research Centre
Harrow
Middlesex, HA1 3UJ
United Kingdom

Carlo Clô
Istituto di Chimica Biologica
Facolta di Medicina e Chirurgia
Universita di Bologna
40126 Bologna, Italy

Ephraim Cohen
Department of Biology
Ben-Gurion University of the Negev
P.O. Box 1025
Beer-Sheva 84110, Israel

Seymour S. Cohen
Department of Pharmacological Sciences
State University of New York at Stony Brook
Stony Brook, New York 11794

Y. Cohen
Department of Botany
The George S. Wise Faculty of Life Sciences
Tel-Aviv University
Tel-Aviv, Israel

Wayne E. Criss
Departments of Human Nutrition and Food
and School of Human Ecology
Cancer Center
Howard University
Washington, D.C. 20059

Y-r. Dai
Department of Biology
Peking University
Beijing, People's Republic of China

Dennis F. Deen
Department of Radiation Oncology
School of Medicine
University of California
San Francisco, California 94143

A. De Matteis
Section of Endocrinology
Istituto dei Tumori
80131 Naples, Italy

A. De Santis
Department of Biochemistry
First Medical School
University of Naples
80138 Naples, Italy

M. A. Desiderio
Institute of General Pathology
and CNR Center for Research in Cell Pathology
University of Milan
Via Mangiagalli 31
20133 Milan, Italy

Hans Desser
The Ludwig Boltzmann Institute
for Leukemia Research and Hematology
The 3rd Medical Department
Hanusch Hospital
1140 Vienna, Austria

B. Deu
School of Human Ecology
Cancer Center
Howard University
Washington, D.C. 20059

P. P. Di Fiore
Institute of General Pathology
Chair of Viral Oncology
II Medical School
University of Naples
80131 Naples, Italy

Ruth Dvir
Endocrine Laboratory
Medical Center and Sackler School of Medicine
Tel-Aviv 64239, Israel

Peter Ebert
Institut fur Physiologische Chemie
Universitat Erlangen-Nurnberg
Fahrstrasse 17
D 8520 Erlangen, Federal Republic of Germany

T. Eloranta
Department of Biochemistry
University of Kuopio
P.O. Box 138
SF-70101 Kuopio 10, Finland

M. Carteni-Farina
Department of Biochemistry
II Chair, First Medical School
University of Naples
80138 Naples, Italy

M. E. Ferioli
Institute of General Pathology
and CNR Centre for Research in Cell Pathology
University of Milan
20133 Milan, Italy

H. Fischler
Department of Membrane Research
The Weizmann Institute of Science
Rehovot 76100, Israel

F. Flamigni
Istituto di Chimica Biologica
e Centro Studi e Ricerche sul Metabolismo Cardiaco
Facolta di Medicina e Chirurgia
Universita di Bologna
Via Irnerio, 48
40126 Bologna, Italy

T. Flatmark
Department of Biochemistry
Arstadveien 19
5000 Bergen, Norway

H. E. Flores
Department of Biology
Yale University
New Haven, Connecticut 06511

Donatella Serafini Fracassini
Istituto Botanico
Universita di Bologna
40126 Bologna, Italy

Ra'anan Friedman
Department of Horticulture
Hebrew University of Jerusalem
P.O. Box 12
Rehovot, Israel

B. Frydman
Facultad de Farmacia y Bioquimica
Universidad de Buenos Aires
Buenos Aires, Argentina

K. Fujita
Department of Nutrition
Jikei University School of Medicine
Minato-ku
Tokyo 105, Japan

Arthur W. Galston
Department of Biology
Yale University
New Haven, Connecticut 06511

D. O. Alonso Garrido
Facultad de Farmacia y Bioquimica
Universidad de Buenos Aires
Buenos Aires, Argentina

Yair Gazitt
The Lautenberg Center for General and Tumor Immunology
The Hebrew University-Hadassah Medical School
Jerusalem, Israel

V. Gentile
Department of Biochemistry
First Medical School
University of Naples
Naples, Italy

Gad M. Gilad
Department of Isotope Research
The Weizmann Institute of Science
Rehovot, Israel

Varda H. Gilad
Department of Isotope Research
The Weizmann Institute of Science
Rehovot 76100, Israel

Sara Goldberg
Endocrine Laboratory
Medical Center and Sackler School of Medicine
Tel-Aviv 64239, Israel

Sara H. Goldemberg
Instituto de Investigaciones Bioquimicas "Fundacion Campomar"
and Facultad de Ciencias Exactas y Naturales
1428 Buenos Aires, Argentina

M. Grieco
Institute of General Pathology
Chair of Viral Oncology
II Medical School
University of Naples
80131 Naples, Italy

Maria A. Grillo
Istituto di Chimica Biologica
Universita di Torino
Torino, Italy

Carlo Guarnieri
Istituto di Chimica Biologica
Facolta di Medicina e Chirurgia
Universita di Bologna
40126 Bologna, Italy

L. Hammar
Institute of Biochemistry
University of Uppsala
Box 576
S-751 23 Uppsala, Sweden

Arieh Harell
Institute of Endocrinology
Medical Center and Sackler School of Medicine
Tel-Aviv 64239, Israel

Shin-ichi Hayashi
Department of Nutrition
Jikei University School of Medicine
Minato-ku
Tokyo 105, Japan

Olle Heby
Department of Zoophysiology
University of Lund
Helgonavagen 3
S-223 62 Lund, Sweden

Y. M. Heimer
Nuclear Research Center - Negev
P.O. Box 9001
Beer-Sheva 84190, Israel

Jacob Hochman
Department of Zoology
Hebrew University
Jerusalem, Israel

Anne-Charlotte Henningsson
Department of Physiology and Biophysics
University of Lund
S-223 62 Lund, Sweden

Stig Henningsson
Department of Physiology and Biophysics
University of Lund
S-223 62 Lund, Sweden

Lenna Alhonen-Hongisto
Department of Biochemistry
University of Helsinki
SF-00170 Helsinki 17, Finland

Yoav Horn
Department of Oncology
Assaf Harofeh Hospital
Sackler School of Medicine
Zerifin, Israel

Yu-Wan Hu
Department of Biochemistry
Memorial University of Newfoundland
St. John's, Newfoundland
Canada A1B 3X9

David T. Hung
Brain Tumor Research Center of the Department of Neurological Surgery
School of Medicine
University of California
San Francisco, California 94143

Seymour H. Hutner
Haskins Laboratories and Biology Department
Pace University
New York, New York 10038

T. Hyvönen
Department of Biochemistry
University of Kuopio
P.O. Box 138
SF-70101 Kuopio 10, Finland

Isaac Icekson
Division of Fruit and Vegetable Storage
ARO
The Volcani Center
P.O. Box 6
Bet Dagan 50250, Israel

Samson T. Jacob
Department of Pharmacology
The Milton S. Hershey Medical Center
The Pennsylvania State University College of Medicine
Hershey, Pennsylvania 17033

Hieronim Jakubowski
Institute of Biochemistry
Agricultural University Wolynska 35
60-637 Poznan, Poland

Juhani Jänne
Department of Biochemistry
University of Helsinki
SF-00170 Helsinki 17, Finland

Arja Kallio
Department of Biochemistry
University of Helsinki
Unioninkatu 35
00170 Helsinki 17, Finland

T. Kameji
Department of Nutrition
Jikei University School of Medicine
Minato-ku
Tokyo 105, Japan

Kazimierz Kamiński
Department of Biochemistry
Silesian Medical Academy
41-808 Zabrze
K. Marksa 19, Poland

R. Kanamoto
Department of Nutrition
Jikei University School of Medicine
Minato-ku
Tokyo 105, Japan

Talya Kapitolnik
Department of Microbiology
Tel-Aviv University
Ramat-Aviv
Tel-Aviv, Israel

Kirsti Käpyaho
Department of Biochemistry
University of Helsinki
SF-00170 Helsinki 17, Finland

Alvin M. Kaye
Department of Hormone Research
The Weizmann Institute of Science
Rehovot 76100, Israel

Helga Kersten
Institut fur Physiologische Chemie
Universitat Erlangen-Nurnberg
Fahrstrasse 17
D 8520 Erlangen, Federal Republic of Germany

Walter Kersten
Institut fur Physiologische Chemie
Universitat Erlangen-Nurnberg
Fahrstrasse 17
D 8520 Erlangen, Federal Republic of Germany

Bracha Kimchi
Department of Microbiology
The George S. Wise Faculty of Life Sciences
Tel-Aviv University
Ramat-Aviv
69978 Tel-Aviv, Israel

W. Kläring
The Ludwig Boltzmann Institute for Leukemia Research and Hematology
The 3rd Medical Department
Hanusch Hospital
1140 Vienna, Austria

B. Knödgen
Centre de Recherche Merrell International
16, rue d'Ankara
67084 Strasbourg Cedex, France

R. Korenstein
Department of Membrane Research
The Weizmann Institute of Science
Rehovot 76100, Israel

Gil Korner
Department of Molecular Biology
Hebrew University-Hadassah Medical School
Jerusalem, Israel

Glenn D. Kuehn
Department of Chemistry
New Mexico State University
Las Cruces, New Mexico 88003

Dimitrios A. Kyriakidis
Laboratory of Biochemistry
Aristotelian University of Thessaloniki
School of Science
Thessaloniki, Greece

T. LaBerge
Departments of Pediatrics and Anatomical Pathology
Oregon Health Sciences University
Portland, Oregon 97201

E. Långström
Department of Oncology
University of Lund
Lasarettet
S-221 85 Lund, Sweden

Lisa J. Leiderman
Laboratory of Biochemistry and Metabolism
National Institute of Arthritis, Diabetes, Digestive Diseases and Kidney Diseases
National Institutes of Health
Bethesda, Maryland 20205

Nitsa Levin
Department of Horticulture
Hebrew University of Jerusalem
P.O. Box 12
Rehovot, Israel

Dan Lichtenstein
Department of Surgery C
Medical Center and Sackler School of Medicine
Tel-Aviv 64239, Israel

M. Linden
Department of Zoophysiology
University of Lund
Helgonavagen 3
S-223 62 Lund, Sweden

Alice Y-C. Liu
Department of Pharmacology
Harvard Medical School
Boston, Massachusetts 02115

G. Löwendahl
Department of Physiology and Biophysics
University of Lund
S-223 62 Lund, Sweden

Warren P. Lubich
Brain Tumor Research Center of the Department of Neurological Surgery
University of California
School of Medicine
San Francisco, California 94143

D. Lutz
The Ludwig Boltzmann Institute for Leukemia Research and Hematology
The 3rd Medical Department
Hanusch Hospital
1140 Vienna, Austria

M. Mach
Institut fur Physiologische Chemie
Universitat Erlangen-Nurnberg
Fahrstrasse 17
D-8520 Erlangen, Federal Republic of Germany

Barbara Maiss
Institut fur Biotechnologie
Kernforschungsanlage Julich
Postfach 1913
D-5170 Julich, Germany

G. S. Marcozzi
Institute of Applied Biochemistry and CNR Centre of Molecular Biology
University of Rome
Rome, Italy

Laurence J. Marton
Brain Tumor Research Center of the Department of Neurological Surgery and the Department of Laboratory Medicine
University of California
School of Medicine
San Francisco, California 94143

S. Matsufuji
Department of Nutrition
Jikei University School of Medicine
Minato-ku
Tokyo 105, Japan

Peter P. McCann
Merrell Dow Research Center
2110 E. Galbraith Road
Cincinnati, Ohio 45215

Gabriele Mezzetti
Istituto di Chimica Biologica
Universita di Modena
41100 Modena, Italy

Shulamit Michaeli
Department of Microbiology
Tel-Aviv University
Tel-Aviv, Israel

J. J. Miret
Instituto de Investigaciones Bioquimicas "Fundacion Campomar"
and Facultad de Ciencias Exactas y Naturales
1428 Buenos Aires, Argentina

Yosef Mizrahi
Applied Research Institute
Ben-Gurion University of the Negev
P.O. Box 1025
Beer-Sheva 84110, Israel

D. Modena
Institute of General Pathology and CNR Centre for Research in Cell Pathology
University of Milan
20133 Milan, Italy

Bruno Mondovi
Institute of Applied Biochemistry and CNR Centre of Molecular Biology
University of Rome
Rome, Italy

M. G. Monti
Istituto di Chimica Biologica
Universita di Modena
41100 Modena, Italy

D. M. L. Morgan
Division of Perinatal Medicine
MRC Clinical Research Centre
Harrow, Middlesex, HA1 3UJ
United Kingdom

Y. Morishita
Departments of Human Nutrition and Food
and School of Human Ecology, Cancer Center
Howard University
Washington, D.C. 20059

G. Moruzzi
Istituto di Chimica Biologica
Universita di Modena
41100 Modena, Italy

M. S. Moruzzi
Istituto di Chimica Biologica
Universita di Modena
41100 Modena, Italy

Y. Murakami
Department of Nutrition
Jikei University School of Medicine
Minato-ku
Tokyo 105, Japan

C. Muscari
Istituto di Chimica Biologica e Centro Studi e Ricerche sul Metabolismo Cardiaco
Facolta di Medicina e Chirurgia
Universita di Bologna
Via Irnerio, 48
40126 Bologna, Italy

Helen M. Naim
Department of Medical Microbiology
St. Mary's Hospital Medical School
Paddington
London, W2 1PG, England

Henry C. Nathan
Haskins Laboratories and Biology Department
Pace University
New York, New York 10038

Jonathan Negrel
Long Ashton Research Station
University of Bristol
Long Ashton
Bristol, BS18 9AF, England

O. Nilsson
Department of Anatomy
University of Uppsala
Box 571
S-751 23 Uppsala, Sweden

Takami Oka
Section on Intermediary Metabolism
National Institute of Arthritis, Metabolism and Digestive Diseases
National Institutes of Health
Bethesda, Maryland 20205

Stina M. Oredsson
Brain Tumor Research Center of the Department of Neurological Surgery
School of Medicine
University of California
San Francisco, California 94143

Tairo Oshima
Mitsubishi-Kasei Institute of Life Sciences
Machida
Tokyo 194, Japan

R-L. Pajula
Department of Biochemistry
University of Kuopio
P.O. Box 138
SF-70101 Kuopio 10, Finland

Lars Paulin
Research Laboratories of the State Alcohol Monopoly
POB 350
SF-00101 Helsinki 10, Finland

Anthony E. Pegg
Department of Physiology
The Milton S. Hershey Medical Center
The Pennsylvania State University
Hershey, Pennsylvania 17033

Antonio Perin
Institute of General Pathology and CNR Center for Research in Cell Pathology
University of Milan
Via Mangiagalli 31
20133 Milan, Italy

Lo Persson
Department of Physiology
University of Lund
Solvegatan 19
S-223 62 Lund, Sweden

G. Piccinini
Istituto di Chimica Biologica
Universita di Modena
41100 Modena, Italy

Carla Pignatti
Istituto di Chimica Biologica
Facolta di Medicina e Chirurgia
Universita di Bologna
40126 Bologna, Italy

M. Porcelli
Department of Biochemsitry
II Chair, First Medical School
University of Naples
80138 Naples, Italy

Raffaele Porta
Department of Biochemistry
First Medical School
University of Naples
80138 Naples, Italy

Hannu Pöso
Research Laboratories of the State Alcohol Monopoly
POB 350
SF-00101 Helsinki 10, Finland

Arne Raina
Department of Biochemistry
University of Kuopio
P.O. Box 138
SF-70101 Kuopio 10, Finland

Bandhana K. Rawal
Department of Medical Microbiology
St. Mary's Hospital Medical School
Paddington
London, W2 1PG, England

A. Riccio
Regina Elena Institute for Cancer Research
Department of Neurosurgery
Rome, Italy

P. Riccio
Institute of Applied Biochemistry
and CNR Centre of Molecular Biology
University of Rome
Rome, Italy

Marta Romano
Department of Biochemistry
Istituto dei Tumori
80131 Naples, Italy

Eliora Z. Ron
Department of Microbiology
Tel-Aviv University
Tel-Aviv, Israel

Kathleen M. Rose
Department of Pharmacology
The Milton S. Hershey Medical Center
The Pennsylvania State University College of Medicine
Hershey, Pennsylvania 17033

Elsa Rosengren
Department of Physiology
University of Lund
Solvegatan 19
S-223 62 Lund, Sweden

Sonia Rozenhak
Department of Microbiology
Tel-Aviv University
Tel-Aviv, Israel

Diane Haddock Russell
Department of Pharmacology
University of Arizona College of Medicine
Tucson, Arizona 85724

A. Sahai
School of Human Ecology, Cancer Center
Howard University
Washington, D.C. 20059

S. Sarhan
Centre de Recherche Merrell International
16, rue d'Ankara
67084 Strasbourg Cedex, France

Giuseppe Scalabrino
Institute of General Pathology
and CNR Centre for Research in Cell Pathology
University of Milan
20133 Milan, Italy

Immo E. Scheffler
Department of Biology, B-022
University of California, San Diego
La Jolla, California 92093

Martha Schwartz
Department of Botany
The George S. Wise Faculty of Life Sciences
Tel-Aviv University
Tel-Aviv, Israel

James E. Seely
Department of Physiology
The Milton S. Hershey Medical Center
The Pennsylvania State University
Hershey, Pennsylvania 17033

Nikolaus Seiler
Centre de Recherche Merrell International
16, rue d'Ankara
67084 Strasbourg Cedex, France

O. Z. Sellinger
Laboratory of Neurochemistry
Mental Health Research Institute
University of Michigan
Ann Arbor, Michigan

P. Seppänen
Department of Biochemistry
University of Helsinki
SF-00170 Helsinki 17, Finland

A. Sessa
Institute of General Pathology and CNR Centre for Research in Cell Pathology
University of Milan
Via Mangiagalli 31
20133 Milan, Italy

J. J. Shaw
MRC Unit Clinical Oncology and Radiotherapeutics
The Medical School
Cambridge University
Cambridge CB2 2QH, England

Ram K. Sindhu
Department of Pharmacological Sciences
State University of New York at Stony Brook
Stony Brook, New York 11794

Albert Sjoerdsma
Merrell Dow Research Center
2110 E. Galbraith Road
Cincinnati, Ohio 45215

Terence A. Smith
Long Ashton Research Station
University of Bristol
Long Ashton
Bristol, BS18 9AF, England

Dalia Sömjen
Hard Tissue Unit
Ichilov Hospital
64239 Tel-Aviv, Israel

Lina Spigel
Department of Oncology
Assaf Harofeh Hospital
Sackler School of Medicine
Zerifin, Israel

Mendel Stavorovsky
Department of Medicine C
Medical Center and Sackler School of Medicine
Tel-Aviv 64239, Israel

Carolyn Steglich
Department of Biology, B-022
University of California, San Diego
La Jolla, California 92093

Frank Sundler
Department of Histology
University of Lund
Biskopsgatan 5
S-223 62 Lund, Sweden

Celia White Tabor
National Institute of Arthritis, Diabetes, and Digestive and Kidney Diseases
National Institutes of Health
Building 4, Room 116
Bethesda, Maryland 20205

Herbert Tabor
National Institute of Arthritis, Diabetes, and Digestive and Kidney Diseases
National Institutes of Health
Building 4, Room 116
Bethesda, Maryland 20205

Y. B. Talwalkar
Departments of Pediatrics and Anatomical Pathology
Oregon Health Sciences University
Portland, Oregon 97201

B. Tantini
Istituto di Chimica Biologica
Facolta di Medicina e Chirurgia
Universita di Bologna
40126 Bologna, Italy

Daniel Teitel
Department of Biology
Ben-Gurion University of the Negev
P.O. Box 1025
Beer-Sheva 84110, Israel

H. Tjälve
Department of Toxicology
University of Uppsala
Box 573
S-751 23 Uppsala, Sweden

P. Torrigiani
Istituto Botanico
Universita di Bologna
40126 Bologna, Italy

E. Turchetto
Istituto di Chimica Biologica
Facolta di Medicina e Chirurgia
Universita di Bologna
40126 Bologna, Italy

Anil K. Tyagi
National Institute of Arthritis, Diabetes, and Digestive and Kidney Diseases
National Institutes of Health
Building 4, Room 116
Bethesda, Maryland 20205

A. Stanley Tyms
Department of Medical Microbiology
St. Mary's Hospital Medical School
Paddington
London W2 1PG, England

Rolf Uddman
Department of Histology
University of Lund
Biskopsgatan 5
S-223 62 Lund, Sweden

K. Utsunomiya
Department of Nutrition
Jikei University School of Medicine
Minato-ku
Tokyo 105, Japan

Victor R. Villanueva
Institut de Chimie des Substances Naturelles
CNRS
91190 Gif sur Yvette, France

Natalio Walach
Department of Oncology
Assaf Harofeh Hospital
Sackler School of Medicine
Zerifin, Israel

R. Waldner
The Ludwig Boltzmann Institute for Leukemia Research and Hematology
The 3rd Medical Department
Hanusch Hospital
1140 Vienna, Austria

John D. Williamson
Department of Medical Microbiology
St. Mary's Hospital Medical School
Paddington
London W2 1PG, England

J. R. Womble
Department of Surgery
University of Arizona College of Medicine
Tucson, Arizona 85724

Q. Watanabe
Department of Pharmacology
Howard University
Washington, D.C. 20059

M. Yariv
Department of Hormone Research
The Weizmann Institute of Science
Rehovot 76100, Israel

N. D. Young
Department of Biology
Yale University
New Haven, Connecticut 06511

Ian S. Zagon
Department of Anatomy
The Milton S. Hershey Medical Center
The Pennsylvania State University
Hershey, Pennsylvania 17033

Vincenzo Zappia
Department of Biochemistry
II Chair, First Medical School
University of Naples
80138 Naples, Italy

Advances in Polyamine Research, Vol. 4, edited by U. Bachrach, A. Kaye, and R. Chayen. Raven Press, New York © 1983.

Roles of Amines in the Action of Antibiotics — Bleomycin, Spergualin, and Aminoglycosides

Hamao Umezawa

Institute of Microbial Chemistry 3-14-23 Kamiosaki, Shinagawa-ku, Tokyo 141, Japan

I have been studying antibiotics since 1944 and have expanded this study to that of low molecular weight enzyme inhibitors. In the course of this study, I have discovered about 45 inhibitors of various enzymes (1-3). Low molecular weight enzyme inhibitors are released into culture media, and almost all of them have no significant antimicrobial activity. Antibiotics and low molecular weight enzyme inhibitors are microbial secondary metabolites which are not involved in the growth function of microbial cells. Many secondary metabolites contain amines which play important roles in their actions.

In this paper, I will discuss the principles involved in the biosynthesis of microbial secondary metabolites, report on bleomycin and spergualin containing polyamines, and show the predominant role that amino groups play in the antibacterial action of aminoglycosides.

GENERAL PRINCIPLE OF BIOSYNTHESIS OF MICROBIAL SECONDARY METABOLITES

On the basis of their structures, antibiotics can be divided into various groups; each group has a common structural part. For instance, there is a 2-deoxystreptamine-containing group which includes kanamycin, neomycin, gentamicin, butirosin, etc. The common structural part of each antibiotic group has no cytotoxicity and is biosynthesized and modified to the final products which are rapidly released extracellularly.

I have discovered various protease inhibitors as shown in Fig. 1: leupeptin (1,3), antipain (1,3), chymostatin (1), elastatinal (1,3), pepstatin (1,3), phosphoramidon (3), bestatin (4,5), amastatin (5,6). We studied the biosynthesis of leupeptin (7) and were successful in the isolation of three enzymes involved in its synthesis. In the first step, acetyl-L-leucine is formed by leucine acetyltransferase. In the next step, leupeptin acid, acetylleucylleucylarginine, is produced by leupeptin acid synthetase in the following reaction sequence in the presence of ATP:

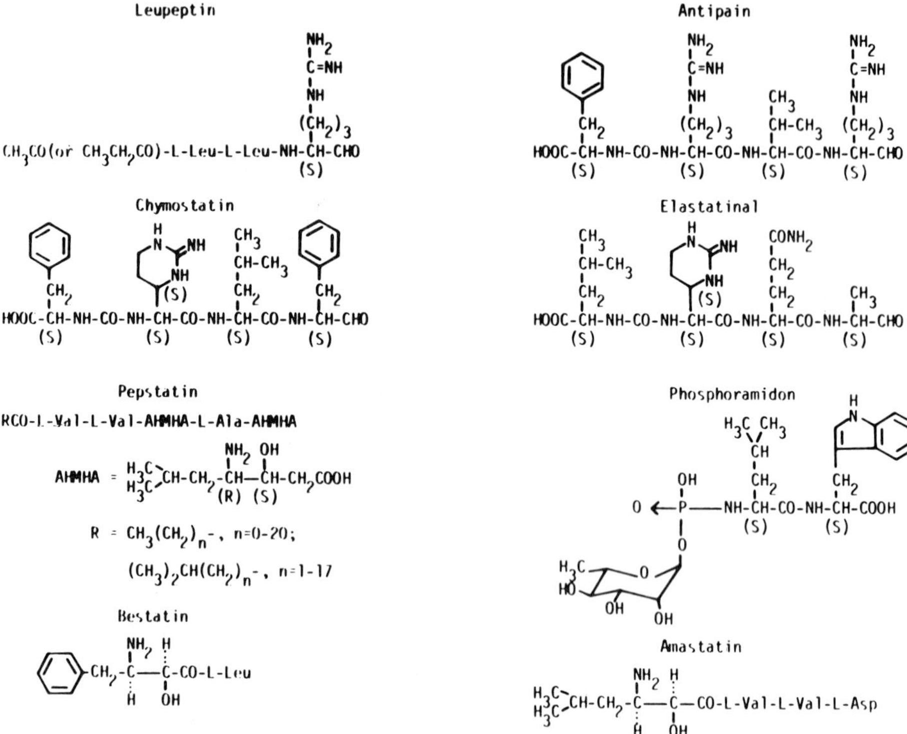

FIG. 1. Protease inhibitors.

acetyl-L-leucine + L-leucine ⟶ acetyl-L-leucyl-L-leucine
⟶ acetyl-L-leucyl-L-leucyl-L-arginine

Leupeptin acid synthetase has been partially purified and is a kind of multifunctional enzyme. Leupeptin acid is reduced to leupeptin by leupeptin acid reductase which requires ATP and NADPH. Leupeptin acid reductase is strongly inhibited by leupeptin. The experimental data (8) indicate that leupeptin is not involved in the growth function of leupeptin-producing cells.

It is possible that genes for the biosynthesis of microbial secondary metabolites have been transferred to other strains (9). In different strains the characteristic structural part could be modified to different final products. It may be concluded that many genes for the synthesis of characteristic or common structural parts of secondary metabolites have been generated and many of them have been transferred to other strains.

BLEOMYCIN-PHLEOMYCIN GROUP ANTIBIOTICS

I have been studying antitumor antibiotics ever since I first reported successful results from the screening procedure in 1953. In the continuation of this study, I found phleomycin. I searched for phleomycin-like antibiotics which had low renal toxicity and found a new one which showed reversible hepatotoxicity in dogs but did not have strong renal toxicity. We named this antibiotic "bleomycin" (10).

Bleomycin has been used in the treatment of Hodgkin's lymphoma, tumors of the testis, and carcinomas of the skin, head and neck (11). By 1972, we had almost elucidated the complete structure of bleomycin (11); and in 1978 we conclusively determined its structure as shown in Fig. 2 (12). Recently, we were successful

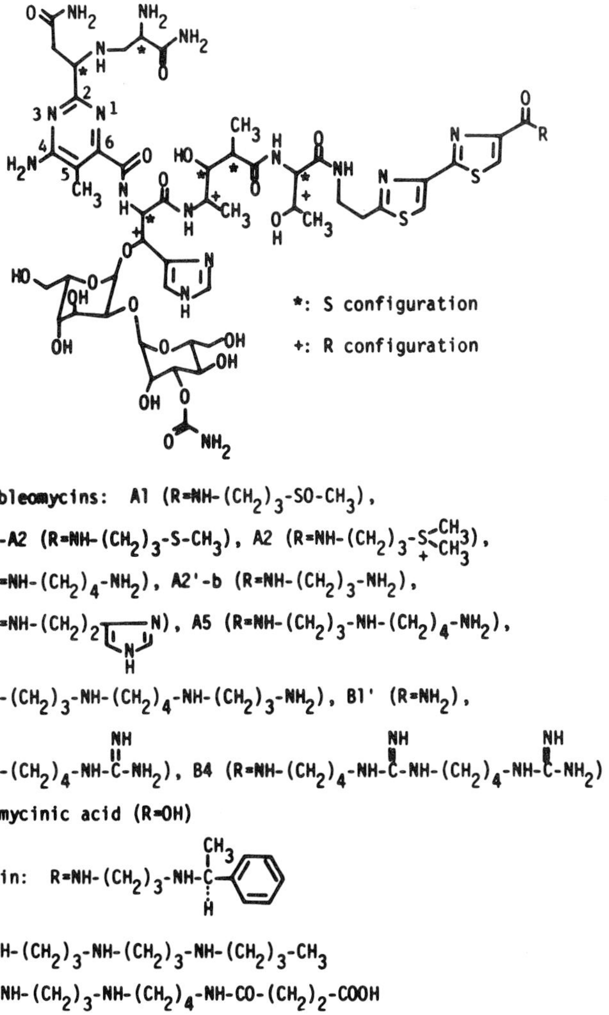

*: S configuration

+: R configuration

Various bleomycins: A1 (R=NH-$(CH_2)_3$-SO-CH_3), Demethyl-A2 (R=NH-$(CH_2)_3$-S-CH_3), A2 (R=NH-$(CH_2)_3$-$S^+(CH_3)_2$), A2'-a (R=NH-$(CH_2)_4$-NH_2), A2'-b (R=NH-$(CH_2)_3$-NH_2), A2'-c (R=NH-$(CH_2)_2$-imidazolyl), A5 (R=NH-$(CH_2)_3$-NH-$(CH_2)_4$-NH_2), A6 (R=NH-$(CH_2)_3$-NH-$(CH_2)_4$-NH-$(CH_2)_3$-NH_2), B1' (R=NH_2), B2 (R=NH-$(CH_2)_4$-NH-C(=NH)-NH_2), B4 (R=NH-$(CH_2)_4$-NH-C(=NH)-NH-$(CH_2)_4$-NH-C(=NH)-NH_2) and Bleomycinic acid (R=OH)

Peplomycin: R=NH-$(CH_2)_3$-NH-CH(CH_3)-C_6H_5

BAPP: NH-$(CH_2)_3$-NH-$(CH_2)_3$-NH-$(CH_2)_3$-CH_3

A5033: NH-$(CH_2)_3$-NH-$(CH_2)_4$-NH-CO-$(CH_2)_2$-COOH

FIG. 2. Bleomycins, bleomycinic acid and three artificial bleomycins (peplomycin, BAPP, A5033).

in the total chemical synthesis of bleomycin A2 (13). Among bleomycins shown in Fig. 2, the terminal amines of bleomycins A5 and A6 are spermidine and spermine, respectively.

One of the amino acids in the bleomycin molecule contains a pyrimidine ring, and we named this new amino acid "pyrimidoblamic acid" (abbreviated as PMB). The following biosynthetic pathway has been suggested by the isolation of the following peptides from fermentation broths: demethylPMB-His → demethylPMB-His-Ala → demethylPMB-His-4-amino-2-hydroxy-4-methylpentanoic (abbreviated as demethylPMB-His-AHP) → demethylPMB-His-AHP-Thr → PMB-His-AHP-Thr → PMB-His-AHP-2'-(2-aminoethyl)-2,4'-bithiazole-4-carboxylic acid (abbreviated as PMB-His-AHP-Thr-BTC) → PMB-β-hydroxyhistidine-AHP-Thr-BTC (abbreviated as PMB-β-OH-His-AHP-Thr-BTC) → PMB-β-OH-His-AHP-Thr-BTC-(3-aminopropyl)-dimethylsulfonium (the peptide part of bleomycin A2).

The methyl group of PMB and the 2-methyl group of AHP are derived from methionine; methylation of the former seems to occur at the demethylPMB-His-AHP-Thr stage. The oxygenation of the β-carbon of the His moiety is suggested to occur at the PMB-His-AHP-Thr-BTC stage.

The study of the incorporation of labeled amino acids into the pyrimidoblamyl moiety suggests that the main molecular part of this amino acid is synthesized from one molecule of L-serine and two molecules of L-asparagine. Labeled BTC is not incorporated into this amino acid moiety of bleomycin. The BTC moiety is produced from β-aminopropionate and 2 molecules of cysteine.

SF-1961 (14), cleomycin (15), tallysomycin (15), phleomycin (11,15), and YA-56X (11,15) which belong to the bleomycin group of antibiotics are different from bleomycin in the parts shown in Table 1. The terminal amines of these bleomycin group antibiotics are shown in Table 2.

Bleomycin binds strongly with Cu(II) and an equimolar bleomycin-Cu(II) complex is formed. We elucidated the structure of this complex shown in Fig. 3 on the basis of the results of X-ray crystallographic analysis of the demethylpyrimidoblamylhistidylalanine-Cu(II) complex (16) and the chemical study of bleomycin-Cu(II) complexes (17). In this structure, the imino group of the amide group linking the PMB moiety to the β-OH-His moiety is deprotonated.

Quenching of fluorescence by DNA indicates the binding of the BTC moiety of both copper-free and copper-bleomycin to the guanine moiety of DNA. The positive charge of the terminal amine facilitates the binding of bleomycin to DNA (18).

In the presence of ferrous ion and oxygen, bleomycin causes double strand scission of DNA. But the bleomycin-Cu(II) complex does not cause strand scission Bleomycin also forms equimolar complexes with Fe(II). ESR studies of pyrimidoblamylhistidine and its model compounds have indicated that the PMB-β-OH-His moiety of the bleomycin molecule is involved in the binding with Fe(II). Although it is not yet conclusive, we have proposed a structure for the bleomycin-Fe(II)-O_2 complex as shown in Fig. 4. Oppenheimer et al. (19) once proposed a structure different from ours for the bleomycin-Fe(II)-CO complex. On the basis of pmr

TABLE 1. Structural parts (except for the terminal amines) of SF-1961, cleomycin, thallysomycin, phleomycin and YA-56X different from bleomycin.

Compound	Structural part	
SF-1961	demethylpyrimidoblamyl	instead of PMB
	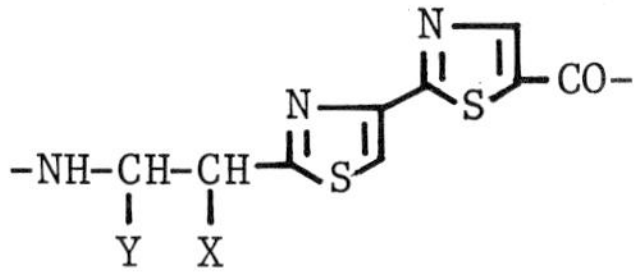 -NH-CH(Y)-CH(X)-[bithiazole]-CO-	instead of BTC
Cleomycin	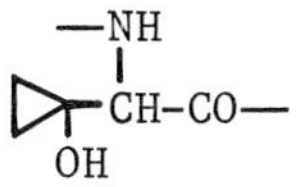 —NH–CH(C(OH)-cyclopropyl)-CO—	instead of Thr
Thallysomycin	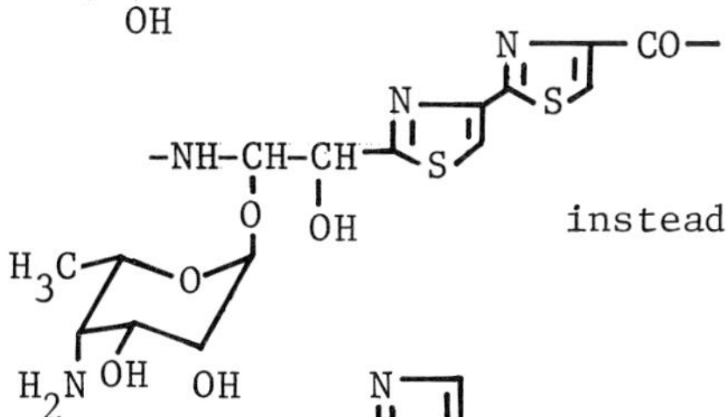 -NH-CH(O-[sugar: H_3C, H_2N, OH, OH])-CH(OH)-[bithiazole]-CO—	instead of BTC
Phleomycin	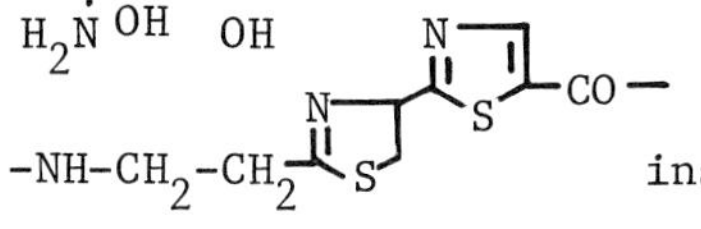 $-NH-CH_2-CH_2$-[thiazoline-thiazole]-CO—	instead of BTC
YA-56X	the same as phleomycin instead of BTC,	
	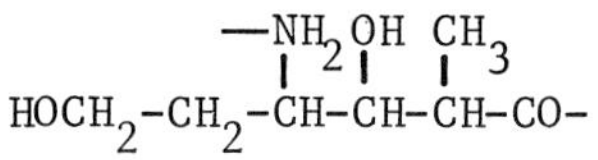 $HOCH_2-CH_2-CH(NH_2)-CH(OH)-CH(CH_3)-CO-$	instead of AHP,
	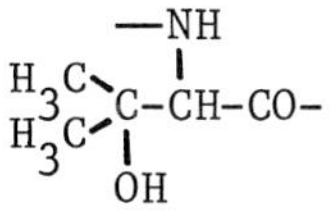 $(H_3C)_2C(OH)-CH(NH-)-CO-$	instead of Thr,
	6-deoxygulose	instead of gulose

analysis, the same authors (20) recently reported that the steric relationship between the α-methin proton of the α-amino-carboxamide part of the PMB moiety and the adjacent methylene protons is different between the demethylpyrimidoblamylhistidyl-alanine-Cu(II) complex crystals (shown by X-ray analysis) and deglycobleomycin-Fe(II)-CO complex. This result does not refute the structure shown in Fig. 4, because the CPK-model study shows that both gauche-trans and gauche-gauche conformers between the methin and methylene protons can exist in the 5-membered chelate-ring constructed as follows: $Fe^{++} \rightarrow NH_2 \rightarrow CH(CONH_2) \rightarrow CH_2 \rightarrow NH(CH—) \rightarrow (Fe^{++})$.

The active species of the bleomycin-iron-O_2 complex involved in the reaction with DNA has been suggested to be the bleomycin-

TABLE 2. The terminal amines of main component of SF-1961, cleomycin, thallysomycin, phleomycin and YA-56X

SF-1961	$-NH_2$
Cleomycin	Same as bleomycin
Thallysomycin A	$-NH(CH_2)_3CH(NH_2)CH_2CONH(CH_2)_3NH(CH_2)_4NH_2$
" B	$-NH(CH_2)_3NH(CH_2)_4NH_2$
Phleomycin D_1	$-NH(CH_2)_4NHC(=NH)NH_2$
" E	$-NH(CH_2)_4NHC(=NH)NH(CH_2)_4NHC(=NH)NH_2$
" G	unknown, containing 2 or more guanidine groups
YA-56X	$-NH(CH_2)_2C(=NH)NH$

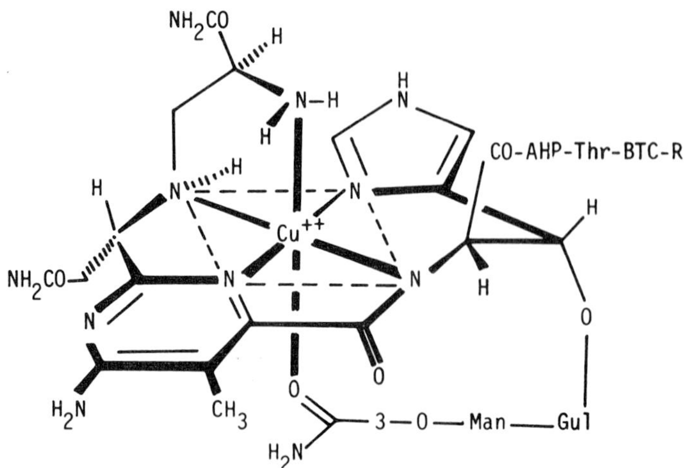

AHP = 4-amino-3-hydroxy-2-methylpentanoyl

Thr = threonyl

BTC = 2'-(2-aminoethyl)-2,4'-bithiazole-4-carboxyl

R = terminal amine

Man-Gul = 2-O-(α-D-mannopyranosyl)-α-L-gulopyranosyl

FIG. 3. Bleomycin copper complex.

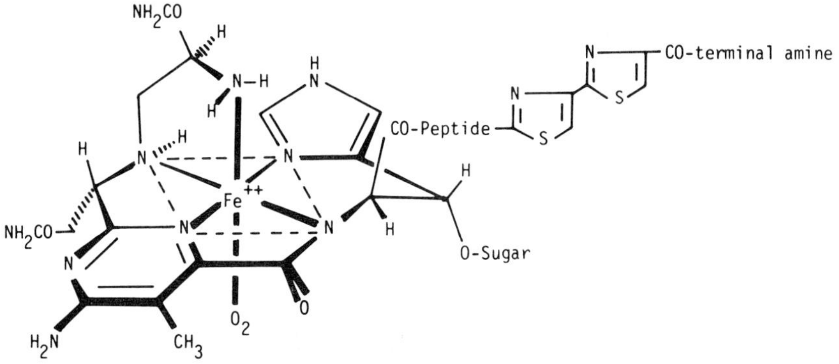

FIG. 4. Bleomycin-Fe^{++} O_2 complex.

Fe(III)-^-O_2H complex (21,22). On the basis of the reaction products of DNA fragmentation (23), it has become possible to show the scheme of the DNA fragmentation reaction caused by bleomycin. The bleomycin-Fe(III)-^-O_2H binding to DNA abstracts the hydrogen radical from the 4-carbon of the 2-deoxyribose moiety, leaving a 4-carbon radical which reacts with molecular oxygen; the 4-carbon is oxygenated; furanose ring is opened between the 4- and 3-carbons, followed by cleavages of phosphate and ester bonds in various ways.

Bleomycin causes double-strand scission. Phleomycin which has the dihydrobithiazole moiety neither intercalates with DNA (24) nor causes double-strand scission. It causes single-strand scission (25).

Injected bleomycin binds with Cu(II) in the blood. After penetrating into cells, the Cu(II) of bleomycin-Cu(II) complex is reduced to Cu(I) by reducing agents such as cysteine, etc. in the cells, and the Cu(I) is then transferred to a cellular protein which can bind selectively to it (26). Cu(II)-free bleomycin thus formed undergoes reaction with bleomycin hydrolase which hydrolyzes the α-aminocarboxamide bond of the pyrimidoblamyl moiety of bleomycin (11,27). This enzyme is widely distributed in human and animal cells. Cu(II)-free bleomycin which escapes this enzymic action reaches nuclei, binds, and reacts with DNA as described above. One of the reasons why bleomycin exhibits a therapeutic effect against squamous cell carcinoma is due to the low content of bleomycin hydrolase in this type of tumor (11).

The degree of renal and pulmonary toxicity is dependent on the terminal amine. We have established fermentation and chemical methods for the preparation of artificial bleomycins which contain various terminal amines (11). With the addition of many kinds of amines to fermentation medium, various bleomycins containing the added amines are produced; and the production of other bleomycins are suppressed. We found acyl agmatine hydrolase in Fusarium sp. which hydrolyzes the terminal peptide bond of bleomycin B2, yielding bleomycinic acid (28). Bleomycinic acid can be obtained also by reaction of cyanogen bromide with bleomycin demethyl A2 (29). Various bleomycins are synthesized from bleomycinic acid.

A method of testing the pulmonary toxicity of bleomycins in mice was developed by Matsuda *et al.* (30). Bleomycin PEP (peplomycin) and BAPP (Fig. 2) were selected as those which have low pulmonary toxicity and strong antitumor activity. The pulmonary toxicity index of PEP was 0.25 and that of BAPP was 0.22. We named bleomycin PEP "pepleomycin"; this name was changed to "peplomycin" (31). Peplomycin was more effective than the present bleomycin against carcinomas of the head, neck, and skin and had lower pulmonary toxicity. Peplomycin has also been confirmed to be effective against prostatic carcinoma.

Radioactive metal complexes of bleomycin are taken up by tumors, especially in lung cancer, and are useful for diagnosis (32). Therefore, the bleomycin analogs and derivatives which are resistant to bleomycin hydrolase can be assumed to have a wider anticancer spectrum than bleomycin. In this paper, I will report two examples of such derivatives, PEP-PEP and DBP-PEP, as shown in Fig. 5. Both are resistant to bleomycin hydrolase. The index

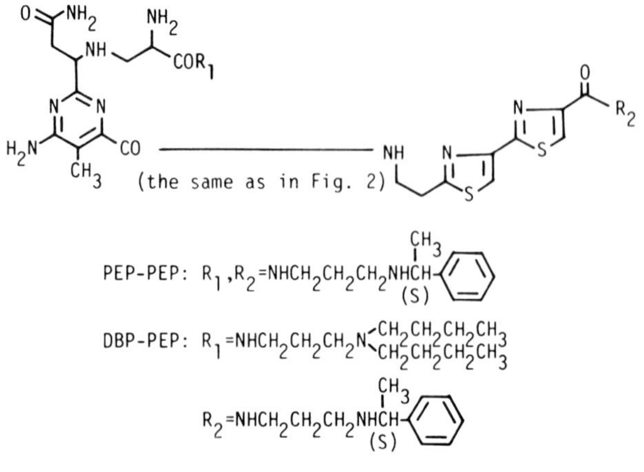

FIG. 5. New bleomycin analogs resistant to bleomycin hydrolase.

of pulmonary toxicity is 0.05 for PEP-PEP and 0 for DBP-PEP. Their therapeutic index against Ehrlich carcinoma is as high as that of bleomycin.

SPERGUALIN, A SPERMIDINE-CONTAINING ANTITUMOR ANTIBIOTIC

As described above, bleomycin A5 and tallysomycins A and B contain spermidine, and bleomycin A6 has spermine. These antibiotics are produced by streptomyces. Several antibiotics containing spermidine have been isolated from bacteria. Edeines shown in Fig. 6 (33) inhibits the growth of bacteria and also has a cytotoxic effect against tumor cells _in vitro_. Laterosporamine

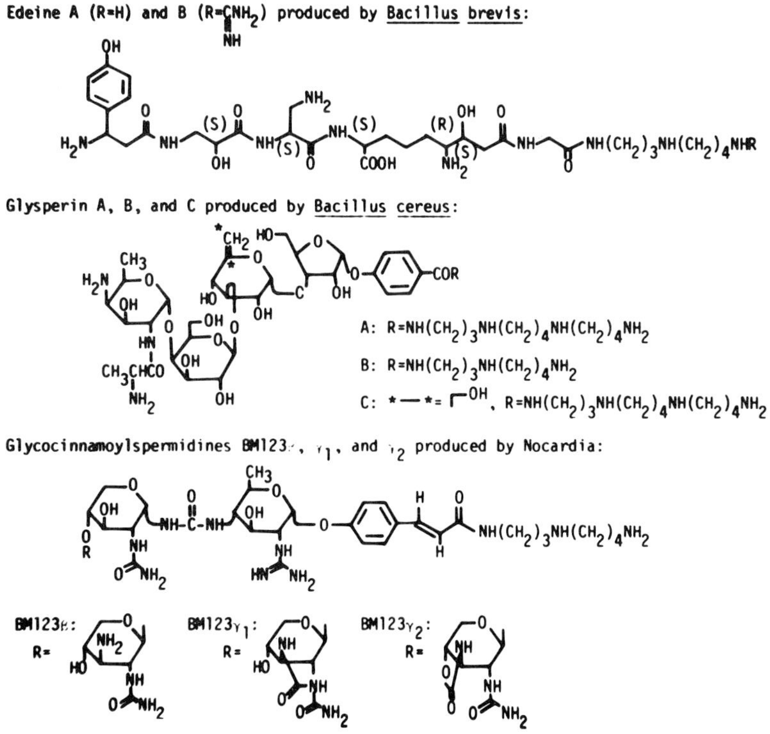

Edeine A (R=H) and B (R=C(=NH)NH$_2$) produced by _Bacillus brevis_:

Glysperin A, B, and C produced by _Bacillus cereus_:

A: R=NH(CH$_2$)$_3$NH(CH$_2$)$_4$NH(CH$_2$)$_4$NH$_2$

B: R=NH(CH$_2$)$_3$NH(CH$_2$)$_4$NH$_2$

C: R=NH(CH$_2$)$_3$NH(CH$_2$)$_4$NH(CH$_2$)$_4$NH$_2$

Glycocinnamoylspermidines BM123β, γ$_1$, and γ$_2$ produced by Nocardia:

FIG. 6. Spermidine-containing antibiotics.

($C_{17}H_{35}N_7O_4$) (34), produced by _Bacillus laterosporus_, and BA-1008 ($C_{16}H_{27}N_4O_7$) (35), produced by _Pseudomonas fluorescens_, have also been reported to contain spermidine. These antibiotics have weak antibacterial activity. Glysperins A, B, C shown in Fig. 6 (36) are produced by _Bacillus cereus_. These antibiotics contain the spermidine moiety. Glycinnamoylspermidine antibiotics, BM123β, BM123γ_1, and BM123γ_2 (Fig. 6) (37) are produced by _Nocardia_. Glycocinnamoylspermidine antibiotics and glysperin have strong antibacterial activity and protect mice from bacterial infections.

Recently we discovered spergualin which has interesting antitumor activity (38-40). The strain producing spergualin was classified as _Bacillus laterosporus_. This antibiotic exhibited a strong inhibitory effect against L-1210 leukemia. We determined the structure of this antibiotic as shown in Fig. 7 (39,40). We have also synthesized spergualin (40).

Spergualin has weak antibacterial activity. The LD_{50} by

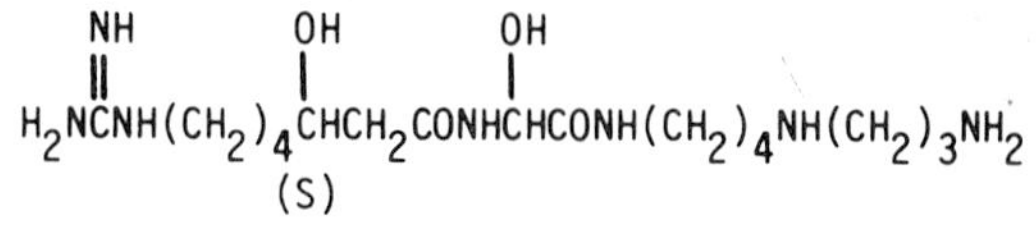

$$H_2N\overset{NH}{\overset{\|}{C}}NH(CH_2)_4\overset{OH}{\overset{|}{C}}HCH_2CONH\overset{OH}{\overset{|}{C}}HCONH(CH_2)_4NH(CH_2)_3NH_2$$

(S)

FIG. 7. Spergualin.

intraperitoneal injection was about 150 mg/kg. It has a very low cumulative toxicity. Spergulain exhibits a very strong inhibition against L-1210 leukemia which had been kept in Institute of Microbial Chemistry as shown in Table 3. The surviving mice rejected a second inoculation of L-1210 cells. However, another

TABLE 3. Antitumor effect of spergualin against 2 lines of L-1210*

Dose (mg/kg/day)	Against L-1210 in IMC		Against L-1210 from NCI	
	T/C (%)	Survivor (60 days)	T/C (%)	Survivor (30 days)
50	295	0/8	250	0/6
25	334	0/8	—	—
12.5	586	4/8	233	0/6
6.25	732	8/8	—	—
3.13	441	3/8	214	0/6
1.56	301	1/8	—	—
0.78	107	0/4	160	0/6

*Indicated dose of spergualin was intraperitoneally injected daily for 9 days starting from day 1 after the intraperitoneal inoculation of 10^5 L-1210 cells to CDF_1 mice.

line of L-1210 which was supplied to the Cancer Institute, Tokyo, by the N.C.I. of the U.S.A. was more resistant to spergualin as shown in Table 3.

In the case of testing for the effect of spergualin against L-1210 in mice kept in the Institute of Microbial Chemistry, none of mice treated with 50 mg/kg or 25 mg/kg daily survived 60 days after inoculation. The death of these mice was due to the delayed growth of the ascites tumor. The recurrent cells were inoculated intraperiotneally into new mice. Then these resulting tumors were resistant to spergualin treatment. However, second generation cells in vitro have completely lost their resistance.

We have synthesized various analogs of spergualin. The analog in which the amino group of the 3-aminobutyl moiety of spermidine is involved in the linkage to the residual part of the molecule has almost no activity against L-1210 *in vivo*. Dehydroxylation of the 7-guanido-3-hydroxyheptanoyl moiety enhances the effect *in vivo*.

THE PREDOMINANT ROLE OF AMINO GROUPS IN THE ACTION OF AMINOGLYCOSIDE ANTIBIOTICS

As described above, the nitrogen atoms of bleomycin and spergualin play important roles in their antitumor actions. In the case of aminoglycoside antibiotics, amino groups, but not hydroxyl groups, also play a predominant role in their antibacterial action.

In 1967, I elucidated the enzymatic mechanism of resistance to aminoglycoside antibiotics (41,42). Enzymes which transferred the terminal phosphate of ATP to the 3'-hydroxyl group of kanamycin, neomycin, and paromomycin were found in cells of strains resistant to these antibiotics (Fig. 8). The involvement of these enzymes in the mechanism of resistance was conclusively established by synthesis of derivatives which lacked the hydroxyl group susceptible to these enzyme reactions; 3'-deoxykanamycin (43) and 3',4'-deoxykanamycin B (Fig. 8) (44) inhibited the

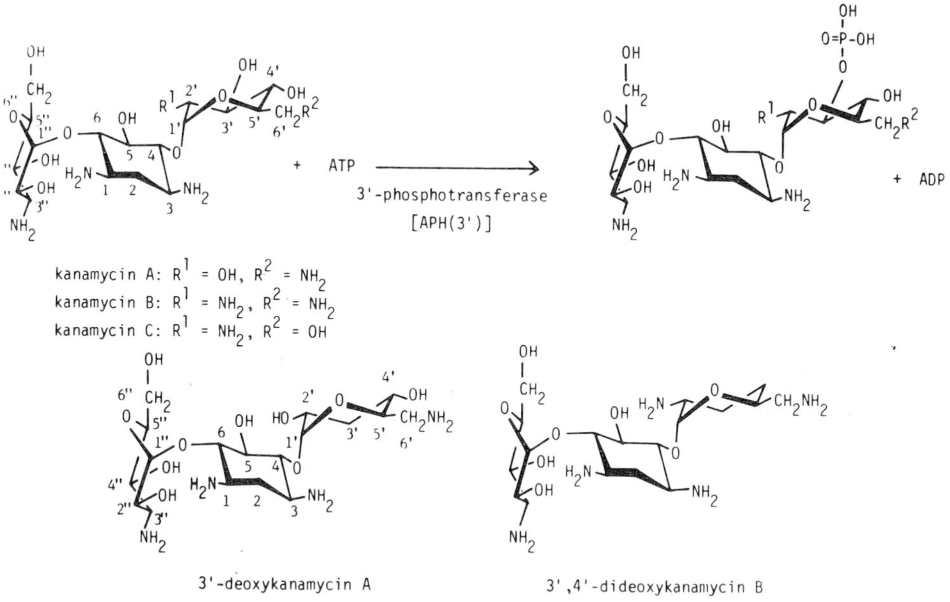

+ ATP → 3'-phosphotransferase [APH(3')] + ADP

kanamycin A: R^1 = OH, R^2 = NH_2
kanamycin B: R^1 = NH_2, R^2 = NH_2
kanamycin C: R^1 = NH_2, R^2 = OH

3'-deoxykanamycin A

3',4'-dideoxykanamycin B

FIG. 8. Reaction of 3'-O-phosphotransferase ((APH(3')) with kanamycins and 3'-deoxykanamycin A and 3',4'-dideoxykanamycin B.

growth of both sensitive and resistant strains which produced 3'-O-phosphotransferase. Thus, it became possible to predict the structures of derivatives which should inhibit the growth of resistant strains, and such derivatives have been synthesized.

At present, 3'-O-phosphotransferase, 2"-O-phosphotransferase, 4'-O-adenylyltransferase, 2"-O-adenylyltransferase, 3-N-acetyltransferase, 2'-N-acetyltransferase, and 6'-N-acetyltransferase are known to be involved in the mechanism of resistance to 2-deoxystreptamine-containing antibiotics (42,45). Resistant strains producing new O-phospho- or O-adenylyltransferases may appear in the future. Therefore, I was interested in the structures of kanamycin derivatives which have strong antibacterial activity and contain the least number of hydroxyl groups. We synthesized deoxy derivatives and found that such derivatives of the structures shown in Fig. 9 contained only one hydroxyl group

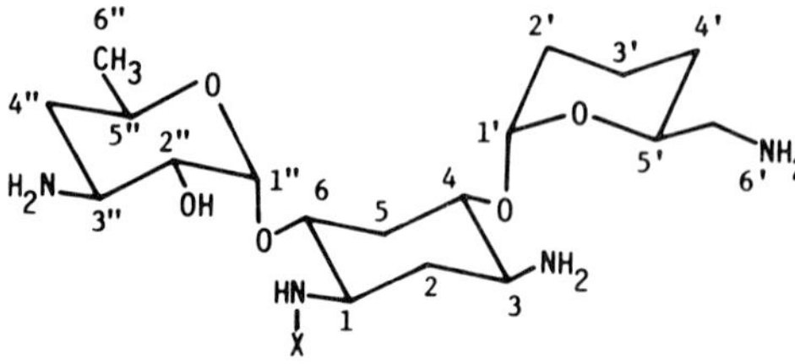

FIG. 9. The structure of kanamycin derivative which has strong antibacterial activity and the least number of hydroxyl group.

and had strong antibacterial activity (46). This indicates that the amino group plays a predominant role in the antibacterial action of aminoglycosides.

From the view point of the predominant role of amino groups in antibacterial action, it can be said that aminoglycoside antibiotics are antibacterial polyamine compounds which have antibacterial activities.

CONCLUSION

Antibiotic research has been expanded to the area of low molecular weight enzyme inhibitors and low molecular weight immuno-modifiers. Many of them contain amines or nitrogen atoms which may be involved in hydrogen bonding or electrostatic binding with nucleotides or proteins or in binding with metal ions; or they may increase the reactivity of neighbouring groups as in the case of β-lactam antibiotics, benzodiazepine antibiotics, etc.

In order to obtain a complete cure in the treatment of cancer, tumors should be macroscopically eliminated; and any minimal residual tumors should be completely eliminated by the action of host defense systems or additional chemotherapy. We screened microbial culture filtrates for inhibitors of enzymes on cell surfaces, and the inhibitors thus found bound to cell surfaces and enhanced immune responses. Among them, bestatin has been suggested to be useful to eliminate minimal residual tumors. It is also necessary to prepare compounds which inhibit the growth or generation of suppressor cells. Such compounds may be found among cytotoxic antibiotics.

Bleomycin and other antibiotics containing polyamines were described in this paper. The polyamine moiety of the bleomycin molecule is involved in the binding to DNA and also in renal and pulmonary toxicity. Spergualin containing spermidine seems to be an interesting antitumor antibiotic. Amino groups but not hydroxyls play predominant role in antibacterial action of aminoglycoside antibiotics. It may be interesting to screen for polyamine compounds in microbial culture filtrates.

REFERENCES

1. Umezawa, H. (1972): Enzyme Inhibitors of Microbial Origin. University of Tokyo Press.
2. Umezawa, H. (1982): In: Ann. Rev. of Microbiology, in press. Annual Reviews Inc.
3. Umezawa, H. (1976): In: Methods in Enzymology, edited by L. Lorand, Vol. 45, pp. 678-695. Academic Press Inc.
4. Umezawa, H., Aoyagi, T., Suda, H., Hamada, M., and Takeuchi, T. (1976): J. Antibiotics, 29: 97-99.
5. Umezawa, H., editor (1981): Small Molecular Immunomodifiers of Microbial Origin - Fundamental and Clinical Studies of Bestatin. Japan Scientific Societies Press, Pergamon Press.
6. Aoyagi, T., Tobe, H., Kojima, F., Hamada, M., Takeuchi, T., and Umezawa, H. (1978): J. Antibiotics, 31: 636-638.
7. Suzukake, K., Hori, M., Tamemasa, O., and Umezawa, H. (1981): Biochim. Biophys. Acta, 661: 175-181.
8. Suzukake, K., Takada, M., Horim, M., and Umezawa, H. (1980): J. Antibiotics, 33: 1172-1176.
9. Umezawa, H., Okami, Y., and Hotta, K. (1978): J. Antibiotics, 31: 99-102.
10. Umezawa, H., Maeda, K., Takeuchi, T., and Okami, Y. (1966): J. Antibiotics, 19: 200-209; 210-215.
11. Umezawa, H. (1976): In: GANN Monograph on Cancer Research, edited by S.K. Carter, G. Mathe, and T. Ichikawa, No. 19, pp. 3-36. University of Tokyo Press, Tokyo.
12. Takita, T., Muraoka, Y., Nakatani, T., Fujii, A., Umezawa, Y., Naganawa, H., and Umezawa, H. (1978): J. Antibiotics, 31: 801-804.
13. Takita, T., Umezawa, Y., Saito, S., Morishima, H., Tsuchiya, T., Miyake, T., Umezawa, H., Muraoka, Y., Suzuki, M., Otsuka, M., and Ohno, M. (1982): Tetrahedron Letters, 23: 521-524.
14. Shomura, T., Omoto, S., Ohba, K., Ogino, H., Kojima, M., and Inouye, S. (1980): J. Antibiotics, 33: 1243-1248.

15. Umezawa, H., and Takita, T. (1980): Structure and Bonding, 40: 73-97.
16. Iitaka, T., Nakamura, H., Nakatani, T., Muraoka, Y., Fujii, A., Takita, T., and Umezawa, H. (1978): J. Antibiotics, 31: 1070-1072.
17. Takita, T., Muraoka, Y., Nakatani, T., Fujii, A., Iitaka, Y., and Umezawa, H. (1978): J. Antibiotics, 31: 1073-1077.
18. Kasai, H., Naganawa, H., Takita, T., and Umezawa, H. (1978): J. Antibiotics, 31: 1316-1320.
19. Oppenheimer, N.S., Rodringuez, L.O., and Hecht, S.M. (1979): Biochemistry, 18: 3439-3445.
20. Oppenheimer, N.J., Chang, C., Chang, L.H., Ehrenfeld, G., Rodriguez, L.O., and Hecht, S.M. (1982): J. Biol. Chem., 257: 1606-1609.
21. Kuramochi, H., Takahashi, K., Takita, T., and Umezawa, H. (1981): J. Antibiotics, 34: 576-582.
22. Burger, R.M., Peisach, J., Horwitz, S.B. (1981): Activated bleomycin. A transient complex of drug, iron and oxygen that degrades DNA. J. Biol. Chem., 256: 11636-11644.
23. Giloni, L., Takeshita, M., Johnson, F., Iden, and Grollman, P. (1981): J. Biol. Chem, 256: 8608-8615.
24. Povirk, L.F. (1981): Biochemistry, 20: 665-670.
25. Okubo, H., Abe, Y., Hori, M., Asakura, H., and Umezawa, H. (1981): J. Antibiotics, 34: 1213-1215.
26. Takahashi, K., Yoshioka, O., Matsuda, A., and Umezawa, H. (1977): J. Antibiotics, 30: 861-869.
27. Umezawa, H., Hori, S., Sawa, T., Yoshioka, T., Takita, T., and Takeuchi, T. (1974): J. Antibiotics, 27: 419-424.
28. Umezawa, H., Takahashi, Y., Fujii, A., Saino, T., Shirai, T., and Takita, T. (1973): J. Antibiotics, 26: 117-119.
29. Takita, T., Fukuoka, T., and Umezawa, H. (1973): J. Antibiotics, 26: 252-254.
30. Matsuda, A., Yoshioka, O., Ebihara, K., Ekimoto, H., Yamashita, T., and Umezawa, H. (1978): In: Bleomycin: Current Status and New Development, edited by S.K. Carter, S.T. Crook, and H. Umezawa, pp. 299-310. Academic Press.
31. Matsuda, A., Yoshioka, O., Takahashi, K., Yamashita, T., Ebihara, K., Eimoto, H., Abe, F., Hashimoto, Y., and Umezawa, H. (1978): In: Bleomycin: Current Status and New Development, edited by S.K. Carter, S.T. Crook, and H. Umezawa, pp. 311-332. Academic Press.
32. Noel, I.P. (1976): In: Fundamental and Clinical Studies of Bleomycin, pp. 301-316. University of Tokyo Press.
33. Hettinger, T.P., and Craig, L.C. (1970): Biochemistry, 9: 1224-1232.
34. Shoji, J., Sakazaki, R., Wakisaka, Y., Koizumi, K., Mayama, M., Matsuura, S., and Matsumoto, K. (1976): J. Antibiotics, 29: 390-393.
35. Kido, Y., Harada, T., Yoshida, T., Umemoto, J., and Motoki, Y. (1980): J. Antibiotics, 33: 791-795.
36. Tsuno, T., Konishi, M., Naito, T., and Kawaguchi, H. (1981): J. Antibiotics, 34: 390-402.

37. Ellestad, G.A., Cosalich, D.B., Broschard, R.W., Martin, J.H., Kunstmann, M.P., Morton, G.O., Lancaster, J.E., Fulmor, W., and Lovell, F.M. (1978): J. Am. Chem. Soc., 100: 2515-2523.
38. Takeuchi, T., Iinuma, H., Kunimoto, S., Masuda, T., Ishizuka, M., Takeuchi, M., Hamada, M., Naganawa, H., Kondo, S., and Umezawa, H. (1981): J. Antibiotics, 34: 1619-1621.
39. Umezawa, H., Kondo, S., Iinuma, H., Kunimoto, S., Ikeda, Y., Iwasawa, H., Ikeda, D., and Takeuchi, T. (1981): J. Antibiotics, 34: 1622-1624.
40. Kondo, S., Iwasawa, H., Ikeda, D., Umeda, Y., Ikeda, Y., Iinuma, H., and Umezawa, H. (1981): J. Antibiotics, 34: 1625-1627.
41. Umezawa, H., Okanishi, M., Kondo, S., Hamana, K., Utahara, R., Maeda, K., and Mitsuhashi, S. (1967): Science, 157: 1559-1561.
42. Umezawa, H. (1974): In: Advances in Carbohydrate Chemistry and Biochemistry, edited by R.S. Tipson, and D. Horton, Vol. 30, pp. 183-225. Academic Press.
43. Umezawa, S., Tsuchiya, T., Muto, R., Nishimura, Y., and Umezawa, H. (1971): J. Antibiotics, 24: 274-275.
44. Umezawa, H., Umezawa, S., Tsuchiya, T., and Okazaki, Y. (1971): J. Antibiotics 24: 485-487.
45. Umezawa, H. (1979): Jap. J. Antibiotics, 32 (Suppl.): S1-S14.
46. Umezawa, H. (1982): In: Microbial Drug Resistance, Proceedings of the 3rd Tokyo Symposium on Microbial Drug Resistance, October, 1981, Vol. 3, in press. Japan Scientific Societies Press.

Advances in Polyamine Research, Vol. 4, edited by U. Bachrach, A. Kaye, and R. Chayen. Raven Press, New York © 1983.

Experimental Approaches to the Systemic and Topical Use of Polyamine Antimetabolites as Antiproliferative Agents

J. Jänne, L. Alhonen-Hongisto, K. Käpyaho, and P. Seppänen

Department of Biochemistry, University of Helsinki, SF-00170 Helsinki 17, Finland

After several decades of intensive basic research, the metabolism of polyamines has apparently become a meaningful target, appreciated even by clinicians, for antiproliferative therapy to be implicated in certain branches of clinical medicine. One of the attractions of this approach is obviously the fact that the therapeutic use of inhibitors of polyamine biosynthesis, i.e. antimetabolites of polyamines, has only emerged from extensive background of basic experimental work, first suggesting (in the 60's), then proving (in the 70's) that polyamines are required for animal cell proliferation, and finally implicating this knowledge experimentally and, to some extent, also clinically in the treatment of proliferative disorders.

The clinicians seem to appreciate the fact that the antimetabolites of polyamines are not just "harvest products" isolated from culture media of some exotic "myces" with little knowledge of their structure and even less of their mode of action. In this respect, we have to admit that polyamine research stands on solid grounds. However, we still face the crucial question, expressed by Seymour Cohen (8) as the title of one of his review articles: "What do the polyamines do?" This surely is the question that has to answered before we can take the full advantage of the experimental phenomenology associated with polyamine research.

In the following few pages we try to show that even though the mode of action of polyamine antimetabolites is known, admittingly with certain reservations, the applications of results obtained with cultured cells, may not be always as simple in "real life", i.e. in experimental animals bearing various tumors, not even touching their relevance to clinical medicine. However, although much has to be done as regards the use polyamine antimetabolites, the combination of differently acting polyamine synthesis inhibitors as to achieve a profound antiproliferative action appears promising in several therapeutic areas.

From the dozens of compounds, which can be classified as polyamine antimetabolites, we will only deal with two of them; DL-2-difluoromethylornithine (DFMO), a specific irrerversible inhibitor of ornithine decarboxylase (33) and methylglyoxal bis(guanylhydrazone) (MGBG), which inhibits the synthesis of spermidine and spermine (55), but also exerts effects that may be unrelated to the biosynthesis of polyamines (39).

Special attention is paid to their combined use, systemic or topical, in experimental murine tumors and in mouse skin undergoing enhanced proliferation. We hope that the reader admits that much has to be done before polyamine antimetabolites can take their place in every day's chemotherapy.

POLYAMINE ANTIMETABOLITES IN EXPERIMENTAL TUMORS

The majority of initial experiments, in which various inhibitors of polyamine biosynthesis were shown to be antiproliferative agents in cell cultures or in animal models were performed in mid seventies (13,24,25,30,37,43), even though MGBG was shown to exhibit profound antileukemic activity, yet without any clues as regards its mode of action, already more than 15 years before that (15). The studies showing that an inhibition of polyamine biosynthesis results in arrest of cell growth are summarized in a recent review article (18). However, as will be shown later, studies with cell cultures and healthy animals are not that simply transferable to tumor-bearing animal models, in which host toxicity and host-dependent compensatory changes are of central importance.

Use of Polyamine Antimetabolites as Single Agents

Methylglyoxal bis(guanylhydrazone)

A marked prolongation of survival time of leukemia L1210-bearing mice by MGBG (and some of its derivatives) was first reported by Freedlander and French (15), and since then substantial evidence has accumulated for the activity of MGBG against various transplantable murine tumors (reviewed in 34 and 35).

Though impressively effective against experimental leukemias, already the early animal experiments indicated that MGBG possessed an extremely narrow therapeutic index. This is exemplified by the fact that the ratio of LD90 to LD10 dose for MGBG in mice was only 1.3, which is far lower than that for other agents commonly used in cancer chemotherapy (35).

However, MGBG apparently has properties that make it unique among other cytostatic compounds: (i) This compound is not transformed or metabolized in animal (35), including human (44), cells. (ii) MGBG is actively concentrated in eukaryotic cells using an uptake system apparently designed for the polyamines

(11); under cell culture conditions, concentration gradients over 1000-fold across the cell membrane are easily achievable (31,48). (iii) The uptake rate of the drug is critically dependent on the growth rate of the cell; rapidly growing tumor cells most effectively accumulate MGBG (23). The latter phenomenon may, in fact, explain the earlier observations indicating that MGBG is better tolerated by tumor-bearing than by healthy animals (35). The preferential accumulation of the drug in actively dividing cells, such as leukemic or ascitic cells, may likewise be the reason for the reported inefficacy of MGBG against experimental solid tumors (35).

Difluoromethylornithine

The data obtained from cell cultures clearly indicated that DFMO acts antiproliferatively, yet exerting cytostatic rather than cytocidal effect (for ref. see 18). In contrast to MGBG, only a few experiments have been done with DFMO employing tumor-bearing animal models. Intraperitoneal administration of the drug to L1210 leukemia-bearing mice produced a moderate (22%) prologation of the survival time (40). In this study the dose of DFMO used was rather low, and, owing to its extremely low host toxicity, almost ten times higher doses have been used later with more pronounced effect on survival (5,7).

Unlike MGBG, DFMO has also shown distinct activity against solid tumors such as murine mammary sarcoma (5,41), glioma (6), rat hepatoma (5) and gliosarcoma (6). Moreover, DFMO is active against several human tumors transplanted into athymic nude mice (14).

Just like in case of MGBG, actively proliferating animal cells are most sensitive to DFMO, as regards the development of polyamine depletion (23).

Combined Use of Polyamine Antimetabolites

After the discovery that all polyamine antimetabolites act antiproliferatively, it was logic to combine differently acting inhibitors of polyamine biosynthesis under various experimental conditions. This was not only an understandable extension of the single-agent studies but a necessity, since it soon became obvious that animal cells brought to a stage of profound polyamine depletion trigger a series of compensatory mechanisms aimed to correct the threatening intracellular deprivation of putrescine, spermidine and spermine. Thus the use of competitive inhibitors of the two biosynthetic decarboxylases (ornithine and S-adenosylmethionine) results in a paradoxic enhancement of the enzyme activities due to stabilization of the otherwise short-lived enzymes by bound inhibitors (9,20,32). The use of irreversible inhibitors produces secondary compensatory mechanisms involving other reactions or processes than the inhibited one. When ornithine decarboxylase activity is inhibited by any of its inhibitors, the activity of adenosylmethionine decarboxylase is greatly enhanced (22), and finally intracellular polyamine depletion as such induces accelerated uptake of extracellular polyamines or almost any organic material that sufficiently resembles the natural polyamines (3,46).

It thus appears that an inhibition of a single reaction of polyamine biosynthetic pathway is doomed to be unsuccesful for antipoliferative purposes.

The first combination experiments were apparently done with unphysiological diamines, such as 1,3-diaminopropane or 1,3-diamino-2-propanol, which are indirect inhibitors of ornithine decarboxylase (42), and with 1,1'-((methylethanediylidene)-dinitrilo) bis(3-aminoguanidine) (MBAG), a close derivative of MGBG. This combination produced an impressive antiproliferative synergism in regenerating liver (54) and in Ehrlich ascites carcinoma-bearing mice (1), but apparently at the cost of severe host toxicity. In all likelihood, the potentiation of the antiproliferative action of the diamines was attributable to the fact that MBAG (38), just like the parent compound MGBG (20), is a powerful inhibitor of diamine oxidase.

Simultaneous addition of MGBG and methylornithine (36) or DFMO (21) into activated lymphocyte cultures, produced an additive antiproliferative action.

Sequential Use of DFMO and MGBG

Even though the strikingly enhanced transport of exogenous polyamines into cells severly depleted of natural polyamines odviously is a mechanism designed for the correction of the intracellular polyamine deprivation and hence to restore normal growth, this phenomenon can be utilized as to obtain just the opposite result. During polyamine depletion, not only are the natural polyamines transported into the cell at greatly accelerated rate, but also MGBG's uptake is stimulated to the same extent (3) by virtue of the fact that the drug uses the same transport system as do polyamines. These two compounds, DFMO and MGBG, being chemically unrelated and intervening entirely different reactions of the polyamine biosynthetic pathway, produce a unique synergism that is most probably based on the altered pharmacokinetics of MGBG, produced by DFMO, rather than a combined inhibition of polyamine biosynthesis.

Under cell culture conditions, a sequential use of DFMO and MGBG produces a profound and rapid antiproliferative effect that appears to be irreversible (47). In every instance the synergistic antiproliferative effect produced by a sequential combination of DFMO and MGBG seems to be related to the enhanced accumulation of MGBG in response to the previous treatment with DFMO. Thus the final cell death is in all likelihood based on the cytocidal action of high intracellular MGBG concentrations, which may not act only through alterations of polyamine metabolism. What is then the rationale of the sequential administration of DFMO and MGBG ? Would it be equivalent to a higher dose of MGBG also automatically resulting in more pronounced host toxicity ? In fact, there is some experimental evidence indicating that the priming effect of DFMO is not just equivalent to a higher MGBG dose. The DFMO-induced enhanced uptake of MGBG is directly related to the extent of putrescine and spermidine depletion produced by the inhibitor of ornithine decarboxylase (23). DFMO, in turn, has quantitatively different effects, as regards the polyamine deprivation, in different tissues (10), and, as it appears, the extent of polyamine depletion produced by DFMO is critically dependent on the growth rate of the target cells (23,47).

We hope that Figs.1 and 2 could be considered as supportive experimental evidence indicating that a prior treatment with DFMO enhances the uptake of MGBG preferentially into rapidly proliferating cells, especially into malignant cells of tumor-bearing animals. In Fig. 1, mice, inoculated with Ehrlich ascites carcinoma, had been primed with DFMO (in drinking water) whereafter MGBG was administered. As seen, DFMO only slightly and insignificantly stimulated the MGBG uptake by liver and small intestine, but more than doubled the cellular concentration of the latter drug in the carcinoma cells.

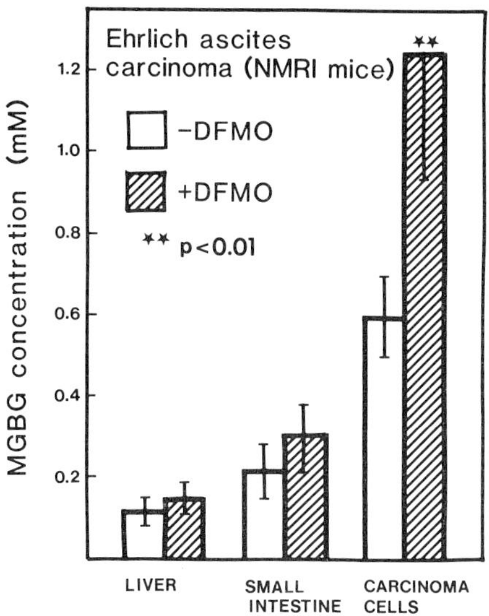

FIG. 1. Effect of DFMO priming on tissue MGBG accumulation in Ehrlich ascites carcinoma-bearing mice. Data from (48).

The picture is even more dramatic when L1210 leukemia-bearing mice were first primed with DFMO (starting on the day of inoculation) and then treated with MGBG. The uptake of MGBG, which was only slightly more rapid in the leukemia cells than in intestinal tissue, was strikingly stimulated in the leukemic cells in response to the treatment with DFMO (Fig. 2). DFMO did not enhance the uptake of MGBG in the normal tissues, including liver (where the uptake was actually decreased by DFMO), bone marrow and small intestine, but did so in spleen. It is, however, well known that L1210 cells, when inoculated intraperitoneally, extremely rapidly infiltrate speen. The view that the enhanced accumulation of MGBG in response to DFMO in spleen was caused by invading leukemia cells was supported by the finding showing that no stimulation of MGBG accumulation was seen in spleens of mice not inoculated with L1210 leukemia cells. In both of these cases, the combined sequential use of DFMO and MGBG also resulted in a more pronounced antiproliferative action not achievable by either of the drugs alone.

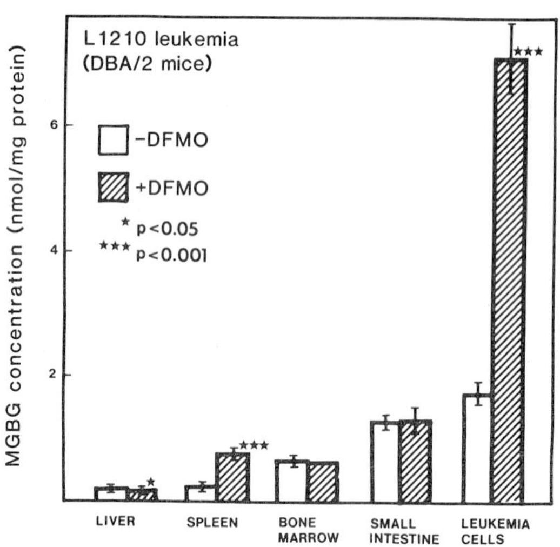

FIG. 2. Effect of DFMO priming on tissue accumulation of MGBG in L1210 leukemia-bearing mice.

Accordingly, the sequential use of DFMO and MGBG produces an enhanced and selective accumulation of the latter drug in malignant cells resulting in a more pronounced antiproliferative action than exerted by MGBG alone. Moreover, when DFMO was added to various MGBG regimens, the life span of L1210-bearing mice was substantially prolonged over that achieved with MGBG as single agent (5,7).

However, although the priming concept is pharmacokinetically sound and even shows pharmacodynamic advantages regarding the tumor selectivity, many, mainly host-dependent, factors still exist that greatly complicate the chemotherapeutic use of polyamine antimetabolites. In fact, there are reports indicating that a combination of MGBG (5) or its close derivative MBAG (41) with DFMO even abolishes the growth-inhibitory action exerted by DFMO on a murine mammary sarcoma. Interestinly, the diguanidines also appeared to correct the DFMO-induced putrescine and spermidine depletion (5,41). This highly unexpected observation has no apparent explanation; both the biosynthetic decarboxylases have been blocked, yet putrescine and spermidine accumulate as in the absence of the inhibitors. One of the compensatory mechanisms triggered by polyamine depletion is the enhanced uptake of exogenous polyamines by the polyamine-deprived cells. In fact, the phenomenon of enhanced amine uptake is not limited to DFMO (3), but as we recently found (4), also treatment with MGBG alone induces the polyamine transport system. The uptake system is not only utilized by the natural polyamines, but a variety amine compounds (46), not necessarily found in animal cells, are transported through the cell membrane via the same transporter. The latter compounds include cadaverine which is occasionally found in murine tissues. Moreover, putrescine, spermidine and spermine can be virtually totally replaced by cadaverine and its aminopropyl derivatives, and this new polyamine pattern still maintains slow cellular growth (2).

Lets have a closer look at the practical implications of the enhanced amine uptake during treatment with DFMO and MGBG. Fig. 3 presents results from an experiment, in which the DFMO priming concept has been applied, side by side, in cell cultures and in leukemia-bearing mice. In both cases (L1210 cells growing in culture or inoculated in mice), the treatments involved DFMO and MGBG as single agents or in combination. As regards DFMO, this drug produced roughly similar depletion of putrescine and spermidine in the tumor cells growing either in culture or in the peritoneal cavity of mice (Fig. 3). Practically the only difference was that small amounts of aminopropylcadaverine (not shown) appeared in response to DFMO in the inoculated L1210 cells. MGBG effectively reduced spermidine and spermine content in the cultured cells, but was practically inactive in this respect in inoculated leukemia cells (Fig. 3). Substantial amounts of cadaverine now appeared in the latter cells. The most striking difference was seen after the combined treatment. In cultured cells, the profound putrescine and spermidine depletion produced by DFMO persisted when MGBG was added to the regimen (Fig. 3). However, in inoculated cells, the putrescine and spermidine depletion, developed in response to the previous DFMO treatment, was virtually abolished upon the administration of MGBG.

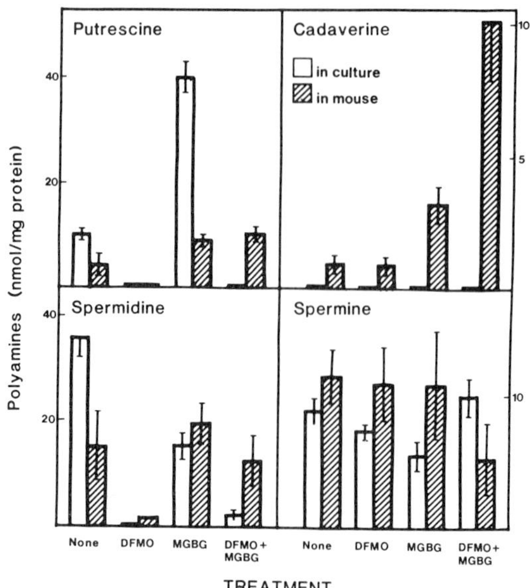

FIG. 3. Effect of DFMO, MGBG or their combination on the polyamine pattern of L1210 leukemia cells grown in culture or inoculated into mice. The cells or the animals were first treated with DFMO for 3 days (3 mM or 3% in the drinking water) whereafter MGBG was added (5 µM or 50 mg/kg/d). The cells were harvested on day 4 (cultured cells) or day 5 (inoculated cells).

One may also notice that after the combined treatment, cadaverine now represented a major intracellular amine (Fig. 3). In both systems, the priming with DFMO resulted in comparable enhancement of MGBG accumulation in the tumor cells (results not shown).

It is thus possible, even probable, that the enhanced uptake of extracellular polyamines during the combined treatment with DFMO and MGBG, rather than any leakage of the biosynthetic block, was responsible for the normalization of cellular polyamine pattern and for the appearance of cadaverine based polyamines. In other words, in some animal species (mouse seems to be an exceptional cadaverine producer) the enhanced MGBG-dependent cytocidal action (polyamine-dependent or independent), as resulted from the priming concept, is compromized by the disappearance of the polyamine depletion-dependent cytostatic action exerted by DFMO. Whether this is a disadvantage or an advantage is extremely difficult to judge on present experimental evidence. Full experimental exploration, including human patients, is still awaited. The fact, however, remains that if tumor specificity is achieved by the combined treatment, the mechanism of MGBG action (polyamine-dependent or polyamine-independent) is of secondary importance. The action of the drug is related to the metabolism or functions of the natural polyamines at least by in the sence that it uses or "misuses" the polyamine transport system.

USE OF POLYAMINE ANTIMETABOLITES IN HYPERPROLIFERATIVE MOUSE SKIN

In addition to various neoplasms, the probable potential use of polyamine antimetabolites includes certain skin conditions, most notably psoriasis. This is a nonmalignant skin disease that has much in common with neoplasms. Psoriasis is characterized by hyperproliferative epidermal growth. In the therapy of psoriasis, many of the principles of cytostatic treatment are used, yet an entirely different risk benefit evaluation is implied; psoriasis, unlike malignant tumors, seldomly is life-threatening. Thus the systemic use of any polyamine antimetabolite, especially of MGBG, may not be justified in uncomplicated psoriasis. This simply means that if used for the treatment of skin diseases, topical formulas of the drugs have to be available. This requirement, however, creates a major chemical problem; all the polyamine antimetabolites are polar, some of them extremely polar as strongly cationic MGBG.

UV-Irradiated Mouse Skin as Model for Epidermal Proliferation

Ultraviolet irradiation of hairless mouse skin not only enhances epidermal proliferation (as indicated by increased mitosis, stimulated DNA synthesis etc.), but also produces an early stimulation of ornithine decarboxylase activity (29) followed by marked elevation of epidermal putrescine and spermidine levels (45). The rise of putrescine concentration precedes and that of spermidine occurs concomitantly with the stimulation epidermal DNA synthesis (45).

Thus UV-induced epidermal proliferation can be used as a convenient model for the screening of polyamine antimebolites as antiproliferative agents in hyperproliferative disorders of skin.

Systemic and Topical Use of Polyamine Antimetabolites in Mouse Skin

Seiler and Knödgen (45) found that the high epidermal concentration of putrescine induced by UV-irradiation could be effectively lowered with systemic or topical DFMO. Despite the prevention of irradiation-induced changes in polyamine biosynthesis the drug did not abolish the UV-induced enhancement of DNA synthesis (45).

After the establishment of the DFMO priming concept in tumor-bearing animals, it was logic to test this concept also in mouse skin model. Using the sequential treatment schedule we found that either systemic or topical DFMO treatment greatly lowered epidermal putrescine and spermidine levels and enhanced the accumulation of systemically administered MGBG (28). However, as discussed earlier, the clinical applicability of the use of systemically administered MGBG (but possibly not of the much better tolerated DFMO) in psoriasis is questionable owing to the toxicity of the compound. Thus all-topical treatment schedule for DFMO and MGBG is not only desirable, but a necessity for succesful clinical use of these polyamine antimetabolites. The problem of poor percutaneous absorption of strongly charged MGBG was partly overcome by using diacetate, instead of dihydrochoride, salt of the compound with suitable commercial cream base (27).

As shown in Fig. 4A, readily detectable epidermal concentrations of MGBG were obtained also after topical administration of the drug onto UV-irradiated mouse skin. Moreover, a preceding topical application of DFMO for a few days strikingly enhanced the epidermal accumulation of subsequently applied MGBG (Fig. 4A).

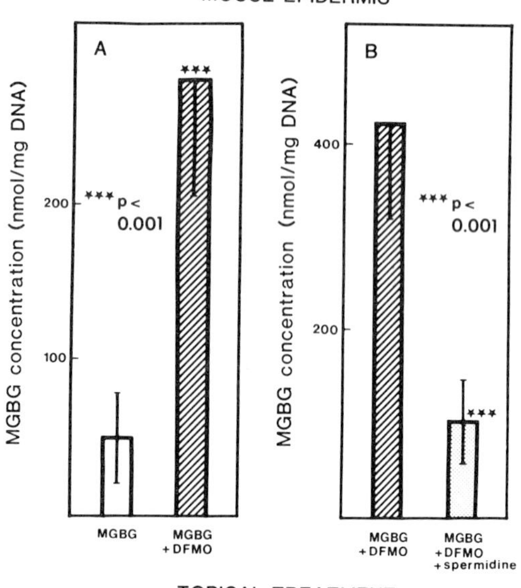

FIG. 4. A. Effect of topical DFMO (20%) on the accumulation of topical MGBG (50mM) in UV-irradiated mouse skin. B. Reduction of epidermal MGBG concentration by topical spermidine (50mM). Data obtained from (27).

As shown in Fig. 4B, topical spermidine can be used "a rescuing means", i.e. high epidermal concentrations of MGBG can be effectively lowered by applications of spermidine ointment. This possibility may be of importance in case of MGBG-induced skin toxicity (providing that spermidine itself does not cause cutaneous irritation).

The effect of topical DFMO and MGBG and their combination on epidermal macromolecular synthesis in UV-irradiated mouse skin is depicted in Fig. 5. DFMO alone depressed thymidine incorporation, yet this difference failed to reach the level of statistical significance. On the other hand, the incoporation of both leucine and histidine into protein was significatly inhibited (Fig. 5). The incorporation of histidine into epidermal proteins largely reflects the activity of keratinization (process of differentiation) in the granular layer of epidermis, whereas leucine incoporation represents more closely general protein synthesis in the lower epidermal layers (16).

MGBG significantly inhibited the incorporation of all precursors into their respective macromolecules, and the

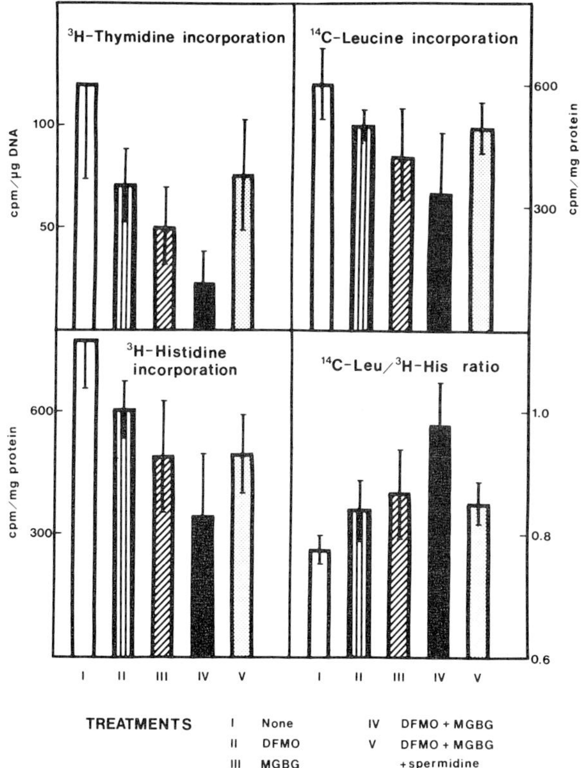

FIG. 5. Effect of topical DFMO and MGBG on the epidermal macromolecule synthesis and the reversal of antimetabolite-induced depression by topical spermidine. Treatment details as in Fig. 4A and B.

combined treatment resulted in more than an additive effect, at least on DNA synthesis (Fig. 5). The ratio of leucine incorporation to histidine incorporation was changed in response to the polyamine antimetabolites in a manner suggesting that the process of keratinization may have been inhibited more than the general protein synthesis.

It is also clearly shown in Fig. 5 that the inhibitory effect exerted by polyamine antimetabolites on macromolecular synthesis in proliferating mouse epidermis could be reversed, at least partly, by topical spermidine. The latter finding further supports the feasibility of the spermidine rescue concept.

It thus appears that topical treatment of hyperproliferative skin diseases with polyamine antimetabolites is an approach worth of further development.

CLINICAL RELEVANCE OF POLYAMINE ANTIMETABOLITES

Use of Polyamine Antimetabolites in the Treatment of Cancer

The main purpose of this chapter is to highlight the experimental approaches of the use of polyamine antimetabolites in different animal models and therefore the clinical activities, as reflected by recent studies in man, are only briefly summarized.

As regards the use of MGBG as single agent in various types of human cancer, the weekly treatment schedule (instead of earlier daily regimens) recently introduced by the Southwest Oncology Group in USA (26), has been widely accepted, and the usefullness the compound is currently being reevaluated. Although most of the recent clinical activity seems to be concentrated to the treatment of solid tumors (for ref. see 23), really promising results are so far obtained from the treatment of lymphomas (52). The empirically designed weekly (26) or even biweekly (52) treatment schedules are fully supported, at least in leukemia, by pharmacokinetic studies involving the determination of MGBG in the leukemic cells (50).

The published information concerning DFMO, as single agent, in the treatment of human malignancies is non-existing. DFMO has been used in a clinical trial for the treatment of nonsuppurative chronic prostatitis, yielding therapeutic results that are not especially impressive (12). The latter study, however, demonstrated the remarkable tolerability of DFMO (for an antiproliferative agent) also in man.

The combined (sequential) use of DFMO and MGBG has so far limited to one pilot clinical trial, primarily designed as a pharmacokinetic study, in childhood leukemia (51). Some single case reports also exist (23). At least pharmacokinetically the priming concept is clinically applicable and also showed therapeutic synergism (23,51). More clinical trials are obviously needed to evaluate the value of this combination.

Use of Polyamine Antimetabolites in the Treatment of Psoriasis

Topical DFMO has been used in the treatment of psoriasis in small scale clinical trials (17,19). Even though a distinct epidermal polyamine depletion was achieved, the therapeutic efficacy was only modest (17,19).

Topical MGBG likewise has shown activity against psoriasis, yet provoking cu taneus irritation preceding the clearing of psoriatic lesions (53).

The published clinical data so far obtained, does not include any combined use of DFMO and MGBG in psoriasis.

FUTURE TRENDS

Although sufficient experimental evidence appears to exist justifying an embarkation to large scale clinical trials with polyamine antimetabolites, much has still to be done at purely experimental level before the therapeutic use of polyamine antimetabolites is fully optimized. One of the critical approaches is the development of a systemic "polyamine rescue concept" that can be empoyed in case of unacceptable host toxicity, especially in connection of MGBG use.

More systematic studies are likewise needed to disclose and possibly also to circumvent all host- or tumor-dependent compensatory mechanisms produced by the treatment with various polyamine antimetabolites.

The combination of polyamine antimetabolites with conventional anti-cancer drugs is an area that requires systematic exploration and may have enormous therapeutic potential.

Apart from these clinical applications and clinically oriented experimental studies, there still is room for basic studies, especially as regards the physiological functions of the polyamines. Let us refer back to the beginning of this chapter and repeat the crucial question: "What do the polyamines do ?"

ACKNOWLEDGEMENTS

The excellent technical assistance of Ms. Riitta Sinervirta, Ms. Merja Roivainen and Ms. Raija Laine at all stages of the experimental work is gratefully acknowledged. The authors are also grateful to Ms. Päivi Jänne for her secretarial help. Studies of the authors' laboratory have received financial support from the National Research Council for Natural Sciences (Academy of Finland) and from the Finnish Foundation for Cancer Research.

REFERENCES

1. Alhonen-Hongisto, L., Kallio, A., Pösö, H. and Jänne, J. (1979): Acta Chem. Scand., B33:559-566.
2. Alhonen-Hongisto, L., Seppänen, P., Hölttä, E. and Jänne, J. (1982): Biochem. Biophys. Res. Commun. 106:291-297.
3. Alhonen-Hongisto, L., Seppänen, P. and Jänne, J. (1980): Biochem. J., 192:941-945.
4. Alhonen-Hongisto, L., Seppänen, P. and Jänne, J. (1982): FEBS Lett., in press.
5. Bartholeyns, J. and Koch-Weser, J. (1981): Cancer Res., 41:5158-5161.
6. Bartholeyns, J., Prakash, N.J., Schechter, P., Mamont, P., Fozard, J.R. and Koch-Weser, J. (1981): Abstr. 3rd NCI-EORTC Symposium on New Drugs in Cancer Therapy, 77.(abstr.).
7. Burchenal, J.H., Lokys, L., Smith, R., Cartmell, S. and Warrell, R. (1981): Proc. Am. Ass. Cancer Res., 22:204.(abstr.).
8. Cohen, S. S. (1978): Nature, 274:209-210.
9. Corti, A., Dave, C., Williams-Ashman, H. G. and Mihich, E. (1974): Biochem. J., 139:351-357.
10. Danzin, C., Jung, M. J., Grove, J. and Bey, P. (1979): Life Sci., 24: 519-524.
11. Dave, C. and Caballes, L. (1973): Fedn. Proc., 32:736.(abstr.)
12. Dunzendorfer, U. (1981): Arzneim.-Forsch., 31:382-385.
13. Fillingame, R. H., Jorstad, C. M. and Morris, D. R. (1975): Proc. Natl. Acad. Sci (USA), 72:4042-4045.
14. Fortmeyer, H. P., Bartholeyns, J., Grove, J. and Schechter, P. (1981): Abstr. 3rd NCI-EORTC Symposium on New Drugs in Cancer Therapy, 78.(abstr.).
15. Freedlander, B. L. and French, F. A. (1958): Cancer Res., 18:360-363.
16. Fukuyama, K. and Epstein, W.L. (1966): J. Invest. Dermatol., 47:551-560.
17. Grosshans, E., Henry, M., Tell, G., Schechter, P., Böhlen, P., Grove, J. and Koch-Weser, J. (1980): Ann. Dermatol. Venereol. (Paris), 107:377-387.
18. Heby, O. and Jänne, J. (1981): in: Polyamines in Biology and Medicine, edited by D.R. Morris and L.J. Marton, pp.243-310. Marcel Dekker, Inc. New York.
19. Henry, M., Juillard, J., Tell, G., Schechter, P.J.S., Grove, J., Koch-Weser, J. and Grosshans, E. (1981): Br. J. Dermatol., 105, Suppl. 20: 33-34.
20. Hölttä, E., Hannonen, P., Pispa, J. and Jänne, J. (1973): Biochem. J., 136:669-676.
21. Hölttä, E., Jänne, J. and Hovi, T. (1979): Biochem. J., 178:109-117.
22. Jänne, J., Alhonen-Hongisto, L., Seppänen, P. and Hölttä, E. (1981): in: Adv. Polyamine Res., vol. 3, edited by C.M. Caldarera, V. Zappia and U. Bachrach, pp.85-95. Raven Press, New York.
23. Jänne, J., Alhonen-Hongisto, L., Seppänen, P. and Siimes, M. (1981): Med. Biol., 59:448-457.
24. Kato, Y., Inoue, H., Gohda, E., Tamada, F. and Takeda, Y. (1976): Gann 67:569-576.
25. Kay, J. E. and Pegg, A. E. (1973): FEBS Lett., 29:301-304.

26. Knight III, W.A., Livingston, R.B., Fabian, C. and Costanzi, J. (1979): Cancer Treat. Rep., 63:1933-1937.
27. Käpyaho, K., Lauharanta, J. and Jänne, J. (1982): Br. J. Dermatol., in press.
28. Käpyaho, K., Linnamaa, K. and Jänne, J. (1982): J. Invest. Dermatol., 78:391-394.
29. Lowe, N., Verma, A. K. and Boutwell, R. K. (1978): J. Invest. Dermatol., 71:417-418.
30. Mamont, P. S., Böhlen, P., McCann, P. P., Bey, P., Schuber, F. and Tardif, C. (1976): Proc. Natl. Acad. Sci. (USA), 73:1626-1630.
31. Mandel, J.-L. and Flintoff, W.F. (1978): J. Cell. Physiol., 97:335-344.
32. McCann, P.P., Tardif, C., Duchesne, M.-C. and Mamont, P.S. (1977): Biochem. Biophys. Res. Commun., 76:893-899.
33. Metcalf, B.W., Bey, P., Danzin, C., Jung, M.J., Casara, J. and Vevert, J.P. (1978): J. Am. Chem. Soc., 100:2551-2553.
34. Mihich, E. (1963): Cancer Res., 23:1375-1389.
35. Mihich, E. (1975): Handbook of Experimental Pharmacology, vol. 37, pp. 766-788. Springer Verlag, New York.
36. Morris, D. R., Jorstad, C. M. and Seyfried, C. E. (1977): Cancer Res., 37:3169-3172.
37. Otani, S., Mizoguchi, Y., Matsui, I. and Morisawa, S. (1974): Molec. Biol. Rep., 1:431-436.
38. Pegg, A. E. and McGill, S. M. (1978): Biochem. Pharmacol., 27:1625-1629.
39. Porter, C.W., Dave, C. and Mihich, E. (1981): in: Polyamines in Biology and Medicine, edited by D.R. Morris and L.J. Marton, pp.407-436. Marcel Dekker, Inc. New York.
40. Prakash, N. J., Schechter, P. J., Grove, J. and Koch-Weser, J. (1978): Cancer Res., 38:3059-3062.
41. Prakash, N.K., Schechter, P.J., Mamont, P.S., Grove, J., Koch-Weser, J. and Sjoerdsma, A. (1980): Life Sci., 26:181-194.
42. Pösö, H. and Jänne, J. (1976): Biochem. Biophys. Res. Commun., 69:885-892.
43. Pösö,H. and Jänne, J. (1976): Biochem. J., 158:485-488.
44. Rosenblum, M. G., Keating, M. J., Yap, B. S. and Loo, T.L. (1981): Cancer Res., 41:1748-1750.
45. Seiler, N. and Knödgen, B. (1979): Biochem. Medicine, 21:168-181.
46. Seppänen, P. (1981): Acta Chem Scand., B35:731-736.
47. Seppänen, P., Alhonen-Hongisto, L. and Jänne, J. (1981): Eur. J. Biochem., 118:571-576.
48. Seppänen, P., Alhonen-Hongisto, L. and Jänne, J. (1981): Biochim. Biophys. Acta, 674:169-177.
49. Seppänen, P., Alhonen-Hongisto, L., Pösö, H. and Jänne, J. (1980): FEBS Lett., 111:99-103.
50. Seppänen, P., Alhonen-Hongisto, L., Siimes, M. and Jänne, J. (1980): Int. J. Cancer, 26:571-576.
51. Siimes, M., Seppänen, P., Alhonen-Hongisto, L. and Jänne, J. (1981):Int. J. Cancer, 28:567-570.
52. Warrell, R.P., Lee, B.J., Kempin, S.J., Lacher, M.J., Straus, D.J. and Young, C.W. (1981): Blood, 57:1011-1014.

53. Weinstein, G.D., McCullough, J.L., Eaglstein, W.H., Golub, A., Cornell, R.C., Stoughton R.B., Clendenning, W., Zackheim, H., Maibach, H., Kulp, K.R., King,L., Baden, H.P., Taylor, J.S. and Deneau, D.D. (1981): Arch. Dermatol., 117:388-393.
54. Wiegand, L. and Pegg, A. E. (1978): Biochim. Biophys. Acta, 517:169-180.
55. Williams-Ashman, H. G. and Schenone, A. (1972): Biochem. Biophys. Res. Commun., 46:288-295.

Advances in Polyamine Research, Vol. 4, edited by
U. Bachrach, A. Kaye, and R. Chayen. Raven Press,
New York © 1983.

Effects of Polyamine Depletion on the Cytotoxicity of Cancer Chemotherapeutic Agents

*†Laurence J. Marton, *Stina M. Oredsson, *David T. Hung, and **Dennis F. Deen

**Brain Tumor Research Center of the Department of Neurological Surgery, and the Departments of †Laboratory Medicine and **Radiation Oncology, School of Medicine, University of California, San Francisco, California 94143*

The effect of polyamines on the structure and stability of cell-free DNA is well-documented. Polyamines stabilize DNA to enzymatic degradation (3), denaturation by X rays (6), and thermal denaturation (55). Using viscoelastometry, we have shown that DNA from 9L rat brain tumor cells depleted of polyamines is either changed in conformation or is considerably more susceptible to shear than DNA from untreated cells (unpublished results). X-ray diffraction studies suggest that primary and secondary amine groups of spermidine and spermine bind ionically to adjacent phosphate groups on one strand of DNA, while the four-carbon chain stretches across the minor groove of the double helix to form a crossbridge between phosphate groups on opposite strands (28). In contrast, Bloomfield and Wilson (5) postulate that polyamine-mediated stabilization of DNA results from relatively nonspecific electrostatic interactions between polyanionic DNA and cationic polyamines. In Z DNA, the structure of which has been solved recently (57), it has been shown that spermine is located not only adjacent to phosphate groups but also adjacent to DNA bases (A. Rich, personal communication).

Many cancer therapeutic agents including radiation and drugs act directly on DNA. Because of the availability of a variety of agents capable of inhibiting polyamine biosynthesis and depleting cells of polyamines (20), we investigated the effect of polyamine depletion on the cytotoxicity of cancer therapeutic agents.

X-IRRADIATION

Our initial study showed that polyamine depletion of 9L rat brain tumor cells in vitro with α-methylornithine, a competitive inhibitor of ornithine decarboxylase (ODC) (the enzyme that converts ornithine to putrescine (Pu)), α-difluoromethylornithine (DFMO), an enzyme-activated, irreversible inhibitor of ODC, methylglyoxal bis(guanylhydrazone) (MGBG), an inhibitor of S-adenosylmethionine (AdoMet) decarboxylase (an enzyme involved in the synthesis of both spermidine (Sd) and spermine (Sp)), or a combination of DFMO and MGBG did not alter the dose response curve for X-irradiated cells compared to X rays alone (45). If DNA is destabilized by polyamine depletion, it was not more susceptible to X-ray damage, possibly because any change in structure would not affect the access to DNA of small hydroxyl radicals produced by X rays.

CHLOROETHYLNITROSOUREAS: IN VITRO

The cytotoxicity of 1,3-bis (2-chloroethyl)-1-nitrosourea (BCNU) against polyamine-depleted 9L cells was enhanced compared to untreated 9L cells (21). The potentiation was reversed when Pu was added back to the culture medium before BCNU treatment, which confirmed that the potentiation was polyamine-related.

In similar experiments we showed that the cytotoxicity of the chloroethylnitrosoureas (CENUs) 1-(2-chloroethyl)-3-trans-4-methylcyclohexyl-1-nitrosourea (MeCCNU) (37) and 2-[3-(2-chloroethyl)-3-nitrosoureido]-D-glucopyranose (chlorozotocin) (36, S.M. Oredsson, D.F. Deen, and L.J. Marton elsewhere in this volume) was also potentiated by DFMO treatment. The dose enhancement ratio for the potentiation of all three CENUs was approximately 1.3.

Most CENUs produce several decomposition products that can alkylate and carbamoylate biomolecules (9, 34, 59). We studied a series of nitrosoureas and found that DFMO pretreatment potentiates exclusively the cytotoxicity of CENUs that alkylate and subsequently form cross-links (36, S.M. Oredsson, D.F. Deen, and L.J. Marton elsewhere in this volume). We also studied DNA monoadduct formation in polyamine-depleted and control 9L cells treated with ^{14}C-MeCCNU; while we expected to find an increase in DFMO pretreated cells, monoadduct formation was the same in DFMO pretreated and control cells (37). The cytotoxicities of the strong alkylating agents N-ethyl-N-nitrosourea and N-methyl-N-nitrosourea, which form only monoadducts and no cross-links, were not potentiated by DFMO pretreatment (36, S.M. Oredsson, D.F. Deen, and L.J. Marton elsewhere in this volume), a fact consistent with the finding that monoadduct formation did not increase in MeCCNU treated cells. CENUs are thought to induce cell kill by alkylation of DNA and subsequent formation of DNA interstrand cross-links (13, 16, 23, 58). Our results show that even though the initial alkylation event is apparently unaffected by polyamine depletion, cross-linking apparently is.

DNA REPAIR ENZYMES

A number of enzymes have been reported to repair alkylation damage to DNA (8, 27, 31, 48). A correlation has been found between the ability to remove alkylation damage from DNA and resistance to alkylating agents (10, 13-15, 47, 49, 53). Inhibition of the activity of these enzymes might increase the cytotoxicity of alkylating agents. We studied the effect of DFMO treatment on O^6-methylguanine-DNA demethylase, 7-methylguanine-DNA glycosylase, and 3-methyladenine-DNA glycosylase in 9L cells in vitro, and found that polyamine depletion did not effect the activity of these enzymes (37).

DECARBOXYLATED ADOMET

When Pu production is inhibited in several cell lines, including 9L cells, by treatment with DFMO, there is a striking accumulation of decarboxylated AdoMet (29, 37, 39). Addition of Pu to the culture medium replenishes intracellular polyamine levels and suppresses decarboxylated AdoMet levels. Pegg et al. (39) have suggested that decarboxylated

AdoMet is a weak methylating agent and that the accumulation of high intracellular levels of this compound might contribute to potentiation of CENU cytotoxicity caused by DFMO. However, DFMO treatment alone and the accumulation of decarboxylated AdoMet is nontoxic to 9L cells in monolayer culture. Thus, if decarboxylated AdoMet contributes to CENU cytotoxicity in DFMO-treated cells, it must be a synergistic interaction.

MGBG-BCNU

To further investigate decarboxylated AdoMet's possible contribution to CENU cytoxocity, we studied the effect of MGBG-induced polyamine depletion in 9L cells (22). MGBG treatment does not cause decarboxylated AdoMet to accumulate but does reduce intracellular Sd and Sp levels significantly. Not only did MGBG pretreatment potentiate BCNU cytotoxicity in 9L cells, the DER of 1.3 is virtually identical to that found for DFMO pretreatment. While it is possible that MGBG and DFMO potentiate BCNU cytotoxicity by different mechanisms, this is probably not the case. The DER's are the same and the potentiation phenomenon for both inhibitors is at least partially reversible by the addition of exogenous polyamines. Although DFMO-mediated potentiation of cytotoxicity can be totally reversed by the addition of Pu, it should be noted that Sd administered to MGBG-treated 9L cultures is somewhat cytotoxic even in the presence of the oxidase inhibitor aminoguanidine, and intracellular polyamine levels cannot be restored sufficiently to reverse potentiation totally.

DFMO-BCNU CELL CYCLE AGE RESPONSE

Even though BCNU is classified as a cell cycle nonspecific agent, it kills more cells in G_1- and G_2-phase than in S-phase (4). If DFMO potentiated the cytotoxicity of BCNU only in S-phase, it would be impractical to use this combination against slow growing tumors because few cells traverse S-phase at any given time. However, we found that DFMO pretreatment of 9L cells in vitro potentiates the cytotoxic effect of BCNU in all phases of the cell cycle, particularly in G_1- and G_2/M-phases (4). Addition of Pu to the cultures reversed DFMO potentiation throughout the cell cycle.

DFMO-BCNU IN VIVO

We studied the in vivo potentiation of BCNU cytotoxicity by DFMO pretreatment in the murine glioma 26 and the 9L rat intracerebral tumors (32). Administration of DFMO in the drinking water of experimental animals for a number of days before intraperitoneal injection of a single dose of BCNU produced a significant increase in survival compared to untreated animals. We also found that administration of DFMO after a single dose of BCNU significantly increased survival. This was expected because DFMO should slow tumor growth and delay repopulation of the tumor after BCNU treatment. However, we were not able to show conclusively that combining DFMO's ability to potentiate BCNU cytotoxicity and delay repopulation of the tumor, by treating animals with DFMO both before and after an intraperitoneal dose of BCNU, was additive in terms of survival.

DFMO-BCNU AGAINST 9L SPHEROIDS

The effects of multiple courses of DFMO-BCNU on tumor growth must be evaluated. Such studies are virtually impossible to conduct in monolayer culture and extremely difficult to conduct in animal models. With the 9L spheroid model (11), which can be used for long-term studies of drug effects, we found that DFMO alone produces a minimal but reproducible cytotoxic effect (unpublished results), in contrast to the total lack of cytotoxicity for 9L monolayer cells. The mechanism of this cytotoxicity is unclear. It is possible that polyamine depletion may interfere with the ability of spheroid cells to adhere to one another, which would disrupt the integrity of the spheroid structure. However, potentiation of BCNU cytotoxocity in 9L spheroids is very similar to that seen in 9L monolayer culture. When DFMO-BCNU treatment is given in two courses to 9L spheroids, growth is delayed dramatically.

CENU-RESISTANT CELLS

A frequent cause of treatment failure in cancer patients is the development of drug resistance (19, 44). CENU-resistant 9L cell lines are available (44); we studied the effect of DFMO pretreatment on BCNU cytotoxicity in a 9L line resistant to CENUs (38). DFMO did not potentiate BCNU cytotoxicity in resistant cells even though DFMO depletes polyamine levels to similar extents in both CENU-resistant and sensitive 9L cells.

Intracellular Pu levels measured during unperturbed growth of 9L CENU-sensitive and resistant lines are different (38). Sensitive 9L cells have elevated levels of Pu during exponential growth, which decrease as growth slows. In contrast, intracellular levels of Pu in resistant cells increase continuously during growth in vitro. Intracellular Pu levels in a BCNU-resistant human glioblastoma cell line also increase continuously during growth in vitro, and DFMO pretreatment did not increase the cytotoxicity of BCNU against these cells (unpublished results). Moreover, cultures of other human glioblastoma cells that were sensitive to BCNU did not have a continuous increase in Pu levels, and pretreatment with DFMO sensitized these cells to BCNU. Although DFMO pretreatment does not overcome resistance to BCNU, the fact that other human brain tumor cell lines are potentiated is encouraging.

CIS-PLATINUM

It is of obvious importance to investigate the possibility that the cytotoxicity of chemotherapeutic agents other than CENUs might be affected by pretreatment with DFMO. We found that the cytotoxicity of cis-diamminedichloroplatinum (II) (cis-platinum) is decreased in 9L cells pretreated with DFMO (35). Platinum complexes interact with DNA by monofunctional binding to bases or by bifunctional cross-linking to bases in the helix on the same or opposite strands (1, 12, 17, 40-42). Cytotoxic lesions induced by cis-platinum are probably the result of DNA cross-linking (18). The two chlorine leaving groups of cis-platinum are separated by 3.3 Å (26, 33), which is close to the interplanar distance (3.4 Å) between DNA bases (25); a number of nucleophilic groups in native DNA are separated by 3.4 Å (17, 30, 43, 51). Presumably, platinum-DNA complexes are formed when these nucleophilic groups displace cis-platinum chlorines and either intra- or interstrand links are formed. An alteration in DNA structure caused by intracellular polyamine depletion possibly increases the distance between and the spatial configuration of nucleophiles, which could make the cross-linking reaction with cis-platinum mechanistically unfavorable. Thus, the results of cytotoxic treatment of cells in which DNA has been destabilized by polyamine depletion may be agent-dependent and may have different but important implications for a variety of drugs that act on DNA.

SISTER CHROMATID EXCHANGE

If the increased or decreased drug-induced cytotoxicities caused by DFMO are indeed the result of DNA destabilization secondary to polyamine depletion, then changes might be detectable on a chromosomal level. The sister chromatid exchange (SCE) assay is a sensitive method for the measurement of damage to chromosomes (50); induction of SCEs by compounds that damage DNA has been clearly established (60). We examined the ability of DFMO to modify the induction of SCEs in 9L monolayer cultures treated with BCNU and cis-platinum (54).

Treatment with DFMO alone did not affect the number of SCEs, while BCNU increased SCEs compared to control levels. In cells pretreated with DFMO and then treated with BCNU, the number of SCEs increased approximately two-fold over the number of SCEs in cells treated with BCNU alone. Addition of Pu to DFMO-pretreated cells before BCNU treatment reversed the increase in SCEs caused by DFMO.

Cis-platinum efficiently induces SCEs in 9L cells. However, if DFMO treatment preceded cis-platinum administration, SCEs were reduced compared to treatment with cis-platinum alone; this effect was also eliminated by addition of Pu.

OTHER CHEMOTHERAPEUTIC AGENTS

A number of chemotherapeutic agents intercalate DNA, and polyamine depletion-related DNA destabilization might effect interactions of these agents. 5, 11-Dimethyl-(6H)-pyrido[4,3-b]carbazole (ellipticine) is an intercalating agent (24) undergoing clinical investigation. We found that the cytotoxicity of this agent was totally unaffected by polyamine depletion of 9L cells.

Bleomycins are chemotherapeutic agents that incorporate polyamines or polyamine analogues as part of their structure (56). It is thought that this part of their structure participates in the attachment of bleomycins to DNA, after which other interactions cause cytotoxicity (7). Bacchi et al. (2) reported that the ability of bleomycin to cure trypanosomal infections in mice is greatly enhanced by small doses of DFMO. Because of these facts, we investigated bleomycin cytotoxicity to DFMO pre-treated 9L monolayer cells and found no effect.

Sunkara et al. (52) showed that HeLa cells treated with DFMO and cytosine arabinoside (ara-C) were killed more effectively than HeLa cells treated with ara-C alone. They found that DFMO blocked HeLa cells in S-phase, where the cytotoxic effects of ara-C, an S-phase specific agent, were most pronounced. We found that DFMO pretreatment of 9L cells significantly decreases the cytotoxicity of ara-C; this effect is reversible by Pu (unpublished results). We have previously shown that 9L cells are not blocked specifically in any phase of the cell cycle, although there might be some tendency towards G_1-phase accumulation (46). However, DFMO slows the growth of 9L cells. Because of this growth inhibitory effect, fewer 9L cells are traversing S-phase at any given time, and the cytotoxicity of ara-C is reduced. Treating 9L cells with DFMO and methotrexate produced similar results (unpublished data.) Differences in the results of treating of 9L and HeLa cells with the same combination of drugs point to the importance of investigating multiple model systems to identify possible contrasting modes of action, particularly with regard to cell cycle kinetic phenomena.

CONCLUSIONS

Polyamine biosynthesis inhibitors are currently undergoing investigation as biological response modifiers for use in cancer chemotherapy. Laboratory studies are of obvious importance at this time to determine optimum combinations of and administration schedules for antineoplastic drugs and polyamine biosynthesis inhibitors. These studies may also lead to a more complete understanding of the function of polyamines and the mode of action of a variety of anticancer drugs.

ACKNOWLEDGMENTS

Supported by ACS Grant RD-137, NIH Program Project Grant CA-13525, the Morris Stulsaft Foundation, and travel grants (to S.M.O.) from the Swedish Natural Science Research Council (R-RA 4685-100) and the Swedish Medical Research Council (B81-04R-6065-504106065). We thank Neil Buckley for expert editorial assistance.

REFERENCES

1. Alvarez, M.V., Cobreros, G., Heras, A., and Lopez Zumel, C. (1978): Br. J. Cancer, 37:68-72.
2. Bacchi, C.J., Nathan, H.C., Hutner, S.H., McCann, P.P., and Sjoerdsma, A. Biochem. Pharmacol. (in press).
3. Bachrach, U., and Eilon, G. (1969): Biochim. Biophys. Acta, 179:494-496.
4. Bjerkvig, R., Oredsson, S.M., Marton, L.J., Linden, M., and Deen, D.F.: Submitted.
5. Bloomfield, V.A., and Wilson R.W. (1981): In: Polyamines in Biology and Medicine, edited by D.R. Morris and L.J. Marton, pp. 183-206. Marcel Dekker, New York.
6. Brown, P.E. (1968): Radiat. Res., 34:24-35.
7. Burger, R.M., Peisach, J., and Horwitz, S.B. (1981): Life Sci., 28:715-727.
8. Cathcart, R., and Goldthwait, D.A. (1981): Biochemistry, 20:273-280.

9. Cheng, C.J., Fujimura, S., Grunberger, D., and Weinstein, I.B. (1972): Cancer Res., 32:22-27.
10. Day, R.S., Ziolkowsky, C.H.S., Scudiero, D.A., Meyer, S.A., Lubiniecki, A.S., Girardi, A.J., Galloway, S.M., and Bynum, G.D. (1980): Nature (Lond.), 288:724-727.
11. Deen, D.F., Hoshino, T., Williams, M.E., Muraoka, I., Knebel, K.D., and Barker, M. (1980): J. Natl. Cancer Inst., 64:1373-1382.
12. Douple, E.B., and Richmond, R.C. (1979): Int. J. Radiat. Oncol. Biol. Phys., 5:1335-1339.
13. Erickson, L.C., Bradley, M.O., Ducore, J.M., Ewig, R.A.G., and Kohn, K.W. (1980): Proc. Natl. Acad. Sci. (USA), 77:467-471.
14. Erickson, L.C., Laurent, G., Sharkey, N.A., and Kohn, K.W. (1980): Nature (Lond.), 288:727-729.
15. Erickson, L.C., Osieka, R., and Kohn, K.W. (1978): Cancer Res., 38:802-808.
16. Ewig, R.A.G., and Kohn, K.W. (1978): Cancer Res., 38:3197-3203.
17. Goodgame, D.M.L., Jeeves, I., Phillips, F.L., and Skapski, A.C. (1975): Biochim. Biophys. Acta, 378:153-157.
18. Harder, H.C., and Rosenberg, B. (1970): Int. J. Cancer, 6:207-216.
19. Harrarp, K.R., and Jackson, R.C. (1978): Antibiot. Chemother., 23:228-237.
20. Heby, O., and Jänne, J. (1981): In: Polyamines in Biology and Medicine, edited by D.R. Morris and L.J. Marton, pp. 243-310. Marcel Dekker, New York.
21. Hung, D.T., Deen, D.F., Seidenfeld, J., and Marton, LJ. (1981): Cancer Res., 41:2783-2785.
22. Hung, D.T., Oredsson, S.M., Deen, D.F., and Marton, L.J.: Submitted.
23. Kohn, K.W. (1977): Cancer Res., 37:1450-1454.
24. Kohn, K.W., Waring, M.J., Glaubiger, D., and Friedman, C.A. (1975): Cancer Res., 35:71-76.
25. Langridge, R., Wilson, L.D., Hooper, C.W., Wilkins, M.H.F., and Hamilton, L.D. (1960): J. Mol. Biol., 2:19-37.
26. Leh, F.K.V., and Wolf, W. (1976): J. Pharm. Sci., 65:315-328.
27. Lindahl, T. (1982): Ann. Rev. Biochem., 51:61-87.
28. Liquori, A.M., Constantino, L., Crescenzi, V., Elia, V., Giglio, E., Puliti, R., DeSantis Savino, M., and Vitagliano, V. (1967): J. Mol. Biol., 24:113-122.
29. Mamont, P.S., Danzin, C., and Wagner, J.: In: Proceedings of the Transmethylation Meeting, edited by E. Usdin, R.T. Borslordt, and C.R. Crevling, in press. MacMillan, New York.
30. Mansy, S., Rosenberg, G., and Thomson, A.J. (1973): J. Am. Chem. Soc., 95:1633-1640.
31. Margison, G.P., and Pegg, A.E. (1981): Proc. Natl. Acad. Sci. (USA), 78:861-865.
32. Marton, L.J., Levin, V.A., Hervatin, S.J., Koch-Weser, J., McCann, P.P., and Sjoerdsma, A. (1981): Cancer Res., 41:4426-4431.
33. Milburn, G.H.W., and Truter, M.R. (1966): J. Chem. Soc. (Lond.), (A), 1609-1616.
34. Montgomery, J.A., James, R., McCaleb, G.S., and Johnston, T.P. (1967): J. Med. Chem., 10:668-674.

35. Oredsson, S.M., Deen, D.F., and Marton, L.J. (1982): Cancer Res., 42:1296-1299.
36. Oredsson, S.M., Deen, D.F., and Marton, L.J.: Submitted.
37. Oredsson, S.M., Pegg, A.E., Deen, D.F., and Marton, L.J.: Submitted.
38. Oredsson, S.M., Tofilon, P.J., Deen, D.F., Rosenblum, M.L., and Marton, L.J.: Submitted.
39. Pegg, A.E., Poso, H., Shuttleworth, K., and Bennett, R.A. (1982): Biochem. J., 202:519-526.
40. Roberts, J.J., and Pascoe, J.M. (1972): Nature (Lond.), 235:282-284.
41. Roberts, J.J., and Thomson, A.J. (1979): Prog. Nucleic Acid Res. Mol. Biol., 22:71-133.
42. Roos, I.A.G. (1977): Chem. Biol. Interact., 16:39-55.
43. Roos, I.A.G., Thomson, A.J., and Mansy, S. (1974): J. Am. Chem. Soc., 96:6484-6491.
44. Rosenblum, M.L., Deen, D.F., Levin, V.A., Dougherty, D.V., Williams, M.E., Weizsaecker, M., Gerosa, M., and Wilson, C.B.: Modern Neurosurg., In Press.
45. Seidenfeld, J., Deen, D.F., and Marton, L.J. (1980): Int. J. Radiat. Biol., 38:223-230.
46. Seidenfeld, J., Gray, J.W., and Marton, L.J. (1981): Exp. Cell Res., 131:209-216.
47. Shiloh, Y., and Becker, Y. (1981): Cancer Res., 41:5114-5120.
48. Singer, B., and Brent, T.P. (1981): Proc. Natl. Acad. Sci. (USA), 78:856-860.
49. Sklar, R., and Strauss, B. (1981): Nature (Lond.), 289:417-419.
50. Solomon, E., and Bobrow, M. (1975): Mut. Res., 30:273-277.
51. Stone, P.J., Kelman, A.D., and Sinex, F.M. (1976): J. Mol. Biol., 104:793-801.
52. Sunkara, P.S., Fowler, S.K., Nishioka, K., and Rao, P.N. (1980): Biochem. Biophys. Res. Comm., 95:423-430.
53. Thomas, C.B., Osieka, R., and Kohn, K.W. (1978): Cancer Res., 38:2448-2454.
54. Tofilon, P.J., Oredsson, S.M., Deen, D.F., and Marton, L.J.: Submitted.
55. Tsuboi, M. (1964): Bull. Chem. Soc. Jpn., 37:1514-1522.
56. Umezawa, H. (1974): Fed. Proc., 33:2296-2302.
57. Wang, A.H.-J., Quigley, G.J., Kolpak, F.J., Crawford, J.L., van Boom, J.H., van der Marel, G., and Rich, A. (1979): Nature (Lond.), 282:680-686.
58. Wheeler, G.P. (1975): In: Antineoplastic and Immunosuppressive Agents. Handbook of Experimental Pharmacology, Vol. 38 (2), edited by A.C. Sartorelli and D.G. Johns, pp. 65-79. Springer-Verlag, Berlin.
59. Wheeler, G.P. (1976): Am. Chem. Soc. Symp. Ser., 30:87-119.
60. Wolff, S. (1977): Ann. Rev. Genet., 11:183-201.

Advances in Polyamine Research, Vol. 4, edited by U. Bachrach, A. Kaye, and R. Chayen. Raven Press, New York © 1983.

The Influence of Polyamine Depletion on Alkylation- and Carbamoylation-Induced Cytotoxicity

*Stina M. Oredsson, *†Dennis F. Deen, and **Laurence J. Marton

**Brain Tumor Research Center of the Department of Neurological Surgery, and the Departments of †Radiation Oncology and **Laboratory Medicine, School of Medicine, University of California, San Francisco, California 94143*

The nitrosoureas are cancer chemotherapeutic agents used to treat a variety of human tumors (6). Under physiological conditions, nitrosoureas decompose to produce two major classes of products, isocyanates and carbonium ions (7, 21). Although isocyanates react in several ways with macromolecules (3, 10, 12, 15, 16, 25, 29, 33), the primary reaction being carbamoylation of proteins, their role in nitrosourea-induced cytotoxicity is not understood. The very reactive carbonium ion intermediates alkylate available nucleophiles in both nucleic acids and proteins (7, 21, 31, 32); the cytotoxicity of many nitrosoureas is thought to be the result of these alkylation reactions. The cytotoxicity of chloroethylnitrosoureas probably is the result of a sequence of reactions in which an initial alkylation of DNA by chloroalkyl carbonium ion intermediates is followed by nucleophilic displacement of a primary chlorine group to form DNA interstrand or DNA-protein cross-links (8, 11, 17, 31).

EFFECTS OF POLYAMINE DEPLETION ON CYTOTOXICITY

We have shown that the in vitro cytotoxicities of the chloroethylnitrosoureas 1,3-bis (2-chloroethyl)-1-nitrosourea (BCNU) (13) and 1-(2-chloroethyl)-3-trans-4-methylcylohexyl-1-nitrosourea (MeCCNU) (23), agents that alkylate and carbamoylate (32), are significantly increased in 9L rat brain tumor cells depleted of polyamines by pretreatment with a nontoxic dose of α-difluoromethylornithine (DFMO), an enzyme-activated, irreversible inhibitor of ornithine decarboxylase (19). Because addition of putrescine to DFMO pretreated cells before treatment with BCNU or MeCCNU reversed this effect, polyamines are implicated in this enhancement. The stabilization of DNA by polyamines is well-documented (2, 4, 5, 30). Although we have postulated that the depletion of polyamine levels caused by DFMO leads to the destabilization of DNA and an increased susceptibility of nucleophilic sites to initial alkylation and formation of cross-links than conformationally stable DNA (23), carbamoylation reactions cannot be excluded from the mechanisms of enhancement.

To evaluate the relative importance of alkylation and carbamoylation in DFMO-mediated increased cytotoxicity, we compared the in vitro cytotoxicities against 9L cells of four nitrosoureas that decompose to give several reactive intermediates. 2-[3-(2-chloroethyl)-3-nitrosoureido]-D-glucopyranose (chlorozotocin) is an alkylating agent (8, 10, 32) and a potent antitumor agent (1). However, because of intramolecular inactivation of the isocyanate, this agent does not carbamoylate (20). Chlorozotocin can cause DNA-interstrand and DNA-protein cross-links and DNA strand breaks (8, 10); 1, 3-bis(trans-4-hydroxy-cyclohexyl)-1-nitrosourea (BHCNU) carbamoylates only (32); N-ethyl-N-nitrosourea (ENU) and N-methyl-N-nitrosourea (NMU) alkylate DNA oxygen and nitrogen, although the degree to which they alkylate these sites differs (18, 24, 26-28). The latter two compounds ethylate and methylate, respectively, and therefore cannot form cross-links; however, both cause a greater frequency of DNA strand breaks than chloroethylnitrosoureas (8, 10). NMU and ENU decompose to give isocyanates that can carbamoylate intracellular molecules (14, 32).

Chlorozotocin, the only cross-linking agent studied here, was significantly more cytotoxic to untreated 9L cells than the other three compounds; DFMO-mediated polyamine depletion potentiated the cytotoxicity of chlorozotocin but not of the other compounds (22). The dose enhancement ratio of 1.3 found for chlorozotocin in these studies is the same as found for BCNU (13) and MeCCNU (23) in DFMO pretreated 9L cells. In contrast, the cytotoxicity of BHCNU, which carbamoylates only, was reduced in polyamine-depleted 9L cells.

The cytotoxicities of ENU and NMU were not affected by DFMO pretreatment (22). Erickson et al. (9) showed that a dose of NMU that is equitoxic to a dose of chlorozotocin produced almost 90 times the number of apparent DNA strand breaks. Both ENU and NMU are monoalkylating agents incapable of cross-linking. Thus DFMO pretreatment does not appear to potentiate the cytotoxicity cause by either DNA strand breaks or by monoalkylation. Our study of MeCCNU, which clearly showed that monoadduct formation was not increased by DFMO pretreatment, supports the second half of this conclusion (23).

DNA is considered to be the probable major target for the cytotoxic action of alkylating agents. Because the availability of a second leaving group on the alkylating moiety greatly increases the cytotoxicity of alkylating agents, cross-linking must be an essential reaction. We found that the cytotoxicity of cross-linking agents is potentiated by DFMO-induced polyamine depletion, while agents that do not form cross-links are either unaffected or made less cytotoxic by polyamine depletion.

ACKNOWLEDGMENTS

Supported by ACS Grant RD-137, NIH Program Project Grant CA-13525, the Morris Stulsaft Foundation, and travel grants (to S.M.O.) from the Swedish Natural Science Research Council (R-RA 4685-100) and the Swedish Medical Research Council (B81-04R-6065-504106065). We thank Neil Buckley for expert editorial assistance.

REFERENCES

1. Anderson, T., McMenamin, M.G., and Schein, P.S. (1975): Cancer Res., 35:761-765.
2. Bachrach, U., and Eilon, G. (1969): Biochim. Biophys. Acta, 179:494-496.
3. Baril, B.D., Baril, E.F., Laszlo, J., and Wheeler, G.P. (1975): Cancer

Res., 35:1-5.
4. Bloomfield, V.A., and Wilson, R.W. (1981): In: Polyamines in Biology and Medicine, edited by D.R. Morris and L.J. Marton, pp. 183-206. Marcel Dekker, New York.
5. Brown, P.E. (1968): Radiat. Res., 34:24-35.
6. Carter, S.K., Schabel, F.M., Jr., Broder, L.E., and Johnston, T.P. (1972): Advan. Cancer Res., 16:273-332.
7. Cheng, C.J., Fujimura, S., Grunberger, D., and Weinstein, I.B. (1972): Cancer Res., 32:22-27.
8. Erickson, L.C., Bradley, M.O., Ducore, J.M., Ewig, R.A.G., and Kohn, K.W. (1980): Proc. Natl. Acad. Sci. (USA), 77:467-471.
9. Erickson, L.C., Bradley, M.O., and Kohn, K.W. (1978): Cancer Res., 38:3379-3384.
10. Ewig, R.A.G., and Kohn, K.W. (1977): Cancer Res., 37:2114-2122.
11. Ewig, R.A.G., and Kohn, K.W. (1978): Cancer Res., 38:3197-3203.
12. Fornace, A.J., Jr., Kohn, K.W., and Kann, H.E., Jr. (1978): Cancer Res., 38:1064-1069.
13. Hung, D.T., Deen, D.F., Seidenfeld, J., and Marton, L.J. (1981): Cancer Res., 41:2783-2785.
14. Jump, D.B., Sudhakar, S., Tew, K.D., and Smulson, M. (1980): Chem. Biol. Interactions, 30:35-51.
15. Kann, H.E., Jr. (1978): Cancer Res., 38:2363-2366.
16. Kann, H.E., Jr., Schott, M.A., and Petkas, A. (1980): Cancer Res., 40:50-55.
17. Kohn, K.W. (1977): Cancer Res., 37:1450-1454.
18. Lawley, P.D., and Warren, W. (1976): Chem. Biol. Interactions, 12:211-220.
19. Metcalf, B.W., Bey, P., Danzin, C., Jung, M.J., Casara, P., and Vevert, J.P. (1978): J. Am. Chem. Soc., 100:2551-2553.
20. Montgomery, J.A. (1976): Cancer Treat. Rep., 60:651-664.
21. Montgomery, J.A., James, R., McCaleb, G.S., and Johnston, T.P. (1967): J. Med. Chem., 10:668-674;
22. Oredsson, S.M., Deen, D.F., and Marton, L.J.: Submitted.
23. Oredsson, S.M., Pegg, A.E., Deen, D.F., and Marton, L.J.: Submitted.
24. Pegg, A.E. (1977): Advan. Cancer Res., 25:195-269.
25. Schmall, B., Cheng, C.J., Fujimura, S., Gersten, N., Grunberger, D., and Weinstein, I.B. (1973): Cancer Res., 33:1921-1924.
26. Singer, B., Bodell, W.J., Cleaver, J.E., Thomas, G.H., Rajewsky, M.F., and Thon, W. (1978): Nature (Lond.), 276:85-88.
27. Singer, B., Spengel, S., and Bodell, W.J. (1981): Carcinogenesis, 2:1069-1073.
28. Sun, L., and Singer, B. (1973): Biochem., 12:1795-1802.
29. Tarantino, A., and Thompson, P. (1979): Cancer Biochem. Biophys., 4:33-36.
30. Tsuboi, M. (1964): Bull. Chem. Soc. Jpn. 37:1514-1522.
31. Wheeler, G.P. (1975): In: Antineoplastic and Immunosuppressive Agents. Handbook of Experimental Pharmacology, Vol. 38 (2), edited by A.C. Sartorelli and D.G. Johns, pp. 65-79. Springer-Verlag, Berlin.
32. Wheeler, G.P. (1976): Am. Chem. Soc. Symp. Ser., 30:87-119.
33. Wheeler, G.P., Bowdon, B.J., and Shuck, R.F. (1975): Cancer Res., 35:2974-2984.

Advances in Polyamine Research, Vol. 4, edited by
U. Bachrach, A. Kaye, and R. Chayen. Raven Press,
New York © 1983.

Polyamines as Biochemical Markers in Cancer

*Yoav Horn, †Stuart L. Beal, *Natalio Walach, §Warren P. Lubich, *Lina Spigel, and §Laurence J. Marton

**Department of Oncology, Assaf Harofeh Hospital, Sackler School of Medicine, Zerifin, Israel; †Department of Laboratory Medicine and the §Brain Tumor Research Center of the Department of Neurological Surgery, School of Medicine, University of California, San Francisco, California 94143*

In 1971 Russell et al. (8) showed that elevated polyamine levels were present in the urine of patients harboring a variety of tumors. The original presumption that urinary polyamine levels could be used to "screen" patients for the presence of tumor was never realized (3,5). Therefore, this assay was investigated as an adjunct to existing diagnostic methods, as a means of monitoring the long-term progression or regression of tumor, and as means of evaluating short-term efficacy of therapy.

A recent review of polyamines as cancer markers points out that despite the considerable effort that has gone into clinical studies since 1971, absolute clinical utility (defined as adding a new variable to existing diagnostic techniques or, at least, as replacing an older diagnostic method with a more efficient one) has been demonstrated thus far only for monitoring the recurrence of medulloblastoma (5), though the demonstration of clinical utility for other tumor types may be forthcoming (6).

Many published studies have not correlated clinical data (tumor staging, activity, therapy, etc.) completely with polyamine levels, and have not obtained serial samples from individual patients. In order to overcome these shortcomings, we initiated stdies (4) using a more controlled clinical setting and a statistical analysis that takes into consideration random biological interpatient variability in polyamine levels and that accounts for serial samples in the same patient (1,2,4,7).

PRELIMINARY STUDIES

All patients included in the study were evaluated initially with a physical examination, blood tests (CBC, urea, electrolytes liver function tests), relevant x-rays, and radionuclide scans; a 24-hour urine was collected for polyamine analysis. Chemotherapy, radiation therapy, or followup without therapy proceeded according to established protocols for the treatment of each disease. All patients were evaluated periodically (1-3 months) by the same medical team at Assaf Harofeh Hospital; followup examinations included routine tests and repeat urine collection for polyamine analysis. At each evaluation, based on all clinical and laboratory findings except for the polyamine assay, patients were judged to have either "active" or "nonactive disease.

A total of 503 polyamine determinations from 192 patients were evaluated. Patients with tumors of the breast, stomach, prostate, female genitalia, and metastatic carcinoma of unknown origin whose disease was classified as active had statistically significant increases in urine polyamine levels compared to patients with the same tumors whose disease was classified as inactive. No such increases in urine polyamine levels was found for patients with tumors of the testes, lung, head and neck, bladder, colon, and rectum. This may have been the result, in part, of the small number of patients with tumors of the testes, lung, bladder, and head and neck. In addition, most bladder tumors were superficial and noninvasive. However, patients with colorectal carcinoma appeared to have been adequately sampled.

Using the groups that demonstrated statistical significance between active and nonactive disease, we have initiated the long-term serial studies necessary to define true utility for the polyamine assay. Evaluation of patients by the same medical team over time is important in obtaining an unequivocal answer regarding this utility for each tumor type. Our initial impression is that such utility might be forthcoming when the assay is used to assist in answering the appropriate questions in the appropriate setting.

ACKNOWLEDGMENT

This work was supported in part by NIH Program Project Grant CA-13525.

REFERENCES

1. Beal, S., and Sheiner, L. (1980): Amer. Statist., 34:118-119.
2. Beal, S., and Sheiner, L.: NONMEM Users Guides Part I-III. Technical Report of the Division of Clinical Pharmacology, University of California, San Francisco.
3. Galen, R.S., and Gambino, S.R. (1975): Beyond Normality: The Predictive Value of Efficiency of Medical Diagnosis. John Wiley & Sons. Inc., New York.
4. Horn, Y., Beal, S.L., Walach, N., Lubich, W.P., Spigel, L., and Marton, L.J. (1982): Cancer Res., in press.
5. Marton, L.J., and Seidenfeld, I. (1981)In: Polyamines in Biology and Medicine, edited by D.R. Morris and L.J. Marton, pp. 337-347. Marcel Dekker, Inc., New York.
6. Milano, G., Cassuto, J.P., Schneider, M., Viguier, E., Lesbats, G., Cambon, P., and Lalanne, C.M. (1980): Nouv. Rev.

Fr. Hematol., 22:249-255.
7. Rao, C.R. (1965): Linear Statistical Inference and Its Applications. John Wiley & Sons, Inc., New York.
8. Russell, D.H., Levy, C.C., Schimpff, S.C., and Hawk, I.A. (1971): Cancer Res., 32:1555-1558.

Advances in Polyamine Research, Vol. 4, edited by U. Bachrach, A. Kaye, and R. Chayen. Raven Press, New York © 1983.

Polyamines and Histamine in Serum from Patients with Hematological Diseases

H. Desser, R. Waldner, W. Kläring, and D. Lutz

The Ludwig Boltzmann Institute for Leukemia Research and Hematology, the 3rd Medical Department, Hanusch Hospital, 1140 Vienna, and the Computer Center of the Veterinary Medical University, 1030 Vienna, Austria

The biosynthesis of the polyamines, spermidine (SPD) and spermine (SPM), and of the diamine putrescine (PUT) is not only closely linked with cell growth but also with cell differentiation (22). PUT is formed by the splitting off of CO_2 by the enzymic action of ornithine decarboxylase from the amino acid, ornithine. This diamine serves as an acceptor for a propylamine group, that is generated over several steps from the amino acid methionine, and produces SPD. A further transfer of a propylamine group results in the biosynthesis of SPM. Both these polyamines, as well as PUT showed intra- and extracellular regulatory influences on cell proliferation. High levels of polyamines were found in the urine, serum and plasma of patients with malignancies (1,22). In addition, high polyamine values in erythrocytes of tumor patients were found (25-27, 29), since these substances are bound to the cell surface (2,3,18,20). The importance of these substances as potential tumor markers and their value in the observation of disease progress during therapy has been shown in clinical studies and various animal models (11,13,20,23,24).

There is little known on the polyamine levels in serum of patients with hematological disorders. In part, contradictory results from patients with acute leukemia, malignant melanoma, pernicious anemia, sickle cell anemia, and polycythemia vera have been found, most likely due to varying methodology (4,10,15,17, 21,28). In contrast, the increase in histamine (HIM) concentration in whole blood of patients with chronic myelocytic leukemia (CML) has long been demonstrated (9).

In previous studies, the polyamine content in the body fluids was mainly determined after acidic hydrolysis that frees the covalently bound amines so that they appear in ionic form. Particulary urine samples were analysed, whereas serum and plasma samples were only occassionally examined. Measurement of the polyamines without releasing covalently bound forms, was made possible after the development and introduction of a sensitive fluorometric method. Thereafter, polyamine concentrations in serum and urine could be determined in the picomole range (14,19).

METHODS AND MATERIALS

140 patients were examined, 62 with CML, 23 with acute leukemia, 21 with malignant lymphoma, and the rest with different preleukemic states or myeloproliferative syndromes. Selection of patients was random. Due to differential diagnostic reasons, patients with CML and myeloproliferative syndromes were preferentially investigated. Most of the patients underwent multiple examinations. Sera of 38 healthy persons between the ages 18-61 were used as controls.

Details of sample preparation and analysis is explicitly described elsewhere (5). The serum was deproteinized with 5-sulfosalicylic acid, the supernatant was centrifuged and applied to a coloum chromatograph. The free ions of SPD, SPM, PUT, and HIM found in the extracts were separated on an adapted amino acid analyzer (Biotronik, München) equipped with an ion exchange resin. The quantative determination was achieved fluorometrically, after reaction of amines with o-phthalaldehyde (OPA), comparing peak height with an external standard.

STATISTICS (16)

Normal distribution was testd by the Kolmogorov-Smirnov method. Calculation of statistical differences between patient group and control population were made by the Mann-Whitney-Wilcoxon rank summation test, necessary due to the absence of normal distribution in some patient populations.Differences were accepted as significant if $p < 0.05$. Polyamine values of individual patients were regarded as abnormally elevated if they deviated from the mean of the control group by more than twice the standard deviation. All amines from the control population were found to have a normal distribution of their values, thus allowing the comparisons to be made with the mean in the case of the controls.

Statistical results for the patient groups are presented as the medians.

RESULTS

Table 1 shows the median and the range for the investigated amines for each disease group and for the controls. The results of the assessment for the existence of a normal distribution is also indicated. The percent frequency of the polyamine elevations versus the healthy control group is given in table 2.

The study of the polyamine profiles with regard to the various disease groups observed displays the following picture:

In untreated CML, SPM and HIM were characteristically high. In 8 out of 10 cases an icreased SPD value before therapy was also observed. Only one patient had an additional increased PUT value.

In the course of the investigation, a total of 30 patients were treated with dibrommanitol (Myelobromol) during the chronic phase of the disease. In this group only about two thirds of the cases showed higher than normal values in the HIM, SPM, and SPD levels. The reoccurrence of elevated levels in these 3 parameters after repeated testing throughout the duration of the disease was correlated with the persistance of the chronic phase. A markedly high HIM value in serum (> 100 nMol/l) was observed in 11 of these patients. In 4 of them, who either had additionally normal SPD and SPM values or a distinct, high PUT value, malignant transformation occured within a short time (blast crisis). Transition to the blast crisis could be determined within a few weeks in another 10 patients who had no increased HIM level during the chronic phase and who showed a decreasing HIM value after repeated measurement during the course of their disease. In contrast to untreated patients and to the chronic phase of CML, increases in the serum level of SPM, SPD and HIM during blast crisis were seldom.

An increase in SPM was seen in 8 out of 23 patients with acute leukemia and 6 cases exhibited higher SPD levels. All but one of these acute leukemia patients suffered from acute myelocytic leukemia. An elevated HIM value was determinded in 3 patients all suffering from acute lymphocytic leukemia. High SPM values were observed in 2 cases of acute monoblastic leukemia. Normal polyamine levels were seen in most of the 12 patients with preleukemic syndromes. Only 2 cases showed a slightly increased SPD value and 5 a minimal increase in the SPM value. In 2 cases PUT and HIM levels were elevated respectively. Distinct increases

TABLE 1. Median and range of serum polyamine concentrations

	CONTROLS	Myeloproliferative syndromes		Chronic myelocytic leukemia			Acute leukemia	Pre-leukemia	Malignant lymphoma	
		Osteomyelo-fibrosis and sclerosis	Poly-cythemia vera	Un-treated	Chronic phase	Blast crisis			Morbus Hodgkin	Non-Hodgkin
PUT	94.5(38) a 33-233 b *	113(15) 56-256 *	118(7) 53-301 *	46.1(10) 0-182 *	139(30) 24-363 *	100(22) 10-391 *	91.4(23) 0-817 --	62.4(12) 29-194 *	80.6(6) 32-346 *	72.8(15) 39-766 --
SPD	75.0(38) 28-191 *	176(15) 34-482 *	247(7) 185-507 *	204(10) 86-504 *	172(30) 39-2160 --	144(22) 38-606 *	113(23) 29-2098 --	66.2(12) 34-275 *	119(6) 49-215 *	101(15) 27-370 *
SPM	22.5(38) 0-128 *	137(15) 8.8-1780 *	147(7) 77-556 *	365(10) 189-980 *	135(30) 33-6110 --	309(22) 22-1170 *	65.7(23) 13-2760 --	43.3(12) 18-270 *	123(6) 10-350 *	51(15) 0-131 *
HIM	not found	4.1(15) 0-156 --	10(7) 0-179 *	128(10) 36-693 *	11(30) 0-1340 --	8.6(22) 0-315 --	0.79(23) 0-28 --	0.5(12) 0-10 --	0(6) -- --	0.77(15) 0-22 --

a: Median (number of cases) in nMol/l.
b: Minimum - maximum.
*: Normal distribution proven by method of Kolmogorov-Smirnov (16).
Underligned: significant difference ($p < 0.05$) versus control group; Mann-Whitney-Wilcoxcon-test (16).

TABLE 2. Percent frequency for polyamine elevations in patients versus normal controls

	Myeloproliferative syndromes		Chronic myelocytic leukemia			Acute leukemia	Pre-leukemia	Malignant lymphoma	
	Osteomyelo-fibrosis and sclerosis	Poly-cythemia vera	Un-treated	Chronic phase	Blast crisis			Morbus Hodgkin	Non-Hodgkin lymphoma
PUT	3/15 a (20) b	1/7 (14)	1/10 (10)	12/30 (40)	3/22 (14)	1/23 (4)	1/12 (8)	1/6 (17)	3/15 (20)
SPD	9/15 (60)	7/7 (100)	8/10 (80)	17/30 (57)	8/22 (36)	6/23 (26)	2/12 (17)	2/6 (33)	6/15 (40)
SPM	14/15 (93)	7/7 (100)	10/10 (100)	21/30 (70)	15/22 (68)	8/23 (36)	5/12 (42)	4/6 (67)	5/15 (33)
HIM	3/15 (20)	4/7 (57)	10/10 (100)	20/30 (67)	11/22 (50)	3/23 (14)	1/12 (8)	0/6 (0)	2/15 (13)

a: Number of patients with elevations / number of cases.
b: Percentage of patients with elevations.

in SPD (16 cases) and especially SPM (21 cases) were observed in 22 cases with myeloproliferative diseases. Over 50% of the cases with polycythemia vera displayed an additional high HIM level. Occassionally an increased PUT level was determined in osteomyelofibrosis (3 cases) and in polycythemia vera (1 case). Patients with malignant lymphoma had high SPM levels in 9 out of 21 cases and 8 had high SPD levels. Of these, patients with Morbus Hodgkin demonstrated especially high increases in their SPM levels. In 4 cases of all lymphoma patients' PUT was high and in 2 cases slight HIM elevations were encountered.

Regarding the behaviour of particular polyamines in the serum of all hematologically diseased patients studied PUT is found to be increased in only 20% of the cases. Abnormal values were observed most frequently in the chronic phase of CML, osteomyelofibrosis (OMF), and in maligant lymphoma. In these diseases the highest single values were also observed. PUT had the highest median value in patients with CML in the chronic phase. This value was found to be statistically significantly higher in comparison to the control group. The median value for PUT in untreated chronic myelocytic leukemia was the lowest in all groups. The statistical test showed it to be significantly lower than the PUT value of the control group.

More frequently, increased SPD and SPM values in the serum of hematologically diseased patients were detected, especially in all 7 patients with polycythemia vera. Equally we found increases in SPD in most of the patients with chronic myelocytic leukemia before treatment. During treatment only about 50% of the patients showed this increase. More than 50% of the cases with OMF also exhibited an increased SPD value. On the other hand, patients with acute leukemia, in blast crisis of CML, with malignant lymphoma, and especially with preleukemia demonstrated abnormally high values less frequently. The median of the serum SPD was highest in myeloproliferative syndromes and untreated CML. Next to polycythemia vera, untreated patients with CML and most OMF cases showed an abnormal elevation of SPM in their serum. In contrast, less than half of the cases with malignant lymphoma, acute leukemia and preleukemia displayed irregularly high SPM values. The median for SPM in CML before treatment and in blast crisis was significantly higher than in the other disease groups. The average SPD values of the myeloproliferative diseases and those of all three differentiated groups of CML were significantly elevated compared to the normal values. Significant differences were also calculated for the SPM values in these diseases, and furthermore in all lymphatic syndromes and leukemias studied.

HIM, which is undetectable in healthy persons, was most frequently found increased in patients with chronic myelocytic leukemia and polycythemia vera. Solely in some individual instances elevated serum histamine was seen in the other disease groups. The median HIM values for chronic myelocytic leukemia and polycythemia vera compared to normal were statistically distinguishable.

DISCUSSION

The biosynthesis and/or aquisition of polyamines is considered a functional requirement for cell proliferation. The secretion of accumulated polyamines to the surrounding body fluids makes it feasible to use the polyamine levels in serum as an indicator for enhanced cell growth. Obviously therefore, it was important to assess the use of polyamine determinations in the serum from patients with malignant diseases as tumor markers (22). Recent results for hematological diseases (4,10,15,17,20,21,28) have shown increased polyamine values for urine and serum of patients with pernicious anemia, chronic myelocytic leukemia, acute leukemia and malignant lymphoma. Patients in our department suffering from myeloproliferative syndromes and chronic myelocytic leukemia displayed increased SPD and SPM levels. Strikingly, polyamine levels of SPD and SPM in patients with acute transformation (blast crisis of CML) were less frequently elevated compared to the chronic phase of this disease. However, as we also encountered only rarely elevated SPD and SPM serum levels in acute leukemia it seems more likely that in cell systems of varying degrees of differentiation, biosynthesis and/or secretion of polyamines are quantitatively distinct.

Based on our results, it seems that the growth of myelocytic cells is coupled with a staggered polyamine synthesis. Abnormally high SPD values were found in three quarters of the cases of myeloproliferative syndromes or chronic myelocytic leukemia. Nearly all patients showed abnormally high SPM levels as well. Elevated polyamine values were less frequently observed under cytostatic treatment which inhibits growth of myelocytic cells. Though we found elevated serum polyamine levels in illnesses with increased growth rates such as acute leukemia and malignant lymphoma they appeared less frequently than in the myeloproliferative disorders (5,6,7).

As opposed to SPD and SPM, PUT was markedly less frequently increased in serum from patients with hematological illness. In our patients none of their hematological diseases under study was specially characterized by a high serum level of PUT. According to our experience we have the impression that patients with infiltration of the liver, independent of thir hematological illness, seem to have increased PUT values in serum more often than other patients. This preliminary impression is compatible with the occurrance of high PUT values found in patients with severe liver damage (post hepatitic, toxic) (8). Since PUT is an important intermediate of polyamine metabolism, it appears probable that the liver plays a predominant role in polyamine metabolism. The detailed relationship is still to be disclosed by specific experiments.

Even though HIM can not be considered to be a polyamine and no metabolic relationship exists, we have nonetheless measured HIM levels - mainly for methodological reasons - in the serum of our patients. HIM in whole blood has been described as a specific marker for chronic myelocytic leukemia. Based on our results, however, it appears that serum HIM is high in other diseases as well, especially in polycythemia vera. Therefore, the diagnostic importance of an elevated HIM level in chronic myelocytic leukemia must be questioned (9). Just as for the polyamines SPD and SPM, HIM was seldomly found to be elevated during treatment in the chronic phase,i.e. during blast crisis in comparison to untreated CML patients.

High serum polyamine concentrations have been observed in divers maladies, even in non-malignant diseases (4,8,20). Therefore, the determination of polyamine levels in serum does not allow a distinct diagnostic statement. Polyamine determination is of particular value in biochemically characterizing disease progress of the individual and in measuring the success or failure of applied therapy (15,22).

SUMMARY

Spermidine, spermine, putrescine, and histamine have been determined by a sensitive fluorometric method in serum samples from patients suffering from myeloproliferative diseases, chronic myelocytic leukeia, acute leukemia, preleukemia, or malignant lymphoma. In the observed groups spermidine concentrations were found to be elevated in 17% of the cases with preleukemia and 100% in polycythemia vera. Elevation of spermine concentrations ranged from 33% in the non-Hodgkin-lymphoma group to 100% in patients with polycythemia vera and untreated chronic myelocytic leukemia. Putrescine levels were rarely elevated, namely between 4% of the cases with acute leukemia and maximally in 40% of chemotherapy treated patients with chronic myelocytic leukemia. The histamine level was elevated especially in cases of chronic phase myelocytic leukemia (100%). In general, serum polyamine concentrations were found to be frequently elevated in the chronic phase of myelocytic leukemia, considerably more frequently than in acute transformation.

ACKNOWLEDGEMENTS

This study was made possible through financial support of the Austrian National Bank (Jubiläumsfondsprojekt Nr. 1057). We also thank Ms. Eva Malek and Mr. Herbert Nentwich for their excellent technical assistance.

REFERENCES

1. Bachrach, U. (1976): Ital. J. Biochem. 25: 77-93.
2. Chun, P.W., Rennert, O.M., Saffen, E.E., and Taylor, W.J. (1976): Biochem.Biophys.Res.Comm. 69:1095-1101.
3. Chun, P.W., Saffen, E.E., Bitore, R.J., Rennert, O.M., and Weinstein, N.H.(1977): Biophys. Chem. 6: 321-335.
4. Desser, H. (1980): In: Polyamines in Biomedical Research, edited by J.H. Gaugas, pp. 415-433.John Wiley, New York.
5. Desser, H., Lutz, D., Krieger, O., and Stierer, M. (1980): Oncology 37: 376-380.
6. Desser, H., Frass, M., Kuzmits, R., Aiginger, P., Müller, M.M., and Kläring, W.J. (1981): In: Adv. Polyamine Res. Vol.3, edited by C.M. Caldarera, V. Zappia, and U. Bachrach, pp. 431-440.
7. Desser, H., Kläring, W.J., Luger, T., and Gebhart, W. (1982): Onkologie 5: 36-41.
8. Desser, H., Kleinberger, G., and Kläring, J. (1981): J.Clin.Chem.Clin.Biochem. 19: 159-164.
9. Gingold, N. (1979): Onkologie 2: 70-74.
10. Hospatta, A.V., Advani, S.H., Vaidya, N.R. Electric, S.E., and Braganca, B.M. (1980): Int.J.Cancer 25: 463-466.
11. Jänne, J., Pösö, H., asnd Raina, A. (1978): Biochem Biophys.Acta 473: 241-293.

12. Lutz, D., Waldner, R., and Desser, H. (1982): in preparation.
13. Marton, L.J. (1978): In: Adv. Polyamine Res. Vol. 2, edited by R.A. Campbell, D.R. Morris, D. Bartos, G.D. Daves, and F. Bartos, pp. 257-263. Raven Press, New York.
14. Marton, L.J., and Lee, P.L.Y. (1975): Clin.Chem. 21: 1721-1724.
15. Milano, G., Cassuto, J.P., Schneider, Ml, Viguier, E., Lesbats, G., Cambon, P., and Lalanne, C.M. (1980): Nouv.Rev.Fr.Hematol. 22: 249-255.
16. Nie, N.H., Hull, C.H., and Jenkins, J.G. (1976): Statistical Package for the Social Sciences. MacGraw Hill, New York.
17. Nishioka, K., Ezaki, K., and Hart, J.S. (1980): Clin.Chim.Acta 107: 59-66.
18. Quash, G., Bonnefoy-Roch, A., Gazzolo, L., and Niveleau, A. (1978): In: Adv. Polyamine Res. Vol. 2, edited by R.A. Campbell, D.R. Morris, D. Bartos, G.D. Daves, and F. Bartos, pp. 85-92. Raven Press, New York.
19. Rennert, O.M., Frias, J., and Shukla, J.B. (1976): Texas Reports Biol. Med. 34: 187-197.
20. Rennert, O.M., and Shukla, J.B. (1978): In: Adv. Polyamine Res. Vol. 2, edited by R.A. Campbell, D.R. Morris, D. Bartos, G.D. Daves, and F. Bartos, pp. 195-211. Raven Press, New York.
21. Romano, M., Cecco, L., Cerra, M., Nontuori, R., and Derosa, C. (1980): Tumori 66: 677-687.
22. Russell, D.H., and Durie, B.G.M. (1978): In: Polyamines as Biochemical Markers of Normal and Malignant Growth, pp. 127-133. Raven Press, New York.
23. Russell, D.H., and Durie, B.G.M. (1978): In: Polyamines as Biochemiocal Markers of Normal and Malignant Growth. p. 134. Raven Press, New York.
24. Russell, D.H., Corney, W.B., Kovacs, C.J., Hopkins, H.A., Marton, L.J., Lebendre, S.M., and Morris, H.P. (1976): Cancer Res. 36: 420-423.
25. Savory, J., Shipe, J.R., and Wills, M.R. (1979): Lancet 2: 1136-1137.
26. Shipe, J.R., and Savory, J. (1980): Ann.Clin.Lab. Sci. 10: 128-136.
27. Takami, H., Romsdahl, M.M., and Nishioka, K. (1979): Lancet 2: 912.
28. Uehara, N., Kita, K., Shirakaw, S., Uchino, H., and Saeki, Y. (1980): Gann 71: 393-397.
29. Uehara, N., Shirakaw, S., Uchino, H., and Saeki, Y. (1980): Cancer 45: 108-111.

Advances in Polyamine Research, Vol. 4, edited by U. Bachrach, A. Kaye, and R. Chayen. Raven Press, New York © 1983.

Polyamine Excretion as Prognostic Marker in Radiologic Therapy of Breast Carcinoma in Metastatic Phase

M. Romano, L. Cecco, M. Cerra, and *A. De Matteis

*Department of Biochemistry and *Section of Endocrinology, Istituto dei Tumori, 80131 Naples, Italy*

The problem generally faced is whether the therapy (surgery, radiation or drugs) totally or partially eliminates the tumor. A biochemical test to estimate the state of the tumor before and during the therapy would be useful and render easier the task of monitoring the therapy. It has been suggested that polyamines may be markers of the disease activity and of the response following therapy (3). Reports have described the accumulation of polyamines in neoplastic tissue (8,10), in cells transformed by tumor viruses (1) or treated with tumor-promoting agents (9). An important observation is that polyamine levels in 24-h urine vary during therapy and may hence be reliable markers of the efficacity of the therapy (3,2,4,5,7).

Aim of this paper was to learn whether the levels of urinary polyamines in breast cancer are tumor markers and whether polyamine excretion during radiotherapy may be useful in evaluating the state of the disease.

METHODS

The methods used have already been reported (7).

RESULTS

Table 1 shows urinary polyamine excretion in normal control subjects and in patients with histologically proven ductal infiltrating breast carcinoma. The patients were divided into two groups: one with and the other without metastases. Patients' urine was analyzed before start of the therapy. The result showed that in the urine of non-metastatic patients the polyamine levels were not different from the controls

(30-50-year-old healthy women). In contrast, the levels in metastatic patients were significantly higher. The increase was observed in about 65% of the patients.

Table 1. URINARY POLYAMINES IN CONTROLS AND IN PATIENTS

	Polyamine content (μg/mg creatinine)					
	Putrescine		Spermidine		Spermine	
	Mean	SD	Mean	SD	Mean	SD
Controls:10	1.90	0.35	1.81	0.30	0.25	0.08
Patients:34						
Non-metastatic (17)	2.01	0.55	1.97	0.36	0.29	0.09
Metastatic (17)	3.78	1.86*	3.38	1.73*	0.72	0.79*

*Values statistically different from the controls at the 0.001 significance level.

The polyamine levels greatly varied ($Cv \geq 50\%$) in the metastatic group. This finding led to examining whether the variability might depend on metastase localization. This group was hence divided into two further groups: one with bone metastases and one with visceral metastases. The results (Table 2) indicated that putrescine significantly increased in both groups, whereas spermidine and spermine increased only in the group with visceral metastases.

Urinary polyamine patterns were determined in patients with bone metastases before and during the first week of radiotherapy. Fig. 1 reports the results observed in 3 patients (A-B-C) with different polyamine patterns at start of the therapy. "A" had normal polyamine values; "B" had a high putrescine and normal spermidine and spermine values; "C" had high putrescine, spermidine and spermine values. After the first dose of radiation "A" (with limited metastases) showed no substantial polyamine variations; "B" (with advanced disease) showed a significant fall of putrescine but no variations in spermidine and spermine; "C" (with advanced disease) showed an immediate increase of the three polyamines, followed by a slight decrease the next day, after which the levels were high throughout.

Radiologic checks revealed a partial regression in "A", a slight improvement in "B", a severe worsening in "C".

Table 2. URINARY POLYAMINES IN CONTROLS AND IN PATIENTS WITH BONE AND VISCERAL METASTASES

	Polyamine content (μg/mg creatinine)					
	Putrescine		Spermidine		Spermine	
	Mean	SD	Mean	SD	Mean	SD
Controls: 10	1.90	0.35	1.81	0.30	0.25	0.08
Metastatic patients: 17						
Bone (9)	3.83	2.37 (o)	2.33	0.61	0.39	0.18
Visceral (8)	3.73	1.22 (*)	4.56	1.85 (*)	1.08	1.06 (*)

Values statistically different from the controls at the 0.005 (o) and 0.001 (*) significance levels.

DISCUSSION

This initial clinical study leads to the conclusion that urinary polyamines do not increase in all breast-cancer patients but only in those with histologically proven advanced cancer, particularly in those with visceral metastases, having, as is known, a very rapid evolution. The results confirm that:
1) polyamine levels are closely linked to the tumor kinetics, hence high polyamine levels may indicate a rapid cell turnover and call for a more aggressive therapy; 2) the polyamine assay may reveal the presence of metastases and hence indicate the therapy required.

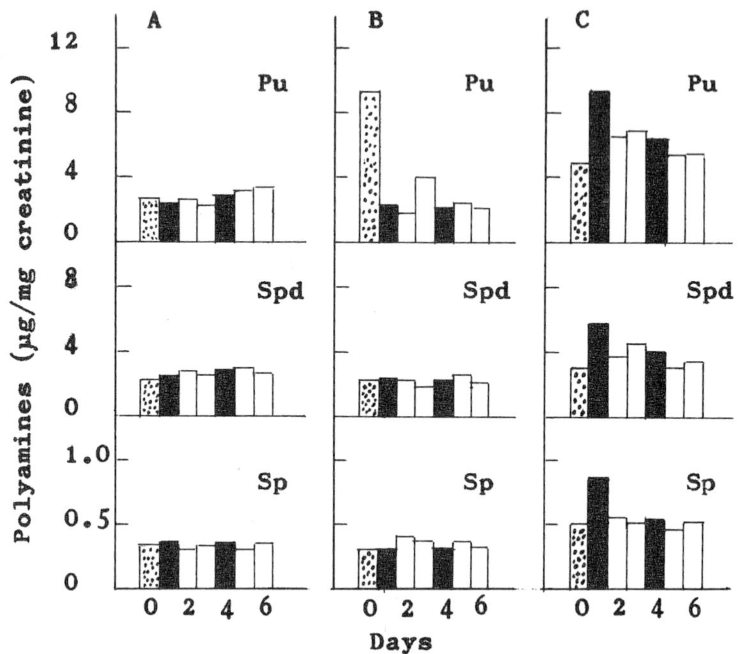

FIG.1. Urinary polyamine patterns in patients with bone metastases during radiotherapy (TCT; 400 RAD twice weekly). Polyamine levels: (dotted) before start of radiotherapy; (black) on the day of radiation and (white) on the 2nd and 3rd days after radiation. Pt=Putrescine; Spd=Spermidine; Sp=Spermine. TCT= telecobalttherapy.

Urinary putrescine may reflect the "cell growth fraction" of a tumor whereas spermidine the "cell death fraction" (8,6). If this is the case, the polyamines patterns observed at the start of the therapy in "B" and "C" should indicate that the tumor kinetics in "B" are different from those of "C". The high putrescine and normal spermidine levels in "B" suggest the presence of a tumor with a slow turnover (high "cell growth fraction" and low "cell death fraction"); the high putrescine and spermidine levels in "C" suggest the presence of a tumor with a rapid turnover (high "cell growth and death fractions").

The variations resulting from local radiation may

foretell that the therapy for "B" will be effective because putrescine dramatically decreases. This is confirmed by the observed regression of metastases and by a slight but generalized improvement of the patient's condition. As for "C", the therapy is not expected to be effective because the polyamine levels further increase (probably due to radiologic stimulation) and remain high throughout. This is confirmed by the observed severe worsening of the disease and the increase of metastase diffusion.

The behaviour of "A" indicates that polyamine assay cannot be useful in patients with limited metastases.

In conclusion, the data show that putrescine and spermidine in the urine of advanced breast cancer patients may be markers of tumor kinetics and useful markers of the radiation therapy.

The positive results obtained are an encouragment to further investigate the possibilities of polyamine assays in monitoring the therapy and evaluating the tumor kinetics.

ACKNOWLEDGMENTS

We are grateful to Mr. Esposito Domenico, Giovanni Menna and Antonio Granata for their competent technical assistence. We are also thankful to the surgeons and pathologists of the Istituto dei Tumori of Naples for the tissue collections and for histological diagnoses.

This research was partially supported by a grant from the Consiglio Nazionale delle Ricerche "Progetto Finalizzato Controllo della Crescita Neoplastica". Contract N° 81.01346. 96, Rome, Italy.

REFERENCES

1. Bachrach, U. (1976): Ital. J. Biochem., 25: 77-93.
2. Dreyfuss, F., Chayen, R., Dreyfuss, G., Dvir, R., and Ratan,J.(1975): Israel J. Med. Sci., 11:785-792.
3. Durie, B.G.M., Salmon, S.E., Russell, D.H. (1977): Cancer Res., 37: 214-221.
4. Fair, W.R., Wehner, N. and Brorsson, U. (1975): J. Urol., 114: 88-95.
5. Heby, O. and Russell, D.H. (1973): In: Polyamines in Normal and Neoplastic growth, edited by D.H. Russell, pp. 221-237. Raven Press, New York.
6. Heby, O., Andersson, G. (1978): Acta Path Microbiol. Scand., 86: 17-20.

7. Romano, M., Cecco, L., Cerra, M., Montuori, R. and De Rosa, C. (1980): Tumori, 66: 677-687.
8. Romano, M., Cecco, L. and Cerra, M. (1981): Tumori, 67: 431-435.
9. Russell, D.H., Durie, B.G.M. and Salmon, S.E. (1975): Lancet, 2: 797-800.
10. Russell, D.H. and Durie, B.G.M. (1978): Polyamines as biochemical markers of normal malignant growth. Raven Press, New York.

Advances in Polyamine Research, Vol. 4, edited by U. Bachrach, A. Kaye, and R. Chayen. Raven Press, New York © 1983.

Urinary Excretion of Spermine in Breast Cancer

*D. Chayen, R. Chayen, R. Dvir, S. Goldberg, A. Harell, *D. Lichtenstein, and *M. Stavorovsky

*Institute of Endocrinology and *Department of Surgery C, Medical Center and Sackler School of Medicine, Tel-Aviv 64239, Israel*

In recent years, there has been increasing interest in the search for biochemical markers of cancer. D.H. Russell (8,9) demonstrated in 1971 the increased excretion of polyamines in all kinds of cancers, and two major studies on breast cancer (3,10) have been concerned with polyamine excretion among many other possible markers. These latter studies indicated that urinary polyamines were infrequently elevated in cases of breast cancer.

We wish to report on the increased urinary excretion of spermine found in 50% of patients suffering from breast cancer. Evidence will be given in support of our belief that improvement in analytical techniques for the measurement of spermine may result in this marker being positive in an even larger percentage of cases.

PATIENTS AND METHODS

The women studied in this investigation were seen in the outpatient clinic for the early detection of breast cancer. The benign tumors were confirmed at biopsy and the malignant tumors at mastectomy. A 24-hour urine was collected before surgery, and was analysed for the polyamines putrescine, cadaverine, spermidine and spermine by our previously published method (4). In most cases, polyamine analysis was repeated one week after mastectomy. Labelled spermine (N,N'-bis-3(aminopropyl)-(1,4-^{14}C) tetramethylene-1,4-diamine tetrahydrochloride), from the Radiochemical Centre, Amersham, was added to urine prior to analysis in some cases, in order to check the recovery. The final stage of the analysis was thin-layer chromatography of dansyl-derivatives. When using labelled spermine, not only was the spot of dansyl-spermine eluted for counting, but also other areas of the chromatogram were counted.

RESULTS

Normal people excrete less than 5 mg putrescine, 2 mg cadaverine, 2 mg spermidine and 0.7 mg spermine per day. In a previous study of polyamine excretion (5), elevated spermine excretion in women was confined to cancer patients. In the present series of breast tumors, elevated spermine excretion was found in 15 out of 30 cases of breast cancer (Tables 1-4), but not in any of 30 patients with benign growths. Increased excretion of other polyamines in the breast cancer and benign growth series, respectively, were as follows:- putrescine, 6 and 1; cadaverine, 9 and 6; spermidine, 3 and 3 cases.

One week after mastectomy, elevated spermine excretions fell to normal in 7 out of 11 patients studied. Neither in these cases, nor in other pre-operative values of spermine excretion, was there any correlation with the presence of metastases in the axillary lymph nodes. There was also no correlation with the histological appearance of the tumor or its size.

Recovery of labelled spermine was usually in the range of 40-50%, in good agreement with our previous observations. However, in some of the cancer cases in which spermine values were normal, the recovery was much lower, in one case being less than 2%. Radioactivity in these samples was found at various points on the chromatogram.

DISCUSSION

The finding of elevated spermine excretion in 15 out of 30 (50%) women suffering from breast cancer is of interest, especially since those with benign tumors did not show this effect. The question that arises is not so much why there should be increased output of spermine as why spermine is not raised in all the cases. With regard to the latter question, it is known that the activity of the enzyme diamine oxidase is raised in certain tumors. These include medullary carcinoma of the thyroid gland (1), small-cell carcinoma of the lung (2) and ovarian cancer (6). Diamine oxidase degrades putrescine, and might therefore prevent the formation of polyamines in the latter part of the biosynthetic pathway (spermidine and spermine). More research needs to be performed in this direction, namely investigation of the diamine oxidase content of breast cancers.

The other obvious possibility for missing elevated spermine output is a fault in the analytical technique. We have noticed the dispersion of radioactivity from labelled spermine added to some urine samples from cancer patients. This was demonstrated in the thin-layer chromatogram of dansyl-polyamines, in which there was occasionally only a fraction of the radioactivity remaining in the region of dansyl-spermine, though considerable amounts in other parts of the chromatogram. This may be explained by breakdown of the spermine prior to dansylation or

TABLE 1. Polyamine excretion in breast carcinoma

Case Number	Lymph node metastases	PUT mg/24h	CAD mg/24h	SPD mg/24h	SPM mg/24h	Day of sample
1	+	11.1	trace	2.5	1.2	Before operation
		2.5	0.5	0.7	0.6	7 days after operation
2	+	2.0	0.4	0.6	0.2	Before operation
		0.2	0.4	0.6	trace	11 days after operation
3	+	1.2	0.1	0.8	0.3	Before operation
4	+	2.0	2.0	0.7	0.7	Before operation
5	-	2.1	0.7	1.6	0.4	Before operation
		4.6	2.0	4.2	0.6	10 days after operation
6	+	2.2	0.8	1.1	0.4	Before operation
		5.2	1.2	2.6	0.4	10 days after operation

TABLE 2. Polyamine excretion in duct carcinoma of the breast

Case Number	Lymph node metastases	PUT mg/24h	CAD mg/24h	SPD mg/24h	SPM mg/24h	Day of sample
1	-	1.9	0.7	0.7	1.5	Before operation
		3.2	trace	2.5	1.3	5 days after operation
2	+	3.0	trace	0.7	trace	Before operation
		5.5	1.7	1.5	1.5	10 days after operation
3	-	4.5	2.6	1.3	trace	Before operation
		3.5	2.1	1.0	trace	7 days after operation
4	+	2.1	1.5	1.4	0.9	Before operation
		4.8	1.5	1.5	0.7	12 days after operation
5	-	4.4	0.8	1.8	1.6	Before operation
		7.6	4.5	1.8	0.5	11 days after operation
6	+	7.2	14.0	1.8	1.0	Before operation
		5.7	67	2.1	0.9	10 days after operation
7	+	4.2	2.3	1.8	1.0	Before operation
		4.2	3.0	2.7	0.5	8 days after operation
8	-	5.1	2.7	1.7	0.8	Before operation
9	-	2.0	1.4	1.0	trace	Before operation

TABLE 3. Polyamine excretion in infiltrating carcinoma of the breast

Case Number	Lymph node metastases	PUT mg/24h	CAD mg/24h	SPD mg/24h	SPM mg/24h	Day of sample
1	-	1.6	10.0	0.5	0.4	Before operation
		6.1	42.0	2.5	1.0	8 days after operation
2	-	6.9	8.3	1.4	0.8	Before operation
		2.4	10.0	1.4	0.5	11 days after operation
3	+	3.7	1.8	1.3	0.5	Before operation
		3.0	0.9	1.0	0.2	7 days after operation
4	-	1.1	0.7	1.2	0.5	Before operation
		3.2	1.7	1.1	0.5	12 days after operation
5	-	1.7	0.6	1.3	trace	Before operation
6	-	2.4	0.8	1.7	0.8	Before operation
		2.1	0.9	0.7	0.7	10 days after operation

TABLE 4. Polyamine excretion in miscellaneous tumors of the breast

Tumor type	Lymph node metastases	PUT mg/24h	CAD mg/24h	SPD mg/24h	SPM mg/24h	Day of sample
Lobular Ca *in situ*	-	4.5	35	3.3	1.1	Before operation
		2.2	9.0	0.8	0.4	12 days after operation
Lobular Ca *in situ*	-	7.0	0.9	2.6	0.4	Before operation
Lobular Ca *in situ*	-	7.4	10.0	1.9	0.4	Before operation
Undifferentiated Ca	-	3.6	0.3	0.5	0.8	Before operation
		3.7	1.2	2.4	1.5	2 days after operation
Anaplastic Ca	+	0.4	trace	0.9	1.4	Before operation
		not measured	1.9	0.6	0.3	8 days after operation
Medullary Ca	-	2.8	0.8	1.2	0.2	Before operation
		not measured	0.8	1.0	0.7	8 days after operation
Medullary Ca	-	2.1	1.1	1.2	1.0	Before operation
Mucoid Ca	-	1.9	0.7	1.9	0.9	Before operation
		4.9	6.0	2.0	0.3	21 days after operation

of dansyl-spermine during chromatography, but may also be a function of multi-derivative (dansyl) formation. Whatever the reason for low recovery of spermine, the result obtained by measuring the fluorescence of the spot of dansyl-spermine can give only a minimum value for spermine excretion. Our finding of elevated output of spermine in 50% of breast cancer patients is therefore probably an underestimation of the true incidence.

Our findings appear to be at variance with those of Coombes et al (3) and of Tormey et al (10) and even with the more recent findings of Tormey et al (11). Both of these groups were unimpressed with the value of polyamines as markers for breast cancer. However, in the work of Coombes et al (3), spermine was not measured at all. The low incidence of high-spermine excretors in the results of Tormey et al (10,11) is more difficult to explain.

In conclusion, it is obvious that much work remains to be done on the metabolism of polyamines, especially in tumors, to explain why spermine in particular should appear in elevated amount in the urine of breast-cancer patients. Likewise, an improvement in analytical technique for the determination of spermine in biological fluids is also essential. Only then may it be possible to use excessive excretion of spermine as a marker for the detection of breast cancer.

ACKNOWLEDGEMENT

This work was supported in part by a grant from the Israel Cancer Association.

REFERENCES

1. Baylin, S.B., Beaven, M.A., Engelman, K., and Sjoerdsma, A. (1970) : New Engl. J. Med., 283:1239-1244.
2. Baylin, S.B., Abeloff, M.D., Wieman, K.C., Tomford, J.W., and Ettinger, D.S. (1975) : New Engl. J. Med.,293:1286-1290.
3. Coombes, R.C., Powles, T.J., Gazet, J.C., Ford, H.T., Sloane, J.P., Laurence, D.J.R., and Neville, A.M. (1977) : Lancet, 1:132-134.
4. Dreyfuss, G., Dvir, R., Harell, A., and Chayen, R. (1973) : Clin. Chim. Acta, 49:65-72.
5. Dreyfuss, F., Chayen, R., Dreyfuss, G., Dvir, R., and Ratan,J. (1975) : Israel J. Med. Sci., 11:785-795.
6. Lin, C.W., Orcutt, M.L., and Stolbach, L.L. (1975) : Cancer Res., 35:2762-2765.
7. Marton, L.J., Russell, D.H., and Levy, C.C. (1973) : Clin. Chem., 19:923-926.
8. Russell, D.H. (1971) : Nature (New Biology) 233:144-145.
9. Russell, D.H., Levy, C.C., Schimpff, S.G., and Hawk, I.A. (1971) : Cancer Res., 31:1555-1558.

10. Tormey, D.C., Waalkes, T.P., Ahmann, D., Gehrke, C.W., Zumwatt, R.W., Snyder, J., and Hansen, H. (1975) : Cancer, 35:1095-1100.
11. Tormey, D.C., Waalkes, T.P., Kuo, K.C., and Gehrke, C.W. (1980) : Cancer, 46:741-747.

Advances in Polyamine Research, Vol. 4, edited by U. Bachrach, A. Kaye, and R. Chayen. Raven Press, New York © 1983.

Determination of Free Polyamines in Blood and Urine During Choriocarcinoma Chemotherapy

Kazimierz Kamiński

Department of Biochemistry, Silesian Medical Academy, 41-808 Zabrze, Poland

There is increasing interest in defining biochemical markers, among others polyamines, that will allow the clinician to diagnose cancer or to evaluate the effects of various regimens upon tumor cells. The important role of polyamines in normal and abnormal biological processes has gained recognition in recent years (2,4,9,13,14).

It has been previously hypothesized that an elevation of polyamines in the serum of cancer patients reflects tumor cell loss (5,9,10,11,12). Changes in serum putrescine and spermidine levels after chemotherapy and radiation therapy of various tumors of rats are mainly the result of polyamine release from the tumor tissue and will be useful markers of the efficacy of the therapeutic regime (5,10,11,12).

It appears, moreover, that polyamine metabolism is related to growth and hormonal stimulation. Elevated putrescine formation in the ovary of rat after the administration of luteinizing hormone (LH) or human choronic gonadotrophin (HCG) was first reported by Kobayashi et al. (6). Recently, Andersson et al. (1) and Rosengren et al. (8) have shown that increased formation of cadaverine and putrescine in the ovary of the pregnant rat might be caused by the hormonal status prevailing during pregnancy, i.e. the effect of human HCG on polyamine metabolism.

In the present study, changes in polyamine concentrations were examined during the chemotherapy of chorion epithelioma (choriocarcinoma). Choriocarcinoma provides a model system in which there is a biological induction of a very intensive growth process by means of the hormone HCG secreted by the cancer itself. In addition, it is of interest to follow the influence of the antimitotic agent Methotrexate used in this case.

MATERIALS AND METHODS

The investigation was performed on five patients aged 21-26 years with clinically and histopathologically diagnosed disease. Samples from the patients were taken for polyamine analysis in serum and 24-h urine completed before and after each chemotherapy cycle, i.e., 48 hours before and 48 hours after the finished treatment. Methotrexate was administered typically for 5 days, 5 mg daily. Blood was collected in the morning, from fasting patients. Urine was preserved by adding 10 ml of 10 M HCl. Samples were frozen at -20°C.

The control group consisted of 20 women volunteers aged 23-25 years, in apparently good health who did not take any drugs. The HCG level in serum of the treated patients was monitored by radioimmunoassay (7).

GLC analysis of polyamines was performed using a Perkin-Elmer gas chromatograph, model 900, with a flame ionization detector. SGI (Australia) glass columns of 1.8 mm interior diameter were used. A 0.9 m column was filled with 3% Dexil 400 on Gas Chrom Q (80-100 mesh) and a 0.6 m column with 5% Poly-S176 on chromsorb G-AW (80-100 mesh).

Standards of putrescine, spermine and spermidine, all as hydrochlorides, were from SIGMA (U.S.A.) n-Dodecylamine (BDH) was used as internal standard. Trifluoroacetic anhydride (TFA) was used for the preparation of N-TFA derivatives of the amines. Acetonitrile was from Merck (W. Germany) and chloroform from POCH (Poland).

The sample, usually 5 c.c. serum or 100 c.c. of a 24-h urine collection, was brought to pH ≤ 2 by the addition of 2 ml 1N-HCl, which converts the polyamines to the hydrochlorides and precipitates proteins. After centrifugation, the supernatant solution was subjected to repeated extractions with chloroform to remove interfering compounds. The aqueous solution was evaporated under reduced pressure and the polyamines were transformed into N-TFA derivatives by adding 4 ml of a mixture of acetonitrile and TFA (1:2) and heating at 100°C for 10 min. Non-reacted TFA and free trifluoracetic acid were removed in a stream of argon and the N-TFA derivative of n-dodecylamine was added as internal standard.

Gas chromatography was performed with the injector temperature at 240°C and detector at 320°C. The Dexil 400 column was held for 3 min at 120°C and the temperature then programmed to 300°C at a rate of 6°C/min. The Poly-S176 was kept at 130°C for 3 min. initially and then programmed to 300°C at a rate of 8°C/min. The flow of the carrier gas (argon) was 15 ml/min for both columns. 2 ul samples were injected.

The Dexil 400 column was used for determination of putrescine and the Poly-S176 for spermidine and spermine.

RESULTS

Free polyamine concentrations in serum and urine of 20 control patients were measured twice a day for 3 days, without noticing large daily fluctuations (Table 1).

Polyamine values in serum and urine of 4 patients, because of the relatively similar dynamics of the observed changes, have been presented in the form of mean values in Table I. These patients presented with bleeding from the reproductive tract, loss of weight, and poor psychological state. In two of these patients, hydatid mole was present, in one - spontaneous abortion, and in the fourth patient - hyperplastic change was diagnosed during abortion. Thus, in all these patients, conditions have developed allowing for the occurrence of choriocarcinoma. All of the 4 patients were in generally good condition.

In the above 4 cases, curetage only was carried out, followed by chemotherapy. Tumor changes did not extend beyond the uterine muscle. HCG concentrations were 350-400 IU/ml. The first determinations of free polyamines ("0", pre-therapy) were made in samples collected just before curetage and 48 hours before the first methotrexate (MTx) treatment. At this stage the levels of polyamines are observed to be many times higher than the control group; putrescine increases 480% in serum, spermine 210%, spermidine 320%. After the first treatment, there is a considerable decrease of free polyamines in serum and this tendency prevails after the consecutive chemotherapeutic cycles. Only the level of putrescine after the 2nd and 3rd treatments does not change, though HCG in serum is undetected. Free putrescine and spermine concentrations in urine increased greatly in choriocarcinoma, while spermidine concentrations decrease. The free putrescine level in urine increases and remains the same after consecutive treatments despite the fact that HCG in urine is undetected. During the whole period of therapy, spermine is increased greatly but it does not undergo any characteristic fluctuations.

The fifth patient, W.A., aged 26, has been presented separately because of the particularly difficult clinical course and longer observation period (Table 2). The patient, with metastases in both lungs, was received in poor condition with haemoptysis and respiratory insufficiency. The free polyamine levels before chemotherapy are very high, all three free both in serum and in urine. After the first treatment, there was an increase of all three free in serum but no changes in urine. In the following therapeutic cycles, there was a gradual decrease of the polyamine levels in serum. After the 3rd cycle, the uterus and appendages were removed. After the operation, a decrease of polyamines in serum and spermidine in urine was observed. Subsequently, the patient's condition worsened with the appearance of liver enlargement, lung X-ray shadows and an increased HCG level prior to death. The above clinical changes were preceeded by an increase of free polyamines in serum. There were lesser changes in urine.

TABLE 1. Free Polyamine Concentrations in 20 Control Patients and in 4 Patients Without Metatasis During Chemotherapy[a]

	Serum (nmol/ml)			Urine (nmol/24 h)		
	PU	SD	SP	PU	SD	SP
Controls	2.6±0.2	2.0±0.2	1.5±0.2	81.4±7.3	21.0±2.4	25.0±3.3
Patients pre therapy	12.2±2.1	6.1±0.8	4.3±0.9	116.4±31.0	78.6±9.9	151.6±24.4
Patients post therapy 1	9.7±1.0	4.1±0.6	1.6±0.3	151.6±15.2	36.0±4.7	169.1±22.1
Patients post therapy 2	5.2±0.4	4.6±0.6	1.5±0.1	162.3±17.8	20.1±1.9	128.6±16.6
Patients post therapy 3	5.4±0.5	2.6±0.4	1.5±0.3	150.3±28.7	21.4±2.8	161.0±29.8

a) Results are expressed as means ± standard deviations, HCG decreases during therapy to non detectable levels

TABLE 2. Free Polyamine Concentrations in a Patient During Methotrexate Treatment for Chloriocarcinoma[a]

		Serum (nmol/ml)			Urine (nmol/24 h)		
Chemotherapy cycle	HCG Conc. (IU/ml)	PU	SD	SP	PU	SD	SP
0	750	13.9	9.3	7.9	306	84	154
1	1250	41.5	13.2	13.4	288	90	161
2	320	25.1	5.4	6.7	254	83	169
3	115	22.5	6.7	4.1	300	33	154
4	150	14.9	4.5	3.4	287	102	153
5	1950	10.6	2.4	1.1	214	79	165
6	2000	19.2	4.6	2.4	225	76	163

a) Between treatments 3 and 4 the uterus and appendages were extirpated

DISCUSSION

Choriocarcinoma is a relatively rare tumor (in Poland 1 per 2,000 pregnancies.) The small number of cases of this hormonally active tumor does not allow for wide-ranging comparisons or conclusions. The high rate of proliferation in this tumor and the responsiveness of the trophoblast to the hormones LH and HCG make the interpretation of changes difficult. High levels of free polyamines before chemotherapy both in serum and in urine may be connected with the high cell loss of tumors (5,10,11,12). It is interesting that a high concentration of free polyamines persisted after the first methotrexate cycle in the patient (W.A.) with highly developed tumor metastases to liver and lungs; this type of change did not appear in the four cases where the tumor process was extremely limited.

The fact that a surge in free polyamines was observed after the first chemotherapeutic treatment in patient W.A. can be accounted for by the fact that there is greater opportunity for cell kill in the above circumstances according to Bakowski and Skipper (3) who hypothesize that a given dose of drugs kills a constant fraction of cells and not a constant number. Thus, in the first treatment the greatest number of cells will be killed and at this point one can expect to detect the release of polyamines into the extracellular fluid. This hypothesis can also explain the occurrence of a clinical response without a rise in extracellular polyamines.

It is difficult to connect the changes found with some unspecific influence of methotrexate. Also the influence of HCG on polyamines in serum and urine seems to be rather negligible - in constrast to this hormone's influence on polyamine metabolism in tissues (1,8).

Thus, it seems that changes in the free polyamine content of biological fluids after chemotherapy may be connected with different intensities of cell killing. However, in the case of choriocarcinoma, at the moment it is difficult to define the usefulness of polyamine determinations for monitoring a patient's condition, the more so as we can make use of HCG determinations.

SUMMARY

Determination of free polyamines (putrescine, spermidine, spermine) in blood and urine can be useful for evaluation of the efficacy of cancer chemotherapy. Free polyamines were estimated for five patients with choriocarcinoma, prior to, during and after chemotherapy with Methotrexate. Gas chromatography with flame ionization detection was used. High concentrations of polyamines before chemotherapy, both in serum and in urine, may be connected with high cell loss in tumors. A high concentration of polyamines was found after the first Methotrexate cycle in a patient with metastases to liver and lungs, but this type of change was not seen in four cases where the tumor process was limited.

REFERENCES

1. Andersson, A. Ch., and Henningsson, S. (1980): Acta Endocrinologica, 95:237-243.

2. Bachrach, U. (1976): Ital. J. Biochem., 25:77-93.

3. Bakowski, M.T., Toseland, P.A., Wicks, R.C., and Trounce, J.R. (1981): Clin. Chim. Acta; 110:273-286.

4. Cohen, S.S. (1971): Introduction to the Polyamines. Prentice-Hall, Inc., Englewood Cliffs, N.J.

5. Desser, H., Frass, M., Kuzmits, R., Aiginger, M., Muller, M., and Klaring, W.J. (1981): In: Advances in Polyamine Research, vol. 3, edited by C.M. Calderera, pp.431-440.

6. Kobayashi, Y., Kupelian, J., Maudsley, D.V. (1971): Science 172:379-380.

7. Kokot, F., and Stupnicki, P. (1979): Metody radioimunologiczne i radiokompetycyjne stosowane w klinice. PZWL, Warszawa.

8. Rosengren, E., Henningsson, A.-Ch., Henningsson, S., and Persson, L. (1981): Med. Biol., 59:320-326.

9. Russell, D.H. (1973): In: Polyamines in Growth: Normal and Neoplastic Growth, edited by D.H. Russell, pp.1-13. Raven Press, New York.

10. Russell, D.H., Looney, W.B., Kovacs, Ch. J., Hopkins, H.A., Marton, L.J., LeGendre, S.M., and Morris, H.P. (1974): Cancer Res., 34:2382-2385.

11. Russell, D.H., Gullino, P.M., Marton, L.J., and LeGendre, S.M. (1974): Cancer Res., 34:2378-2381.

12. Russell, D.H., Looney, W.B., Kovacs, Ch. J., Hopkins, H.A., Datilo, J.W., and Morris, H.P. (1976): Cancer Res., 36:420-423.

13. Tabor, H., and Tabor, C.W. (1972): In: Advances in Enzymology and Related Areas of Molecular Biology, edited by A. Meister, vol. 36, pp.203-268. Interscience Publishers, New York.

14. Williams-Ashman, H.G., Coppoc, G.C., and Weber, G. (1972): Cancer Res., 32:1924-1932.

Advances in Polyamine Research, Vol. 4, edited by U. Bachrach, A. Kaye, and R. Chayen. Raven Press, New York © 1983.

Decreased Activity of Ornithine Decarboxylase Antizyme During Hyperplastic and Neoplastic Growth of Rat Liver

G. Scalabrino, M. E. Ferioli, and D. Modena

Institute of General Pathology and C.N.R. Centre for Research in Cell Pathology, University of Milan, 20133 Milan, Italy

> Distat tamen ab aeternorum contemplatione actio qua bene utimur temporalibus rebus, et illa sapientiae, haec scientiae deputatur...... In haec differentia intellegendum est ad contemplationem sapientiam, ad actionem scientiam pertinere.
> Augustine of Hippo, De Trinitate, XII, XIV.

Sensitive modulation of polyamine biosynthesis in mammalian tissues in relation to intracellular and/or extracellular stimuli can be achieved through rapid fluctuations in the activity of L-ornithine carboxy--lyase (L-ornithine decarboxylase, EC 4.1.1.17; ODC), that is the initial and rate-limiting enzyme in the polyamine biosynthetic pathway and is the only route in eukaryotic cells for synthesis of putrescine (14,20,40, 43). Moreover, even more important, ODC appears to play an important role in the regulation of proliferation and differentiation as well as in growth, both controlled and uncontrolled, of mammalian cells (14,20, 40,43). Substantial data have been accumulated with whole animal tissue, perfused organs and cultured cell lines, whether or not neoplastic, demonstrating that striking increases in ODC activity occur at the same time as the onset of proliferation and/or development and that, conversely, ODC activity markedly decreases when either of these biological processes ceases (20, 43). Therefore, the regulation of ODC activity and polyamine biosynthesis and the regulation of the proliferative capacity of the eukaryotic cells appear to be intimately connected.

In addition to a great variety of molecules, some naturally occurring, some not, that stimulate ODC activity in mammalian tissues (20,43), the activity of eukaryotic ODC is also regulated negatively, mainly by amines, such as the three chief naturally occurring polyamines and some diamines not normally found in living cells (14,20). Many mechanisms have been proposed to explain the repression of cell ODC activity by this variety of amines (5,28). On the basis of the different experimental data, this negative regulation has been considered by different investigators to occur at the transcriptional level, at the post-transcriptional level, at the translational level and at the post--translational level. However, the mechanism of inhibition of ODC activity by various diamines was elucidated when Canellakis and coworkers reported that putrescine and other di- and poly-amines also regulate intracellular ODC levels by an indirect mechanism, inducing the synthesis of an inhibitory protein (named "ODC-antizyme"), under some experimental conditions both in vivo and in cultured cells (8,15-18). This has been widely confirmed (3,9,12,21,24,25,29,30,36,38,47). It is therefore apparent that ODC is regulated by its product putrescine by many different mechanisms, not necessarily mutually exclusive.

Rat liver and mouse skin have received considerable attention in recent years as models for chemical carcinogenesis. The induction of epidermal ODC activity is considered to be an essential event in the formation of mouse skin tumors (34) and, similarly, a prolonged elevation of hepatic ODC activity has been believed to be a necessary event for the development of neoplastic foci and eventual formation of rat hepatomas (37). Furthermore, it is widely known that elevation of ODC activity is one of the earliest and more impressive biochemical phenomena observed during rat liver regeneration (20,40). Although it is clear that enhanced ODC activity very frequently occurs in response to differing stimuli that promote cell growth of differing types, both controlled and uncontrolled, the detailed mechanism of this increase in ODC activity is not yet fully known. Theoretically, intracellular enhancement of ODC activity could be accomplished by several mechanisms, not necessarily mutually exclusive. One mechanism has been demonstrated to be the appearance in some cell types when stimulated to grow, of different forms of ODC with increased affinity for the substrate or for pyridoxal-5-phosphate (10,26,31-33, 49). Another mechanism, only hypothetical at the present, would be a decreased capacity of growing cells to form the ODC-antizyme in response to polyamines.

In this regard, it seems worthwhile to recall here that some rapidly proliferating tissues (e.g., regenerating rat liver (20,40), mitogen-activated lymphocytes (7,22) and rat liver undergoing chemical carcinogenesis (41)) during some phases of the growth process show the apparently paradoxical coexistence of high levels of ODC activity and high levels of putrescine. This biological inconsistency hints at the possibility that there is some failure in the control of ODC activity by putrescine and other polyamines in proliferating cells. Recently, studies carried with cultures of virally- or chemically-transformed cells have provided further support for this hypothesis, since they demonstrated that tumour cells have a decreased sensitivity to the control of ODC activity by exogenously added polyamines (1,2,13,35,36). On the contrary, data obtained in normal perfused rat liver suggest that ODC activity is tightly regulated in this organ by putrescine levels, the first increasing when the latter are decreasing and _vice versa_ (21).

With all this in mind, we undertook the present study to investigate possible differences in ODC regulation by its antizyme in models of controlled hepatic growth (i.e., in rat livers receiving stimuli that enhance ODC levels, followed by normal cell proliferation) and in models of uncontrolled hepatic growth (i.e., in rat livers receiving a stimulus that enhances ODC levels but followed by neoplastic transformation). In more detail, we measured the activity of ODC-antizyme, induced by the _in vivo_ administration of exogenous putrescine, in normal quiescent rat liver; in regenerating rat liver, monitored at various times after partial hepatectomy; in livers of rats fed 3'-methyl--4-(dimethylamino)azobenzene (3'-Me-DAB, a well known hepatocarcinogen) or α-naphthylisothiocyanate (α-NIT, which causes hepatic hyperplasia, but never hepatomas (27)) for fixed time periods.

MATERIALS AND METHODS

Chemicals. L-[1-^{14}C] Ornithine (specific activity 59 mCi/mmol) was purchased from the Radiochemical Centre (Amersham, Bucks., U.K.) and was diluted to the required specific activities by the addition of unlabeled L-ornithine obtained from Sigma Chemical Co. (St. Louis, MO, U.S.A.). Dexamethasone 21-acetate was obtained from Sigma Chemical Co. (St. Louis, MO, U.S.A.) and cycloheximide from Calbiochem (San Diego, CA, U.S.A.). The source of 3'-Me-DAB was Koch Light (Colnbrook, U.K.) and that of α-NIT was Aldrich-Europe (Beerse, Belgium).

Animals. Adult male rats of the Sprague-Dawley strain, initially weighing 150-170 g, were used in all experiments.

Treatments of animals. The composition of the basal diet was essentially that reported by Sneider and Potter (46). For the oncogenic diet 0.06% 3'-Me-DAB was added. The diet containing α-NIT was the basal diet with 1 g α-NIT/kg added (27). Rats were maintained on an automatic daily cycle of 12 hrs of light and 12 hrs of darkness (lights on 06:00 A.M. to 06:00 P.M.). During both the entrainment period and the experiments, the nocturnal feeding habit of the rats was reinforced by removal of the food when the lights came on and by replacement of the food just before the lights were turned off. Partial hepatectomy was performed under light ether anaesthesia by the method of Higgins and Anderson (19). One intraperitoneal injection of dexamethasone (1 mg/100 g body weight) was given to rats to induce hepatic ODC within 3 hours.

Preparation of liver extracts for the enzyme assay and for the ODC-antizyme assay. After decapitation of the rats, the livers were rapidly removed, and weighed and homogenized separately in 2 vol. of a medium whose composition has been previously described (42). All the successive steps were performed at 0°-2°C. The homogenates were centrifuged at 20,000 x g for 20 min and the supernatant was centrifuged in a Spinco ultracentrifuge at 100,000 x g for 60 min. An aliquot of the 20,000 x g supernatant was used for the assay of ODC activity and an aliquot of the 100,000 x g supernatant was used for the assay of ODC-antizyme activity.

ODC assay. ODC activity was assayed by the method routinely used in our laboratory, in the presence of a saturating concentration (2mM) of L-ornithine (42). The ODC activity was determined by measuring the release of $^{14}CO_2$ from carboxyl-labeled L-ornithine. In the standard assay, amounts of 3-5 mg protein were used. Enzyme assays were done in triplicate and the enzyme activity was expressed as pmoles of $^{14}CO_2$ liberated from carboxyl-labeled L-ornithine per mg protein per 30 min incubation.

Assay of ODC-antizyme activity. Hepatic ODC, induced as described above, was partially purified (about 10 fold) essentially as described by Fong et al. (8) before assay. This ODC preparation had a specific activity of 10 nmoles $^{14}CO_2$ released per mg protein per 60 min incubation. The ODC-antizyme activity in the homogenates obtained from the livers taken from putrescine-treated rats was assayed by adding known amounts of the partially purified ODC (approximately 1.3 nmol of ODC, corresponding to about 5000 dpm $^{14}CO_2$ per 60 min incubation) to aliquots of the above-mentioned homogenates and measuring the amount of this added activity remaining, as suggested by Heller et al. (17). The extent by which the homogenates reduced the activity of the purified enzyme was calculated by subtracting the radioactivity observed when both the purified enzyme and the homogenate were present from that released by the purified enzyme alone. One unit of inhibitor is defined as the amount that inhibited the release of one nmol of $^{14}CO_2$ in the assay in a 60 min incubation. The activity of ODC-antizyme can also be expressed as the percentage of inhibition of the ODC activity, with the radioactivity liberated by the purified enzyme alone designated as 100%. Liver homogenates were not dialyzed because there was no difference between the antizyme activity in undialyzed and dialyzed extracts (results not reported here), confirming similar reports by Heller and Canellakis (18) and by McCann et al. (29). Separation of the ODC--antizyme complex was not considered to be necessary in this study and therefore it was not carried out. In addition to the above mentioned amount of purified ODC, the incubation mixture contained, in a final volume of 0.25 ml: 100 mM Tris-HCl buffer (pH 7.1 at 37°C); 4 mM EDTA; 4 mM dithiothreitol; 0.4 mM pyridoxal phosphate; 0.42 mM L-[1-^{14}C] ornithine (2 Ci/mol) and 0.05 ml tissue extract, in which the protein content ranged from 1.5 to 2.1 mg.

The radioactivity of the samples obtained from the ODC assays and the ODC-antizyme assays was counted in a Tri-Carb liquid scintillation spectrometer (Packard model 300 C). Counting efficiency was determined by both internal and external standardization and was 80 ± 5% by both methods. Non-enzymatic release of $^{14}CO_2$ was subtracted from the sample counts.

Protein concentrations were estimated by the method of Geiger and Bessman (11) using crystalline bovine serum albumin as standard.

Data are presented as the means of a given number of observations ± SEM. The statistical significance of differences between means was calculated by appropriate statistical tests. Differences for which the P values were greater than 0.05 were not considered to be significant.

RESULTS

The data in Table 1 show that the activity of the ODC-antizyme (measured 3 hr after the first i.p. injection of putrescine) is markedly lower in 2.5 hr-regenerating liver and still lower 4 hr postoperatively than in livers of unoperated or sham-operated rats. It is worthwhile to point out that in the 2.5 hr-regenerating liver experiments, the first injection of putrescine was made 30 min before partial hepatectomy. Parenthetically, it is also apparent from Table 1 that the sham-operation slightly reduces ODC-antizyme activity, probably owing to the stress. At later times of the early phase of liver regeneration, namely at

TABLE 1. Activities of ODC and of ODC-antizyme during the early phases of rat liver regeneration.

Hours after P.H.	Treatment	Times of i.p. treatment (min before killing) 1st	2nd	ODC activity (pmol $^{14}CO_2$/mg prot/ /30 min)	ODC-antizyme activity (% inhibition)
0	SAL	180	90	35.5 ± 3.33 (4)	-
0	PUT	180	90	4.1 ± 1.00••• (4)	82.0 (4)
0(S.O.)	PUT	180	90	21.0 ± 3.54••• (4)	60.0 (4)
2.5	SAL	180	90	572.2 ± 52.76 (4)	-
2.5	PUT	180	90	17.6 ± 2.09••• (4)	23.6 (4)
4.0	SAL	180	90	1240.7 ± 257.14 (4)	-
4.0	PUT	180	90	43.1 ± 6.72••• (4)	17.8 (4)
6.0	SAL	180	90	908.5 ± 143.34 (4)	-
6.0	PUT	180	90	11.6 ± 1.73••• (4)	53.9 (4)
8.0	SAL	180	90	368.5 ± 54.53 (4)	-
8.0	PUT	180	90	5.7 ± 1.30••• (4)	50.0 (4)

The results for ODC are expressed as means ± SEM. P.H.=Partial Hepatectomy. SAL=Saline (0.5 ml for each injection). PUT=Putrescine (200 μmol/100 g BW for each injection). S.O.=Sham-Operation. i.p.=Intraperitoneal. Number of animals in parentheses. Within each experimental group the difference between the two means was tested for statistical significance by Student *t* test. ••• $P<0.001$ vs. saline-treated rats. The differences between the means for regenerating controls at various hours were tested for statistical significance by Duncan's multiple range test (6) and the P values always indicated statistical significance ($P<0.05$).

TABLE 2. Activities of ODC and of ODC-antizyme in the rat liver that has regenerated 20 or 24 hours.

Time after P.H. (hrs)	Treatment	Times of i.p. treatment (min before killing)		ODC activity (pmol $^{14}CO_2$/mg prot/ /30 min)	ODC-antizyme activity (% inhibition)
		1st	2nd		
20	SAL	180	90	426.2 ± 40.40 (4)	-
20	PUT	180	90	7.6 ± 1.42••• (4)	78.3 (4)
24	SAL	180	90	295.2 ± 34.91 (4)	-
24	PUT	180	90	3.4 ± 0.85••• (4)	70.6 (4)

The results for ODC activity are expressed as means ± SEM. PUT = Putrescine (200 μmol per 100 g BW for each injection). SAL = Saline (0.5 ml for each injection). i.p. = Intraperitoneal. P.H. = Partial Hepatectomy. Number of animals in parentheses. Within each experimental time group the difference between the two means was tested for statistical significance by Student *t* test. •••$P < 0.001$ vs. saline-treated rats.

6- and 8-hr postoperatively, ODC-antizyme activity is restored almost completely, as is demonstrated by the same levels of percentage of inhibition observed in the livers of sham-operated rats (see Table 1). This restoration of ODC-antizyme activity is coordinated in time with the phase in which the ODC activity declines after its first peak, while the decrease in ODC-antizyme activity is coordinated in time with the phase during which ODC activity is progressively increasing, until the first peak at 4 hr after the operation. Table 1 also lists the effects of injections of putrescine on ODC activity in regenerating rat livers at the various times tested. The decreases in ODC activity brought about by this diamine are always very great, regardless of the time during liver regeneration when putrescine was injected. This is in complete agreement with earlier reports from other laboratories (20). But, what is much more important, these decreases in ODC levels are nearly always of the same order of magnitude, whether the activity of the ODC-antizyme is maximal or minimal.

In recent years it has became increasingly evident that in regenerating rat liver the ODC activity at the first peak, occurring nearly at 4 hr postoperatively, is regulated in a different manner, at least partly, from that at the second peak, which occurs toward the end of the first day of liver regeneration (23,39). Therefore, we also determined the activity of ODC-antizyme in rat livers that had regenerated from 20 hr to 24 hr. Table 2 shows that the ODC-antizyme activities, expressed as percentage of the ODC inhibition, both in 20 hr-regenerating liver (i.e., when the second peak of ODC activity is observed under our experimental conditions) and in 24 hr-regenerating liver (i.e., when the ODC activity begins to decline once again) are very similar to that observed in unoperated controls and reported in Table 1.

Data presented in Table 3 confirm the previously reported results that both 3'-Me-DAB and α-NIT feeding enhance hepatic ODC activity at the end of the first month of experimentation and that the oncogenic drug enhances ODC activity more than the hyperplasia-inducing drug (44). Additionally, Table 3 shows that putrescine greatly depresses hepatic ODC activity, re-

TABLE 3. ODC activity in livers of rats fed different diets for different time periods.

Type of diet	Duration of feeding (months)	ODC activity (pmol of $^{14}CO_2$/mg prot/30 min) SAL	PUT
Basal	1	37.8 ± 6.14 (12)	10.4 + 1.08 (6)
Oncogenic (3′-Me-DAB)	1	194.4 ± 20.10•• (9)	15.4 ± 2.97 (5)
Hyperplaseogenic (α-NIT)	1	*129.6 ± 18.94•• (5)	8.3 ± 1.77 (5)
Basal	2	30.7 ± 8.19 (6)	8.6 ± 1.78 (5)
Oncogenic (3′-Me-DAB)	2	73.6 ± 6.20•••(10)	4.2 ± 1.03 (5)

The results are expressed as means ± SEM. Number of animals in parentheses. SAL = Saline (0.5 ml for each of the two i.p. injections). PUT = Putrescine (200 μmol/100 g BW, for each of the two i.p. injections). Saline or putrescine was administered at 180 and 90 min before killing. i.p. = Intraperitoneal. Within the 1-month group, the differences between the means were tested for statistical significance by Duncan's multiple range test (6): ••$P < 0.01$ vs. controls; * $P < 0.05$ vs. 3′-Me-DAB. In the 2-month group, the difference between the two means was tested for statistical significance by Student *t* test: •••$P < 0.001$. The differences between the means of saline-treated rats and those of putrescine-treated rats were always statistically highly significant ($P < 0.001$).

TABLE 4. Activity of ODC-antizyme in livers of rats fed different diets for different time periods.

Type of diet	Duration of feeding (months)	ODC-antizyme activity (nmol $^{14}CO_2$/mg prot/60 min)	(% inhibition)
Basal	1	0.52 ± 0.03 (5)	88.0 (5)
Oncogenic (3′-Me-DAB)	1	0.39 ± 0.02•• (5)	48.4 (5)
Hyperplaseogenic (α-NIT)	1	*0.14 ± 0.01•• (5)	29.8 (5)
Basal	2	0.44 ± 0.03 (5)	85.5 (5)
Oncogenic (3′-Me-DAB)	2	0.10 ± 0.01••• (5)	21.1 (5)

Hepatic ODC-antizyme was induced by intraperitoneal administration of putrescine (200 μmol/100 g BW for each of the two injections, at 180 and 90 min before killing). Number of animals in parentheses. The results, when expressed as nmoles of labeled carbon dioxide per mg protein per a 60 min incubation, are presented as means ± SEM. Within the 1-month group, the differences between the means were tested for statistical significance by Duncan's multiple range test (6): ••$P < 0.01$ vs. controls; * $P < 0.01$ vs. 3′-Me-DAB. In the 2-month group, the difference between the two means was tested for statistical significance by Student *t* test: •••$P < 0.001$.

gardless of the type of diet and regardless of the length of the treatment. As for ODC-antizyme, Table 4 shows that the percentage of inhibition of ODC activity at the end of the first month of the oncogenic treatment is reduced nearly to half that in the controls while after the same time of the α-NIT treatment the percentage of enzyme inhibition is reduced to nearly one third. Furthermore, when we compare the ODC activity and the ODC-antizyme activity at the end of the second month of oncogenic feeding with those at the end of the first month, there is a decrease both in ODC activity (which is in agreement with a previous report (44)) (see Table 3) and a further decrease in ODC-antizyme activity (see Table 4). From Table 4 it is also evident that the trend in ODC-antizyme activity in livers of rats fed either 3′-Me-DAB or α-NIT is the same, whether when it is expressed as percentage of inhibition of ODC activity or as nmoles of labeled carbon-dioxide liberated per mg protein per 60 min. And this agreement between these two ways of expressing ODC-antizyme activity also was true for regenerating livers at the various times of testing, although the results in nmoles are not given in Tables 1 and 2.

Finally, when cycloheximide (1.5 mg/kg body weight) was injected with the first injection of putrescine, no ODC-antizyme activity was observed in supernatants obtained from livers of rats fed either the basal or the oncogenic diet (results not reported here). Furthermore, Heller and Canellakis (18) have recently demonstrated that in cultured neoplastic cells the higher the level of ODC activity when putrescine was added, the longer the delay in appearance of the antizyme. Therefore, we measured the hepatic ODC-antizyme activity of rats fed 3′-Me-DAB for one month (i.e., when the greatest increase in ODC activity is observed) which were injected twice with putrescine, at 240 min and at 150 min before killing. This means that the time for ODC-antizyme formation was prolonged to 4 hr, instead of the routinely used 3 hr. No difference in ODC-antizyme activity in rats injected 3 hr before killing and those injected 4 hr before killing was noted (results not reported here).

DISCUSSION

In this paper evidence has been presented that there is a failure of ODC control by its antizyme in rat liver stimulated to proliferate, either surgically or chemically. In the experimental models of hepatic growth we used, we have demonstrated that the inhibition of ODC activity caused by putrescine and the activity of ODC-antizyme induced simultaneously by the same amount of the diamine can differ in magnitude and that, consequently, they are not mandatorily directly correlated. In other words, the deep repression of ODC activity by putrescine is not always accompanied by and does not necessarily entail contemporaneous high ODC-antizyme activity. This has been demonstrated for different types of growth in rat liver, such as several stages of the early phase of the regeneration, hyperplasia caused by α-NIT, and during carcinogenesis by 3'-Me-DAB. In rat liver regeneration, the coexistence of a large decrease in ODC activity by putrescine and relatively small ODC-antizyme activity induced by the diamine is transient, since it is apparent only at 2.5 and 4 hr postoperatively, i.e., at the times when ODC activity first increases. These results strongly indicate that during rat liver regeneration the control of the first peak of ODC activity by its antizyme differs from that of the second peak. Not only when different inhibitors of ODC activity are used (23,39), but also when the activity of ODC-antizyme is measured, it further highlights that during rat liver regeneration the ODC increases in the early stages and those at the end of the first day are regulated by mechanisms which differ at least partially. In this regard, it is well known that two phases are recognized during the first 24 hr of liver regeneration in the rat: hypertrophy, lasting approximately 12-16 hr, and subsequent hyperplasia, characterized by a peak in DNA synthesis at about 24 hr, followed 6-8 hr later by mitosis (4).

Admittedly, the trend in hepatic ODC-antizyme activity during the early months of rat liver carcinogenesis by 3'-Me-DAB is quite different from that observed during rat liver regeneration. In fact, as in the hepatic compensatory growth process, there is a marked and progressive decrease in ODC-antizyme activity at the end of the first and the second month of the oncogenic treatment, but, unlike during liver regeneration, this decrease is a steady feature of the transforming hepatocytes, since it is also present at the end of the third month 3'-Me-DAB feeding (results not reported here). As for α-NIT feeding, the relatively short period of time of the experiment does not allow to us to say whether or not the observed decrease in ODC-antizyme activity is permanent.

Since it is widely known that in rat liver that is either restoring its mass or undergoing chemical carcinogenesis there is an intracellular accumulation of putrescine (20,40,41), it is tempting to speculate that what we have shown to occur with exogenous putrescine could also happen with the augmented amount of endogenous putrescine in these two types of proliferating hepatocytes. In other words, it is conceivable that these two types of growing hepatocytes are less able to modulate the synthesis of ODC-antizyme in relation to their intracellular putrescine content than the quiescent hepatocytes. This relative insensitivity is transient in regenerating liver and stable in chemically transformed liver and it could allow both these types of hepatocytes to continue to synthesize polyamines in the presence of high intracellular putrescine levels. Therefore, ODC in rat liver undergoing chemical carcinogenesis by 3'-Me-DAB appears to be insensitive to several of the main controlling mechanisms, the intracellular putrescine content, the light-darkness and the food-starvation cycles. In fact, we have demonstrated that the hepatic ODC circadian rhythm disappears from rat livers continuously stimulated to grow into hepatoma by 3'-Me-DAB feeding (44). Therefore, hepatic ODC during neoplastic transformation is less sensitive or even insensitive to some repressor located both inside the cell and outside the cell. In keeping with this are reports indicating that a cultured transformed cell line failed to synthesize ODC-antizyme in the presence of high concentrations of putrescine (36) and that the ODC activity of some cultured neoplastic cells is less sensitive than the normal cells to repression by putrescine (1, 2,13). Interestingly enough, experimental data indicate that some lines of cultured neoplastic cells are insensitive to polyamine depletion or restriction and continue to progress through the cell cycle phases, unlike the majority of cultured normal cells, in which inhibition of putrescine and spermidine synthesis results in a G_1 arrest (48).

Finally, we realize that the decreased ODC-antizyme activity we found in the present investigation might have been caused by several derangements in more than one step of polyamine biosynthesis and/or catabolism. This decrease could be due to a lesser uptake of putrescine by rat liver stimulated to grow chemically or surgically, or to increased activity of diamine-oxidase or of polyamine-oxidase in proliferating rat livers, so that the intracellular amount of putrescine for eliciting formation of ODC-antizyme would be quickly reduced. A modification of the interconversion of the polyamines, which is a physiological process and is a part of the polyamine turnover (45), could also somehow interfere in determining indirectly the activity of ODC-antizyme in regenerating rat liver and in rat liver undergoing chemical carcinogenesis. Experiments to investigate some of these points are now in progress.

ACKNOWLEDGEMENTS

We thank Prof. E.S. Canellakis (New Haven) and his coworkers for their help in this research.

This work was financially supported by the National Research Council, project "Control of Neoplastic Growth" (grant N° 80.01479.96), Rome, Italy.

REFERENCES

1. Bethell, D.R., and Pegg, A.E. (1979): Biochem. J., 180:87-94.
2. Bethell, D.R., and Pegg, A.E. (1981): J. Cell. Physiol., 109:461-468.
3. Branca, A.A., and Herbst, E.J. (1980): Biochem. J., 186:925-931.
4. Bucher, N.L.R., and Malt, R.A. (1971): Regeneration of Liver and Kidney. Little, Brown and Co., Boston.

5. Canellakis, E.S., Viceps-Madore, D., Kyriakidis, D.A., and Heller, J.S. (1979): Curr. Top. Cell. Regul., 15:155-202.
6. Duncan, D.B. (1955): Biometrics, 11:1-42.
7. Fillingame, R.H., and Morris, D.R. (1973): Biochemistry, 12:4479-4487.
8. Fong, W.F., Heller, J.S., and Canellakis, E.S. (1976): Biochim. Biophys. Acta, 428:456-465.
9. Friedman, Y., Park, S., Levasseur, S., and Burke, G. (1977): Biochim. Biophys. Acta, 500:291-303.
10. Fuller, J.L., Greenspan, S., and Clark, J.L. (1978): Adv. Polyamine Res., 1:31-38.
11. Geiger, P.J., and Bessman, S.P. (1972): Anal Biochem., 49:467-473.
12. Grillo, M.A., Bedino, S., and Testore, G. (1980): Int. J. Biochem., 11:37-42.
13. Haddox, M.K., Magun, B.E., and Russell, D.H. (1980): Cancer Res., 40:604-608.
14. Heby, O. (1981): Differentiation, 19:1-20.
15. Heller, J.S., Fong, W.F., and Canellakis, E.S. (1976): Proc. Natl. Acad. Sci. USA, 73:1858-1862.
16. Heller, J.S., Kyriakidis, D., Fong, W.F., and Canellakis, E.S. (1977): Eur. J. Biochem., 81:545--550.
17. Heller, J.S., Chen, K.Y., Kyriakidis, D.A., Fong, W.F., and Canellakis, E.S. (1978): J. Cell. Physiol., 96:225-234.
18. Heller, J.S., and Canellakis, E.S. (1981): J. Cell. Physiol., 107:209-217.
19. Higgins, G.M., and Anderson, R.M. (1931): Arch. Path., 12:186-202.
20. Jänne, J., Pösö, H., and Raina, A. (1978): Biochim. Biophys. Acta, 473:241-293.
21. Jefferson, L.S., and Pegg, A.E. (1977): Biochim. Biophys. Acta, 484:177-187.
22. Kay, J.E., and Cooke, A. (1971): FEBS Lett., 16:9-12.
23. Kallio, A., Pösö, H., Scalabrino, G., and Jänne, J. (1977): FEBS Lett., 73:229-234.
24. Kallio, A., Löfman, M., Pösö, H., and Jänne, J. (1979): Biochem. J., 177:63-69.
25. Klingensmith, M.R., Freifeld, A.G., Pegg, A.E., and Jefferson, L.S. (1980): Endocrinology, 106:125-132.

26. Lau, C., and Slotkin, T.A. (1982): Eur. J. Pharmacol., 78:99-105.
27. Lopez, M., and Mazzanti, L. (1955): J. Pathol. Bacteriol., 69:243-250.
28. McCann, P.P. (1980): In: Polyamines in Biomedical Research, edited by J.M. Gaugas, pp. 109-123. J. Wiley & Sons, Chichester.
29. McCann, P.P., Tardif, C., and Mamont, P.S. (1977): Biochem. Biophys. Res. Commun., 75:948-954.
30. Minaga, T., Marton, L.J., Piper, W.N., and Kun, E. (1978): Eur. J. Biochem., 91:577-585.
31. Mitchell, J.L.A. (1981): Adv. Polyamine Res., 3: 15-26.
32. Mitchell, J.L.A., and Mitchell, G.K. (1982): Biochem. Biophys. Res. Commun., 105:1189-1197.
33. Obenrader, M.F., and Prouty, W.F. (1977): J. Biol. Chem., 252:2860-2865.
34. O'Brien, T.G. (1976): Cancer Res., 36:2644-2653.
35. O'Brien, T.G., and Diamond, L. (1978): In: Carcinogenesis, vol. 2, edited by T.J. Slaga, A. Sivak, and R.K. Boutwell, pp. 273-287. Raven Press, New York.
36. O'Brien, T.G., Saladik, D., and Diamond, L. (1980): Biochim. Biophys. Acta, 632:270-283.
37. Olson, J.W., and Russell, D.H. (1980): Cancer Res., 40:4373-4380.
38. Pegg, A.E., Conover, C., and Wrona, A. (1978): Biochem. J., 170:651-660.
39. Pösö, H., Kallio, A., Scalabrino, G., and Jänne, J. (1977): Biochim. Biophys. Acta, 497:288-297.
40. Russell, D.H., and Durie, B.G.M. (1978): Polyamines as Biochemical Marker of Normal and Malignant Growth. Progr. Cancer Res. Ther., VIII. Raven Press, New York.
41. Scalabrino, G., Pösö, H., Hölttä, E., Hannonen, P., Kallio, A., and Jänne, J. (1978): Int. J. Cancer, 21:239-245.
42. Scalabrino, G., Pigatto, P., Ferioli, M.E., Modena, D., Puerari, M., and Carù, A. (1980): J. Invest. Dermatol., 74:122-124.
43. Scalabrino, G., and Ferioli, M.E. (1981): Adv. Cancer Res., 35:151-268.
44. Scalabrino,G., Ferioli, M.E., Puerari, M., Modena, D., Fraschini, F., and Majorino, G. (1981):

J. Natl. Cancer Inst., 66:697-702.

45. Seiler, N. (1981): In: Polyamines in Biology and Medicine, edited by D.R. Morris, and L.J. Marton, pp. 169-180. M. Dekker, Inc., New York.
46. Sneider, T.W., and Potter, V.R. (1969): Adv. Enzyme Regul., 7:375-394.
47. Stoscheck, C.M., Florini, J.R., and Richman, R.A. (1980): J. Cell. Physiol., 102:11-18.
48. Sunkara, P.S., and Rao, P.N. (1981): In: The Transformed Cell, edited by I.L. Cameron, and T.B., Pool, pp. 267-291. Academic Press, New York.
49. Weiss, J.M., Lembach, K.J., and Boucek, R.J.Jr. (1981): Biochem. J., 194:229-239.

Advances in Polyamine Research, Vol. 4, edited by U. Bachrach, A. Kaye, and R. Chayen. Raven Press, New York © 1983.

Enhancement of Thermal Killing by Spermine in EMT6 Multicellular Tumour Spheroids Versus Monolayer Cells

*E. Ben-Hur, J. J. Shaw, and N. M. Bleehen

**Nuclear Research Center-Negev, Beer-Sheva 84190, Israel; MRC Unit of Clinical Oncology and Radiotherapeutics, The Medical School, Cambridge University, Cambridge CB2 2QH, United Kingdom*

Hyperthermia at temperatures above 41°C has recently gained increased interest as a potentially useful modality in cancer treatment (8). Much of the rationale for considering the use of hyperthermia in cancer therapy has come from results in vitro. Thus, hyperthermia sensitizes mammalian cells to radiation (2) and drugs (1,13) and preferentially kills S-phase cells which are normally more resistant to X-rays than other cell cycle stages (19). Thermal sensitivity is affected greatly by environmental factors. Nutritional deficiency (12) lowered pH in the growth medium (11) and the naturally occurring polyamines when supplied exogenously (3,9), all enhance thermal response. The mechanism(s) leading to the enhanced response to polyamines are still obscure, although both the chromatin structure (4-6) and the plasma membrane (10) have been implicated.

Multicellular tumour spheroids are an *in vitro* model system representing an intermediate level of complexity between monolayer cultures and solid tumours *in vivo* (16). This system has been used to study the response to hyperthermia and chemotherapy of various tumour cell lines. Thus in EMT6 tumour spheroids, a marked resistance to adriamycin was shown when compared with exponentially growing monolayer cells (17). We have described in a similar system a synergistic interaction between hyperthermia and the cytotoxic drugs bleomycin and adriamycin as measured by growth delay and cell survival (14).

This paper demonstrates that, compared with monolayer cells, the EMT6 multicellular tumour spheroids are resistant to the combination of hyperthermia and spermine. The latter is the most effective polyamine in enhancing thermal response. Poor penetration of spermine to the inner cells of the spheroid appears to be part of the reason for this resistance.

MATERIALS AND METHODS

The methods used for growth and assay of the EMT6/Ca/VJAC spheroid system were described previously (14,18). Briefly, spheroids were grown from a single-cell suspension in culture flasks base coated with agar to prevent cell adhesion to the surface (20). Spheroids were used in experiments on day 6, by which time their average diameter was 220 ± 20 μm. Heat treatment was in a waterbath with the temperature controlled to ±0.1°C.

Spermine tetrahydrochloride was dissolved in water and stored at -20°C in small aliquots as a stock 0.1M solution. Prior to use it was diluted in Hanks' balanced salt solution to a concentration 50-fold higher than the final concentration required in the growth medium. To each 4.9 mℓ spheroids suspension to be treated was added 0.1 mℓ of the diluted spermine. At the same time aminoguanidine was added to the growth medium to a final concentration of 10^{-5}M in order to inhibit spermine degradation by polyamine oxidases (15).

After treatment the spheroids were transferred into fresh medium. Twenty four representative spheroids were selected from each treatment group and placed in individual wells on plastic multidishes for regrowth studies. The remaining spheroids were trypsinized and the cell suspension was then assayed for cell survival (18). Exponentially growing cells in a monolayer were treated as above, trypsinized and cell survival was determined. Uptake of labelled spermine into cells was measured as described previously (4). $[1,4-^{14}C]$ - spermine tetrahydrochloride (New England Nuclear, 77 mCi/mM) was added to 6 day spheroids suspended in a 5 mℓ medium at a final concentration of 1 mM and 0.1 µCi/mℓ. After various incubation times the spheroids were rinsed with buffer and extracted with 5% cold trichloracetic acid. The radioactivity in the acid-soluble fraction was counted using a liquid scintillation spectrometer. For treatment of monolayer cells the final concentration of the labelled spermine was 0.05 mM and 0.02 µCi/mℓ. About 2×10^6 cells were used for each determination in both spheroids and monolayer cells.

RESULTS

Fig. 1 shows the survival curves for spheroids disaggregated immediately after exposure to spermine for 1 h at 42°C and 43°C. Spheroids were apparently very resistant to heat enhanced killing by spermine as compared to cells in log-phase treated similarly as a monolayer. Thus 0.1 mM spermine killed ≈99% of the cells in a monolayer at 42°C or 43°C. Under the same conditions only 18% to 50% of the cells in spheroids were killed (Fig. 1).

Since most of the cells in spheroids are kinetically equivalent to plateau-phase cells, we have also tested the effect of spermine at 42°C on plateau-phase cells in a monolayer. Indeed, there was an increased resistance in such cells compared to cells in log-phase, due to a larger shoulder (data not shown). However, at 0.4 mM spermine there were $< 1 \times 10^{-4}$ survivors while spheroids displayed resistance to spermine at concentrations > 1mM. These results suggest that a low spermine concentration the resistance of spheroids could be due to a large fraction of non-cycling cells. This however, can not explain the relative lack of effect at spermine concentrations > 0.4 mM. It should be noted that the effect of increasing the temperature from 42°C to 43°C is to eliminate the shoulder on the survival curve of log-phase cells in monolayer. In spheroids the main effect is to reduce the

effect of increasing the temperature from 42°C to 43°C is to eliminate the shoulder on the survival curve of log-phase cells in monolayer. In spheroids the main effect is to reduce the level at which the survival levels off.

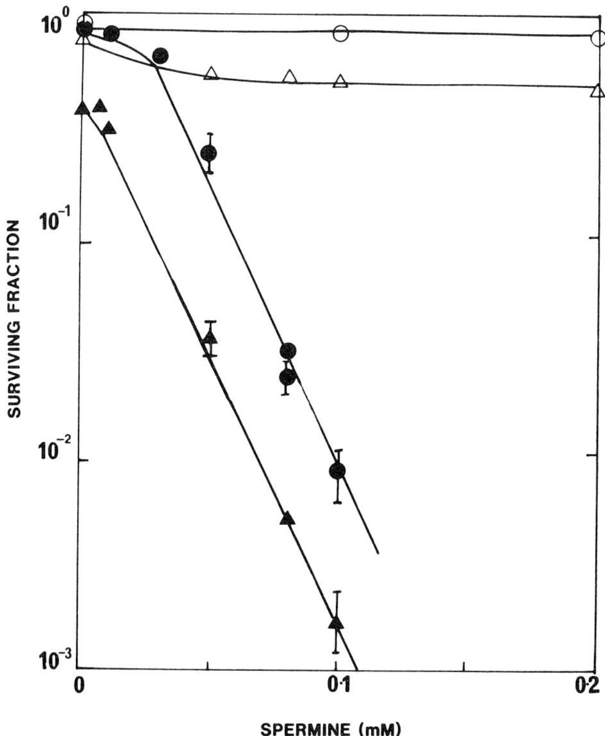

FIG. 1. Survival of EMT6 tumour cells to heat in the presence of spermine. Cells were exposed for 1 h either as a monolayer (closed symbols) or as spheroids (open symbols) at 42°C (circles) or 43°C (triangles) in the presence of various spermine concentrations. Error bars are shown when larger than the symbols.

The resistance of spheroids to heat plus spermine could be due to poor penetration of the polyamine to the inner cells. Prolonged exposure would presumably allow better penetration. Fig. 2 shows that prolonged exposure to 1 mM spermine at 42°C caused progressive cell killing. Heat by itself produced the usual sigmoidal survival curve with a tail beyond 4 h, suggesting development of thermotolerance. Spermine by itself became significantly toxic only at exposure times longer than 4 h.

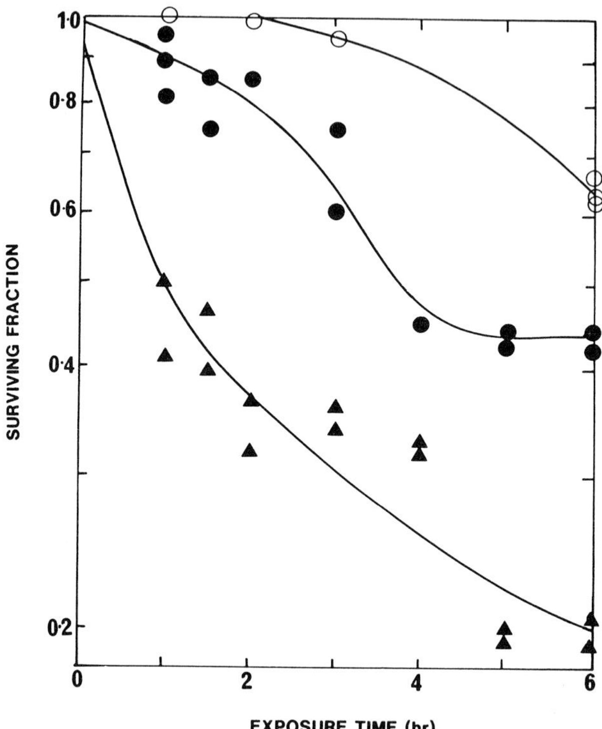

FIG. 2. Survival of EMT6 multicellular spheroids exposed for various times to 1 mM spermine at 37°C (o) or 42°C (▲) Closed circles denote exposure at 42°C in the absence of spermine.

The effect of heat plus spermine on regrowth of spheroids has been studied using plotd of mean spheroid diameter against days after treatment (Fig. 3). A progressive decrease in the rate of regrowth can be seen with increasing exposure time to 1 mM spermine at 42°C.

These results are consistent with the survival data shown in Fig. 2. A more quantitative analysis of the regrowth curves is based on the growth delay obtained from such data (18). This is shown in Fig. 4. Exposure at 42°C produced a small growth delay which was proportional to time and reached 1.9 days after 6 h. In the presence of 1 mM spermine heat was more cytotoxic, resulting in a growth delay of 5.8 days after 6 h treatment. At 37°C spermine had only very small effect on the regrowth of spheroids.

The uptake of labelled spermine into spheroids and monolayer cells

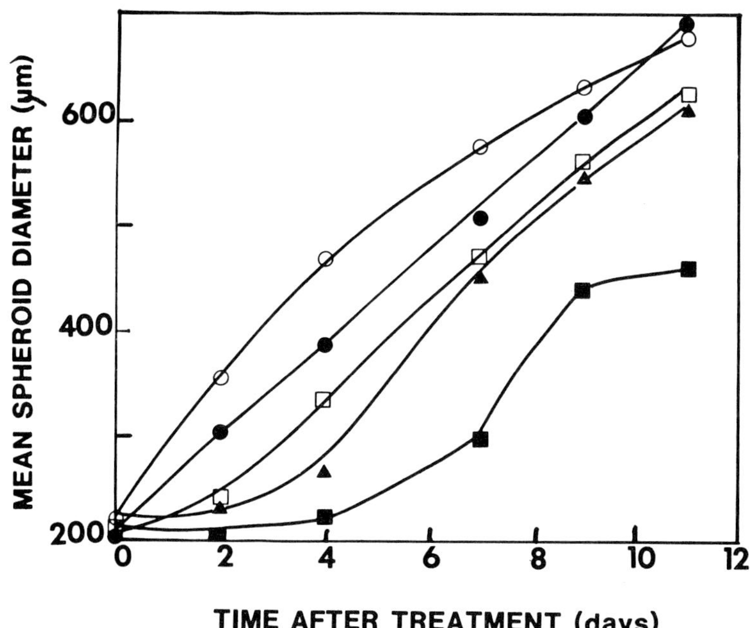

FIG. 3. Growth curves of EMT6 spheroids treated on day o for various times at 42°C in the presence of 1 mM spermine. S.E. of the mean (24 spheroids per each datum point) were smaller than 10% and are not shown. o, control; ●, 2 h at 42°C + spermine; ▲ 5 h at 42°C + spermine; ■, 6 h at 42°C + spermine; □, 6 h at 42°C.

following increasing incubation times at 37°C and 42°C is shown in Fig. 5. In monolayer, uptake was initially faster at 42°C but reached a plateau after 2 h. At 37°C uptake continued up to 4 h before levelling off. In spheroids uptake was again faster at 42°C than at 37°C but at both temperatures it was slower than in monolayer cells. However, in spheroids uptake continued up to 6 h without any indication of levelling off. Considering that the concentration of spermine in the growth medium for spheroids was 20-fold higher than for the experiments with monolayer cells, the relative ease of uptake in the latter is even more pronounced than is apparent from the data in Fig. 5.

DISCUSSION

The main conclusion from this work is that EMT6 multicellular spheroids are strikingly resistant to the enhancement of thermal killing by spermine when compared with monolayer cells. This is particularly evident at spermine concentrations $>1\times10^{-5}$M where survival of monolayer cells begins to drop sharply while that of spheroids does not (Fig. 1). Survival of the latter levels off at ≈5×10^{-5}M spermine, leaving ≈80% and 50% of the cells viable at 42°C and 43°C respectively. However, prolonging the exposure time up to 6 h and using 1 mM spermine at 42°C can lead to a progressive reduction in the surviving fraction of cells in spheroids (Fig. 2). This is consistent with the increased growth delay of spheroids observed after a similar treatment schedule (Fig. 4). Both end-points demonstrate syngergism heat and spermine in spheroids althought it is much smaller than in monolayer cells.

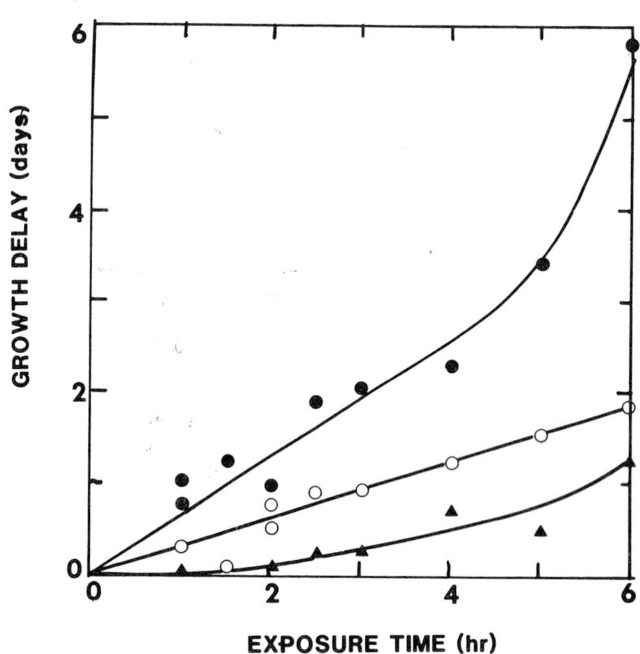

FIG. 4. Growth delay calculated from the results of growth experiments (e.g. see Fig. 3) in which EMT6 spheroids were exposed at 42°C (circles) with (●) or without (o) 1 mM spermine for various times. ▲, 1 mM spermine at 37°C.

The results of uptake experiments (Fig. 5) showed that spermine penetrates less easily into spheroids than in monolayer cells and that heat facilitates this process. These results, as well as the kinetics of cell survival and growth delay, suggest that the resistance of spheroids may at least in part be due to a difficulty of spermine in penetrating to the inner cells. This will make the use of spermine combined with heat treatment *in vivo* of doubtful value, considering the long treatment time required for a marked effect (up to 6 h).

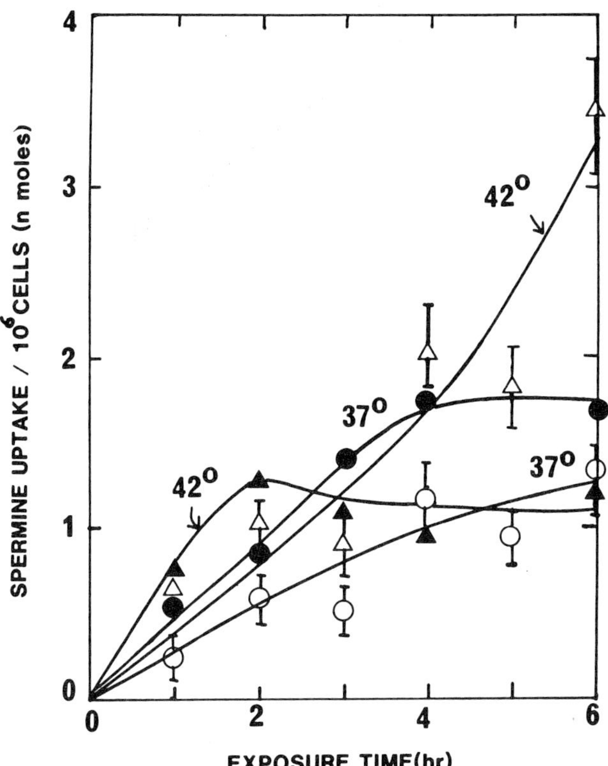

FIG. 5. Uptake of labelled spermine by multicellular spheroids (open symbols) and monolayer cells (closed symbols) as described in materials and Methods. Exposure to spermine was either at 37°C (circles) or at 42°C (triangles). S.E.M. are shown only for spheroids. For monolayer cells they were less than 10%

In spite of these observations, polyamines may be of value for cancer chemotherapy after appropriate modification. Thus, the N-acetyl derivatives of polyamines (7) penetrate more easily into cells, presumably due to the reduced positive charge. Since one of these derivatives was shown to retain some of the capacity to enhance thermal killing of monolayer cells (our unpublished observations) it is now being tested also in spheroids. If efficacy in spheroids is retained it will then be tested *in vivo*.
While the present work did not clarify the mechanism by which polyamines enhance thermal killing, it suggested that uptake into the cells may be an essential step. If confirmed by results using

N-acetyl derivatives, this would argue against a role for the plasma membrane as a primary target.

ACKNOWLEDGEMENT

We whish to thank Dr. P.R. Twentyman for helpful discussions. One of us (EBH) was supported in part by funds provided by the International Cancer Research Data Bank Programme of the National Cancer Institute, National Institute of Health (US), under contract No. N01-CO-65341 (International Cancer Research Technology Transfer - ICRETT) with the International Union Against Cancer.

REFERENCES

1. Ben-Hur, E. and Elkind, M.M. (1974): *Cancer Biochem. Biophys.* 1: 23-32.
2. Ben-Hur, E., Elkind, M.M. and Bronk, B.V. (1974): *Radiat. Res.* 58: 38-51.
3. Ben-Hur, E., Prager, A. and Riklis, E. (1978): *Int. J. Cancer* 22: 602-606.
4. Ben-Hur, E. and Riklis, E. (1978): *Int. J. Cancer,* 22: 607-610.
5. Ben-Hur, E. and Riklis, E. (1979): *Radiat. Res.,* 78: 321-328.
6. Ben-Hur, E. and Riklis, E. (1979): *Cancer Biochem. Biophys.,* 4: 25-31.
7. Blankenship, J. and Walle, T. (1978): Adv. Polyamines Res., 2: 97-110.
8. Field, S.B. and Bleehen, N.M. (1979): Cancer Treat. Rev., 6: 63-95.
9. Gerner, E.W., Cress, A.E., Stickney, D.G., Holmes, D.K., and Culver, P.S. (1980): *Ann. N.Y. Acad. Sci.* 334: 215-226.
10. Gerner, E. W. Holmes, D.K. Stickney, D.G. Noterman, J.A. and Fuller, D.M. (1980): *Cancer Res.,* 40: 432-438.
11. Gerweck, L., and Rottinger, E. (1976): *Radiat. Res.,* 67: 508-511.
12. Hahn, G.M. (1974): *Cancer Res.,* 34: 3117-3123.
13. Hahn, G.M. Braun, J., and Har-Kedar, I. (1975): *Proc. Natl. Acad. Sci. U.S.A.,* 72: 937-940.
14. Morgan, J.E. and Bleehen, N.M. (1981): *Br. J. Cancer,* 43: 384-391.
15. Shore, P.A., and Cohn, V.H.Jr. (1960): *Biochem. Pharmacol.* 5: 91-95.
16. Sutherland, R.M., and Durand, R.E. (1976): *Curr. Topics Radiat. Res.,* 11: 87-139.
17. Sutherland, R.M. Eddy, H.A., Bareham, B., Reich, K. and Vanantwerp, D. (1979): *Int. J. Radiat. Oncol. Biol. Phys.,* 5: 1225-1231.

18. Twentyman, P.R. (1980): *Br. J. Cancer,* 42: 297-304.
19. Westra, A., and Dewey, W.C. (1971): *Int. J. Radiat. Biol.,* 19: 467-477.
20. Yuhas, J.M., Li, A.P., Martinez, A.O., and Ladman, A.J. (1977): *Cancer Res.,* 37: 3639-3645.

Advances in Polyamine Research, Vol. 4, edited by U. Bachrach, A. Kaye, and R. Chayen. Raven Press, New York © 1983.

Myeloid Body Formation Under Conditions of Polyamine Stress

R. A. Campbell, T. LaBerge, *M. Campbell-Boswell, R. E. Brooks, and Y. B. Talwalkar

*Departments of Pediatrics and Anatomical Pathology, Oregon Health Sciences University, Portland, Oregon 97201; *Institute of Pathology, School of Medicine, Case Western Reserve University, Cleveland, Ohio 44100*

When one considers the chemical structures of polyamines, as well as their remarkable number of biological functions, many of which are critical to cell and host survival, it is not difficult to imagine unregulated alterations in intracellular and extracellular polyamine levels might result in adverse effects (13,2,7). Intracellular levels of polyamines are regulated by hormones, blood gases, nutrients, the ionic milieu and polyamine constituents of the extracellular space (6,21,38). Polyamines are taken up from the extracellular space by cells when there is an appropriate need (28,42). When one is dealing with a complex collection of highly specialized cells, as is found in higher organisms, certain organs appear to take over a large measure of responsibility for maintaining polyamines at acceptable levels outside of the cells. Based on present knowledge, the liver and the kidney probably share joint responsibility for preventing extracellular polymine excess (8,48,49). When extracellular polyamine homeostasis is disturbed as in the uncontrolled cellular proliferation of cancer or when kidney function is compromised, it behooves one to compare these clinical hyperpolyaminemic disorders of diverse etiology to determine if they share common symptoms, signs and biochemical derangements.

Cancer and uremic patients have weight loss and tissue wasting. They also have carbohydrate intolerance, disturbed cellular immunity, normocytic-normochromic anemias with shortened red cell survival time, similar bone marrow pictures and a tendency to peripheral neuropathy. One has to distinguish between acute, chronic and acute-on-chronic states in evaluating pathophysiological patterns, and hyperpolyaminemic disturbances are no exception. The presenting manifestations of cancer are often chronic ones of weight loss and fatigue. Cancer may be converted to an acute problem following tissue damage resulting from radiation or chemotherapy when symptoms such as pernicious vomiting typically occur. The same obtains with chronic uremia. The patient may complain of fatigue and weight loss. Chronic uremia becomes acute with intercurrent problems such as infection or terminal deterioration of kidney function. Intractible vomiting is the rule (56). Likewise, simple acute renal failure can be a highly symptomatic disturbance (46). Other hyperpolyaminemic conditions in which there is associated acute systemic toxicity include hyperemesis gravidarum and Reye's syndrome (53,9). It is no surprise that the parenteral administration of polyamines in man resulted in pernicious vomiting as well as other toxic signs (47). The symptoms, signs and metabolic disturbances of uremia and comparison of them with effects of polyamine administration to man, animals, cells and tissue culture has recently been reviewed (10). The anemia of uremia now appears to be due, in large part, to the associated disorders in polyamine metabolism (45). It is evident there is a remarkable concordance between the pathophysiological and chemical disturbances in uremia and those produced by the polyamines.

Very little work has been carried out to determine the effect of administration of polyamines on intact mammals. Studies suggest that the nervous system, the kidney, and possibly the liver are organs particularly susceptible to toxic concentrations of polyamines (50,20,59). Polyamines parenterally injected into humans caused pernicious vomiting, hematuria and proteinuria. Large parenteral doses in experimental animals caused convulsions, coma and death. Lower doses had a diuretic action on the kidney and also produced hypothermia and adypsia. Intermediate doses over several days caused acute renal failure as well as mild focal hepatic necrosis. In all experiments, spermine (SPM) was consistently more toxic than spermidine (SPD). Cadaverine (CDR) and putrescine (PTC) showed very little effect unless massive doses were employed.

The purpose of this study was to demonstrate the effects of sequential parenteral administration of SPM on subcellular morphological events occurring in the rat nephron.

METHODS

Pre-adolescent Sprague-Dawley rats received daily intraperitoneal injections of spermine tetrahydrochloride (Sigma Chemical Co., St. Louis, Mo.), 40 mg (23 mg of free base) per kg of body weight as a 0.27% solution in normal saline. Control animals, matched by age and sex, received like volumes of saline. Tissue preparations were initiated four hours after daily scheduled injections. Animals were ether anesthetized and immediately received prefixation tissue perfusion according to Maunsbach (36). Tissue fixation was accomplished by perfusion with a solution containing 1% gluteraldehyde in 0.1 molar phosphate buffer. One millimeter cubes of perfusion fixed kidney were stored in gluteraldehyde fixative at 4°C overnight. Tissues were then rinsed twice for 30 minutes in 0.1 molar cacodylate buffer (pH 7.4 in 6% sucrose solution), post-fixed for two hours in 3% osmium tetroxide (3% sucrose and 0.05 molar cacodylate buffer), dehydrated in graded ethanol solution and embedded in Araldite 502 (Ciba Products Co., Fairlawn, N.J.). Thin sections cut with glass knives on an LKB ultramicrotome were stained with uranyl acetate and lead citrate and examined on a Philips 200 electron microscope at 60 kv.

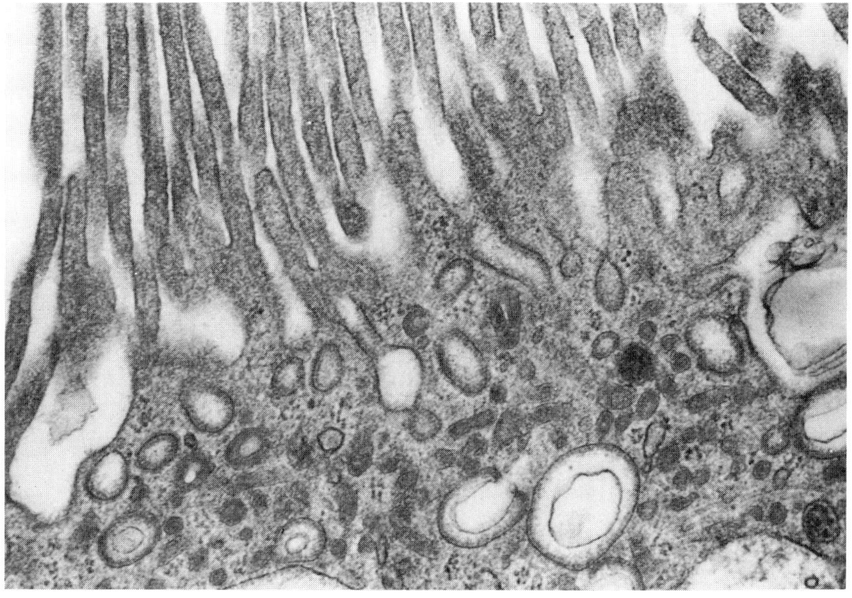

FIG. 1. Brush border endocytotic processes.

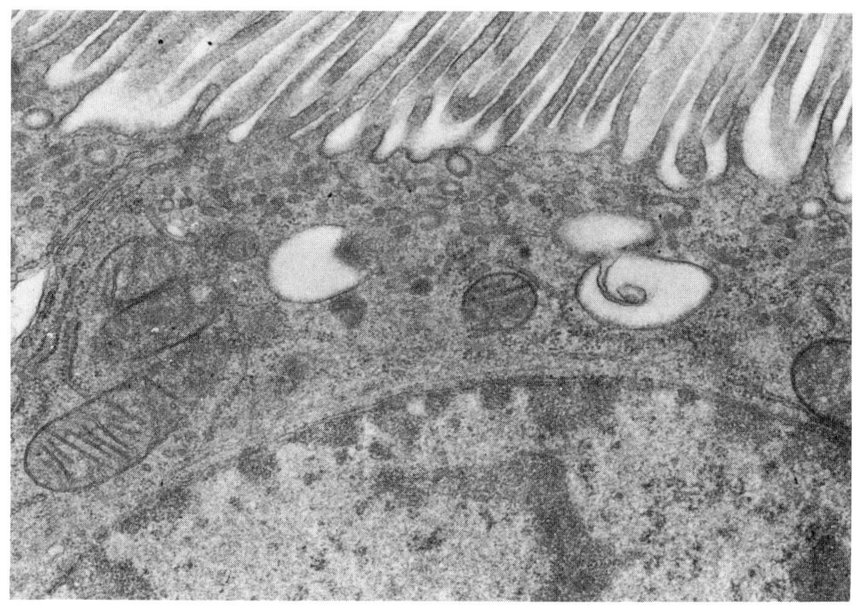

FIG. 2. Tubular vesicle-vacuole fusion and apparent transfer processes.

RESULTS

By 48 hours of single daily SPM injections some experimental animals began to show signs of ataxia and hind leg paralysis.

Figure 1 is an electron micrograph of the proximal tubule of a female rat treated for 24 hours. The brush border is intact and subapical endocytotic activity is prominent. The endocytotic vesicle which is composed of a fluid-filled space surrounded by a plasma membrane is shown to be undergoing change as it proceeds in an antiluminal migration into the cytoplasm. The vesicle increases in size and assumes a spherical shape. These enlarging vesicles show two membranous structures emerging as separate entities from the formerly unified single plasma membrane. The innermost of these is quite discrete and electron-dense, somewhat irregular in configuration and is derived from the outer surface of the plasma membrane which has become internalized. The boundary membrane for the vesicle is different in that its outer surface is sharp and quite electron-dense, but attached to its inner surface are structural elements which may be seen as highly organized striations poised at 90 degree angles over the entire inner aspect. This outer membrane originally derived from the inner component of the plasma membrane. It would appear that we are dealing with the evolution of a sphere within a sphere as membrane dehiscence takes place. In contrast to these endocytotic structures which are increasing in size, an alternate pattern of change in vesicles may occur with downward migration. As one examines the array of endocytotic vesicles displayed in the subapical region, one observes that the vesicles are generally larger immediately beneath the brush border, and as one proceeds in a more antiluminal direction, in fact, they are smaller, more compact and without visible vacuolization. The clear-cut double membrane gradually disappears. While treatment with SPM seemed to rapidly increase the number of endocytotic vesicles present in the proximal tubule cells of the experimental animals when compared to controls, morphometric analysis was not carried out to verify this impression. Thus, below the erect microvillus structures one sees the long tubular elements from which the endocytotic vesicles continuously pinch off, and these endocytotic vesicles, immediately after achieving discontinuity with the plasma membrane, appear destined to follow one of two kinds of behavior. The first path is that of swelling of the vesicle with dehiscence of membrane elements during the downward migration and the second, collapse and obliteration of the vesicle lumen as it progresses into the cytoplasm.

Figure 2. This tissue is from a SPM treated female rat, day 2. Endocytotic vesicles which were undergoing swelling and dehiscence have now assumed the proportion of subapical vacuoles. On the right side of this electron micrograph the process of endocytotic vesicle transfer of membranous material into a larger vacuole during the process of fusion is displayed. On the left, one sees accumulation of electron-dense material

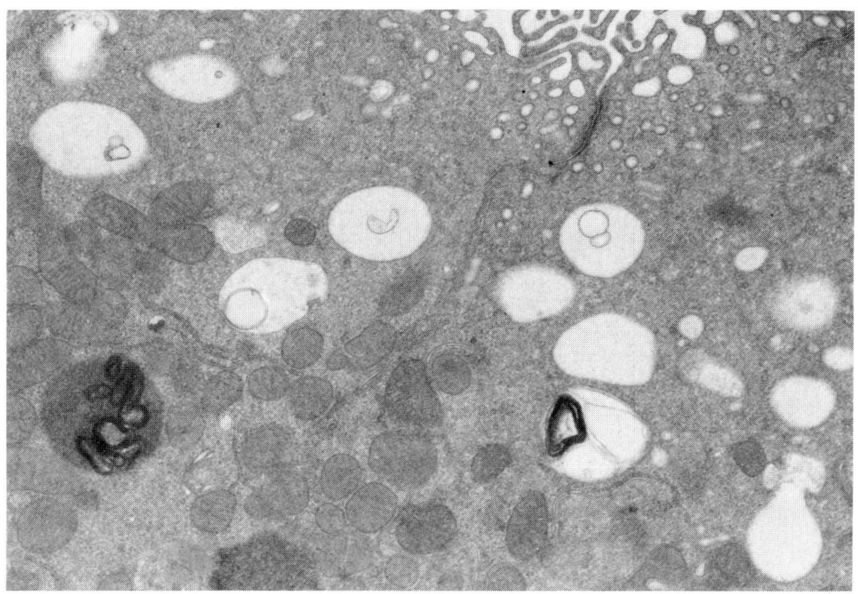

FIG. 3. Vacuolization of proximal tubule; vacuolar and lysosomal myeloid bodies.

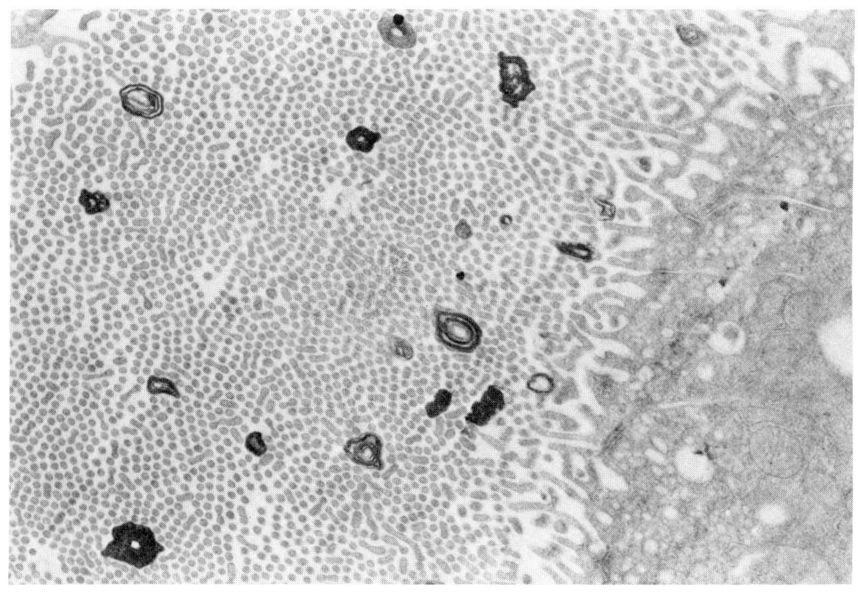

FIG. 4. Intermicrovillar myeloid bodies.

typically situated in a marginated position against the wall of the vacuole. Endocytotic vesicle-vacuolar fusions and vacuole-vacuole fusions with mixing of their contents are frequently seen at this stage of vacuolar development.

Figure 3. This proximal tubule cell from a female rat after three days of SPM treatment is quite typical of the morphological changes in the proximal tubular cells after continued SPM injections. The ringlet-like remnants of dehiscent outer plasma membrane accumulating in the enlarging vacuoles clearly indicates the endocytotic vesicle origin of these elements. Some of the vacuoles contain quantities of what appears to be proteinaceous detritus, while others are quite clear except for the membrane remnants. Other vacuoles (not shown) were completely filled with granular, moderately electron-dense material. In the lower right-hand corner of this figure typical vacuole-vacuole fusion is occurring. The prominent intravacuolar triangular shaped electron-dense laminated body with membranous streamers extending across the entire vacuole represents what will become a mature myeloid body (residual body) of the unicentric type. Immediately above these vacuoles one may observe downward streaming of endocytotic vesicles committed to swelling and ultimate fusion with larger vacuoles. A secondary lysosome is present in the lower left corner of the electron micrograph. It contains clusters of unicentric myeloid bodies. We have identified myeloid bodies in the secondary lysosomes four hours after the first injection of SPM. This occurs ordinarily before vacuoles of significant size or number develop. One may see immediately above this isolated lysosome a small vacuolar remnant with a marginated myeloid body within its lumen. Margination is the rule.

Figure 4 is a tangential cut across the proximal tubular brush border revealing the transverse anatomy of the microvillae and the immediate subapical regions of the cell in a female rat treated three days with SPM. What is striking here is the presence of a shower of myeloid bodies in the process of being delivered from the cell into the luminal fluid by exocytosis for passage into the urine. Characteristic electron-dense membranous laminations are present; however, please note, like snow flakes, the configuration of each myeloid body is unique.

Figure 5 has several morphological elements within it. In this female rat treated with SPM for three days, the prominent laminated figure-eight myeloid body is being extruded into the glomerular capillary lumen from the cytoplasm of the glomerular endothelial cell. The endothelial cell surrounds interposing spheroidal mesangial substance. This picture captures what one might suspect to be an improbable event, that of a myeloid body being expressed into the blood stream. We have not only observed free-floating myeloid bodies (not shown) within the lumina of capillaries but also collections of free-floating exocytotic ringlets. In contrast to myeloid bodies found in the brush border, those in the capillary endothelium of the glomerulus tend to have disproportionately large, clear unicenters.

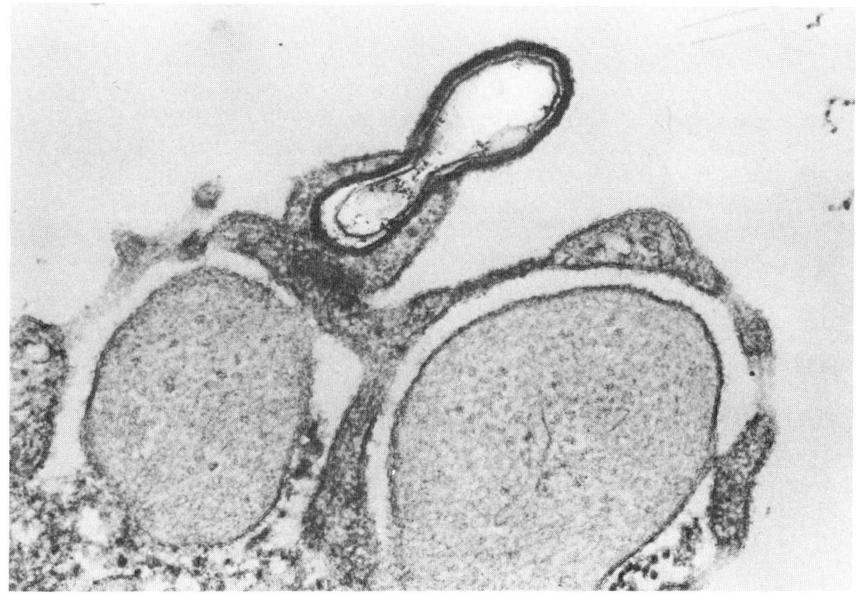

FIG. 5. Glomerular capillary endothelium with partially extruded myeloid body.

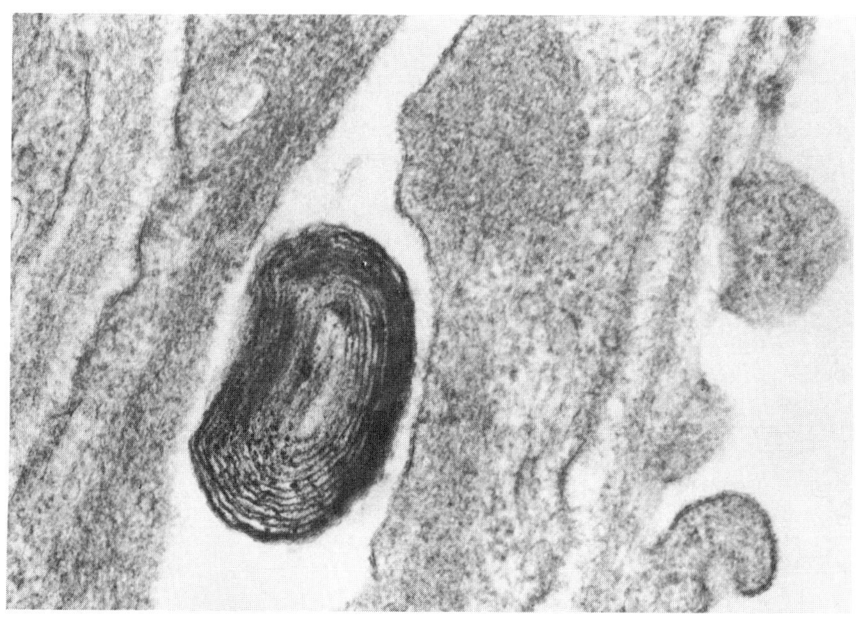

FIG. 6. Mature glomerular myeloid body in uriniferous space.

Figure 6. Here a myeloid body is wedged in juxtaposition to two visceral epithelial surfaces of the glomerulus. The capillary lumen may be seen on the right. The two prominent vertical stripes represent glomerular basement membranes. This large, mature, well-organized myeloid body with little vacuolar remnant in its center is floating within the uriniferous space. Source of it is unknown.

Continued treatment of the experimental animals with SPM for four or five days resulted in further packing of the subapical spaces with vacuoles, the breakdown of the brush border microvillae, loss of cytoplasmic integrity and discharge of the cellular contents into the tubular lumen. Preservation of the tubular basement membrane was a uniform finding.

DISCUSSION

Because of the kidney's homeostatic role in polyamine metabolism and the fact that urine is readily available for examination, the renal excretory system provides a remarkable opportunity for understanding the toxic effects of polyamines.

Perhaps the first clue to the possibility that the kidney played a role in polyamine metabolism in man dates back almost one hundred years to the discovery by Von Udranszky and Bauman that cystinuric patients' urine contained quantities of PTC and CDR (68). Twenty-five years ago the kidney was demonstrated to have a propensity for accumulation of polyamines (51). Somewhat later it was shown that a number of days were required for the complete release of injected polyamine by involved renal parenchyma (60). While the characteristic defects in the dibasic amino acid pathway were being defined in classical cystinuria and Lowe's syndrome, it was not until the last decade that the polyamines were again reported to be present in the urine of patients with these inborn errors of metabolism (15,27,5). A number of sensitive methods have since been developed which enable investigators to consistently identify and quantitate polyamines in body fluids (11). With the report of elevated serum polyamine levels in end-stage renal disease (uremia), a specific role for the kidney in PA homeostasis appeared secure (8). Evidence for the importance of the kidney in polyamine biotransformation was gleaned from the observations that not only was there excessive urinary excretion of PTC, CDR and SPD in patients with cystinuria, Lowe's and Fanconi's syndromes, but, contrary to the normal condition, there was also a deficiency in the acetylation of the urinary polyamine, SPD (4). Since polyamine acetylating enzymes had been isolated from renal cortical tissue, one could easily be persuaded to assert that these disorders of tubular transport resulted not only in the failure of normal biotransformation of the polyamines in the process of excretion, but that the polyamine burden derived from the extracellular pool and delivered to the kidney might, in fact, be much larger than previously supposed.

A binding site for SPM exists on the brush border of the proximal tubule of the rat kidney. Quantities of SPM required to produce 50% inhibition of the binding of the cationic peptide, aprotinin, were far less than those of a variety of organic polycations found in nature (29). Several aminoglycoside antibiotics showed inhibitory capacity of the same order of magnitude as SPM. Polyamines, through a mechanism of competitive inhibition, were shown to prevent the brush border membrane vesicle uptake of the aminoglycoside, gentamicin (31). Aminoglycosides are selectively accumulated in the rat renal cortex, and their gradual release over a period of days parallels the observations made with exogenously administered polyamines (34,18). The nephrotoxicity of the aminoglycoside antibiotics can also be ranked according to the number of primary and secondary amine groups present on the individual molecules (21). Thus, the available evidence suggests the nephrotoxic character of SPM, directed to the proximal tubule, results from the existence of closely juxtaposed primary and secondary amines within the molecule, a particular affinity for an existing transepithelial organic and cationic transport system and the availability of SPM in quantity for delivery to brush border binding sites.

The distinguishing features of SPM toxicity as seen in our experimental model, that is vacuolization, vacuolar and lysosomal myeloid body formation and the exocytosis of the myeloid bodies, may be seen occasionally in normal cells. These morphological changes are, however, isolated and modest events. A small subapical vacuole may be identified in a normal tubular epithelial cell. A secondary lysosome may be found, after examining a number of cells, to contain a myeloid body and we have occasionally caught a myeloid body undergoing extrusion out through the brush border into the tubular lumen. These observations suggest that what we observe are normal cellular mechanisms which only become prominent under appropriate metabolic stress. We never see in normal cells the massive vacuolization or the waves of exocytosis seen in the SPM treated animals. On the other hand, myeloid bodies found in the glomerular capillary endothelial cells and in the uriniferous space of the glomerulus to our knowledge have never before been described in experimental nephropathology.

The Swiss cheese effect resulting from the accumulation of the large cytoplasmic vacuoles in the brush border of SPM treated animals is associated with an increased number of downward migrating endocytotic vesicles of intermediate size. Whether or not this represents an increased rate of endocytosis or an increased number of endocytotic vesicles committed to undergoing fusion with the large vacuoles cannot be ascertained from these static electron micrographs. It is conceivable that downward migration has been slowed by the metabolic effects of SPM and, thus, one gets what appears to be accumulation of these vesicles. However, if that were the case, one would not expect to see such dramatic accumulation of the large vacuoles and great quantities of outer plasma membrane and detritus within their lumina.

Amines have been reported to alter endocytotic activity, but confirmation of our impressions would require rapid sequence marker and morphometric analysis to determine the precise sequence of endocytotic events. We did note the production of a few lysosomal myeloid bodies within four hours after the parenteral injection of SPM and this occurred before significant vacuolization. The polyamine antibiotic, gentamicin, has been shown to enter lysosomes in a matter of minutes (67). This suggests these endocytotic related macromolecular events happen with great rapidity and that endocytotic vesicle migration and ultimate disposition may evolve through more than one anatomical pathway. Since it is established that cells recycle a significant portion of the internalized membrane, the inordinate sequestration of membranous material in lysosomes and in large vacuoles seems neither appropriate nor efficient (54,55). Further, the accumulation of intravacuolar ringlets from endocytotic activity as seen in Figure 3 indicates that this membranous material is being isolated from normal recycling processes. At times, vacuoles are crowded wall-to-wall with endocytotic membrane remnants in the form of clustered ringlets (not shown).

Vesicle-vacuole and vacuole-to-vacuole fusions must be very active processes in the cell under stress of SPM excess. Based on current biophysical concepts of fusion, the small endocytotic vesicle contact with a larger vacuole may readily take place because the radius of curvature of the endocytotic vesicle is so small it can approach the vacuolar membrane with point charge resolution by London-Van der Waal forces (43). Thus, the hydrophobic lipid interaction of the components of the two bodies can overcome the tendency for electrostatic repulsion and membrane fusion is followed by formation of a single larger vacuole. With respect to vacuole-vacuole interactions in the tubule cell, the vacuoles typically achieve proximity, and one vacuole, often assuming an elipsoid shape as it increases in size in the subapical region, will begin pointing at one end to form an hyperacute angulation directed to intimate contact with the other vacuole. Results of this process can be seen in the flattened surface of the fusion process taking place in the vacuoles in the lower right-hand corner of Figure 3. We have noted the smaller vacuoles commonly deliver their vacuolar substance into a larger vacuole following membranous continuity as seen in Figure 2. By contrast, other endocytotic vesicles in a deeper antiluminal position have obliterated lumina. With apparent inward progression into the deeper regions of the cytoplasm, the vesicles can no longer be identified. Thus, those structural elements positioned most deeply must ultimately disperse. In certain cell lines this type of inward migration has been described as saltatory and is thought to result in committed endocytotic vesicles entering the GERL region of the cell (40).

In contrast to the vacuolar production of myeloid bodies clearly resulting from the accretion of membranes derived from inner membranes of the endocytotic vesicles as seen in Figure 3, the lysosomal myeloid bodies seem to arise de novo, at least in the proximal tubule of the kidney, rather than by fusion processes. Despite systematic examination, we have never observed fusion of endocytotic vesicles or vacuoles with lysosomal bodies. The dispersed substrates may present to the secondary lysosome following entry into the GERL complex for processing. While in some cells we noted both an increase in number and size of lysosomes, this was not always the case and we made no attempts to quantitate lysosomes. However, following the administration of SPM for two to three days, many lysosomes had at least one-third to one-half of their apparent volume composed of unicentric myeloid bodies. By day 4 of SPM administration, bizarre, extremely electron-dense cytosegresomes of giant proportions (not shown) were seen in the tubular epithelial cytoplasm.

Aminated compounds have been described as being lysosomotrophic (14). These are proposed to enter the lysosome by diffusion and because of the acid lysosomal pH to ionize and become trapped. A number of these compounds are psychotropic agents, amphipathic in nature, and have been demonstrated to induce myeloid body formation. Gentamicin, the most notable of the nephrotoxic aminoglycosides, produces striking numbers of myeloid bodies in lysosomes within the kidney tubule in animals and man (26,1). Lysosomal myeloid bodies induced by gentamicin have been detected in the human kidney several weeks after the administration of the antibiotic has stopped (3).

We have now demonstrated that under the stress of SPM excess, the proximal tubule cell is able to sequester plasma membrane in the form of myeloid bodies in the vacuolar system and also to induce myeloid body formation in lysosomes. We have identified similar bodies in glomerular capillary endothelial cells (please see Figure 5); these are of the vacuolar type. Free-floating myeloid bodies have been readily identified in the proximal tubular lumen, the interstitial space beneath the tubular basement membrane, the lumina of the blood vessels, and in the uriniferous space within the Malphigian corpuscle (please see Figure 6). Based on these findings, it would appear that normal enzymatic digestion by the SPM stressed cell of membranous electron-dense substances is prevented in both the subapical vacuoles and in the lysosomes. Myeloid bodies have a characteristic structural lamination and electron density and are considered to be derived primarily from accumulating phospholipid membrane, and, indeed, analysis of myeloid bodies indicates their chemical composition to be primarily membrane type structural phospholipids (1,35). Polyamines are well-known to be membrane stabilizers (32,62,63). Therefore, it is not inappropriate to suggest that the polyamines, particularly SPM with its strong cationic charge at biological pH, might bind electrostatically with negatively charged phospholipid membranes being endocytosed into the cytoplasma of the cell. This view is supported by the observation of high binding affinity for SPM in brush border vesicles. Moreover, the polyamines *in vitro* have the demonstrated capacity not only to aggregate purified phospholipids, but also to inhibit enzymes such as phospholipases and phosphodiesterase essential for the breakdown of these substances (58,16). Thus, the accumulation of phospholipids within the vacuoles may presently be rationalized by the explanation that SPM shields the substrate from enzymatic attack through ionic interaction with the polar head groups.

The characteristically unicentric nature of the myeloid body with its clear center suggests the membranes accumulating in the vacuoles undergo aggregation around a single initiating endocytotic ringlet. We would propose, based on present evidence, that the polyamines would neutralize the charge of the exposed phospholipids on the inner surface of the ringlet. This would then facilitate the flattening of the stabilized and coalescing ringlets as their outer hydrophobic surfaces come into contact. The result would be the laying down of a series of concentric layers of phospholipid around the central core of the myeloid body as seen in Figure 3. Following compaction and the gradual release of the vacuolar fluid surrounding the myeloid body, these residual structures begin to migrate luminally to the brush border where they undergo exocytosis (please see Figure 4). The same sequence of events take place in cells lining the blood vessel walls except, in this instance, the extruded myeloid bodies would, perforce, go to the spleen or liver for removal from the blood. Since we note exocytosis of myeloid bodies by the second and third day of SPM treatment in our experimental animals, this would suggest the process might serve as an acute mechanism for ridding the cell of excess polyamines. The protective lysosomal sequestration and digestion of the polyamine phospholipid complexes would result in a much slower release of polyamines from the cell. This is consistent with the findings of retention of SPM and SPD, as previously described, in the renal parenchyma for a number of days and the findings of retention of the antibiotic gentamicin.

The reasons for the incremental toxic vacuolization of the cells with continued administration of polyamines is not entirely clear. It is reasonable to conclude that a significant proportion of the fluid of the vacuoles results from the demonstrated accretion of endocytotic vesicles. This being the case, the vacuolar fluid should be rich in sodium, chloride and calcium inasmuch as the proximal tubule brush border fluid is similar in composition to that of plasma. Because of their enlarging size, water is certainly not being rapidly transported out of these vacuoles. However, exchange of water and ions could be taking place. Several forces at work preventing vacuolar regression may be operating. Vacuoles, which have also been called receptosomes or heterophagic vacuoles, bring with them, at least in several cell systems, a number of enzymes which are present in the plasma membrane of the cell (39,40). In addition, over time it would appear that the lysosomal marker enzyme, acid phosphatase, accumulates in these vacuoles indicating they have a digestive function similar in some respects to secondary lysosomes (61). The exo-enzymes with associated receptor markers to appropriate plasma membrane receptors may be incorporated into the vesicle. Digestion as well as transport may be major functions normally carried out by these organelles so that dissolution of membrane and disappearance of the vesicles into the regions of the GERL and secondary lysosomes would be the more usual course of events in an unstressed cell. Contemplating these organelles accumulating in great numbers, the question of a normal process being slowed or completely inhibited by SPM metabolic stress has to be seriously entertained. Inhibition of the normal breakdown of phospholipids to lysophospholipids might reduce the permeability of the vesicle-vacuole system (52,33) and prevent disassembly. The suggestion that the integrity of these vacuolar outer membranes (inner plasma membranes) is intact is consistent with the general finding that myeloid bodies tend to marginate and remain inside in the vacuoles. In addition, the apparent digestion (or depolymerization) and dispersion of the striated material, i.e., the striations disappear and the outer membrane of the vacuole becomes attenuated and discrete (Figures 1 and 3), suggest the vacuole is partially successful in carrying out some digestive functions. The possibility that increased osmotic forces resulting from substrate breakdown may play a role in preserving the size of these vacuoles, even when they are great distances from the incoming stream of endocytotic vesicles, must be considered. In addition, under normal circumstances one would not expect to see the tremendous accumulation of extracellular fluid as one sees during this toxic vacuolization. It is well established that polyamines may inhibit transmembranous transport (8,10). Polyamines do inhibit _in vitro_ Na^+-K^+ ATPase activity (44). What mechanisms are operating to prevent the gradual transfer of fluid into the cytosol remains to be determined. Finally, the accumulation of numerous large vacuoles within the cell might seriously compromise normal tubular function. This appears to be the case, for when we treat an experimental animal with SPM long enough the proximal tubule cells undergo complete necrosis. Certainly, as may be noted in Figure 4, the massive exodus of myeloid bodies into the tubular lumen is not without purpose. Inasmuch as it is well established that the concentration of polyamines within cells are precisely regulated in order to carry out vital functions, it would not be surprising that vigorous attempts might be made to either remove excess polyamines by exocytosis as is summarized in Fig. 7, or by sequestration within the secondary lysosomes (24).

The phenomenon of tolerance has been reported following the chronic parenteral administration of both polyamine and aminoglycoside antibodies (51,22). In the case of polyamines, a number of well-defined enzymatic systems might be activated to promote peptidization, acetylation and oxidative biodegradation. Thus, larger quantities of spermine could be processed (25,37, 57,58). Further, the visible and large scale elimination of outer plasma membrane does suggest the likelihood that both the cationic intoxicant and its membrane receptors are participating in the exocytotic process. Such events would reduce the number of receptors available for recycling into renewed plasma membrane and, perforce, would decrease the rate of intracellular polyamine accumulation. Since the nephrotoxic aminoglycosides do not undergo biotransformation, selective changes in plasma membrane phospholipid constituents of functioning residual and regenerating tubular epithelial cells provide, with our present knowledge, a reasonable explanation for the development of aminoglycoside tolerance. In effect, the consequences of exocytosis and sequestration would also permit a defensive form of cellular transport down-regulation. Such a mechanism may very well have evolved over time to protect organisms against noxious environmental as well as endogenous amines. Chloroquin, an amphiphilic cation which induces myeloid body formation, has recently been reported to impair receptor recycling by both macrophages and fibroblasts (23,64,65).

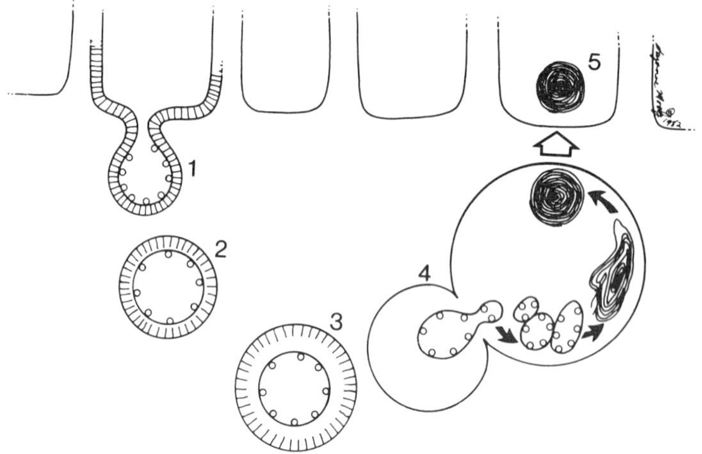

FIG. 7. Initiation of endocytosis (1), separated vesicle (2), dehiscence of membranes (3), vesicle fusion with subapical vacuole, transfer and myeloid body formation (4), myeloid body exocytosis (5).

Myeloid bodies have been described in the kidneys of normal animals and man (17,66). Remarkably enough, there seems to be a sex difference in the number of myeloid bodies present in animal kidneys (30). Male mice have increased numbers of myeloid bodies in their proximal tubular cells. This intracellular sex dimorphism is eliminated by castration in the male mouse. Production of myeloid bodies in the kidney occurs following the administration of testosterone. Of interest are the observations that animals treated with testosterone or other anabolic agents have increased quantities of polyamines in their urine (41).

This would suggest that the anabolic effects of androgenic substances, even under normal conditions of sexual differentiation, increase the total number of polyamines delivered to the kidney in the male animal as a result of growth stimulation and increasing body muscle mass. While testosterone is known to cause renal hypertrophy, induction of myeloid bodies as a direct normal target organ effect would seem highly improbable. We have found SPM to be more toxic clinically in our male adolescent Sprague-Dawley rats but have not found this to be the case in SPM treated adolescent B Albino C mice.

In making systematic comparisons between SPM induced tissue changes in experimental animals and those from biopsies of patients with Reye's syndrome and toxemia of pregnancy, we note remarkable concordance in the disordered configurations of intracellular organelles. Since these conditions have associated hyperpolyaminemia, the possibility of a cause and effect relationship between disturbed polyamine metabolism and the pathological changes noted in these clinical disorders must be raised (9,12). Studies concerning altered polyamine metabolism and associated tissue changes in these clinical syndromes will be the subject of separate publications.

Summary

SPM induces characteristic nephrotoxic fine-structure changes in the cytoplasma of several glomerulo-tubular cell types. These morphological alterations are most striking in the epithelial cells of the proximal convoluted tubule. We have documented the following cellular processes: (1) dehiscence of endocytotic plasma membrane components, (2) vesicle-vacuolar transfer of outer membrane phospholipid, (3) the accumulation of fluid vacuoles through vesicle-vacuole and vacuole-vacuole fusions, (4) the formation of myeloid bodies in both subapical vacuoles and secondary lysosomes, and (5) the exocytosis of vacuolar myeloid bodies. These quantitative deviations from normal reveal visible macromolecular mechanisms suitable for elimination and sequestration of noxious amines. Large accumulations of sequestered outer plasma membrane phospholipids suggest receptor down-regulation with consequent reduction in polyamine transport may be operating.

The strategic position of the kidney in the transport, metabolism and excretion of polyamines helps explain the particular susceptibility of the organ to experimental intoxication. The involvement of not only proximal tubule cells but also glomerular capillary endothelial and visceral epithelial cells point to a more general, albeit less dramatic, cellular response to polyamine stress. Inasmuch as SPM is a normal endogenous metabolic product, the observer encountering the presence of the cytopathological stigmata herein described in such study preparations as human biopsy material, specimens from intact experimental animals or cells in tissue culture should entertain the possibility of co-existing polyamine stress.

ACKNOWLEDGMENTS

We wish to thank Donna Abbott, Norma Fritz, Mary C. Campbell and Ann Childers for help in the preparation of the figures and manuscript.

We gratefully acknowledge support through gifts from the Selwyn Bingham family and a grant from the Chiles Foundation.

REFERENCES

1. Aubert-Tulkens, G., Van Hoof, F., and Tulkens, P. (1979): Lab. Invest., 40:481-491.
2. Bachrach, U. (1973): Functions of Naturally Occurring Polyamines. Academic Press, Inc., New York and London.
3. Bennett, W.M., Gilbert, D.N., Houghton, D., and Porter, G.A. (1977): West. J. Med., 126(1):65-68.
4. Berry, H.K., Glazer, H.S., Denton, M.D., and Fogelson, M.H. (1978): In: Advances in Polyamine Research Volume 2, edited by R.A. Campbell, D.R. Morris, D. Bartos, G.D. Daves, Jr., and F. Bartos, pp. 313-318. Raven Press, New York.
5. Bremer, H.J., Kohne, E., and Enders, W. (1971): Clin. Chim. Acta, 32:407.
6. Caldarera, C.M., Zappia, V., and Bachrach, U., editors (1981): Advances in Polyamine Research Volume 3. Raven Press, New York.
7. Campbell, R.A., Morris, D.R., Bartos, D., Daves, G.D., Jr., and Bartos, F., editors (1978): Advances in Polyamine Research Volumes 1 and 2. Raven Press, New York.
8. Campbell, R., Talwalkar, Y., Bartos, D., Bartos, F., Musgrave, J., Harner, M., Duri, H., Grettie, D., Dolney, A.M., and Loggan, B. (1978): In: Advances in Polyamine Research Volume 2, edited by R.A. Campbell, D.R. Morris, D. Bartos, G.D. Daves, Jr., and F. Bartos, pp. 319-343. Raven Press, New York.
9. Campbell, R.A., Isom, J.B., Bartos, D., and Bartos, F. (1979): Clin. Res., 27(1):116A.
10. Campbell, R.A., Bartos, F., and Bartos, D. (1981): In: Pediatric Nephrology, edited by A.B. Gruskin and M.E. Norman, pp. 484-493. Martinus Nijhoff Publishers, The Hague, Boston, London.
11. Campbell, R.A., Grettie, D.P., Bartos, D., and Bartos, F. (1981): In: Advances in Polyamine Research Volume 3, edited by C.M. Caldarera, V. Zappia and U. Bachrach, pp. 409-423. Raven Press, New York.
12. Casti, A., Reali, N., Orlandini, G., Ferrari, B., Salvadori, B., and Bernasconi, S. (1981): In: Advances in Polyamine Research Volume 3, edited by C.M. Caldarera, V. Zappia, and U. Bachrach, pp. 463-472. Raven Press, New York.
13. Cohen, S.S. (1971): Introduction to the Polyamines. Prentice Hall, Inc., New Jersey.

14. deDuve, C., DeBarsy, T., Poole, B., Trouet, A., Tulkens, P., and Van Hoof, F. (1974): Biochem. Pharmacol., 23:2495-2531.
15. Dent, C.E., and Rose, G.A. (1951): Q.J. Med., 20:205.
16. Eichberg, J., Zetusky, W.J., Bell, M.E., and Cavanaugh, E. (1981): J. Neurochem., 36(5):1868-1871.
17. Ericsson, J.L.E., Trump, B.E., and Weibal, J. (1965): Lab. Invest., 14:1341.
18. Fabre, J., Rudhart, M., Blanchard, P., and Regamay, C. (1976): Kidney Int., 10:444-449.
19. Falco, F.G., Smith, H.M., and Arcieri, G.M. (1969): J. Infect. Dis., 119:406-409.
20. Fisher, E.R., and Rosenthal, S.M. (1954): Arch. Pathol., 57:244-253.
21. Gaugas, J.M., editor (1980): Polyamines in Biomedical Research. John Wiley and Sons, Chichester, New York.
22. Gilbert, D.N., Houghton, D.C., Bennett, W.M., Plamp, C.E., Reger, K., and Porter, G.A. (1979): Proc. Soc., Exp. Biol. Med., 160:99-103.
23. Gonzalez-Noriega, A., Grubb, J.H., Talkad, V., and Sly, W.S. (1980): J. Cell Biol., 85:839-852.
24. Heby, O., and Andersson, G. (1980): In: Polyamines in Medical Research, edited by J.M. Gaugas, pp. 17-34. John Wiley and Sons, Chichester, New York.
25. Hölltä E. (1977): Biochemistry, 16(1):91-100.
26. Houghton, D.C., Campbell-Boswell, M.V., Bennett, W.M. Porter, G.A., and Brooks, R.E. (1977): Clin. Nephrol., 10(4):140-145.
27. Jagenburg, O.R. (1959): Scand. J. Clin. Lab. Invest., (Suppl 43), 11:5.
28. Jänne, J., Alhonen-Honigsto, L., Seppänen, P., and Höllta, E. (1981): In: Advances in Polyamine Research Volume 3, edited by C.M. Caldarera, V. Zappia and U. Bachrach, pp. 85-95. Raven Press, New York.
29. Just, M., and Habermann, E. (1977): Arch. Pharm., 300:67-76.
30. Koenig, H., Goldstone, A., and Hughes, C. (1978): Lab. Invest., 39(4):329.
31. Lipsky, J.J., Chang, L., Saktor, B., and Lietman, P.S. (1980): J. Pharmacol. Exp. Ther., 215:390.
32. Little, J.R. (1962): Proc. Soc. Exp. Biol. Med., 109:910-912.
33. Lucy, J.A. (1970): Nature, 227:814.
34. Luft, F.C., and Kleit, S.A. (1974): J. Infect. Dis., 130: 656-659.
35. Lüllman-Rauch, R. (1979): In: Lysosomes in Applied Biology and Therapeutics, edited by J.T. Dingle, P.S. Jaques, and I.H. Shaw, pp. 49-130. North-Holland Publishing Company, Amsterdam, New York, Oxford.
36. Maunsbach, A.B. (1966): J. Ultrastruct. Res., 15:283-309.
37. Morgan, D.M.L. (1980): In: Polyamines in Biomedical Research, edited by J.M. Gaugus, pp. 285-305. John Wiley, Chichester, New York.

38. Morris, D.R., Marton, L.J., editors (1981): Polyamines in Biology and Medicine. Marcel Dekker, Inc., New York and Basel.
39. Neufeld, E.F., Sando, G.N., Garvin, A.J. and Rome, L.H. (1977): J. Supramol. Struct., 6:95-101.
40. Pastin, I.H., and Willingham, M.C. (1981): Science, 214: 504-509.
41. Pearson, J.T., and Buttery, P.J. (1979): Proc. Nutr. Soc., 38:91A.
42. Pohjanpelto, P. (1976): J. Cell Biol., 68:512-520.
43. Poste, G., and Allison, A.C. (1971): J. Theor. Biol., 32:165.
44. Quarfoth, G., Ahmed, K., and Foster, D. (1978): Biochem. Biophys. Acta, 526:580-590.
45. Radtke, W.H., Rege, A.B., LaMarche, M.B., Bartos, D., Bartos, F., Campbell, R.A., and Fisher, J.W. (1981): J. Clin. Invest., 67:1623-1629.
46. Raskin, N.H., and Fishman, R.A. (1976): N. Engl. J. Med., 294:143-148.
47. Risetti, U., and Mancini, G. (1954): Acta. Neurol. (Napoli), 8:911-915.
48. Rosenblum, M.G., and Russell, D.H. (1977): Cancer Res., 37:47-51.
49. Rosenblum, M.G., Durie, B.G.M., Salmon, S.E., and Russell, D.H. (1978): Cancer Res., 38(10):3161-3163.
50. Rosenthal, S.M., Fisher, E.R., and Stohlman, E.F. (1952): Proc. Soc. Exp. Biol. Med., 80:432-434.
51. Rosenthal, S.M., and Tabor, C.W. (1956): J. Pharm. Exp. Ther., 116:131-138.
52. Rossi, C.R., Sartorelli, L., Tato, L., and Siliprandi, N. (1964): Arch. Biochem. Biophys., 107:170-175.
53. Russell, D.H., and Durie, B.G.M. (1978): Polyamines as Biochemical Markers of Normal and Malignant Growth. Raven Press, New York.
54. Schneider, Y-J., Tulkens, P., deDuve, C., and Trouet, A. (1979): J. Cell Biol., 82:449-465.
55. Schneider, Y-J., Tulkens, P., deDuve, C., and Trouet, A. (1979): J. Cell Biol., 82:466-474.
56. Schreiner, G.E., and Maher, J.F. (1961): Uremia: Biochemistry, Pathogenesis, and Treatment. Charles C. Thomas, Springfield, Ill.
57. Seale, T.W., Chan, Y.W., Shulka, J.R., and Rennert, O.M. (1979): Arch. Biochem. Biophys., 198(1):164-174.
58. Sechi, A.M., Cabrini, L., Pasquali, P. and Lenaz, G. (1978): Arch. Biochem. Biophys., 186(2):248-254.
59. Shaw, G.G. (1979): Biochem. Pharmacol., 28:1-6.
60. Siimes, M. (1967): Acta Physiol. Scand. (Suppl), 298:1-66.
61. Straus, W. (1964): J. Cell Biol., 20:497-507.
62. Tabor, C.W. (1960): Biochem. Biophys. Res. Comm., 2:117-120.
63. Tabor, C.W. (1962): J. Bacteriol., 83:1101-1111.
64. Tietze, C., Schlesinger, P., and Stahl, P. (1980): Biochem. Biophys. Res. Commun., 93:1-8.

65. Tietze, C., Schlesinger, P., and Stahl, P. (1982): J. Cell Biol., 92:417-424.
66. Tisher, C.C., Bulger, R.E., and Trump, B.F. (1966): Lab. Invest., 15:1357.
67. Vera-Roman, J., Krishnakantha, T.P., and Cuppage, F.E. (1975): Lab. Invest., 33:412-417.
68. Von Udranszky, L., and Baumann, E. (1889): Z. Physiol. Chem., 13:562.

Advances in Polyamine Research, Vol. 4, edited by U. Bachrach, A. Kaye, and R. Chayen. Raven Press, New York © 1983.

Evidence That an Elevated Level of Ornithine Decarboxylase Activity is an Essential Component of Tumor Promotion

R. K. Boutwell

McArdle Laboratory for Cancer Research, University of Wisconsin, Madison, Wisconsin 53706

PRINCIPLES OF SKIN TUMOR PROMOTION

Skin tumor formation in mice is readily divisible into two components that are named initiation and promotion. Initiation is accomplished by treating the skin with a carcinogen at a dose that is sufficiently low that tumors develop only rarely during the animals life. Skin so treated appears to contain some cells that are irreversibly altered (initiated) so that application of a second agent will elicit many tumors, both benign and malignant. The second agent has the unique capability of eliciting tumors in initiated skin but is not an effective carcinogen when applied to the skin of mice that are not initiated. The second agent, called a promoting agent, is best exemplified by the diterpene, 12-O-tetradecanoylphorbol-13-acetate (TPA), which was isolated from croton oil. TPA promotes skin tumors in mice when applied twice weekly to the skin of initiated mice in the dose range of about 1 to 10 nmol per 0.2 ml of acetone, depending on the genetic background of the mice. It causes responses in cultured cells at levels of 10^{-7} to 10^{-10} nM. A number of reviews on the subject of tumor promotion exist (6-8,11,21).

The initiation-promotion model system in mouse skin has been very useful for studying the biochemical mechanism of tumor promotion utilizing TPA and its congeners. The response of mouse epidermis to TPA is a typical pleiotrophic response. Within 2 minutes, the level of cyclic adenosine-3':5'-monophosphate and cyclic guanosine 3':5'-monophosphate in epidermis begins to increase (18). Elevated levels of prostaglandins E and F are detected within 30 mintues (1,10) and the turnover of phosphatidyl choline is increased (19). A many-fold induction of ornithine decarboxylase activity reaches a peak by 5 to 6 hr and returns to normal levels with a half-life of 15-20 min (16,17). These responses are followed by increased incorporation of appropriate precursors into the RNA, protein, and DNA of epidermal cells (4). After 3 to 5 days, the epidermis has increased from a thickness of about 2 cell layers to many cell layers (7). These responses are all reversible in skin that has not been initiated and after about 3 weeks there is little remaining evidence of the fact that TPA had been applied to the skin. A number of other biochemical and morphological effects have been described (8,11,21).

It must be emphasized that many of the responses may be accomplisehd by agents that do not promote tumors. For example, mezerein is as potent as TPA for the induction of ODC yet mezerein does not promote tumors (14). Thus, although the evidence for an essential role for an induced level of ODC activity will be presented in this report, an elevated level of ODC is not sufficient to account for skin tumor promotion. In addition to an elevated level of ODC, it is likely that TPA causes several other responses which will eventually be proven to be essential for promotion. These include increased DNA synthesis and hyperplasia (4,7), elimination of metabolic cooperation (9,33), inhibition of maturation of the epidermal stem cell (5,11,21), and the production of dark cells (13).

The mouse skin system is a uniquely valuable model for studying tumor promotion. First, TPA is remarkably effectve at dose-levels of less than 10 nmol per application and derivatives of TPA varying greatly in activity are available (3,11,21). Furthermore, TPA is effective upon direct application to the skin without the necessity for metabolic activation and because it acts directly (21), the complications of systemic administration are avoided. Tumors arise as early as 5 to 6 weeks after TPA treatment of initiated mice, their presence is readily measured in quantitative terms and the biochemical responses to TPA are readily correlated with the tumor response because identical treatments can be given to elicit both the biochemical response and the tumor response (11,21). Finally, by the use of appropriate agents, one is able to ascertain if both the biochemical response and the tumor response are altered in a parallel fashion by the test agent. This last point will constitute the major theme of this report.

There is one final set of facts about the initiation-promotion model that aids ones conception of the model. First, there are initiators that have minimal biochemical and perhaps no morphological effects on the skin, for example, urethan and 7,12-dimethylbenz[a]anthracene at minimal initiating doses (20). In contrast the promoter, TPA at promoting dose levels, causes mouse skin to exhibit many of the biochemical and morphological characteristics of neoplastic cells (6,11,21). Moreover, exposure of cells in culture to TPA causes the cells to mimic the transformed state (32). Thus, one may hypothesize that in those few cells that are initiated, the changes accomplished by TPA are not reversible but rather locked-in due to a subtle defect introduced during the process of initiation (16). These few cells may be responsible for the clonogenic origin of the tumors that become grossly detectable.

ODC ACTIVITY AND TUMOR PROMOTION

The first evidence that the induction of ODC by TPA and the resultant elevation of putrescine and spermidine may be an essential component of the process of tumor formation was reported by O'Brien et al. (16,17). TPA induces ODC activity as much as 300-fold at 4 to 5 hours after a promtoing dose of TPA. The stimulation of S-adenosyl methionine activity is about 6-to 7-fold with a peak of activity at 9-12 hours. The stimulation of activity of both enzymes is dependent on the dose of TPA and correlates well with the promoting ability of these doses on mouse skin. Cycloheximide pretreatment abolishes the induction of these enzymes by TPA and furthermore reveals that the half-life of the induced activity is about 17 and 41 minutes for ODC and S-adenosylmethionine decarboxylase, respectively. Further studies revealed that the inductive ability of a series of structurally varied phorbol esters for the activity of these two enzymes correlates well with the promoting potency of these phorbol derivatives. Furthermore, structurally unrelated tumor promoters including iodoacetic acid, anthralin, and Tween 60 also induce these same enzymes. In addition, while initiating levels of 7,12-dimethylbenz[a]anthracene is not capable of inducing the polyamine biosynthetic enzymes, completely carcinogenic dose levels are effective. In addition, ultraviolet light at dose levels that are carcinogenic for mouse skin induces polyamine biosynthetic activity (27).

In addition to the fact that skin carcinogens and promoters cause the induction of ODC, the resulting tumors consistently show high levels of ODC activity but variable levels of S-adenosylmethionine decarboxylase activity (2,15). Determination of the half-life of the two enzymes in induced skin tumors by means of cycloheximide gave interesting results. The half-life of S-adenosylmethionine decarboxylase in squamous papillomas was found to be 45 minutes, a value comparable to the induced level in non-neoplastic skin. In contrast, the half-life of ODC activity in mouse skin papillomas declines at a rate similar to that in TPA-treated epidermis for only the first 15 to 20 minutes after cycloheximide injection. Thereafter, at times when protein synthesis was 90% inhibited, the ODC activity actually reverts to high levels. It appears that the level of ODC activity is stabilized at high levels in papillomas, suggesting that the control mechanism for ODC activity is lost or deranged in the tumors. It will be of interest to determine if this loss of control is critical to the altered differentiation and/or growth rate of the tumor cells.

MODIFYING FACTORS

Additional evidence for a role for deranged control of ODC activity as one of the essential components of tumor promotion (and presumably of carcinogenesis) is based on the ability of certain agents to inhibit both the induction of ornithine decarboxylase activity and the formation of tumors. The molecular mechanism of the induction of ODC activity by TPA is not understood. We know, for example, that TPA increases the level of cyclic nucleotides in the epidermis within minutes and that the rate of turnover of phosphatidyl choline is accelerated. The latter serves as a source of arachidonic acid, a precursor of prostaglandins. As a part of the investigation into the mechanism of induction of ODC by TPA, we found that indomethacin and other inhibitors of the cyclooxygenase system inhibited the induction of ODC activity by TPA (23,28). The inhibition was dependent on the dose of indomethacin applied to the skin two hours before TPA. The threshold for inhibition was approximately 10 nmol and inhibition was almost complete at 2800 nmol indomethacin. Indomethacin was most effective when applied two to five hours prior to TPA; the inhibition was nearly eliminated when applied 2 hours after TPA or 12 to 24 hours prior to TPA. Furthermore, the inhibition could be reversed in a dose-dependent manner by prostaglandin E_2 applied at the same time as TPA; as little as 2 nmol of prostaglandin E had activity and the inhibition of ODC induction by 280 nmol of indomethacin was completely eliminated by 40 nmol of prostaglandin E_2. Indomethacin did not inhibit the induction of S-adenosylmethionine decarboxylase by TPA.

Just as the effect of alterations in the dose and structure of TPA correlated with respect to the induction of ODC and tumor promotion, the correlation of the two endpoints is observed following indomethacin treatment. Furthermore, treatment of mouse skin with prostaglandin E_2 alone was capable neither of inducing ornithine decarboxylase activity nor of promoting tumor formation (23).

These data on the effect of indomethacin on the responses of mouse skin to TPA strengthen the possibility that the induction of ODC is an essential phenotypic change that is essential for tumor promotion.

Analogous results have been obtained with a number of retinoids (compounds related to vitamin A). Namely, certain retinoids inhibit ODC induction as well as the promotion of both benign and malignant tumors by TPA (24,26,29,30). In contrast, retinoic acid is unable to inhibit the induction of ODC and tumor formation by carcinogenic doses of 7,12-

dimethylbenz[a]anthracene; often a small increment in tumor incidence is observed in mice treated with both a retinoid and dimethylbenzanthracene (26). In contrast, 7,8-benzoflavone, which is incapable of inhibiting the induction of ODC activity by TPA, inhibits both the induction of ODC and neoplasia caused by 7,12-dimethylbenz[a]anthracene (26). Because of these facts one concludes, first, that the mechanism for the induction of ODC activity and for tumor formation by dimethylbenzanthracene is not the same as by TPA. Second, nonetheless, an induced level of ODC may be an essential component of the mechanism of dimethylbenzanthracene carcinogenesis. Third, the protective effect of retinoids on skin carcingenesis is restricted to specific promoting agents and it should not be used for prophylaxis in man at the present time (12,26).

The evidence that the inhibition of tumor promotion by retinoids involves their ability to inhibit the induction of ODC and the accumulation of putrescine will be reviewed briefly.

The application of 1.7 nmol of retinoic acid in 0.2 ml of acetone to the skin of the back of the mouse 1 hour before the application of 17 nmol of TPA in 0.2 ml of acetone to the same area of the back inhibits the induction of ODC and the accumulation of putrescine in the treated epidermis. The time course of the induction is not altered. However, the induction of S-adenosylmethionine decarboxylase is not inhibited. The inhibition of ODC is dependent on the dose of retinoic acid through the range of about 0.017 nmol to 3.4 nmol; at the higher dose, inhibition of induction is essentially complete. Furthermore, the time of application of retinoic acid is critical. Inhibition is maximal if the retinoid is applied between 1 hour before and 1 hour after TPA and by 3 hours after TPA the effect is negligible (the mice are killed for determination of ODC acctivity in an epidermal preparation 4 1/2 hours after TPA; retinoic acid has no effect on the assay). If retinoic acid is applied 12 or more hours before TPA, the inhibitory effect is lost indicative of the short half-life of retinoic acid in the skin at these low dose-levels. Finally, the inhibitory effect varies greatly with the structure of the retinoid; over 30 have been tested (24,25,29,30).

Under all the conditions of dose, time of application, and structure, the ability of retinoids to inhibit tumor promotion by TPA has been found to parallel the effect on ODC induction. Of particular importance is the fact that retinoic acid applied 24 hours after each repetitive treatment with TPA not only did not inhibit the induction of ODC, it did not inhibit tumor promotion (30). Furthermore, the ability of retinoic acid to inhibit ODC induction is retained when applied 1 hour before each TPA application in a tumor induction experiment (30).

These effects of retinoids are compatible with an essential role for an induced level of ODC activity and/or an elevated level of putrescine as an essential component of tumor promotion by TPA.

Further evidence in support of a role for the induction of ODC activity in tumor promotion arose from the observation that the induction is inhibited by the application of putrescine or other amines to mouse skin (31). The degree of inhibition depended on both the dose and time of putrescine application; application of 20 μmol of putrescine 2 hours after TPA treatment inhibited the induction of ODC activity by 50% whereas the induction of S-adenosylmethionine decarboxylase activity was unaffected. Spermidine, spermine, and 1,7-diaminoheptane were the most effective of a number of amines that were tested, causing an inhibition of 90% at the 20 μmol dose.

At the 20 μmol dose level, putrescine had no effect on the ODC assay system, although higher levels added directly to the assay medium inhibited the activity of the enzyme.

Under the conditions that putrescine inhibited the induction of ODC activity, it inhibited the formation of tumors by TPA. At doses of 100 μmol of putrescine applied to the skin 2 hours after each application of TPA, papilloma formation was only 20% that of the control mice (31).

These data strengthen the probability that a continuous high level of ODC activity and/or putrescine may be one of the essential components of tumor promotion. By administering putrescine to the mice at the appropriate time relative to TPA treatment, ODC induction is precluded and therefore a high level of ODC activity is not achieved. The time-course data for inhibition of ODC induction by topical application of putrescine provides evidence that the half-life of the administered putrescine is only a few hours and is incapable of replacing a continuous high level of endogenously formed putrescine which is proposed to occur in the initiated cells because of a defect in the control mechanism for ODC.

Most recently, we have found that tumor promotion by TPA was inhibited by α-difluoromethylornithine (DFMO), an irreversible inhibitor of ODC activity (22). Because of the specificity of DFMO for the inhibition of ODC activity and the synthesis of putrescine, it is clear that the enzyme and/or its products (most likely putrescine or its metabolites) play an essential role in tumor promotion by TPA. Precluding the synthesis of putrescine in response to TPA treatment inhibited tumor formation.

The inhibition of the induction of ODC is dependent on the amount of DFMO applied to mouse skin 1 hour after TPA treatment. As little as 0.046 μmol of DFMO caused inhibition and the inhibition amounted to 59 and 90% by 0.137 μmol and 1.37 μmol, respectively, when applied in 0.2 ml of solvent mixture. Likewise, a reduction in both the basal and induced levels of putrescine in the epidermis occur following DFMO treatment (22).

SUMMARY AND CONCLUSIONS

Treatment of mouse skin with carcinogens and tumor promoting agents induces ornithine decarboxylase activity. In order to determine the relevance of the induction to the biochemical mechanism of skin tumor formation, a number of experiments were done. These included ascertaining the effect of a number of experimental protocols on the induction of the enzyme, the tissue level of its products, and the incidence of tumors. The variables incorporated into the protocols included: 1) the structure of the agent responsible for eliciting tumors, 2) the dose of that agent, 3) the effect of substances capable of modifying the induction of the enzyme and the incidence of tumors. The latter included prostaglandin synthesis inhibitors, retinoids, certain polyamines, and α-difluoromethylornithine. In all cases, alterations in ornithine decarboxylase activity and putrescine level in the skin paralleled alterations in tumor incidence.

It is concluded that elevated levels of ornithine decarboxylase and/or its product putrescine or its metabolites play an essential role in tumor promotion; precluding the synthesis of putrescine in response to treatment with a tumor promoter inhibited tumor formation. The specificity of α-difluoromethylornithine for the inhibition of ornithine decarboxylase activity and its ability to inhibit tumor promotion under identical protocols provides very strong evidence to support the conclusion.

References

1. Ashendel, C.L., and Boutwell, R.K. (1979): Biochem. Biophys. Res. Commun., 90:623-627.
2. Astrup, E.G., and Boutwell, R.K. (1982): Carcinogenesis, 3:303-308.
3. Baird, W.M., and Boutwell, R.K. (1971): Cancer Res., 31:1074-1079.
4. Baird, W.M., Sedgwick, S.A., and Boutwell, R.K. (1971): Cancer Res., 31:1434-1439.
5. Berenblum, I. (1954): Cancer Res., 14:471-477.
6. Boutwell, R.K. (1964): Prog. Exp. Tumor Res., 4:207-250.
7. Boutwell, R.K. (1974): CRC Crit. Rev. Toxicol., 2:419-443.
8. Diamond, L., O'Brien, T.G., and Baird, W.M. (1980): Adv. Cancer Res., 32:1-74.
9. Fitzgerald, D.J., and Murray, A.W. (1980): Cancer Res., 40:2935-2937.
10. Furstenberger, G., and Marks, F. (1980): Biochem. Biophys. Res. Commun., 92:749-756.
11. Hecker, E.H., Fusenig, N.E., Kunz, W., and Marks, F., editors (1981): Carcinogenesis, A Comprehensive

Survey, Vol. 7: Cocarcinogenesis and Biological Effects of Tumor Promoters, Raven Press, New York.
12. Hennings, H., Wenk, M.L., and Donahoe, R. (1982): Cancer Lett., 16:1-5.
13. Klein-Szanto, A.J.P., Major, S.K., and Slaga, T.J. (1980): Carcinogenesis, 1:399-406.
14. Mufson, R.A., Fischer, S.M., Verma, A.K., Gleason, G.L., Slaga, T.J., and Boutwell, R.K. (1979): Cancer Res., 39:4791-4795.
15. O'Brien, T.G. (1976): Cancer Res., 36:2644-2653.
16. O'Brien, T.G., Simsiman, R.C., and Boutwell, R.K. (1975): Cancer Res., 35:2426-2433.
17. O'Brien, T.G., Simsiman, R.C., and Boutwell, R.K. (1975): Cancer Res., 35:1662-1670.
18. Perchellet, J.-P., and Boutwell, R.K. (1980): Cancer Res., 40:2653-2660.
19. Rohrschneider, L.R., O'Brien, D.H., and Boutwell, R.K. (1972): Biochim. Biophys. Acta, 280:57-70.
20. Slaga, T.J., Bowden, G.T., Shapas, B.G., and Boutwell, R.K. (1974): Cancer Res., 34:771-777.
21. Slaga, T.J., Sivak, A., and Boutwell, R.K., editors (1978): Carcinogenesis, A Comprehensive Survey, Vol. 2: Mechanisms of Tumor Promotion and Cocarcinogenesis, Raven Press, New York.
22. Takigawa, M., Verma, A.K., Simsiman, R.C., and Boutwell, R.K. (1982): Biochem. Biophys. Res. Commun., 105:969-976.
23. Verma, A.K., Ashendel, C.L., and Boutwell, R.K. (1980): Cancer Res., 40:308-315.
24. Verma, A.K., and Boutwell, R.K. (1980): Carcinogenesis, 1:271-276.
25. Verma, A.K., and Boutwell, R.K. (1980): In: Polyamines in Biomedical Research, edited by J.M. Gaugas, pp. 185-202. John Wiley and Sons, Chichester.
26. Verma, A.K., Conrad, E.A., and Boutwell, R.K. (1980): Carcinogenesis, 1:607-611.
27. Verma, A.K., Lowe, N.J., and Boutwell, R.K. (1979): Cancer Res., 39:1035-1040.
28. Verma, A.K., Rice, H.M., and Boutwell, R.K. (1977): Biochim. Biophys. Res. Commun., 79:1160-1166.
29. Verma, A.K., Rice, H.M., Shapas, B.G., and Boutwell, R.K. (1978): Cancer Res., 38:793-801.
30. Verma, A.K., Shapas, B.G., Rice, H.M., and Boutwell, R.K. (1979): Cancer Res., 39:419-425.
31. Weekes, R.G., Verma, A.K., and Boutwell, R.K. (1980): Cancer Res., 40:4013-4018.
32. Weinstein, I.B., Lee, L.-S., Fisher, P.B., Mufson, A., and Yamasaki, H. (1979): J. Supramol. Struct., 12:195-208.
33. Yotti, L.P., Chang, C.C., and Trosko, J.E. (1979): Science, 206:1089-1091.

Advances in Polyamine Research, Vol. 4, edited by U. Bachrach, A. Kaye, and R. Chayen. Raven Press, New York © 1983.

Diamine Oxidase and Polyamine Catabolism

N. Seiler, B. Knödgen, G. Bink, S. Sarhan, and F. Bolkenius

Centre de Recherche Merrell International, 67084 Strasbourg Cedex, France

A number of related copper containing amine oxidases exist in the mammalian organism with partially overlapping substrate patterns (5,6,31). The classic hog kidney diamine oxidase (E.C.1.4.3.6) attacks among others putrescine and cadaverine, but spermidine less, and spermine hardly at all (30,32). The reverse is true for serum spermine oxidase (29).

Based on this differential substrate preference one can imagine that these enzymes could play an important role in polyamine catabolism. However, relatively little is known in this regard as one may see from recent reviews (3,4,13,20,21).

Evidence for a role of diamine oxidases in polyamine catabolism is provided by the finding (17) that rats treated with aminoguanidine, a well known inhibitor of diamine oxidase (19), excrete not only increased amounts of putrescine, but also of spermidine. A similar observation, which will be reported in this paper, stimulated the initiation of the work presented here. We shall provide evidence for the diamine oxidase catalyzed formation of putreanine in vivo.

MATERIALS AND METHODS

Laboratory animals

Male albino CD1 mice (weighing 43 $\pm$ 2 g) and Sprague-Dawley rats (300 $\pm$ 100 g) were used. They were from Charles River, Saint Aubin-les-Elboeuf, France.

All animals had access to standard diet and water ad libitum throughout the experiment. Usually, the experiments were carried out during the natural 12 hr light period of the animals.

Urine was collected from rats which were adapted for at least three days to the stainless steel metabolic cages before beginning the actual experiment. Daily weighings allowed us to establish when the animals had resumed normal eating habits.

Sample preparation

Extracts of brain with 0.2 N perchloric acid were separated by HPLC after dilution with 2-3 volumes of 0.2 M acetic acid. Chromatographic separation was improved and samples corresponding to larger tissue amounts were in some cases analyzed using the recently reported cleanup procedure (26). The method is briefly as follows:

0.3 ml aliquots of the tissue extracts prepared by homogenization with 10 volumes of 0.2 N $HClO_4$ are run through 0.3 ml columns of Dowex 50 W x 8 (200-400 mesh, H^+ form) (Serva, Heidelberg, G.F.R.). The columns are washed with 1.5 ml of 2N HCl. Putreanine, isoputreanine and compounds of related polarity are eluted with 1.5 ml of 4 N HCl. This fraction is evaporated to dryness and the residue is dissolved in 0.3 ml of 0.2 N $HClO_4$. Before separation samples are diluted appropriately with 0.2 M acetic acid.

Determination of putreanine and isoputreanine by high performance liquid chromatography. (26)

For the separation of putreanine and isoputreanine from other tissue constituents ion-exchange chromatography using a high performance column (Whatman, Partisil-10/25 SCXD;250 x 4.6 mm) was used.

The elution system consisted of a linear gradient which was prepared by mixing of 0.2 M acetic acid and 0.2 M sodium acetate adjusted with acetic acid to pH 4.50. Buffer flow rate was 1 $ml.min^{-1}$.

Detection was achieved by post-column derivatization using the o-phthalaldehyde/2-mercaptoethanol reagent (18) and recording of fluorescence intensity.

Determination of polyamines in tissues and urine.

For polyamine determinations our previously published HPLC method was used (24). Perchloric acid tissue extracts were directly chromatographed after appropriate dilution with 0.2 N $HClO_4$. Urine samples were run through a cleanup procedure (23) prior to analysis.

Measurement of time-dependent inhibition of serum spermine oxidase by aminoguanidine.

Increasing concentrations of aminoguanidine sulfate were incubated at 37°C with 300 µl newborn calf serum and 300 µl of 0.06 µM Soerensen phosphate buffer pH 7.4 under constant shaking. (The calf serum was extensively dialyzed against the phosphate buffer before it was used). At 0,5,10,15,20 and 30 min 40 µl aliquots were removed, and serum spermine oxidase activity was determined in these samples, using the method of Snyder and Hendley (28) with 0.5 mM spermidine as substrate.

The incubation buffer for serum spermine oxidase activity determination contained 8 mg homovanillic acid and 2 mg horse radish peroxidase in 100 ml of the above mentioned phosphate buffer. Two ml of the incubation buffer were mixed with 0.2 ml spermidine.3HCl solution, and 0.76 ml of phosphate buffer. To this mixture the 40 μl aliquots of the preincubates were added and the samples were shaken for 60 min at 37°C. Fluorescence intensity was measured at 425 nm (fluorescence activation at 315 nm). Blanks contained no substrate.

Determination of diamine oxidase activity in small intestine of mice.

Mice received 0.5 $mmol.kg^{-1}$ of aminoacetonitrile intraperitoneally. At various times after administration of this drug, the animals were decapitated and small intestines were quickly isolated and extensively washed with ice-cold physiol. NaCl. Intestine samples were homogenized with 10 vol. of 0.06 M Soerensen phosphate buffer pH 7.4.

Diamine oxidase activity was assayed in 50μl aliquots of the homogenates by determination of hydroperoxide formation. The experimental procedure was the same as that used for serum spermine oxidase activity measurements, except that 0.5 mM putrescine.2HCl was used as substrate.

Chemicals

Usual laboratory chemicals were of A grade and were purchased either from E. Merck, Darmstadt, G.F.R. or from Baker Chemicals, Deventer, the Netherlands. Spermidine.3 HCl was from Fluka, Buchs, Switzerland and putreanine. $H_2SO_4.\frac{1}{2}H_2O$ was purchased from CalBiochem. Los Angeles, Ca, U.S.A.. Isoputreanine.2HCl was prepared from N-(3-aminopropyl)-2-pyrrolidinone (Aldrich, Beerse, Belgium) by hydrolysis with conc. H_2SO_4. It was purified by chromatography on Dowex 50W x 8 (200-400 mesh H^+ form) using an HCl-gradient (between 0 and 3 N HCl). The elemental analysis corresponded to the expected composition. Aminoguanidine sulfate was from Schuchardt, München, G.F.R., aminoacetonitrile was from Aldrich, Beerse, Belgium. The commercial product was purified by chromatography on Dowex 50Wx8, using an HCl gradient, and the hydrochloride was recrystallized from water/propanol-2. Newborn-calf serum was from Gibco (Europe) Glasgow, Scotland U.K.

Homovanillic acid was from Fluka, Buchs, Switzerland, horse radish peroxidase (grade II) from Boehringer, Mannheim, G.F.R.

RESULTS

In vivo evidence for polyamine catabolism by diamine oxidase.

Treatment of mice and rats with 2 ml/kg of carbon tetrachloride causes the enhanced formation of N^1-acetylspermidine in the liver (1,22) and the occurence of N^1-acetylspermine could be demonstrated (22). Putrescine concentrations also increase, whereas spermidine and spermine concentrations decrease during the first 24 hr.

In agreement with the polyamine changes in liver, an enhancement of putrescine and N^1-acetylspermidine excretion in the urine was observed during the first 24 hr after CCl_4-administration, which gradually returned to normal during the following days (Table 1). Spermidine, spermine or N^8-acetylspermidine excretion was not significantly altered by this treatment.

Since N^8-acetylspermidine is a good substrate of serum spermine oxidase, in contrast to N^1-acetylspermidine (9,11) we expected to see enhanced excretion of N^8-acetylspermidine after administration of aminoguanidine in doses sufficiently high to inhibit serum spermine oxidase. However, treatment with 100 $mg.kg^{-1}$ of the inhibitor for three consecutive days enhanced putrescine and spermidine excretion but had only a marginal effect on N^1- or N^8-acetylspermidine excretion (Table 1). This finding is in agreement with that of Rojansky et al. (17) as far as spermidine and putrescine are concerned, with the difference, however, that these authors determined total putrescine and spermidine, i.e. after hydrolytic cleavage, whereas the present data represent the original urinary pattern. (N-Acetylputrescine has not been included here because of technical difficulties with its assay).

Very significant elevations of urinary putrescine, spermidine and also of N^1-acetylspermidine were observed, in aminoguanidine- treated rats which in addition received a single dose of 2 ml/kg of CCl_4 (Table 1). Since N^8-acetylspermidine excretion was not changed, the ratio of N^1-acetylspermidine/-N^8-acetylspermidine increased on the first day from 2.8 (controls) to 9.5 and remained significantly elevated in the urine collected on the third day.

Taking the mentioned substrate characteristics of N^1-acetylspermidine into account these results allow one to conclude that under physiological rates of polyamine metabolism only a minor proportion of N^1-acetylspermidine is exposed to oxidation by diamine oxidases, whereas in CCl_4-intoxicated animals with unphysiologically high rates of N^1-acetylspermidine formation, a significant proportion of this compound is oxidized by diamine oxidases.

TABLE 1. The effect of treatment with CCl_4 and aminoguanidine sulfate* on the urinary excretion of polyamines (μmol/24 hr; mean values ± S.D.)

Treatment	No. of Animals	day of Treatment	Putrescine	Spermidine	Spermine	N^1-Acetyl-spermidine	N^8-Acetyl-spermidine	N^1-AcSpd / N^8-AcSpd *
**	9	–	2.6 ± 0.5	0.6 ± 0.2	0.08 ± 0.03	0.35 ± 0.08	0.13 ± 0.02	2.8 ± 0.5
2 $ml.kg^{-1}$	3	1	5.8 ± 0.2	0.5 ± 0.2	0.06 ± 0.02	0.83 ± 0.27	0.10 ± 0.02	8.2 ± 2.6
CCl_4	3	2	3.7 ± 1.1	0.7 ± 0.2	0.06 ± 0.02	0.50 ± 0.17	0.10 ± 0.002	5.1 ± 1.7
(1 dose)	3	3	3.6 ± 0.2	0.4 ± 0.3	0.11 ± 0.05	0.39 ± 0.08	0.12 ± 0.02	3.3 ± 0.8
100 $mg.kg^{-1}$	3	1	7.7 ± 1.1	0.7 ± 0.2	0.07 ± 0.03	0.41 ± 0.06	0.14 ± 0.02	3.0 ± 0.1
AG i.p.	3	2	7.5 ± 0.5	1.4 ± 0.5	0.13 ± 0.09	0.42 ± 0.07	0.14 ± 0.01	3.0 ± 0.4
(daily)	3	3	8.9 ± 1.0	1.5 ± 0.2	0.2 ± 0.1	0.45 ± 0.04	0.16 ± 0.01	2.9 ± 0.2
2 $ml.kg^{-1}$	3	1	22.4 ± 3.0	1.3 ± 0.5	0.4 ± 0.4	1.6 ± 0.2	0.17 ± 0.02	9.5 ± 0.2
CCl_4 (1 dose)	2	2	15.2 ± 3.4	3.1 ± 0.6	0.1 ± 0.01	1.1 ± 0.1	0.19 ± 0.01	5.8 ± 0.1
+ 100 $mg.kg^{-1}$ AG ip (daily)	2	3	10.0 ± 0.8	1.0 ± 0.2	0.1 ± 0.02	0.68 ± 0.02	0.16 ± 0.01	4.4 ± 0.06

* Ac-Spd = acetylspermidine ; * AG = Aminoguanidine sulfate

** One group of 3 animals received physiol. NaCl, two groups received 8 $ml.kg^{-1}$ of maize oil, which was used as vehicle for CCl_4.

Inhibition of in vivo formation of putreanine from spermidine by aminoacetonitrile and aminoguanidine.

In a recent paper (26) we reported on the in vivo rate of putreanine formation from exogenous spermidine in mice after a single intraperitoneal dose of 1 $mmol.kg^{-1}$ of spermidine. The measured in vivo rates of putreanine and isoputreanine formation presumably do not reflect the maximal capacity of mice to perform the transformation reaction. Because of unacceptable toxicity it could not be established, whether the rate-limiting step of the reaction sequence is saturated with its substrate. As is shown in Fig. 1, there is however, a linear relationship between time and putreanine content in mouse liver, if 1 $mmol.kg^{-1}$ of spermidine is administered. Inhibition of the oxidative deamination reaction should be reflected by the putreanine content of liver 30 min after spermidine administration if the enzyme responsible for this process is inhibited to an extent that it becomes rate-limiting. This approach was used, therefore, in the following experiments which were designed to assist in clarifying the role of diamine oxidases in the transformation process.

Two inhibitors were considered for this study, aminoguanidine sulfate, a well known inhibitor of diamine oxidase (19) and aminoacetonitrile. This latter compound inhibits Cu^{2+}-containing oxidases and according to Abeles (2) preferentially inhibits serum spermine oxidase. Fig. 2 shows the effect of different doses of aminoguanidine.½.H_2SO_4 and aminoacetonitrile.HCl on the accumulation of putreanine in mouse liver 1 hr after the intraperitoneal administration of the inhibitors. Using aminoacetonitrile, 50 percent inhibition of putreanine formation was achieved with about 0.1 $mmol.kg^{-1}$ (9.3 $mg.kg^{-1}$), whereas only about 0.16 $\mu mol.kg^{-1}$ (27.5 $\mu g.kg^{-1}$) of aminoguanidine was needed to achieve comparable inhibition.

Recovery of the capacity to form putreanine from exogenous spermidine was tested using doses of the inhibitors which were sufficient to completely prevent putreanine formation at 1 hr after injection. After a single intraperitoneal dose of 0.2 $mmol.kg^{-1}$ of aminoguanidine or of 0.5 $mmol.kg^{-1}$ of aminoacetonitrile in vivo formation of putreanine from spermidine was blocked by 92 percent or more at 24 hr. After this time the putreanine forming capacity recovered exponentially.

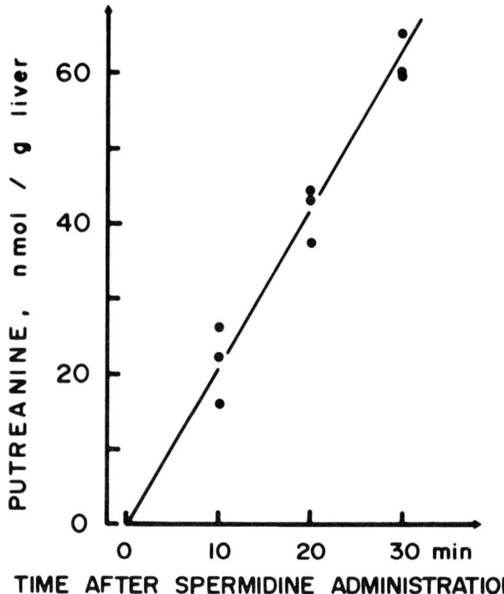

Fig. 1. Putreanine content in mouse liver after the intraperitoneal administration of 1 $mmol.kg^{-1}$ spermidine. Each point represents the experimental value obtained from a separate animal.

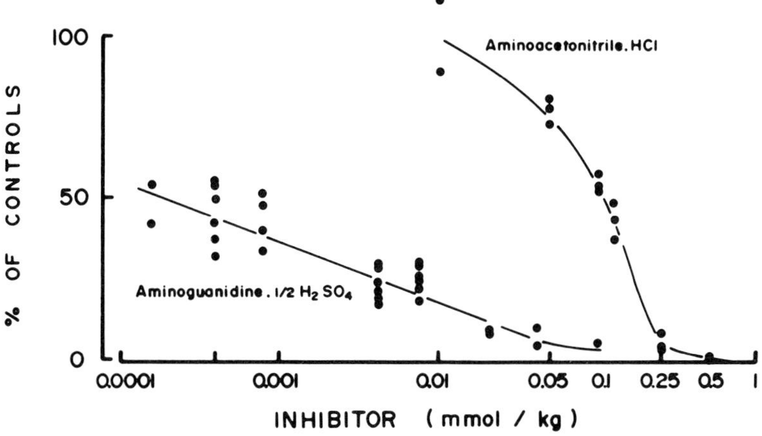

Fig. 2. Effect of aminoguanidine sulfate and aminoacetonitrile.HCl on the in vivo formation of putreanine from spermidine.3HCl in mice; dose effect. The inhibitors were given 1 hr before the administration of 1 $mmol.kg^{-1}$ of spermidine.

The times for recovery of 50 percent of the control rate were of the same order for both inhibitors, namely 25 hr for aminoguanidine and 19 hr for aminoacetonitrile (Fig. 3).

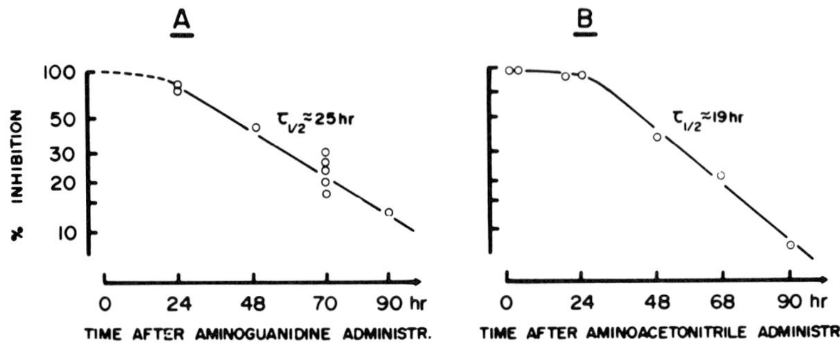

Fig. 3. Recovery of the capacity to form putreanine from intraperitoneally injected spermidine.3HCl after treatment with A: aminoguanidine sulfate or B: aminoacetonitrile hydrochloride. Each point is the mean value of three animals which were treated together.

Fig. 4 gives an impression of the efficiency of the analytical method used in this work.

Polyamine concentrations were measured in liver, kidney and small intestine of mice treated with 0.2 $mmol.kg^{-1}$ of aminoguanidine or 0.5 $mmol.kg^{-1}$, aminoacetonitrile respectively at 1,4 and 24 hr after administration of the inhibitors. The results are summarized in Table 2. Only in small intestine were putrescine and spermidine concentrations significantly elevated. In this organ diamine oxidase obviously plays a prominent role in the regulation of tissue polyamine levels. In kidney, putrescine levels were decreased at 24 hr after treatment with either inhibitor. We have presently no explanation for this latter finding.

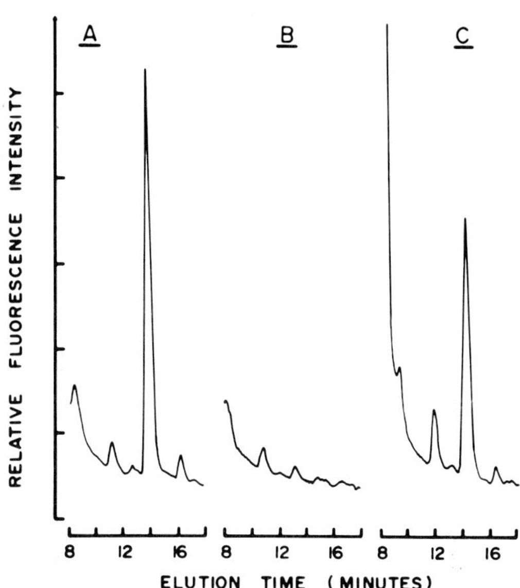

Fig. 4. HPLC of perchloric acid mouse liver extracts.
A: Extract prepared 30 min after the administration of 1 $mmol.kg^{-1}$ of spermidine.

B: Liver extract from an animal which had received 0.5 mmol. kg^{-1} of aminoacetonitrile 1 hr before the administration of spermidine.

C: Extract prepared from the liver of an animal which had received 0.5 $mmol.kg^{-1}$ of aminoacetonitrile 48 hr before administration of spermidine.

Chromatogram B is not different from separations obtained from liver extracts of untreated control animals.

Ex vivo activity of diamine oxidase in small intestine of mice after administration of aminoacetonitrile.

Fig. 5 demonstrates that a single intraperitoneal dose of 0.5 $mmol.kg^{-1}$ of aminoacetonitrile has a long-lasting effect on intestinal diamine oxidase activity.

Enzyme activity was measured in small intestines removed at various times after the administration of the inhibitor, using 0.5 mM putrescine as substrate. (For details see the Methods Section).

Inhibition was not complete 4 hr after aminoacetonitrile

TABLE 2. Polyamines in tissues of mice after a single intraperitoneal dose of aminoguanidine or aminoacetonitrile (mean values ± S.D.; N = 3)

Drug	Tissue	Time after drug admin.(hr)	Putrescine nmol/g	Spermidine μmol/g	Spermine μmol/g	Spermidine / Spermine
Aminoguanidine	Liver	-	29 ± 10	1.89 ± 0.34	1.48 ± 0.22	1.28
		1	31 ± 10	1.56 ± 0.60	1.78 ± 0.52	0.88
		4	37 ± 17	1.40 ± 0.28	1.64 ± 0.38	0.85
		24	29 ± 1	1.65 ± 0.35	1.69 ± 0.23	0.98
0.2 mmol.kg^{-1}	Kidney	-	85 ± 17	0.38 ± 0.06	0.93 ± 0.11	0.41
		1	110 ± 45	0.40 ± 0.01	0.92 ± 0.06	0.43
		4	93 ± 25	0.47 ± 0.02	0.97 ± 0.08	0.48
		24	22 ± 7 *	0.36 ± 0.03	0.76 ± 0.16	0.47
	Small Intestine	-	181 ± 10	1.90 ± 0.21	0.81 ± 0.08	2.35
		1	275 ± 66 *	1.73 ± 0.05	0.75 ± 0.04	2.31
		4	300 ± 27 *	1.68 ± 0.17	0.73 ± 0.05	2.30
		24	322 ± 58 *	2.57 ± 0.09 *	1.02 ± 0.09	2.52

TABLE 2. Cont'd

Drug	Tissue	Time after drug admin. (hr)	Putrescine nmol/g	Spermidine µmol/g	Spermine µmol/g	Spermidine/Spermine
Aminoaceto-nitrile	Liver	-	29 ± 10	1.89 ± 0.34	1.48 ± 0.22	1.28
		1	22 ± 5	1.45 ± 0.27	1.78 ± 0.30	0.81
		4	26 ± 15	1.51 ± 0.68	1.62 ± 0.28	0.93
0.5 $mmol.kg^{-1}$		24	33 ± 12	1.53 ± 0.47	1.66 ± 0.53	0.92
	Kidney	-	85 ± 17	0.38 ± 0.06	0.93 ± 0.11	0.41
		1	89 ± 15	0.40 ± 0.07	0.94 ± 0.11	0.43
		4	73 ± 12	0.52 ± 0.13	1.04 ± 0.15	0.50
		24	35 ± 8 *	0.40 ± 0.09	0.84 ± 0.19	0.48
	Small Intestine	-	181 ± 10	1.90 ± 0.21	0.81 ± 0.08	2.35
		1	263 ± 31 *	1.87 ± 0.28	0.76 ± 0.02	2.46
		4	315 ± 45 *	1.72 ± 0.20	0.76 ± 0.06	2.26
		24	228 ± 36 *	2.18 ± 0.42	0.92 ± 0.03	2.37

* Statistically significant difference as compared to the value from non-treated controls ($p \leq 0.001$)

administration, but recovery of enzyme activity was considerably slower ($\tau_{\frac{1}{2}} \approx 44$ hr) than the recovery of the ability to transform exogenous spermidine into putreanine (Fig. 4).

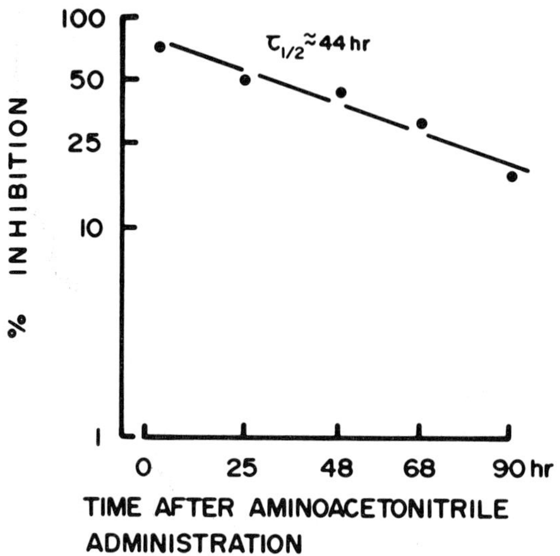

Fig. 5. Diamine oxidase activity in the small intestine of mice after a single intraperitoneal dose of 0.5 $mmol.kg^{-1}$ aminoacetonitrile. Diamine oxidase was determined in small intestine homogenates using 0.5 mM putrescine.2HCl as substrate. Each point represents the mean value of duplicate determinations from three animals.

Inhibition by aminoguanidine of spermidine oxidation by newborn calf serum.

Incubation of newborn calf serum with aminoguanidine in the absence of substrate inactivates serum spermine oxidase in a time-dependent manner. The rate of inactivation increases with inhibitor concentration and seems not to be saturable. As can be seen in Fig. 6A 50 μM aminoguanidine is necessary to inactivate within 30 min, 50 percent of serum spermine oxidase of dialyzed newborn calf serum in the absence of substrate. The ten-fold higher inhibitor concentration reduces $\tau_{\frac{1}{2}}$ to about 3 min.

The presence of spermidine or spermine in the incubation mixtures diminishes the initial rate of inactivation, showing that inhibitor and substrate compete for some site of the enzyme. With progressive removal of substrate from the incubation mixture the rate of enzyme inactivation increases (Fig. 6B).

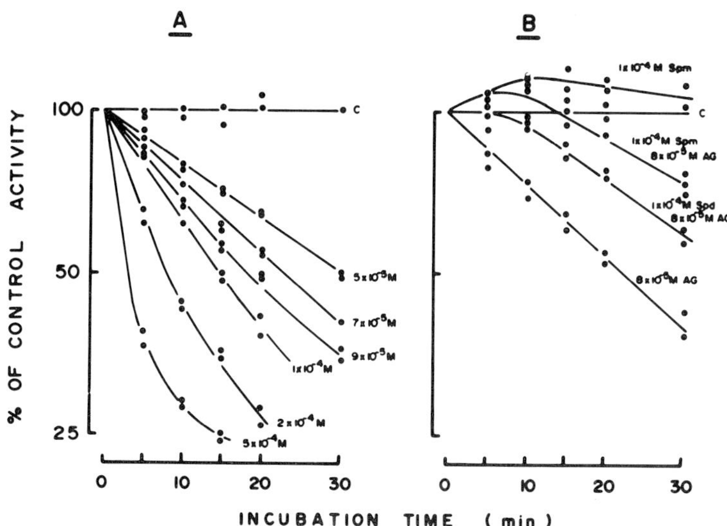

Fig. 6. Time-dependent inhibition of newborn-calf serum spermine oxidase by aminoguanidine sulfate (AG).
A: Incubations in the absence of substrate with various concentrations of aminoguanidine sulfate.
B: Incubations with aminoguanidine sulfate in the presence of spermidine.3HCl (Spd) or spermine.4HCl (Spm). C = control incubations in the absence of inhibitor or substrate.
Each point represents the result of a separate incubation.

Brain putreanine in mice and treatment with amine oxidase inhibitors.

Groups of 3 to 5 mice were administered daily intraperitoneal doses of 0.8 $mmol.kg^{-1}$ of aminoguanidine, for 3 or 5 days. Similarly aminoacetonitrile was administered at a dose of 0.5 $mmol.kg^{-1}$ for up to 5 days, and the concentration of brain putreanine was determined after these treatments. Table 3 summarizes the results. Treatment with aminoguanidine did not produce a significant change in the endogenous brain levels of putreanine. Administration of 0.5 $mmol.kg^{-1}$ aminoacetonitrile, resulted in a minor decrease of brain putreanine level after 3 and 5 days of treatment.

TABLE 3. Putreanine in the brain of mice treated with amine oxidase inhibitors (mean values ± S.D). (To each treated group belonged a control group which was analysed in parallel).

Inhibitor	Dose mmol.kg^{-1}	Time after the last dose (h)	No. of animals	Putreanine nmol.g^{-1}
Amino-guanidine	-	-	5	17.9 ± 1.9
	3 x 0.8 i.p. 24 h intervals	6	5	17.7 ± 1.7
	-	-	5	16.0 ± 2.9
	5 x 0.8 i.p. 24 h intervals	6	5	18.6 ± 2.5
Amino-acetonitrile	-	-	3	16.0 ± 0.23
	3 x 0.5 i.p. 24 h intervals	6	3	14.7 ± 0.15
	5 x 0.5 i.p. 24 h intervals	6	3	14.6 ± 0.35

DISCUSSION

In a recent publication we showed that intraperitoneally injected spermidine and spermine were transformed to a significant degree into the corresponding amino acids putreanine and N^8-(2-carboxyethyl)spermidine. In vitro this transformation process could be mimicked by incubation of the polyamines with mouse and rat liver homogenates or slices in the presence of newborn calf serum. Neither serum alone, nor liver homogenates alone were capable of forming singificant amounts of putreanine or N^8-(2-carboxyethyl)spermidine from spermidine and spermine. Both in vivo and in vitro the rate of transformation of exogenous polyamines into the corresponding amino acids was limited by aminoguanidine and pargyline (25).

These findings were consistent with the assumption that the transformation takes place by a two-step reaction, the first being the oxidative deamination of the polyamines to the corresponding monoaldehydes by serum spermine oxidase, or an enzyme with similar properties, the second being the oxidation of the aldehydes to the corresponding amino acids. The latter step is necessarily intracellular because of the absence of adequate aldehyde dehydrogenases in the circulation.

In further support of this interpretation are the following facts:

(a) According to Zeller et al (30,32) spermidine, and even more spermine are poor substrates for diamine oxidase.

(b) Mouse serum contains measurable serum spermine oxidase activity (13).

(c) Substitution of newborn calf serum by mouse serum produced in the presence of mouse or rat liver homogenate small, but detectable amounts of N^8-(2-carboxyethyl)spermidine from spermine (25).

Nevertheless it was decided to reevaluate our previous findings using two inhibitors, aminoacetonitrile and aminoguanidine.

Aminoacetonitrile inactivates Cu^{2+}-containing oxidases; serum spermine oxidase is inactivated irreversibly at 2.5. 10^{-5}M. The $\tau_{\frac{1}{2}}$ for inhibition being approximately 1 min (12). When administered to rabbits intravenously at 12 mg.kg^{-1} (0.13 mmol.kg^{-1}) it showed significant reduction of the rate of mescaline oxidation in serum, kidney, lung and also some in brain (16).

Aminoguanidine is a highly selective and potent inhibitor of diamine oxidase (6,19). A dose of 0.07 μmol.kg^{-1} is sufficient to reduce intestinal diamine oxidase of rats by 50 percent, and 0.4 μmol.kg^{-1} to achieve the same effect in cat kidney in vivo, whereas only 0.00007 μM of aminoguanidine was needed to produce 50 percent inhibition of rat intestinal or pig kidney diamine oxidase in vitro, using cadaverine as substrate (7,8). The selectivity of this compound is underlined by our finding that 50 μM aminoguanidine was needed to inactivate 50 percent of serum spermine oxidase within 30 min. On the other hand, about four times the dose of aminoacetonitrile considered to be adequate for the in vivo inhibition of serum spermine oxidase in rabbits (16) produced a very considerable, long-lasting inactivation of diamine oxidase in small intestine of mice, demonstrating a poor selectivity of this inhibitor for serum spermine oxidase.

The fact that administration of only 0.16 μmol.kg^{-1} of aminoguanidine was necessary to inhibit putreanine formation by 50 percent argues in favor of a prominent role for a diamine oxidase-type enzyme in the oxidative deamination of spermidine in mice in vivo. This figure is close to the value of 0.07 μmol.kg^{-1} reported by Burkard et al. (7) to be necessary for the inactivation of rat intestinal diamine oxidase by 50 percent.

No conclusive result with respect to the enzyme could be drawn from the fact that about 0.1 mmol.kg^{-1} of aminoacetonitrile was necessary for 50 percent inhibition of putreanine formation. There is, however, a discrepancy between observations made with this inhibitor which should be emphasized.

A single intraperitoneal dose of 0.5 mmol.kg^{-1} of acetonitrile caused 97-92 percent inhibition of putreanine formation from spermidine during the first 24 hr after drug administration, $\tau_{\frac{1}{2}}$ being about 19 hr for the recovery of this capacity. In contrast, diamine oxidase activity, as measured with 0.5 mM putrescine as substrate in small intestine homogenates after administration of the same dose of the inhibitor was inhibited by only 70 percent at 4 hr, $\tau_{\frac{1}{2}}$ being about 44 hr for the recovery of diamine oxidase activity. One possible explanation for this finding is that the diamine oxidase activity measured under these conditions with putrescine as substrate is not identical with the spermidine oxidizing enzyme.While the capacity to oxidize diamines and spermidine is clearly related with two distinct proteins in adult bovine serum (9), it has been recently reported (10) that a diamine oxidase exists in human pregnancy serum which has a high affinity for spermidine (Km 10.9 μM) and aminoguanidine (Ki 0.8 μM). It is tempting to speculate that enzymes with similar characteristics might occur in certain tissues and be responsible for putreanine formation. But oxidases with considerably lower affinity for spermidine should also be capable of reacting with polyamines in conditions where dramatically elevated cellular levels exist.

In rodents, the highest diamine oxidase activity disregarding the placenta, is found in the small intestine with considerably lower activities in all other tissues (27). We found, in agreement with this distribution pattern, putreanine and N-(3-aminopropyl)-2-pyrrolidinone formation by incubation of homogenates of small intestine with spermidine. 3HCl, but no significant amounts of these compounds were found by analogous incubations with homogenates of mouse kidney and liver (N. Seiler and B. Knödgen, unpublished observations). The fact that only in small intestine, but not in kidney or liver were putrescine and spermidine levels significantly elevated after treatment with aminoguanidine or aminoacetonitrile, is also in agreement with the above-mentioned distribution of diamine oxidase activity in rodent tissues.

Changes of liver polyamines after administration of aminoguanidine were found after partial hepatectomy (14), but these results are difficult to interpret.

It is not inconceivable that the putreanine found in liver and kidney after spermidine administration was actually formed in the intestines and has been distributed into other organs by the circulation. But local formation of putreanine in diamine oxidase containing organs other than small intestine and oxidative deamination of polyamines in the circulation according to our previous hypothesis (25) cannot be completely ruled out, especially not for endogenous spermidine which is released from cells into the circulation.

A role for aminoguanidine-sensitive oxidases in the degradation of endogenous spermidine and N^1-acetyl spermidine has clearly been shown by the demonstration of elevation of urinary putrescine, spermidine and N^1-acetyl spermidine in aminoguanidine and carbon tetrachloride-treated rats. Since the effect on urinary polyamines after treatment with aminoguanidine alone was small it is suggested that aminoguanidine-sensitive oxidases play mainly a role in situations of excessive release of polyamines from tissues.

Endogenous putreanine levels in brain are relatively high (15). It was speculated that if brain putreanine was formed outside the brain and only stored in brain cells after its uptake from the circulation, or if the corresponding aminoaldehyde was formed within the circulation and only the final step of the reaction sequence, namely reaction with the aldehyde dehydrogenase took place inside the brain, according to our hypothesis (25), it should be possible to decrease brain putreanine levels by inhibition of diamine oxidase and/or serum spermine oxidase in the periphery. However, as was shown here, treatment with doses of aminoguanidine higher than needed to completely inhibit metabolism of exogenous spermidine to putreanine, did not change endogenous levels of putreanine in mouse brain, even if aminoguanidine was given on five consecutive days. However, treatment with aminoacetonitrile produced after three days of treatment with 0.5 mmol. kg^{-1} doses per day a slight decrease of brain putreanine, which was not further decreased after 5 days of treatment.

There are two possible explanations for these findings:

(a) Putreanine or its precursor aldehyde is formed outside the brain, but its turnover rate in brain is so slow, that even complete inhibition of its formation does not cause a significant loss after 5 days.

(b) Putreanine is formed inside the brain. Aminoguanidine does not inhibit the responsible oxidase, but due to its penetration into brain (16) amino-acetonitrile inhibits the brain oxidase to some extent and causes a slight decrease of putreanine levels.

Shaff and Beaven (27) succeeded in demonstrating very low but measurable diamine oxidase activities in thalamus, hypothalamus and pons medulla of rat brain. If indeed this low diamine oxidase activity is responsible for the formation of putreanine in brain one has to assume that it is stored in a very stable form, i.e. there is only a very slow elimination rate from brain in view of the relatively high concentration of putreanine in this organ. In agreement with this is the finding that intraventricular injections of 100 μg of aminoacetonitrile.HCl to mice did not decrease brain levels of putreanine within 24 hr (results not shown), although this amount of the drug should have been sufficient to inhibit brain diamine oxidase.

From these findings it is unlikely that putreanine plays a role in dynamic processes of brain function.

SUMMARY

The effects of two inhibitors of Cu^{2+} containing amine oxidases, aminoguanidine sulfate and aminoacetonitrile.hydrochloride were studied with respect to their effect on putreanine formation from intraperitoneally injected spermidine in mice. Fifty percent inhibition of the transformation process at 1 hr after administration of the inhibitors was obtained with 0.1 $mmol.kg^{-1}$ of aminoacetonitrile, and 0.16 $\mu mol.kg^{-1}$ of aminoguanidine. The time for recovery of 50 percent of the capacity to form putreanine from spermidine was 19 hr and 25 hr for aminoacetonitrile and aminoguanidine, respectively.

The low effective dose of aminoguanidine is consistent with the assumption that a diamine oxidase-type enzyme is responsible for spermidine oxidation under the experimental conditions. Evidence is also presented for the role of a diamine oxidase in the degradation of endogenously formed spermidine and N^1-acetylspermidine.

REFERENCES

1. Abdel-Monem, M.M., and Merdink, J.L. (1981): Life Sci., 28:2017-2023.

2. Abeles, R. (1976): Accounts Chem. Res., 9:313-319.

3. Andersson, A.C., Henningsson, S, and Rosengren, E. (1980): In: Polyamines in Biomedical Research, edited by J.M. Gaugas pp. 273-283. Wiley, Chichester, New York, Brisbane, Toronto.

4. Bachrach, U. (1981): In: Polyamines in Biology and Medicine, edited by D.R. Morris, and L.J. Marton, pp. 152-168. Marcel Dekker, New York, Basel.

5. Blaschko, H. (1974): Rev. Physiol. Biochem. Pharmacol., 70:83-148.

6. Buffoni, F. (1966): Pharmacol. Rev., 18:1163-1198.

7. Burkard, W.P., Gey, K.F., and Pletscher, A. (1960): Biochem. Phamacol., 3:249-255.

8. Burkard, W.P., Gey, K.F., and Pletscher, A. (1962): Biochem. Pharmacol., 11:177-182.

9. Gahl, W.A., and Pitot, H.C. (1982): Biochem. J., 202: 603-611.

10. Gahl, W.A., Vale, A.M., and Pitot, H.C. (1982): Biochem.

J., 201:161-166.

11. Mamont, P.S., Seiler, N., Siat, M., Joder-Ohlenbusch, A.-M., and Knödgen, B. (1981): Med. Biol., 59:347-353.

12. Maycock, A.L., Sura, R.H., and Abeles, R.H. (1975): J. Amer. Chem. Soc., 97:5613-5614.

13. Morgan, D.M.L. (1980): In: Polyamines in Biomedical Research, edited by J.M. Gaugas, pp. 285-302. Wiley, Chichester, New York, Brisbane, Toronto.

14. Moulinoux, J.-P., Quemener, V., and Chambon, Y. (1981): C.R. Soc. Biol., 175:828-834.

15. Nakajima, T. and Matsuoka, Y. (1971): J. Neurochem., 18: 2547-2548.

16. Riceberg, L.J., Simon, M., Van Vunakis, M., and Abeles, R.H. (1975): Biochem. Pharmacol., 24:119-125.

17. Rojansky, N., Neufeld, E., and Chayen, R. (1979): Biochim. Biophys. Acta, 586:1-9.

18. Roth, M., and Hampai, A. (1973): J. Chromatog., 83:353-356.

19. Schuler, W. (1952): Experientia, 8:230-232.

20. Seiler, N. (1980): Physiol. Chem. and Physics, 12:411-429.

21. Seiler, N., Bolkenius, F.N., and Rennert, O.M. (1981): Med. Biol., 59:334-346.

22. Seiler, N., Bolkenius, F.N., Knödgen, B., and Haegele K. (1980): Biochim. Biophys. Acta, 676:1-7.

23. Seiler, N., and Knödgen, B. (1979): J. Chromatog., 164: 155-168.

24. Seiler, N., and Knödgen, B. (1980): J. Chromatog., 221: 227-235.

25. Seiler, N., Knödgen, B., Gittos, M.W., Chan, W.-Y., Griesmann, G., and Rennert, O.M. (1981): Biochem. J., 200:123-132.

26. Seiler, N., Knödgen, B., and Haegele, K. (1982): Biochem. J., in press.

27. Shaff, R.E., and Beaven, M.A. (1976): Biochem. Pharmacol., 25:1057-1062.

28. Snyder, H., and Hendley, E.D. (1968): J. Pharm. Exptl. Therap., 63:386-392.

29. Tabor, C.W., Tabor, H., and Rosenthal, S.M. (1954): J. Biol. Chem., 208:645-661 .

30. Zeller, E.A. (1938): Helv. Chim. Acta, 21:1645-1665.

31. Zeller, E.A. (1963): In: Metabolic Inhibitors, vol. 2, edited by R.M. Hochster and J.H. Quastel, pp. 53-78. Academic Press, New York.

32. Zeller, E.A., Schär, B., and Staehlin, S. (1939): Helv. Chim. Acta, 22:837-850.

Advances in Polyamine Research, Vol. 4, edited by U. Bachrach, A. Kaye, and R. Chayen. Raven Press, New York © 1983.

Polyamine Oxidation and Human Pregnancy

D. M. L. Morgan

Division of Perinatal Medicine, Medical Research Council, Clinical Research Centre, Harrow, Middlesex, HA1 3UJ United Kingdom

An association between maternal peripheral blood amine oxidase activity and pregnancy has been recognised for more than 40 years. Histaminase (1,16,34,54), diamine (8,44-46,52,56,57), spermidine (18) and polyamine (spermine, 23) oxidases all show a progressive increase in activity with increase in gestational age to 21 weeks or beyond, and decline to vanishingly low levels 3-4 days after delivery (benzylamine oxidase is excluded from this dicussion as it behaves quite differently, 30). Only traces of activity have been found in male sera or in sera from non-pregnant women of reproductive age (1,8,18,23,27,28,54,57). The placenta has long been regarded as the source of these enzymes but there is increasing evidence that they may be decidual in origin (24,31,51,54,57, 58).

PLACENTAL AMINE OXIDASES

Several attempts have been made to purify the human placental and serum amine oxidases (4,6,14,41,49). Smith (49), who fractionated a placental homogenate with ammonium sulphate followed by sequential chromatography on DEAE-cellulose and cellulose phosphate obtained two active fractions from the latter column but discarded the early, or breakthrough peak. Gel filtration on Sephadex G-200 of the second peak gave a fraction with a 500-fold increase in specific activity compared to the starting material. The rates of oxidation of a number of amines by this preparation were compared by measuring hydrogen peroxide production (Table 1). An apparent K_m of 2×10^{-4}M was obtained for putrescine and cadaverine and 6×10^{-6}M for histamine. Smith (49) also examined the effects of a number of inhibitors on his preparation (Table 2) and commented that in substrate and inhibitor specificities the enzyme resembled the diamine oxidases of hog kidney and pea seedling. He noted that placental histaminase oxidised several aliphatic diamines faster than histamine and suggested that the enzyme might be more concerned with the metabolism of these substances rather than histamine.

TABLE 1. Relative rates of oxidation of amines by placental histaminase (49)

Substrate	Comparative rate[a] (cadaverine = 100)
1,3-diaminopropane	23
1,4-diaminobutane (putrescine)	54
Histamine	23
Benzylamine	5

[a]Measured by rate of decolorization of indigodisulphonate

Paolucci and co-workers (41) used acetone and alcohol precipitation followed by chromatography on DEAE-cellulose, manganese phosphate and Bio-Gel A-5M to obtain a 9,600 fold purification of human placental diamine oxidase. Gel filtration gave four active fractions with molecular masses that were multiples of 125 000 $\pm$ 5 000. Apparent Michaelis constants (K_m) for histamine, putrescine and cadaverine were 0.6×10^{-5}M, 3.3×10^{-4}M, and 3.3×10^{-6} M respectively.

TABLE 2. Inhibition of oxidation of putrescine by placental histaminase (49)

Concentration of Inhibitor	Inhibition (%)[a]		
	10^{-3}M	10^{-5}M	10^{-7}M
Inhibitor			
Aminoguanidine	100	100	63
Isoniazid	44	24	0
Diethytdithiocarbamate	100	44	13
Semicarbazide	100	56	2

[a]Measured by rate of decolorization of indigodisulphonate

Substrate specificity (Table 3) and effects of inhibitors (Table 4) were also examined. The most extensive study of placental diamine oxidase is that carried out by Bardsley and co-workers (4,11-14), who used ethanol-chloroform fractionation followed by separation on DEAE-Sephadex and G-200 to isolate placental diamine oxidase with a 4 600-fold purification. Ammonium sulphate and ethanol-chloroform fractionation, then chromatography on CM-Sephadex followed by gel-filtration gave placental monoamine oxidase with a purification of 22,400-fold. Enzyme activity was monitored spectrophotometrically using *p*-dimethylaminomethylbenzylamine as substrate for the diamine oxidase (5), and benzylamine (55) or *p*-dimethylaminobenzylamine (15) for the monoamine oxidase. Some of their results are given in Tables 5 and 6; interestingly both preparations show some activity towards polyamines.

TABLE 3. Relative rates of oxidation of amines by placental diamine oxidase (41)

Substrate	Comparative rate[a] (putrescine = 100)
Histamine	85
1,3-Diaminopropane	13
1,5-Diaminopentane (cadaverine)	69
Spermine	6
Benzylamine	0
Tyramine	0
Tryptamine	0
Lysine	0

[a]Measured by rate of decolorization of indigodisulphonate

Modification of the procedure for diamine oxidase purification by inclusion of affinity chromatography on Concanavalin A-Sepharose increased the purification factor to 30 800 (14). The enzyme had

TABLE 4. Inhibition of oxidation of putrescine and cadaverine by placental diamine oxidase (41)

Inhibitor (10 ^{5}M)	Inhibition (%)[a]	
	Putrescine	Cadaverine
Aminoguanidine	100	100
Semicarbazide	27	49

[a]Measured colorimetrically following condensation of reaction product with *o*-aminobenzaldehyde (21).

a molecular weight of approximately 70 000 and contained both copper and manganese. Tropocollagen, collagenase-digested collagen, {8-arginine}vasopressin and {8-lysine}vasopressin (K_m 1.8×10^{-5}M) were oxidised at rates which were 50, 85, 50 and 32% of those obtained using *p*-dimethylaminomethylbenzylamine (K_m 1×10^{-5}m), measured by oxygen consumption or hydrogen peroxide formation). The oxidation of tropocollagen was inhibited by β-aminopropionitrile (9), known to inhibit lysyl oxidase, the enzyme catalysing the first step in the cross-linking of collagen (40). Placental diamine oxidase thus shows a stronger resemblence to lysyl oxidase (48) than to the 'classical' hog kidney diamine oxidase.

PREGNANCY PLASMA DIAMINE OXIDASE

Diamine oxidase from pregnancy plasma has been purified 3000-fold by affinity chromatography on cadaverine-Sepharose and elution with sodium heparin (6). The molecular weight was 90 000 and the purified enzyme oxidised histamine and putrescine

TABLE 5. Relative rates of oxidation of amines by placental diamine oxidase (DAO) monoamine oxidase (MAO) and pregnancy plasma (4)

Substrate	Comparative rate[a] (putrescine = 100)		
	DAO	MAO	plasma
p-Dimethylaminomethyl-benzylamine	100	100	9.5
Benzylamine	0	67	0
1,3-Diaminopropane	14	6	-
1,5-Diaminopentane	147	84	31
Histamine	30	49	45
3-Methylhistamine	18	270	37.5
Tyramine	0	185	24
Tryptamine	0	14	0
Adrenaline	0	1300	0
Nonadrenaline	0	450	31.5
Spermine	32	50	0
Spermidine	0	140	106

[a]Measured by coupled oxidation of *o*-dianisidine by peroxidase and H_2O_2 (32).

TABLE 6. Effect of inhibitors on placental diamine oxidase (DAO) monoamine oxidase (MAO) and pregnancy plasma (4)

Inhibitor[a]	Inhibition (%)[b]		
	DAO	MAO	plasma
MAO Inhibitors			
Iproniazid	66	0	45
Isoniazid	40	0	41
trans-Phenylcyclopropylamine	14	84	23
Pargyline	-	-	51
Carbonyl Group Reagents			
Aminoguanidine	100	0	58
Semicarbazide	50	0	37
SH-group Reagents			
Iodoacetamide	6	100	0
p-chloromercuribenzoate	0	50	0
Copper chelators			
Cuprizone	> 50	> 50	30
Diethyldithiocarbamate	40	100	92

[a]Inhibitor concentration 10^{-4}M or 10^{-5}M.

[b]Measured by aldehyde formation (5).

(relative rates 100 and 75, respectively, measured by estimation of {^{3}H}water formed from tritiated substrates) but not benzylamine or tryptamine. The effect of inhibitors is shown in Table 7.

TABLE 7. Inhibition of oxidation of {β-^{3}H}histamine by diamine oxidase from human pregnancy plasma (6)

Inhibitor	Concentration (M)	Inhibition (%)[a]
trans-Phenyl-cyclopropylamine	10^{-4}	24
Iproniazid	10^{-4}	68
Semicarbazide	10^{-5}	8
	10^{-4}	99
Aminoguanidine	2×10^{-5}	100

[a]Measured by formation of {^{3}H} water.

PREGNANCY SERUM POLYAMINE OXIDASES

Recently Gahl and co-workers (19) using {^{14}C}spermidine have shown the presence of spermidine oxidase activity in pregnancy serum. This activity co-purified with diamine oxidase activity (determined using {^{14}C}putrescine) when pregnancy serum was subjected to cadaverine-Sepharose and hydroxyapatite chromatography and gel-filtration. A 400-fold purification was achieved with a near constant ratio of diamine oxidase to spermidine oxidase activity at all stages. The preparation showed an apparent K_m of 2.5×10^{-6}M for putrescine and 10.9×10^{-6}M for spermidine. Both enzyme activities were competitively inhibited by aminoguanidine. Spermidine, spermine and putrescine were oxidised at similar rates (100: 102: 112, measured by hydrogen peroxide production), but N^1-acetylspermidine was metabolised more rapidly than the N^8-acetyl derivative (in the ratio 157: 101, ref 17), a feature in common with rat liver polyamine oxidase (22).

Illei and Morgan (23), using {^{14}C}spermine, demonstrated spermine oxidase activity in maternal peripheral blood serum and showed that the activity in retroplacental serum (derived mainly from intervillous blood but containing some decidual and placental interstitial fluid) was 30-40 times greater (24). Ammonium sulphate fractionation followed by gel-filtration gave a single peak of enzyme activity representing a 400-fold increase in specific activity but the preparation was contaminated by haemproteins (Fig. 1). When retroplacental serum was subjected to affinity chromatography on Cibacron blue-Sepharose (gift of Dr. P.G.H. Byfield, ref 29a), two peaks of spermine oxidase activity were obtained (Fig. 2) and the separation was confirmed by polyacrylamide gel electrophoresis (Fig. 3). A similar distribution of activity was obtained by both methods, about 80% in the major

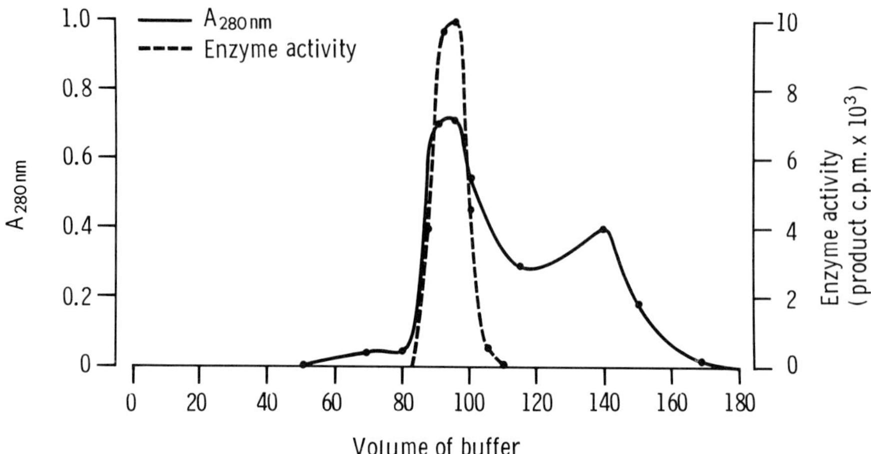

FIG. 1. Elution pattern of spermine oxidase activity obtained by gel-filtration on Sephadex G-100 (10 mM Tris/HCl, pH 7.4) of the material precipitated from retroplacental serum by 55% saturation with ammonium sulphate. Enzyme activity was assayed using {^{14}C} spermine (38).

---

peak and 10-15% in the minor peak. Both fractions oxidised spermine, spermidine and putrescine in the ratio 1:3:25, measured by the formation of labelled products from {^{14}C}-labelled substrates. Spermine appears to inhibit the oxidation of both spermidine and putrescine, as only a low putrescine peak and even smaller peak of a putrescine metabolite was found when the reaction products were separated by ion-exchange chromatography (Fig. 4). Similarly, with spermidine as substrate, putrescine oxidase activity was inhibited since only trace amounts of the putrescine metabolite(s) were formed. When putrescine was the substrate then the reaction products eluted with 0.25-0.5M HCl, while putrescine was retained until the HCl concentration reached 0.75-1.0M (Morgan, unpublished data). Recovery of radiolabel was approximately 98% in each case.

Pargyline and *trans*-phenylcyclopropylamine failed to inhibit the spermidine oxidase activity in either fraction; the results obtained with other inhibitors are shown in Table 8. An interesting feature of the effect of aminoguanidine is the constant degree of inhibition obtained (50% for peak 1, 70% for peak 2) over a wide range of concentration, suggesting the presence of at least two species of enzyme in each fraction, one sensitive to aminoguanidine, the other not. The inhibition by Quinacrin confirms the earlier reports of its effects on the spermine oxidase activity in unfractionated retroplacental serum, which was also inhibited by β-aminopropionitrile (23,36).

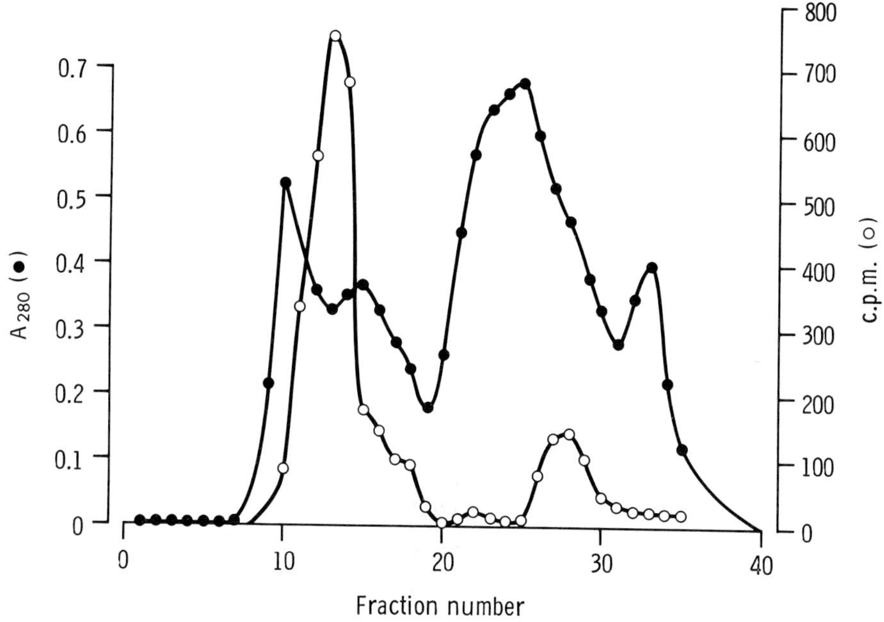

FIG. 2. Elution pattern of spermine oxidase activity obtained by chromatography of pooled retroplacental serum on Cibacron blue-Sepharose (10 mM Tris/CHl, pH 7.4, containing 0.1 mM disodium EDTA and 0.1 mM β-mercaptoethylamine).

BIOLOGICAL ROLE OF POLYAMINE OXIDASES IN PREGNANCY

The apparent lack of specificity shown by the pregnancy-associated amine oxidases towards a variety of substrates and inhibitors has led to confusion in the definition and classification of amine oxidases in the literature (35); indeed as Gahl has remarked (17) the name used may reflect currently available assays rather than the functional importance of the enzyme. Confusion is further compounded by the tendency of some authors to use inappropriate names or, when referring to the same enzyme, to use different names in different publications. Furthermore, in many cases the reaction products have not been adequately identified, if at all. The hitherto accepted specificity of many inhibitors is also open to question. Certain common features can be discerned, however, in the reported data for pregnancy-associated polyamine oxidases summarised here. Cadaverine (1,5-diaminopentane), putrescine (1,4-diaminobutane) and 1,3-diaminopropane are oxidised by all preparations, as are spermine and/or spermidine where these have been tested; all preparations are inhibited by low concentrations of aminoguanidine and the enzyme(s)

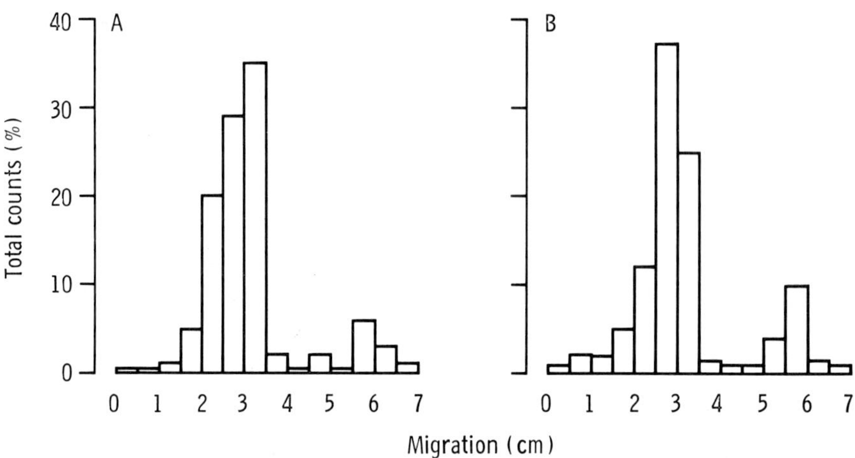

FIG. 3. Distribution of spermidine oxidase activity obtained by gradient (4%-27% acrylamide) slab gel electrophoresis (0.1 M Tris/ glycine, pH 8.9, for 20 h at 4°C) of individual samples of retroplacental serum. Enzyme activity was estimated by slicing the gel transversely into 0.5 cm strips which were incubated with buffer and substrate in the usual way (38). Sample A has been stored at -20°C for several months, sample B was freshly obtained.

TABLE 8. Inhibition of oxidation of spermidine by two fractions of polyamine oxidase from retroplacental serum

Inhibitor	Concentration (M)	Inhibition (%)[a] Peak 1	Peak 2
Aminoguanidine	10^{-8}	0	0
	10^{-7}	37	70
	10^{-6}	43	71
	10^{-5}	53	68
	10^{-4}	51	71
Methylglyoxal bis(guanylhydrazone)			
	10^{-8}	10	24
	10^{-7}	38	29
	10^{-6}	45	66
	10^{-5}	57	71
	10^{-4}	74	75
Quinacrin	10^{-8}	0	0
	10^{-7}	46	10
	10^{-6}	55	49
	10^{-5}	67	69
	10^{-4}	71	87

[a] Enzyme activity was measured as described elsewhere (38)

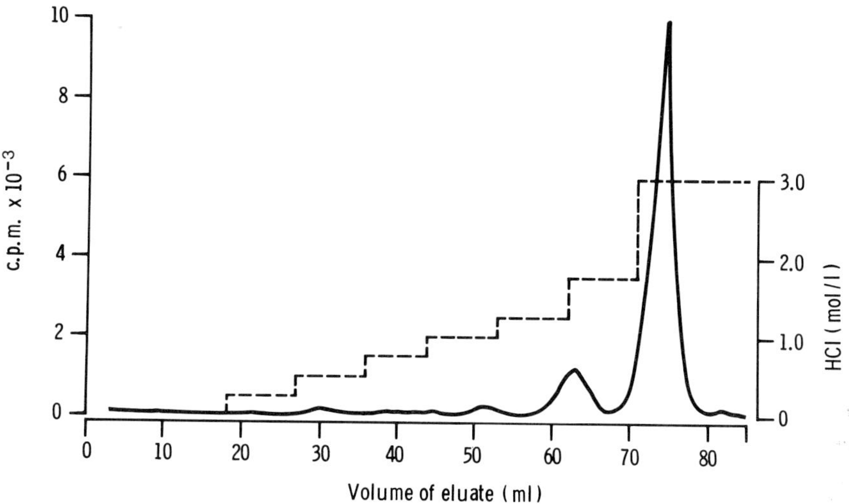

FIG. 4. Elution pattern of {^{14}C}-labelled compounds from an assay of polyamine oxidase activity in retroplacental serum. Separation was by HCl gradient on Dowex-50 resin. 1 ml fractions were collected. The major peak is spermine (unreacted substrate), the two minor peaks to the left comigrate with spermine and putrescine both on ion-exchange chromatography and paper electrophoresis (38). The trace of activity eluted with 0.5M HCl is a metabolite of putrescine.

bind to DEAE. It seems probable, to quote Crabbe (10), that what we are dealing with is "a range of catalytic proteins with amine-oxidising activity, some preferring monoamines to diamines, some vice-versa, and some polyamines or protein-bound lysyl groups". The data of Table 8 and Fig. 4 support this view. The apparent lack of specificity may well be a consequence of our ignorance of the conditions in which these enzymes function *in vivo*. The biological effect produced will, however, depend upon the relative local concentrations of a particular enzyme and a particular substrate at a given point of time.

It seems to be common ground among workers in this area that the amine oxidases exert a protective function in the physiology of pregnancy, although opinions differ on the form this takes. It has been suggested that raised amine oxidase levels protect mother and fetus from elevated concentrations of biogenic amines (3). Another possibility (14) is that the human placental enzyme may not only influence placental biogenic amine metabolism, but also placental collagen synthesis and possibly placental growth by affecting polyamine concentrations which in turn affect the rate of cell division. An alternative hypothesis arises from the suggestion (7) that the placenta liberates an immunosuppressive factor which "switches off" potentially harmful maternal lymphocytes. Studies *in vitro* have shown that the action of polyamine oxidase on polyamines produces inhibitors of cell proliferation, which are not cytotoxic (7); human pregnancy serum also will inhibit lymphoproliferation *in vitro* in the presence of added spermine (20). Morgan and Illei (39) obtained similar results using human lymphoid cells and human retroplacental serum and showed that the effect is proportional to the polyamine oxidase content. These profoundly suppressive effects of spermine and spermidine on *in vitro* parameters of immunity, together with the observation that the placenta is rich in spermine (43), the increase in circulating polyamine in maternal blood (8a) and the very high levels of polyamine oxidase activity in retroplacental blood (24) suggest that the products of the interaction of polyamines and polyamine-degrading enzymes contribute to the protection of the fetus against maternal immune rejection (29,36,37).

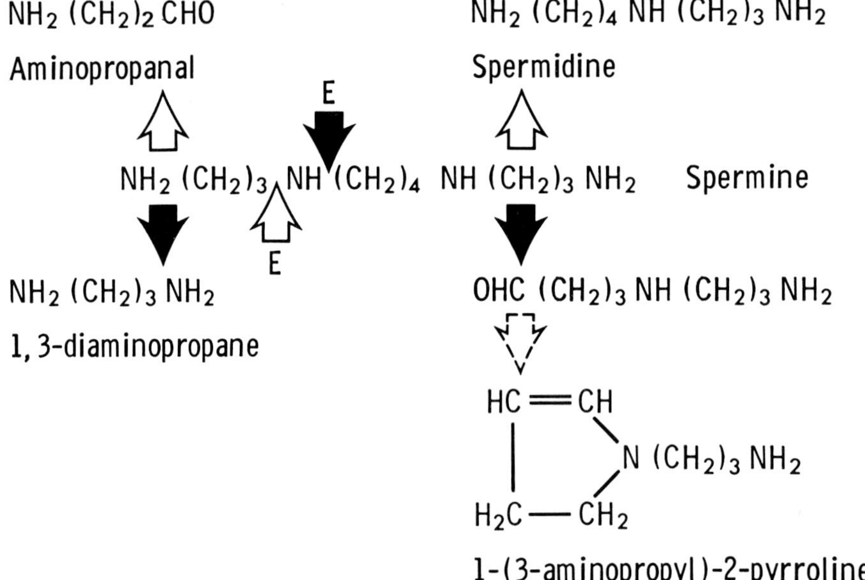

Oxidases that cleave polyamines at a secondary amino group to give aminopropanal or 1,3-diaminopropane and an aldehyde that cyclises to a pyrrol are found in plants, bacteria, fungi and protozoa (2,25,26,33,35,50,53) as well as in many mammalian cells (47). This widespread distribution indicates that the products of the polyamine oxidases (most probably the aminoaldehydes) may have a generalised growth regulatory function (2,36,42). The available data suggest that in pregnancy, in addition to the postulated immunoregulatory function, the polyamine-polyamine oxidase system might also regulate placental growth and limit the extent of placental invasion of the myometrium.

ACKNOWLEDGEMENTS

I thank Mr. P. Gough of this Department for carrying out the gel electrophoresis and Mrs. R.C. White for her help in the preparation of the manuscript.

REFERENCES

1. Ahlmark, A. (1944): Acta Physiol. Scand., 9 suppl. 28: 1-107.
2. Bachrach, U. (1981): In: Polyamines in Biology and Medicine, edited by D.R. Morris, and L.J. Marton, pp. 151-168. Marcel Dekker, New York and Basel.
3. Bardsley, W.G., and Crabbe, M.J.C. (1973): Biochem. Soc. Trans., 6: 494-496.
4. Bardsley, W.G., Crabbe, M.J.C., and Scott, I.V. (1974): Biochem. J. 139: 169-181.
5. Bardsley, W.G., Crabbe, M.J.C., Shindler, J.S., and Ashford, J.S. (1972): Biochem. J., 127: 875-879.
6. Baylin, S.B., and Margolis, S. (1975): Biochim. Biophys, Acta., 397: 294: 306.
7. Byrd, W.J., Jacobs, D.M., and Amoss, M.S. (1978): Adv. Polyamine Res., 2: 71-83.
8. Carrington, E.R., Frishmuth, G.J., Oesterling, M.J., Adams, F.M., and Cox, S.E. (1972): Obstet. Gynaecol., 39: 426-430.
8a. Casti, A., Reali, N., Orlandini, G., Ferrari, B., Salvadori, B., and Bernasconi, S. (1981): Adv. Polyamine Res., 3: 463-472.
9. Crabbe, M.J.C. (1979): Agents and Actions, 9: 41-42.
10. Crabbe, J. (1981): Agents and Actions, 11: 3-8.
11. Crabbe, M.J.C., and Bardsley, W.G. (1974a): Biochem J., 139: 183-189.
12. Crabbe, M.J.C., and Bardsley, W.G. (1974b): Biochem. Pharmacol., 23: 2983-2990.
13. Crabbe, M.J.C., Childs, R.E., and Bardsley, W.G. (1975): Eur J. Biochem., 60: 325-333.
14. Crabbe, M.J.C., Waight, R.D., Bardsley, W.G., Barker, R.W., Kelly, I.D., and Knowles, P.F. (1976): Biochem. J., 155: 679-687.
15. Deitrich, R.A., and Erwin, V.G. (1969): Anal. Biochem., 30: 395-402.
16. Dodge, A.C. (1952): Am. J. Obstet. Gynecol., 63: 1213-1222.
17. Gahl, W.A., and Pitot, H.C. (1981): Life Sci., 29: 2177-2179.
18. Gahl, W.A., Raubertos, R.F., Vale, A.M., and Golubjatnikov, R. (1982a): Br. J. Obstet. Gynaecol., 89: 202-207.
19. Gahl, W.A., Vale, A.M., and Pitot, H.C. (1982b): Biochem. J., 201: 161-164.
20. Gaugas, J.M., and Curzen, P. (1978): Lancet, 1: 18-20.
21. Holmstedt, B., and Tham, R. (1959): Acta Physiol. Scand., 45: 152-163.
22. Holtta, E. (1977): Biochemistry, 16: 91-100.

23. Illei, G., and Morgan, D.M.L. (1979a): Br. J. Obstet. Gynaecol., 86: 878-881.
24. Illei, G., and Morgan, D.M.L. (1979b): Br. J. Obstet. Gynaecol., 86: 873-877.
25. Isobe, K., Tani, Y., and Yamada, H. (1980a): Agric. Biol. Chem., 44: 2651-2658.
26. Isobe, K., Tani. Y., and Yamada, H. (1980b): Agric. Biol. Chem., 44: 2749-2751.
27. Kapeller-Adler, R. (1944): Biochem. J. 38: 270-274.
28. Kobayashi, Y. (1963): J. Lab. Clin. Med., 62: 699-702.
29. Labib, R.S., and Tomasi, T.B. (1981): Eur. J. Immunol., 11: 266-269.
29a. Land, M., and Byfield, P.G.H. (1979): Int. J. Biolog. Macromolecules, 1: 223-226.
30. Lewinsohn, R., and Sandler, M. (1982): Clin. Chim. Acta., 120: 301-312.
31. Lin, C-W., Chapman, C.M., DeLellis, R.A., and Kinley, S. (1978): J. Histochem. Cytochem., 26: 1021-1025.
32. McEwen, C.M., Jr. (1965): J. Biol. Chem., 240: 2003-2010.
33. Mach, M., Kersten, H., and Kersten, W. (1982): Biochem, J., 202: 153-162.
34. Marcou, I., Athanasiu-Vergu, E., Chiricéance, D., Cosma, G., Gingold, N., and Panhon, C-C. (1938): Presse Med., 46: 371-374.
35. Morgan, D.M.L. (1980): In: Polyamines in Biomedical Research, edited by J.M. Gaugas, pp. 285-302. Wiley, Chichester, and New York.
36. Morgan, D.M.L. (1981a): Adv. Polyamine Res., 3: 65-73.
37. Morgan, D.M.L. (1981b): Biochem. Soc. Trans., 9: 400-401.
38. Morgan, D.M.L. and Illei, G. (1980a): Med. Lab. Sciences., 38: 49-56.
39. Morgan, D.M.L. and Illei, G. (1980b): Br. Med. J. 280: 1295-1297.
40. Narayanan, A.S., Siegel, R.C., and Martin, G.R. (1972): Biochem. Biophys. Res. Comm., 46: 745-751.
41. Paolucci, F., Cronenberger, L., Plan, R., and Pachou, H. (1971): Biochimie, 53: 735-749.
42. Patt, L.M., and Houck, J.C. (1980): FEBS Lett., 120: 163-170.
43. Porta, R., Servillo, L., Abbruzzese, A., and Pietra, G.D. (1978): Biochem. Med., 19: 143-147.
44. Resnik, R., and Levine, R.J. (1969): Am. J. Obstet. Gynecol., 104: 1061-1066.
45. Scott, I.V., Bardsley, W.G., Gregory, H.H., and Tindall, V.R. (1979): In: Clinical Enzymology Symposia, 2, edited by A. Burlina and L. Galzigna, pp. 563-578. Piccin Medical Books, Padua and London.
46. Scott, I.V., Childs, R.E., Crabbe, M.J.C., Tindall, V.R. and Bardsley, W.G. (1977): In: Clinical Enzymology Symposia 1, edited by A. Burlina and L. Galazigna, pp. 97-112. Piccin Medical Books, Padua and London.

47. Seiler, N., Bolkenius, F.N., Knodgen, B., and Mamont, P. (1980): Biochim. Biophys, Acta., 615: 480-488.
48. Siegel, R.C., and Fu, J.C.C. (1976): J. Biol. Chem., 251: 5779-5785.
49. Smith, J.K. (1967): Biochem. J. 103: 110-119.
50. Smith, T.A. (1977): Progr. Phytochem., 4: 27-81.
51. Southren, A.L., Kobayashi, Y., Brenner, P., and Weingold, A.B. (1965): J. Appl. Physiol., 20: 1048-1051.
52. Southren, A.L., Kobayashi, Y., Carmody, N.C., and Weingold, A.B. (1966): Am. J. Obstet. Gynecol., 95: 615-620.
53. Suzuki, Y., and Yanagisawa, H. (1980): Plant and Cell Physiol., 21: 1085-1094.
54. Swanberg, H. (1950): Acta Physiol. Scand., 23, Suppl. 79: 1-69.
55. Tabor, C.W., Tabor, H., and Rosenthal, S.M. (1954): J. Biol. Chem., 208: 645-661.
56. Ward, H., Whyley, G.A. and Millar, M.D. (1972): J. Obstet. Gynaecol. Br. Comm., 79: 216-221.
57. Weingold, A.B., and Southren, A.L. (1968): Obstet. Gynaecol., 32: 693-606.
58. Weisburger, W.R., Mendelsohn, G., Eggleston, J.C. and Baylin, S.B. (1978): Lab. Invest., 38: 703-706.

Advances in Polyamine Research, Vol. 4, edited by U. Bachrach, A. Kaye, and R. Chayen. Raven Press, New York © 1983.

Polyamine Oxidation and the Killing of Intracellular Parasites

D. M. L. Morgan and *J. Christensen

*Divisions of Perinatal Medicine and *Clinical Cell Biology, Medical Research Council Clinical Research Centre, Harrow, Middlesex HA1 3UJ, United Kingdom*

In humans and domestic animals certain single-celled parasites multiply in red blood cells and produce widespread and important diseases. One of these is malaria, caused by several species of *Plasmodium* transmitted by anopheline mosquitos. Another is babesiosis, an infection of domestic and other animals, transmitted by ticks and associated with the appearance of various species of *Babesia* parasites in red blood cells. When *B. microti*, a European species adapted to the mouse (11), is injected into normal mice it produces a transient parasitaemia; the animals recover and are then resistant to further challenge with the same organism, whereas nude (athymic) mice maintain a parasitaemia of about 50% for the rest of their lives.

Precisely how intra-erythrocytic protozoa are killed as the host overcomes the infection and acquires immunity is not understood. Primary infection or rechallenge with *B. microti* is terminated normally by death of the parasites within the red cells (8). Reported experiments (4,5,7,10), indicate that a soluble non-antibody factor(s) is responsible for the intracellular death of the parasites without the direct assistance of other cells, and that premature lysis of the infected cells is not involved. Clark (4) suggested that in normal mice cells of the mononuclear phagocyte system sensitised with a specific T-cell factor, could release soluble products on combining with parasite antigen, since the releasing cell did not carry the theta antigen and appeared to be present in spleen, rare in lymphnodes and absent from thoracic lymph.

Oxidised polyamines are known to be toxic to a variety of cell types (1,19) including trypanosomes (18). The possibility that polyamine oxidase or one of its products might be the soluble factor reported by Clark seemed worth investigation, particularly since activated macrophages have increased polyamine oxidase activity (15).

EXPERIMENTAL

High parasitaemia blood collected in heparin from mice infected with *Babesia rodhaini* (normally fatal to mice) was diluted 1 + 1 with dimethyl sulphoxide, the dimethylsulphoxide being added to the blood. Aliquots (200 μl) were dispensed into nylon vials, frozen and stored in liquid nitrogen. Frozen blood was rapidly thawed, as required, by immersion in a 37° water bath immediately prior to intraperitoneal injection into a mouse. When its parasitaemia was ascending this mouse was used as a source of parasites to infect further experimental mice. In this way the parasite was maintained in continuous passage. No account was taken of the number of parasites per cell so where the dose is given as 10^6 parasites this means 10^6 parasitised red cells.

Newborn calf serum (20) and human retroplacental serum (13) were used as sources of polyamine oxidase. In addition amine oxidase (from bovine plasma; lot no. 7018) was obtained from Miles Laboratories Ltd., Stoke Podges, Bucks, U.K. The mice used were male or female Balb/C or C57 Bl strains from the Clinical Research Centre colony.

Blood taken from mice showing a rising parasitaemia of 30-50% was added to solutions of β-D-glucose (4 mM, final concentration; haematocrit 0.5-5%) with additions to give the final concentrations shown (Table 1) and incubated with occasional shaking for 2 hours at 37°C. Equal quantities of treated blood were given by intraperitoneal injection to experimental mice (10^7 - 2×10^8 parasites/animal) usually in groups of 3. Blood smears were taken from the tail, at about day 4, air-dried, fixed for a few seconds with methanol, then stained for 45 minutes with Giemsa's stain diluted 1 + 9 with 0.067M phosphate buffer pH 6.8, washed with tap water and air-dried. A minimum of 200 cells per smear were counted, and the parasitaemia expressed as a percentage of the total number of red cells counted.

RESULTS

From a total of 23 mice injected with *B. rodhaini* infected cells that had been treated with 5 mM spermine and 10% newborn calf serum 21 did not develop a parasitaemia by day 4 and survived beyond day 60, as compared with those given cells treated with 5 mM spermine alone, 10% bovine serum alone, or controls given cells treated with neither, all of which developed a parasitaemia by day 4 and subsequently died by about day 8 (Table 1).

Similarly mice given cells that had been incubated with commercially purified amine oxidase and spermine did not develop a parasitaemia, whereas those given cells treated with amine oxidase only, did, even when the enzyme concentration was increased 100-fold. The data clearly show that the protective effect is dependant on both enzyme and substrate concentration. Mice given cells treated with retroplacental serum and spermine, did develop a parasitaemia.

TABLE 1. Survival of mice infected with blood infected with *B. rodhaini* that had been incubated in 5 mM glucose with additions at the final concentrations shown (AO is amino oxidase, RPS is retroplacental serum)

{Spermine} (mM)	{Newborn Calf serum} (%)	Other additions	No. Survivors/No in group (no. of expt)
5	5	-	0/7 (2 experiments)
5	1	-	0/3
1	5	-	0/3
0.1	10	-	0/3
5	0	-	0/10 (3 experiments)
0	10	-	0/10 (3 experiments)
0	0	-	0/26 (8 experiments)
5	10	-	21/23 (7 experiments)
1	10	-	5/6 (2 experiments)
5	-	0.012 mg/ml AO	12/12 (4 experiments)
0	-	0.012 mg/ml AO	0/12 (4 experiments)
0	-	1.12 mg/ml AO	0/3
5	-	10% RPS	0/8 (3 experiments)
0	-	10% RPS	0/9 (3 experiments)

Experiments with *Plasmodium chaubaudi* in C57Bl mice showed that treatment of infected blood with amine oxidase and spermine prevented the appearance of a parasitaemia in recipient mice. (Table 2). *Plasmodium yoelii* infection also failed to develop after similar treatment. As before, treatment with amine oxidase or spermine alone, or retroplacental serum with or without spermine, gave rise to a parasitaemia in the recipients. Further experiments using *B. microti* were inconclusive, with 7 out of 12 mice showing no parasitaemia after being injected with cells that had been treated with calf serum and spermine, and 3 out of 6 mice given cells treated with amine oxidase and spermine also showing no parasitaemia.

Incubation of parasitised red cells with amine oxidase and spermine results in a considerable amount of lysis (Table 3), but infected cells were no more fragile than controls. However, 10^7-10^8 parasites were injected in the *B. rodhaini* experiments and 10^4 parasites will produce a lethal infection (J. Christensen, unpublished data). In addition it has been shown that parasites freed from the erythrocyte are just as infective as those in the host cell (14).

As has already been noted (4) mice which are recovering from infection with *B. microti* have greatly enlarged spleens. Estimation of the polyamine oxidase activity in spleens from *B. microti* infected mice showed a 3-fold increase in specific activity, which, coupled with a nearly 4-fold increase in size (Table 4) greatly increased the total enzyme activity per spleen.

TABLE 2. Effect of spermine and polyamine oxidase treatment of parasitised cells on the development of a parasitaemia in infected C57B1 mice (RPS is retroplacental serum, NBCS is newborn calf serum)

Parasite	{Spermine} (5 mM)	{Amine oxidase} (12 μg/ml)	Other additions	No. developing parasitaemia/ No. in group
Plasmodium chaubaudi	O	O	10% RPS	3/3
	+	O	10% RPS	3/3
	O	+		3/3
	+	+		0/3
Plasmodium yoelii	+	O		3/3
	O	+		3/3
	+	+		0/3
Babesia microti	O	O		9/9
	+	O		6/6
	O	+		9/9
	+	+		3/6
	O	O	10% NBCS	12/12
	+	O	10% NBCS	5/12

Infection with *B. rodhaini* does not cause spleen enlargement or an increase in polyamine oxidase activity.

TABLE 3. Lysis of *B. rodhaini* infected cells following incubation with amine oxidase and spermine

	Lysis (%)		
Experiment	Amine oxidase + spermine	Spermine only	Amine oxidase only
1	49.6	6.1	4.6
2	24.9	8.6	5.4
3	46.9	7.4	5.6

DISCUSSION

Evidence for the involvement of a soluble factor(s) in the mechanism by which mice recover from *B. microti* infections has been provided by earlier reports from this laboratory (4). Serum from recovered mice failed to protect other mice when given as single large doses. Daily small doses (150 μl ip) reduced the parasitaemia by 50% when given from the day of infection (day 0), and by 80% when given from day 12, indicating a short half-life for the protecting factor. Parasitised red cells in diffusion chambers (2) implanted intraperitoneally were killed more rapidly

TABLE 4. Polyamine oxidase activity in, and mean weights of, spleens from mice infected with *Babesia microti*.

	Polyamine oxidase activity (n mol/min/mg protein)	Wet weight (g)
B. microti spleens	0.098 ± 0.018	0.590 ± 0.045
Controls	0.036 ± 0.009	0.157 ± 0.021

in immune mice than in controls, implying that soluble factors in the immune mice can either kill or inhibit parasites without assistance from immune cells.

Intraperitoneal chambers containing immune spleen cells reduced the parasitaemia in nude mice, again suggesting protection by the release of soluble factors. Intraperitoneal administration of immune spleen cells also transferred immunity, the effect being significant if given at the same time as the parasite and absolute if given 4-8 weeks earlier. Neither incubation of immune spleen cells on nylon wool columns (T-enrichment) nor treatment with anti-theta serum and complement (B-enrichment) greatly altered their ability to transfer immunity to nude mice. Immune lymphnode cells had no protective effect. Removal of the spleen before administration of *B. microti* caused a prolonged parasitaemia; splenectomy on day 18 resulted in recrudescence of the parasitaemia and temporary loss of immunity.

Mice given intravenous or intraperitoneal injections of live *Mycobacterium bovis* (strain Bacillus Calmette-Guerin; BCG, 9), or *Brucellas abortis* (strain 19) (12) or killed *Corynebacterium parvum* (8), 2-4 weeks before infection were strongly protected against *B. microti* and even against the highly virulent *B. rodhaini*.

Spleens, livers and lungs of mice given BCG contain small tubercules (3) which are most obvious 10-14 days after injection and then become less active. Protection against a final challenge with *B. microti* is retained for at least 3 months after *C. parvum* has been given. Over this period these mice retained the epitheliod cells of the tubercules, although most of the rest had gone. The observation that the full protective effect that BCG and *C. parvum* can exert takes some weeks to develop (8), apart from being evidence against adjuventicity, is consistent with the gradual development, and then persistance, of epithelioid cells in these mice. Epithelioid cells are thought to be derived from the mononuclear phagocyte series (16) and have been shown to be secretory (17). Spleens, livers and lungs of mice given BCG subcutaneously, which did not protect against parasites but could be expected to express powerful adjuventicity, contained no epithelioid cells (7). Alveolar macrophages from rabbits given BCG subcutaneously or intravenously have been shown to contain elevated levels of polyamine oxidase (15).

From the data presented here (Table 1) it would appear that the polyamine oxidase in calf serum was the agent responsible for initiating the events that lead to the death of the intracellular parasites used in this study. Polyamine oxidase would also seem to be a candidate for the soluble factor from spleen cells observed by Clark (4,6). The lack of effect with human polyamine oxidase suggests that the products formed must differ from those of the bovine enzyme.

REFERENCES

1. Bachrach, U. (1970): Ann. Rev. Microbiol., 24: 109-134.
2. Berman, I., and Kaplan, H.S. (1959): Blood, 14: 1040-1046.
3. Blanden, R.W., Lefford, M.J., and Mackaness, G.B. (1969): J. Exp. Med., 129: 1079-1106.
4. Clark, I.A. (1976): PhD Thesis, London.
5. Clark, I.A. and Allison, A.C. (1976a): New Scientist, 69: 668-669.
6. Clark, I.A., and Allison, A.C. (1976b): Nature, 259, 309-311.
7. Clark, I.A., Cox, F.E.G., and Allison, A.C. (1977): Parasitology, 74: 9-17.
8. Clark, I.A., Richmond, J.E., Wills, E.J., and Allison, A.C. (1975): Lancet, ii: 1128-1129.
9. Clark, I.A., Wills, E.J., Richmond, J.E. and Allison, A.C. (1977): Infect. Immun., 17: 430-438.
10. Cox, F.E.G. (1978): Parasitology, 76: 55-60.
11. Cox, F.E.G., and Young, A.S. (1969): Parasitology, 59: 257-268.
12. Herod, E., Clark, I.A., and Allison, A.C. (1978): Clin. Exp. Immunol., 31: 518-523.
13. Illei, G., and Morgan, D.M.L. (1979): Br. J. Obstet. Gynaecol. 86: 873-877.
14. Jack, R.M., and Ward, P.A. (1980): J. Immunol., 124: 1566-1573.
15. Morgan, D.M.L., Ferluga, J., and Allison, A.C. (1980): In: Polyamines in Biomedical Research, edited by J.M. Gaugas, pp. 303-308. John Wiley, Chichester.
16. Payling-Wright, G. and Heard, B.E. (1976): In: Systemic Pathology, edited by G. Payling-Wright, and W. St.C. Symmers, pp. 333-426. Longmans, London.
17. Spector, W.G. (1976): Ann. N.Y. Acad. Sci., 278: 3-6.
18. Tabor, C.W. and Rosenthal, S.M. (1956): J. Pharmacol. Exper. Therap., 116: 139-155.
19. Tabor, H., and Tabor, C.W. (1964): Pharmacol. Rev., 16: 245-300.
20. Tabor, H., and Tabor, C.W. (1972): Adv. Enzymol., 36: 203-268.

Advances in Polyamine Research, Vol. 4, edited by U. Bachrach, A. Kaye, and R. Chayen. Raven Press, New York © 1983.

Induction of Diamine Oxidase Activity in Some Processes of Growth

A. Perin, A. Sessa, and M. A. Desiderio

Institute of General Pathology and Consiglio Nazionale delle Ricerche, Center for Research in Cell Pathology, University of Milan, 20133 Milan, Italy

In contrast to the large number of investigations on putrescine biosynthesis, few studies have been carried out on the regulation of catabolism of this diamine in cellular growth. Putrescine, as other aliphatic diamines with 3 to 6 carbon atoms, and histamine, which can be considered a substituted cyclic diamine, are substrates of diamine oxidase (EC 1.4.3.6), which catalyzes their oxidative deamination to corresponding amino aldehydes (7, 41). In tissues with a high diamine oxidase activity, such as intestine, thymus, placenta (14, 21, 36) and serum during pregnancy (19, 40), putrescine degradation seems to take place through diamine oxidase. In tissues with a low diamine oxidase activity, putrescine seems to be mainly acetylated to monoacetylputrescine, which is excreted as such or oxidized by a monoamine oxidase (EC 1.4.3.4) (32).

The occurrence, purification, and structure of diamine oxidase have been the object of many studies (4, 23, 41), but its in vivo function as a rate-limiting enzyme in the oxidative metabolism of putrescine has only recently been emphasized. In fact, in normal (22) and pregnant rats (1) as well as in normal mice (34), exogenous putrescine is mainly degraded by diamine oxidase, since aminoguanidine, a strong and irreversible inhibitor of this enzyme (24), reduces almost completely the oxidation of this substance to respiratory CO_2.

Studies carried out in this laboratory have demonstrated that diamine oxidase activity increases in different processes of growth (9, 25, 35). Here we discuss some problems concerning diamine oxidase activity determination in tissue homogenates and we report data demonstrating that the growth stimulus induces a de novo synthesis of enzyme in tissues.

STUDIES ON DIAMINE OXIDASE ACTIVITY DETERMINATION IN TISSUE HOMOGENATES

Among the various methods available for diamine oxidase activity determination, the radiochemical assay of Okuyama and Kobayashi (24), which measures the [^{14}C]Δ^1-pyrroline using as substrate [^{14}C]putrescine, is considered the most specific and sensitive (20, 24). Our studies (9, 35), in agreement with those of Fogel et al. (11, 12) and Andersson et al. (2), have demonstrated that aldehyde-metabolizing enzymes present in tissue homogenates may interfere with diamine oxidase activity determination by this method, since a fraction of γ-aminobutyraldehyde, the first product of putrescine oxidation, instead of cyclizing to Δ^1-pyrroline may be further oxidized to γ-aminobutyric acid (34) or reduced to 4-amino-1-butanol by aldehyde reductase and/or alcohol dehydrogenase (10). For this reason, the use of inhibitors of aldehyde-metabolizing enzymes may be a better approximation of the actual values of diamine oxidase activity in homogenates. As reported in Fig. 1, acetaldehyde, a competitive inhibitor of aldehyde dehydrogenase, enhanced Δ^1-pyrroline formation in rat liver and kidney without modifying diamine oxidase activity, since this substance did not affect the activity of a commercial preparation of hog kidney diamine oxidase. In contrast, chloral hydrate, disulfiram, pyrazole, and phenobarbital showed a direct influence on diamine oxidase because they enhanced or diminished the enzyme activity. On the basis of these results, we measured diamine oxidase activity in tissue homogenates in the presence of 10 mM acetaldehyde, which gave the highest yield in Δ^1-pyrroline (9, 35).

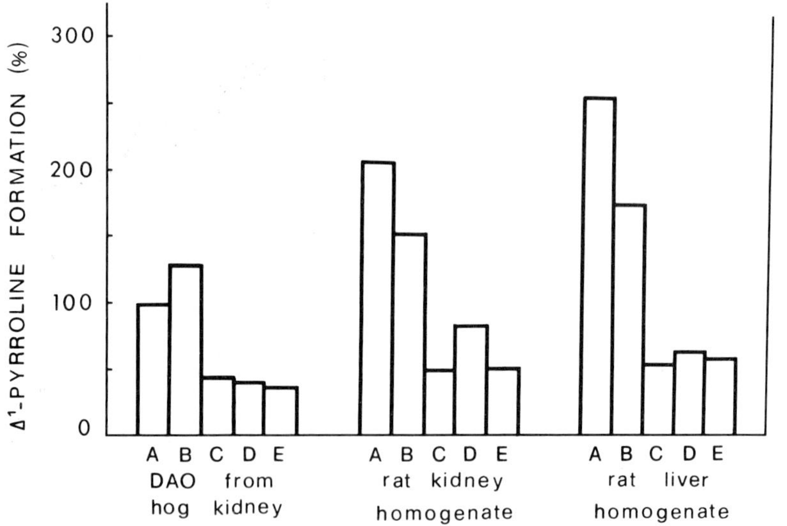

Fig. 1. Effects of inhibitors of aldehyde-metabolizing enzymes on Δ^1-pyrroline formation from putrescine by hog kidney diamine oxidase as well as rat kidney and liver homogenates. 10 mM Acetaldehyde (A); 1 mM chloral hydrate (B); 0.1 mM disulfiram (C); 2 mM pyrazole (D); 10 mM phenobarbital (E). The values are expressed as the percentage of activity measured without any addition.

Another problem for diamine oxidase determination in tissue homogenates concerns the specificity of the method for this enzyme. In this regard, we considered as diamine oxidase activity only that sensitive to aminoguanidine, which corresponds to the amount of Δ^1-pyrroline obtained by the difference between the values measured in the absence and in the presence of aminoguanidine (9, 35).

DIAMINE OXIDASE ACTIVITY IN GROWTH PROCESSES

In human and rat gestation, in addition to profound modifications in diamine (histamine, putrescine, and cadaverine) formation related to the biochemical changes and extensive cell proliferation of fetal (38) and reproductive tissues (3, 14, 21, 30), there is a striking enhancement of blood diamine oxidase activity (19, 40). The uterus, where this enzyme activity is normally undetectable, and maternal placenta, the principal source of blood enzyme (14, 19), exhibit high diamine oxidase activity starting from the first stages of pregnancy (14, 21). In rat kidney (27) and glioma (6) cultured cells, diamine oxidase activity increases in the early phases of growth, and in glioma cells the increase is associated with an enhancement in ornithine decarboxylase activity (6).

The close involvement of polyamines in cell growth and proliferation (16) and the possible relation between their levels and diamine oxidase activity led us to study the behavior of this enzyme activity in regenerating rat liver after partial hepatectomy as well as in rat kidney undergoing compensatory hypertrophy after unilateral nephrectomy. As shown in Fig. 2, diamine oxidase activity significantly increased in the first 24 and 48 h after the application of growth stimulus, in liver and kidney, respectively.

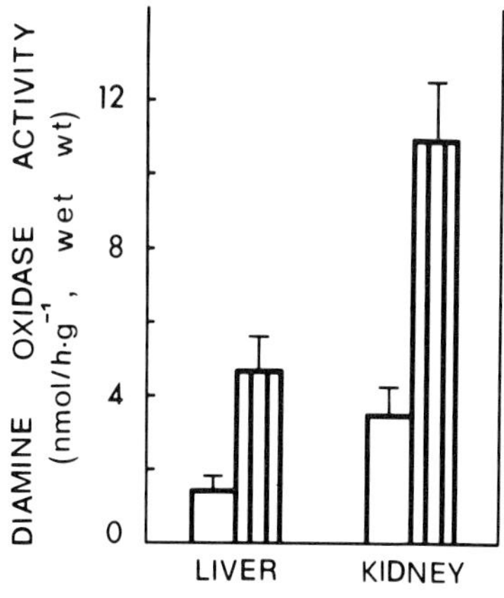

Fig. 2. Diamine oxidase activity in regenerating liver and kidney undergoing compensatory hypertrophy of rat. Values indicate the enzymatic activity measured 24 and 48 h after the operation in liver and kidney, respectively (dashed columns) and in corresponding normal tissues (blank columns).

The study of the time course of this enzyme activity and of putrescine concentration showed that in the early phases of hepatic regeneration and renal hypertrophy, diamine oxidase activity followed the increase in putrescine levels (9, 25, 35). In the same experimental models of growth, various authors have found a sudden increase in ornithine decarboxylase activity (8, 17, 31), the rate-limiting enzyme in putrescine biosynthesis, followed by an enhanced formation of putrescine. The course of the events herein described suggests that the growth stimulus per se or through the enhancement in putrescine biosynthesis may regulate diamine oxidase activity.

Experiments with cycloheximide given in vivo (Fig. 3) demonstrated that in the studied models there is an induction of diamine oxidase activity with synthesis of new enzyme molecules. In fact, this drug, which is an inhibitor of protein synthesis (26), prevented the increase in enzyme activity observed in regenerating liver and hypertrophic kidney. Experiments with actinomycin D, an inhibitor of DNA-dependent RNA synthesis (18), indicated that after the growth stimulus new genetic transcription is involved in diamine oxidase induction, since this drug, administered in the early phases of growth, prevented the increase in diamine oxidase activity in regenerating liver and hypertrophic kidney.

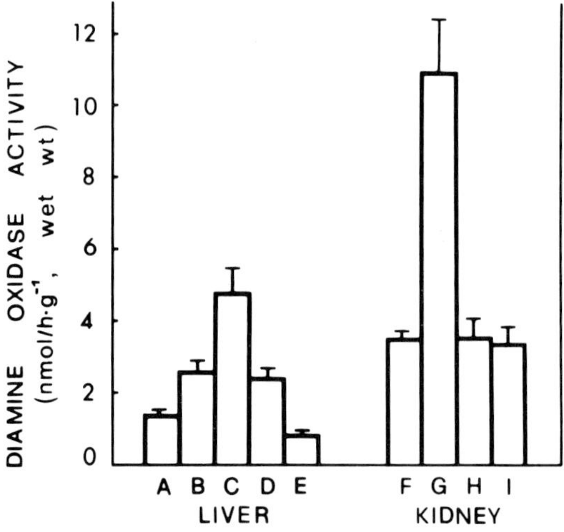

Fig. 3. Effects of cycloheximide and actinomycin D on diamine oxidase activity of regenerating liver and kidney undergoing compensatory hypertrophy of rat. Liver: normal (A); 8 h-regenerating (B); 16 h-regenerating (C); 16 h-regenerating plus cycloheximide given 8 h after hepatectomy (D); 16 h-regenerating plus actinomycin D given at the time of hepatectomy (E). Kidney: normal (F); 48 h after unilateral nephrectomy (G); 48 h after unilateral nephrectomy plus cycloheximide given at the time of operation and 24 h later (H); 48 h after unilateral nephrectomy plus actinomycin D given at the time of operation (I).

The study of the apparent half-life of diamine oxidase in normal and growing tissues using cycloheximide also confirms the enzyme induction, since it permits us to exclude that the increased activity is the result of a slowed degradation of enzyme. In fact, the half-life of diamine oxidase was 14 – 15 h in hypertrophic kidney and regenerating liver as well as in corresponding normal tissues of rat. A half-life for diamine oxidase of about 10 h has been found by Shaff and Beaven (36) in rat small intestine. It is interesting to note that after administration to rats of aminoguanidine or heparin, which releases diamine oxidase from intestine, the half time recovery of putrescine-catabolizing ability in vivo has been found to be 15 – 18 h (22, 36, 37).

CONCLUDING REMARKS

In regenerating rat liver and in rat kidney undergoing compensatory hypertrophy after unilateral nephrectomy, the increase in ornithine decarboxylase activity (8, 17, 31) and in putrescine level (8, 17, 31) is associated with the enhancement in diamine oxidase activity (9, 25, 35). This enzyme has been shown to be inducible in response to growth stimulus, since its enhancement is mainly due to a de novo synthesis of enzyme as demonstrated in experiments with inhibitors of protein and RNA syntheses. Our data also demonstrate that the enhancement in diamine oxidase activity does not depend on a slowing of enzyme degradation, and suggest that the release of inactive enzyme from subcellular cryptic structures, if any occurs, is an event of minor importance in determining the increase in enzyme activity.

The parallel distribution of diamine oxidase activity, ornithine decarboxylase activity, and putrescine in growing tissues may be significant for the role of diamine oxidase in the metabolism of putrescine in growth processes. Diamine oxidase activity induction, together with other mechanisms such as conversion to spermidine (39), production of an ornithine decarboxylase inhibitor (13, 15), transamination or conjugation (33), and diffusion in extracellular fluids (29), may contribute to regulate the levels of putrescine, which is considered a critical regulatory molecule for cell growth.

The mechanism that regulates the induction of diamine oxidase in growing cells is still unknown. The synthesis of ornithine decarboxylase, which in many growing tissues precedes or parallels that of diamine oxidase, seems to be regulated by a cyclic AMP-dependent protein kinase (5, 28). Whether this mechanism may also operate in diamine oxidase induction in growing tissues needs to be verified.

ACKNOWLEDGEMENT

This work was supported by a grant from the Consiglio Nazionale delle Ricerche, Rome (Target Project "Control of Tumor Growth," no. 80.01479.96).

REFERENCES

1. Andersson, A.Ch., and Henningsson, S. (1981): Acta Physiol. Scand., 113:525-534.
2. Andersson, A.Ch., Henningsson, S., Persson, L., and Rosengren, E. (1978): Acta Physiol. Scand., 102:159-166.
3. Andersson, A.Ch., Henningsson, S., and Rosengren, E. (1979): Acta Physiol. Scand., 105:508-512.
4. Argento-Cerù, M.P., Sartori, C., and Autuori, F. (1973): Eur. J. Biochem., 34:369-374.
5. Bachrach, U. (1975): Proc. Natl. Acad. Sci. U.S.A., 72:3087-3091.
6. Bachrach, U. (1980): Biochem. J., 188:387-392.
7. Bardsley, W.G., and Ashford, J.S. (1972): Biochem. J., 128:253-263.
8. Brandt, J.T., Pierce, D.A., and Fausto, N. (1972): Biochim. Biophys. Acta, 279:184-193.
9. Desiderio, M.A., Sessa, A., and Perin, A. (1982): Biochim. Biophys. Acta, 714:243-249.
10. Dupré, S., Granata, F., Dederici, G., and Cavallini, D. (1975): Physiol. Chem. Physics, 7:517-522.
11. Fogel, W.A., Bieganski, T., and Maśliński, C. (1979): Agents Actions, 9:42-44.
12. Fogel, W.A., Bieganski, T., Woźniak, J., and Maśliński, C. (1978): Biochem. Pharmacol., 27:1159-1162.
13. Fong, W.F., Heller, J.S., and Canellakis, E.S. (1976): Biochim. Biophys. Acta, 428:456-465.
14. Guha, S.K., and Jänne, J. (1976): Biochim. Biophys. Acta, 437:244-252.
15. Heller, J.S., Fong, W.F., and Canellakis, E.S. (1976): Proc. Natl. Acad. Sci. U.S.A., 73:1858-1862.
16. Jänne, J., Pösö, H., and Raina, A. (1978): Biochim. Biophys. Acta, 473:241-293.
17. Jänne, J., and Raina, A. (1968): Acta Chem. Scand., 22:1349-1351.
18. Kahan, E., Kahan, F.M., and Hurwitz, J. (1963): J. Biol. Chem., 238:2491-2497.
19. Kobayashi, Y. (1964): Nature, 203:146-147.
20. Kusche, J., Richter, H., Hesterberg, R., Schmidt, J., and Lorenz, W. (1973): Agents Actions, 3:148-156.
21. Maudsley, D.V., and Kobayashi, Y. (1977): Biochem. Pharmacol., 26:121-124.
22. Missala, K., and Sourkes, T.L. (1980): Eur. J. Pharmacol., 64:307-311.
23. Mondovì, B., Guerrieri, P., Costa, M.T., and Sabatini, S. (1981): In: Advances in Polyamine Research, edited by C.M. Caldarera, V. Zappia, and U. Bachrach, Vol. 3, pp. 75-84, Raven Press, New York.
24. Okuyama, T., and Kobayashi, Y. (1961): Arch. Biochem. Biophys., 95:242-250.

25. Perin, A., Sessa, A., and Desiderio, M.A. (1981): In: Advances in Polyamine Research, edited by C.M. Caldarera, V. Zappia, and U. Bachrach, Vol. 3, pp. 397-407, Raven Press, New York.
26. Piestka, S. (1971): Annu. Rev. Biochem., 41:697-710.
27. Quash, G., Keolouangkhot, T., Gazzolo, L., Ripoll, H., and Saez, S. (1979): Biochem. J., 177:275-282.
28. Russell, D.H., Byus, C.V., and Manen, C.A. (1976): Life Sci., 19:1297-1306.
29. Russell, D.H., and Durie, B.G.M. (1978): Polyamines as Biochemical Markers of Normal and Malignant Growth, Raven Press, New York.
30. Russell, D.H., and McVicker, T.A. (1972): Biochim. Biophys. Acta, 259:247-251.
31. Russell, D.H., and Snyder, S.H. (1968): Proc. Natl. Acad. Sci. U.S.A., 60:1420-1427.
32. Seiler, N. (1980): Physiol. Chem. Physics, 12:411-429.
33. Seiler, N. (1980): In: Polyamines in Biology and Medicine, edited by D.R. Morris, and L.H. Marton, pp. 127-150, Marcel Dekker, New York.
34. Seiler, N., and Eichentopf, B. (1975): Biochem. J., 152: 201-210.
35. Sessa, A., Desiderio, M.A., Baizini, M., and Perin, A. (1981): Cancer Res., 41:1929-1934.
36. Shaff, R.E., and Beaven, M.A. (1976): Biochem. Pharmacol., 25:1057-1062.
37. Sourkes, T.L., and Missala, K. (1981): Agents Actions, 11: 20-27.
38. Sturman, J.A., and Gaull, G.E. (1974): Pediatr. Res., 8: 231-237.
39. Tabor, C.W., and Tabor, H. (1976): Annu. Rev. Biochem., 45: 285-306.
40. Tryding, N., and Willert, B. (1968): Scand. J. Clin. Lab. Invest., 22:29-32.
41. Zeller, E.A. (1963): In: The Enzymes, edited by P.D. Boyer, H. Lardy, and K. Myrbäck, Vol. 8, pp. 313-335, 2nd edition, Academic Press, New York and London.

Advances in Polyamine Research, Vol. 4, edited by
U. Bachrach, A. Kaye, and R. Chayen. Raven Press,
New York © 1983.

Amine Oxidase Activity in Malignant Human Brain Tumors

B. Mondovì, P. Riccio, *A. Riccio, and G. S. Marcozzi

*Institute of Applied Biochemistry and CNR Centre of Molecular Biology, University of Rome, Rome, Italy; *Department of Neurosurgery, Regina Elena Institute for Cancer Research, Rome, Italy*

As reported in recent literature (16) the activity of enzymes involved in polyamine metabolism is higher in rapid-growing tissues, and particularly in tumors, as compared with normal tissues.

Quash et al (14) demonstrated that diamine oxidase (DAO) activity rose rapidly during lag phase in coltures of normal and transformed rat kidney cells. Both normal and transformed cells showed maximum DAO activity 24 hr after seeding, the latter being 50% greater than that of normal cells at this time. DAO and polyamine oxidase activities seemed to correlate with the cells growth. As cells approached confluence, the normal ones showed about twice the activity of their transformed counterparts.

Sessa et al (19) observed that DAO activity increased rapidly in regenerating rat liver and reached a maximum 16-18 hr after partial hepatectomy. A noteworthy increase in DAO activity was also observed in 4-dimethylaminoazobenzene-induced and Yoshida ascites hepatomas.

This work was supported by the Italian National Research Council, Contract No. 810141696 of the Finalized Project "Control of Tumor Growth".
This paper is dedicated to Professor Alessandro Rossi Fanelli on his 75th birthday.

It was suggested that the high DAO activity ascertained in some malignant thyroid cells could be employed as an useful marker to distinguish between hyperplasia and malignancy in thyroids with early proliferative C-cells disorder (2).

The study of amine oxidase (AO) activity[1] in growing tissues and especially in tumors appears to be important for the understanding of the regulatory process of cell multiplication: actually amine aldehydes, which are the product of polyamine oxidative deamination, have a strong antiproliferative effect on cells (1, 5, 11, 12) because of their bindings to DNA (1) with consequent inhibition of DNA synthesis (1, 11).

Within this framework, the balance between the biosynthesis of polyamine and the degradation of them and their metabolic products, is very important for the growth of cells.

In other words, the increase in polyamines goes together with proliferation, while if the balance is shifted towards the intracellular aldehyde concentration, proliferation should be apparently interrupted.

An increase in the activity of polyamine biosynthetic enzymes ornithine decarboxylase (ODC) and S-adenosyl-L-methionine decarboxylase (SAMD) has recently been demonstrated in some human brain tumors (17). Furthermore, high polyamine levels were found in the cerebrospinal fluid of patients with medulloblastoma (10) and glioblastoma multiforme (4).

In order to ascertain the possible involvement of AO in the oxidation of polyamines produced in malignant human brain tumors, AO activity has been determined both in the benign and the malignant forms of some human tumors of the Central Nervous System.

EXPERIMENTAL METHODS

All chemicals used were of the highest purity available. Tissue samples obtained immediately after surgery from patients of both sexes were stored at -20°C until used.

---

[1]The term "amine oxidase (AO) activity" is used to indicate the activity of amine oxidizing enzymes regardless to substrate specificity: i.e., it may refer to monoamine oxidase, diamine oxidase or polyamine oxidase.

Samples were dissected and washed with ice-cold 0.9% NaCl solution, weighed and homogenized (10% w/v) with a Potter apparatus in ice-cold 0.1 M potassium phosphate buffer at pH 7.4.

AO activity was detected by using ^{14}C putrescine according to Okuyama and Kobayashi (13) as modified by Quash et al (14), except for the fact that the test tubes contained 1.5 ml homogenate in a final volume of 2 ml. Blanks were performed by heating samples at 100°C for 20 min in boiling water.

Incubation was carried out for 1 hr at 37°C and the reaction was stopped by placing test tubes in an ice-cold bath. 0.2 ml of a 2% Na_2CO_3 solution were immediately added to each tube to allow best extraction efficency; 10 ml toluol containing 6 g/l PPO plus 0.1 g/l POPOP were added, the mixture was vigorously shaken for 30 sec and then centrifuged in order to separate the organic solvent which was transferred into counting vials.

Radioactivity of Δ' ^{14}C pyrroline was detected by a liquid scintillation counter.

Enzyme activity is expressed in units/mg protein. 1 unit corresponds to 1 μmole of Δ' pyrroline formed in 1 min.

AO inhibitors were added to the reaction mixture and incubated at 37°C for 15 min or 60 min before the addition of substrate.

Tumors were classified according to the degree of malignancy as suggested by Zülch (22). Anaplastic astrocytoma and glioblastoma multiforme correspond, respectively, to astrocytomas degree III and IV.

RESULTS AND DISCUSSION

Table 1 shows AO activity in typical and atypical meningiomas, astrocytomas and medulloblastoma.

No enzyme activity has been detected in the typical forms of meningiomas (tumors without a malignant character), whereas in the atypical forms a high AO activity was present.

Medulloblastoma, which is one of the most malignant brain tumors, showed low AO activity.

The best correlation between degree of tumor malignancy and levels of AO activity has been obtained in the astrocytomas, where the activity is proportional to the degree of malignancy (see also Fig.1, where activity values are reported in logarithmic scale).

TABLE 1 - AO activity in some brain tumors

Tumors	AO µunits/mg protein/min
Meningiomas (typical form)	no detectable activity
Meningiomas (atypical form)	4.5
Astrocytoma degree I	0.7
Astrocytoma degree II	1.0
Astrocytoma degree III (anaplastic astrocytoma)	3.3
Astrocytoma degree IV (glioblastoma multiforme)	4.6
Medulloblastoma	0.4

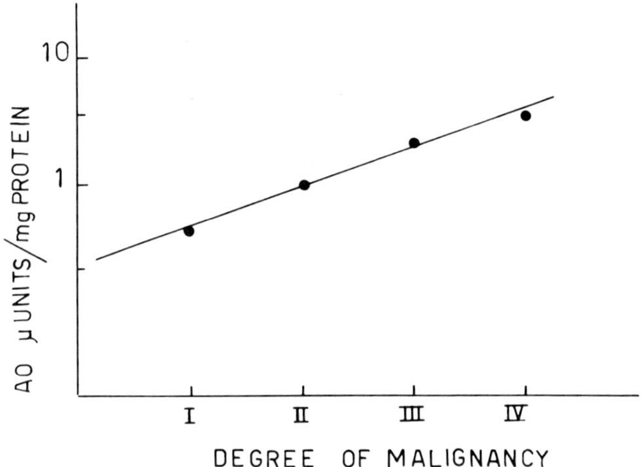

FIG.1. Relationship between AO activity and degree of malignancy in astrocytomas.
Enzyme activity is expressed in logarithmic values.

A good correlation has also been observed between polyamine biosynthetic and degradative enzymes. Fig. 2 shows the relationship between AO, ODC and SAMD activities: values represent a mean obtained between the benign (degree I-II) and the malignant (degree III - IV) astrocytomas. AO activity appears to be over 4 times higher in the malignant forms than in the benign ones.

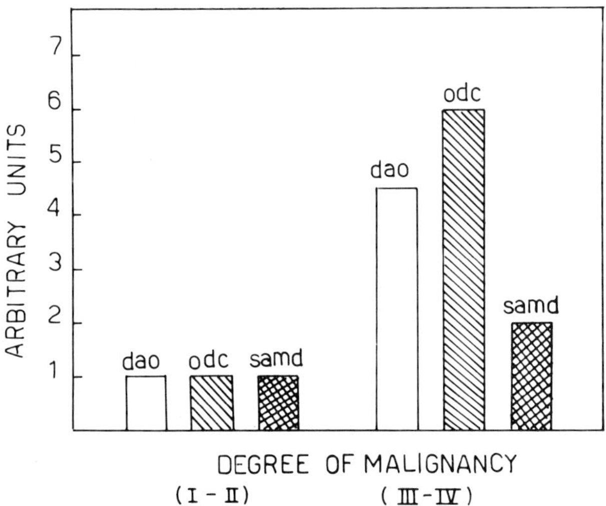

FIG. 2. Relationship between AO, ODC and SAMD activities in benign and malignant astrocytomas. AO values are those obtained from the present research. ODC and SAMD values are obtained from Scalabrino et al (17).

It is interesting to note that in peritumoral tissue, AO activity was in some cases found to be higher than that found in the tumor itself (see Table 2). Similar results were obtained by Maslinski (personal communication) in the peritumoral tissue of the rat intestinal mucosa.

In the case of medulloblastoma, which showed a low AO activity as compared with other malignant tumors, values of the same order of magnitude as glioblastoma were found in peritumoral tissue.

TABLE 2 - AO activity in tumoral and peritumoral tissue

	tumoral tissue AO (μunit/mg prot.)	peritumoral tissue AO (μunit/mg prot.)
Astrocytoma (degree II)	1.0	1.7
Medulloblastoma	0.4	5.5

It is interesting to note that an excellent correlation was assessed between liquoral polyamine levels and tumor recurrence in patients with medulloblastoma (9), while polyamine levels in the cerebrospinal fluid of patients with glioblastoma multiforme were not well correlated with the tumor progression. This might be attributed to the propinquity of medulloblastoma to the ventricular ependyma and/or to the fact that polyamine levels ought to be lower in tumors having a high AO activity such as glioblastoma.

As reported by Seiler and Al Therib (18), the degradative pathway for putrescine in mammalian brain comprises its acetylation followed by oxidative deamination of monoacetyl putrescine to acetyl-γ-aminobutyraldehyde, which in turn is further oxidized to N-acetyl-γ-aminobutyrate, this being then deacetylated to γ-aminobutyric acid.

Icekson and Bachrach (7) demonstrated that in rat glioma cell putrescine and spermidine are converted into acetyl putrescine and acetylspermidine, respectively. Monoacetyl-cadaverine and monopropionylcadaverine were also found to be substrates for both monoamine oxidase (MAO) in rat liver mitochondria and hog kidney DAO (20).

In order to obtain more information on the type of AO involved in the tumors analyzed, experiments were performed by preincubating tumor homogenates in the presence of aminoguanidine, a specific inhibitor of DAO (21), pargyline, a specific inhibitor of both type A and type B mitochondrial MAO (15) and phenylhydrazine as inhibitor of both copper and FAD-dependent amine oxidases (8, 15).

As reported in Table 3, about 20% inhibition was obtained by aminoguanidine and pargyline, and 80% by phenylhydrazine;

no substantial differences were obtained on preincubation of samples in the presence of these inhibitors for 15 min or 60 min at 37°C.

TABLE 3 - Effect of some AO Inhibitors [a]

Inhibitor	Concentration (M)		
	$2 \cdot 10^{-5}$	$2 \cdot 10^{-4}$	$2 \cdot 10^{-3}$
Aminoguanidine	15	25	-
Phenylhydrazine	-	-	80
Pargyline	-	-	20

[a]Values are expressed as percentage decrease of AO activity with respect to that obtained in the absence of inhibitors. Preincubation time: 15 min. At 60 min, results were similar.

Because of the absence of severe inhibition observed in tumor samples pre-treated with aminoguanidine or pargyline, further experimenting is needed to obtain more adequate information on the type of AO involved in human brain tumors.

In order to minimize the effect of aldehyde metabolizing enzymes, tumor samples were incubated in the presence of chloral hydrate, acetaldehyde or butyraldehyde as inhibitors of aldehyde dehydrogenase and phenobarbital as inhibitor of alcohol dehydrogenase (3, 19). Controls were effected by the addition of acetaldehyde, chloral hydrate and butyraldehyde at the end of incubation to blanks performed with boiled samples, in order to avoid any incorrect result attributable to the reaction between the aldehyde group of the inhibitor and radioactive putrescine with the formation of a Schiff base (Marcozzi, personal communication).

Table 4 shows the increase of AO activity in the presence of inhibitors.

The cerebrospinal fluid of patient affected by malignant brain tumors was also tested for AO activity: the enzyme is present, but no correlation has been obtained so far between the liquoral enzyme activity and that of the corresponding tumor. The different localization of each tumor with respect to the ventricular system or basal cisterns should be taken into account (4).

TABLE 4 - Effect of aldehyde dehydrogenase and alcohol dehydrogenase inhibitors[a]

Inhibitor	Concentration (M)			
	2.10^{-4}	2.10^{-3}	5.10^{-3}	2.10^{-2}
Chloral hydrate	-	4	15	50
Acetaldehyde	-	50	-	250
Butyraldehyde	20	50	-	150
Phenobarbital	-	40	-	30

[a]Values are expressed as percentage increase of AO activity with respect to that obtained in the absence of inhibitors.

Experiments are in progress to employ AO activity in cerebrospinal fluid as a marker of malignant brain tumors.

REFERENCES

1. Bachrach, U., Persky, S. (1969): Biochim. Biophys. Acta, 179: 484-493.
2. Baylin, B.S., Mendelsohn, G., Weisburger, W.R., Gann, D.S. and Eggleston, J.C. (1979): Cancer, 44: 1315-1321.
3. Fogel, W.A., Bieganski, T., Wozniak, J. and Maslinski, C., (1978): Biochem. Pharmac., 27: 1159-1162.
4. Fulton, D.S., Levin, V.A., Lubich, W.P., Wilson, C.B. and Marton, L.J. (1980): Cancer Res., 40: 3293-3296
5. Gaugas, J.M., Dewey, D.L. (1981): J. Pathol., 134: 243-252.
6. Goa, J. (1953): Scand. J. Chim. Invest., 5: 218-222.
7. Icekson, I. and Bachrach, U. (1980): Jerusalem, 13th FEBS Meeting Abstr., S3-P15: 58.

8. Lindström, A. and Petterson, G. (1973): Eur. J. Biochem., 34: 564-568.
9. Marton, L.J., Edwards, M.S., Levin, V.A., Lubich, W.P. and Wilson, C.B. (1979): Cancer Res., 39: 993-997
10. Marton, L.J., Edwards, M.S., Levin, V.A., Lubich, W.P. and Wilson, C.B. (1981): Cancer, 47: 757-760.
11. Mondovì, B., Guerrieri, P., Costa, M.T. and Sabatini, S. (1981): Adv. Polyamine Res., 3: 75-84, Raven Press, New York.
12. Mondovì, B., Gerosa, P., Cavaliere, R.: Agents and Actions (in press).
13. Okuyama, T., Kobayashi, Y. (1961): Arch. Biochem. Biophys., 95: 242-250.
14. Quash, G., Keolouangkhot, T., Gazzolo, L., Ripoll, H. and Saez, S. (1979): Biochem. J., 177: 275-282.
15. Roth, J.A., Eddy, B.J., Pearce, B. and Mulder, K.M. (1981) Biochem. Pharmacol., 30: 945-950.
16. Scalabrino, G. and Ferioli, M.E. (1981): Adv. Cancer Res., 35: 151-268.
17. Scalabrino, G., Ferioli, M.E., Modena, D., Puerari, M. and Luccarelli, G. (1982): It. J. Biochem., 31: 60-62
18. Seiler, N. and Al-Therib, M.J. (1974): Biochem.J., 144: 29-35.
19. Sessa, A., Desiderio, M.A., Baizini, M. and Perin, A. (1981): Cancer Res., 41: 1929-1934.
20. Suzuki, O., Matsumoto, T., Oya, M., Katsumata, Y. and Stepita Klauco, M. (1980): Experimentia, 36: 535-537.
21. Webb, J.L. (1966): Enzyme and Metabolic Inhibitors, Vol.2, Acad. Press, New York and London.
22. Zülch, K.J. and Woof, A.L. (1964): Classification of Brain Tumors, Springer, Vienna.
23. Zülch, K.J. (1980): Neuroradiology, 19: 59-66.

Advances in Polyamine Research, Vol. 4, edited by
U. Bachrach, A. Kaye, and R. Chayen. Raven Press,
New York © 1983.

Diamines and Polyamines in Mammalian Reproduction

A.-Ch. Henningsson, S. Henningsson, and *O. Nilsson

Department of Physiology and Biophysics, University of Lund, S-223 62 Lund, Sweden;
**Department of Anatomy, University of Uppsala, S-751 23 Uppsala, Sweden*

In the nineteen-thirties pregnancy was found to be a physiological condition closely related to elevated diamine oxidase levels as the placenta (12) and the blood (41) of pregnant women were found to possess a strong histaminolytic property. These findings were followed by numerous investigations where the plasma diamine oxidase level during the course of pregnancy was measured and correlations between the plasma level and the diagnosis of pregnancy, the prognosis of threatening abortion as well as other obstetrical conditions were made (20,33,61). The hope was excited that measurements of plasma diamine oxidase levels would be clinically valuable as an index of fetoplacental functionality and that fetal salvage programmes might be based on assays of the enzyme (60). Some authors, however have presented critical views on the value of plasma diamine oxidase determinations in pregnancy as they found that the diamine oxidase values during normal pregnancy could vary over a rather wide range (48). Measurements of diamine oxidase activity in plasma during early pregnancy for prediction of date of confinement and in vaginal fluid for the diagnosis of ruptured fetal membranes were found to be techniques suitable for routine clinical use (14,15).

In the early reports on diamine oxidase and reproduction elevated levels of this enzyme was thought to occur only in human pregnancy. However, Ahlmark (1) showed a strongly increased histaminolytic power of placental tissue and plasma of several species, e.g. rat and guinea pig, during pregnancy. These findings in the rat were later confirmed and extended (37,49) and elevated diamine oxidase levels were also shown in several other tissues of the pregnant rat (36).

In the measurement of diamine oxidase activity in vitro various methods and substrates have been used. With putrescine as substrate for diamine oxidase, the product of the reaction, γ-aminobutyraldehyde, is non-enzymatically transformed into the cyclic compound Δ^1-pyrroline (58) which, if radioactively labelled, is extracted and used as a measure of diamine oxidase activity (44). However, in vivo studies involving the administration of ^{14}C-putrescine to rats and mice, have shown the occurrence of a further metabolism of γ-aminobutyraldehyde. Thus radiolabelled putrescine carbon has been shown both to be incorporated into γ-aminobutyric acid (GABA) (27,56) and excreted as $^{14}CO_2$ (27,28,55).

The pregnant rat, in spite of the elevated levels of diamine oxidase in plasma and various organs excreted large amounts of histamine in the urine during the last third of pregnancy. The excreted amounts were shown to be larger the greater the number of young in the litter, suggesting an importance of the presence of the fetuses for the elevated amounts of histamine appearing in the urine of the mother (30). This view was supported by experiments showing that after removal of the fetuses before term the increased urinary excretion of histamine was rapidly abolished. Further, in vitro experiments revealed a high rate of fetal histamine formation (31). Particularly the livers of the fetuses exhibited a high ability of forming histamine (32). The increased histamine formation of the fetuses as reflected in the urinary excretion of the mother roughly corresponded to the period of rapid increase in weight of the fetuses occurring after the 15th day of pregnancy.

The female hamster, likewise excreted elevated amounts of histamine in the urine during the last third of pregnancy (51). However, the histamine formation of the fetuses was rather low. In this species the placenta was likely to account for the excessive amounts of urinary histamine occurring during pregnancy as it was found to possess a high ability to form histamine in vitro.

At variance with the situation in the rat and hamster the mouse excreted increased amounts of histamine in the urine from the very first day of pregnancy, ascribed to an increased formation of the amine in the kidneys of the mother. Further, the fetuses of the mouse, like the fetuses of the rat formed histamine in vitro, though the main capacity of histamine formation of the mouse fetus was located to the skin (50).

The physiological significance of histamine during pregnancy was suggested to be related to processes of rapid growth (29,32). However, the role of histamine as a growth regulating factor has remained elusive and the interest in this respect has been focused on the diamine putrescine and the polyamines spermidine and spermine, low molecular aliphatic bases which appear to be ubiquitously present in animal, microbial as well as plant material.

As to the regulation of polyamine biosynthesis in mammalian tissue the important discovery of the putrescine forming enzyme, i.e. ornithine decarboxylase was made in 1968 in among other tissues the chick embryo (54). The ornithine decarboxylase activity which was found elevated in the developing chick embryo was correlated to the earlier observations (10,46) of changes in the diamine and polyamine content of this tissue. These changes were closely parallelled by changes in protein and nucleic acid concentrations (10,43). Also in the fetal rat putrescine and polyamine content increased on a parallel with ornithine decarboxylase activity (52) followed by an increased concentration of RNA (53).

During the last decade a considerable amount of work has confirmed a correlation between the ability of various hormones related to female reproduction to stimulate growth and differentiation and their ability to affect a change in diamine metabolism at an early stage of their action. Systems with such a response include the effects of LH on the ovaries of immature or adult rats (34,38). 17β-oestradiol on the chick oviduct (11) or the uterus of ovariectomized or immature rats (11,35).

In contrast to the great number of investigations of histamine, putrescine, spermidine and spermine in various physiological connections, our knowledge of cadaverine in mammalian tissue is scarce. Cadaverine was, together with putrescine early found to be excreted in the urine of patients with cysteinuria (59). This excretion was believed to be due to an impaired absorption of the precursor amino acids ornithine and lysine, from the intestinal lumen, leading to an increased diamine formation by intestinal bacteria followed by absorption of the diamines and excretion by the kidneys (42). Cadaverine has also been described to be present in the urine of healthy humans (45) as well as in the urine of normal rats (9), but it was still thought to be of intestinal origin. Evidence for biosynthesis of cadaverine from lysine in mammalian tissue cells was originally reported in the mouse kidney stimulated to growth by an anabolic steroid (26).

URINARY EXCRETION OF DIAMINES AND POLYAMINES IN THE PREGNANT RAT

The urinary excretion of histamine, methylhistamine, putrescine, cadaverine, spermidine and spermine was studied in rats before, during and after pregnancy (6). The relation between the urinary content of the different amines during the course of pregnancy in one of the rats is shown in FIG. 1A. The pattern of urinary excretion of histamine, methylhistamine, putrescine and cadaverine during pregnancy was very similar, i.e. no noteworthy change during the first two weeks of pregnancy and a highly elevated excretion during the last week. Peak values were obtained a few days before parturition. After that the urinary excretion fell towards the control levels which after the birth of the young were rapidly resumed. The urinary excretion of spermidine displayed a biphasic course with one peak before parturition and another occurring after delivery. The content of spermine in the urine was much lower than that of any of the other amines studied and no change in its excretion was obtained during the course of pregnancy.

In view of the profound elevations in diamine oxidase activity in plasma and various tissues during pregnancy in the rat, the urinary excretion of the diamines and polyamines was studied under the influence of the diamine oxidase inhibitor aminoguanidine. FIG. 1B depicts the results obtained in two of the rats. It is obvious that essentially the same pattern of excretion of diamines and polyamines occurred in the pregnant rat treated with aminoguanidine though the control levels as well as the peak values were highly elevated as compared to the values obtained in undisturbed pregnancy (FIG. 1). Administration of aminoguanidine did not however affect the urinary excretion of spermine.

The distinct changes in the excretion of diamines and their derivatives during the last third of rat pregnancy suggest a relation between changes in diamine and polyamine metabolism and the hormonal state of pregnancy and related processes of growth. The

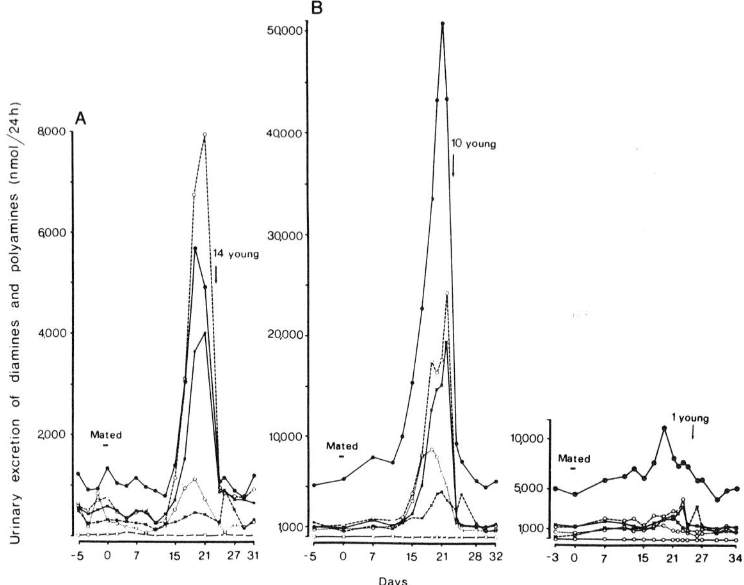

Fig. 1. Urinary excretion (nmol/24 h) of histamine (o--o), methylhistamine (x—x), putrescine (●—●), cadaverine (o····o), spermidine (x--x) and spermine (o—o) in rats before, during and after pregnancy. A: Undisturbed pregnancy in a rat giving birth to 14 young on the 23rd day of pregnancy. B: rats under the influence of aminoguanidine; one rat gave birth to 10 young on the 24th day of pregnancy and the other was delivered of one young by Caesarian section on the 25th day of pregnancy.

lactating rat also excretes increased amounts of diamines and polyamines in the urine (to be published) furhter strengthening this relation.

To investigate the origin of these amines various ectomizing operations were performed on the 19th day of pregnancy (4). The urinary amine content was determined in samples from the 7th day of pregnancy, the day before surgery and for three days postoperatively. Removal of the fetuses without dislodging the placentae reduced the excretion of histamine as well as methylhistamine towards the level of non-pregnant rats (FIG. 2B), indicating that the fetuses might be capable not only to form large amounts of histamine but also to methylate a large portion of the newly formed amine. The possibility of methylation by maternal tissues must however also be considered. The increased urinary excretion of putrescine, cadaverine and spermidine was not much affected by fetectomy (FIG. 2B) while removal of also the placentae reduced the urinary excretion of these amines, indicating a crucial importance of the presence of the placentae but not the fetuses for the elevated output of these amines.

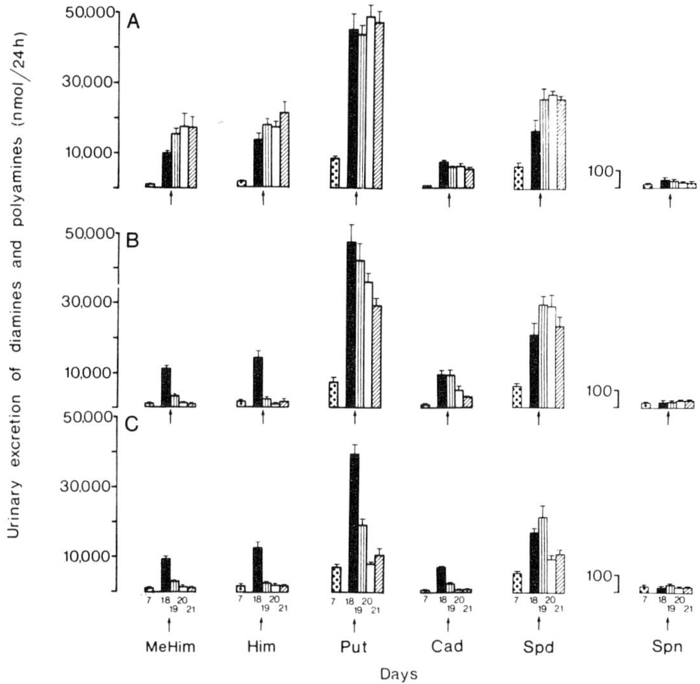

Fig. 2. Urinary excretion (nmol/24 h) of methylhistamine (MeHim), histamine (Him), putrescine (Put), cadaverine (Cad), spermidine (Spd) and spermine (Spn) in aminoguanidine treated rats subjected to various ectomizing operations. The arrow denotes the time of surgery. A: sham-operation, B: fetectomy, C: fetectomy + placentectomy. Each column represents the mean value ± S.E. of mean of determinations from four rats.

FORMATION AND CONTENT OF CADAVERINE IN THE PREGNANT RAT

As to the origin of the increased amount of cadaverine excreted in the urine during the last third of rat pregnancy the ability of various tissues to form cadaverine in vitro was investigated (7). Of the tissues examined the fetal placenta displayed the highest in vitro formation of cadaverine, suggesting that the increased amounts of urinary cadaverine during rat pregnancy being of placental origin. The ovary of the pregnant rat also formed cadaverine at a high rate (Table 1). The cadaverine concentration in tissues of pregnant rats was also determined. In untreated pregnant rats most cadaverine was found in the ovary and the placenta. Aminoguanidine treatment of the animals doubled the concentration of cadaverine in the maternal part of the placenta which is very rich in diamine oxidase. In the fetal part with a low diamine oxidase activity the change in cadaverine content was only slight. It appears conceivable that diamine oxidase could function as a regula-

TABLE 1. Formation of cadaverine (nmol/g x h) in different tissues of pregnant and non-pregnant rats. The tissues of the pregnant rats were excised on the 19th day of pregnancy. The figures are mean values ± S.E. of mean of 8 observations.

Tissue	Non-pregnant	Pregnant
Ovary	25 ± 7.47	110 ± 16.3 ***
Uterus	22 ± 3.40	9.0 ± 1.49 **
Kidney	2.1 ± 1.03	3.9 ± 1.38
Liver	2.7 ± 2.11	1.8 ± 1.01
Fetus		5.9 ± 1.94
Whole placenta		250 ± 20.9
Fetal placenta		350 ± 29.0
Maternal placenta		52 ± 10.0

** = $p < 0.01$ and *** = $p < 0.001$ versus non-pregnant rats.

tor of cadaverine content in certain tissues rich in the enzyme.

The elevated in vitro formation of cadaverine in the ovary during rat pregnancy was thus consistent with an increased endogenous cadaverine content of the organ indicating that cadaverine can be formed in vivo by decarboxylation of lysine.

As to the physiological significance of cadaverine it is notable that in microorganisms cadaverine has been suggested to possess some of the biological activity of putrescine and spermidine, possibly by the formation of a spermidine analog (13). Thus, parallel studies of the metabolism of cadaverine and putrescine would benefit a better understanding of diamine metabolism not only in reproductive physiology but also in other physiological connections.

DIAMINE AND POLYAMINE METABOLISM IN RAT OVARY AFTER ADMINISTRATION OF HUMAN CHORIONIC GONADOTROPHIN

As the significantly increased formation of cadaverine in the ovary of the pregnant rat (7) might be caused by the hormonal state prevailing during pregnancy, the effect of human chorionic gonadotrophin (hCG) on diamine and polyamine metabolism in the rat ovary was investigated (3). Results concerning the in vitro formation of cadaverine and putrescine in the ovaries at various intervals after the administration of 50 IU hCG are shown in FIG. 3. Peak values of the formation of both diamines were obtained 5 h after the hormone injection. The formation of cadaverine and putrescine in the ovary was also determined 5 h after the injection of hCG in the dose-range of 1.5-200 IU. Maximum diamine formation was observed by doses between 25 and 50 IU. The content of cadaverine, putrescine, spermidine and spermine was determined in the ovaries 1-12 h after the injection of 50 IU hCG. The administration of hCG increased the content of the diamines, while there was a decrease in the level of the polyamines.

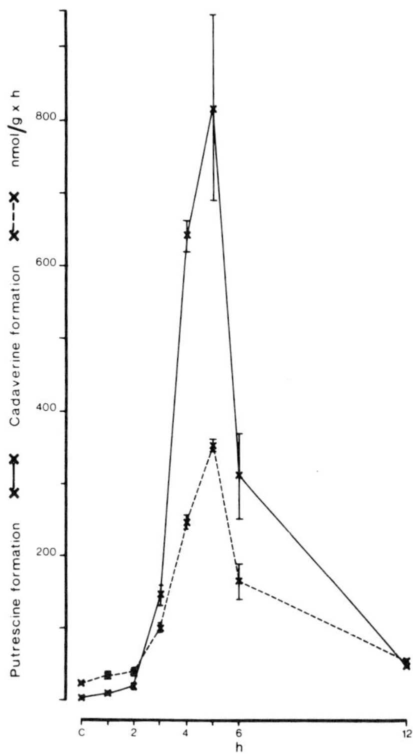

Fig. 3. Formation of cadaverine and putrescine (nmol/g x h) in the ovaries of rats injected with hCG (50 IU) at 28 days of age. The rats were sacrificed at different times after the injection. Control rats (C) received saline. Each point represents the mean value ± S.E. of mean. Controls: n=7, hCG-treated: n=4.

Regarding putrescine, the conspicuous stimulation of the formation of this amine in the ovary in the evening of pro-estrus, i.e. between the surge of LH and ovulation, appeared to be closely related to the ovulatory process (38). The finding of an increased formation of cadaverine in response to a gonadotrophin made it tempting to suggest a functional role also for this diamine in connection with ovulation (3). However, in rats in which the increased putrescine formation was completely inhibited by treatment of the animals with α-difluoromethyl ornithine, ovulation took place normally but the number of uterine implantation sites following mating at the next cycle was increased indicating that the putrescine generated might be inhibitory on the development of follicles for ovulation in the succeeding cycle (19).

DETERMINATION OF TISSUE DIAMINE OXIDASE ACTIVITY

When putrescine is incubated in the presence of purified diamine oxidase γ-aminobutyraldehyde is formed which is non-enzymatically transformed into the cyclic compound Δ^1-pyrroline. In vitro measurements of diamine oxidase activity have been based on the determination of ^{14}C-Δ^1-pyrrolone formed from ^{14}C-putrescine. As in vivo experiments had shown a further metabolism of γ-aminobutyraldehyde to GABA and carbon dioxide it appeared relevant to examine whether these metabolites were formed also in vitro from ^{14}C-putrescine. If so, determination of only Δ^1-pyrroline could not constitute a valid measure of tissue diamine oxidase activity. Tissues of pregnant and non-pregnant rats were incubated with ^{14}C-putrescine and the nature of the ^{14}C-metabolites formed in the incubates was determined (5). In addition to ^{14}C-Δ^1-pyrroline radioactive CO_2, GABA and some not identified compound(s) were found to be metabolites formed via the diamine oxidase pathway. In the pregnant rat the highest diamine oxidase activity per unit of weight was found in the maternal part of the placenta, followed by the uterus and the distal ileum (FIG. 4). The main metabolite formed from putrescine by these tissues was Δ^1-pyrroline, in fact it constituted about 94% of the total amount of metabolites formed. In the other tissues of the pregnant rat examined, the pattern of metabolites was different. In the ovary the formation of Δ^1-pyrroline and GABA was equal in magnitude, while in the fetal placenta and the liver GABA was the metabolite found in the highest concen-

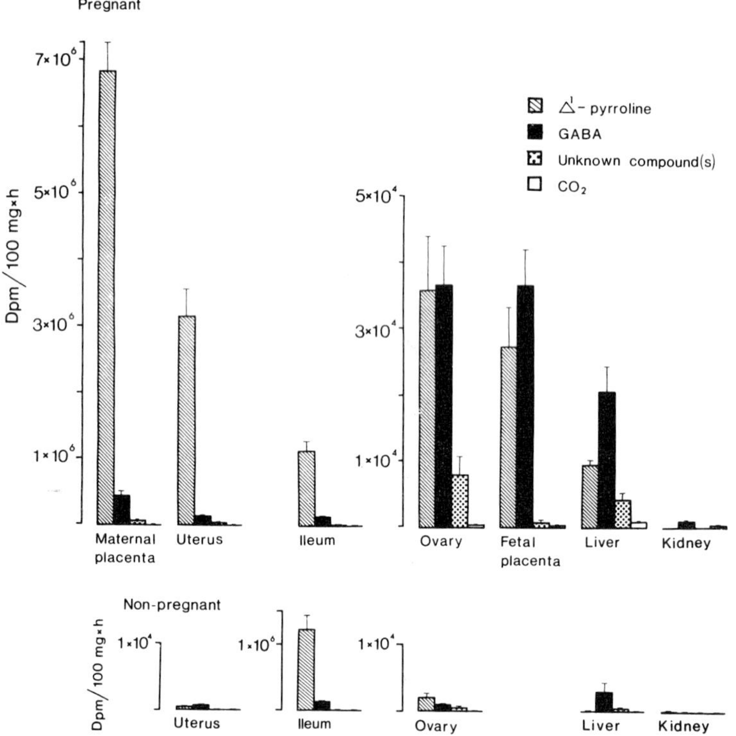

Fig. 4. Rate of in vitrc formation of various metabolites from ^{14}C-putrescine in tissues of pregnant and non-pregnant rats. The tissues of the pregnant rats were excised on the 19th day of gestation. Each column is the mean value ± S.E. of mean of 4 determinations.

tration constituting an average of 57% of the total amount of metabolites formed. In the kidney the diamine oxidase activity was low; only small amounts of GABA and carbon dioxide were formed. In the non-pregnant rat all the tissues examined showed a low diamine oxidase activity; the only exception was the ileum where the diamine oxidase activity was of the same magnitude as in the pregnant rat (FIG. 4). Thus, Δ^1-pyrroline alone can be used for determination of diamine oxidase activity in certain tissues but measurements of the total amount of metabolites formed from putrescine are necessary in other tissues to obtain a valid measure of the diamine oxidase activity (5). Similar conclusions were made using ^{14}C-cadaverine as substrate for diamine oxidase (24).

IN VIVO METABOLISM OF PUTRESCINE IN THE PREGNANT RAT

In the pregnant rat with its highly elevated diamine oxidase activity in plasma and several tissues, it was of interest to investigate the metabolic fate of injected radiolabelled putrescine. 1,4-(^{14}C)-Putrescine (2.5 µCi; 100 µg) was injected to untreated rats before pregnancy, on the 19th day of pregnancy and ten days after parturition (5). Immediately after the injection of the isotope radiorespirometric recordings of the expired $^{14}CO_2$ were made for 10 x 30 min in one group of rats and urine was collected for 3 x 24 h for analysis of its radioactivity in another group of rats. In FIG. 5 the excretion of $^{14}CO_2$ following the injection of ^{14}C-putrescine is shown. The putrescine was rapidly metabolized; within 30 min of the administraiton of the isotope 9% of the injected radioactivity was recovered as $^{14}CO_2$ in the rats examined before and after pregnancy. During pregnancy this fraction was significantly increased, constituting 12% of the injected radioactivity. However, during the whole period of 5 h there was no significant increase in the $^{14}CO_2$ excretion of the pregnant rats. Forty-eight per cent of the given radioactivity was expired within 2.5 h and 60% within the whole recording period of 5 h. As seen in FIG. 5B administration of aminoguanidine almost completely inhibited the excretion of $^{14}CO_2$. In the pregnant rats treated with the inhibitor 0.8% of the injected radioactivity was recovered as $^{14}CO_2$ during the 2.5 recording period.

In the urine of the pregnant rats, collected for three days following the administration of ^{14}C-putrescine, 16% of the injected radioactivity was recovered. Corresponding figures for radioactivity in the urine of the rats before and after pregnancy were 22% and 19%, respectively. The major part of the radioactivity excreted in the urine was found during the first day following the administration of ^{14}C-putrescine. In rats not given aminoguanidine GABA, the unidentified compound(s) and unmetabolized putrescine were excreted in the urine collected for 24 h following the ^{14}C-putrescine administration. GABA and the unidentified compound(s) were excreted in almost equal amounts by rats before, during and after pregnancy while ^{14}C-putrescine was excreted in significantly less amounts by the pregnant rats. Under the influence of aminoguanidine the pregnant rats excreted 8 times more unmetabolized

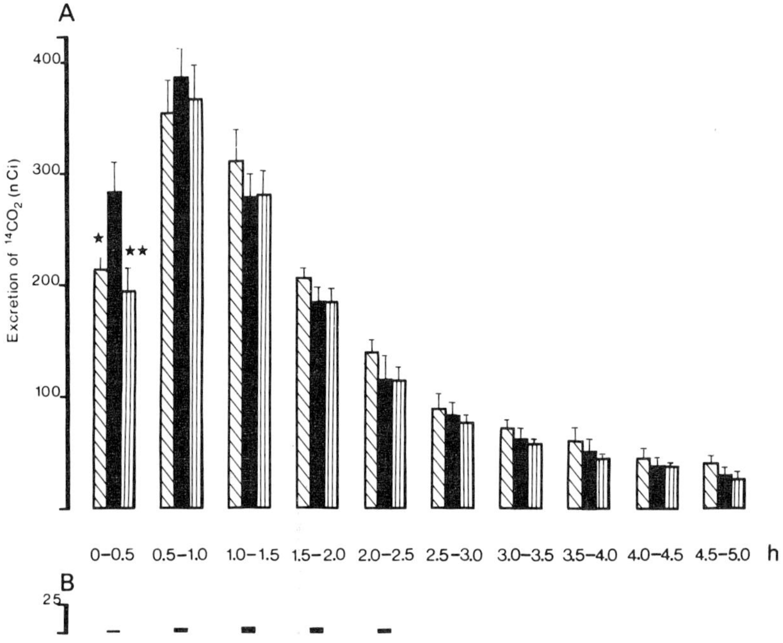

Fig. 5. Excretion of $^{14}CO_2$ (nCi) after the injection of ^{14}C-putrescine (2.5 µCi; 100 µg) to rats. ▧ : rats before pregnancy, ■ : rats on the 19th day of pregnancy, ▥ : rats 10 days after delivery. A: untreated rats, B: rats treated with aminoguanidine. The columns are mean values ± S.E. of mean. Untreated rats: n=4, rats treated with aminoguanidine: n=2. ★ = $p < 0.05$, ★★ = $p < 0.01$ versus pregnant rats.

putrescine in the urine, while the excretion of the unidentified compound(s) and GABA was significantly decreased. In addition, in the urine of the aminoguanidine-treated rats, small amounts of ^{14}C-spermidine appeared; no urinary ^{14}C-polyamines were found in the urine of the untreated rats.

In most mammalian tissues putrescine has been considered mainly to serve as a precursor of the polyamines. Formation of polyamines from ^{14}C-putrescine was not detectable in the tissues investigated in the pregnant rat nor in the urine following injection of ^{14}C-putrescine, indicating a preference for the metabolism of putrescine via oxidative deamination rather than conversion to polyamines. The biological significance of diamine oxidase is yet not settled but a better understanding of its metabolic function may provide a further insight not only into the physiological implication of this enzyme but also into that of diamines and polyamines in pregnancy as well as other biological conditions.

EFFECTS OF INHIBITORS OF POLYAMINE SYNTHESIS ON EARLY STAGES OF MAMMALIAN EMBRYOGENESIS

Early embryogenesis is characterized by rapid cellular proliferation and differentiation and several studies have described marked increases in the putrescine forming enzyme, i.e. ornithine decarboxylase, and polyamine concentrations in association with this process in mammalian and non-mammalian species (For ref. see 18). Using inhibitors of polyamine synthesis in non-mammalian embryogenesis a requirement of diamines and polyamines has been revealed for oogenesis (23), cleavage (39), blastulation (8) and gastrulation (16,22,40).

In mammals the employment of DFMO in murine pregnancy (17) has provided strong evidence that changes in diamine and polyamine metabolism are essential for the continuation of normal mammalian embryogenesis. Administration of DFMO to mice during a critical period after implantation, i.e. days 5-8 of gestation resulted in a greatly retarded fetal development followed by resorption and/or expulsion of the fetuses. Corresponding results were obtained in the rat (18,47) where no young pups were delivered by animals treated with DFMO during the critical period. Administration of DFMO before this period did not cause any notable effects on the number of young per litter or on the number of implantation sites (17,47). From the experiments referred to above it could be concluded that DFMO administration to mice and rats in vivo did not seem to affect preimplantation or implantation processes. However, in vitro, Alexandre (2), using methylglyoxal bis(guanylhydrazone) (MGBG) showed an inhibitory effect on cavitation, i.e. blastocyst formation, arresting development at the morula stage.

The early blastocyst is surrounded by a mucopolysaccharide membrane, named the zona pellucida. The mouse blastocyst reaches the uterine cavity four days after fertilization. It then loses its zona pellucida, i.e. hatching of the blastocyst, attaches to and invades the endometrium, a process known as implantation (For ref. see 57). In experiments on mouse blastocysts in vitro (25) we have studied two stages of early mammalian embryogenesis not earlier shown to be affected by inhibitors of polyamine synthesis, i.e. the disappearance of zona pellucida and the outgrowth of the blastocyst, the latter process corresponding to implantation (21). In cultures of mouse blastocysts obtained on the 4th day of pregnancy low doses of DFMO (10 μM) or MGBG (0.2 μM) completely inhibited the disappearance of zona pellucida (FIG. 6). Using a hundredfold higher concentration of DFMO or MGBG the outgrowth of the blastocysts was inhibited. Thus, a clear difference in sensitivity to polyamine inhibitors was found between the two processes studied.

In view of the knowledge of the apparent lack of effects of inhibition of polyamine synthesis on preimplantation and implantation processes after in vivo administration of DFMO the dramatic influence in vitro of low concentrations of DFMO or MGBG on these processes seemed notable. Possible explanations of this discrepancy may be that DFMO administered in vivo is secreted into the uterine cavity in concentrations too low to affect the blastocysts or that the blastocysts in vivo are more resistent to polyamine limitation.

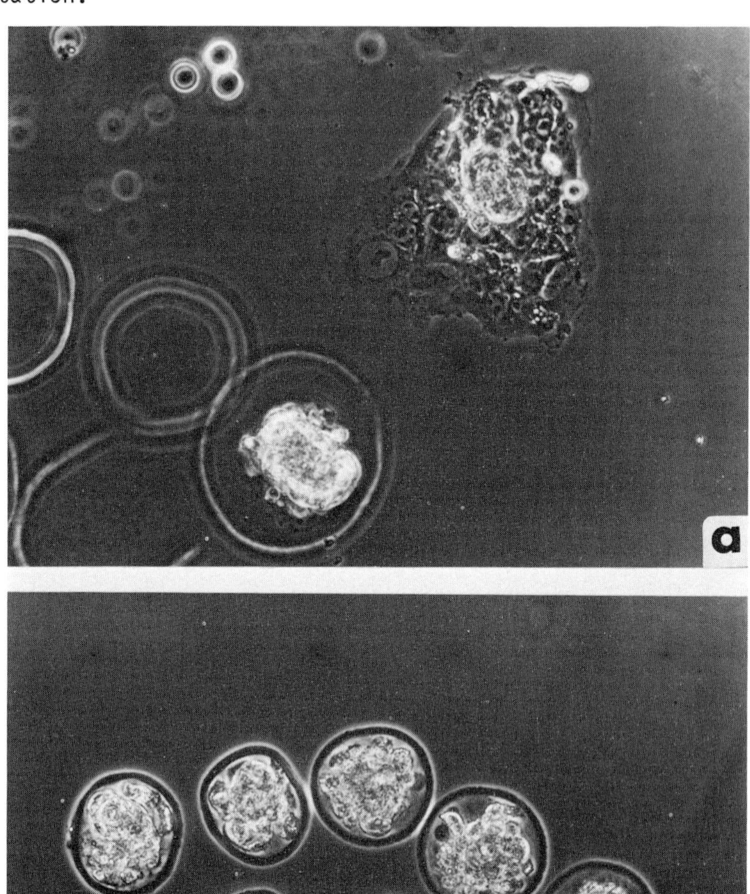

FIG. 6. Mouse blastocysts flushed from the uterus on the fourth day of pregnancy and cultured in vitro for four days in (a) control medium showing one blastocyst surrounded by its zona, one blastocyst free of its zona and broken and empty zonae. In (b) where the culture medium contained 10 μM DFMO all the blastocysts were zona-enclosed. In a medium containing 0.2 μM MGBG the hatching of the blastocysts was also inhibited.

ACKNOWLEDGEMENTS

Financial support was received from the Swedish Medical Research Council, the Faculty of Medicine, University of Lund, Torsten and Ragnar Söderbergs´ Foundations, the Royal Physiographic Society and from the Swedish Society for Medical Research. DFMO was kindly donated by Centre de Recherche Merrell International, Strasbourg (France).

REFERENCES

1. Ahlmark, A. (1944): Acta Physiol. Scand., 9:suppl. 28:1-107.
2. Alexandre, H. (1979): J. Embryol. exp. Morph., 53:145-162.
3. Andersson, A.-Ch., and Henningsson, S. (1980): Acta Endocrinol., 95:237-243.
4. Andersson, A.-Ch., and Henningsson, S. (1981): Acta Endocrinol., 98:456-463.
5. Andersson, A.-Ch., and Henningsson, S. (1981). Acta Physiol. Scand., 113:523-532.
6. Andersson, A.-Ch., Henningsson, S., and Rosengren, E. (1978): J. Physiol., 285:311-324.
7. Andersson, A.-Ch., Henningsson, S., and Rosengren, E. (1979): Acta Physiol. Scand., 105:508-512.
8. Brachet, J., Mamont, P., Boloukhère, M., Baltus, Ē., and Hanocq-Quertier, J. (1978): C.R. Acad. Sc. (Paris), 287:1289-1292.
9. Bremer, H.J., and Kohne, E. (1971): Clin. chim. Acta, 32:407-418.
10. Caldarera, C.M., Barbiroli, B., and Moruzzi, G. (1965): Biochem. J., 97:84-88.
11. Cohen, S., O´Malley, B.W., and Stastny, M. (1970): Science, 170:336-338.
12. Danforth, D.N., and Gorham, F. (1937): Am. J. Physiol., 119:294-295.
13. Dion, A.S., and Cohen, S.S. (1972): J. Virol., 9:423-430.
14. Elmfors, B., and Tryding, N. (1976): Br. J. Obstet. Gyneac., 83:6-10.
15. Elmfors, B., Tryding, N., and Tufvesson, G. (1974): J. Obstet. Gynaec. Br. Commonw., 81:361-362.
16. Emanuelsson, H., and Heby, O. (1978): Proc. Natl. Acad. Sci. USA, 75:1039-1042.
17. Fozard, J.R., Part, M.-L., Prakash, N.J., and Grove, J. (1980): Eur. J. Pharmac., 65:379-391.
18. Fozard, J.R., Part, M.-L., Prakash, N.J., Grove, J., Schechter, P.J., Sjoerdsma, A., and Koch-Weser, J. (1980): Science, 208:505-508.
19. Fozard, J.R., Prakash. N.J., and Grove, J. (1980): Life Sci., 27:2277-2283.
20. Genell, S. (1950): Acta obstet. gynec. scand., 30:suppl. 7:413-421.
21. Gwatkin, R.B.L. (1966): J. Cell. Physiol., 68:335-344.
22. Heby, O., and Emanuelsson, H. (1980): Biochem. Biophys. Res.

Commun., 92:75-79.
23. Heby, O., and Emanuelsson, H. (1981): Med. Biol. 59:417-422.
24. Henningsson, A.-Ch., and Henningsson, S. (1982): Abstract European Histamine Research Society, Bled 1982.
25. Henningsson, A.-Ch., Henningsson, S., and Nilsson, O. (1982): Manuscript.
26. Henningsson, S., Persson, L., and Rosengren, E. (1976): Acta Physiol. Scand., 98:445-449.
27. Henningsson, S., and Rosengren, E. (1976): Br. J. Pharmac., 58:401-406.
28. Jänne, J. (1967): Acta Physiol. Scand., suppl. 300:1-71.
29. Kahlson, G. (1962): Perspect. Biol. Med., 5:179-197.
30. Kahlson, G., Rosengren, E., Westling, H. (1958): J. Physiol., 143:91-103.
31. Kahlson, G., Rosengren, E., Westling, H., and White, T. (1958): J. Physiol., 144:337-348.
32. Kahlson, G., Rosengren, E., and White, T. (1960): J. Physiol., 151:131-138.
33. Kapeller-Adler, R. (1944): Biochem. J., 38:270-274.
34. Kaye, A.M., Icekson, I., Lamprecht, S.A., Gruss, R., Tsafriri, A., and Lindner, H.R. (1973): Biochemsitry, 12:3072-3076.
35. Kaye, A.M., Icekson, I., and Lindner, H.R. (1971): Biochim. biophys. Acta, 252:150-159.
36. Kim, K.S., and Harris, M.E. (1973): In: Histamine int. Congr. Physiol. Sci., edited by Cz. Maślinśki, pp.312-326.
37. Kobayashi, Y. (1964): Nature (Lond.), 203:146-147.
38. Kobayashi, Y., Kupelian, J., and Maudsley, D.V. (1971): Science, 172:379-380.
39. Kusonoki, S., and Yasumasu, I. (1978): Devl. Biol., 67:336-345.
40. Löwkvist, B., Heby, O., and Emanuelsson, H. (1980): J. Embryol. exp. Morph., 60:83-92.
41. Marcou, I., Athanasiu-Vergu, E., Chiricéanu, D., Cosma, G., Gingold, N., and Parhon, C.-C. (1938): Presse méd., 46:371-374.
42. Milne, M.D., Asatoor, A.M., Edwards, K.D.G., and Loughridge, L.W. (1961): Gut, 2:323-337.
43. Moruzzi, G., Barbiroli, B., and Caldarera, C.M. (1968): Biochem. J., 107:609-613.
44. Okuyama, T., and Kobayashi, Y. (1961): Archs Biochem. Biophys., 95:242-250.
45. Perry, T.L., and Schröder, W.A. (1963): J. Chromatogr., 12: 358-373.
46. Raina, A. (1963): Acta Physiol. Scand., 60:suppl. 218:1-81.
47. Reddy, P.R.K., and Rukmini, V. (1981): Contraception, 24:215-221.
48. Resnik, R., and Levine, R.J. (1969): Am. J. Obstet. Gynec., 104:1061-1066.
49. Roberts, M., and Robson, J.M. (1953): J. Physiol.,119:286-291.
50. Rosengren, E. (1963): J. Physiol., 169:499-512.
51. Rosengren, E. (1965): Proc. Soc. exp. Biol. Med., 118:884-888.
52. Russell, D.H., Gregory, J.L., Medina, V.J., and Snyder, S.H.

(1969): Physiologist, 12:344.
53. Russell, D.H., and McVicker, T.A. (1972): Biochim. biophys. Acta, 259:247-258.
54. Russell, D., and Snyder, S.H. (1968): Proc. natn. Acad. Sci. USA, 60:1420-1427.
55. Seiler, N., and Eichentopf, B. (1975): Biochem. J., 152:201-210.
56. Seiler, N., Wiechmann, M., Fischer, H.A., and Werner, G. (1971): Brain Res., 28:317-325.
57. Sherman, M.I., and Wudl, L.R. (1976): In: The Cell Surface in Animal Embryogenesis and Development, edited by G. Poste, and G.L. Nicolson, pp.81-125. North Holland, Amsterdam.
58. Tabor, H. (1951): J. biol. Chem., 188:125-136.
59. Udränszky, L.v., and Baumann, E. (1889): Hoppe Seylers Z. Physiol. Chem., 13:562-594.
60. Weingold, A.B., and Southren, A.L. (1968). Obstet. Gynec., 32:593-606.
61. Zeller, E.A., and Birkhäuser, H. (1940): Schweiz. med. Wschr., 70:975-976.

Advances in Polyamine Research, Vol. 4, edited by U. Bachrach, A. Kaye, and R. Chayen. Raven Press, New York © 1983.

Polyamines and Methionine Sulfoximine-Induced Seizures

R. Porta, M. Camardella, A. De Santis, V. Gentile, and *O. Z. Sellinger

*Department of Biochemistry, First Medical School, University of Naples, Naples, Italy; *Laboratory of Neurochemistry, Mental Health Research Institute, University of Michigan, Ann Arbor, Michigan 48109*

The possibility that spermidine (Spd) and/or spermine (Spm) may play a specific functional role in the brain has been frequently suggested. However, most of the knowledge linking cerebral polyamine metabolism to behavior changes derives only from pharmacological experiments. Thus, it is well established that intraventricularly injected Spd and Spm produce convulsions in mice (2), rabbits (1) and chicks (7), and are effective in producing insomnia also suppressing feeding behavior in rats (12). Noteworthy in this regard is that Spd is very much less active than Spm. Such a notion, in addition to the described presence of high affinity uptake systems for Spm in rat cerebral cortex slices (4), supports the hypothesis that Spm is the polyamine playing some neuromodulatory role. On the other hand "as a general feature it appears that Spm concentrations are relativly higher in structures rich in nerve cells, whereas white matter, spinal cord and peripheral nerves are extremely rich in Spd" (17).

The analog of the amino-acid L-methionine (Met), L--methionine-d,l-sulfoximine (MSO), produces an epileptiform syndrome that has a long latency of onset, resembling human epilepsy for the recurrent bouts of seizures interspersed with quiescent interictal periods (11). The seizure activity of the drug varies, depending on dose, from clonus

methionine "Met"	methionine sulfoximine "MSO"
COOH	COOH
H-C-NH$_2$	H-C-NH$_2$
CH$_2$	CH$_2$
CH$_2$	CH$_2$
S	O=S=NH
CH$_3$	CH$_3$

to generalized tonic-clonic convulsions.

As concerns the biochemical effects of MSO an approach to its mechanism of action is based on the observed increase of S-adenosylmethionine (AdoMet) utilization in the brain after i.p. injection of the drug (3,10,13,14,15,16). This contribution is devoted to the effects of MSO on AdoMet utilization in the polyamine biosynthesis, pointing out the possible specific role of increased cerebral production of Spm in the MSO-elicited seizures.

EXPERIMENTAL PROCEDURES

Equipment. A Varian Mod. 5020 liquid chromatograph with UV-5 selectable five wavelength single-beam detector was used for HPLC analyses. The chromatographic runs were recorded on a Varian Mod. 9176 strip chart recorder and the peak areas were measured by a Varian Mod. CDS-111L data system. Radioactivity contained in the collected HPLC eluates was measured in a Packard Tri-Carb 460C liquid spectrometer. Spectral analyses were carried out with a Varian Mod. Cary 219 spectrophotometer.

Animals. Adult male mice (Swiss strain) and 21-day-old rats (Wistar strain) were housed (7:00 A.M. to 7:00 P.M. light cycle followed by a 12 h dark cycle) in stainless steel cages (one family of rats/cage and 6 mice/cage) for at least 48 h before use. The mice were fasted overnight but were allowed water ad libitum prior to the experimental procedures. All the injections were carried out between 8:00 and 11:00 A.M.. After treatment each animal was transferred to an individual cage. Animals were always sacrificed by immersion (head first) into liquid nitrogen and were stored frozen overnight at -20°C. Whole brains were rapidly removed, immediately refrozen in liquid nitrogen and kept at -20°C until assayed.

Determination of polyamine levels and specific radioactivity. The preparation of brain samples for polyamine determination and the high-performance liquid chromatographic (HPLC) separation of Spd and Spm were essentially performed as previously described (9). Some modifications concerning the UV detection of benzoyl (BZ)-polyamines were carried out in order to improve the sensitivity of the method. Thus, we isolated the BZ-putrescine (Pt), -Spd and -Spm (prepared from commercially available polyamines, Sigma, St. Louis, MO, U.S.A.)according to the reported HPLC procedure (9) by elution of the peaks with spectroscopic grade methanol solution. Spectral analyses were carried out on the collected peaks by using the HPLC eluant solution as blank. As reported in Fig. 1 the individual UV spectra of the three BZ-polyamines show absorption maxima centered between 220 and 225 nm. The absence of derivatized secondary amino groups could be responsible of the slight difference observed in the spectral behavior of Pt derivative with respect to the BZ-Spd and BZ-Spm ones. These findings suggested the possibility to improve the sensitivity in the detection of the BZ-polyamines following their HPLC separation. In fact, by using the selectable UV wavelength detector at 215 nm we obtained an increase in the sensitivity of the method 3.5 times for BZ-Pt, 5 times for BZ-Spd and 5.6 times for BZ-Spm. A major improvement appears theoretically realizable by providing the liquid chromatograph with a variable wavelength detector selected at 224 nm for BZ-Pt and 221 nm for BZ-Spd and BZ-Spm detection.

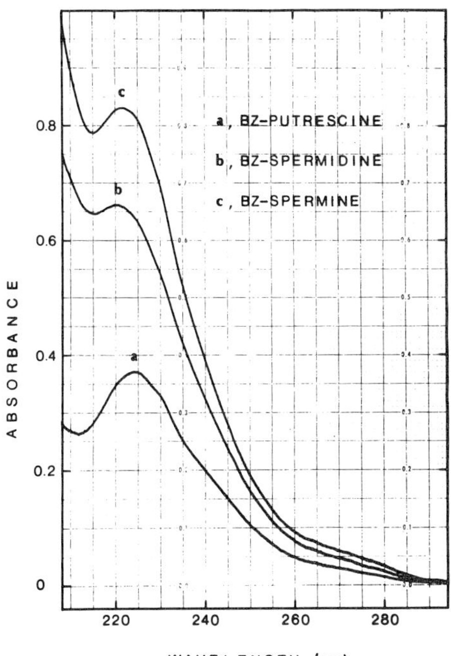

FIG.1. *Ultraviolet light absorption spectra of BZ-polyamines.*

RESULTS

In vitro inhibition of brain Spd synthase by alicyclic amines.

Cerebral aminopropyltransferase activities were assayed as previously described (8) by using an HPLC procedure in order to separate Pt from Spd (Spd synthase assay) or Spd from Spm (Spm synthase assay). Since the BZ-derivatives of the alicyclic amines, contained in the assay mixtures, interfered with the typical HPLC elution pattern of BZ-polyamines, the radiometric evaluation of the BZ-derivatives of the reaction products (Spd or Spm) was preferred to their quantitation by UV absorbance. In Fig. 2 is reported a typical HPLC elution pattern of BZ-(^{14}C)polyamines as derived from aminopropyltransferase assays.

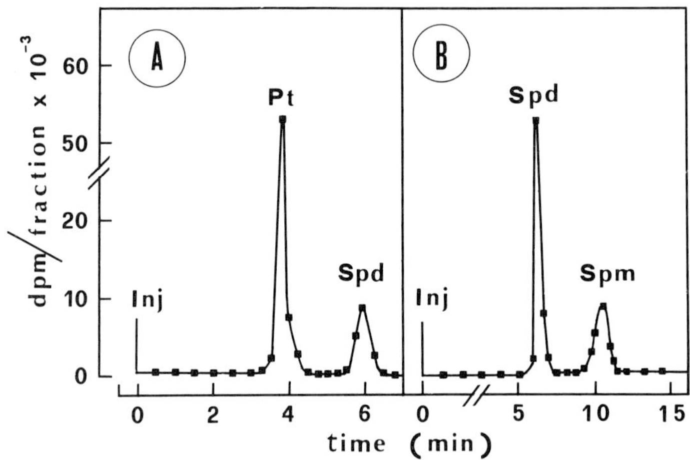

FIG. 2. *HPLC patterns of BZ-(^{14}C)polyamines as derived from Spd synthase (A) and Spm synthase assay (B) samples.* The reaction mixtures (0.5 ml) contained 0.1 M sodium phosphate buffer (pH 7.4), 0.06 mM decarboxylated AdoMet, 2 mg of enzyme protein (supernatant at 100,000 x *g* of mouse brain homogenate) and 0.05 mM (1,4-^{14}C) Pt (207,000 dpm) for Spd synthase assay (A) or 0.05 mM (^{14}C)Spd (230,000 dpm) for Spm synthase assay (B). Blanks contained no decarboxylated AdoMet. The reactions were stopped by perchloric acid addition after 90 min of incubation at 37° C. The supernatants recovered by centrifugation were subjected to the benzoylation procedure. An aliquot (50 μl) of the final methanol solution (100 μl) were injected onto MicroPak MCH-10 column (4 mm i.d. x 30 cm) in order to separate the BZ-polyamines. Elution was with 60% methanol (v/v) at a flow rate of 1.5 ml/min. The HPLC eluates were collected into scintillation vials and counted after addition of Pico-Fluor 30 (Packard Instr. Co. Inc., ILL).

Fig. 3 shows the inhibition of Spd synthase activity assayed *in vitro* in the presence of different concentrations of alicyclic amines. It is worthy to note that the inhibitory effectiveness of these compounds seems to be in strict relation with the size of their aliphatic rings. In fact dicyclohexylamine (DCHA) and cyclopentylamine (CPA) result to possess the most effective inhibiting properties. Conversely, none of the mentioned alicyclic amines was able to inhibit *in vitro* cerebral Spm synthase activity at the maximal concentration tested (1 mM).

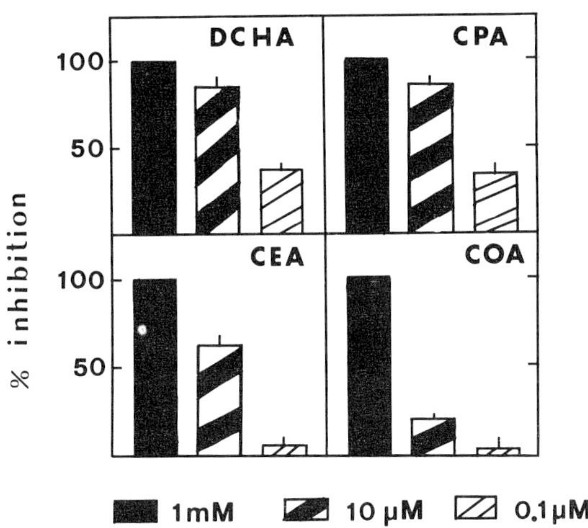

FIG. 3. *In vitro inhibition of mouse brain Spd synthase activity by dicyclohexylamine (DCHA), cyclopentylamine (CPA), cycloheptylamine (CEA) and cyclooctylamine (COA).* The assays were performed as described in Fig. 2 in the presence of the indicated concentrations of inhibitors.

Our findings are in complete agreement with the recently published data of Hibasami *et al.* (5,6). In fact, these authors noted both the specific inhibition of rat ventral prostate Spd synthase by cyclohexylamine and DCHA, a.d an antitumor effect following depletion of Spd levels in the tumor cells caused by DCHA treatment.

Effect of MSO and DCHA on the conversion of polyamine (^{14}C)precursors to (^{14}C)Spd and (^{14}C)Spm in the brain.

As shown in Table 1, the specific radioactivity (s.r.a.) values of (^{14}C)Spd were found significantly higher in the brains of the MSO-treated animals following a 60 min pulse of L-(3,4-^{14}C)Met, and significantly lower in the brains of the DCHA-treated animals

TABLE 1

The effect of MSO and DCHA on the conversion of polyamine labeled precursors into (^{14}C)spermidine.[a]

(^{14}C)PRECURSOR	PULSE TIME (min)	SPERMIDINE (s.r.a., dpm/nmol)	
		Control	MSO
(^{14}C)Met, 2 μCi	60	28.4±1.0 (7)	42.7±3.9 (11)[b]
(^{14}C)Met, 1 μCi	180	22.4±1.6 (6)	25.4±0.8 (12)
		Control	DCHA
(^{14}C)Pt, 0.2 μCi	180	228.3±15.0 (6)	60.6±6.5 (8)[c]

[a] Adult mice were given an i.p. injection of MSO (150 mg/Kg) or saline 180 min before killing. 21-day-old rats were i.p. injected with DCHA (300 mg/Kg) or saline 9 and 6 h before killing. The labeled polyamine precursors dissolved in 10 μl of Merlis solution were intracerebrally injected at the indicated times before animal killing. Values are expressed in dpm/nmol as means ± S.E.M.. Number in parentheses represents the number of animals in each experimental group.

[b] Significantly different from control value: $p<0.005$ (Student's t-test).

[c] Significantly different from control value: $p<0.001$ (Student's t-test).

TABLE 2

The effect of MSO and DCHA on the conversion of polyamine labeled precursors into (^{14}C)spermine.[a]

(^{14}C)PRECURSOR	PULSE TIME (min)	SPERMINE (s.r.a., dpm/nmol)	
		Control	MSO
(^{14}C)Met, 2 μCi	60	40.1±1.2 (7)	60.6±4.3 (11)[b]
(^{14}C)Met, 1 μCi	180	38.8±2.4 (6)	40.6±1.1 (8)
		Control	DCHA
(^{14}C)Spd, 1 μCi	180	418.4±50.3 (6)	544.7±39.8 (6)

[a]For details see Table 1.

[b]Significantly different from control value: $p<0.001$ (Student's t-test).

following a 180 min pulse of (^{14}C)Pt. Conversely, in Table 2 is reported the significant increase of Spm s.r.a. values in the brain of animals treated with MSO, whereas it is possible to note that the conversion of (^{14}C)Spd to (^{14}C)Spm resulted to be not affected by DCHA treatment. The opposite effects of MSO and DCHA only toward Spd production reflect both the more active utilization of AdoMet, caused by MSO, in the Spd and Spm biosynthetic pathway, and the specific *in vivo* inhibition of Spd synthase by DCHA.

Production of convulsions in mice by the combination of DCHA and subconvulsive dosages of MSO.

DCHA injected to adult albino mice 3 h before treatment with subconvulsive dosages of MSO was effective in promoting convulsions (57% of the animals seizured). The concurrent administration of the two drugs induced convulsions only in 10% of the treated animals (Fig. 4). The separate administrations of either 400 mg/kg of

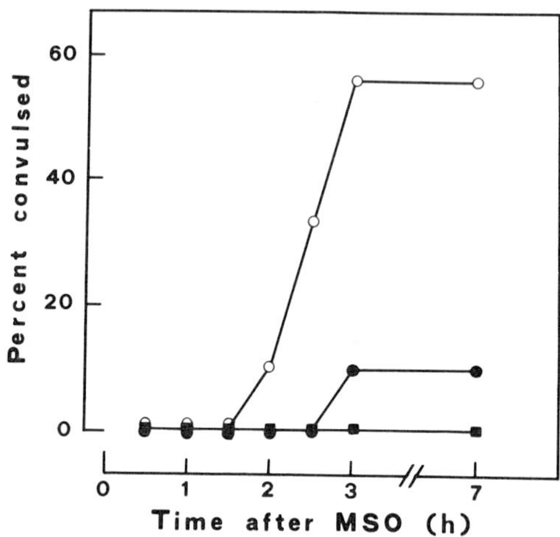

FIG. 4. *Effect of DCHA (400 mg/Kg, i.p.) on the production of tonic-clonic convulsions in mice treated with subconvulsive dosages of MSO (130 mg/kg, i.p.).* Each point represents the percent of convulsed mice at different times after MSO administration. Eighteen animals per group were used. Animals injected with MSO (control group) (■-■); animals injected with DCHA at the same time (●-●) and 3 h before MSO administration (o-o).

DCHA or 130 mg/kg of MSO were without marked behavioral effects. DCHA thus induces convulsions only under certain conditions. Our explanation of the present results is the possible iperproduction of Spm caused by the concurrent effects of MSO and DCHA. In fact, the MSO--induced increased utilization of AdoMet could be specifically directed toward Spm production under Spd synthase inhibition by DCHA (Fig. 5 C). This speculation appears in line with the previously reported data showing the effectiveness of methylglyoxal bis(guanylhydrazone) (MGBG), a well known inhibitor of decarboxylated AdoMet synthesis, in preventing MSO seizures (10) (Fig. 5 B).

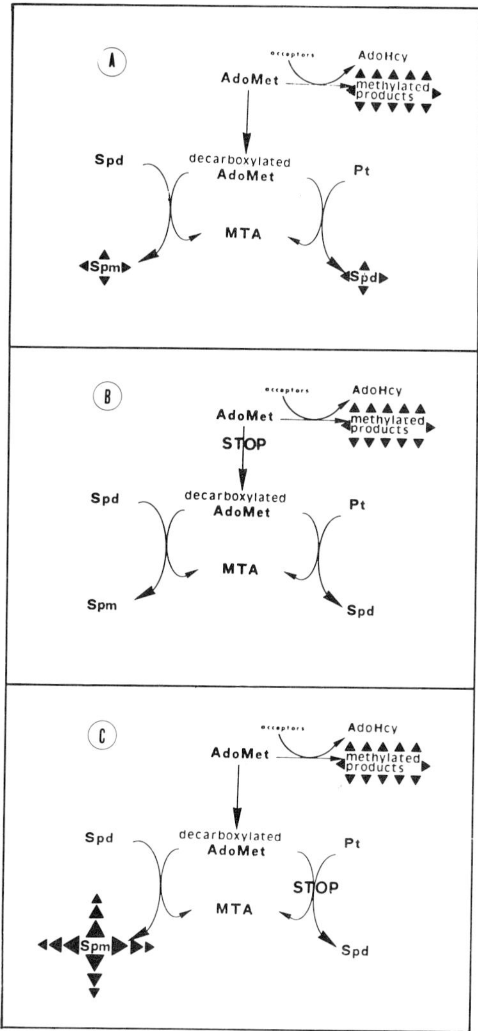

FIG. 5. *Proposed mechanism of the modulation of MSO-induced seizures by MGBG and DCHA.* (A) Increase of Spd and Spm cerebral levels in MSO-treated animals. (B) Hypothesized capacity of MGBG to prevent MSO-induced increase of Spd and Spm levels. (C) Hypothesized capacity of DCHA to contribute to the MSO-induced increase of Spm levels by preventing the increase of Spd levels.

The increased transmethylation flux, a general characteristics of the MSO-epileptogenic brain (A), should not be affected by MGBG (B) and DCHA (C).

ACKNOWLEDGEMENTS

The authors would like to express their appreciation to Dr. L. Servillo for his excellent technical assistance and valuable suggestions in the attempts to improve the sensitivity of polyamine analysis. Portions of this work were supported by a grant from U.S.Public Health Service, NINCDS 06294 (OZS).

REFERENCES

1. Anderson, D.J., and Shaw, G.G. (1974): Br. J. Pharmac., 52:205-211.

2. Anderson, D.J., Crossland, J., and Shaw, G.G. (1975): Neuropharmacology, 14:571-577.

3. Dainat, J., Salas, C.E., and Sellinger, O.Z. (1978): Biochem. Pharmacol., 27:2655-2658.

4. Harman, R.J., and Shaw, G.G. (1981): J. Neurochem., 36:1609-1615.

5. Hibasami, H., Tanaka, M., Nagai, J., and Ikeda, T. (1980): FEBS Lett., 116:99-101.

6. Hibasami, H., and Ito, H. (1981): Gann, 72:512-516.

7. Nisticò, G., Ientile, R., Rotiroti, D., and Di Giorgio, R.M. (1980): Biochem. Pharmacol., 29:954-957.

8. Porta, R., Esposito, C., and Sellinger, O.Z. (1981): J. Chromatogr., 226:208-212.

9. Porta, R., Doyle, R.L., Tatter, S.B., Wilens, T.E., Schatz, R.A., and Sellinger, O.Z. (1981): J. Neurochem., 37:723-729.

10. Porta, R., Schatz, R.A., and Sellinger, O.Z. (1982): In: The Biochemistry of S-Adenosylmethionine and Related Compounds, edited by R.T. Borchardt, E. Usdin, and C.R. Creveling, (in press) Macmillan Publishers.

11. Proler, M., and Kellaway, P. (1962): Epilepsia, 3: 117-130.

12. Sakurada, T., and Kisara, K. (1978): Japan. J. Pharmacol., 28:125-132.

13. Salas, C.E., Ohlsson, W.G., and Sellinger, O.Z. (1977): Biochem. Biophys. Res. Commun., 76:1107-1115.

14. Schatz, R.A., and Sellinger, O.Z. (1975): J. Neurochem., 24:63-66.

15. Schatz, R.A., and Sellinger, O.Z. (1975): J. Neurochem., 25:73-78.

16. Schatz, R.A., Frye, K., and Sellinger, O.Z. (1978): J. Pharmacol. Exp. Ther., 207:794-800.

17. Seiler, N. (1981): Neurochem. Int., 3:95-110.

Advances in Polyamine Research, Vol. 4, edited by U. Bachrach, A. Kaye, and R. Chayen. Raven Press, New York © 1983.

Antagonism of Polyamine Metabolism—A Critical Factor in Chemotherapy of African Trypanosomiasis

Cyrus J. Bacchi, *Peter P. McCann, Henry C. Nathan, Seymour H. Hutner, and *Albert Sjoerdsma

*Haskins Laboratories and Biology Department, Pace University, New York, New York 10038; *Merrell Dow Research Center, Cincinnati, Ohio 45215*

Function and metabolism of polyamines in hemoflagellates has been the subject of intense study for less than 8 years; however, these ubiquitous cell constituents are emerging as important - possibly indispensable - factors for normal hemoflagellate metabolism and reproduction. These parasites are highly vulnerable to agents which block polyamine synthesis or the functional sites of polyamines. In this report, some aspects of polyamine metabolism in trypanosomes and its relationship to chemotherapy will be examined.

BACKGROUND: THE ORGANISMS

The African trypanosomes have plagued man and animals for centuries and are a continuing burden to animal husbandry and human well-being on the continent. Although there are many important species of trypanosomes including major animal pathogens such as *Trypanosoma congolense* and *Trypanosoma vivax* the organisms most often studied biochemically are those of the *Trypanosoma (Trypanozoon) brucei* species complex which includes *T. brucei brucei*, a parasite of cattle and horses, and *T. b. gambiense* and *T. b. rhodesiense*, agents of west African and east African human sleeping sickness, respectively. These organisms cause serious problems in livestock: in cattle, milk production is lowered and there is a general wasting effect with decreased usable muscle tissue, while horses are rendered useless for transportation. Large land areas become unfit for human habitation where the human parasites and the *Glossina* sp. vectors are endemic (31).

Control of disease by a vaccine seems remote since surface coat variation within a single strain is enormous (37,44). Efforts to control the tsetse fly vector and game and human reservoirs have been only sporadically successful (50). No novel practical drugs have been developed since the introduction of Melarsoprol over 35 years ago, and the toxicity of this and the other available drugs makes large-scale treatment impractical and dangerous (1). This is especially true for the late, central nervous system stages of the disease wherein conventional drugs clear trypanosomes from peripheral blood but, except for Melarsoprol, fail to destroy the cryptic CNS infection, which soon repopulates the blood.

Polyamine Content and Synthesis

Our early studies with insect trypanosomatids (*Crithidia fasciculata*, *Leptomonas seymouri*), as well as *T. b. brucei*, demonstrated that putrescine and spermidine were present in these hemoflagellates and spermine was not detectable (4). Although polyamine quantitation in this study was crude, it was evident that bloodstream forms of *T. b. brucei* had considerably more spermidine than putrescine and that the total polyamine content was higher than in the *Crithidia* and *Leptomonas* species. Use of the amino acid analyzer for polyamine analysis confirmed an overall spermidine to putrescine ratio of more than 4:1 in bloodstream *T. b. brucei*; minute quantities of spermine were also detectable under certain biological conditions (27). From use of radiolabeled precursors we determined that synthetic pathways for polyamines followed established routes from ornithine to spermidine, with S-adenosylmethionine forming the aminopropyl group donor for spermidine formation. Synthesis of spermine in normal bloodstream *T. b. brucei* also seems to occur at a very low rate (5). Unusual polyamines, e.g. norspermidine or norspermine, as found in Euglena (18,22,47), may also be present in bloodstream stages (4,5) but have not been positively identified.

Ornithine decarboxylase (ODC) is as active in rapidly growing bloodstream trypanosomes (16) as in most mammalian cell lines or regenerating tissues (25), with rates approaching 40 nmole CO_2 evolved/h/mg protein in a standard reaction system (16). These rates also fluctuate during culture growth of other trypanosomatids including *Leishmania* sp. (10,41); *Blastocrithidia* (13), *Leptomonas* (4) and *T. b. brucei* culture forms (C.J. Bacchi, unpublished observations) with peak activity at early-log phase. It appears, therefore, that *T. b. brucei* has, except for spermine synthesis, a mammalian-like pattern of polyamine metabolism.

Anti-polyamine Agents as Trypanocides

a. D,L-α-Difluoromethylornithine (DFMO)

A major advance in polyamine research and antipolyamine chemotherapy was the synthesis of the enzyme-activated (suicide) inhibitor DL-α-difluoromethylornithine (DFMO), a potent inhibitor of ornithine decarboxylase (29,43). This agent arrested a fulminating short-term (5-7 day) mouse infection of T. b. brucei EATRO 110 and a 1% or 2% solution in drinking water, given for 3 days beginning 24 h after infection, was invariably curative (6). The bloodstream was, in fact, cleared of parasites 72 hours after cessation of treatment (33). Co-administration of putrescine, spermidine or spermine, but not diaminopropane or cadaverine antagonized the DFMO effect and annulled the cures (33). DFMO was later shown to cure fulminating mouse infections of T. b. rhodesiense, a human pathogen (26), as well as rodent infections of T. b. gambiense, another human parasite (27). Recent data has also demonstrated that DFMO will cure fatal infections of T. congolense in mice (21).

DFMO has its rapid suppressive effect on T. b. brucei via direct inhibition of polyamine biosynthesis. Within 1 hour of incubation of intact bloodstream forms with 50 μM DFMO, putrescine production from exogenous ornithine was reduced more than 50% (6). ODC from the parasite is greater than 60 times more sensitive to DFMO than rat liver ODC with a K_i of 0.6 μM (16). Furthermore, in a study of DFMO action on a model infection in rats we found that DFMO acts rapidly in vivo, i.e. within 12 h of initiation of treatment, to cause a variety of important changes in the parasite including: a) 50-90% reduction in putrescine and spermidine levels; b) complete inhibition of ODC activity; c) more than 50% inhibition of DNA and RNA synthesis; d) a 1500 fold increase in decarboxylated S-adenosylmethionine (dSAM), a substrate, along with putrescine, for spermidine synthetase (C.J. Bacchi, unpublished observations).

In separate related studies we discovered that the antitumor antibiotic Blenoxane (bleomycin) also cures the short-term infection of T. b. brucei in mice, and that this action could be blocked by spermidine or spermine, but not putrescine (34). In view of the unexpected link between DFMO and bleomycin revealed by mutual polyamine reversal, we tested the efficacy of combination therapy in such infections. We found striking synergism: minimum curative levels of DFMO could be lowered by a factor of 4 (to 0.25% in drinking water for 3 days) and bleomycin by a factor of 12 (0.25 mg/kg i.p., once daily for 3 days) (8,27). As might be predicted, these cures were blocked by co-administration of putrescine, spermidine or spermine (8). There was a synergistic effect only with concurrent treatment or if DFMO was given before bleomycin; administration of bleomycin before DFMO did not cure (27). It thus appeared that polyamine depletion by priming with DFMO was an absolute condition for the bleomycin-DFMO synergism. Of greater potential import was the finding that, used at higher concentrations, DFMO + bleomycin in combination cured a T. b. brucei long-term central nervous system infection (27).

DFMO at 2% (drinking water) for 1 week + bleomycin 7 mg/kg (i.p. once daily) for 1 week cured all animals infected with the neurotropic TREU 667 isolate. Neither drug used alone was curative for this chronic central nervous sytem infection. This model infection is characterized by waves of parasitemia, and eventual death of the animals occurs 30-80 days after inoculation. The model mimics a human central nervous system infection and is resistant to standard trypanocides (27).

b. Methylglyoxal-bis (guanylhydrazone) (MGBG) and related compounds

MGBG was initially used as an antitumor agent, but fell into clinical disuse because of toxicity (30). Recently, encouraging clinical results have been obtained in human cancers by the use of less rigorous chemotherapeutic regimens (23,39,45,49). MGBG has also been tried in combination with DFMO against human leukemia (20) and with trypanocidal agents against murine leukemia L1210 (11).

Early preliminary reports had noted that MGBG apparently cured some trypanosome infections, i.e. *Trypanosoma equiperdum* (19) and *T. b. brucei*, *T. b. gambiense* and *T. b. rhodesiense* (38), in rodents. Growth of the insect trypanosomatid *Crithidia fasciculata* in defined media was also moderately inhibited by MGBG (30). Chang *et al.* (12), however, studying the biochemical and pharmacological effects of MGBG on bloodstream *T. b. brucei* and *T. congolense*, found it non-curative for either infection although it suppressed the parasitemia for several days. We have also found that MGBG prolongs the life of infected animals but does not cure (32; Table 1).

One specific interaction of DFMO with MGBG *in vivo* in human leukemia depends on depletion of polyamines by DFMO resulting in an accelerated uptake of MGBG as a spermidine analog via an induced polyamine transport system (20,42). Although DFMO treatment does cause 2-4 fold increase in *T. b. brucei* polyamine transport (C.J. Bacchi, unpublished observations), we found that MGBG and DFMO do not act synergistically in this system. Table 1 shows attempts to cure the EATRO 110 isolate of *T. b. brucei* with a combination of MGBG and DFMO. MGBG alone at 25-75 mg/kg was not curative. Combinations of MGBG with 0.25% DFMO were apparently antagonistic, since the survival time was shorter in the combination group, and animals were heavily infected at the time of death. As mentioned previously, DFMO at 0.25% is highly active in combination with bleomycin (8,27).

Ulrich and co-workers have synthesized a series of MGBG related aromatic guanylhydrazones and tested their activity against *T. b. brucei* and *T. congolense* infections. Two derivatives - 2,6 diacetylpyridine bisguanylhydrazone and 1,3 diacetylbenzene bisguanylhydrazone (8-40 mg/kg) cured mice infected with T. b. brucei (46). These guanylhydrazones were more active than MGBG seemingly because of their aromatic nature. We have also studied another aromatic derivative, 4,4'diaminodiphenylurea-bis(guanylhydrazone) (DDUG). This potent inhibitor of L1210 leukemia was not thought to be of great clinical utility because of its toxicity (30). Nevertheless, as shown in Table 1, DDUG was active against the acute EATRO 110 T. b. brucei infection at 10 mg/kg. Although the drug is not thought to antagonize polyamine synthesis (28) spermine, but not spermidine or putrescine, blocked cures when co-administered with the drug.

TABLE 1. Effect of bis-guanylhydrazones on T. b. brucei EATRO 110 in mice[a]

Drug	Dose (mg/kg)	Co-administered Polyamine (mg/kg)	Average Survival[c]
MGBG	25	--	3.4
	50	--	7.5
	75	--	13.6
DFMO	0.25% (d.w.)[b]	--	0
DFMO/MGBG	0.25%/25	--	1.2
	0.25%/50	--	3.1
	0.25%/75	--	1.6
DDUG	0.1	--	1.0
	1.0	--	4.0
	5.0	--	13.5
	10.0	--	>30
	100.0	--	>30
	10.0	putrescine 500	>30
	10.0	spermidine 300	>30
	10.0	spermine 50	0

[a] Groups of 5 mice were infected with 5 X 10^6 trypanosomes as in Nathan et al. (32). Treatment, begun 24 h after infection, consisted of single doses (i.p.) of drug daily for 3 days. Polyamines were administered concurrently as separate injections.

[b] DFMO was administered as a 0.25% solution in drinking water for 3 days, 24 h post-infection.

[c] Average survival in days beyond control deaths. Cures are indicated by >30. Results are the average of at least 2 separate experiments (10 animals).

c. Diamidines and Other Cationic Trypanocides

Several strongly cationic agents are used in human and animal trypanosomiases, including the diamidines pentamidine and Berenil (diminazine aceturate), quinapyramine (Antrycide) and the phenanthridine prothidium. The mode(s) of action of these drugs are unclear. Nevertheless, there are strong indications from the literature that some of these agents may be amine antagonists. Newton (35,36) found that quinapyramine at 10^{-4} M displaced Mg^{++} and polyamines from isolated ribosomes of an insect trypanosomatid. Our work had determined that a critical trypanosome respiratory enzyme, α-glycerophosphate dehydrogenase (NAD^+-linked) required spermidine or spermine for full activity (3,5), and that 10^{-5} M quinapyramine or ethidium (24) or the diamidines pentamidine and imidocarb (C.J. Bacchi, unpublished observations) replaced this requirement. Cohen and co-workers have also demonstrated ethidium displacement of polyamines in nucleic acids (14,40). Wallis (48) reported that diamidines released polyamines and Mg^{++} from insect trypanosomatid ribosomal preparations, and that pentamidine bound to ribosomes in treated whole cells.

The babesicides amicarbalide and imidocarb cured *T. b. brucei* infections in mice (32) while spermidine or spermine, but not putrescine, administered concurrently with either drug blocked cures (7). Since both drugs resemble diamidines, we also attempted to block the action of other trypanocides by concurrent polyamine administration *in vivo*. Table 2 indicates that the action of a number of drugs including pentamidine, quinapyrimine, prothidium and Berenil can be blocked by spermidine and/or spermine. Since this response to polyamine administration seemed so general, and because the drug structures contained amine groups whose spacing suggested the physiological polyamines (2), we believed that there could be a uniform mechanism of polyamine interference with drug action related to drug transport. Bachrach *et al.* (9) had, in fact, suggested that pentamidine interfered with transport of radioactive putrescine and its subsequent conversion to spermidine in *Leishmania* sp., and Damper & Patton (15) had demonstrated transport of [^{3}H]-pentamidine through a specific transport site in *T. b. brucei*, the natural function of which was not identified. Using [^{3}H]-pentamidine we have begun to examine the nature of diamidine uptake in bloodstream *T. b. brucei*. We used DFMO-treated and untreated control cells to ascertain whether uptake was greater in the polyamine depleted cells. As shown in Table 3, *T. b. brucei* treated with DFMO for 12 hours took up [^{3}H]-pentamidine (10 μM) at the same rate as control cells. Furthermore, the addition of polyamines at 1000 times the level of the drug did not influence drug uptake. These preliminary studies may thus indicate that diamidines are not taken up via a polyamine transport site and that polyamine reversal of *in vivo* action occurs by a more complex intracellular interaction of polyamine and drug.

TABLE 2. Effects of polyamines on the activity of trypanocidal agents in mice infected with T. b. brucei (EATRO 110)[a]

Drug	Dose (mg/kg)	Co-administered Polyamine	(mg/kg)	Average Survival[b]
Pentamidine	0.5	--		>30
	0.5	spermidine	100[c]	>30
	0.5	spermine	50	0
	1.0	--		>30
	1.0	spermidine	100	>30
	1.0	spermine	50	5.7
Berenil	0.1	--		4.6
	0.3	--		>30
	0.3	spermidine	30	16.8
	0.3	spermine	10	>30
	1.0	--		>30
	1.0	spermidine	300	25
	1.0	spermine	10	>30
	1.0		25	>30
	1.0		50	22.5
	1.0		100	0
Antrycide	0.5	--		>30
	0.5	spermidine	100	4.3
	0.5	spermine	50	5.8
	2.5	--		>30
	2.5	spermidine	300	20
	2.5	spermine	100	0
Prothidium	0.5	--		20.2
	0.5	spermidine	100	0
	0.5	spermine	50	0
	1.0	--		27.1
	1.0	spermidine	100	1.0
	1.0	spermine	50	0
	2.0	--		>30
	2.0	spermidine	300	0
	2.0	spermine	25	>30
	2.0	spermine	50	>30
	2.0	spermine	100	2.6

[a] Groups of 5 mice were infected with 2 X 10^5 trypanosomes as in Nathan et al. (32). Treatment, begun 24 h after infection, consisted of single doses (i.p.) of drug daily for 3 days. Polyamines were administered concurrently as separate injections.

[b] Average survival in days beyond control deaths. Cures are indicated by >30. Results are the average of at least 2 separate experiments (10 animals).

[c] Spermidine 300 mg/kg and spermine 100 mg/kg had no influence on the course of infection.

TABLE 3. Uptake of [^{3}H]pentamidine by bloodstream *T. b. brucei*[a]

Experiment	Cell Type	Addition	cpm x 10^{-4}/mg protein/h
1	Control	--	1.69
		10 mM putrescine	1.73
		10 mM spermidine	1.71
		10 mM spermine	1.69
2	Control	--	1.86
		10 mM spermidine	1.69
		10 mM spermine	1.67
	12 h DFMO treated	--	1.71
		10 mM spermidine	1.70
		10 mM spermine	1.72

[a] EATRO 110 strain was used. Trypanosomes were obtained from a 72 h infection as described (5), washed and incubated (37 °C) in 3 ml aliquots (1 X 10^{8} cells/ml) in pH 8.0 phosphate-saline-glucose buffer + 1% bovine serum albumen + penicillin/streptomycin. Ten µM [^{3}H]pentamidine (New England Nuclear special synthesis; 600 mCi/mM; final concentration 0.2 µCi/ml) was added to start the reaction. After 1 h, duplicate 3 ml samples were filtered (Whatman GS/B) and the filters washed 3X (10 ml/wash) in buffer containing 10 µM pentamidine. Filters were air dried overnight and counted.

CONCLUSION

It is apparent that polyamine metabolism is a crucial aspect of the physiology of trypanosomes. Activity of many standard trypanocides as well as novel agents such as bis(guanylhydrazones) are influenced by co-administration of polyamines. DFMO, a direct inhibitor of polyamine biosynthesis, is successful in blocking division of the parasites and as a result can cure infections *in vivo*. There is compelling evidence that polyamine depletion via DFMO is the means by which DFMO is able to act synergistically with bleomycin. Thus, DFMO may be valuable with a number of antitumor agents as well as known trypanocides to effect treatment of the various forms of African sleeping sickness.

ACKNOWLEDGEMENTS

The authors are grateful to Dr. Curtis Patton of Yale University for his gift of [^{3}H]-pentamidine, and to Dr. Carl Porter of the Roswell Park Memorial Institute for his gift of

4,4′diaminodiphenylurea bis(guanylhydrazone). This research was supported by N.I.H. grant AI 17340, a grant from the UNDP/World Bank/WHO Special Programme for Research and Training in Tropical Diseases, and a Pace University Scholarly Research Award.

REFERENCES

1. Apted, F.I.C. (1980): Pharmacol. Ther., 11:391-413.
2. Bacchi, C.J. (1981): J. Protozool., 28:20-27.
3. Bacchi, C.J., Marcus, S.L., Lambros, C., Goldberg, B., Messina, L., and Hutner, S.H. (1974): Biochem. Biophys. Res. Commun., 58:778-786.
4. Bacchi, C.J., Lipschik, G.Y., and Nathan, H.C. (1977): J. Bacteriol., 131:657-661.
5. Bacchi, C.J., Vergara, C., Garofalo, J., Lipschik, G.Y., and Hutner, S.H. (1979): J. Protozool., 26:484-488.
6. Bacchi, C.J., Nathan, H.C., Hutner, S.H., McCann, P.P., and Sjoerdsma, A. (1980): Science, 210:332-334.
7. Bacchi, C.J., Nathan, H.C., Hutner, S.H., Duch, D.S., and Nichol, C.A. (1981): Biochem. Pharmacol., 30:883-886.
8. Bacchi, C.J., Nathan, H.C., Hutner, S.H., McCann, P.P., and Sjoerdsma, A. (1982): Biochem. Pharmacol., in press.
9. Bachrach, U., Brem, S., Wertman, S.B., Schnur, L.F., and Greenblatt, C.L. (1979): Exp. Parasitol., 48:464-470.
10. Bachrach, U., Abu-Elheiga, L., Talmi, M., Schnur, L.F., El-On, J., and Greenblatt, C.L. (1981): Med. Biol., 59:441-447.
11. Burchenal, J.H., Lokys, L., Smith, R., Cartmell, S., and Warrell, R. (1981): Proc. Am. Assoc, Cancer Res., 22:230.
12. Chang, K.-P., Steiger, R.F., Dave, C., and Cheng, Y.-C. (1978): J. Protozool., 25:145-149.
13. Chang, K.-P., and Dave, C. (1980): In: Proc. Int. Colloquium, Endosymbiosis and Cell Research, edited by W. Schwemmler, and H. Schlenk, pp. 349-359. Walter de Gruyter, Berlin and New York.
14. Cohen, S.S. (1978): Nature, 274:209-210.
15. Damper, D., and Patton, C.L. (1976): Biochem. Pharmacol., 25:271-276.
16. Garofalo, J., Bacchi, C.J., McLaughlin, S.D., Mockenhaupt, D., Trueba, G., and Hutner, S.H. (1982): J. Protozool., 29:389-394.
17. Gashumba, J.K., and Mwambu, P.M. (1981): Tropical Doctor, 11:175-178.
18. Hegewald, E., and Kneifel, H. (1981): Arch. Hydrobiol. (Suppl. 60.3), 313-323.
19. Jaffee, J.J. (1965): Fed. Proc., Fed. Am. Soc. Exp. Biol., 24:455.

20. Jänne, J., Alhonen-Hongisto, L., Seppänen, P., and Siimes, M. (1981): Med. Biol., 59:448-451.
21. Karbe, E., Bottger, M., McCann, P.P., Sjoerdsma, A., and Freitas, E.K. (1982): Tropenmed. Parasit., 33: in press.
22. Kneifel, H., Schuber, F., Aleksijevic, A., and Grove, J. (1978): Biochem. Biophys. Res. Commun., 85:42-46.
23. Knight, W.A. III, Livingston, R.B., Fabian, C., and Costanzi, J. (1979): Cancer Treat. Rep., 63:1933-1937.
24. Lambros, C., Bacchi, C.J., Marcus, S.L., and Hutner, S.H. (1977): Biochem. Biophys. Res. Commun., 84:1227-1233.
25. McCann, P.P. (1980): In: Polyamines in Biomedical Research, edited by J.M. Gaugas, pp. 109-123. John Wiley & Sons, New York.
26. McCann, P.P., Bacchi, C.J., Hanson, W.L., Cain, G.D., Nathan, H.C. Hutner, S.H., and Sjoerdsma, A. (1981): In: Advances in Polyamine Research, Vol. 3, edited by C.M. Caldarera, V. Zappia, and U. Bachrach, pp. 97-110. Raven Press, New York.
27. McCann, P.P., Bacchi, C.J., Clarkson, A.B., Jr., Seed, J.R., Nathan, H.C., Amole, B.O., Hutner, S.H., and Sjoerdsma, A. (1981): Med. Biol., 59:434-440.
28. Mikles-Robertson, F., Dave, C., and Porter, C.W. (1980): Cancer Res., 40:1054-1061.
29. Metcalf, B.W., Bey, P., Danzin, C., Jung, M.J., Casara, P., and Vevert, J.P. (1978): J. Am. Chem. Soc., 100:2551-2553.
30. Mihich, E. (1975): In: Antineoplastic and Immunosuppressive Agents. Handbook of Experimental Pharmacology, vol. 38/2, edited by A.C. Sartorelli and D.G. Johns, pp. 766-788, Springer-Verlag, Berlin.
31. Mulligan, H.W., editor (1970): The African Trypanosomiases, Allen and Unwin, London.
32. Nathan, H.C., Soto, K.V.M., Moreira, R., Chunosoff, L., Hutner, S.H., and Bacchi, C.J. (1979): J. Protozool., 26:657-660.
33. Nathan, H.C., Bacchi, C.J., Hutner, S.H., Rescigno, D., McCann, P.P., and Sjoerdsma, A. (1981): Biochem. Pharmacol., 30:3010-3013.
34. Nathan, H.C., Bacchi, C.J., Sakai, T.T., Rescigno, D., Stumpf, D., and Hutner, S.H. (1981): Trans. R. Soc. Trop. Med. Hyg., 75:394-398.
35. Newton, B.A. (1963): Biochem. J., 89:93.
36. Newton, B.A. (1966): Symp. Soc. Gen. Microbiol., 16:213-234.
37. Pays, E., Lheureux, M., Vervoort, T., and Steinert, M. (1981): Mol. Biochem. Parasitol., 4:349-357.
38. Risby, E.L. (1976): J. Protozool., 23 (Suppl.), 23:27A.
39. Rosenblum, M.G., Stewart, D.J., Yap, B.S., Leavens, M., Benjamin, R.S., and Loo, T.L. (1981): Cancer Res., 4:459-462.

40. Sakai, T.T., Torget, R.I.J., Freda, C.E., and Cohen, S.S. (1975): Nucl. Acids Res., 2:1005-1022.
41. Schnur, L.F., Bachrach, U., Greenblatt, C.L., and Ben Joseph, M. (1979): FEBS Lett., 106:202-206.
42. Seppänen, P. (1981): Acta Chem. Scand., B25:731-736.
43. Sjoerdsma, A. (1981): Clin. Pharmacol. Ther., 30:3-22.
44. Strickler, J.E., and Patton, C.L. (1982): Mol. Biochem. Parasitol., 5:117-131.
45. Todd, R.F., III, Garnick, M.B., Canellos, G.P., Richie, J.P., Gittes, R.F., Mayer, R.J., and Skarin, A.T. (1981): Cancer Treat. Rep., 65:17-20.
46. Ulrich, P.C., Grady, R.W., and Cerami, A. (1982): Drug Dev. Res., 2:219-228.
47. Villanueva, V.R. (1981): In: Adv. Polyamine Res., Vol. 3, edited by C.M. Caldarera, V. Zappia, and U. Bachrach, pp. 389-395. Raven Press, New York.
48. Wallis, O.C. (1966): J. Protozool., 13:234-239.
49. Warrell, R.P., Jr., Lee, B.J., Kempin, S.J., Lacher, M.J., Straus, D.J., and Young, C.W. (1981): Blood, 57:1011-1014.
50. W.H.O. (1981): Annual Rep., African Trypanosomiases (Special Programme Res. Training Trop. Dis.). Geneva, Switzerland.

Advances in Polyamine Research, Vol. 4, edited by U. Bachrach, A. Kaye, and R. Chayen. Raven Press, New York © 1983.

Effect of Putrescine Derivatives on Growth and Protein Synthesis

Sara H. Goldemberg, I. D. Algranati, J. J. Miret, *D. O. Alonso Garrido, and *B. Frydman

*Instituto de Investigaciones Bioquímicas "Fundación Campomar" and Facultad de Ciencias Exactas y Naturales, 1428 Buenos Aires, Argentina; *Facultad de Farmacia y Bioquímica, Universidad de Buenos Aires, Buenos Aires, Argentina*

Putrescine, spermidine and spermine are the most widely distributed polyamines in living cells. They are strongly cationic substances with a hydrophobic backbone of considerable conformational flexibility. Due to these structural characteristics polyamines are able to interact with nucleic acids, proteins and phospholipids through ionic forces and hydrogen or hydrophobic bonds (2-5, 20, 21).

Although a great number of investigations have shown the essential role of polyamines in many biological processes, such as cell proliferation, DNA replication, RNA synthesis, translation, transport and membrane reactions (2, 3, 20), many aspects of polyamine action at the molecular level and the relation with their chemical structure are still poorly understood.

Some attempts to explain the structural features which give the polyamines their capacity to influence so many processes have been carried out. Heller et al. (15) examined the effect of spermidine, spermine and a series of diamines of general structure $H_2N(CH_2)_nNH_2$ on the induction of ornithine decarboxylase and its antienzyme. Igarashi and coworkers (16) tested the addition of some macrocyclic polyamines and polymethylenediamines on various reactions influenced by polyamines. Morris et al. (11, 12, 17) synthesized spermidine analogues of general structure $H_2N(CH_2)_3NH(CH_2)_nNH_2$ and tested their action on cell growth and macromolecular synthesis. The above mentioned authors tried to change the distance between nitrogen atoms by altering the number of methylene groups between them. More recently Karpetsky et al. (18) studied the effect of some other features of the polyamine molecule; they analysed the action of certain compounds containing polyamine substructures on the poly(A)-induced inhibition of ribonuclease activity. In addition Mamont and Danzin studied the effect of several changes in putrescine structure on the activation of eukaryotic S-adenosyl-L-methionine decarboxylase (19).

In this paper we report the results obtained following a somewhat different approach. We have synthesized several putrescine derivatives in which the distance between the amino groups was unaltered, but which were N-substituted in order to modify the basicity or the hydrogen-bonding capacity of the polyamine. These substances were tested on some putrescine-requiring mutants of *Escherichia coli*, analysing the effect on growth, protein synthesis and ribosomal distribution patterns.

MATERIALS AND METHODS

Putrescine dihydrochloride, spermidine trihydrochloride, spermine tetrahydrochloride, cadaverine dihydrochloride, 1,3diaminopropane dihydrochloride and arcaine sulfate were purchased from Sigma Chemical Co; dansyl chloride was obtained from Pierce.

The N-monoalkylputrescines (compounds VI-IX, see below) were obtained by condensation of 4-bromobutyronitrile with the corresponding benzylalkylamines in the presence of triethylamine. Hydrogenolysis and reduction of the resulting N,N-benzylalkyl-4-aminobutyronitriles over Pd/C afforded the expected N-alkylputrescines in good overall yields. Condensation of 4-bromobutyronitrile with dimethylamine gave the N,N-dimethyl-4-aminobutyronitrile, which was reduced to N,N-dimethylputrescine (X). Diacetylputrescine (V) was prepared by acetylation of putrescine with an excess of acetic anhydride, while monoacetylputrescine (III) obtained by partial acetylation of the former (22), had to be purified from contaminant putrescine by TLC on cellulose. Diformylputrescine (IV) was obtained by heating putrescine formate at 200° C for 1 h. Reduction of IV, V and N,N'-dipropionylputrescine with diborane in tetrahydrofuran afforded the compounds XI, XII and XIII.

Reductive alkylation of putrescine with CH_2O/H_2 over Pd/C gave substance XIV. Succinodiamidine dihydrochloride (II) was obtained from succinodinitrile following an established procedure (8).

Escherichia coli BGA 8 (thi leu thr spe B spe C), a strain obtained in our laboratory (14), and MA 255 (6) have been used throughout. These bacterial strains were grown in minimal medium and starved for polyamines as previously reported (13). Several aliquots of an exponential culture were grown in the absence or presence of different polyamines as indicated in each case. Cell growth was followed by the A_{490} values obtained at different times.

In vivo protein synthesis was measured on aliquots of the cultures by the incorporation of $[^{14}C]$ phenylalanine (0.05 μC/ml, specific activity 532 Ci/mol) into TCA-insoluble material.

In order to analyse the ribosomal distribution patterns corresponding to different growth conditions, cells were collected after slow cooling of exponentially growing cultures, disrupted by sonication and centrifuged. The resulting S_{30} extracts were layered on top of 15-45% linear sucrose gradients containing 10 mM Tris-HCl buffer, pH 7.5, 5 mM magnesium acetate and 50 mM KCl, and then centrifuged for 135 min at 45,000 rpm in a Spinco SW 50.1 rotor. Ribosomal profiles were obtained by monitoring at 254 nm with an ISCO ultraviolet analyzer. Polyamine intracellular pools were determined after extraction of cells with 0.2 M perchloric acid followed by dansylation of extracts according to Dion and Herbst (7) and TLC of the dansyl derivatives (13).

RESULTS

Several authors have reported that increasing the distance between the amino groups of putrescine and spermidine leads to a gradual reduction of the stimulation of growth of a polyamine-requiring mutant of E. coli (16, 17). We decided therefore to test other changes of putrescine structure, trying to keep the hydrophobic flexible backbone untouched. The following modifications were carried out:

1) Changes in the basicity of the molecule
   a. Increase in basicity, keeping a high hydrogen bonding capacity. The compounds tested were arcaine (I) and succinodiamidine (II)

$HN{=}C(NH_2){-}HN{-}(CH_2)_4{-}NH{-}C(NH_2){=}NH$ (I)

$H_2N{-}C({=}NH){-}(CH_2)_2{-}C({=}NH){-}NH_2$ (II)

b. Decrease in basicity and hydrogen-bonding capacity

The following compounds were assayed: N-acetylputrescine (III), N,N'-diformylputrescine (IV) and N,N'-diacetylputrescine (V)

$H_2N-(CH_2)_4-NH \cdot COCH_3$ (III)

$OHC \cdot HN-(CH_2)_4-NH \cdot CHO$ (IV)

$CH_3CO \cdot HN-(CH_2)_4-NH \cdot COCH_3$ (V)

2) Modifications of the hydrogen-bonding capacity without major changes in basicity.

Alkyl residues of varying length were introduced in one or both amino groups, resulting in a decreased hydrogen-bonding capacity with a concomitant enhanced hydrophobicity of the molecule. The substances tested are shown below:

$$\begin{matrix}R_1\\R_2\end{matrix}\!>N \cdot (CH_2)_4 \cdot N<\!\begin{matrix}R_3\\R_4\end{matrix}$$

	R_1	R_2	R_3	R_4	Compound
VI	CH_3	H	H	H	N-methylputrescine
VII	C_2H_5	H	H	H	N-ethylputrescine
VIII	$n-C_3H_7$	H	H	H	N-propylputrescine
IX	$n-C_4H_9$	H	H	H	N-butylputrescine
X	CH_3	CH_3	H	H	N,N-dimethylputrescine
XI	CH_3	H	CH_3	H	N,N'-dimethylputrescine
XII	C_2H_5	H	C_2H_5	H	N,N'-diethylputrescine
XIII	$n-C_3H_7$	H	$n-C_3H_7$	H	N,N'-dipropylputrescine
XIV	CH_3	CH_3	CH_3	CH_3	N,N'-tetramethylputrescine

Bacterial Growth, "in vivo" Protein Synthesis and Structural Changes of the Putrescine Molecule

Fig. 1 shows the effect of putrescine and different polyamines on the growth rate of putrescine-depleted cultures of E. coli BGA 8. Similar results were obtained with the strain MA 255.

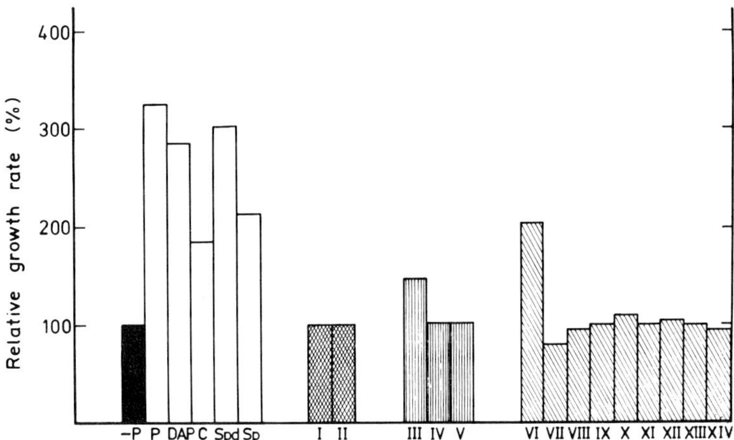

FIG. 1. Growth rate of E. coli BGA 8 cultures in the absence and presence of different polyamines. Growth was expressed as mass doublings per hour; values in the absence of putrescine were taken as 100%. All polyamines were used at concentrations of 100 μg/ml.

It can be seen that changes giving higher or lower basicity of the molecule decreased or abolished the stimulatory effect of putrescine on growth. N-acetylputrescine was able to enhance the growth of polyamine-starved bacteria, although this effect was observed only after a lag period of several hours. When alkyl groups were introduced altering the hydrogen-bonding capacity without modifying the basicity, the following results were observed: substitutions in both amino groups or in both hydrogen atoms of one amino group suppressed the activating effect of putrescine on bacterial growth. Partial stimulation was obtained when one hydrogen atom was replaced by a methyl group. If the hydrogen was replaced by an ethyl group, there was a small but reproducible inhibition of the basal growth rate of the putrescine-depleted culture. Substitution with a propyl or butyl residue abolished all effect on the basal growth rate. It is interesting to remark that while spermidine was almost fully active in the stimulation of bacterial growth, N-propylputrescine, lacking only one amino terminal group, was unable to enhance growth of the polyamine-requiring mutant of *E. coli*.

Fig. 2 depicts the effect of putrescine and some of the above mentioned derivatives on *in vivo* protein synthesis. In this case a similar pattern could be observed: N-methylputrescine showed a stimulatory action of about 50 per cent compared to the value obtained with putrescine; N-ethylputrescine gave a slight inhibition and all other modifications of the putrescine molecule abolished the stimulation of amino acid incorporation into polypeptides.

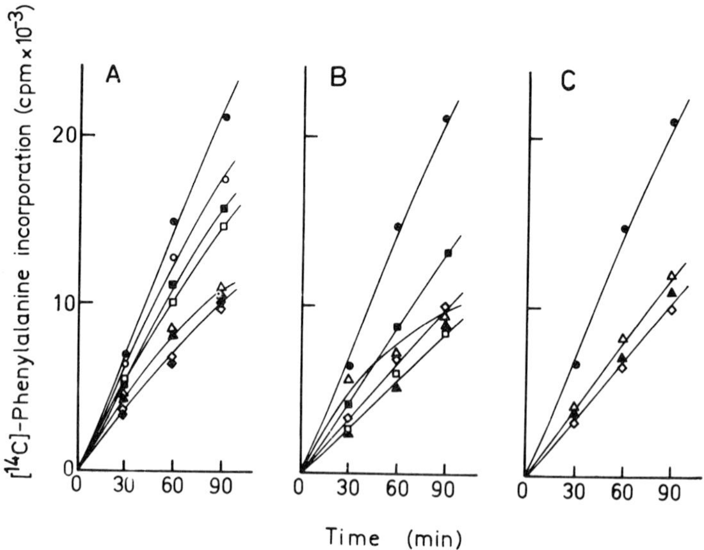

FIG. 2. *In vivo* protein synthesis of *E. coli* BGA 8 grown in the absence and presence of different polyamines. Symbols: A) culture without putrescine (◇), with arcaine (◆), N,N'-diacetylputrescine (△), N-acetylputrescine (▲), 1,3 diaminopropane (□), cadaverine (■), spermidine (○) and putrescine (●). B) culture without (◇) or with (●) putrescine; with N-methylputrescine (◆), N-ethylputrescine (▲), N-propylputrescine (△) and N-butylputrescine (□). C) culture without (◇) or with (●) putrescine; with N,N-dimethylputrescine, N,N'-dimethylputrescine or N,N'-tetramethylputrescine (▲) and with N,N'-dipropylputrescine (△). All polyamines were used at 100 μg/ml.

Since most of the putrescine analogues assayed did not show any effect either on cell growth or protein synthesis, it could be argued that these compounds were not taken up by the bacteria. Analyses of intracellular polyamine pools have shown that this was not the case, at least for most of the N-alkyl putrescine derivatives used in our experiments (data not shown). We have detected only traces of N-acetylputrescine inside the cells. In some occasions after several hours incubation in the presence of this compound we have found putrescine in the intracellular pool. This result seems to indicate that N-acetylputrescine can be hydrolysed either in the growth medium or immediately after being taken up by the bacteria.

Effect of N-Methylputrescine

All the modifications carried out in the putrescine molecule, except the introduction of a methyl residue in only one hydrogen atom of the amino group, led to a loss of stimulatory activity on growth and protein synthesis. Therefore, we decided to study the effect of N-methylputrescine in further detail, in order to improve our understanding of the molecular basis of polyamine functions and its correlation with some structural characteristics.

Fig. 3 shows the time course of amino acid incorporation into polypeptides in a putrescine-starved culture of E. coli BGA 8, using different doses of putrescine (A) or N-methylputrescine (B). It can be seen that very small concentrations (0.01 µg/ml or less) of either diamine had no effect on protein synthesis; 0.1 µg/ml putrescine gave some stimulation after a lag period, while the same amount of N-methylputrescine did not alter basal incorporation. Higher concentrations of the diamines led to shorter lag periods for both compounds. This result might indicate that at low concentrations of these substances their uptake by the cells or the readjustment of the protein synthetic machinery would take longer periods of time. Further differences were observed in the maximal values attained and in the shape of the curves: N-methylputrescine promoted lower incorporations and the curves levelled off after a 90 minutes incubation period. This different behavior indicates that the decreased effect of N-methylputrescine should be attributed to the structural change of the polyamine molecule rather than to a reduced uptake of the analogue by the microorganism.

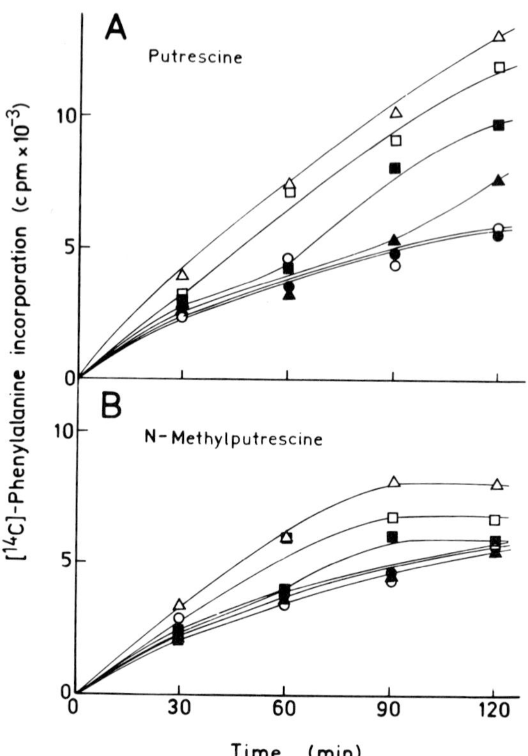

FIG. 3. Effect of different levels of putrescine (A) and N-methylputrescine (B) on amino acid incorporation into proteins. Cultures of E. coli BGA 8 were carried out in the absence (○) or presence of 0.01 μg/ml (●), 0.1 μg/ml (▲), 1 μg/ml (■), 10 μg/ml (□) and 100 μg/ml (△) polyamine.

Ribosomal Distribution Patterns in Bacteria Grown in the Absence or Presence of Putrescine or N-Alkyl Putrescines

We have previously shown that in polyamine-starved E. coli the equilibrium between 70S monomers and ribosomal subunits was shifted towards the subparticles and that this change correlated with decreased growth rate and protein synthesis (1, 9, 10). The results described above have demonstrated that the use of putrescine derivatives instead of putrescine in cultures of polyamine-requiring bacteria caused a reduction of cell growth and in vivo protein synthesis. In order to investigate whether these changes were also accompanied by modifications in the equilibrium pattern of ribosomal particles, we have analysed the ribosomal distribution profiles in E. coli BGA 8 cells grown in the absence and presence of putrescine or several N-alkyl putrescine derivatives.

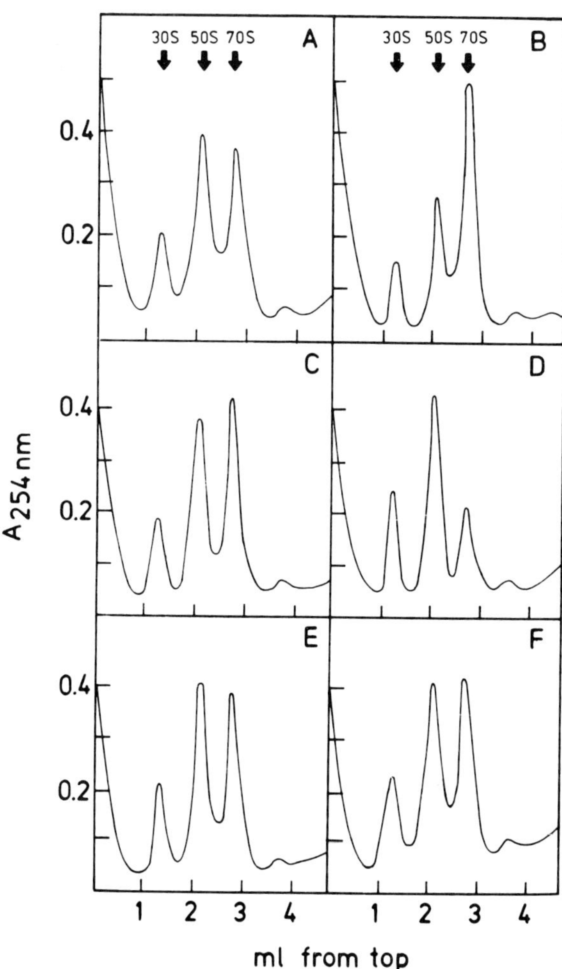

FIG. 4. Ribosomal profiles of extracts obtained from E. coli BGA 8 grown in the absence (A) or presence of putrescine (B), N-methylputrescine (C), N-ethylputrescine (D), N-propylputrescine (E) and N-butylputrescine (F). All polyamines were used at a concentration of 100 μg/ml.

Fig. 4 shows that while N-methylputrescine increased the association of ribosomal subparticles (Fig. 4 C), N-ethyl putrescine gave a marked dissociation (Fig. 4 D). These results are in accordance with the effects of both compounds on cell growth and protein synthesis. On the other hand, N-propyl and N-butylputrescine gave ribosomal distribution profiles (Fig. 4 E and 4 F) almost identical to those observed in extracts from bacteria cultivated in the absence of putrescine (Fig. 4 A), and concomitantly these alkyl derivatives did not produce any effect on bacterial growth and protein synthesis.

CONCLUDING REMARKS

In order to elucidate the structural specificity of polyamines necessary for their physiological functions in bacteria, we have studied the effects of several putrescine derivatives on growth and in vivo protein synthesis. The substituents were chosen so as to modify mainly the basicity and/or the hydrogen-bonding capacity of the molecule, because these structural characteristics seem to be essential for the interactions between polyamines and nucleic acids or proteins (2-5, 20, 21). Our results led us to the conclusion that even maintaining the same tetramethylene backbone of the molecule, any change of the basicity or the hydrogen-bonding capacity caused a marked reduction or completely abolished the stimulatory activity of putrescine on growth and protein biosynthesis of polyamine-depleted cultures. These facts strongly support the idea that two primary amino groups are required in the polyamine structure to be fully active.

The experiments with a series of N-monoalkyl derivatives of putrescine with almost identical basicity and hydrogen-bonding capacity, but differing in the hydrophobicity of the N-alkyl residue have indicated that only N-methylputrescine was partially active. While N-propyl and N-butylputrescine did not cause any effect, N-ethylputrescine inhibited bacterial growth and protein synthesis and at the same time induced a further shift of the equilibrium between 70S monomers and ribosomal subparticles towards the subunits.

All the results described above demonstrate that the structural requirements of functional polyamines are indeed rather strict. Further investigations with other polyamine derivatives may prove very useful for the elucidation of the optimal structural characteristics necessary for the appropriate interaction at the intracellular sites of action. With this approach new insights of the molecular basis of polyamine function might be obtained.

Acknowledgements. The authors thank Ms Marta Eirin and Ms Susana Raffo for their skillful technical assistance. S.H.G., I.D.A. and B.F. are career investigators of the Consejo Nacional de Investigaciones Científicas y Técnicas (Argentina).

REFERENCES

1. Algranati, I.D., Echandi, G., Goldemberg, S.H., Cunningham-Rundles, S. and Maas, W.K. (1975):J.Bacteriol., 124:1122-1127.
2. Algranati, I.D., and Goldemberg, S.H. (1977):Trends Biochem.Sci., 2:272-274.
3. Bachrach, U. (1973):Function of Naturally Occurring Polyamines, Academic Press, New York.
4. Cohen, S.S. (1971): Introduction to the Polyamines, Prentice-Hall, Englewood Cliffs, N.J.
5. Cohen, S.S., and McCormick, F.P. (1979): Adv.Virus Res., 24:331-387.
6. Cunningham-Rundles, S., and Maas, W.K. (1975): J.Bacteriol., 124:791-799.
7. Dion, A.S., and Herbst, E.J. (1970): Ann.N.Y.Acad. Sci., 171:723-734.
8. Dox, A.W. (1958): Org.Synth.Coll., vol. 1:5-7.
9. Echandi, G., and Algranati, I.D. (1975): Biochem. Biophys.Res.Commun., 62:313-319.
10. Echandi, G., and Algranati, I.D. (1975): Biochem. Biophys.Res.Commun., 67:1185-1191.
11. Geiger, L.E., and Morris, D.R. (1980): Biochim.Biophys.Acta, 609:264-271.
12. Geiger, L.E., and Morris, D.R. (1980): J.Bacteriol. 141:1192-1198.
13. Goldemberg, S.H., and Algranati, I.D. (1977): Mol. Cell.Biochem., 16:71-77.
14. Goldemberg, S.H., and Algranati, I.D. (1981): Eur. J.Biochem., 117:251-255.
15. Heller, J.S., Chen, K.Y., Kyriakidis, D.A., Fong, W.F., and Canellakis, E.S. (1978): J.Cell.Physiol., 96:225-234.
16. Igarashi, K., Kashiwagi, K., Kakegawa, T., Hirose, S., Yatsumani, T., and Kimura, E. (1980): Biochim. Biophys.Acta, 633:457-464.
17. Jorstad, C.M., Harada, J.J., and Morris, D.R. (1980) J.Bacteriol., 141:456-463.
18. Karpetsky, T.P., Shriver, K.K., and Levy. C.C. (1981): Biochem.J., 193:325-337.
19. Mamont, P.S., and Danzin, C. (1981): In: Adv.Polyamine Res., edited by C.C. Caldarera, V. Zappia, and U. Bachrach, vol. 3, pp. 123-135, Raven Press.

20. Scalabrino, G., and Ferioli, M.E. (1981): Adv.Cancer.Res., 35:151-268.
21. Tabor, C.W., and Tabor, H. (1976): Ann.Rev.Biochem. 45:285-306.
22. Tabor, C.W., Tabor, H., and Bachrach, U. (1964): J.Biol.Chem., 239:2194-2203.

Advances in Polyamine Research, Vol. 4, edited by U. Bachrach, A. Kaye, and R. Chayen. Raven Press, New York © 1983.

Mammalian Propylamine Transferases

A. Raina, T. Eloranta, T. Hyvönen, and R. -L. Pajula

Department of Biochemistry, University of Kuopio, SF-70101 Kuopio 10, Finland

Enzymic synthesis of spermidine and spermine from decarboxylated S-adenosylmethionine (decarboxy-AdoMet; DeSAM) and the appropriate propylamine acceptor was first demonstrated by Pegg and Williams-Ashman (20,21) using crude extracts of rat ventral prostate as enzyme source. Soon thereafter it was demonstrated that two separate enzymes were involved in the transfer of the propylamine group to either putrescine (spermidine synthase) or spermidine (spermine synthase) in rat brain (3,24), liver (2,23) and prostate (9). However, only recently the development of rapid and sensitive assay methods (6,22,26), and the application of affinity chromatography for the purification of spermine synthase (14,15,27) and spermidine synthase (30,31) have made it possible to characterize these enzymes from mammalian tissues in greater detail. In this paper we describe some properties of spermidine and spermine synthases, with special emphasis on the propylamine transferases purified from bovine brain.

ASSAY OF SPERMIDINE AND SPERMINE SYNTHASES

Two methods, based on the use of ^{14}C-decarboxy-AdoMet as a substrate, have recently been described for a rapid assay of propylamine transferase activities (6,22,26). The radioactive product, ^{14}C-labeled polyamine or (methyl-^{14}C)methylthioadenosine, is isolated by a simple chromatographic procedure.

Labeled S-adenosylmethionine was synthesized from DL-(2-^{14}C)methionine or from L-(methyl-^{14}C)methionine as described by Pegg and Williams-Ashman (20). Both types of radioactive S-adenosylmethionine of high specific activity are now commercially available, too. Radioactive decarboxy-AdoMet was prepared from labeled S-adenosylmethionine using S-adenosylmethionine decarboxylase from E.coli as the enzyme. The product was purified on a Dowex 50-H^+ column, followed by preparative paper electrophoresis as previously described (24). (Methyl-^{14}C)Decarboxy-AdoMet was also purified by chromatography on a phosphocellulose column (Cellex-P, Bio-Rad).

The assay procedure using (propylamine-1-^{14}C)decarboxy-AdoMet as the labeled substrate has been previously described in detail (14,26,27). With (methyl- ^{14}C)decarboxy-AdoMet as the labeled substrate the original method developed by Hibasami and Pegg (6) was modified as follows. Dowex 50-H^+ cation exchanger was replaced by a weak cation exchange material (Cellex-P). This modification made the elution of labeled methylthioadenosine (and its degradation products) easier and also increased the counting efficiency. In kinetic studies with the purified enzymes bovine serum albumin was added to a final concentration of 1 mg/ml to prevent denaturation of the very dilute enzyme. To prevent adsorption of decarboxy-AdoMet to glass at low substrate concentrations, the appropriate dilutions of the substrate were prepared in 0.15 M NaCl. At the end of incubation 0.5 ml of 25 mM HCl was added to the reaction mixture (0.1 ml). An aliquot of 0.5 ml of the acid solution (pH slightly below 3) was applied to a phosphocellulose column (containing approximately 0.5 ml of the cation exchanger Cellex-P; the void volume was about 0.8 ml) previously equilibrated with 25 mM HCl. The effluent (0.5 ml) was discarded and the radioactive methylthioadenosine (plus degradation products) was eluted directly into a scintillation vial with 1.8 ml of 25 mM HCl. Ten ml of scintillation solution (ACS; Amersham) was added and mixed thoroughly. Counting efficiency was about 82 %. The column was regenerated with 2 x 2 ml of 0.1 M NaOH and/or 2 x 2 ml of 0.5 M HCl.

One unit of enzyme activity represents the formation of 1 nmol of spermidine or spermine in 1 min under standard assay conditions.

PURIFICATION OF PROPYLAMINE TRANSFERASES FROM MAMMALIAN TISSUES

The activities of spermidine and spermine synthases in most rat tissues are considerably higher (in terms of specific activity) than those of ornithine and S-adenosylmethionine decarboxylases (26). The highest activities of spermidine synthase were found in the pancreas and prostate, whereas the spermine synthase activity was remarkably high in the brain (Table 1). In most cases prostatic and brain tissues have been used as starting material for the purification of propylamine transferases from mammalian sources.

Spermine synthase was first purified to an apparent homogeneity (about 6000-fold) from bovine brain using spermine-Sepharose affinity chromatography (14,15). We have recently applied the same method for the purification of spermine synthase from rat ventral prostate. A summary of the purification procedure is presented in Table 2. The final preparation (purified about 2300-fold) was more than 90 % pure as judged by polyacrylamide gel electrophoresis. It co-migrated with spermine synthase purified from bovine brain at two different gel concentrations. It therefore appears that the molecular weight of spermine synthase isolated from these two different sources is approximately the same.

TABLE 1. Propylamine transferase activities in some tissues of male rats at the age of 10 weeks[a]

Tissue	Spermidine synthase	Spermine synthase
	(pmol/mg protein per min)	
Brain	103	107
Prostate	603	91
Thymus	190	36
Spleen	135	30
Pancreas	1260	19
Kidney	48	17
Liver	107	9

[a]Data taken from (26).

Spermidine synthase has been purified to an apparent homogeneity from rat prostate (30,31) and from bovine brain (30). The design of a new affinity chromatographic adsorbent, S-adenosyl(5')-3-thiopropylamine linked to Sepharose (31), has greatly facilitated the purification procedure. The kinetic experiments described in this paper have been performed using spermidine synthase purified about 4500-fold from bovine brain.

PROPERTIES OF MAMMALIAN PROPYLAMINE TRANSFERASES

Previous studies on the enzymology of spermidine and spermine synthesis have recently been reviewed (25, 34). Some physical and catalytic properties of spermidine and spermine synthases purified from bovine brain are summarized in Table 3. It appears that both enzymes are composed of two subunits of equal size. The apparent molecular weight of spermidine synthase purified from bovine brain (70 000) is approximately the same as that (73 000) reported for the rat prostate enzyme (31). An apparent molecular weight of 88 000 has been obtained for spermine synthase from bovine brain (14) and from rat ventral prostate (vide supra).

TABLE 2. Purification of spermine synthase from rat prostate

Step	Total protein (mg)	Specific activity (units/mg)	Total activity (units)	Yield (%)
Ammonium sulfate	3 220	0.09	283	100
DEAE-cellulose	1 000	0.25	254	90
First sp-Sepharose	3.30	42.7	141	50
Second sp-Sepharose	0.33	206.0	68	24

TABLE 3. Some properties of propylamine transferases from bovine brain

	Spermidine synthase	Spermine synthase
Molecular wt.	70 000	88 000
Subunit molecular wt.	35 800	45 000
K_m (DeSAM) (μM)	0.2	0.1
K_m (putrescine/spermidine) (μM)	30	50
Substrate inhibition (DeSAM)	yes	no
Substrate specifity for		
DeSAM	less strict	strict
propylamine acceptor	less strict	strict
Inhibition by MTA		
K_i (DeSAM variable) (μM)	15	0.3
K_i (amine variable) (μM)	80	10

Both propylamine transferases have a remarkably high affinity for decarboxy-AdoMet. It appears that the K_m -values of the brain enzymes are of the order of 10^{-7} M. Somewhat higher values, varying from 0.6 to 50 μM (3,9,13,14,21,30), have previously been reported for mammalian propylamine transferases from various sources. Although there might be some tissue and species differences in the affinities of the enzymes for decarboxy-AdoMet, it seems likely that the discrepancy is mostly due to insensitive assay methods used previously. It is interesting that the concentration of decarboxy-AdoMet in several rat tissues has been reported to be about 1 μM (5,11,18).

It has been found that prostatic spermidine synthase is inhibited by its substrate, decarboxy-AdoMet (1,4). We have confirmed this observation with purified brain enzyme. This substrate inhibition, observed at concentrations as low as 10-20 μM, interferes with kinetic studies and complicates interpretation of the results. No substrate inhibition at concentrations up to 0.4 mM of decarboxy-AdoMet was observed with purified spermine synthase, whereas this enzyme was strongly inhibited by methylthioadenosine, one of the reaction products (13,14). It is not known whether methylthioadenosine has any regulatory functions in the synthesis of polyamines in vivo. Its concentration in rat tissues has been reported to be about 10 μM (32). Although methylthioadenosine is rapidly degraded in vivo (see 34), it may contribute to the preferential accumulation of spermidine in tissues undergoing rapid growth, such as regenerating liver. It is worth noting that at least in some human leukemic cell lines methylthioadenosine is mostly, if not solely, produced by the polyamine-biosynthetic pathway (10), although additional routes may operate in other tissues (see 34).

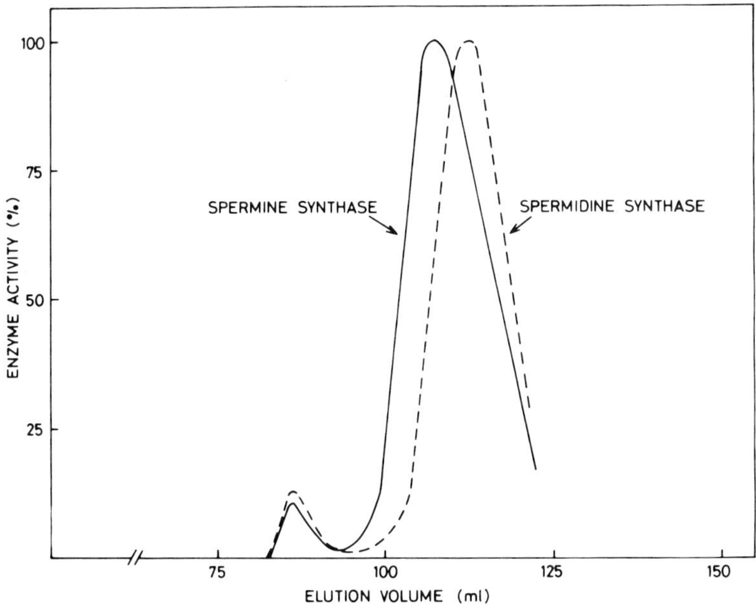

FIG. 1. Separation of propylamine transferase activities of the soluble fraction of rat brain homogenate by gel filtration. Brains were homogenized in one vol. of ice-cold 10 mM potassium phosphate buffer, pH 7.4, containing 140 mM KCl and 1 mM 2-mercaptoethanol. An aliquot of 0.8 ml of the 100 000 g supernatant fraction was applied to a Sephacryl S-200 Superfine (Pharmacia) column (1.6 x 91 cm). Elution was performed with the homogenization buffer at a rate of 6 ml/h. Enzyme activities in each fraction (1.4 ml) are expressed as percent of the peak value.

Substrate specificity of mammalian propylamine transferases has mainly been studied using prostatic and brain enzymes. Several diamines, notably cadaverine, have been shown to serve as propylamine acceptors in vitro in the spermidine synthase reaction (6,7,19,29,30), although the reaction rates were only a few percent of that obtained with putrescine. Using purified spermine synthase from bovine brain we could not demonstrate any propylamine acceptor activity with several diamines and spermidine homologs (14,27). However, it appears that also spermine synthase can catalyse the propylamine transfer to acceptors other than spermidine, as demonstrated by Pegg and coworkers with the prostatic enzyme (19).

Spermidine synthase from rat prostate was able to use several analogs of decarboxy-AdoMet, carrying an ethyl, propyl or butyl group in place of the methyl group, as the propylamine donor (29). Aminoethyl and aminobutyl analogs of decarboxy-AdoMet were inactive as substrates (29). Also 7-deaza analog of decarboxy-AdoMet has been shown to act as a substrate in the spermidine synthase reaction (1). S-Adenosyl(5')-1-methyl-3-(methylthio)propylamine has been shown to be a substrate for spermidine synthases from bovine brain and rat ventral prostate but not a substrate for the spermine synthases from the same sources (16).

The half-lives of spermidine and spermine synthases are at least several hours, perhaps days. In fact, no changes in the activities of these enzymes have been observed within 4-12 hours after treatment with cycloheximide (12,25).

It has been suggested that even if the four enzymes involved in the biosynthesis of polyamines in mammalian tissues are readily separable from one other, they might engage in physical and functional associations in the cell cytoplasm (34). We have tried to explore this possibility by preparing a soluble extract from rat brain at physiological ionic strength and by applying it to gel filtration. Excepting the small peak of aggregated enzyme close to the void volume, spermidine synthase and spermine synthase emerged from the column as separate peaks, closely corresponding to their molecular weights (Fig. 1.). The activity of S-adenosylmethionine decarboxylase in the effluent was too low to be accurately measured. The result does not support the existence of a physical complex between the two propylamine transferases. Further work is, however, needed to resolve this problem unambiguously.

INHIBITORS OF PROPYLAMINE TRANSFERASES

The observation that methylthioadenosine strongly inhibited purified spermine synthase (13,14) readily suggested that nucleosidase-resistant analogs of methylthioadenosine, such as alkylthiotubercidins, should be tested as inhibitors of polyamine synthesis. In fact, a number of nucleoside analogs, particularly methylthiotubercidin and derivatives of S-adenosylhomocysteine have been shown to inhibit spermidine and spermine synthases in vitro (4,17,28). Methylthiotubercidin also reduced the concentration of spermidine in cultured cells and produced growth inhibition at fairly low concentrations (17,28). However, it appears that methylthiotubercidin and other nucleoside derivatives have effects on cellular metabolism unrelated to biosynthesis of polyamines (28) which limit their value as inhibitors.

Dicyclohexylamine (8) and some analogs of decarboxy-AdoMet (29) have been shown to inhibit spermidine synthase in vitro. Recently Pegg and coworkers (18,33) have reported that prostatic spermidine synthase is powerfully inhibited by so-called transition-state analogs, by S-adenosyl-3-thio-1,8-diaminooctane in particular. Evidently these compounds are valuable for studies on the mechanism of the propylamine transfer reaction. It would be interesting to know whether they are able to specifically inhibit polyamine synthesis in vivo.

ENZYMIC MECHANISM OF THE PROPYLAMINE TRANSFERASE REACTION

The high affinity of propylamine transferases for one of their substrates, decarboxy-AdoMet, has made it difficult to perform adequate kinetic analysis of the propylamine transferase reaction. Availability of radioactive decarboxy-AdoMet of high specific activity and development of feasible assay methods (see above) have considerably facilitated such studies.

We have recently performed initial velocity as well as product inhibition studies with purified spermidine and spermine synthases from bovine brain. The details of these studies will be published elsewhere. Some of the kinetic data are presented in Table 3. Our results with both spermidine synthase and spermine synthase speak for a sequential reaction mechanism rather than a Ping Pong mechanism suggested to operate in the propylamine transferase reaction catalysed by the E.coli enzyme (35,36). Our data are in agreement with the mechanism proposed by Pegg and coworkers (18,33,34), involving a direct nucleophilic attack by putrescine (or spermidine) on the aminopropyl group of decarboxy-AdoMet, without any formation of covalent adducts between the aminopropyl group and the active site of the enzyme. This mechanism is supported by studies with transition-state analogs (33). On the basis of product inhibition studies it is not possible to decide whether the reaction mechanism is ordered or random. Further work, for example binding studies are definitely needed for understanding the mechanism(s) of the propylamine transferase reaction.

REFERENCES

1. Coward, J.K., Motola, N.C., and Moyer, J.D. (1977): J. Med. Chem., 20: 500-505.

2. Hannonen, P., Jänne, J., and Raina, A. (1972): Biochem. Biophys. Res. Commun., 46: 341-348.

3. Hannonen, P., Jänne, J., and Raina, A. (1972): Biochim. Biophys. Acta, 289: 225-231.

4. Hibasami, H., Borchardt, R.T., Chen, S.Y., Coward, J.K., and Pegg, A.E. (1980): Biochem. J., 187: 419-428.

5. Hibasami, H., Hoffman, J.L., and Pegg, A.E. (1980): J. Biol. Chem., 255: 6675-6678.

6. Hibasami, H., and Pegg, A.E. (1978): Biochem. J., 169: 709-712.

7. Hibasami, H., and Pegg, A.E. (1978): Biochem. Biophys. Res. Commun., 81: 1398-1405.

8. Hibasami, H., Tanaka, M., Nagai, J., and Ikeda, T. (1980): FEBS Lett., 116: 99-101.

9. Jänne, J., Schenone, A., and Williams-Ashman, H.G. (1971): Biochem. Biophys. Res. Commun., 42: 758-764.

10. Kamatani, N., and Carson, D.A. (1980): Cancer Res., 40: 4178-4182.

11. Mamont, P.S., Danzin, C., Wagner, J., Siat, M., Joder-Ohlenbusch, A-M., and Claverie, N. (1982): Eur. J. Biochem., 123: 499-504.

12. Oka, T., Perry, J.W., and Kano, K. (1977): Biochem. Biophys. Res. Commun., 79: 979-986.

13. Pajula, R-L., and Raina, A. (1979): FEBS Lett., 99: 343-345.

14. Pajula, R-L., Raina, A., and Eloranta, T. (1979): Eur. J. Biochem., 101: 619-626.

15. Pajula, R-L., Raina, A., and Kekoni, J. (1978): FEBS Lett., 90: 153-156.

16. Pankaskie, M.C., Abdel-Monem, M.M., Raina, A., Wang. T., and Foker, J.E. (1981): J. Med. Chem., 24: 549-553.

17. Pegg, A.E., Borchardt, R.T., and Coward, J.K. (1981): Biochem. J., 194: 79-89.

18. Pegg, A.E., and Coward, J.K. (1981): In: Advances in Polyamine Research, vol. 3, edited by C.M. Caldarera, V. Zappia, and U. Bachrach, pp. 153-161. Raven Press, New York.

19. Pegg, A.E., Shuttleworth, K., and Hibasami, H. (1981): Biochem. J., 197: 315-320.

20. Pegg, A.E., and Williams-Ashman, H.G. (1969): J. Biol. Chem., 244: 682-693.

21. Pegg, A.E., and Williams-Ashman, H.G. (1970): Arch. Biochem. Biophys., 137: 156-165.

22. Raina, A., Eloranta, T., and Pajula, R-L. In: Methods Enzymol., in press.

23. Raina, A., and Hannonen, P. (1970): Acta Chem. Scand., 24: 3061-3064.

24. Raina, A., and Hannonen, P. (1971): FEBS Lett., 16, 1-4.

25. Raina, A., and Jänne, J. (1975): Med. Biol., 53: 121-147.

26. Raina, A., Pajula, R-L., and Eloranta, T. (1976): FEBS Lett., 67: 252-255.

27. Raina, A., Pajula, R-L., and Eloranta, T. In: Methods Enzymol., in press.

28. Raina, A., Tuomi, K., and Pajula, R-L. (1982): Biochem. J., 204: 697-703.

29. Samejima, K., and Nakazawa, Y. (1980): Arch. Biochem. Biophys., 201: 241-246.

30. Samejima, K., Raina, A., Yamanoha, B., and Eloranta, T. In: Methods Enzymol., in press.

31. Samejima, K., and Yamanoha, B. (1982): Arch. Biochem. Biophys., in press.

32. Seidenfeld, J., Wilson, J., and Williams-Ashman, H.G. (1980): Biochem. Biophys. Res. Commun., 95: 1861-1868.

33. Tang, K-C., Pegg, A.E., and Coward, J.K. (1980): Biochem. Biophys. Res. Commun., 96: 1371-1377.

34. Williams-Ashman, H.G., and Pegg, A.E. (1981): In: Polyamines in Biology and Medicine, edited by D.R. Morris, and L.J. Marton, pp. 43-73. Marcel Dekker, New York.

35. Zappia, V., Cacciapuoti, G., Pontoni, G., Della Ragione, F., and Carteni-Farina, M. (1981): In: Advances in Polyamine Research, vol. 3, edited by C.M. Caldarera, V. Zappia, and U. Bachrach, pp. 39-53. Raven Press, New York.

36. Zappia, V., Cacciapuoti, G., Pontoni, G., and Oliva, A. (1980): J. Biol. Chem., 255: 7276-7280.

Advances in Polyamine Research, Vol. 4, edited by U. Bachrach, A. Kaye, and R. Chayen. Raven Press, New York

Effect of Magnesium on the Interaction of Polyamines with Nucleic Acids

A. K. Abraham and T. Flatmark

Department of Biochemistry, Årstadveien 19, 5000 Bergen, Norway

A large number of studies have demonstrated that polyamines can influence the rate and accuracy of protein synthesis in vitro (1,9,7,3). The positive effects of these polyvalent cations depend to a large extent on assay conditions (3). Further, the requirements for polyamines in the translational assay has varied significantly from one cell-free system to another and for the same cell-free system in different laboratories (6,3,7). An obvious explanation for these variations could be that during the preparation of the cell free system, polyamines associated with components of the cell-free system have been lost to varying extents. We have investigated this possibility in the studies reported here. Our data indicate that the cell-free systems that did not require addition of polyamines for maximal protein synthesis, become stimulatable by polyamines if these extracts are gel-filtrated in the presence of Mg^{2+}. These extracts also showed increased accuracy of protein synthesis when polyamines were added back.

The effect of Mg^{2+}, pH and sonication on the efficiency of perchloric acid extraction of polyamines from the nucleus and cytoplasm of rat liver was also studied. Under optimal conditions, Mg^{2+} slightly improved the yield of polyamines. The data indicate that the combined effect of Mg^{2+}, pH and sonication can produce significant variations upon the extractability of polyamines.

MATERIALS AND METHODS

Preparation of Rabbit Reticulocytic Lysate

Crude rabbit reticulocyte lysate, and nuclease

treated lysate were prepared as described earlier (12). Gel filtration of lysate was carried out on a Sephadex G-50 (fine) column (0.9 x 40 cm) equilibrated with 20 mM Tris-HCl buffer (pH 7.5) containing 100 mM KCl 1 mM DTT and Mg^{2+} as indicated.

Protein Synthesis in Reticulocyte Lysate

Protein synthesis in reticulocyte lysate was measured as described earlier (12). Polyphenylalanine synthesis was measured in the nuclease treated lysate using poly(U), and (^{14}C)-phenylalanine.

Fidelity of Translation

Fidelity of translation was monitored in the poly(U) system, using (^{3}H)-leucine (23 Ci/mmole) and (^{14}C)-phenylalanine (430 mCi/mmole). Misreading is expressed as the number of leucine residues incorporated as a percentage of the phenylalanine incorporated.

The Wheat Germ System

Preparation of wheat germ extract and assay conditions were as described earlier (2). Wheat germ ribosomes were isolated by gel filtration on a Sephacryl S-500 column. The concentration of ribosomes was calculated from absorption at 260 nm (1 A_{260}unit = 19 pmoles ribosomes).

Separation of Ribosomal Subunits

Wheat germ ribosomes were incubated in the buffer used for binding studies with 1 mM puromycin and 1 mM GTP at 30 °C for 5 min. The KCl concentration was increased to 0.5 M. Subunits were separated on a 10-30 % sucrose gradient at 40,000 rpm in a SW-40 rotor, for 5 hrs.

Extraction of polyamines from Rat Liver

Liver was homogenised in Hepes-Sucrose buffer and nuclear, mitochondrial and post-mitochondrial fractions were prepared as described earlier (15). Nuclear and mitochondrial pellets were resuspended in the same buffer. Mg^{2+} concentration and pH were adjusted as indicated. Sonication was performed in a Branson sonifier B-12 at 50 watts output. Perchloric acid (70 %) was added with thorough mixing to each of the subcellular fractions, to a final concentration of 10 %. After 30 min the samples were centrifuged at 8000 rpm for 10 min in a Sorvall HB-4 rotor. The supernatant was collected and the pH was adjusted to 4.2 with 0.72 N KOH containing 0.36 M $KHCO_3$. After 15 min at 2 °C, $KClO_4$ was removed by centrifugation and the clear supernatant was collected for high pressure liquid chromatographic separation of polyamines.

Assay of Polyamines by High-performance Liquid Chromatography

The high-performance liquid chromatograph with fluorescence detector and integrator (Hewlett-Packard Model 3390) has previously been described in detail (5,8,15). The chromatographic separation was achieved at 20 °C on a reversed-phase silica support (C18, 5 μm particles from Supelco, Inc., PENN, U.S.A., packed in a 50 mm x 4.6 mm I.D. stainless-steel tube) with a theoretical plate number of 72 140/m. A precolumn (same dimension as the analytical column) packed with a pellicular packing material (LC-18 from Supelco) was used.

The mobile phase, consisting of 0.157 M Na-acetate (pH 4.5) with 40 % (v/v) acetonitrile and 10 mM hexanesulfonate (Regis Chemical Co., U.S.A.) was pumped at a flow rate of 1 ml/min.

Column effluent and *o*-phthalaldehyde reagent were mixed in a 2:1 ratio using the URS 050 post column derivatization system of Kratos Analytical Instruments (New Jersey, U.S.A.). The *o*-phthalaldehyde reagent was prepared as described (14). Fluorescence intensity was continuously recorded using 230 nm and 455 nm as the excitation and emission wavelength, respectively.

20 μl of the perchloric acid extracts were diluted with 80 μl of the mobile phase and the total volume was injected into the liquid chromatograph. Spermidine and spermine were recovered with retention times of 1.9 min and 2.7-2.9 min, respectively.

RESULTS AND DISCUSSION

Effect of Spermidine on the Rate and Accuracy of Protein Synthesis by Reticulocyte Lysate

The stimulatory effect of polyamines on protein synthesis has been demonstrated clearly in the Wheat germ system (6,1,3) and the purified ascites system (4). In contrast, protein synthesis in rabbit reticulocyte lysate, the system most widely used for *in vitro* translation, is not stimulated by polyamines.

We have therefore studied the factors that produce polyamine requirement in reticulocyte lysate.

Table 1 shows the effect of gel filtration of lysates in the absence or presence of different concentrations of Mg^{2+}. Under identical conditions a concentration of spermidine that stimulated protein synthesis in a lysate gel-filtered in the presence of 5 mM Mg^{2+}, inhibited protein synthesis in crude extracts. A lower degree of stimulation was observed in a lysate gel-filtered in the presence of low Mg^{2+} (0.5 mM). The activity of lysate gel-filtered in the absence of Mg^{2+} could not be fully restored by the addition of polyamines and Mg^{2+}. The apparent increase of amino acid incorporation in gel-filtered lysates (Table 1) is due to the removal of cold amino acids from these lysates.

TABLE 1. Effect of Spermidine on Protein Synthesis in Gel-filtered Lysates

Treatment of Lysate	Mg^{2+} in Gel-filtration buffer (mM)	(^{3}H)-Leucine Incorporated ÷ spd (cpm)	+ spd (cpm)	stimulation (%)
None		68580	65370*	÷4.5
	0	41966	45218	+8
Gel-filtered	0.5	50024	61730	+2 3
	5.0	51045	78002	+5 3

*Concentration of spermidine 10 µM. Lysates containing 1 mg protein (50 µg) were incubated at 37 °C as described earlier (12) with 1 µM (3H)-leucine (2 Ci/mmole). The concentration of Mg^{2+} in all samples was 1.5 mM, and spermidine, when present, was 0.66 mM, except in lysate that was not gel-filtered. 10 µl samples were withdrawn in duplicate after 10 min and Trichloroacetic acid insoluble radioactivity was estimated.

The effect of spermidine on the fidelity of poly(U) translation is shown in Figure 1. Spermidine reduced misreading in gel-filtered lysates. The low level of misreading seen in the crude lysate is not reduced further upon the addition of spermidine. These results indicate that polyamines bound to the components of cell-free system contribute to the rate and accuracy of protein synthesis. Removal of bound polyamines during gel filtration is most

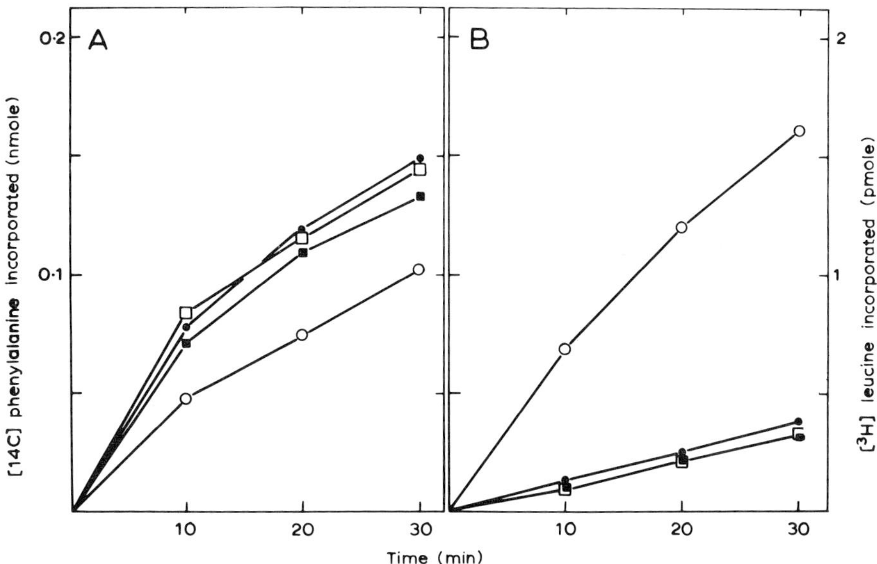

Fig. 1. Effect of Spermidine on the Fidelity of Poly(U) Translation. Polyphenylalanine synthesis was carried out in a nuclease treated rabbit reticulocyte lysate as described earlier (12). The values given are nmole phenylalanine and pmole leucine incorporated per 10 nmole of ribosomes. Background incorporation in the absence of poly(U) was substracted. Squares, crude nuclease treated lysate; circles, nuclease-treated lysate gel-filtered in the presence of 5 mM Mg^{2+}. Open circles and squares, no spermidine added. Filled circles and squares spermidine (0.66 mM)

effective in the presence of Mg^{2+}, and produces a polyamine requirement in the gel-filtered lysate.

Effect of Mg^{2+} on the Binding of Spermidine and Spermine to Ribosomes

Effect of Mg^{2+} on the binding of (^{14}C)-spermidine and spermine to ribosomes isolated from gel-filtered lysate is demonstrated in Figure 2. A fraction of (^{14}C)-spermidine added to the system is bound to ribosomes and co-sediments in the sucrose gradient. The ribosome bound fraction of spermidine was signifi-

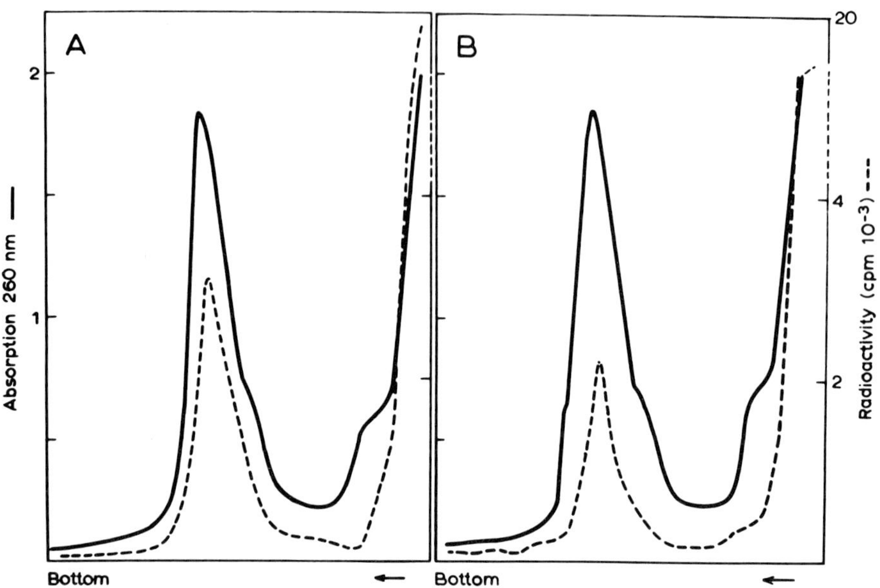

Fig. 2. Effect of Mg^{2+} on the Binding of Spermidine to Ribosomes. Ribosomes (10 A_{260}units) were incubated in the gradient buffer (100 mM KCl, 20 mM Tris-HCl pH 7.5, 1 mM DTT) with ^{14}C) spermidine (100 μM) for 5 min at 20 °C, in a final volume of 200 μl. Concentrations of Mg^{2+} were 0.5 mM (A) and 5 mM (B). Samples were layered on top of a 10-30 % sucrose gradient and centrifuged for 2 hrs at 40,000 rpm in a Spinco SW 40 rotor. The gradient was harvested from the bottom. Samples (0.5 ml) were collected into scintillation vials for radioactivity measurement.

cantly higher at low Mg^{2+} (0.5 mM) concentration. Since polyamines alone as the only polyvalent cation are not able to support protein synthesis in cell-free systems (3) the binding experiments were not performed in the absence of Mg^{2+}.

Binding of polyamines to dissociated ribosomes is shown in Fig. 3. Ribosomal subunits bound less spermine and spermidine than intact ribosomes. The stoichiometry of polyamine binding under different conditions was calculated from A_{260} of peak fractions

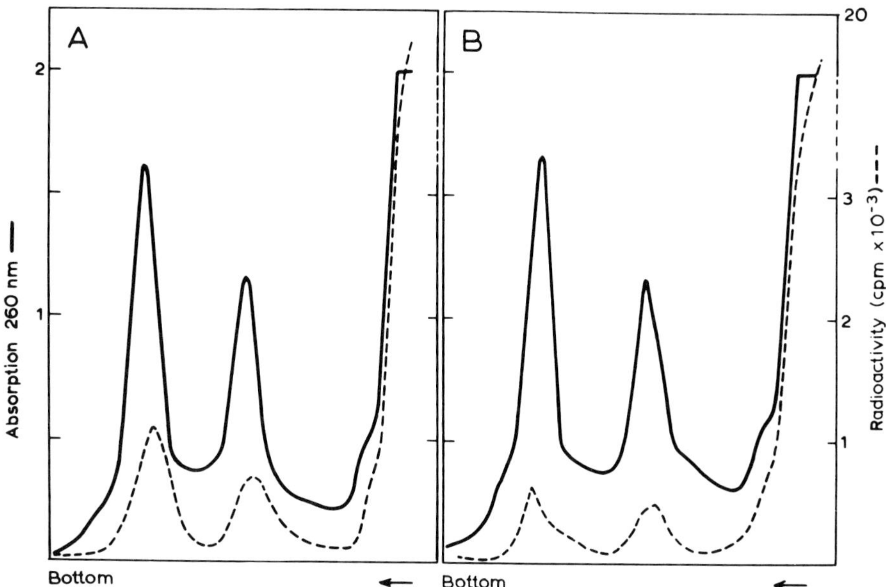

Fig. 3. Spermidine Binding to Dissociated Ribosomes
Ribosomes (10 A_{260} Units) were incubated with 1 mM puromycin, 1 mM GTP and 100 μM (^{14}C)-spermidine in the gradient buffer (500 mM KCl, 20 mM Tris-HCl pH 7.5, 1 mM DTT) in a final volume of 200 μl. Concentrations of Mg^{2+} were 0.5 mM (A) and 5 mM (B) respectively. Samples were layered on top of a 10-30 % sucrose gradient and centrifuged for 5 hrs at 40,000 rpm in a Beckman SW 40.1 rotor at 40,000 rpm. The gradients were harvested from the bottom. Samples (0.5 ml) were collected into scintillation vials for radioactivity measurement.

collected from the gradient and their radioactivity. Assuming that all bound polyamines are removed from ribosomes during gel-filtration in the presence of Mg^{2+}, roughly 9 moles of spermidine or 3 moles of spermine are bound per mole of ribosomes. (Table 2). These binding sites are not highly specific, since very little radioactive spermidine was bound to ribosomes in the presence of non-radioactive spermine and vice versa (Table 2).

TABLE 2. Quantitation of Spermidine and Spermine Bound to Ribosomes

Conditions of Incubation		Radioactive Polyamine Bound	
Mg^{2+} (mM)	Non-radioactive Polyamine added (mM)	Spermidine (mole/ribosome)	Spermine
0.5	none	8.7	3.2
0.5	spermine (0.6)	0.9	-
0.5	spermidine (0.6)	-	0.4
5.0	none	6.1	2.5
5.0	spermine (0.6)	0.7	-
5.0	spermidine	-	0.3

Ribosomes were incubated in the gradient buffer under conditions described in Figure 2 with either (^{14}C)-spermidine (60 μM) or (^{14}C)-spermine (60 μM). After sentrifugation peak ribosome fraction was collected. Ribosome concentration was calculated from A_{260}. Radioactivity was determined in a scintillation counter.

Factors Influencing the Extraction of Polyamines with Perchloric Acid

In contrast to the significant effect of Mg^{2+} on the removal of polyamines during gel-filtration at physiological pH, only a marginal effect of Mg^{2+} was observed on the recovery of polyamines during perchloric acid extraction (Figure 4). The pH of the extraction medium and sonication have also a slight effect on the extractability of polyamines. In our experiments extractability of polyamines in the Hepes-sucrose medium (pH 7.4), is superior to that in Tris-HCl pH 8.5. The combined effect of these 3 parameters can produce up to 50 % variation in the extractability of polyamines from the nucleus. A similar but less significant effect was also observed with the cytosol fraction (data not shown). The concentration of polyamines in the mitochondrial fraction was too small to enable any definite conclusion. The content of spermidine and putrescine calculated per gm wet weight of liver is of the same order as reported earlier (11). A major portion of spermine was found in the nuclear fraction, while spermidine was distributed more or less equally

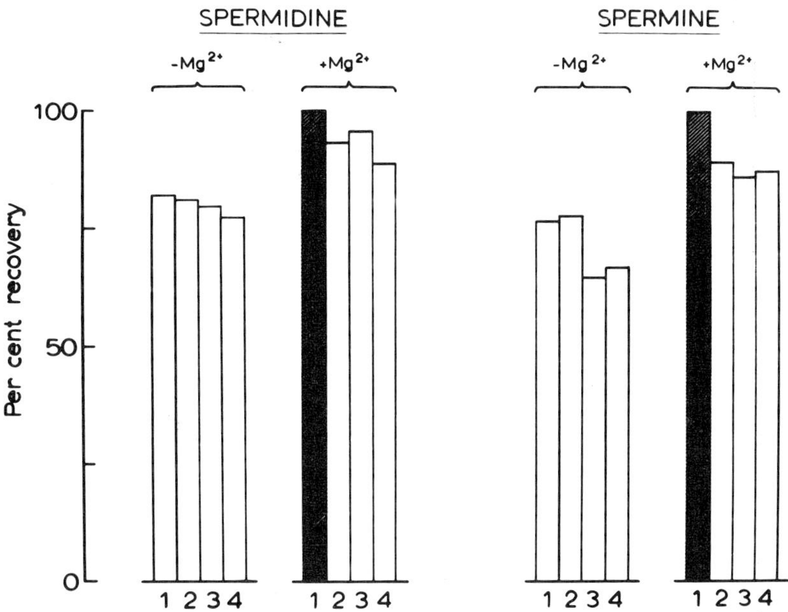

Fig. 4. Factors Influencing the Extraction of Polyamines from the Nuclear Fraction of Rat Liver Homogenates. For experimental details, see Methods section. The maximum recoveries of both spermidine (262 pmol/µg DNA) and spermine (273 pmol/µg DNA) were obtained by extraction with Hepes-sucrose medium pH 7.4 containing 10 mM Mg^{2+}, and these values were used as reference values (100 %-values). (1) Hepes-sucrose medium, pH 7.4; (2) Tris-sucrose medium, pH 8.5; (3) Hepes-sucrose medium, pH 7.4 and sonication; (4) Tris-sucrose medium, pH 8.5 and sonication.

between nuclear and cytoplasmic fractions. Almost all putrescine was present in the cytoplasmic fraction (data not shown). These results are not in agreement with the subcellular distribution of polyamines in BSC-1 cells (10).

CONCLUSION

The results of the present study indicate that polyamines have essentially the same effect on protein synthesis in a rabbit reticulocyte lysate as in the wheat germ cell-free system. The apparent lack of stimulation of crude reticulocyte lysate by added polyamines is due to the presence of ribosome-bound polyamines in such lysates. Gel filtration of crude lysates in the presence of Mg^{2+} can lead to removal of most of the bound polyamines from ribosomes. Polyamines influence both the rate and accuracy of protein synthesis in gel-filtered reticulocyte lysates.

Spermidine binds to ribosomes and ribosomal subunits more effectively than spermine. These binding sites are, however, not specific. Some of the polyamines bound to ribosomes are removed upon dissociation of ribosomes. The affinity of ribosomes for polyamines is reduced in the presence of Mg^{2+}. In contrast to the effect of Mg^{2+} on the interaction of polyamines with nucleic acid-containing structures at physiological pH, Mg^{2+} has only a marginal effect at the low pH used for the extraction of polyamines from tissues. However, the combined effect of Mg^{2+}, pH and sonication on the extractability of polyamines is significant and should be taken into consideration when performing polyamine extraction.

ACKNOWLEDGEMENT

This work was supported by the Norwegian Cancer Society. The skilful technical assistance of Mrs. Teresa Cicplinska and Sissel Riise is greatfully acknowledged.

REFERENCES

1. Abraham, A.K., Olsnes, S. and Pihl, A. (1979): FEBS Lett., 101: 93-96

2. Abraham, A.K., and Pihl, A. (1980): Eur. J. Biochem. 106: 257-262

3. Abraham, A.K., (1981): Med.Biol. 59: 368-373.

4. Atkins, J.F., Lewis, J.B., Anderson, C.W., and Gesteland, R.F. (1975): J.Biol.Chem. 250: 5688-5695

5. Flatmark, T., Skotland, T., Ljones, T. and Ingebretsen, O.C. (1978): J.Chromat, 146: 433-438

6. Hunter, A.R., Farrell, P.J., Jackson, R.J. and Hunt, T. (1977): Eur.J.Biochem., 75: 149 157

7. Igarashi, K., Kashiwagi, K. Aoki, R., Kojima, M. and Hirose, S. Biochem.Biophys.Res.Commun.(1981): 91: 440-448

8. Ingebretsen, O.C., and Flatmark, T. (1979): J.Biol.Chem., 254: 3833-3839.

9. Jelenc, P.C. and Kurland, C.G. (1979): Proc. Natl.Acad.Sci., 76: 3174-3178

10. Mach, M., Ebert, P., Popp, R. and Ogilvie, A. (1982): Biochem.Biophys.Res.Commun. 104: 1327-1334

11. Pegg, A.E., Matsui, I., Seely, J.E., Pitchard, M.L. and Pösö, H. (1981): Med.Biol. 59: 32-333

12. Pelham, H.R.B., and Jackson, R.J. (1976): Eur.J.Biochem. 67: 247-256

13. Romslo, I. and Flatmark, T. (1973): Biochim. Biophys.Acta 305: 29-35

14. Seiler, N., and Knödgen, B. (198): J.Chromat. 221: 227-235

15. Terland, O., Flatmark, T., and Kryvi, H. (1979): Biochim.Biophys.Acta 553: 460-468

Advances in Polyamine Research, Vol. 4, edited by U. Bachrach, A. Kaye, and R. Chayen. Raven Press, New York © 1983.

Effect of 1,25-Dihydroxycholecalciferol on Intestinal Spermine-Binding Proteins in Chicks

G. Mezzetti, M. S. Moruzzi, M. G. Monti, G. Moruzzi, and B. Barbiroli

Istituto di Chimica Biologica, Università di Modena, 41100 Modena, Italy

One unanswered and critical question of actuality is whether polyamines are "free" within the cells, or rather if the regulation of their intracellular concentrations and localization may be mediated by the binding with specific structures and/or macromolecules. In fact polyamine-conjugated proteins have been detected in several tissues and fluids, i.e. chick duodenal mucosa (1), rat ventral prostate (2), human lymphocytes (3), human amniotic fluid (4), human and rabbit serum (5, 6).

During the last few years we focused our attention on the study of specific polyamine-proteins interaction in the chick intestine (1, 7, 8). This tissue, quite rich in polyamines, is a primary target organ of 1,25-dihydroxycholecalciferol, the hormonal form of vitamin D_3 (9).

This chapter will describe studies on the relationship between the duodenal spermine-binding protein and the vitamin D status of the animal.

MATERIALS AND METHODS

$[^3H]$ Spermine tetrahydrochloride (44.3 Ci/mmole) was obtained from New England Nuclear. Electrophoretically pure nucleases purchased from Worthington were always desalted before use. Cycloheximide, Actinomycin D and cold spermine were obtained from Sigma. 1,25-Dihydroxycholecalciferol (1,25-$(OH)_2D_3$) was a generous gift of Prodotti Roche, Milano (Italy). Animals used in all experiments were White Leghorn cockerels that were raised for 3-4 weeks on a vitamin D-deficient diet (10). Experiments were carried out in duplicate and/or were repeated at least four times to assure reliability. Vitamin D dosed chicks received, unless stated otherwise, 0.5 nmole of 1,25-$(OH)_2D_3$ intracardially in 0.1 ml of ethanol/propylene glycol (1:9 v/v) 4 hrs before being killed. The control animals received injection of the same amount of vehicle. All chicks were starved for 16 hrs before death.

All manipulations were carried out at 0-2°C. Duodenum was removed, rinsed in 40 mM Tris-HCl, pH 7.5, and the mucosal layer separated from the underlying muscle layers using a glass slide. The mucosa was homogenized in 2 vol of the same buffer with a Potter-Elvehjem homogenizer fitted with a Teflon pestle. The homogenate was centrifuged at 105,000 x g for 1 h. The upper 3/4 of the supernatant were taken and used as the cytosol fraction.

The cytosol fraction (200-400 mg of protein) was then chromatographed on a DEAE-cellulose (Whatman DE-52) column (1.5 x 13 cm) that was equilibrated with the same buffer. The column was eluted stepwise with buffer containing 0.1, 0.2, and 0.3 M KCl. The protein fractions were dialyzed extensively against 40 mM glycine buffer, pH 8.7, and assayed for their spermine-binding activity.

The spermine-binding activity of the freshly fractionated duodenal protein was analyzed by gel filtration on Sephadex G-25 as described previously (1). Since RNAse treatment of the protein sample does not affect spermine-binding activity (1) in all the experiments the sample was digested with a protease-free solution of RNAse (pancreatic A + T_1, 20:1) to minimize the error of measurements due to RNA molecules, eventually contaminating the preparation. The protein sample (150 µg) was incubated with 0.2 µCi of 15 µM $[^3H]$ spermine in 0.35 ml of 40 mM glycine buffer, pH 8.7, at 0°C for 10 min. After incubation the reaction mixture was passed through a Sephadex G-25 column (0.7 x 16 cm) to measure the amount of radioactive spermine bound to the macromolecules in the void volume of the effluent. 0.5 ml fractions were collected and the binding activity was estimated from the area of the peak of radioactivity (1).

Unless stated otherwise we studied the 1,25-$(OH)_2D_3$ effect by comparing the spermine-binding activities of the protein fraction which are retained by a DEAE-cellulose column and which could be eluted from the column by 0.3 M KCl but not by 0.2 M KCl.

The preparation of duodenal nuclei was performed as originally described by Yu (11).

For the preparation of nucleoplasm the purified nuclei were lysed for one hour in the cold with a sufficent volume of a medium containing 1 mM Tris-HCl (pH 8) 0.1 mM EDTA, 0.5 mM dithiothreitol, 12.5% glycerol. The nucleoplasm was separated from chromatin by centrifugating at 15,000 x g for 15 min (12).

Protein concentration was measured by the method of Warburg and Christian (13).

RESULTS AND DISCUSSION

When the duodenal mucosa cytosol is incubated with micromolar concentration of [^{3}H]-spermine and the mixture is analyzed by gel-filtration on Sephadex G-25 column a distinct radioactive peak is eluted in the void volume of the effluent (1). Trypsin and other proteolytic enzymes but not RNAse or DNAse could abolish the formation of the macromolecular radioactive complex.

The spermine-binding protein can be purified by DEAE-cellulose column chromatography (Fig. 1). More than 90% of cytosol pro-

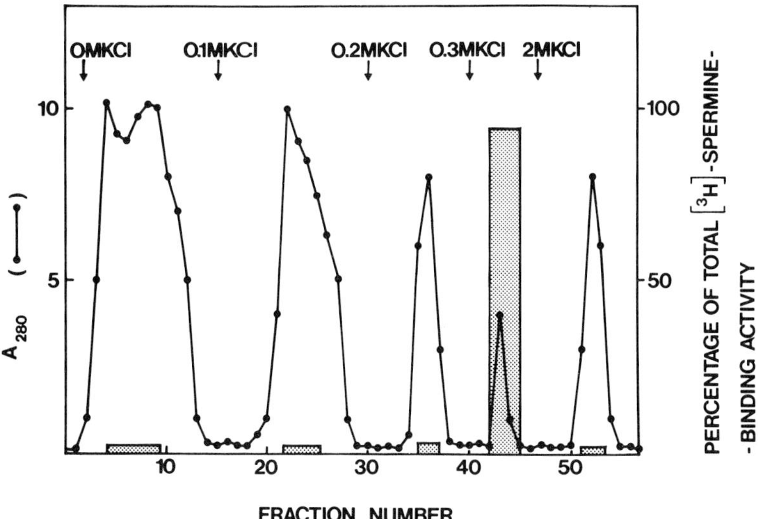

FIG. 1. DEAE-cellulose column chromatography of cytosolic spermine-binding protein. 400 mg of cytosol protein were applied to a DEAE-cellulose column (1.5 x 13 cm) that was equilibrated with 40 mM Tris-HCl, pH 7.5. Details of elution are given under the methods section. Aliquots from each dialyzed fraction (0.14 O.D_{280}) were assayed for spermine-binding activity after extensive RNAse treatment (see methods section). Vertical bars represent the distribution of the total cytoplasmic binding activity into the different elution peaks.

Reprinted by permission of Biochem. Biophys. Res. Commun. (7).

tein is eluted from the column by 0.2 M KCl, but these proteins do not have significant spermine-binding activity. Almost all the spermine binding-protein is eluted from the column between 0.2 and 0.3 M KCl whereas the major contaminating RNA is eluted at higher KCl concentrations. Further purification can be achieved by ammonium sulphate precipitation, spermine-Sepharose phosphocellulose, and preparative polyacrilamide gel-electrophoresis. The purified protein behaves like an acidic protein and appears to have a molecular weight about 32,000 (8). The protein-spermine interaction has been studied also by mean of competitive radiospermine-binding assay using chelex 100 (14) and carboxymethylcellulose (see Mezzetti et al., elsewhere, this volume) as exchanger resins.

From these studies it appears that only spermine is bound with high affinity and with a stoichiometry of 1:1, whereas other polyamines and calcium ions are not efficiently bound. We have estimated the dissociation constant for spermine to be approximatively 1 μM which is much lower than the spermine concentration in duodenal mucosa (millimolar)(8).

Sensitivity to 1,25-dihydroxycholecalciferol

1,25-Dihydroxycholecalciferol is now considered to be a hormone that mediates calcium transport in target tissues and acts in protein synthesis like other steroid hormones (15). This is thought to involve extranuclear receptor(s) for 1,25-$(OH)_2D_3$, subsequent association of the receptor complex with the chromatin fraction of the nucleus, and the ultimate increase in protein synthesis as a response to the hormonal message.

Several cellular effects induced by this hormone have been described in chick duodenal mucosa, including the induction of the synthesis of various proteins thought to be correlated with calcium transport (9).

Recently we have reported that the spermine-binding activity from chick duodenal mucosa changes in relation to the circulating level of 1,25-dihydroxycholecalciferol (7).

Fig. 2 reports typical patterns of spermine-binding activity obtained by gel filtration on Sephadex-G 25 of partially purified cytosol proteins. Spermine-binding activity has been studied in three weeks old chicks maintained since hatching on a vitamin-D deficient diet or on a standard laboratory diet. At the end of the feeding period spermine-binding activity has been found to be dramatically decreased in rachitic chicks as compared with the values found in the control chicks. In the experiment reported rachitic chicks (dashed line) show a 21% residual spermine-binding activity of their

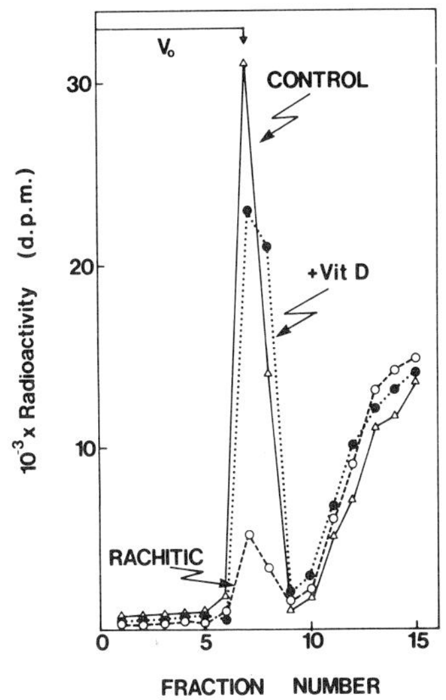

FIG. 2. Stimulation of spermine-binding activity from purified duodenal protein by 1,25-$(OH)_2D_3$ in vivo. The spermine-binding activity of the DEAE-cellulose fractionated cytosol proteins from normal animals (△—△), rachitic chicks (o--o), and rachitic chicks injected with 1,25-$(OH)_2D_3$ (●···●) were analyzed by gel filtration method as described under the methods section.

corresponding controls (continuous line). Fig. 2 reports also the effect of 1,25-$(OH)_2D_3$ administration to the vitamin D-deficient chicks on duodenal spermine-binding activity (dotted line). In the experiment reported 4 hrs after a single intracardial dose of 0.5 n-mole of the active form of vitamin D_3 the spermine-binding activity was increased by 365% above the values obtained from the vehicle-dosed rachitic chicks and reached 97% of the control animals.

Spermine-binding activity could be detected also in many different organs from adult chick. Nevertheless injection of 1,25--$(OH)_2D_3$ resulted in the stimulation of spermine-binding activity only in vitamin D sensitive tissues (unpublished results).

A marked dose dependence in the stimulation of spermine-binding activity was observed after a single intracardial dose of the active form of vitamin D_3 (Fig. 3). A maximal stimulation of binding activity was obtained with a dosage between 0.5 and 1 nmole of 1,25-$(OH)_2D_3$, and its enhanced activity was almost four times higher than the control.

In Table I is reported that the increase of activity of the spermine-binding protein is sensitive to actinomycin D or cycloheximide administration. In fact the large enhancement of binding activity found 3 hrs after 1,25-$(OH)_2D_3$ injection to rachitic chicks was almost completely prevented when the animals received a prior injection of 50 μg of either actinomycin D or cycloheximide.

From the above experiments it appears that 1,25-$(OH)_2D_3$ is able to regulate the activity of the spermine-binding protein pos-

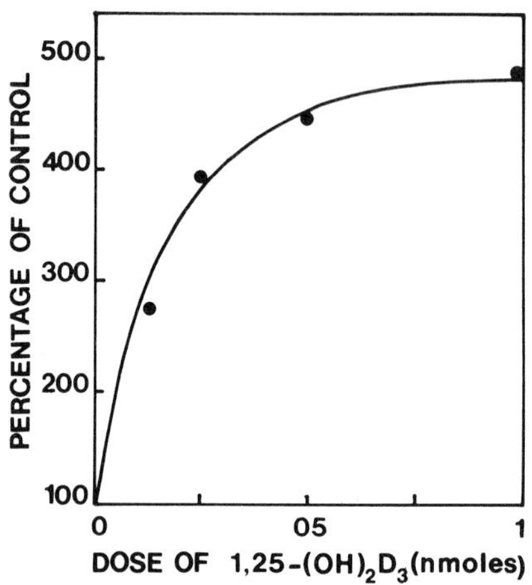

FIG. 3. Dose dependency in the stimulation of cytoplasmic spermine-binding protein. 1,25$(OH)_2D_3$ was administered to groups of rachitic chicks and the animals were killed 4 hrs later. The DEAE cellulose chromatographed binding proteins were prepared from each group and aliquots (140 μg) assayed for their spermine-binding activities as described under the methods section. Each point represents the mean of at least four determinations.

TABLE I. Prevention of stimulating effect of 1,25-$(OH)_2D_3$ on the spermine-binding activity from duodenal mucosa of rachitic chick by actinomycin D or cycloheximide injection[a]

Treatment to rachitic chicks	Spermine-binding activity	
	dpm/150 μg of protein	% of control value
None	9,300	100
1,25-$(OH)_2D_3$	40,000	433
Actinomycin D	8,922	96
Actinomycin D + 1,25-$(OH)_2D_3$	9,669	104
Cycloheximide	8,754	94
Cycloheximide + 1,25-$(OH)_2D_3$	8,782	94

[a] 1,25-$(OH)_2D_3$ (0.5 nmole) was injected into three weeks old rachitic chicks 3 hrs before sacrifice. Inhibitors: actinomycin D (50 μg/chick) and cycloheximide (50 μg/chick) were injected i.p. 1 hr before steroid administration. The conditions of spermine-binding assay were as described under the Methods section.

sibly by inducing the synthesis of new protein molecules. Although the cycloheximide sensitivity suggests that 1,25-$(OH)_2D_3$ may induce the spermine-binding protein it is also possible that the active form of vitamin D induces another protein or enzyme which modifies the binding protein and enhances its spermine-binding activity.

Fig. 4 shows the time course of change in duodenal spermine-

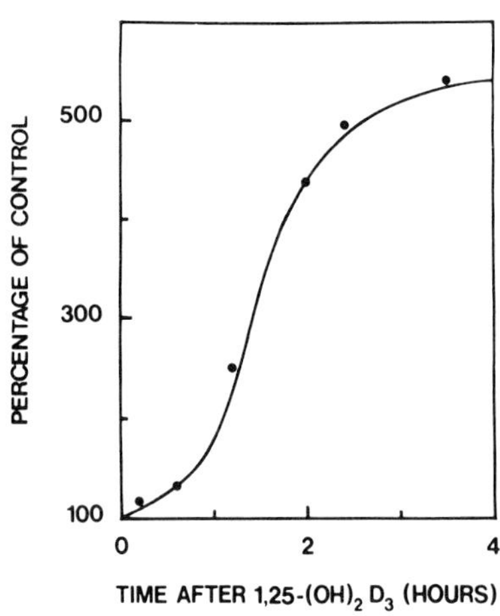

FIG. 4. Rapid effect of $1,25(OH)_2 \cdot D_3$ on the spermine-binding activity of the cytosolic protein fraction. Groups of rachitic chicks were dosed with 0.34 nmole of $1,25(OH)_2 \cdot D_3$ or vehicle and killed at the time indicated. DEAE-cellulose chromatographed binding proteins were prepared from each group and assayed for their spermine-binding activity as described under the methods section. Each point represents the mean of at least four determinations.
Reprinted by permission of Biochem. Biophys. Res. Commun. (7).

binding activity in response to a single intracardial dose of 0.34 nmole of the active form of vitamin D_3. A significant and reproducible stimulation of the binding activity was observed as early as 50 min after $1,25-(OH)_2D_3$ dosing. This phase is followed by a period up to 2.5 hrs of rapid and remarkable increase in activity. Thereafter the activity increased slower and eventually reached a plateau value. The rapidity of stimulation of spermine-binding protein is striking so that this event is among the earliest detected in the duodenal cell after exposure to the hormone.

On the Functional Significance of Spermine-Binding Protein

It is not yet clear whether the spermine-binding protein plays a specific biological role in the cell. In fact it is possible that by means of the binding to spermine we are simply assaying a vitamin D sensitive protein(s) and that the binding with spermine does not represent any biological important function for the cell. Yet, the high affinity of the spermine-protein complex (in the micromolar range), the high polyamine concentration in the duodenal cell (millimolar) and the rapidity of induction of the binding activity by the hormone give strong evidence against an accidental binding within the cell.

The spermine-binding protein may be an enzyme or a regulatory part of it; in this case, its activity may be regulated by spermine. Alternatively the spermine-binding protein may be involved in concert with other polyamine-binding proteins or conjugators in the regulation of intracellular and extracellular levels of polyamines.

In fact it is very conceivable that both intracellular and extracellular levels of these polications may be precisely regulated in response to the physiological status of the cell. Woo et al. have shown that in cell culture polyamines are released by and taken up from the cells in relation to the phase of the cell cycle (16).

The spermine-binding protein may play a major role in the intracellular communication of hormone message since the protein and/or spermine may be translocated to specific cellular structure such as ribosomes or chromatin to fulfill jointly or separately their functional roles in mediating the response to 1,25-$(OH)_2D_3$.

Recently Mach et al. using a rapid method for the separation of cytoplasm and nuclei of cultured BSC-1 cells measured the concentration of polyamines in these cellular comparts (17). The technique used was able to prevent leakage of polyamines from the organelles and redistribution during the preparation. Strikingly they found a different polyamine distribution in the cell compartments. In particular, putrescine and spermine were found almost in equal amounts in cytoplasm, whereas spermine is preferentially accumulated in the nucleus, where it was found 5 fold more concentrated than other polyamines tested. This evidence prompted us to search a spermine-binding activity in the nuclei from duodenal mucosa. We prepared nuclei from this tissue using a method preventing nuclear loss and that could provide good yield and reproducibility (11). Nucleoplasm was then gently separated from chromatin (12) and treated extensively with DNAse I and RNAse mixture.

Fig. 5 shows that duodenal nucleoplasm contains a factor(s)

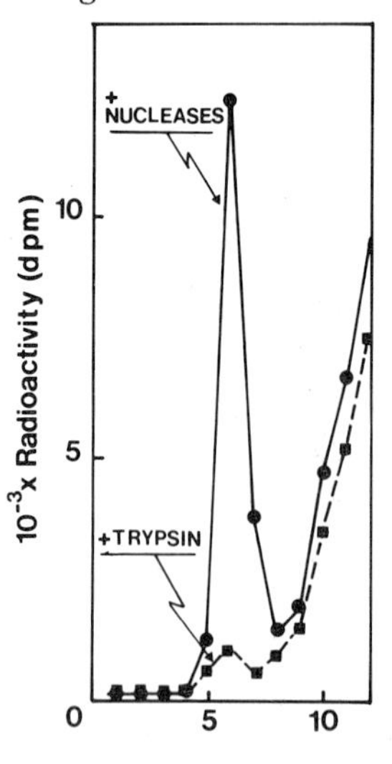

FIG. 5. Radiospermine-binding by a nuclease-resistant and trypsin sensitive factor from chick duodenal nucleoplasm. [^{3}H]spermine was incubated with 0.15 mg of dialyzed nucleoplasm protein pretreated with 20 μg of DNAse I and 50 μg of RNAse mixture undigested with trypsin (●—●) or digested with 100 μg of trypsin for 4 hrs at 25°C (■--■). The mixtures were analyzed by gel filtration as described under the methods section.

binding spermine at micromolar concentration that is resistant to extensive treatment with nuclease and very sensitive to the proteolytic enzyme trypsin.

To test the relative polyamine specificity of the nucleoplasmic binding factor, radioactive amines were compared for their binding to the protein. Similarly to the cytoplasmic spermine-binding protein the nucleoplasmic factor reacts selectively with spermine whereas other polyamines or diamines are not effectively bound. The affinity constant of the cytoplasmic and nuclear spermine complex are comparable (unpublished results).

In order to assess a preliminary purification criterion we have performed a DEAE cellulose column chromatography of the soluble nuclear extract. The elution of the column was performed step-wise essentially as described for cytosolic preparation. The spectroscopic data suggested that the materials eluted with the wash, or 0.1, 0.2 and 0.3 M KCl named respectively K_o, K_1, K_2 and K_3 were mainly proteins whereas the material eluted with 2 M KCl named K_{20} was essentially nucleic acids. The RNAse-resistant spermine-binding activities shown by the DEAE-fractionated nucleoplasmic extracts of rachitic and 1,25-$(OH)_2D_3$ dosed chicks are presented in fig. 6. All fractions from DEAE chromatography of nucleoplasm extracts from rachitic chicks showed a very low spermine-binding activity except for the fraction named K_2 which showed a more significant binding activity.

1,25-$(OH)_2D_3$ administration to rachitic chicks affected only the spermine-binding activity associated with fraction K_3, which was enhanced by to a quite high level. On the contrary the binding activity associated to K_2 and to the other protein fractions did not respond to the active form of vitamin D. From these results it appears that cytosolic and nucleoplasmic spermine-binding activities show different DEAE elution patterns, but either one responds to the injection of the hormone by remarkably increasing the binding activity associated with fraction K_3.

The significance of these preliminary findings of nuclear spermine-binding activity is not clear and more data are needed on the specificity of this radiospermine-protein complex. Nevertheless the possibility of a hypothetical passage of the polyamine-bound protein from cytosol into the nucleus is extremely appealing. To solve such a question precise kinetic data and immunological probes that can retrieve spermine-binding proteins in a mixture of proteins are needed.

In the present article we have tried to integrate the experimental evidence of 1,25-$(OH)_2D_3$ induction of spermine-binding activity both in the cytosol and in the nucleus into some future working perspectives. Since the hypothesis put forward are so far amenable to direct experimental evaluation, we hope that they will stimulate further work toward the elucidation of the cellular mechanism in which polyamine-binding protein is implicated.

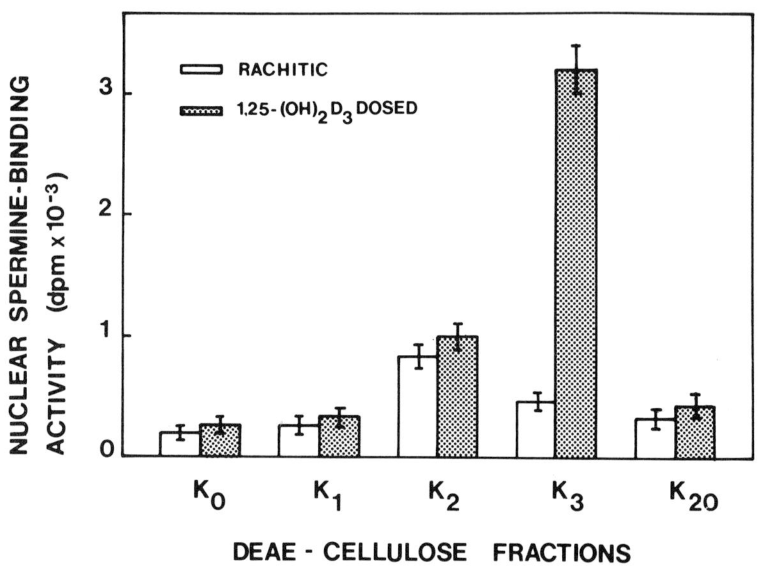

FIG. 6. Effect of 1,25-$(OH)_2D_3$ administration on nucleoplasmic spermine-binding activity after DEAE-cellulose chromatography. 200 mg of protein of crude duodenal nucleoplasm from rachitic chicks injected without and with 0.4 nmole of 1,25-$(OH)_2D_3$ were chromatographed on a DEAE-cellulose column (0.9 x 20 cm) as described under the methods section. Aliquots from each fraction (0.18 $O.D_{280}$) digested with 20 μg of pure DNAse I and 50 μg of RNAse mixture were assayed by gel filtration as described under the methods section. Results are means $\pm$ S.E.M.

ACKNOWLEDGEMENTS

The authors wish to thank Dr. A. Kaiser (Hoffman-La Roche & Co., Ltd, Basel, Switzerland) and Prodotti Roche, Milano, Italy, for the generous gift of 1,25-dihydroxycholecalciferol.

REFERENCES

1. Mezzetti, G., Moruzzi, M.S., Capone, G. and Barbiroli, B. (1980): Biochem. Biophys. Res. Commun., 97:222-229.
2. Mezzetti, G., Loor, R. and Liao, S. (1978): Biochem. J., 184: 431-440.
3. Mezzetti, G., Drusiani, F., Montagnani, G., Capone, G. and Barbiroli, B. (1982): Proc. Soc. Exp. Biol. Med., 169: 280-283.
4. Chan, W.J., Seale, T.W., Shukla, J.B. and Rennert, O.M. (1979): Clinica Chimica Acta, 91:233-241.
5. Rock, A.M., Quash, G.A., Ripoll, J.P. and Saez, S. (1979): Recent Results Cancer Res., 67:56-63.
6. Bartos, D., Bartos, F., Campbell, R.A., Grettie, D.P. and Smejtek, P. (1980): Science, 208:1178-1181.
7. Mezzetti, G., Moruzzi, M.S., and Barbiroli, B. (1981): Biochem. Biophys. Res. Commun., 102:287-294.
8. Mezzetti, G., Moruzzi, M.S., Capone, G., Moruzzi, G. and Barbiroli, B. (1981): In: Advances in Polyamine Research, vol. 3, edited by C.M. Caldarera et al., pp. 237-247, Raven Press, New York.
9. Lawson, D.E.M. (1978): Vitamin D, 167-200, Academic Press, London and New York.
10. McNutt, K.W. and Haussler, M.R. (1973): J. Nutr., 103:681-689.
11. Yu, F.L. (1975): Biochim. Biophys. Acta, 395:329-336.
12. De Pomerai, D.I., Chesterton, C.J. and Butterworth, P.H.W. (1974): Eur. J. Biochem., 46:461-471.
13. Warburg, O. and Christian, W. (1942): Biochem. Z., 310:384-385.
14. Mezzetti, G., Moruzzi, M.S., Capone, G. and Barbiroli, B. (1980): Ital. J. Biochem., 29:185-190.
15. DeLuca, H.F. (1981): In: The Harvey Lectures, 75:333-379.
16. Woo, K.B., Perini, F., Sadow, J., Sullivan, C. and Funkhauser, W. (1979): Cancer Res., 39:2429-2432.
17. Mach, M., Ebert, P., Popp, R. and Ogilvie, A. (1982): Biochem. Biophys. Res. Commun., 104:1327-1334.

Advances in Polyamine Research, Vol. 4, edited by
U. Bachrach, A. Kaye, and R. Chayen. Raven Press,
New York © 1983.

The Use of Carboxymethyl Cellulose Resin in the Competitive Radiospermine-Binding Assay

G. Mezzetti, M. G. Monti, M. S. Moruzzi, G. Piccinini,
and B. Barbiroli

Istituto di Chimica Biologica, Università di Modena, 41100 Modena, Italy

Chick duodenal mucosa contains a cytosol protein that can bind radiospermine selectively (1) and is dependent on the vitamin D status of the animal (2). Properties, partial purification and characterization of the duodenal binding protein have been previously reported (1, 3).

The present article describes a preliminary study on a convenient and rapid method for assaying spermine-binding protein based on the use of a competitor resin for radiospermine in the test system.

MATERIALS AND METHODS

Animals used in all experiments were 3-4 weeks old White Leghorn chicks obtained from a commercial distributor. All manipulations were carried out at 0-2°. The duodenum was removed, rinsed in 40 mM Tris-HCl, pH 7.5, and the mucosal layer separated from the underlying muscle layers using a glass slide. The mucosa was homogenized in two volumes of the same buffer with a Potter-Elvehjem homogenizer and teflon pestle. The homogenate was centrifuged at 105,000 x g for 1 h. The upper three quarters of supernatant were taken and used as the cytosol fraction. In some experiments the cytosol fraction was purified by chromatography on a DEAE-cellulose (Whatman DE-52) column as described previously (2). The protein fraction retained by the column which could be eluted by 0.3 M KCl but not 0.2 M KCl was used as a source of purified spermine-binding protein (2). Duodenal protein fractions to be assayed were dialyzed overnight against 10 mM glycine buffer, pH 8.7. Spermine-binding activity of the protein fractions was analyzed by a modification of the ion exchange assay of Wasserman and Taylor (4). The cation exchange resin carboxymethyl cellulose (Whatman CM-52)(0.15 ml of packed resin) was mixed with a solution of spermine-binding protein in 1 ml of 10 mM glycine buffer, pH 8.7, in the presence of 0.2 μCi of 1 μM [^{3}H] spermine (NEN 44.3 Ci/mmole) until an equilibrium distribution was achieved. The samples were vigorously vortexed for 15 sec and centrifuged at 1,500 rev/min for 10 min. The percentage of radiospermine bound respectively to the soluble binding protein and to the resin was estimated. The distribution coefficients at various protein concentrations in the assay were then calculated using the following expression given by Schubert et al. (5).

$$K_d = \frac{\text{\% Radiospermine in resin x volume of solution (ml)}}{\text{\% Radiospermine in the solution x mass of resin (mg)}}$$

RESULTS AND DISCUSSION

Spermine-binding proteins from different tissue sources and at various degree of purification may be analyzed for their binding ability by an ion-exchange assay using carboxymethyl cellulose resin (CM-52). The resin binds spermine and other cations; in fact, as shown in fig. 1, under our experimental conditions low amounts of CM-52 suspension were very effective in sequestering radiospermine from the supernatant of the test system in the absence of bind-

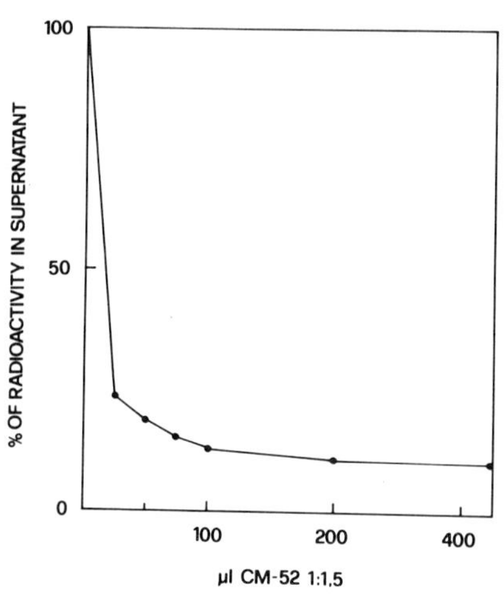

Fig. 1. Effect of increasing concentrations of CM-52 suspension on the percentage of radiospermine in the supernatant of the test system measured in the absence of binding protein. Experimental details are given under the methods section.

ing protein.

When a protein solution containing a spermine-binder is mixed with CM-52 resin a complex equilibrium results between spermine, spermine bound to the soluble binder, and spermine bound to the insoluble resin. At equilibrium two binding reactions can be written to express the possible states of the radioactive polycation

$$*Sp + R \rightleftharpoons SpR$$
$$*Sp + P \rightleftharpoons SpP$$

where R = noninteracting binding sites for spermine on the resin, P = noninteracting binding sites for spermine on the protein, *Sp = concentration of unbound radiospermine, SpR = concentration of spermine-resin complex, SpP = concentration of spermine-protein complex. P and R are assumed not to interact.

Fig. 2 shows that the increase of percentage of radiospermine in supernatant (o--o) measured with progressively increased concentrations of binding proteins rapidly approaches a plateau value.

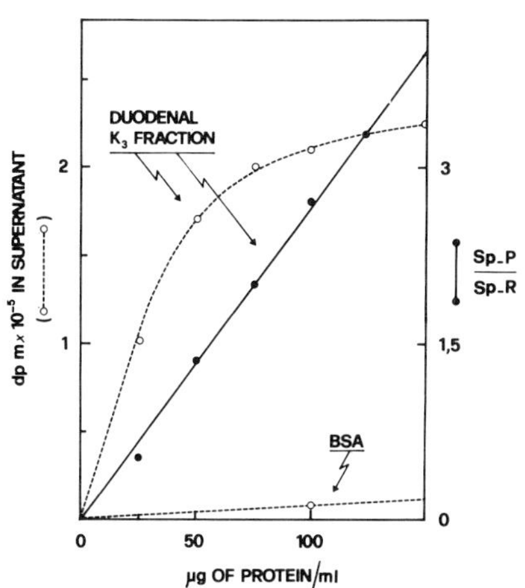

Fig. 2. Relationships of spermine-binding protein concentration to binding activity. Stated amounts of DEAE-cellulose-purified binding proteins were assayed for their binding activity by the CM-52 method. The binding data are expressed in two ways, as the percentage of $[^3H]$ spermine remaining in the supernatant (o---o) and as the ratio of $[^3H]$ spermine associated with protein to that associated with resin (Sp-P /Sp-R) (●—●). For experimental details see methods section and text.

In fact binding data expressed in this way, as one can expect from the above consideration, are linear only over a limited range of protein concentration and they are reasonably suitable only for qualitative assay. When a generic protein such as bovine serum albumine replaced the duodenal preparation the binding activity was always near to background values.

In order to obtain a linear relationship between spermine-binding protein concentration and a parameter representing spermine distribution in the two phases assay system (●—●), we considered the following equation based on considerations originally proposed by Schubert et al. (5) and then formulated for this purpose by Wasserman et al. (6):

$$P = K \frac{*SpP}{*SpR}$$

where P = unbound spermine-binding protein, K = constant including the formation constants of protein and resin with radiospermine and total resin concentration. *SpP and *SpR are respectively protein bound radiospermine in the supernatant and resin bound radiospermine in the solid phase. The above equation states that there is a linear relationship between binding protein concentration and the ratio of radiospermine associated with protein to that associated with resin. As shown in fig. 2, a linear relationship, tested experimentally, does exist between protein concentration and the ratio *Sp-P/*Sp-R (●—●) making this way of expressing the binding data reasonably valid for a quantitative assay.

In fig. 3 is reported a study on the classes of binding sites and on the binding capacity of purified duodenal binding-protein for various polyamines. DEAE chromatographed binding-protein was diluted with buffer to the concentrations stated in fig. 3 and then assayed with 1 μM radioactive spermine (●), spermidine (△), pu-

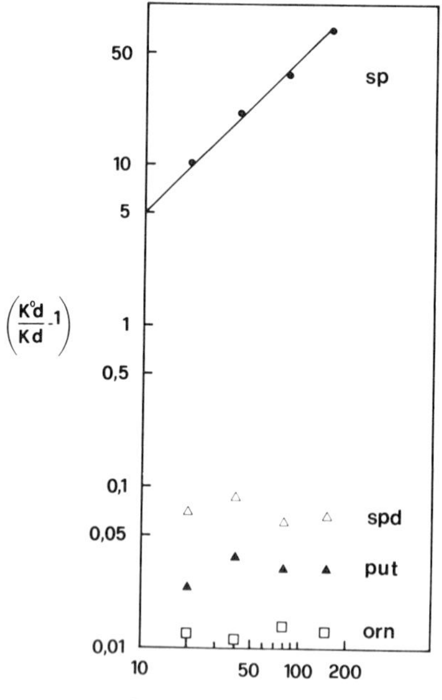

Fig. 3. Variation of $(k^{o}d/Kd)-1$ with concentration of DEAE-cellulose purified binding protein. $[^{3}H]$ Spermine (●), $[^{3}H]$-spermidine (△), $[^{14}C]$ putrescine (▲) or $[^{14}C]$ ornithine (□) bindings was assayed by the CM-52 method. For experimental details and definition of symbols see the methods section and text.

trescine (▲) or ornithine (□). In order to estimate the molar binding capacity of the protein for the various polyamines we calculated their respective distribution coefficients (K_d) in our competition assay system as described under materials and methods. A value of K_d was obtained for each binding protein concentration in the assay and K_d^o was the distribution coefficient calculated in the absence of protein. A log-log plot of $(K_d^o/K_d)-1$ versus the relative binding-protein concentration yielded a straight line with a slope of unity only for spermine.

These data suggest that the molar ratio of spermine to spermine-binding protein is 1:1 and there is mainly one class of binding sites present according to previous results(3) obtained by a different assay system and using crude protein preparations. Fig. 3 shows also that other polyamines do not bind efficiently the protein and the log-log plot of $(K_d^o/K_d)-1$ versus the protein concentrations do not yield a straight line suggesting the aspecific nature of these bindings.

In a previous paper we described that spermine-binding activity of rachitic chicks is rapidly enhanced in the duodenal mucosa by the administration of 1,25-dihydroxycholecalciferol (1,25-$(OH)_2D_3$) (2). In that study the spermine-binding assay was performed by using the more laborious gel filtration technique. In order to ascertain whether the ion exchange techinque described here could provide an alternative and reliable assay method we studied the effect of 1,25-$(OH)_2D_3$ on the intestinal spermine-binding activity of three weeks old chicks raised on a vitamin D-deficient diet (7). Fig. 4 shows the effect of a single intracardial dose of 1,25-$(OH)_2D_3$ (0.5 nmole) (generous gift of Hoffman La Roche) to rachitic chicks on the spermine-binding activity from various segments of the small intestine. The active form of vitamin D rapidly increased the binding activities in all segments of intestine tested but the intensity of effect decreased in the following order: duodenum > jejunum > ileum. The data reported in fig. 4 correlate well with sites at which vitamin D_3 is known to exert its effect (8); in fact similarly to spermine-binding activity, both calcium transport and vitamin D_3 induced calcium binding protein are differentially stimulated in various tracts of small intestine.

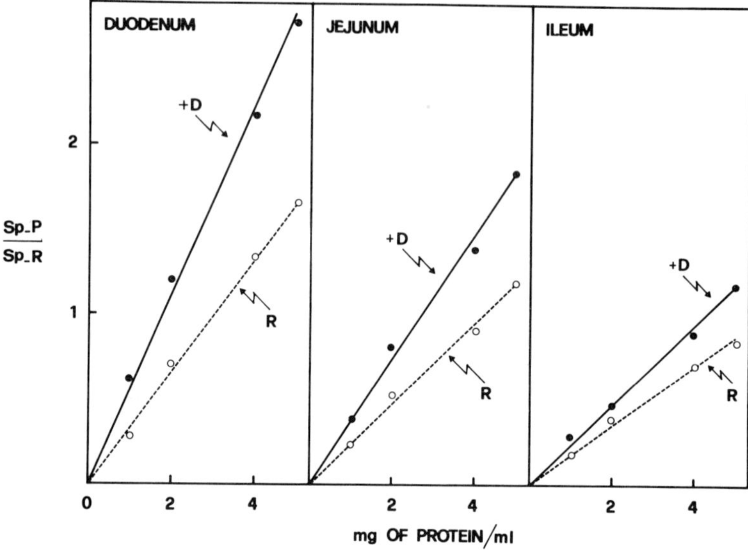

Fig. 4. Effect of a single dose of 0.5 nmole of 1,25-$(OH)_2D_3$ on spermine-binding activity measured in various tracts of small intestine. The spermine bound was assayed by the CM-52 method using crude cytosol as a source of spermine-binding protein. For experimental details see methods section. Binding data are expressed as the ratio of [^{3}H]-spermine associated with protein to that associated with resin (Sp-P/Sp-R).

REFERENCES

1. Mezzetti, G., Moruzzi, M.S., Capone, G. and Barbiroli, B. (1980): Biochem. Biophys. Res. Commun., 97:222-229.
2. Mezzetti, G., Moruzzi, M.S. and Barbiroli, B. (1981): Biochem. Biophys. Res. Commun., 102:287-294.
3. Mezzetti, G., Moruzzi, M.S., Capone, G., Moruzzi, G. and Barbiroli, B. (1981): In Advances in Polyamine Research, vol. 3, edited by C.M. Caldarera et al., pp. 237-247, Raven Press, New York.
4. Wasserman, R.H. and Taylor, A.M. (1966): Science, 152:791-793.
5. Schubert, J., Russel, E.R. and Mayers, L.S. (1950): J.Biol. Chem., 185:387-398.
6. Wasserman, R.H., Corradino, R.A. and Taylor A.N. (1968): J. Biol. Chem., 243:3978-3986.
7. McNutt K.W. and Haussler, M.R.(1973): J. Nutr., 103:681-689.
8. Taylor, A.N. and Wasserman, R.H. (1967): Arch. Biochem. Biophys., 119:536-540.

Advances in Polyamine Research, Vol. 4, edited by
U. Bachrach, A. Kaye, and R. Chayen. Raven Press,
New York © 1983.

Polyamine Compartmentalization and DNA Synthesis in Polyamine Depleted and Restored Mammalian Cells

M. Mach, P. Ebert, and W. Kersten

Institut für Physiologische Chemie, Universität Erlangen-Nürnberg, D 8520 Erlangen, Federal Republic of Germany

Kidney cells from african green monkey can be separated into nuclei and cytosol within seconds. To polyamine depleted cells putrescine, spermidine and spermine were added separately and their cellular distribution and their effects on the restoration of DNA synthesis are compared. Although spermine is the main polyamine in the nuclei, DNA synthesis in polyamine depleted cells is restored by all three polyamines about equally well.

INTRODUCTION

Polyamines are suggested to be involved in DNA synthesis and cell division. However the specificity of polyamine interaction with nucleic acids and the mechanism by which polyamines affect the cell progression through the cell cycle and its subsequent division are still unknown. The obvious means of elucidating whether polyamines are required for cell growth and multiplication, is to specifically deplete the cells of their polyamines. The development of specific inhibitors of polyamine biosynthetic enzymes made it possible to block the increase in polyamine synthesis, normally ensuing upon stimulation of cell proliferation, and, at least partly, to deplete cells of their polyamines.

In studies with bovine lymphocytes activated with concanavalin A, Morris and coworkers (15) showed that methylglyoxal-bis (guanylhydrazone = MGBG) blocks the synthesis of spermidine and spermine and inhibits DNA synthesis by 50%. A combination of MGBG and α -methylornithine, a competitive inhibitor of ornithine decarboxylase (= ODC), inhibited DNA synthesis even more effectively. Knutson and Morris (10) isolated nuclei from bovine lymphocytes depleted of polyamines by MGBG and α -methylornithine and demonstrated that DNA synthesis is inhibited by 75% as compared with DNA synthesis in nuclei from cells cultured in the absence of the inhibitors. Herbst and Elliot (6) reported similar results with nuclei from HeLa-cells. Krokan and Erikson (11) attributed the decreased rate of DNA synthesis to a reduction of the number of replication units active in DNA synthesis. In contrast, Geiger and Morris (3) have presented data indicative of a reduced rate of DNA replication fork movement in polyamine starved E. coli.

Using chinese hamster ovary (CHO) cells depleted of polyamines with α -methylornithine, Harada and Morris (4) showed that the G_1 and the S phase of these cells were lengthened proportionally to the polyamine deficiency within the cells. Studies with other experimental systems suggest that inhibitors of polyamine synthesis, alone or in combination, may block the entry of the cells into the S phase (13,20). Interference with the initiation of DNA synthesis is suggested also by the observation that cells accumulate in the G_1 phase of the cell cycle following MGBG treatment (5,18).

In lymphocytes, inhibition of polyamine accumulation during mitogenesis resulted in a reduced rate of DNA synthesis (15,16). Thus the basal levels of polyamines in unactivated lymphocytes support only a depressed rate of DNA synthesis. In all other cell types (HTC cells, rat embryo fibroblasts), the inhibition of polyamine biosynthesis during the stimulation of growth allowed one round of DNA replication with no apparent inhibition of the rate (13,18). Contrary to the situation in lymphocytes, it appears in this cases that the basal levels of polyamines present in the quiescent cells were sufficient to fulfil the polyamine requirement of DNA synthesis in the first cycle after activation.

Thus a minimum level of polyamines, which in some cells is well below that observed in steady state growth, is required to support optimal rates of DNA replication. The addition of putrescine or spermidine or spermine has usually been found to cause an immediate resumption of DNA synthesis and cell division. As to whether a particular polyamine specifically affects DNA synthesis is however unclear.

We have recently described the compartmentalization of polyamines in a cell line from the kidney of african green monkey (BSC-1) which can be separated into cytoplasm and nuclei within seconds (12). This system has been applied to evaluate the distribution of polyamines in polyamine depleted cells and the restoration of growth and DNA synthesis in cells restored with putrescine or spermidine or spermine.

METHODS

BSC-1 is a cell line derived from the kidney of the african green monkey Ceropithecus aethiops (8). The cells are grown in RPMI 1640 medium, supplemented with 10% of fetal calf serum. Under these conditions the cells grow as monolayers with a doubling time of approximately 24 h. The separation of cytoplasm and nuclei is usually performed as follows:

The culture medium is quickly removed by aspiration and the cells in one petri dish are treated with buffer containing 0.5% Nonidet P-40 as a detergent. Treatment is exactly for 20 sec under gentle shaking. After 20 sec an aliquot of the lysate is removed and pipetted into ice-cold trichloroacetic acid. The remaining solution is aspirated and the nuclei - still attached to the surface of the petri dish - are washed for exactly 5 sec with buffer (12). The wash fluid is treated in the same way as the cytoplasmic lysate. Polyamines of the nuclei are extracted by adding 10% trichloroacetic acid into the petri dish. Further details are described in reference (12).

RESULTS

As other mammalian cells, BSC-1 cells contain the diamine putrescine and the polyamines spermidine and spermine (12). The intracellular levels of the particular polyamines show distinct variations if the cells grow from lower to higher cell densities and exponentially to a density of 3 x 10^5 cells in 35 mm diameter petri dishes. If the cells are seeded with 25.000 per dish then the cells start growing after a lag period of about 8 hours (Fig. 1). Upon the onset of growth the putrescine content decreases continously from 20 nmol/10^6 cells to 9.5 nmol after 48 hours. At this time the cells have a density of 1 x 10^5 per petri dish and still grow exponentially. The levels of spermidine and spermine remain almost constant during the same time period. The

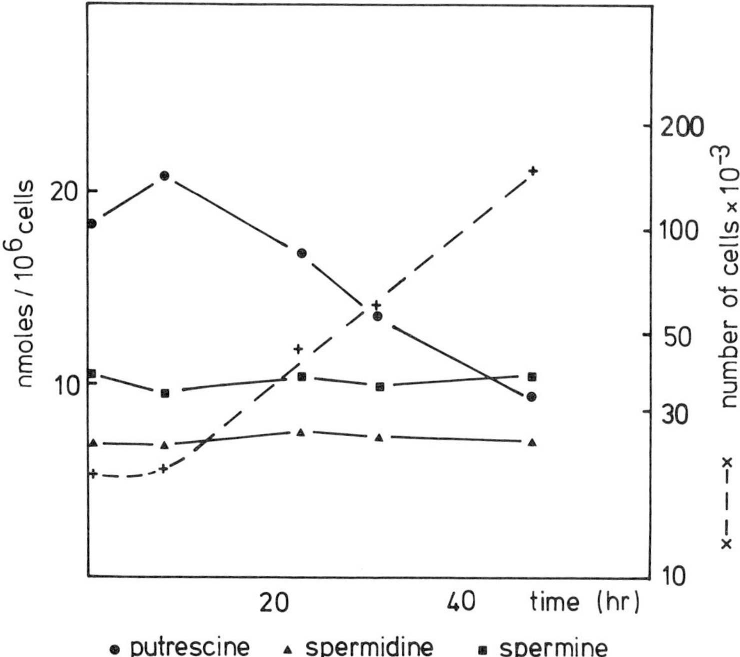

FIGURE 1. Polyamine content and cell number in exponentially growing BSC-1 cells.

cell-density dependent variations in the putrescine content of the BSC-1 cells make it difficult to define the 'physiological' intracellular polyamine concentration. Populations of exponentially growing cells differ from each other depending on the initial cell density, the time after seeding and the actual cell density.

Distribution of Polyamines in Cell Compartments

For the determination of the distribution of the various polyamines in the compartments of the cells we usually analysed log phase cells. The data from a representative experiment are shown in Table 1.

If spermidine is used as a reference and set at 1 the cells contain putrescine, spermidine and spermine at ratios of 1.4 : 1 : 1.8. Approximately the same ratios are found in the cytoplasmic fraction (2.3 : 1 : 1.4). In the nucleus however, the ratios are 1 : 1 : 5.5. In the nucleus spermine represents 74% but in the cytoplasm only 30% of the polyamine content.

To exclude the possibilities that alterations in the distribution of polyamines are caused by preferential loss of the less tightly bound polyamines from the nuclei we have washed the nuclei several times and have analysed the wash fluids for polyamines (Table 2). The polyamines can be washed out easily of the nuclei. However, a preferential loss of one of the polyamines does not occur. Therefore the results support the hypothesis that in eukaryotes spermine accumulates in the nucleus and is associatet preferentially with the DNA or the chromatin.

TABLE 1. Distribution of polyamines in the cytoplasm and nuclei of BSC-1 cells (nmoles/culture)

	Put	Spd	Sp
whole cells	11.5	8.2	14.8
cytoplasm	11.0	4.9	6.7
nuclei	0.58	0.6	3.33

The values are the mean of duplicate cultures. Separation of cytoplasm and nuclei and analyses of polyamines were performed as described (12).

TABLE 2. Polyamines in the cytoplasm, the washing fluids and the three times washed nuclei (nmoles/culture)

	Put	Spd	Sp
cytoplasm	10.3	6.3	7.7
wash fraction I	0.87	1.59	4.46
wash fraction II	-	0.23	1.09
wash fraction III	-	-	0.28
3 x washed nuclei	-	-	0.13

The washing procedure was performed for 5 sec with 2 ml buffer.

Cellular Levels and Distribution of Polyamines upon Treatment with Inhibitors of Polyamine Synthesis

We have used 2 different inhibitors of polyamine synthesis to reduce the amount of polyamines in the BSC-1 cells. Difluoromethylornithine (= DMFO), an irreversible inhibitor of ODC (14) and MGBG an inhibitor of S-adenosyl-methionine-decarboxylase (17). If 1 mM DFMO is added to the culture medium the intracellular level of putrescine decreases rapidly. After 10 - 15 hours the putrescine levels are below the limit of detection (< 0.05 nmol/10^6 cells, Fig. 2 b) and remain low for 50 hours. The spermidine concentration decreases more slowly. Spermidine levels start to decrease at the time when putrescine is already undetectable. After 30 hours treatment the cells contain only 40% of the spermidine present in control cells.

A tremendous increase of the putrescine content is observed upon treatment with 2 μM MGBG in the growth medium (Fig. 2 c). Maximum levels were found between 25 - 30 hours upon addition of the drug. Thereafter the putrescine level starts to decrease with a simultaneous increase of the spermine concentration. The varia-

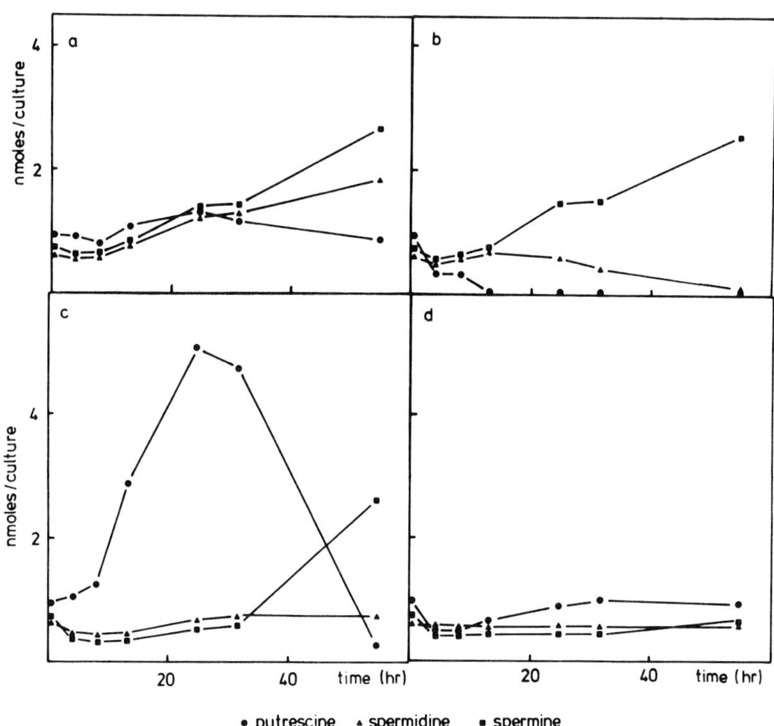

FIGURE 2. Polyamine content in BSC-1 cells during treatment with inhibitors of polyamine synthesis. Cultures were incubated: (a) in the absence of inhibitors or (b) in the presence of 1 mM DFMO, (c) 2 μM MGBG and (d) 1 mM DFMO + 2 μM MGBG.

tions in the polyamine content after 30 hours may be due to a limitation of the inhibitor in the growth medium.

Only the combination of 1 mM DFMO together with 2 μM MGBG leads to a significant decrease in the content of spermidine and spermine, whereas putrescine stays at levels of the untreated cultures (Fig. 2 d).

As can be seen from Table 3 the relative amounts of the polyamines in the nuclei are not disturbed if the inhibitors are present alone, in combination however they lead to a considerable decrease of spermine relative to spermidine. In the cytoplasm the polyamine ratios are not altered profoundly upon depletion of polyamines with both inhibitors.

TABLE 3. Polyamine ratios in the cytoplasm and nuclei of BSC-1 cells after 25 h inhibition of polyamine-synthesis

	Cytoplasm			Nuclei		
	Put	Spd	Sp	Put	Spd	Sp
control	1.28	1	0.66	0.1	1	4.95
DFMO 1 mM	0.36	1	0.74	-	1	5.92
MGBG 2 μM	4.41	1	0.58	1.0	1	6.09
DFMO + MGBG 1 mM 2 μM	1.70	1	0.51	-	1	3.27

^{3}H Thymidine-incorporation into DNA of Polyamine depleted and restored cells

The results of several investigators support the view that the polyamines primarily act on DNA, even though in some experimental systems the inhibition of polyamine synthesis also results in disturbances of protein and RNA synthesis (7,9,11).

To evaluate further as to whether
i: a specific polyamine primarily stimulates proliferation which in consequence gives rise to the initiation of DNA synthesis or
ii: is directly involved in the initiation and/or elongation of DNA synthesis, we have depleted BSC-1 cells of polyamines. To restore DNA synthesis each of the polyamines was added separately and the distribution of the polyamines within the cell compartments and ^{3}H thymidine incorporation into DNA were determined.

DNA synthesis in BSC-1 cells can be inhibited only by the combined action of DFMO and MGBG (Fig. 3). Neither DFMO nor MGBG alone could reduce the amount of ^{3}H thymidine incorporation into acid insoluble material. A complete depletion of the putrescine pool by DFMO or a fivefold excess of putrescine over control values caused by MGBG has no inhibitory effect on DNA synthesis. Contrary, in both cases a stimulation of the incorporation of thymidine into DNA was observed (Fig. 3).

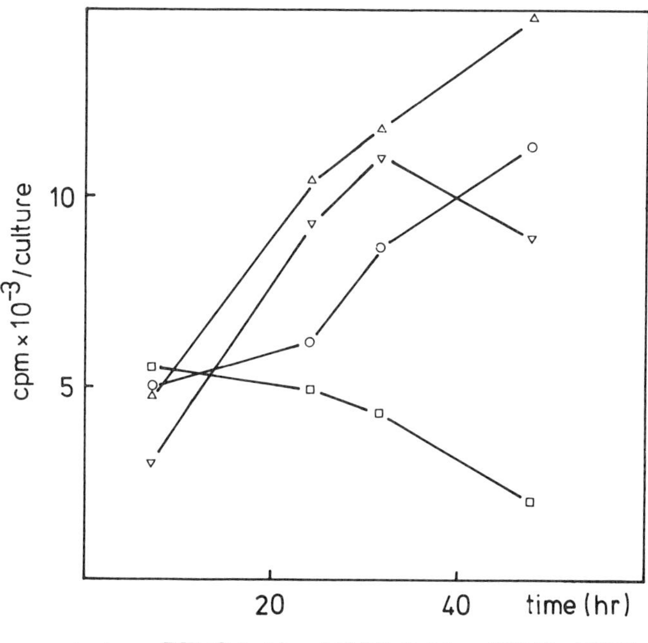

FIGURE 3. ^{3}H-Thymidine incorporation into untreated BSC-1 cells. Time = hours after addition of inhibitor. ^{3}H-Thymidine incorporation was assayed in 35 mm petri-dishes as described (12). (The values are the mean of duplicate cultures).

Treatment of the cells with a combination of DFMO and MGBG results in a putrescine level almost equal to that found in control cells. The spermidine and spermine levels however are drastically reduced (spermidine by 50%, spermine by 60%). This reduction of spermidine and/or spermine leads to a reduced DNA synthesis (see Fig. 2 and 3).

In restoration experiments the polyamines were added separately for 3 hours in a serum free medium and then removed. ^{3}H-Thymidine incorporation was measured for 30 min from 3 to 15 hours after the addition of polyamines. Polyamine analysis has shown that in medium containing 10% fetal calf serum polyamines are degraded within a few hours. To prevent degradation of polyamines and hence formation of toxic products, polyamines were added to cells in serum free medium.

Whereas untreated cells do not take up polyamines to an appreciable extent the polyamine depleted cells show rapid increase in the intracellular polyamine level up to normal values for spermidine and spermine. Putrescine is more easily taken up and accumulated within the cells up to the threefold normal level. The phenomenon that polyamine depleted cells take up exogenous polyamines to a greatly enhanced rate has been described also by Alhonen-Hongisto and Coworkers (1).

Determination of the polyamines in the cell compartments of control cells, polyamine depleted and polyamine restored cells are shown in Table 4.

TABLE 4. Distribution of polyamines in the cytoplasm and nuclei after 25 h inhibition (DFMO + MGBG) and addition of each polyamine for 3 h

	nmoles/culture					
	Cytoplasm			nuclei		
	Put	Spd	Sp	Put	Spd	Sp
control	9.2	6.8	4.9	-	0.12	0.87
DFMO + MGBG 1 mM 2 μM	8.8	3.2	1.6	-	0.11	0.65
DFMO + MGBG + Put 10 μM	49.8	3.5	1.8	0.42	0.08	0.63
DFMO + MGBG + Spd 10 μM	7.7	11.2	2.1	-	0.26	0.86
DFMO + MGBG + Sp 10 μM	5.2	8.5	3.4	-	0.21	1.40

In the cytoplasm the concentration of polyamines are altered profoundly in contrast to the nuclei, were they are rather constant.

Determination of ^{3}H-thymidine incorporation and polyamine levels in depleted and restored cells (Table 5) show in repeated experiments an increase of DNA-synthesis 5 hours after the addition of polyamines. At 11 hours putrescine causes 86% recovery, spermidine 88% and spermine 71%, compared with the control cells (100%). If polyamines were not restored ^{3}H-thymidine incorporation was 47%.

DISCUSSION

Recently we have determined the distribution of polyamines in BSC-1 cells. In the nuclei of BSC-1 cells spermine is the main polyamine, putrescine on the other hand was found almost exclusively in the cytoplasm (12).

The inhibitors of polyamine synthesis DFMO or MGBG, when added alone to the cells have hardly any effect on the growth rate dur-

TABLE 5. ^{3}H-Thymidine incorporation into DNA and polyamine content of BSC-1 cells depleted of polyamines and after readdition of polyamines

	% ^{3}H-Thymidine incorp.			nmol PA/culture (8 hours)		
	5 h	8 h	11 h	Put	Spd	Sp
control	100	100	100	1.15	1.12	1.12
DFMO + MGBG 1 mM 2 μM	54	51	47	0.59	0.86	0.65
DFMO + MGBG + Put (5 μM)	65	66	87	5.23	1.00	0.77
DFMO + MGBG + Spd (5 μM)	76	80	88	0.19	1.56	1.32
DFMO + MGBG + Sp (10 μM)	57	55	71	-	0.42	1.39

Treatment with inhibitors was for 25 h. The polyamines were added subsequently for another 3 h in a serum free RPMI 1640 medium also containing the inhibitors. Other experimental details see previous tables.

ing the first 50 hours. The lowered polyamine levels achieved during the first 50 hours are sufficient to support growth of the cells. Upon further exposure of BSC-1 cells to DFMO or MGBG cells stop to divide. A combination of both compounds inhibits cell growth starting at 30 hours almost completely. Although polyamine levels decrease the nuclei retain the spermine/spermidine ratio almost independent of the inhibitor and the extent of polyamine depletion.

Thymidine incorporation is blocked during the first 50 hours only by a combination of DMFO and MGBG. However, ^{3}H-thymidine incorporation into DNA continues in the presence of only one of the inhibitors. From these experiments one cannot decide unequivocally as to whether this is DNA synthesis or whether the uptake and/or the metabolism of ^{3}H-thymidine is altered. It is known from CHO cells (4) and fibroblasts (18) that polyamine defiency causes a prolongation of the G_1 and the S phase proportionally to the polyamine limitation of the cells, therefore the reduced incorporation of ^{3}H-thymidine probably reflects decreased DNA synthesis.

Putrescine or spermidine or spermine reverse the inhibition of DNA synthesis in polyamine depleted BSC-1 cells. The effect of added putrescine to reverse the inhibition of DNA synthesis is surprising. The experiments were performed under conditions where DFMO and MGBG were present over the whole time period. Polyamine analysis of whole cells showed that there is no conversion of putrescine into higher polyamines in the presence of DFMO plus MGBG. Thus it seems unlikely that the effect of putrescine is caused simply by the conversion to spermidine and spermine. In polyamine depleted and putrescine restored cells relatively high amounts of putrescine are found associated to the nuclear fraction. It is therefore possible that in putrescine restored cells this polyamine penetrates the nuclear membrane and compensates the lack of spermine.

The potency of spermidine and spermine to revers the inhibition of DNA synthesis is well documented (6,10,11). In most systems like HTC cells (13) human lymphocytes (7) and regenerating rat liver (17) the inhibition of DNA synthesis can also be reversed by putrescine. However, in BALB/3T3 fetal mouse cells or WI-38 fetal human cells putrescine cannot reverse the suppression of DNA synthesis (1,19).

When BSC-1 cells are arrested in the G_1 phase by serum deprivation they need about 10 - 12 h to start DNA synthesis. The time course of the resumption of DNA synthesis in polyamine depleted cells after restoration of polyamines resembles that of serum arrested cells. Contrary in cells that are arrested in the G_1 phase with thymidine and hydroxyurea DNA synthesis starts 1 - 2 hours after the removal of the block. This polyamine depletion in BSC-1 cells leads to cellular damages that have to be restored by polyamines before DNA replication is initiated.

The results show that BSC-1 cells are a suitable system to evaluate specific effects of particular polyamines during different phases of the cell cycle.

ACKNOWLEDGEMENT

We thank Mrs. I. Zenger and Miss U. Lösel for skilful technical assistance.

This work was supported by the Deutsche Forschungsgemeinschaft, SFB 118 and the W. Sander-Stiftung.

REFERENCES

1. Alhonen-Hongisto, L., Seppänen, P. and Jänne, J. (1980): Biochem.J. 192:941 - 945
2. Boynton, A.L., Whitfield, J.F. and Walker, P.R. (1980) In: Polyamines in Biomedical Research, J.M. Gaugas (ed.), John Wiley, New York, 63 - 80
3. Geiger, L.E. and Morris, D.R. (1978): Nature, 272:730 - 732
4. Harada, J.J. and Morris, D.R. (1981): Mol.Cell.Biol., 1:594 - 599
5. Heby, O., Marton, L.J., Wilson, C.B. and Gray, J.W. (1977): Europ.J.Cancer, 13:1009 - 1017
6. Herbst, E.J. and Elliot, Q.D. (1981): Medical Biology, 59: 410 - 416
7. Höltä, E., Jänne, J. and Hovi, T. (1979): Biochem.J. 178:109 - 117

8. Hopps, H.E., Bernheim, B.C., Nisalak, A., Tijo, J.H. and Smadel, J.H. (1963): J.Immunol., 91:416 - 424
9. Kay, J.E. and Pegg, A.E. (1973): FEBS Lett., 29:301 - 304
10. Knutson, J.C. and Morris, D.R. (1978):Biochim.Biophys.Acta, 520:291 - 301
11. Krokan, H. and Eriksen, A. (1977): Eur.J.Biochem.72:501 - 508
12. Mach, M., Ebert, P., Popp, R. and Ogilvie, A. (1982): Biochem.Biophys.Res.Comm., 104:1327 - 1334
13. Mamont, P.S., Böhlen, P., McCann, P.P., Bey, P., Schuber, F. and Tardif, C. (1976): Proc.Natl.Acad.Sci., USA, 73:1626-1630
14. Mamont, P.S., Duchesne, M.-C., Grove, J. and Bey, P. (1978): Biochem.Biophys.Res.Comm., 81:58 - 66
15. Morris, D.R., Jorstadt, C.M. and Seyfried, C.E. (1977): Cancer Res., 37:3169 - 3172
16. Otani, S., Mizoguchi, Y., Matsui, I. and Morisawa, S.,(1974): Mol.Biol.Rep., 1:431 - 436
17. Pösö, H. and Pegg, A.E. (1982): Biochim.Biohphys.Acta, 696:179 - 186
18. Rupniak, H.T. and Paul, D. (1978): J.Cell.Physiol., 94:161 - 170
19. Rupniak, H.T. and Paul, D. (1978): In: Adv.Polyamine Res., R.A. Campbell et.al. (ed.) Raven Press, New York, 117 - 126
20. Wiegand, L. and Pegg, A.E. (1978): Biochim.Biophys.Acta, 517:169 - 180
21. Williams-Ashman, H.G. and Schenone, A. (1972): Biochem.Biophys.Res.Comm., 46:288 - 295

Advances in Polyamine Research, Vol. 4, edited by U. Bachrach, A. Kaye, and R. Chayen. Raven Press, New York © 1983.

Polyamines, Carnitine, and Fatty Acid Metabolism in Human Platelets

Victor R. Villanueva

Institut de Chimie des Substances Naturelles, CNRS, 91190 Gif-sur-Yvette, France

Among the different biological effects of polyamines-putrescine (Pu), spermidine (Sd) and spermine (Sm)-on cellular events (1,3,7, 32)evidence has been presented suggesting the possible involvement of these organic polycations on lipid metabolism interrelated process such as glycerolipid biosynthesis regulation (15), inhibition of phospholipase A_2 and C (31), lipolyses supression (20),stimulation of sn-glycerol-3-phosphate acyltransferase and of diacylglycerol acyltransferase and inhibition of palmitoyl hydrolase (13, 14). Of particular interest is the finding that inhibition of ornithine decarboxylase causes a decrease of both polyamines and phospholipid synthesis (24).

In connection with our current studies on platelet aggregation which has been reported to be inhibited by polyamines (27), we were interested to obtain more information on the effect of polyamines on fatty acid oxidation which, together with glycolysis and glucose oxidation,is the main source for the energetic function of platelets (10) on processes such as aggregation.

Herein we report our results on polyamine biosynthesis and also on the effect of polyamines in fatty acid metabolism in human platelets.

POLYAMINE ANALYSIS AND BIOSYNTHESIS

Table 1 shows the ornithine (Orn), arginine (Arg) and polyamine contents of soluble TCA extracts of normal human platelets analysed by automatic ion exchange chromatography (33,35). No significant difference was obtained in the polyamine content of the samples before and after acid hydrolysis (HCl 6N/ 24h/110°C) thereby suggesting that platelets do not contain conjugate polyamines. Similar results have been reported by Rennert and Shukla (28).

Pu, Sd and Sm biosynthesis was assayed by incubating crude homogenates of human platelets for 1 hour in the presence of the appropriate ^{14}C labelled precursors and cofactors using experimental conditions already described (16,17,25,26,29,34).Results are summarized in Table 2. Although in our experimental conditions we were not able to detect ornithine decarboxylase activity, its presence can not be rule out as this enzyme has a very short half life (30)

TABLE 1 . Human platelets content of polyamines and their precursors*

Sample N°	Orn	Arg	Pu	Sd	Sm
1	450	1720	152	303	60
2	540	1450	65	310	40
3	340	940	80	160	42
4	420	600	110	190	46
5	495	1520	40	280	32
$\bar{X}$	449	1246	89	249	44
SD	76	462	43	69	10

*The concentration is given in pmol/mg protein

TABLE 2 . Polyamine biosynthesis in human platelets*

Precursors (specific radioactivity)	Label incorporated into the products formed (dpm)		
	Pu	Sd	Sm
Orn [5-^{14}C] (11.5mCi/mmol)	100		
Pu [1,4-^{14}C] (76.6mCi/mmol)		3500	
Sd [1,4-^{14}C] (120mCi/mmol)			6300

*Label was incorporated into the polyamine after 1 hour incubation at 30°C using 1.1mg protein , 1mM S-adenosyl methionine and 100000 dpm of the corresponding precursor. Polyamine analysis and radioactivity measurements were as previously described (34).

and the purification process, of platelets (19) for these experiments took us the whole day. Nevertheless, with this preparation and in the presence of S-Adenosyl methionine, Pu was converted into Sd and Sd into Sm. Maximal activity for Sd and Sm synthetases was observed at pH 7.4 and reactions were almost linear during the first 60 minutes (fig 1). The specific activity of these enzymes, in the preparation used, was about 20 pmoles/mg protein. From the results obtained it can be inferred that S-adenosyl methionine decarboxylase is present in human platelets.
During the course of these studies the presence of arginase in human platelets was detected for the first time. We have partially purified this enzyme and determined some of its kinetic parameters (36).

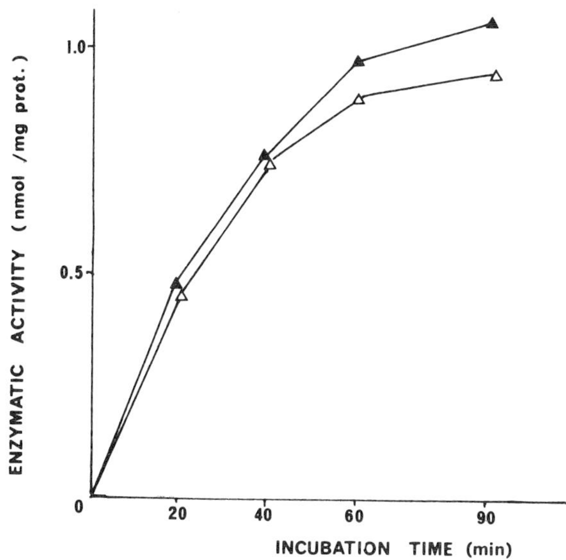

Fig. 1. Formation of Sd (△) and Sm (▲) (from Pu or Sd and S-adenosyl methionine respectively) as a function of time by human platelet homogenate at pH 7.4.

POLYAMINES AND FATTY ACID METABOLISM

"In vitro" studies have shown that polyamines inhibit platelet aggregation (27). This is in agreement with the observation that platelets from psoriatic subjets (which contain higher levels of polyamines compared to normal controls) present a very low capacity of aggregation(28) . Platelet aggregation is a high energy dependent biological process and the oxidative metabolism of fatty acids is one of the sources of the energy needed for platelet function "in vivo" . It was therefore of interest to study the effect of polyamines in fatty acid metabolism in these cells. We now wish to report our results in this area.

These studies were carried out using mitochondria prepared according to previously described methods (12, 19, 36) from fresh-platelets , collected from normal donors in acid citrate-dextrose, provided by the Centre National de Transfusion Sanguine . Owing to the low content of mitochondria of human platelets we used the equivalent of 2.5 L of blood for each set of experiments. The mitochondria preparation obtained was controlled, just prior to use,by the difference spectra of its cytochrome contents (4, 5, 6) . Furthermore, the oxidative capacity of motichondria was checked by using succinate , which was well oxidized in state 2 . Experiments were then carried out either with triton X-100 solubilized mitochondria or with intact mitochondria.

TABLE 3 . Inhibitory effect (%) of polyamines on the oxidation of palmitoyl (^{14}C-1)CoA*

Polyamines concentration	Pu	Sd	Sm
2mM	18	70	89
4mM	46	87	91

* After 15 min preincubation of the solubilized mitochondria (0.5mg protein) with or without (controls) polyamines at the indicated concentrations, cofactors and substrates were added. Final incubation system contains in 0.5 ml : Tris. HCl 0.1 M (pH 7.5), NAD^+ 0.1 mM, FAD 0.1mM, ATP 1mM, $KHPO_4$ 4mM, $MgCl_2$ 1mM, L-carnitine 2mM and palmitoyl [1-^{14}C] CoA 260.000 dpm (59m Ci/mmol). Experimental conditions used were as described by Panter and Mudd (23). The percentage inhibition effect of polyamines was calculated by comparison with the radioactivity, as $^{14}CO_2$, found in the controls (3950 dpm)

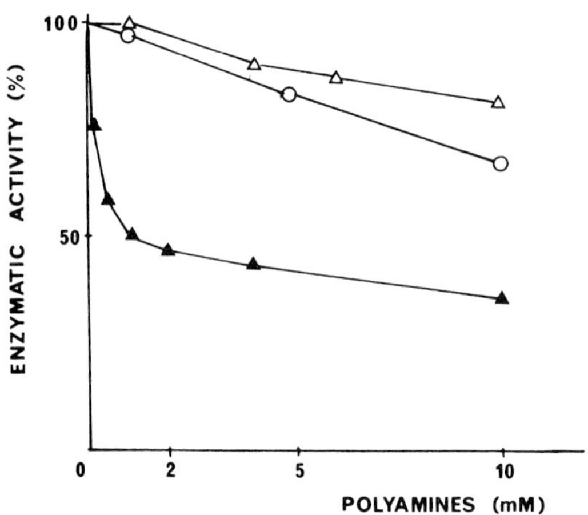

Fig.2. , Effet of polyamines on mitochondrial carnitine palmitoyl transferase activity : putrescine (Δ) spermidine (O) spermine (▲). Duplicate assays contained in a volume of 0.5ml : Tris HCI 0.1M (pH 7.4) palmitoyl CoA 50μM, L-carnitine 2mM and 300 000 dpm of DL-(methyl ^{14}C) carnitine (54.3m Ci/mmol). Reactions were started by addition of 100μl of solubilized mitochondria preincubated during 15 min with different concentrations of polyamines (0,05, 1,2,4 and 10 mM). After 5 min incubation at 37°C the reactions were stopped by addition of 500μl of 1.2 N HCl and the palmitoyl carnitine formed extracted by 500μl of butanol and aliquots counted for its radioactivity contents according to (22).

Experiments with Solubilized Mitochondria

TABLE 3 shows the inhibitory effect of polyamines on the oxidation of palmitoyl (^{14}C-1) CoA as measured by the $^{14}CO_2$ released during incubation with solubilized mitochondria (23). Duplicate assay gave almost 90 % inhibition of β-oxidation by spermine at 2mM.

$$R-\underset{\overset{\|}{O}}{C}-SCoA + HO-\underset{\underset{H_2C-COO^{\ominus}}{|}}{CH}-CH_2-N^{\oplus}(CH_3)_3 \rightleftharpoons R-\underset{\overset{\|}{O}}{C}-O-\underset{\underset{H_2C-COO^{\ominus}}{|}}{CH}-CH_2-N^{\oplus}(CH_3)_3 + HSCoA$$

As carnitine palmitoyl transferase is a key enzyme for the transport of acyl groupsinto the mitochondria matrix , and thereby for the β-oxidation of palmitoyl CoA we examined the effect of polyamines on this enzymatic reaction.
Fig. 2 shows that Pu and Sd, at the concentration of 2mM, have negligible effects, while Sm results in more than 50% inhibition of enzyme activity.

Experiments with intact mitochondria

It is known that during the course of β-oxidation of palmitoyl CoA the FADH and NADH formed are reoxidised at the respiratory chain level (21). It was thus interesting to verify the effect of spermine on the oxygen uptake during this process. Table 4 shows the inhibitory effect of spermine on the oxidation of palmitoyl CoA (82%) and palmitoyl carnitine (50%) as measured by the use of Clark-type electrode (9)

TABLE 4 . Effect of spermine on O_2 uptake during oxidation of palmitoyl derivatives by platelets mitochondria *

Substrates	O_2 uptake (nmol/mg prot/min)		inhibition
	no spermine	spermine 2mM	(%)
Palmitoyl L carnitine	0.82	0.41	50
Palmitoyl CoA	0.43	0.08	82

* Oxygen uptake was measured using a Clark-type polarographic electrode (Beckman oxygen analyser model 777). The reaction chamber was maintained at 25°C and contained in a total volume of 4ml : 300 mM mannitol , 10 mM sodium phosphate buffer (pH 7.4) 10 mM KCl, 5mM $MgCl_2$, 0.1% (w/v) BSA , 0.1 mM NAD^+ , 1mM ATP (1mM L-carnitine for oxidation of palmitoyl CoA) and a suspension of mitochondria . Experimental conditions were as described in (9).

As the reversible transfer of palmitic acid to CoA and to carnitine is one of the first steps in the sequence of mechanisms leading to the β-oxidation we examined the action of spermine on these enzymatic reactions . The aim of these experiments was to determine the enzymatic steps in which carnitine palmitoyl transferase (CPT) inhibition occurs :

(A) Formation of palmitoyl carnitine from carnitine and palmitoyl CoA ;(B) formation in the mitochondrial matrix of palmitoyl CoA from palmitoyl carnitine and CoA as shown in Fig 3. The results show (table 5) that after incubation of a mitochondria suspension with palmitoyl CoA and DL - [methyl- ^{14}C]carnitine the labelled palmitoyl carnitine formed is diminished when spermine (2mM) is added to the incubation medium. Moreover,the lack of accumulation of palmitoyl carnitine and its continued metabolism as a function of time provide presumptive evidence that spermine acts in the first step (A) of the enzymatic sequence.

In conclusion, these studies show the inhibitory effect of polyamines on the β-oxidation of palmitoyl CoA and particulary on the activity of carnitine palmitoyl transferase (in the direction of the formation of palmitoyl carnitine). Spermine inhibits carnitine palmitoyl transferase only by reducing the formation of palmitoyl carnitine thus suggesting that platelets mitochondria could contain two forms of carnitine palmitoyl transferase , as in rat liver cells (2).

TABLE 5. Effect of spermine on the formation of palmitoyl carnitine by platelet mitochondria *

Incubation time (min)	Label extracted palmitoyl carnitine (dpm)	
	no spermine	spermine (2mM)
15	9840	2230
30	5180	1330
60	2700	600

* The incubation medium (1ml) contained : 0,1 M Tris -HCl (pH 7.4), 50 μM palmitoyl CoA, 2mM L-carnitine , 300 000 dpm DL - [methyl- ^{14}C] carnitine (54.3 mCi/mmol) and a suspension of mitochondria (8mg). The reaction was stopped by 1.2 N HCl and extraction of palmitoyl carnitine was performed by butanol as in (22). Similar results were obtained in three separate experiments.

These experiments showed that spermine is more effective on carnitine palmitoyl transferase than spermidine and putrescine.

This difference on the inhibitory effect of polyamines, which has been already observed in other studies (8,18), could be due to a specific spatial separation of their amino-groups. Probably polyamines, owing to their cationic nature, may act on enzymes via interaction with their positively charged amino groups with resultant modification of their catalytic properties. Of course the possibility exists that this inhibitory effect of spermine in the formation of palmitoyl carnitine could be due in part to a stimulatory effect of spermine on other acyl transfer systems which can be present in the mitochondrial preparation used. Our current results would suggest, however, that Sm in mitochondria inhibit the β-oxidation although the enzymatic oxidative degradation step(s) remains to be determined. Finally it is interesting to note that analysis of the polyamine content of the whole platelets showed the presence of putrescine, spermidine and spermine (35) while, within the detection limit of our method, platelets mitochondria contain only putrescine and spermidine ; this absence of spermine in mitochondria could be of regulatory significance for platelet lipid metabolism.
Further work in this area is currently in progress in our laboratory.

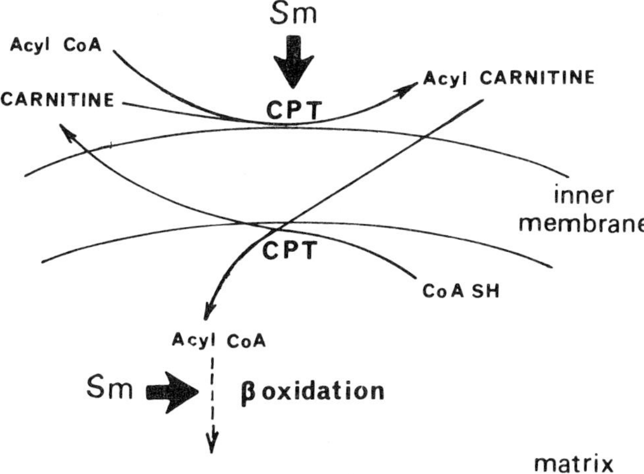

Fig.3. The arrows show the suggested effect of spermine at the outer mitochondria membrane level by inhibiting the activity of CPT, and in the matrix by blocking one or more steps in the β-oxidation sequence.

ACKNOWLEDGEMENTS

We wish to thank Dr.J.Vaur of Centre National de Transfusion Sanguine who kindly arranged for a supply of platelet samples.

REFERENCES

1. Bachrach, U. (1973) : Function of Naturally Occurring Polyamines ., Academic Press, New York.

2. Bergstrom, J.D., and Reitz, R.C. (1980) : Arch. Biochem. Biophys., 204 : 71-79.

3. Campbell, R.A. , Morris, D.R., Bartos, D., Daves, G.D., and Bartos, F., editors (1978) : Advances in Polyamines Research, vol 1 and Vol 2. Raven Press, New York .

4. Chance, B., and Williams, G.R. (1955) : J. Biol. Chem., 217: 383-397.

5. Chance, B., and Williams, G.R. (1955) : J. Biol. Chem.,217 : 395-407.

6. Chance, B., and Williams, G.R. (1956) : Adv. Enzymol.,17 : 65-134.

7. Cohen, S.S. (1971) : Introduction to the Polyamines Prentince Hall, Englewood Cliffs , New Jersey.

8. Cox, R. (1979) : Biochem. Biophys. Res. Comm., 86 : 594-598.

9. Estabrook, R.W. (1967) : Methods in Enzymology, 10 : 41-47.

10. Doery, J.C., Hirsh, J., and Cooper, I. (1970) : Blood , 36 : 159-168.

11. Farstad, M., Aas, M., and Sander, J. (1973) : Scand. J. Clin. Lab. Invest., 31 : 149-155.

12. Giret, M., and Villanueva, V.R., (1981) : Mol. Cell. Biochem., 37: 65-69.

13. Jamdar, S.C. (1977) : In : Advances in Polyamine Research , edited by R.A. Campbell et al. Vol. 2, pp 139-144 Raven Press, New York.

14. Jamdar, S.C. (1979) : Arch. Biochem. Biophys., 186 :248,254.

15. Jamdar, S.C. (1979) : Arch. Biochem. Biophys., 195 : 81-94.

16. Jänne, J., and Williams-Ashman, H.G. (1971) : J. Biol. Chem., 246 : 1725-1732.

17. Jänne, J., and Williams-Ashman, H.G.(1971) : Biochem.,Biophys. Res. Comm., 42 : 222-229.

18. Lapinjoki, S.P., Pajunen, E.I., Hietala, O.A., and Piha,R.S. (1980) : FEBS Lett., 112 : 289-292.

19. Levy Toledane, S., Rendu, F., Besson, P., and Caen, J. (1972): Rev. Europ. Etud. Clin. Biol., 17: 513-518.

20. Locwood, D.H., and East, L.E. (1974) : J. Biol. Chem., 249 : 7717-7722.

21. Martens , M.E., and Lee, C.P. (1980) : Can. J. Biochem., 58 : 549-558.

22. Mc Garry, J.D., Leatherman, G.F., and Foster, D.W. (1978) : J. Biol. Chem., 253 : 4128-4136.

23. Panter, R.A., and Mudd, J.B. (1973) : Biochem J., 134 : 655-658.

24. Peter, H.W., Gunter, T. , and Seiler, N. (1979) : FEMS Microb. Lett., 5: 389-393.

25. Raina, A. (1963) : Acta Physiol. Scand. , 60 : 7-81.

26. Raina, A. and Hannonen, P. (1970) : Acta Chem. Scand. (B) 24 : 3061-3063.

27. Rennert, O., Buehler, B., Miale, T., and Lawson, D. (1976): Life Sci. , 19 : 257-264.

28. Rennert, O., and Shukla, J.B. (1978) : In : Advances in Polyamines Research , edited by R.A. Campbell, et al. Vol 1 pp. 15-211 Raven Press, New York.

29. Russell., D.H., and SNYDER, S.H. (1968) : Proc. Natl. Acad. Sci., 60: 1420-1427.

30. Russell., D.H., and Snyder, S.H. (1969) : Mol. Pharmacol., 5: 253-262

31. Sechi, A.M., Cabrini, L., Landi, L., Pasquali, P., and Lenaz, G., (1978) : Arch. Biochem. Biophys., 186 : 248-254.

32. Tabor, C.W., and Tabor, H., (1976) : Ann. Rev. Biochem.,45: 285-306.

33. Villanueva, V.R., Adlakha, R.C. (1978) : Anal. Biochem. 91 : 264-275.

34. Villanueva, V.R., Adlakha, R.C., and Calvayrac, R. (1980): Phytochemistry , 19: 787-790.

35. Villanueva, V.R., Adlakha, R.C., and Giret, M. (1979) : Feuillets de Biologie , 106 : 141-146.

36. Villanueva, V.R., and Giret, M. (1980) : Mol. Cell. Biochem., 33: 97-100.

Advances in Polyamine Research, Vol. 4, edited by U. Bachrach, A. Kaye, and R. Chayen. Raven Press, New York © 1983.

The Polyamine Specificity of the Valyl-tRNA Synthesis Reaction Catalysed by Procaryotic *(Escherichia Coli)* and Eucaryotic (Plant, *Lupinus Luteus*) Valyl-tRNA Synthetases

Hieronim Jakubowski

Institute of Biochemistry, Agricultural University, 60-637 Poznan, Poland

The polyamines spermine and spermidine and the diamine putrescine occur in essentially all plant and animal cells. Procaryotic cells do not synthesize spermine (1,2). The significance of polyamines in physiological and biochemical processes is well recognized (3) but the molecular basis of their effects are only beginning to be understood (4). One of the puzzles of living systems is the fact that spermine is exclusively an eucaryotic polyamine. It is conceivable that the function of spermine is related to the nucleus. This suggestion is strengthened by the recent finding that spermine is present in the nucleus in 5-fold excess over putrescine and spermidine whereas in cytoplasm the polyamines are present in roughly equimolar concentrations (5). However, so far we do not know any of the biochemical processes in which spermidine could not fulfill the function of spermine in eucaryotic cells. Possible exclusive targets for spermine are spermine-binding proteins of unknown function present in some specialized tissues of mammalian organisms (6- 9).

One approach to the problem is to study the polyamine specificity of defined polyamine-dependent enzymatic systems of procaryotic and eucaryotic origin. It is to be expected that at least some eucaryotic systems, in contrast to procaryotic ones, would exhibit a marked specificity for spermine by effectively discriminating against putrescine and spermidine. Also, it is likely that some unique effects of spermine would be manifested in eucaryotic systems. An enzymatic system of choice for this study is aminoacylation of $tRNA^{Val}$ catalysed by valyl-tRNA synthetase (10). This reaction is stimulated up to 10-fold by polyamines, both in procaryotic (Escherichhia coli) and eucaryotic (plant, Lupinus luteus) systems, in the presence of limiting concentrations of magnesium. Effects of a number of diamines, polyamines and their derivatives on the rate of the tRNA aminoacylation reaction were determined in both procaryotic and eucaryotic systems. The data indicate a relative lack of polyamine specificity of the E. coli aminoacylation system and high specificity for spermine of the plant aminoacylation system. In addition, higher concentrations of spermine inhibit completely the $tRNA^{Val}$ aminoacylation reaction catalysed by the plant valyl-tRNA synthetase, whereas no inhibition of the reaction is observed in the E. coli system. Spermine behaves like a regulatory effector of the $tRNA^{Val}$ aminoacylation reaction catalysed by the plant valyl-tRNA synthetase. This novel function and high preference for spermine observed in the plant system may explain the need for this polyamine in eucaryotes.

MATERIALS AND METHODS

The valyl-tRNA synthetases from Escherichia coli (11) and lupin (Lupinus luteus) seeds (12) were purified to homogeneity by the published procedures. The E. coli valyl-tRNA synthetase was a generous gift from Dr. A.R. Fersht, Imperial College, London. Transfer RNAs from E. coli (Serva, Heidelberg, F.R.G.) and lupin seeds (13) were chromatographed on benzoylated diethylaminoethyl-cellulose (0.1 M acetate, pH 4.5, 5 mM $MgCl_2$, 0.5 mM EDTA, 0.45-0.8 M NaCl) to give enriched preparations of $tRNA_1^{Val}$ of acceptor activity of 400 pmoles/A_{260} (E. coli) and 200 pmoles/A_{260} (lupin). Concentration of the enzymes was determined by active site titration with [^{14}C]valyl adenylate (14). Uniformly labeled [^{14}C]valine (175 Ci/mol; 1 Ci=3.7 X 10^{10} Becquerels) was purchased from the Institute for Research, Production and Application of Radioisotopes (Prague, Czechoslovakia). ATP was from Sigma (St. Louis, MO, U.S.A.). Polyamines (chloride salts) and diamines (free bases) were from Serva (Heidelberg, F.R.G.). Edeine was from Calbiochem (La Jolla, CA, U.S.A.). Higher homologues of spermidine, i.e., N-1-(aminopropyl)-1,5-diaminopentane (AP5), N-1-(aminopropyl)-1,6-diaminohexane (AP6), N-1-(aminopropyl)-1,7-diamino heptane (AP7) and N-1-(aminopropyl)-1,8-diaminooctane (AP8) were a kind gift of Dr. R.D. Morris (Seattle, Washington, U.S.A.)

The rate of tRNA aminoacylation was measured by precipitation of [^{14}C]valyl-tRNA with trichloroacetic acid on Whatman 3 MM filter paper discs (15). Reaction mixtures contained 50 mM Hepes (pH 8.0), 1 mM $MgCl_2$, 1 mM ATP, 0.2 mM EDTA, 5 mM 2-mercaptoethanol, 0.2 mg/ml bovine serum albumin, 11 μm [^{14}C]valine, 1.2 nM valyl-tRNA synthetase, $tRNA^{Val}$, and polyamine or diamine as indicated in Results. After time intervals at 25°C, aliquots were taken in 5% trichloroacetic acid and the acid-insoluble radioactivity was determined by scintilation counting.

RESULTS

Valyl-tRNA synthetases from lupin and E. coli are almost identical with respect to moleculr weight; both are composed of one polypeptide chain (12), exhibit very similar kinetic constants for substrates, (16) and identical amino acid specificity both qualitatively and quantitatively (17,14). However, the two enzymes differ in amino acid composition and in tRNA specificity (not shown). The lupin enzyme aminoacylates heterologous (E. coli) $tRNA_1^{Val}$ but the E. coli enzyme does not aminoacylate lupin $tRNA_1^{Val}$. As will be shown below, the two enzymes also exhibit strikingly different polyamine specificity.

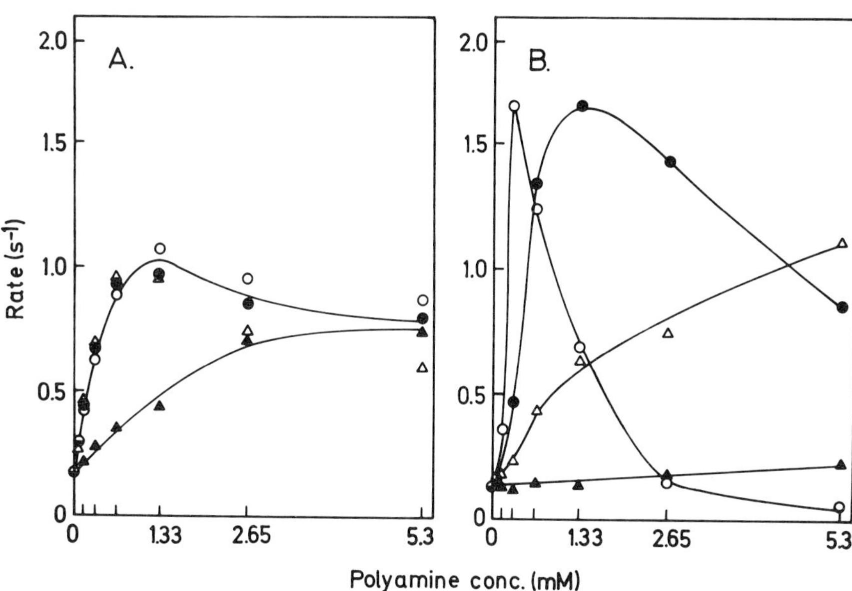

FIG. 1. The rate of $tRNA^{Val}$ aminoacylation as a function of polyamine concentration. A. Escherichia coli $tRNA^{Val}$ (7.5 μM) and coli valyl-tRNA synthetase. B. Lupin $tRNA^{Val}$ (2.2 μM) and lupin valyl-tRNA synthetase. Spermine (o), spermidine (●), AP8 (△), 1,8-diamineoctane (▲).

The aminoacylation of $tRNA^{Val}$ catalysed by valyl-tRNA synthetases from lupin and E. coli is stimulated by spermine and spermidine in the presence of limiting concentrations of magnesium. Carbon analogues of spermine and spermidine, N-1-(aminopropyl)-1,8-diaminooctane (AP8) and 1,8-diaminooctane, respectively, also stimulate the reaction. However, concentration dependence of the effect reveals different discrimination properties of the eucaryotic and procaryotic $tRNA^{Val}$ aminoacylation systems (Fig. 1). The aminoacylation of procaryotic $tRNA^{Val}$ by the procaryotic enzyme is stimulated by spermine, spermidine and AP8 with the same concentration dependence of the effect (Fig. 1A). Apparently, this enzymatic system does not discriminate between the three polyamines. 1,8-diaminooctane is some 4-fold less effective than the polyamines in stimulating $tRNA^{Val}$ aminoacylation catalysed by E. coli valyl-tRNA synthetase (Fig. 1A). Contrary to this, stimulation of the aminoacylation of lupin $tRNA^{Val}$ catalysed by lupin valyl-tRNA synthetase is affected by the three polyamines in a different way (Fig. 1B). Spermine is the best stimulator of the eucaryotic aminoacylation followed by spermidine, AP8, and 1,8-diaminooctane, a very poor stimulator. Only 1.6-fold stimulation was observed at 5.3 mM 1,8-diaminooctane, which is to be compared with 11.5-fold stimulation at 0.33 mM spermine. Thus, the three polyamines (spermine, spermidine and AP8) are distinguished by the eucaryotic aminoacylation system, but not by the procaryotic one. Moreover, excess concentrations of spermine and spermidine inhibit the reaction in the eucaryotic, but not in the procaryotic system. Complete inhibition of lupin $tRNA^{Val}$ aminoacylation catalysed by lupin valyl-tRNA synthetase is observed in the presence of sufficiently high concentrations of spermine (Fig. 1B).

Lupin valyl-tRNA synthetase aminoacylates E. coli $tRNA^{Val}$ at about the same rate as that observed with homologous lupin $tRNA^{Val}$. E. coli valyl-tRNA synthetase does not charge valine into lupin $tRNA^{Val}$. The effects of polyamines were also determined in heterologous system with the expectation of getting information about which component (either enzyme, or tRNA, or both) is the site of action of the polyamines. The aminoacylation of E. coli $tRNA^{Val}$ catalysed by lupin valyl-tRNA synthetase is stimulated by AP8 and 1,8-diamineoctane with the concentration dependence of the effect similar to that observed in the homologous E. coli system (Fig. 2). Low concentrations of spermine, spermidine and AP8 stimulate the heterologous aminoacylation to the same extent, as in the homologous E. coli system. However, higher concentrations of spermine and spermidine inhibit aminoacylation of E. coli $tRNA^{Val}$ catalysed by lupin valyl-tRNA synthetase, just as in the homologous lupin system. It is tempting to propose that the inhibition by spermine and spermidine is a novel function acquired by, and unique to eucaryotic valyl-tRNA synthetase; this is brought about by direct binding of the polyamines to the enzyme. This proposal is supported by the fact that spermine stimulates the ATP-PP_i exchange activity of the enzyme under the same conditions (10). It is to be noted that spermine fulfills the inhibitory function at concentrations at least one order of magnitude lower than spermidine (Fig. 1A and 2).

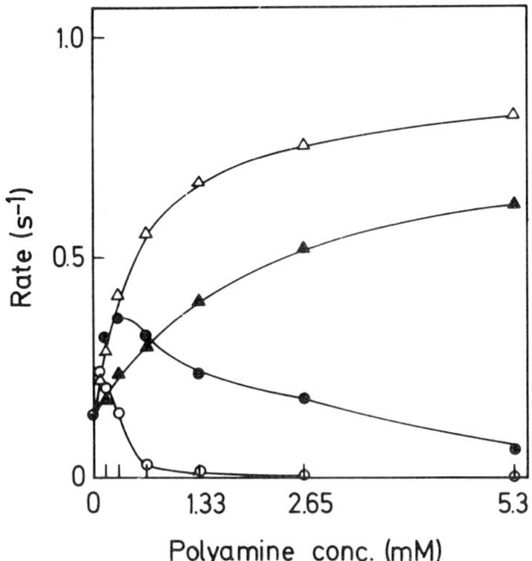

FIG. 2. The rate of coli $tRNA^{Val}$ aminoacylation by lupin valyl-tRNA synthetase as a function of polyamine concentration. Spermine (o), spermidine (●), AP8 (△), 1,8-diamineoctane (▲). The concentration of the tRNA was 7.5 μM.

A number of higher homologues of spermidine and some derivatives of spermidine were tested as effectors in the aminoacylation of lupin $tRNA^{Val}$ catalyzed by lupin valyl-tRNA synthetase (Fig. 3). N-1-substituted spermidine, edeine, stimulates the reaction with a concentration dependence similar to that observed with spermidine (Fig. 3 and 2B). Thus, N-1-substitution of spermidine by a bulky polypeptide substituent (18,19) does not affect its function. Ethidium bromide, formally a cyclic analogue of spermidine (20), is an inhibitor of the reaction. The ATP-PP_i exchange reaction is not affected by ethidium bromide (not shown). Thus, although spermidine and ethidium bromide bind to the same sites on tRNA (20,21), the spermidine-tRNA and ethidium-tRNA complexes exhibit different reactivity in the aminoacylation reaction, the former being active and the latter inactive. N-1-(aminopropyl)-1,n-diaminoalcanes containing n=5-8 carbon atoms are all stimulatory with the stimulating capacity decreasing with increasing number of atoms.

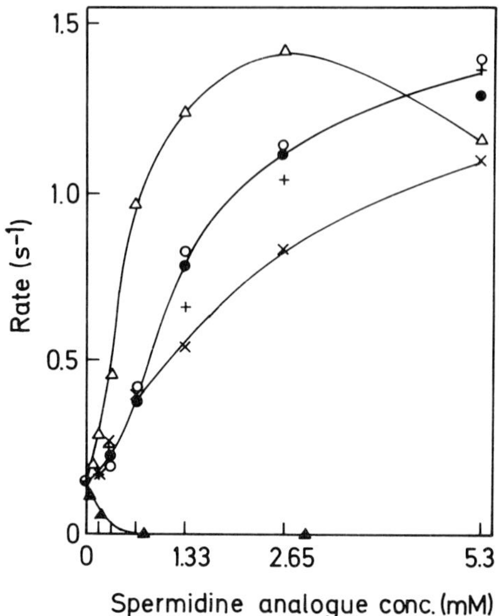

FIG. 3. The rate of lupin $tRNA^{Val}$ aminoacylation by lupin valyl-tRNA synthetase in the presence of spermidine analogues. Edeine (△), AP5 (o), AP6 (●), AP7 (+), AP8 (x), ethidium bromide (▲). The concentration of the tRNA was 2.2 μM.

The aminoacylation of $tRNA^{Val}$ is also stimulated by diamines. However, the diamines are much better stimulators of the reaction in the E. coli system than in the lupin system (Fig. 4). At limiting concentrations, putrescine is only 2-fold less effective than spermidine in the E. coli system. However, putrscine is an almost 100-fold less effective stimulator than spermidine in the lupin system (Fig. 4B). Diaminoethane and diaminohexane are 4-fold less effective stimulators than putrescine, diamineoctane is 8-fold less effective than putrescine and diaminodecane does not affect at all the aminoacylation of lupin $tRNA^{Val}$ catalysed by lupin valyl-tRNA synthetase. Again, this is in marked contrast to the effects of the diamines on the aminoacylation of E. coli $tRNA^{Val}$ catalysed by E. coli valyl-tRNA synthetase. Diaminodecane at limiting concentrations is a 2-fold better stimulator than putrescine and as good as spermine in stimulation of the tRNA aminoacylation in the procaryotic system. Diaminooctane and diaminohexane are only slightly worse stimulators than putrescine. Diaminoethane is 2-fold less effective than putrescine in stimulation of the reaction in procaryotic system. The most striking difference between the procaryotic and eucaryotic aminoacylation systems is their response to diaminodecane. Diaminodecane stimulates the reaction catalysed by the E. coli enzyme as well as spermine and spermidine do, whereas the diamine does not affect the reaction catalysed by the lupin enzyme.

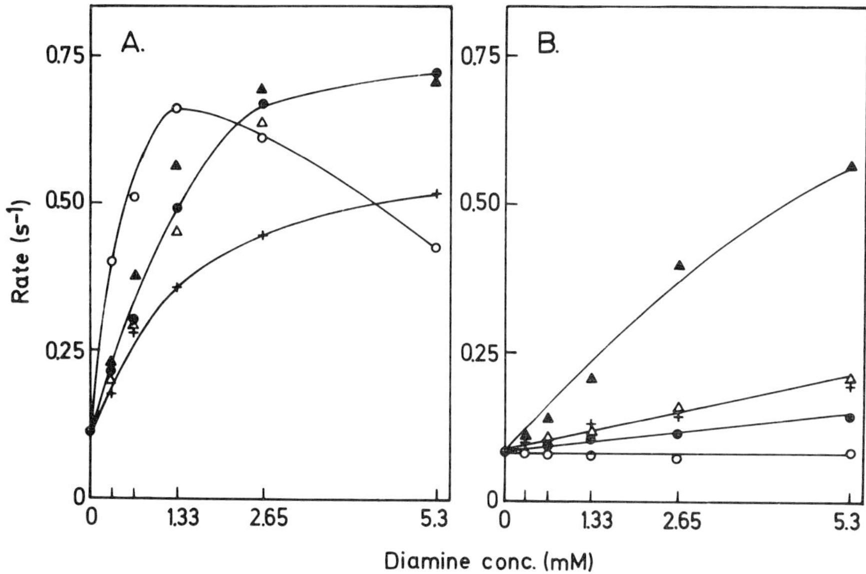

FIG. 4. Rate of tRNA&S.Val. aminoacylation as a function of diamine concentration. A. E. coli $tRNA^{Val}$ (1.5 μM) and E. coli valyl-tRNA synthetase. B. Lupin $tRNA^{Val}$ (2.2 μM) and lupin valyl-tRNA synthetase. 1,10-Diaminedecane (o), 1,8-diaminooctane (●), 1.6-diaminohexane (△), 1,4-diaminobutane (putrescine) (▲), 1,2-diaminoethane (+).

In order to quantitate the observed differences between procaryotic and eucaryotic $tRNA^{Val}$ aminoacylation systems, the effects of spermine, spermidine and putrescine were measured as a function of valine, ATP, Mg^{2+} and $tRNA^{Val}$ concentrations. Changes in the concentrations of either valine or ATP over one order of magnitude (10-100 μM valine; 30-1000 μM ATP) have no effect on the degree of stimulation of tRNA aminoacylation by polyamines, both in procaryotic and eucaryotic systems. Stimulation by polyamines decreases as Mg^{2+} concentration increases and no stimulation is observed in the presence of 5-10 mM Mg^{2+}. However, the inhibition by spermine of the aminoacylation reaction in the eucaryotic system (Fig. 1B) is independent of the

Mg^{2+} concentration in the range from 1-10 mM $MgCl_2$.

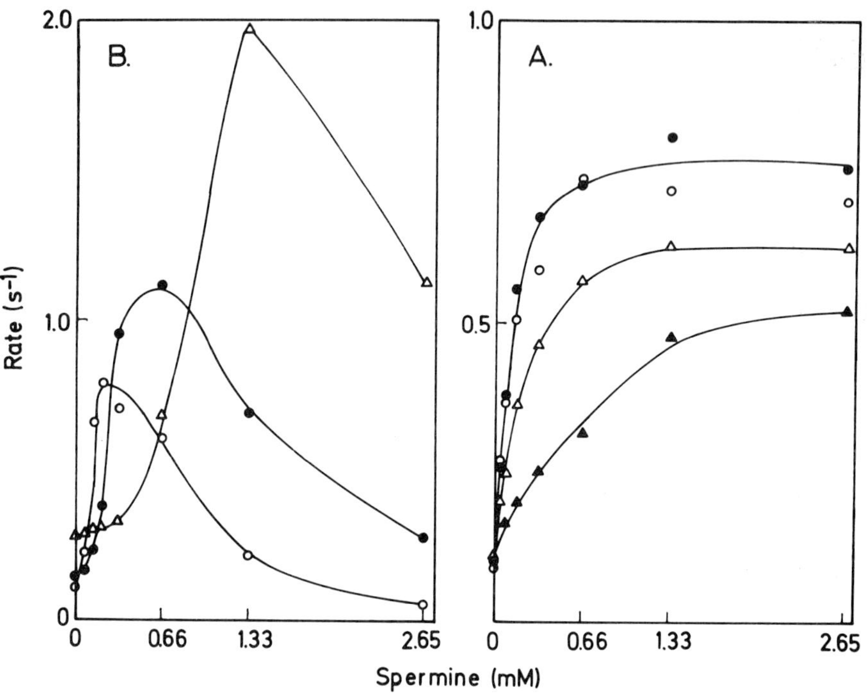

FIG. 5. Rate of $tRNA^{Val}$ aminoacylation as a function of spermine concentration in the presence of fixed concentrations of $tRNA^{Val}$. A. E. coli $tRNA^{Val}$ and E. coli valyl-tRNA synthetase. B. Lupin $tRNA^{Val}$ and lupin valyl-tRNA synthetase. The concentrations of $tRNA^{Val}$ were: 0.6 µM (o), 1.8 µM (●), 7.5 µM (△), 20 µM (▲).

The concentration dependence of the stimulation by spermine of the $tRNA^{Val}$ aminoacylation is markedly affected by changes in concentration of the tRNA in both procaryotic and eucaryotic systems (Fig. 5). Similar effects of the changes in the tRNA concentration have been observed when the rate of the reaction was measured as a function of the concentration of spermidine and putrescine. The stimulation of the reaction by the polyamines became independent of the tRNA (e.g., Fig. 5A). Thus it was possible to determine concentrations of the polyamines at which half of the effect was observed and which do not depend on the concentration of the substrates of the reaction and which can be regarded as a measure of dissociation constants of the polyamine.enzyme.tRNA complex. The relevant values (Table 1) indicate some fundamental differences in the interaction of polyamines with procaryotic and eucaryotic $tRNA^{Val}$ aminoacylation systems. The E. coli system does not discriminate between spermine and spermidine, and the two polyamines are some 6 to 7-fold preferred over putrescine. The procaryotic system seems to have only one kind of binding site for polyamines. The lupin system, on the other hand, exhibits at least two types of binding sites for the polyamines; high affinity sites responsible for stimulation and low affinity sites responsible for inhibition. The eucaryotic system binds spermine 4 to 8-fold better than spermidine and about 100-fold better than putrescine.

Hill plots of the stimulation by polyamines of $tRNA^{Val}$ aminoacylation reveal another difference between the procaryotic and eucaryotic systems. In both systems, stimulation by putrescine is proportional to its concentration (Fig. 6). However, stimulation by spermine and spermidine both are proportional to the 1.0-0.3 power of their concentration in the procaryotic system, and to 1.9 and 2.5 power of concentrationof spermidine and spermine, respectively, in the eucaryotic system. The stimulation by spermine of tRNA-Phe aminoacylation in another eucaryotic system (yeast) was also proportional to the concentration of spermine at higher than the first power (22). The data indicate different modes of interaction of the polyamines spermine and spermidine with the procaryotic and eucaryotic systems. Putrescine seems to interact in the same way with both procaryotic and eucaryotic $tRNA^{Val}$ aminoacylation systems.

Table 1. The $S_{0.5}$ and $I_{0.5}$ values for polyamines in the $tRNA^{Val}$ aminoacylation reaction catalysed by procaryotic (E. coli) and eucaryotic (lupin) valyl-tRNA synthetases. The values are independent of concentrations of substrates in both systems.

Polyamine	E. coli system	Lupin system	
	$S_{0.5}$(mM)	$S_{0.5}$(mM)	$I_{0.5}$(mM)
Spermine	0.18	0.06	0.52
Spermidine	0.18	0.25	4.3
Putrescine	1.2	$\gtrsim$ 5.	-

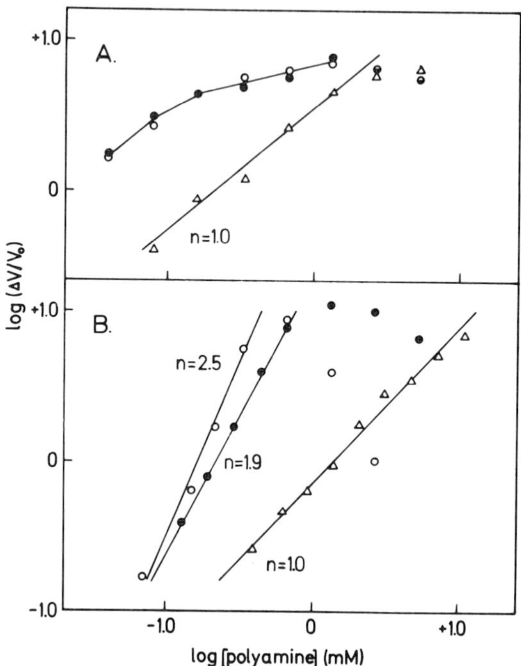

FIG. 6. Rate of $tRNA^{Val}$ aminoacylation as a function of polyamine concentration: Hill plots. A. *coli* $tRNA^{Val}$ (1.5 μM) and *E. coli* valyl-tRNA synthetase. B. Lupin $tRNA^{Val}$ (1.5 μM) and lupin valyl-tRNA synthetase. The logarithm of the relative increment in rate (V/V_o) is plotted as a function of the logarithm of polyamine concentration (mM). Spermine (o), spermidine (●), putrescine (Δ).

DISCUSSION

The aim of this investigation was to determine the specificity of the stimulation by polyamines of aminoacylation of $tRNA^{Val}$ in procaryotic and eucaryotic systems. From the data presented above it is evident that different features of the polyamines are essential for their stimulatory effects in the two aminoacylation systems. The procaryotic system exhibits only very limited specificity for spermidine as it is stimulated as effectively by a number of derivatives of spermidine and spermine. It is remarkable that the procaryotic system does not distinguish spermine and spermidine. In general, even diamines are good stimulators of the $tRNA^{Val}$ aminoacylation catalysed by *E. coli* valyl-tRNA synthetase, being only 2-4-fold less effective than spermidine. On the contrary, the eucaryotic system exhibits a pronounced specificity towards spermine; other polyamines and diamines are some 5-800-fold less effective than spermine. The extreme example of the discriminating ability of the eucaryotic $tRNA^{Val}$ aminoacylation system is the case of diaminodecane. This diamine does not affect the reaction in the eucaryotic system, but it is as good a stimulator as spermidine and spermine in the procaryotic system.

The different effects of spermine and spermidine in eucaryotic $tRNA^{Val}$ aminoacylation systems and identical effects of these polyamines in procaryotic systems indicate that at last a part of the polyamine effect is due to interaction with the enzyme, in addition to interaction with tRNA. It is well-known that polyamines are bound strongly to tRNA (21,20) with spermine being bound over 10-fold better than spermidine (23). Thus, it is to be expected that, if the effect of polyamines were exclusively on tRNA, we should observe different concentration dependencies of the effect for spermine and spermidine with over a 10-fold difference between the concentrations of spermine and spermidine at which half of the stimulation is observed. However, as shown in Table 1, this is not so. Previous data also indicate that polyamines interact with valyl-tRNA synthetase (10).

Another feature which distinguishes the eucaryotic and procaryotic $tRNA^{Val}$ aminoacylation systems is the inhibition of the aminoacylation reaction in the eucaryotic system by spermine and spermidine. Again, spermine is almost a 10-fold better inhibitor than spermidine. The effects of spermine on the eucaryotic $tRNA^{Val}$ aminoacylation system suggest that this polyamine may be a regulator of the $tRNA^{Val}$ aminoacylation reaction in eucaryotic cells. First, the stimulation is proportional to more than the square of spermine concentration (in the procaryotic system it is proportional to the 1.0-0.3 power of spermine concentration). Second, after a certain threshold concentration, spermine is a potent inhibitor of the reaction in the eucaryotic system (no inhibitory effect of polyamines is observed in the procaryotic system). The eucaryotic $tRNA^{Val}$ aminoacylation system exhibits two kinds of spermine binding sites, whereas the procaryotic system has only one.

There is a basic difference in the effect of increase in concentration of Mg^{2+} on the two types of spermine binding sites. Magnesium apparently competes with spermine for the high affinity site(s) responsible for stimulation (no stimulation by spermine is observed in the presence of saturating concentrations of Mg^{+2}). By contrast, the site(s) responsible for inhibition is specific for spermine and is not affected by Mg^{2+} (since the inhibition of $tRNA^{Val}$ aminoacylation by spermine does not depend on the concentration of Mg^{2+}).

CONCLUSION

The plant aminoacylation system has acquired two novel features (high specificity for spermine and ability to be regulated by spermine), which are absent in the E. coli system. Spermidine can perform the same functions but at about one order of magnitude higher concentrations than spermine. This feature may justify the need for spermine in the eucaryotic organism. It is more economical energetically for the eucaryotic cell to make a new polyamine, spermine, than to increase almost 10-fold the concentration of spermidine to achieve the same physiological goal.

REFERENCES

1. Cohen, S.S. (1970): Introduction to the Polyamines. Prentice-Hall, Englewood Cliffs, N.J.

2. Tabor, C.W., and Tabor, H. (1976): Ann. Rev. Biochem., 45:285-306.

3. Jane, J., Poso, H., and Raina, A. (1976): Biochim. Biophys. Acta, 473:241-293.

4. Cohen, S.S. (1978): Nature, 274:209-210.

5. Teeter, M.M., Quigley, G.J., & Rich, A. (1980): In: Nucleic Acid Metal Ion Interactions, edited by T.G. Spiro, Chapter 4. Wiley, New York.

6. Liang, T., Mezetti, G., Chen, C., and Liao, S. (1978): Biochem. Biophys. Acta, 524:430-441.

7. Mezetti, G., Loor, R., and Liao, S. (1979): Biochem. J., 184:431-440.

8. Bartos, D., Bartos, F., Campbel, R.A., Grettie, P.D., and Smejtek, P. (1980): Science, 208:1178-1181.

9. Mezzetti, G., Moruzzi, M.S., Capone, G., Moruzzi, G., and Barbioli, B. (1981): Advances in Polyamine Research, 3:237-248.

10. Jakubowski, H. (1980): FEBS Lett., 109:63-66.

11. Mulvey, R.S., and Fersht, A.R. (1977): Biochemistry, 16:4005-4013.

12. Jakubowski, H., and Pawełkiewicz, J. (1975): Eur. J. Biochem., 52:301-310.

13. Dziegielewski, T., and Jakubowski, H. (1975): J. Chromatogr., 103:364-367.

14. Jakubowski, H., and Fersht, A.R. (1981): Nucl. Acid Res., 9:3105-3117.

15. Mans, R.J., and Novelli, G.D. (1961): Arch. Biochem. Biphys., 94:48-53.

16. Jakubowski, H. (1978): Biochim. Biophys. Acta, 521:584-596.

17. Jakubowski, H. (1980): Biochemistry, 19:5071-5078.

18. Hettinger, T.P., Kurylo-Borowska, Z., and Craig, L.C. (1968): Biochemistry, 7:4153-4160.

19. Hettinger, T.P., and Craig, L.C. (1970): Biochemistry, 9:1224-1232.

20. Sakai, T.T., and Cohen, S.S. (1976): Progr. Nucl. Acid. Res. Mol. Biol., 17:15-42.

21. Sakai, T.T., Torget, R.I.J., Freda, C.F., and Cohen, S.S. (1975): Nucl. Acid. Res., 2:1005-1022.

22. Loftfield, R.B., Eigner, E.A., and Pastuszyn, A. (1981): J. Biol. Chem., 256:6729-6735.

23. Schreier, A.A., and Schimmel, P.R. (1975): J. Mol. Biol., 93:323-329.

Advances in Polyamine Research, Vol. 4, edited by U. Bachrach, A. Kaye, and R. Chayen. Raven Press, New York © 1983.

Acetylation of Spermidine in Chicken Tissues: Effect of Starvation, Insulin, and Glucagon

M. A. Grillo

Instituto di Chimica biologica, Università di Torino, Torino, Italy

The polyamines formed during decarboxylation of ornithine, followed by aminopropyltransfer from decarboxylated S-adenosylmethionine to putrescine and then to spermidine, can be interconverted to form spermidine from spermine and putrescine from spermidine. This transformation includes acetylation of the polyamine and oxidation of its acetyl derivative (16). The first reaction is catalyzed by a spermidine acetyltransferase with low $t_{\frac{1}{2}}$ (40 min in the liver of thioacetamide-treated rats)(12). The enzyme is enhanced in the mouse liver after prolonged fasting (17), with a parallel increase of N^1-acetylspermidine, and also in rat liver after treatment with hepatotoxic agents, such as CCl_4 (10) and thioacetamide (11). Since the maximal rate of conversion of spermidine into putrescine occurs when the acetyltransferase activity is also maximal, and since polyamine oxidase is present in a much greater amount and is not altered by treatment with CCl_4, Matsui et al.(14) hold that acetylation of spermidine is the rate-limiting step in the interconversion of polyamines.

We have previously shown (8) that in chicken pancreas (but not in liver) insulin promotes a decrease in spermidine concentration and an increase in putrescine. These changes are not correlated with the modifications in ornithine decarboxylase and S-adenosylmethionine decarboxylase, usually believed to be responsible for intracellular polyamine concentration. Since they are not dependent on the polyamine synthesis pathway, they could be due to an effect on the interconversion pathway. The results presented here confirm this hypothesis.

It is shown that several chicken tissues contain an enzyme able to acetylate spermidine. This is very active in pancreas and rapidly modified by different stimuli, such as starvation, and the effect of glucagon, insulin and spermine. Insulin acts through different mechanisms: one results in a decrease of the rate of enzyme degradation. The changes previously shown may therefore depend on enhancement of spermidine acetyltransferase activity.

MATERIALS AND METHODS

Groups of fed or 20-hr fasted, 2 to 4 week old chickens received glucagon (0.1 mg/100 g), theophylline (9 mg/100 g), insulin (4.6 I.U./100 g), spermidine or spermine (75 µmoles/100 g) at different times before sacrifice, as shown. In a few cases, actinomycin D (0.5 mg/100 g) was also given. To measure $t_{\frac{1}{2}}$, cycloheximide was given i.p. (1 mg/100 g), and residual enzyme activity was measured at different times.

Tissues rapidly excised were homogenized in 2 volumes of 0.25 M sucrose containing 50 mM Tris, pH 7.5, 25 mM KCl and 5 mM $MgCl_2$. The homogenate was centrifuged 15 minutes at 27.000xg. The supernatant was decanted and used for measuring the enzyme activity of the soluble fraction; the residue was brought to the initial volume with the same solution and used as the corpusculate fraction. In some experiments, enzyme activity was measured on a partially purified preparation, obtained by precipitation with ammonium sulphate (50% saturation) and dialysis.

Spermidine acetyltransferase activity was determined according to Libby (9) by measuring radioactivity incorporated from ^{14}C-acetyl-CoA into acetylspermidine or acetylspermine in 4 minutes incubation at 28° C. Under these conditions, proportionality between the amount of enzyme preparation and incorporated radioactivity is observed. The reaction mixtures contained 0.3 µmoles spermidine (or spermine or putrescine), 10 µmoles Tris, pH 7.8, 40 nCi (0.8 nmoles) ^{14}C-acetyl-CoA, and the enzyme preparation in a total volume of 100 µl. After 4 min at 28° C, 20 µl 1 M NH_2OH.HCl were added and the reaction mixtures were kept in a boiling water bath for 3 min. After centrifugation, 50 µl supernatant were applied to discs of P 81 paper. These were washed several times with water and then with ethanol, dried, and counted in PPO-POPOP-toluene.

S-adenosylmethionine decarboxylase of crude 10,000xg pancreas supernatant was assayed by measuring $^{14}CO_2$ evolution from the carboxyl labelled substrate, as previously described (7).

Insulin, spermidine, spermine, putrescine, cicloheximide, actinomycin D were obtained from Sigma Chem. Co., St. Louis; ^{14}C-acetylCoA from the Radiochemical Centre, Amersham; theophylline from Böhringer, Mannheim.

RESULTS

In 20-hr fasted chickens, spermidine acetyltransferase was active in all tissues tested, particularly in the pancreas and intestinal mucosa, followed in decreasing order by retina, brain, liver, kidney, and heart. Except in brain and heart, activity was much higher in the soluble fraction (Table 1).

TABLE 1. Spermidine acetyltransferase activity of subcellular fractions of chicken tissues

Tissue	Fraction	
	Cytosol	Particulate
	nmol/g tissue/min	
liver	1.14	0.69
kidney	1.14	0.60
heart	0.40	0.39
pancreas	6.47	3.12
intestinal mucosa	4.28	1.45
brain	1.59	1.64
retina	2.54	1.12

Enzyme activity was measured as described under Methods.

As in mammals, spermidine acetyltransferase of pancreas soluble fraction was active on both spermidine and spermine, more so on the former. Acetylation of spermidine in the presence of spermine occurs at a reduced rate, showing that they are substrates of the same enzyme (Table 2). Km measured on the partially purified enzyme preparation was 0.41 and 1.4 mM for spermidine and spermine respectively.

Spermidine acetyltransferase activity is much higher in the pancreas of fasted birds (about 7-fold)

TABLE 2. Acetylation of polyamines by an enzyme preparation of chicken pancreas cytosol

Substrate	Relative activity
3 mM spermidine	100
3 mM spermine	35
3 mM spermidine + 3 mM spermine	69
3 mM putrescine	12

then in the pancreas of fed birds, but it returns to the baseline after short refeeding (4-hr)(Table 3).

TABLE 3. Effect of fasting and refeeding on spermidine acetyltransferase activity of chicken pancreas

	Activity (nmol/g tissue/min)
controls fed	1.14±0.19 (4)
15-hr fasted	8.55±0.13 (3)
15-hr fasted + 4-hr refeeding	1.40±0.12 (4)

Glucagon is known to increase in plasma after brief starvation. It promoted marked enhancement of spermidine acetyltransferase activity to a maximum after 3 hr (Fig. 1). The effect is not additional to

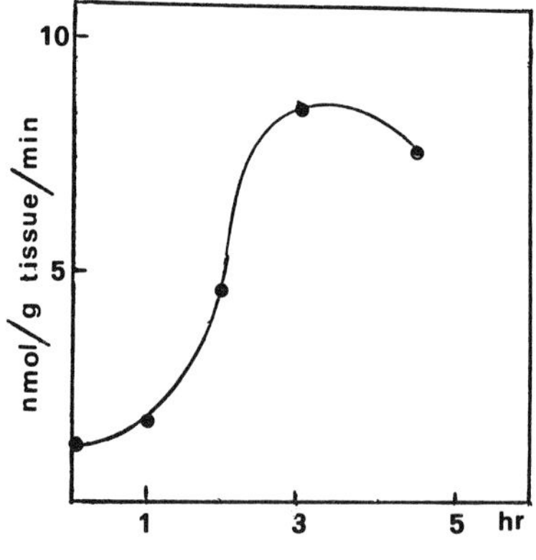

Fig. 1. Effect of glucagon on spermidine acetyltransferase activity in chicken pancreas.
0.1 mg glucagon/100 g were injected i.p. at the time shown before measuring enzyme activity.

that of theophylline (Table 4). In some chicken, as opposed to mammal, tissues insulin enhances cyclic AMP (5). Like glucagon, it increases spermidine acetyl transferase activity, and this effect is maximal after only 5-6 hr (Fig. 2). The increase is much lower in fasted then in fed birds (2-fold as opposed to 7-fold) (Table 5).

TABLE 4. Effect of glucagon and of theophylline on spermidine acetyltransferase activity of chicken pancreas

	Activity (nmol/g tissue/min)
controls	1.33±0.17 (7)
glucagon-treated (3-hr)	9.10±1.40 (6)
theophylline-treated (3-hr)	6.26±0.86 (4)
glucagon + theophylline-treated	6.21±0.76 (4)

0.1 mg glucagon/100 g or 9 mg theophylline/100 g or both were injected i.p. in the chickens 3-hr before measuring enzyme activity in the pancreas.

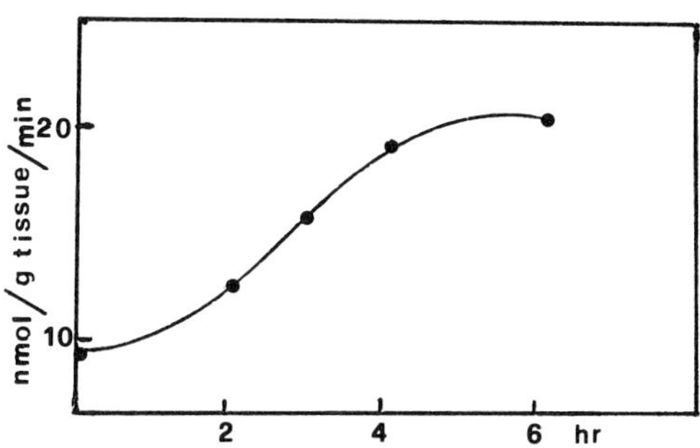

Fig. 2. Effect of insulin on pancreas spermidine acetyl transferase activity of fasted chickens 4.6 I.U. insulin were injected i.m. in the chickens at the time shown before measuring enzyme activity.

TABLE 5. Effect of insulin on spermidine acetyltransferase activity of the pancreas of fed and fasted chickens

	Activity (nmol/g tissue/min)	
	Controls	+ Insulin
Fed	1.33±0.17 (7)	10.57±1.31 (6)
Fasted	8.80±0.55 (8)	20.06±1.73 (6)

4.6 I.U. insulin were injected i.m. in fed or 16-hr fasted chickens 6-hr before measuring enzyme activity.

Insulin should thus have at least two mechanisms. The second may affect the rate of degradation of the enzyme: apparent $t_{\frac{1}{2}}$ of the spermidine acetyltransferase in the pancreas of fasted chickens is 22 min, but it is 45 min after treatment with insulin (Fig. 3). Actinomycin D only partially blocks the increase in spermidine acetyltransferase activity (results not shown), in line with the view that insulin has more than one mechanism of action.

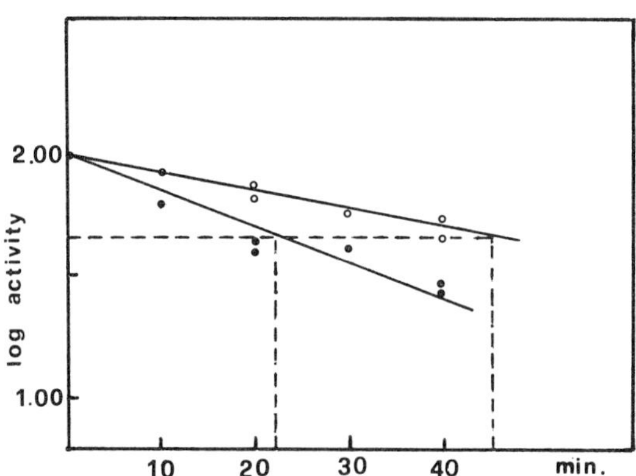

Fig.3. Apparent half-life of spermidine acetyltransferase.
20-hr fasted chickens, treated or not with 4.6 I.U. insulin/100 g (6-hr) received i.p. 1 mg cycloheximide/100 g. At the time shown, the birds were killed and enzyme activity was measured in the pancreas.
•——• , fasted chickens: $t_{\frac{1}{2}}$= 22 min; o——o , fasted chickens + insulin: $t_{\frac{1}{2}}$= 45 min.

Spermine, but not spermidine, also promotes a marked increase of spermidine acetyltransferase activity. Its effect is additional to that of insulin or fasting, and its mechanism must also be different (Table 6).

TABLE 6. Effect of spermidine and of spermine on spermidine acetyltransferase activity in chicken pancreas in the presence and in the absence of insulin

	Activity (nmol/g tissue/min)
controls fasted	8.80±0.55 (8)
spermidine-treated	10.39±1.37 (6)
insulin-treated	20.06±1.73 (6)
spermine-treated	16.13±1.47 (5)
insulin- and spermine-treated	34.25±2.68 (4)

The birds were treated i.m. with 4.6 I.U. insulin/100 g or with the polyamine (75 μmoles/100 g) 6-hr before measuring enzyme activity.

In the pancreas of fasted birds, where spermidine acetyltransferase is increased, S-adenosylmethionine decarboxylase is markedly reduced (65%) (Table 7).

TABLE 7. S-adenosylmethionine decarboxylase activity in the pancreas of fed and fasted chickens

	Activity (nmol/g tissue/30 min)
fed birds	6.93±0.40 (6)
fasted birds	2.45±0.67 (6)

Similar modifications of spermidine acetyltransferase activity are not promoted by insulin in liver. A much lower change is observed in liver after fasting (Table 8).

TABLE 8. Spermidine acetyltransferase activity in the liver of fed and fasted chickens. Effect of insulin.

	Activity (nmol/g tissue/min)
fed birds	0.30
fasted birds	0.85
fasted birds + refeeding	0.22
fasted birds + insulin	0.95

Average activity measured in duplicate on the liver of 2 or 3 birds.

DISCUSSION

Spermidine acetyltransferase has been demonstrated in rat (14,1) and mouse (17) liver, and in rat kidney (13). Our results show that it is also present in other tissues. In the fasted chicken, activity is maximal in the pancreas, followed by the intestinal mucosa. The enzyme acts on both spermidine and spermine, though it has a higher affinity for the former.

Our results suggest that spermidine acetyltransferase may have an important role in the regulation of intracellular polyamine concentration in chicken pancreas: since its $t_{1/2}$ is low (22 min in the fasted bird), its concentration may change following several stimuli, namely glucagon and insulin, and a high concentration of spermine, though not of spermidine, contrary to what occurs in rat kidney (3).

Enhancement of enzyme activity by insulin is followed by a decrease in spermidine and an increase in putrescine concentration (8), showing that spermidine acetyltransferase activity has an effect on polyamine levels.

The effect of glucagon and insulin is mediated by cyclic AMP, which is known to increase in some chicken tissues following their injection (5).However, with insulin a second mechanism is involved, since it modifies the degradation rate of the enzyme, as is shown by the increase of $t_{1/2}$ (from 22 to 45 min). This observation is in line with the action of insulin on tyrosine aminotransferase in rat hepatoma cell cultures (18) and on other proteins (6). However, it is in contrast with the hypothesis that insulin acts in this way on long-lived proteins only (4).

The effect of spermine in increasing spermidine acetyltransferase activity seems to be even different from that of fasting and of insulin. This is confirmed by the fact that no modification of half-life is observed after spermine (data not shown).

The fast modification of spermidine concentration following spermidine acetyltransferase activity variation in chicken pancreas suggests that spermidine has in the pancreas, as in the mammary gland(15) an active role, though this modification occurs, at least partially, through a decrease of the degradation rate in the pancreas, and through an increase of the rate of synthesis (due to enhancement of arginase, ornithine decarboxylase and S-adenosylmethionine decarboxylase) (2) in the mammary gland. In the pancreas, therefore, during fasting, when less proteins - enzymes - must be formed by the gland, spermidine is rapidly acetylated and transformed into putrescine. S-adenosylmethionine decarboxylase, which forms the substrate for reconverting putrescine into spermidine, is also fairly active in the pancreas, and the transformation can occur again when needed. The increase in this activity on refeeding helps the transformation.

In the liver, spermidine acetyltransferase seems not so important for the regulation of spermidine concentration: only after starvation is enzyme activity increased, but much less than in the pancreas. Insulin does not have any effect. This may be due to the different function of the tissue.

In conclusion, these experiments prove that regulation of spermidine concentration is obtained by modification of the synthesis and/or the degradation rate depending on the tissue involved.

REFERENCES

1. Bolkenius, F.N., and Seiler, N.(1981): Int.J.Biochem. 13: 287-292.
2. Brosnan, M.E., Ilic, V., and Williamson, D.H.(1982): Biochem.J., 202: 693-698.
3. Della Ragione, F., Matsui, I., and Pegg, A.E.(1982): Fed.Proc., 41:904.
4. Epstein, D., Elias-Bisko, S., and Hershko, A.(1975): Biochemistry, 14: 5199-5204.
5. Frohlich, A.A., and Marquardt, R.R. (1972): Biochim. Biophys.Acta, 286: 396-405.
6. Goldberg, A.L., and St.John, A.C. (1976): in Annual Rev.Biochem., 45: 747-803.
7. Grillo, M.A., Bedino, S., and Testore, G. (1978): Int.J.Biochem., 9: 185-189.
8. Grillo, M.A., Bedino, S., and Testore, G. (1978): Int.J.Biochem., 9: 673-676.

9. Libby, P.R. (1978): J.Biol.Chem., 253: 233-237.
10. Matsui, I., and Pegg, A.E. (1980): Biochem.Biophys. Res.Commun., 92:1009-1015.
11. Matsui, I., and Pegg, A.E. (1980): Biochim.Biophys. Acta, 633: 87-94.
12. Matsui, I., and Pegg, A.E. (1981): Biochim.Biophys. Acta, 675: 373-378.
13. Matsui, I., and Pegg, A.E. (1982): FEBS Letters, 139: 205-208.
14. Matsui, I., Wiegand, L., and Pegg, A.E. (1981): J.Biol.Chem., 256: 2454-2459.
15. Oka, T., and Perry, J.W. (1976): J.Biol.Chem.,251: 1378-1744.
16. Seiler, N., Bolkenius, F.N., and Rennert, O.M. (1981): Med.Biol.,59: 334-346.
17. Seiler, N., Bolkenius, F.N., and Sarhan, S. (1981): Int.J.Biochem., 13: 1205-1214.
18. Spencer, C.J., Heaton, J.H., Gelehrter, T.D., Richardson, K.I., and Garwin, J.L. (1978): J.Biol. Chem., 253: 7677-7682.

Advances in Polyamine Research, Vol. 4, edited by
U. Bachrach, A. Kaye, and R. Chayen. Raven Press,
New York © 1983.

Involvement of Polyamines in Phospholipid-Induced Changes of Cyclic Nucleotide Contents and Macromolecular Synthesis in Cultured Heart Cells

C. Pignatti, B. Tantini, C. Rossoni Caldarera, E. Turchetto, and C. Clò

Istituto di Chimica Biologica, Facoltà di Medicina e Chirurgia, Università di Bologna, 40126 Bologna , Italy

It is well known that phospholipids (PL), as major constituents of cell membranes, play a primary role in various cellular functions in which membranes partecipate (20,37). Recently, evidence has shown that cellular PL are also involved in the control of genome activity (12) and, in view of their interaction with DNA, (39) they may partecipate in the regulatory mechanisms of duplication, transcription and cell differentiation (40).

The possibility that also exogenous PL might influence some processes related to cell growth is based on their property to affect membrane fluidity and permeability (44), whose importance for normal cell growth and metabolism is well known (3), as well as to regulate the activity of different enzymes, partly membrane-bound (46) and partly soluble enzymes (52,54), most of which are closely involved in the regulation of cell proliferation. Although a causal relationship between the supply of lipids, including PL, and the onset of cell division has been suggested (13), most of the data concerning the effects of exogenous PL on cell metabolic activities related to growth come from experiments performed in vitro, while little evidence is available as far as living cells are concerned. The purpose of this work was to provide a contribute at this regard.

Cultured cells are usually regarded as appropriate model systems in studying the role of naturally occuring substances on specific metabolic processes under controlled conditions. As far as PL are concerned, the use of heart cell cultures provide a further advantage, being their metabolic and functional activities closely dependent on the supply of lipids with medium (24). The proliferative activity of cells in culture is ensured by the presence and amount of serum, which also represents the source of lipids for the cells. The importance of lipids in the growth stimulatory action of serum, is supported by the fact that cells fed with lipid-depleted serum do not proliferate appreciably, and that optimal growth may be restored by the treatment with different lipids (30). Starting from these bases, we have studied in chick embryo heart cell cultures the role of exogenous PL, as normal costituents of serum, in the control of processes leading to cell growth. Two experimental designs have been used in this study. In the first the cells were grown up to confluency in a medium containing a suboptimal concentration of serum (5%) enriched with a complex of PL, and growth activity was assessed by the rate of protein, RNA and DNA synthesis. In the second, the effects of single PL, such as phosphatidylcholine (PC) and phosphatidylethanolamine (PE), has been evaluated on confluent cultures in condition of serum-starvation. Furthermore since changes in cyclic nucleotide (47) and in polyamine contents (10,45) represent early and constant events following a stimulus affecting cell growth, we have investigated a possible involvement of these endogenous factors in the action of PL. The results here reported are consistent with a mediating role of polyamines.

METHODS

Cell Cultures

Primary beating heart cells obtained from 10 days-old chick embryos as previously described (9), were grown as monolayers in 5 ml of Eagle's minimum essential medium (MEM F-15, GIBCO) containing 10% foetal calf serum, 10% triptose phosphate broth, antibiotics and buffered with 25 mM Hepes, pH 7.4. The seeding density was 5.10^6 cells/90 mm Falcon plastic plate.

Experimental procedure.

Starting from the 40th h after seeding (time zero) the cells were daily fed with 4 ml of fresh medium containing 5% foetal calf serum, supplemented or not with different doses (0.1, 1.0, 100 µg/ml of medium) of a purified PL complex from calf heart (48.3% phosphatidylcholine; 11.0% sphingomyelin; 9.3% phosphatidylethanolamine; 5% phosphatidylserine; 4.1% cardiolipin; 3.5% lysophosphatidylcholine and 18.8% plasmalogens + phosphatidylinositol + phosphatidic acid; Aesculapius Co., Italy). Multilamellar liposomes were prepared at 37°C by mechanical stirring of the dried PL complex dispersed in an aqueous solution and diluted in MEM, according to Jonah et al (34). Only fresh stock solutions of PL were used and care was taken to protect them from oxygen and heat. The same lot of serum was used routinely.

Other groups of plates were grown to complete confluency, maintained for 20 h in 4 ml of a serum-free MEM and then added with 20 µg/ml (final concentration) of PC or PE (SIGMA), previously sonicated for 2 min in 1 ml of 10 mM Tris-HCl (pH 7.4) in order to obtain multilamellar liposomes. In the experiments in which DNA synthesis was studied confluent and serum starved cultures were shifted to fresh MEM, supplemented or not with 10% serum, or PC or PE. Group of 3-4 plates were used for each experimental condition.

Isotope labelling procedure. Protein and RNA synthesis were extimated by exposing the cells for 1 h to 0.3 μCi/ml of ^{14}C-leucine (59 mCi/mmol) or to 1 μCi/ml of ^{3}H-uridine (25.3 Ci/mmol) while DNA synthesis was determined after a labeling period of 2 h with 1.5 μCi/ml of ^{3}H-thymidine (5.0 Ci/mmol).
At the end of the incubations the pulsing medium was rapidly decanted and the cells were rinsed three times with ice-cold phosphate-buffered saline (pH 7.4), collected by scraping in cold 0.6 M $HClO_4$, frozen and thawed twice and centrifuged at 15,000 g for 10 min. The precipitate was used for measurement of protein (38), RNA (7) and DNA (42) contents and for radioactivity analysis. Radioactivities were determined by means of a Packard Tri-Carb PLD Spectrometer with Instagel as scintillation fluid. Correction for quenching was made with automatic external standardization.

Polyamine and cyclic nucleotide analysis. For analysis of polyamine,cyclic AMP (cAMP) and cyclic GMP (cGMP)contents the cells were treated and collected as described above. After centrifugation at 15,000 g for 10 min,aliquots of the supernatants were assayed for polyamine and cyclic nucleotide contents, while the pellets were solubilized into 1N NaOH and used for protein determination (38).
Polyamines were quantified by fluorescence determination of their dansylated derivatives separated on silica t.l.c. in ethylacetate: cyclohexane (2:3, v/v) (49), while intracellular cAMP and cGMP were assayed by means of kits from Radiochemical Centre, after purifying the cyclic nucleotides according to Chan and Lin (8). In the experiments in which α-DL-difluoromethylornithine (DFMO) was used, polyamine were determined by high-pressure liquid chromatographic technique (51).

Determination of decarboxylase activities. Ornithine decarboxylase activity was determined according to Bachrach (1) except that the cells were centrifuged at 10,000 g for 10 min and the assay buffer contained 0.1 mM EDTA, 0.02 mM pyridoxal phosphate, 2.5 mM ditiothreitol in 10 mM sodium phosphate buffer, pH 7.2. The reaction mixture contained 0.5 ml of the supernatant fluid, 0.4 μCi of ^{14}C-ornithine (47.2 mCi/mmol) and 20 nmoles of unlabelled ornithine in a final volume of 0.6 ml. S-adenosylmethionine decarboxylase activity was measured as reported by Bachrach (2). Proteins were determined according to Bradford (5).

RESULTS AND DISCUSSION

Since organ specificity has been postulated with exogenous PL (20) in the first set of experiments a complex of purified PL from calf heart was daily given to the cells in a medium containing 5% serum. This suboptimal concentration of serum was able to allow cell proliferation, without masking the effects of exogenous PL. Fig. 1 reports the changes in DNA, RNA and protein specific acti

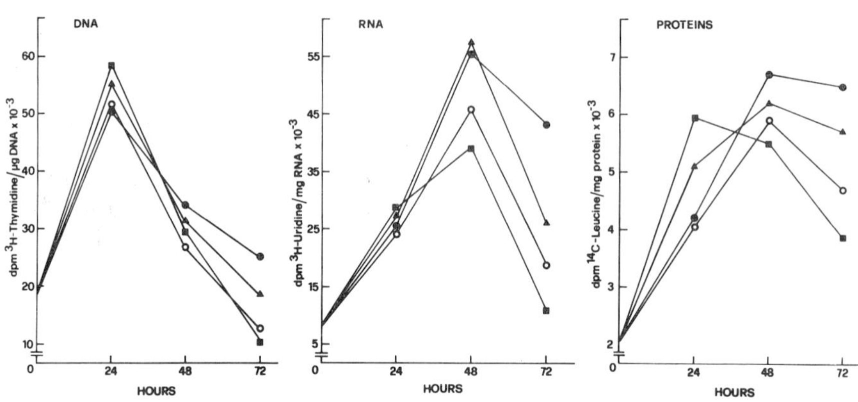

Fig. 1. Effect of exogenous PL on the rate of macromolecular synthesis in growing heart cells. Starting from the 40th h after seeding (time zero) cells were daily fed with fresh MEM containing 5% serum (○) or 5% serum plus PL at the dose of 0.1 µg/ml (●); 1.0 µg/ml (▲) and 100 µg/ml (■). Pulse labeling was as described in Methods. Data are the means of six determinations from three separate experiments, differing by no more than 10%.

vities induced by different doses of PL. It may be seen that PL--untreated cultures (controls) exhibit the maximum rate of DNA synthesis 24 h after the first change of medium, while protein and RNA synthesis progressively increase up to 48 h. Thereafter a significant decline in DNA, RNA and, to a lesser extent, of protein synthesis occurs. This behaviour may be related to the fact that in our experimental conditions at the 72 nd h (i.e. 112 h after seeding) the cultures were approaching confluency, because of the repeated changes of medium. At this stage of growth, it is known that the ability of cells to synthetize macromolecules as well as to take up precursors and nutrients is strongly reduced in comparison with actively dividing cells (15). In particular, as reported for other cell types (21), DNA synthesis is the first to be affected by increasing cell density.

The addition of PL the complex at the dose of 0.1 µg/ml of medium significantly counteracts the decline of DNA, RNA and particularly of protein syntheis, suggesting for a synergistic effect of PL with the component of fresh medium, especially serum. This effect might be due to a better utilization of nutrients and precursors by the cells or to a stronger binding of the growth factors of serum to the cells, as it has been observed with other mitogenic factors (41). Moreover, the possibility exists that the supply of PL, at this concentration might ensure membrane integrity and cell functions, chiefly when cells reduce their capablity to synthetize lipids de-novo because of crowding (31). Our previous results had indicated that in confluent and serum restricted heart cells the addition of the same PL at relative low doses significantly increases in time the rate of ^{3}H-leucine uptake and its incorporation into proteins (9). Conversely the presence of PL at the highest dose (100 µg/ml) causes, within 24 h, a stimulation of DNA and protein synthesis while it has no significant effect on RNA specific activity. However, when cultures approach confluency, the same dose of PL induces a more rapid decrease of the rate of precursor incorporation, when compared to cells treated with serum alone. The intermediate dose of the PL complex (1 µg/ml) exhibits intermediate effects.

Other data (not shown) indicate that in these experimental conditions, exogenous PL influence the rate of macromolecular synthesis of growing heart cells, not only by affecting the uptake of precursors, but primarily by modifying the capability of the cells to utilize them, independently on their endogenous availability, suggesting the possibility that exogenous PL affect macromolecular synthesis also by a membrane independent mechanism. Since exogenous PL are in part taken up as intact molecules by cultured cells (28) and are reported to affect in vitro enzymes of nucleic acid metabolism (29,30), the existence of a direct effect of these compounds on cellular gene activity cannot be excluded.

Moreover, the negative effect elicited by the repeated supply of PL at high dose might be related to an excessive accumulation of lipids inside the cells, leading to an uncoupling (4) and/or to alterations in membrane properties, possibly as a consequence of a detergent action of PL or of peroxidative processes (14, 18).

Alternatively, since it has been reported that exogenous lipids may influence the activity of different enzymes including those of cyclic nucleotide (14, 18, 50) and of polyamine metabolism (35), a possible involvement of these endogenous factors in PL action has been examined in the same experimental conditions.

Fig. 2 reports the effect of the different doses ot the PL complex on cellular cAMP and cGMP contents of the cultures described in Figure 1. It may be noted that in control cultures fed with MEM containing 5% serum, cAMP slightly falls 24 h after the first change of medium, but in time (48 and 72 h) it progressively accumulates towards confluency. The presence of PL in the same medium stimulates the decrease of cAMP within 24 h in a dose dependent manner. Conversely in time PL either counteract (0.1 and 1.0 µg/ml) or enhance (100 µg/ml) cAMP accumulation. A parallel but opposite behaviour is observable with cGMP in both control and PL stimulated cells.

In view of the inverse or direct relationship existing between cAMP or cGMP contents and the rate of macromolecular synthesis and cell division (19, 48), the kinetics of the two nucleotides well correlate with the data of Fig. 1.

Fig. 3 reports the time course of spermine (SPM) and spermidine (SPD) contents. Both polyamines progressively accumulate after repeated changes of medium and further increase in the presence

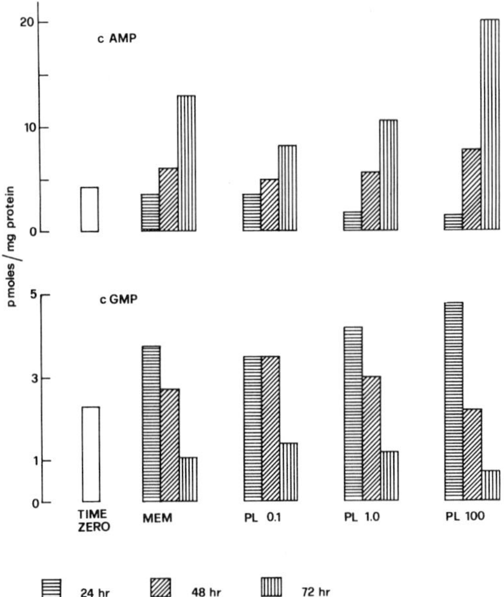

Fig.2. Effect of exogenous PL on cyclic nucleotide contents of growing heart cells. Experimental conditions were as given in Fig.1. Data are the means of six determinations from three separate experiments differing by no more than 10%.

of low doses of PL (0.1 and 1.0 μg/ml). At 100 μg/ml PL, while increasing polyamine contents within 24 h, thereafter bring about their decline, particularly evident in the case of SPD.

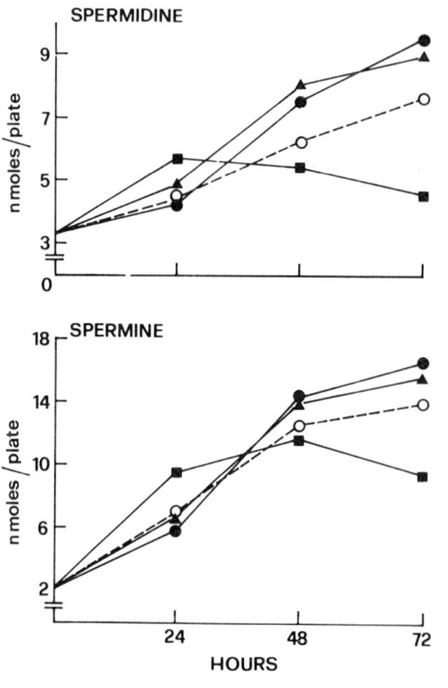

Fig.3. Changes in polyamine contents of growing heart cells treated with PL. Experimental conditions and symbols were as given in Fig.1. Data are the means of six determinations from three separate esperiments differing by no more than 10%.

All together the data of this first set of experiments indicate that exogenous PL stimulate or reduce the rate of macromolecular synthesis with respect to their concentrations and to the lenght of their presence in the medium. These changes are associated with variations in cyclic nucleotide and polyamine contents. Furthermore it appears that the magnitude of the cellular response to PL increases with the age of the cultures. Considering that polyamines positively correlate with the rate of macromolecular synthesis (32, 45) and are able to reduce or increase cAMP or cGMP respectively, as observed by us (11), the behaviour of these polycations under the different experimental conditions might account for the differences observed at the level of macromolecular synthesis and of cyclic nucleotide contents.

The possibility that polyamines may mediate the effect of exogenous PL has been further investigated in another set of experiments with single PL, such as phosphatidylcholine (PC) and phosphatidylethanolamine (PE). These are among the most representative PL of the complex above used and the predominant PL of cell membranes in which, at confluency, undergo a larger turnover (15). At this regard confluent and serum-starved heart cells were utilized in order to have a population of cells all synchronized in early G_1 (17) and to avoid the interferences of lipids present in serum. PC and PE were singularly added to the cultures at the final concentration of 20 µg/ml of medium. This dose is the same (PC) or about 30 fold (PC) that normally found in a medium supplemented with 10% of serum (53).

From Fig. 4 it appears that both PC and PE cause, shortly after

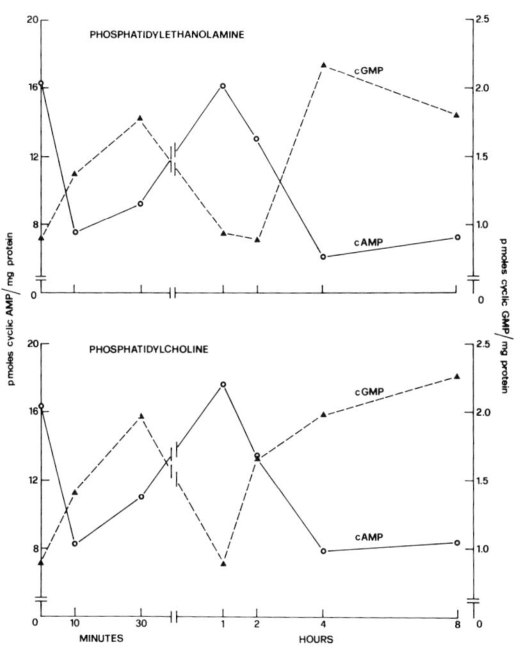

Fig.4. Effect of exogenous PL on cAMP and cGMP levels of heart cells. Confluent and 20 h serum-starved cells were added with 20 µg/ml of PC or PE. Data are the means of six determinations from three separate samples differing by no more than 10%.

their addition, a net decrease of cAMP and increase in cGMP with the maximum within 10 and 30 min, respectively. Both the nucleotides reach the basal values at 1 h, but thereafter a new fall in cAMP and increase in cGMP occur, still evident after 8 h. The early changes of the cyclic nucleotides induced by exogenous PL lead to a dramatic fall (about 70%) of the cAMP/cGMP ratio within 10 min. This is characteristic of confluent and serum-starved cells stimulated by serum, growth factors or nutrients to enter the cell cycle again and to proliferate (47,48).

Also the behaviour of polyamines in the presence of the two PL (Fig. 5) is similar to that observed by us with serum in heart cells (10). Both the PL cause a net accumulation of putrescine (PTC), SPD and SPM, which, particularly in the case of PC, is

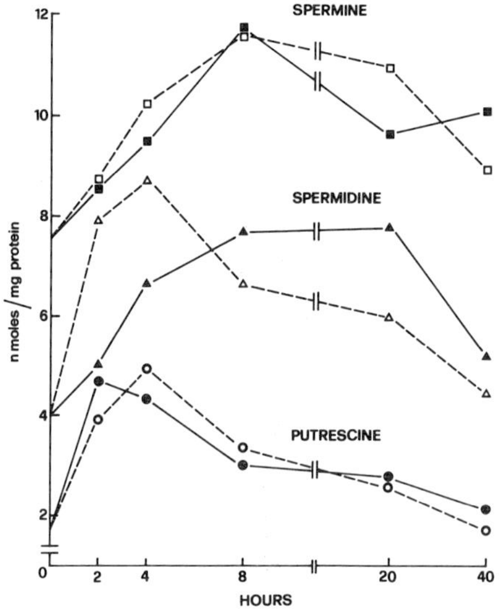

Fig. 5. Changes in polyamine contents of PL-stimulated heart cells. Experimental conditions were as given in Fig. 4. PC: closed symbols; PE: open symbols. Data are the means of six determinations from three separate experiments differing by no more than 10%.

still evident after 40 h. Maximum increase of PTC occurs within 2-4 h, followed by a progressive decline. SPD accumulates earlier with PE but remains more elevated in time with PC. No significant differences are observable in SPM during the first 8 h between the two PL. It should be noted that the increase in polyamine contents is concomitant with the last changes in cAMP and cGMP (see Fig. 4), thus suggesting a cause-effect relationship between the two events.

This hypothesis is supported by the results of experiments performed by using a specific inhibitor of polyamine biosynthesis, such as α-difluoromethylornithine (DFMO). In fact, the presence of DFMO while preventing polyamine accumulation by PC or PE (Table 1), significantly reduce the changes in cAMP and cGMP occurring 1 h after the addition of PC or PE (Fig. 6). The data provide further evidence that polyamines are involved in the control of cellular cyclic nucleotide contents, as elsewhere reported (11).

The rise in polyamine contents induced by PC and PE is related

TABLE 1. Effect of 1 mM DFMO on polyamine accumulation in PL-stimulated heart cell cultures[a]

Addition	Time (h)	PTC	SPD	SPM	PTC + SPD + SPM
		(nmol/mg protein)			(neq N/mg protein)
None (control	0	1.6	4.0	7.5	45.2
PC	4	4.3	6.6	9.5	66.4
	8	3.0	7.7	11.7	75.9
PC + DFMO	4	2.1	4.1	7.8	47.7
	8	1.8	4.0	7.8	46.8
PE	4	4.9	8.7	10.2	76.7
	8	3.3	6.6	11.6	72.8
PE + DFMO	4	2.0	4.2	7.7	49.4
	8	1.5	3.9	7.8	45.9

[a]Experimental conditions were as given in Fig. 4.

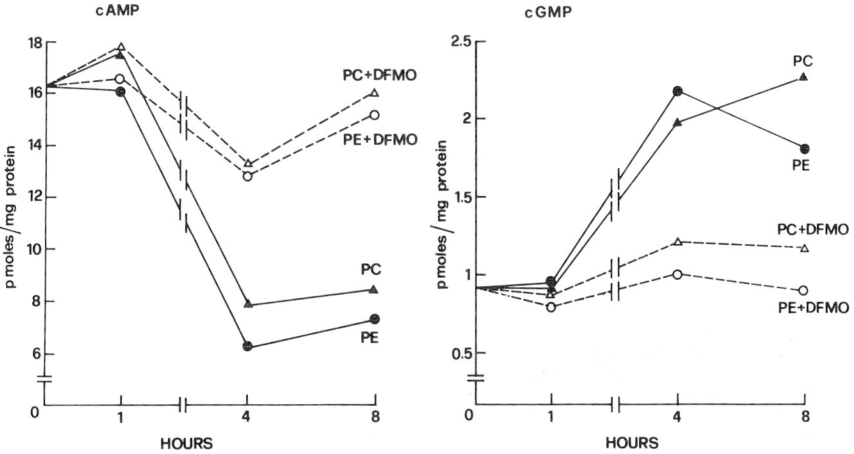

Fig. 6. Effect of 1 mM DFMO on cyclic nucleotides in PL-stimulated heart cells. Experimental conditions were as given in Fig. 4. Data are the means of six determinations from three separate experiments, differing by no more than 10%.

to a significant increase in the activity of the first two enzymes of polyamine biosynthesis, such as ornithine decarboxylase (ODC) and S-adenosylmethionine decarboxylase (SAMD) (45). Fig. 7 shows that ODC activity is maximal 2-3 h after the addition of each PL, particularly with PC.

As far as the mechanisms by which the two PL increase the activity of the enzyme, its independence by cAMP is supported by the

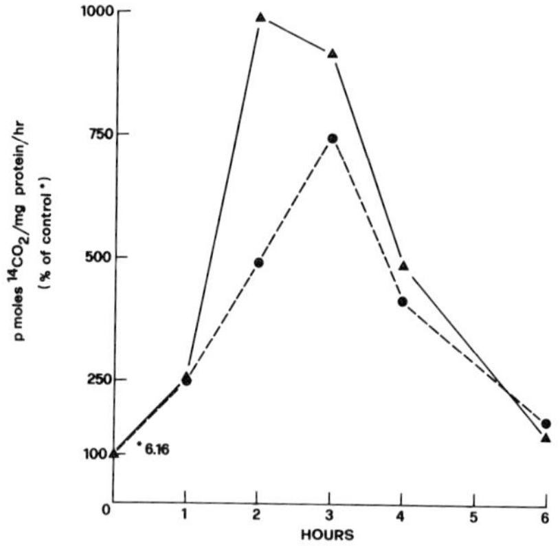

Fig. 7. Effect of exogenous PL on ODC activity heart cells. Experimental conditions were as given in Fig. 4. Data are the means of six determinations from three separate experiments differing by no more than 10%. PC (▲); PE (●).

data of Fig. 4. Like cAMP, it has been observed that cGMP too can induce ODC via a protein-kinase activation (33), however its possible involvement in PL action must be supported by experimental evidence. The failure to prevent cGMP accumulation without affecting cAMP makes this study quite problematic. It has been shown (35) that PC and PE, differently from anionic PL, increase the activity of a highly purified enzyme from rat liver. A direct activation of the enzyme by the two PL, in our conditions, must involve their transfer as intact molecules into the cells. However, experiments with inhibitors of RNA or protein synthesis give support for a de novo synthesis of the enzyme in the presence of the two PL (Table 2). In fact both actinomycin D and cycloheximide completely prevent the increase in ODC activity elicited by PC and PE, measured at the time of the respective peaks. Both the inhibitors did not affect the basal activity (data not shown).

It is quite difficult at the present to understand the mechanism by which PL induce the synthesis of the enzyme. It should be worthwhile to consider whether it might involve an increase in cGMP-dependent protein kinase activity leading to an increase in the phosphorilation of nuclear proteins (33), which represents an important derepression mechanism. The capability of PC and PE to affect protein kinase activity has been reported (16). Alternatively PL, taken up by the cells as intact molecules, might affect the genome activity, as observed in vitro (6,39). Both the possibilities do not exclude each-other and will be the object of our future investigations.

The increase in ODC activity is accompanied by an evident increase, more than 5 fold the basal value, in SAMD activity (Fig.8). While PE causes the maximum increase within 2-4 h, PC progressively stimulates it in time. The different effect of the two PL may account for the observed differences in the behaviour of SPD (see Fig. 5).

In many types of cultured cells the increase in polyamine contents is accompanied by the onset of DNA synthesis (26,29,32). This is especially the case of serum-stimulated quiescent cells.

TABLE 2. Effects of actinomycin D or cycloheximide on ODC activity in PL-stimulated heart cells[a]

Addition	Time (h)	ODC activity (pmoles $^{14}CO_2$/mg protein/h)	%
None (control)	0	7.20	100
PC	2.5	83.37	1158
PC + ACT.D	2.5	7.63	106
PC + CIX	2.5	9.14	127
PE	3	55.72	773
PE + ACT.D	3	7.24	100
PE + CIX	3	8.28	115

[a]Experimental conditions were as given in Fig. 4, except that 5 μg/ml of actynomicin D (ACT.D) or 25 μg/ml of cycloheximide (CIX) were added together with phosphatidylcholine (PC) or phosphatidylethanolamine (PE). Data are the means of six determinations from three separate experiments differing by no more than 10%.

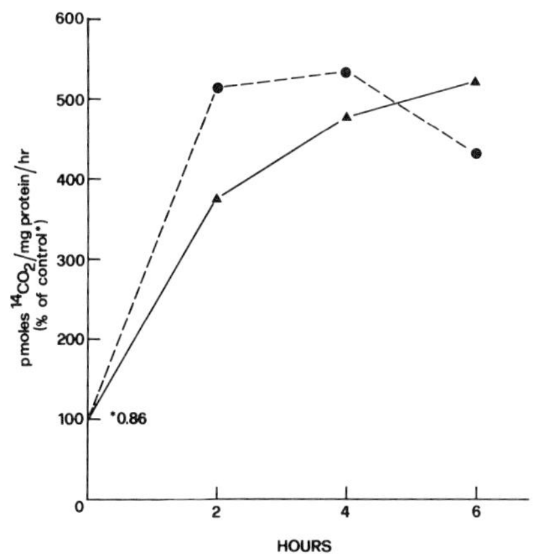

Fig. 8. Time course of SAMD activity in PL-stimulated heart cells. Experimental conditions are as given in Fig. 4. Data are the means of six determinations from three separate experiments, differing by no more than 10%. PC (▲); PE (●).

In order to understand whether both these events take place in PL-stimulated cells and are linked by a cause-effect ralationship, we have evaluated and compared the effects of serum or PL additions on polyamine accumulation and on ^{3}H-thymidine (^{3}H-TdR) incorporation into DNA, in the presence or absence of DFMO. From Table 3 it may be seen that the addition of fresh MEM to confluent and serum-starved cells has no effect on polyamine accumulation and ^{3}H-TdR incorporation during the first 20 h, but after 40 h it leads to their decrease. On the other hand, 10% serum-containing MEM causes a net rise in polyamine content and a parallel increase in ^{3}H-TdR incorporation, particularly evident 20 h after its addition. The presence of DFMO totally abolishes polyamine accumulation by serum, but the synthesis of DNA is not greatly disturbed

TABLE 3. Effect of DFMO on polyamine accumulation and DNA synthesis in confluent heart cells treated with serum or PL[a]

Addition	Time (h)	PTC	SPD	SPM	^{3}H-TdR incorporation (dpm/mg protein x 10^{-3})
		(nmoles/mg protein)			
None (control)	0	1.6	4.0	6.7	9.8
MEM	20	1.7	3.6	6.2	10.6
	40	1.3	3.2	5.0	5.3
MEM+serum	20	6.3	9.8	10.8	292.8
	40	4.3	8.4	9.6	177.4
MEM+serum+DFMO	20	1.0	2.6	7.9	271.6
	40	0.8	2.4	7.7	84.5
MEM+PC	20	2.7	7.7	8.5	28.0
	40	2.1	5.2	8.9	15.6
MEM+PC+DFMO	20	1.0	3.4	6.0	10.0
	40	0.8	2.8	4.8	5.6
MEM+PE	20	2.5	5.9	9.7	20.3
	40	1.7	4.4	7.9	11.6
MEM+PE+DFMO	20	1.0	2.7	5.4	16.3
	40	0.7	2.6	5.2	5.3

[a]Confluent and 20 h serum-starved heart cells were fed with fresh MEM containing 10% foetal calf serum, PC (20 μg/ml) or PE (20 μg/ml) ± 1 mM DFMO. Data are the means of six determinations from three separate samples, differing by no more than 10%.

for up to 40 h. Similar data have been reported with other cell types (27,29) and it has been suggested that the inhibition of polyamine accumulation does not affect the first round of cell division. As in the case of serum readdition, the addition of PC or PE while increasing polyamine contents causes the onset of DNA synthesis. This is particularly evident in the presence of PC and after 20 h of incubation with each PL. However the magnitude of these changes is far from that elicited by 10% serum, which provides other important factors able to act as positive signals for the reinitiation of cell proliferation (28). The possibility that also in the case of PL the increase in polyamine contents correlates with that of DNA synthesis is supported by the fact that addition of DFMO together with the PL, while blocking polyamine accumulation, either counteracts (PE) or prevents (PC) the onset of DNA synthesis at 20 h, but later it prevents in both cases.

CONCLUSION

All together these data provide evidence that in primary beating heart cell cultures exogenous PL are able to affect a variety of biochemical processes notable linked to the rate of cell growth. This has been observed both in the presence and absence of serum and by utilizing single PL or a complex of different PL. The effects of PL are more evident as cells approach confluency or are completely confluent, when their capability to synthesize lipid de-novo or to take up precursors or nutrients is strongly reduced (31), and their requirement for growth stimulating factors is increased (22). In the usual culture medium, growth stimulating activity is supplied by the components of serum, which provides also a lipid source for the cells, including PL. The importance of lipids in the growth stimulating activity of serum is supported by the fact that lipid-depleted serum no longer supports cell growth (25,30). Probably this might also be due to the fact that some enzymatic activities, whose products are closely involved in the growth regulation (32,36), strongly depend upon the lipid environment (46,52) or, as in the case of guanylate cyclase, respond to lipids but are insensitive to hormonal agents (19,52). These and other considerations might account for the capability both of a PL complex to potentiate, in certain experimental conditions, the effect of serum and of single PL to partially substitute for serum. Furthermore the data indicate that, like serum-stimulated heart cells, PL-stimulated cells accumulate polyamines as consequence of an elevated biosynthetic activity. The property of ODC and SAMD activities to early and rapidly increase in response to PL addition, suggests that as in the case of other growth promoting factors, polyamine accumulation might be a very important step in the mechanism of action of exogenous PL. This hypothesis is indirectly supported by the fact that by blocking polyamine synthesis with DFMO the response of the cells to PL addition is strongly reduced or completely prevented, at least as far as the cyclic nucleotide contents and DNA synthesis are concerned. Further experiments are in progress to clarify the role of polyamines in the action of PL on other cellular activities.

ACKNOWLEDGMENTS

This work was supported by a grant from Ministero della Pubblica Istruzione, Rome (Italy).

REFERENCES

1. Bachrach, U. (1976): Biochem. Biophys. Res. Commun., 72: 1008-1013.
2. Bachrach, U. (1976): FEBS Lett., 75:201-204.
3. Barnett, R.E., Furcht, L.T., and Scott, R.E. (1974): Proc. Nat. Acad. Sci. U.S.A., 71:1992-1994.
4. Borst, P., Loos, J.A., and Christ, E.J. (1962): Biochim. Biophys. Acta, 62:509-518.
5. Bradford, M.H. (1976): Anal. Biochem., 72:248-254.
6. Capitani, S., Caramelli, E., Felcao, M., Miscia, S., and Manzoli, F.A. (1981): Physiol. Chem. Phys., 13:153-158.

7. Ceriotti, F. (1955): J. Biol. Chem., 214:59-70.
8. Chan, P.S., and Lin, M.C. (1974): Methods in Enzymology, 38: 38-41.
9. Clô, C., Pignatti, C., Coccolini, M.N., Tantini, B., and Turchetto, E. (1981): Ital. J. Biochem., 30:271-278.
10. Clô, C., Tantini, B., Coccolini, M.N., and Caldarera, C.M. (1981): Adv. Polyamine Res., 3:333-345.
11. Clô, C., Tantini, B., Pignatti, C., Guarnieri, C., and Caldarera, C.M. (1982): Adv. Polyamine Res. This volume.
12. Cocco, L., Maraldi, M.N., and Manzoli, F.A. (1980): Biochem. Biophys. Res. Commun., 96:890-898.
13. Cornell, R., Grove, G.L., Rothblat, G.M., and Horwitz, A.F. (1977): Exp. Cell Res., 109:299-307.
14. Cornwell, D.G., Huttner, J.J., Milo, G.E., Panganamala, R.V., Sharma, H.M., and Geer, J.C. (1979): Lipids, 14:194-207.
15. Cunningham, D.D. (1972): J. Biol. Chem., 247:2464-2470.
16. Endo, T., and Hidaka, H. (1981): Arch. Biochem. Biophys., 211:108-112.
17. Frelin, C., and Padieu, P. (1976): Biochimie, 58:953-959.
18. Gavino, V.C., Miller, J.S.,Ikharebha, S.O., Milo, G.E., and Cornwell, D.G. (1981): J. Lipid Res., 22:763-769.
19. Goldberg, N.D., and Haddox, M.K. (1977): Ann. Rev. Biochem., 46:823-896.
20. Gregoriadis, G., Siliprandi, N., and Turchetto, E. (1977): Life Sci., 29:1773-1786.
21. Griffiths, J.B. (1971): J. Cell Sci., 8:43-52.
22. Griffiths, J.B. (1972): Exp. Cell Res., 75:47-56.
23. Hachmann, M.J., and Lezius, A.G. (1975): Europ. J. Biochem., 50:357-366.
24. Harary, I., Seraydarian, M., and Gershenson, L.E. (1974):In: Factors Influencing Myocardial Contractility, edited by R.D. Tanz, F. Kavalerand, R. Jay, pp. 231-243. Academic Press, New York.
25. Hatten, M.E., Horwitz, A.F., and Burger, M.M. (1977): Exp. Cell Res., 107:31-34.
26. Heby, O., Marton, L.J., Zardi, L., Russell, D.H., and Baserga, R. (1975): Exp. Cell Res., 90:8-14.
27. Heby, O., and Janne, J. (1981): In: Polyamines in Biology and Medicine, edited by D.R. Morris and L.J. Marton, pp. 243-310. Marcel & Dekker, New York.
28. Holley, R.W. (1975): Nature, 258:487-490.
29. Holtta, E., Pohjanpelto, P., and Janne, J. (1979): FEBS Lett., 97:9-13.
30. Horwitz, A.F., Wight, A., Ludwig, P., and Cornell, R. (1978): J. Cell Biol., 77:334-357.
31. Howard, B.U., and Howard, W.J. (1974): Adv. Lipid Res., 12: 51-96.
32. Janne, J., Pösö, H., and Raina, A. (1978): Biochim. Biophys. Acta, 437:241-293.
33. Johnson,E.M., and Hadden, J.W. (1975): Science, 187:1198-1200.

34. Jonah, M.M., Cerny, E.A., and Rahman, Y.E. (1975): Biochim. Biophys. Acta, 401:336-348.
35. Kitani, T., and Fujisawa, H. (1981): FEBS Lett., 132:296-298.
36. Kram, R., Mamont, P., and Tomkins, G.M. (1973): Proc. Nat. Acad. Sci. U.S.A., 70:1432-1436.
37. Lenaz, G. (1977): In: Membrane Proteins and their Interactions with Lipids, edited by R.A. Capaldi, vol. 1, pp.47-49. Marcel Dekker Inc., New York.
38. Lowry, O.H., Rosebrough, N.J., and Farr, A.L. (1951): J. Biol. Chem., 193:265-275.
39. Manzoli, F.A., Muchmore, J.B., Bonora, B., Capitani, S., and Bartoli, S. (1974): Biochim. Biophys. Acta, 340:1-15.
40. Manzoli, F.A., Capitani, S., Maraldi, N., Cocco, L., and Barnabei, O. (1977): Adv. Enzyme Regulation, 17:175-194.
41. Ozato, K., Huang, L., and Pagano, R.E. (1978): Membr.Biochem., 1:27-42.
42. Patterson, M.K. (1979): Methods in Enzymology, 58:141-152.
43. Pagano, R.E. (1978): Ann. Rev. Biophys. Bioeng., 7:435-468.
44. Poste, G. (1980): In: Liposomes in Biological Systems, edited by G. Gregoriadis and A.C. Allison, pp. 101-151. John Wiley & Sons, London.
45. Raina, A., and Janne, J. (1975): Med. Biol., 53:121-147.
46. Sandermann, H.Jr. (1978): Biochim. Biophys. Acta, 515:209-237.
47. Schönhöfer, P.S., and Peters, H.D. (1977): In: Cyclic 3'-5' Nucleotides: Mechanism of action, edited by H. Cramer and J. Schultz, pp. 107-131. John Wiley & Sons, London.
48. Seifert, W., and Rudland, P.S. (1974): Proc. Nat. Acad. Sci. U.S.A., 71:4920-4924.
49. Seiler, N., and Weichman, N. (1965): Experientia, 15:203-204.
50. Shier, W.T., Baldwin, J.M., Hamilton, N.M., Hamilton, R.T., and Thanassi, N.M. (1976): Proc. Nat. Acad. Sci. U.S.A., 73: 1586-1590.
51. Weeks, C.E., and Abdel-Monem, M.M. (1977): J. Pharmac. Sci., 66:1586-1589.
52. White, A.A., Karr, D.B., and Patt, C.S. (1982): Biochem. J., 204:383-392.
53. Witter, B., and Debuch, H. (1982): J. Neurochem., 38:1029-1038.
54. Wolff, D.J., and Brostrom, C.O. (1976): Arch. Biochem. Biophys. 173:720-731.

Advances in Polyamine Research, Vol. 4, edited by
U. Bachrach, A. Kaye, and R. Chayen. Raven Press,
New York © 1983.

The Cinnamic Acid Amides of the Di- and Polyamines

Terence A. Smith, Jonathan Negrel, and Colin R. Bird

Long Ashton Research Station, University of Bristol, Long Ashton, Bristol, BS18 9AF, United Kingdom

Although the sporadic identification of the conjugates formed between amines and cinnamic acids in various plants has been reported over many years, it is only recently that the widespread nature and potential significance of these amides has been recognised. In higher plants loss of ammonia from phenylalanine to give cinnamic acid and from tyrosine to give 4-coumaric acid are major pathways, ultimately leading to the biosynthesis of lignin (35, 106). In animals, cinnamic acid and other analogous phenolic acids are not major metabolites and when these acids are found in the urine, they probably originate principally in the diet. Even so, 4-coumaric acid is known to be a metabolite of tyrosine in the rat and rabbit (18, 75). The aromatic acids which are excreted occur both in the free form and as glycine conjugates (3). Di- and polyamines are found widely as amides of acetic acid in animals (1), although these acetyl derivatives have not hitherto been reported in plants. Both acetyl (81) and, more rarely, cinnamoyl (30) derivatives of the amines are found in micro-organisms.

The di- and polyamine conjugates formed with the cinnamic acids are found in many higher plants. Those occurring in the tobacco plant have been the subject of recent intensive research by groups at Dijon and Braunschweig. In other higher plants the cinnamoyl-amine conjugates form complex alkaloids which are considered later in this review. Amines other than the di- and polyamines are found in combination with cinnamic acids in plants. The di- and polyamines are also found as amides of fatty acids in certain plants and the occurrence of some of these conjugates has been reviewed by Smith (84).

We have made a study of the antifungal hordatines found in barley which are dimers of coumaroylagmatine and we have now investigated the biosynthesis of this amide and purified the enzyme responsible for its formation (16). It seems likely that the related amides found in other plants are synthesized by a basically similar mechanism.

PHENOLIC AMIDES IN THE TOBACCO PLANT

In tobacco, hydroxycinnamic acid amides were first isolated from callus tissue culture by Mizusaki et al. (61) who were working on nicotine biosynthesis. They showed that radioactivity from ^{14}C-ornithine and ^{14}C-putrescine could be incorporated into 4-coumaroylputrescine, caffeoylputrescine and feruloylputrescine (44, 61). Although at that time these workers could not find any hydroxycinnamic acid amides in normal tobacco tissue, caffeoylputrescine was later isolated from the flowers and vegetative apex of tobacco plants (19). More recently the general occurrence and physiology of hydroxycinnamic acid amides in *Nicotiana tabacum* has been studied in France and Germany.

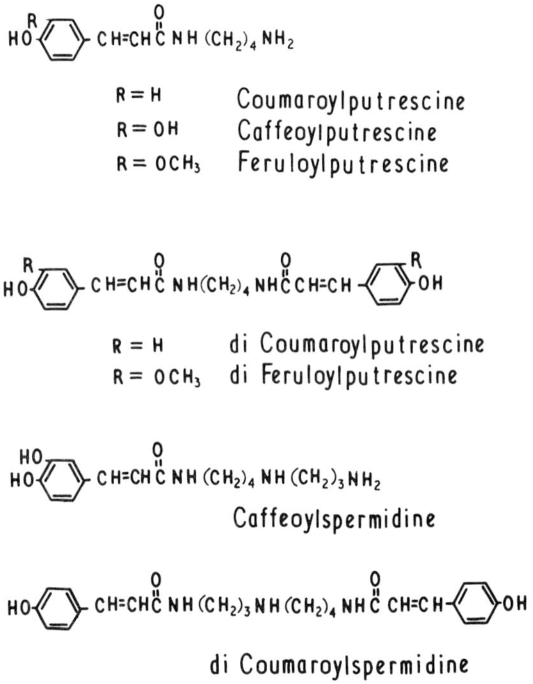

HO-(R)C₆H₃-CH=CHC(=O)NH$(CH_2)_4NH_2$

R = H Coumaroylputrescine
R = OH Caffeoylputrescine
R = OCH_3 Feruloylputrescine

HO-(R)C₆H₃-CH=CHC(=O)NH$(CH_2)_4$NHC(=O)CH=CH-C₆H₃(R)-OH

R = H di Coumaroylputrescine
R = OCH_3 di Feruloylputrescine

(HO)₂C₆H₃-CH=CHC(=O)NH$(CH_2)_4$NH$(CH_2)_3NH_2$

Caffeoylspermidine

HO-C₆H₄-CH=CHC(=O)NH$(CH_2)_3$NH$(CH_2)_4$NHC(=O)CH=CH-C₆H₄-OH

di Coumaroylspermidine

FIG. 1. Amides of hydroxycinnamic acids formed with di- and polyamines which are found in tobacco plants together with coumaroyl- and feruloyltyramine

Coumaric, caffeic and ferulic acids are found conjugated with putrescine, spermidine and tyramine in the tobacco plant in large concentrations and in a variety of combinations (Fig. 1). *Nicotiana tabacum* cv. *Xanthi* n.c. forms large amounts of mono- and di-feruloylputrescine after infection with tobacco mosaic virus (TMV). Similarly *Nicotiana sylvestris* forms di-feruloyl-putrescine on infection with the TMV strain Aucuba. These amides are at their greatest concentration when synthesis of the virus has ended (55). They accumulate especially in the living cells surrounding the hypersensitive zone, together with lesser amounts of coumaroylputrescine and di-coumaroylputrescine. Application to excised leaf discs at ca. 1 mM caused a significant reduction in the number of lesions on infection with TMV by comparison with an untreated control (56).

The concentrations of free putrescine, spermidine and spermine were relatively unchanged with age, although free tyramine increased 5-fold in the flowers (28,72). Potassium, sulphur and phosphorus deficiencies caused putrescine accumulation in tobacco (94,105) and deficiencies of potassium, calcium, magnesium and phosphorus gave rise to the accumulation of caffeoylputrescine and mono- and di-caffeoylspermidine in tobacco leaves (26).

Caffeoylputrescine and several other amides appear in especially large concentrations in tobacco flowers. Ovaries were shown to contain mono-caffeoylputrescine and mono-caffeoyl-spermidine, while the anthers contained di-coumaroylputrescine and di-coumaroylspermidine. In addition anthers contained coumaroyltyramine. The total concentration was about 2 μmol/g fresh weight in the floral spike after 75 days growth. The amides are accumulated in the tops at 20°C after floral initiation. On increasing the temperature, the plant becomes vegetative and the amide concentration is reduced. Only 0.1 μmol/g fresh weight of caffeoylputrescine was found in plants grown at 30°C (20,21). Removal (topping) of the upper flowering spike of plants which are more than 65 days old causes accumulation of amides in the remaining upper leaves, rising to 13 μmol/g fresh weight. On regeneration of the axillary buds, this concentration declines. These experiments in which amide concentration declines indicate that the amides are not always end products of metabolism, but that they could also undergo catabolism (21).

At first it was thought that the amides migrated from the leaves to the flowers acting as a 'florigen'. This was given some support by the fact that the amides were accumulated on floral initiation and well before bud formation. Moreover the experiment in which amide accumulation was found on removal of the tops indicated an accumulation of amides which could no longer be translocated. However, since the key enzyme involved in cinnamic acid formation, phenylalanine ammonia-lyase, is greatly increased in the remaining leaves on topping, the increase is not attributable solely to loss of a sink (22). Even so, the tobacco mutant RMB_7, which cannot flower never synthesizes the amides. Moreover TMV causes formation of local lesions in this mutant, but the multiplication of the TMV becomes systemic (52).

The close association of these amides with fertile reproductive organs in tobacco and in other plants (see next section) suggests the possibility that their formation is functionally linked with reproduction. Moreover viruses are rarely seed-transmitted, and resistance to viruses is often associated with flowering. In addition, since viral multiplication is associated with the formation of amides and appears to be retarded in their presence, a role in virus resistance is implied (56).

The effect of metabolic inhibitors on the formation of the hydroxycinnamic acid amides in tobacco cell cultures has been studied in Germany. Plant cells which are resistant to fluorophenylalanine have often been found to accumulate greater concentrations of phenolic compounds than wild type cells (7). A fluorophenylalanine resistant line of tobacco cells accumulated 6-10 times more phenolic compounds, of which more than 85% were caffeoyl- and feruloylputrescine. This was accompanied by a greatly increased activity of phenylalanine ammonia-lyase (12), and arginine decarboxylase (ADC) and ornithine decarboxylase (ODC) (8, 10, 13). The cinnamoylputrescines increased and ODC activity was considerably enhanced on limitation of phosphate in the medium (46,47). Wild type cells showed a greater response to phosphate deficiency than cells which had been selected for the phenolic compounds (8). In the wild type cell, the ODC inhibitor 2-difluoromethylornithine (DFMO) (0.1 to 0.5 mM) inhibited cell division but caused considerable cell enlargement. Growth of the 4-fluorophenylalanine resistant line was not altered by DFMO. The biosynthesis of the amides in the wild type cell line was increased by 50% on a dry weight basis in the presence of DFMO. Since ODC shows no activity in the DFMO treated cells, the putrescine must therefore be formed via ADC (9). Accumulation of putrescine amides may also be induced in the wild type cell line by adding phenylalanine, ornithine, or 2-fluorophenylalanine to the growth medium. On removal of these inducers, the amide concentration remained high for 3 cell cycles (11).

SIMPLE PHENOLIC AMIDES IN OTHER PLANTS

The hydroxycinnamic acid amides have now been found throughout the plant kingdom. In a survey of 20 species representing 13 families they occurred as the main phenolic constituents of the reproductive organs of all plants studied (53). In fertile maize, amides of ferulic acid with di- and polyamines are present in the anthers, but anthers of a male sterile line F7T having Texas male sterile cytoplasm contained no amides, and other male sterile maize lines also had no amides. Seeds and ovaries of both male fertile and sterile maize contained a wide spectrum of the amides. Moreover in the seeds which were to produce the sterile plants feruloylputrescine was absent (54).

2-Hydroxyputrescine amides of 4-coumaric acid and ferulic acid were detected in the leaves of wheat infected with rust and other pathogens. These amides could also be induced on treatment of wheat leaves with hot water, but they could not be detected in oats or barley. The origin of the hydroxyputrescine is not yet established, although ^{14}C-putrescine was not incorporated into the amides (79,92). It seems possible that the hydroxyputrescine arises from a pathway involving hydroxyarginine, an amino acid which has been demonstrated in legumes (6). Free hydroxyputrescine could not be detected in the wheat plant (79,92).

Coumaroylhydroxyputrescine

$HO-C_6H_4-CH{=}CH-C(=O)-NHCH_2CH(OH)CH_2CH_2NH_2$

Feruloylhydroxyputrescine

$CH_3O(HO)C_6H_3-CH{=}CH-C(=O)-NHCH_2CH(OH)CH_2CH_2NH_2$

Caffeoylputrescine, also known as paucine, (Fig. 1) is found in the seeds of *Pentaclethra macrophylla* (pauco nuts, Leguminosae) in which it is the main ninhydrin positive component, the hydrochloride dihydrate comprising 1.4% of the dry weight (57). Seeds of *P. macrophylla* showed a growth depressant activity for rats, despite amino acid supplementation. This may be attributed to the presence of the caffeoylputrescine (58).

Feruloylputrescine (Fig. 1) was first discovered in *Salsola subaphylla* (Chenopodiaceae) and given the name subaphylline (78). Feruloylputrescine has been identified in grapefruit and oranges but it could not be found in tangerine or lemons. In grapefruit the content ranged from 22 to 34 µg/ml juice (103). Feruloylputrescine and other related compounds have hypotensive activity. In its hypotensive properties and its effects on respiration in the cat feruloylputrescine resembles isoamylenylputrescine, but its action is less pronounced. When used intravenously at 2 mg/kg bodyweight, it causes a rapid drop in blood pressure which however recovers equally rapidly. The LD_{50} for feruloylputrescine is 225 mg/kg bodyweight in white mice (70,71).

Methylated putrescine amides are found in leaves of *Kniphofia flavovirens*. Unlike most other amides which have the natural *trans* form of the cinnamic acids, these *Kniphofia* amides were found mainly in the *cis* form (77). These amides are:-

Coumaroylmethylputrescine

$HO-C_6H_4-CH{=}CH-C(=O)-N(CH_3)(CH_2)_4N(CH_3)_2$

Methoxycinnamoylmethylputrescine

$CH_3O-C_6H_4-CH{=}CH-C(=O)-N(CH_3)(CH_2)_4N(CH_3)_2$

Maytenine which is an antipyretic and vasodilator is obtained from the South American *Maytenus chuchuhuasha* where it occurs in the roots and bark (31,41,59,80).

Maytenine

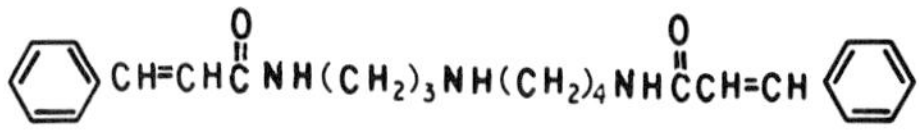

Tricoumaroylspermidine has been found in the hawthorn *Crataegus* sp.) where it occurs only in the flowers. The structure was established by synthesis (42).

Tricoumaroylspermidine

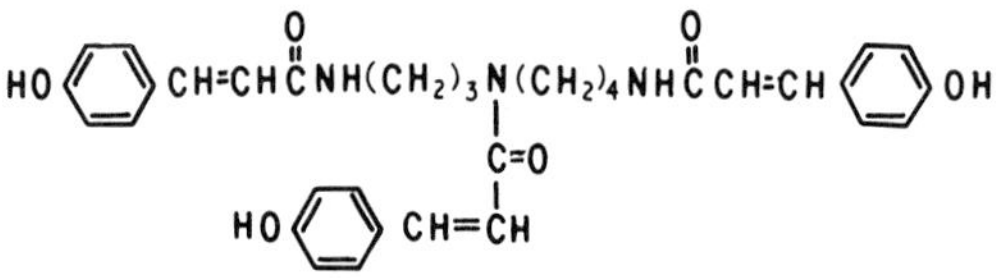

Kukoamine, which is a linear molecule derived from spermine and two coumaric acid moieties with reduced sidechains, is the active constituent of 'jikoppi', an oriental medium used in the treatment of hypertension. This drug is prepared from root barks of *Lycium chinense* (Solanaceae) now cultivated as a garden plant in the western world (23).

Kukoamine

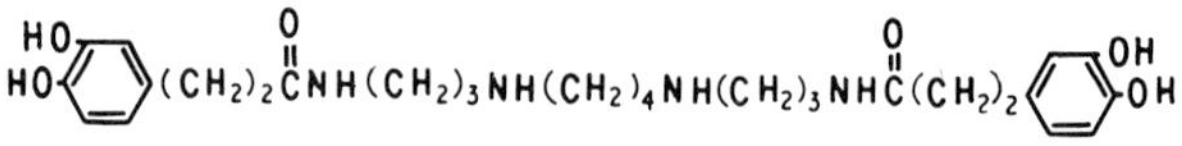

The cinnamic acid amine conjugates occur only rarely outside the plant kingdom. They have been identified in a new class of antibiotics, the glycocinnamoylspermidines, isolated from an unidentified species of *Nocardia*. These are effective against a wide range of gram negative bacteria and they show protective effects against infections in mice (30). Glycocinnamoylspermidine LL-BM 123 β has the following structure:

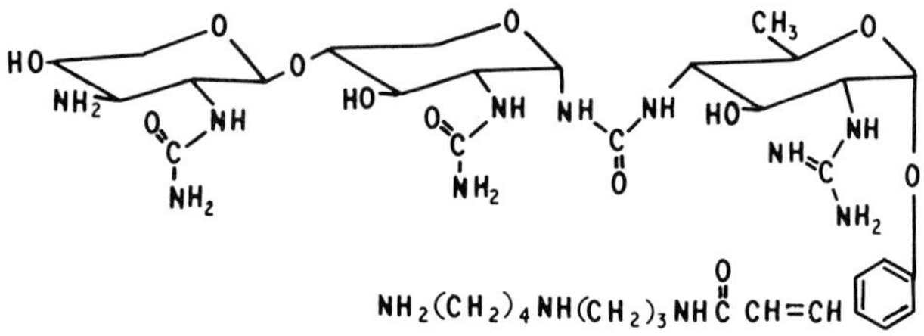

Compounds apparently related to the cinnamic acid amides are probably toxic components in the venom of bird catching spiders. These amides include the N-'(4-hydroxyphenylpyruvoyl)-derivatives of diaminopropane and spermine. Similar derivatives of 4-hydroxybenzoic acid, 4-hydroxyphenylacetic acid and 4-hydroxyphenyllactic acid also occurred (32,34,64).

N^1-(4-hydroxyphenylpyruvoyl)spermine

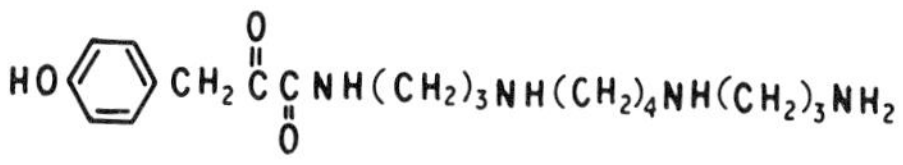

$HO-C_6H_4-CH_2-CO-CO-NH(CH_2)_3NH(CH_2)_4NH(CH_2)_3NH_2$

On feeding benzoic acid and 4-nitrobenzoic acid to the scorpion, these acids were recovered as amides of agmatine. These were derived by decarboxylation of the arginine derivatives which were synthesized first (40). Similar derivatives were formed from these acids in the arthropod *Peripatus* (43).

THE MACROCYCLIC POLYAMINE ALKALOIDS

In the plant kingdom many compounds have been found in which di- or polyamines are bound in amide linkage to cinnamic acid or its derivatives and some of the structures are quite complex. Periphylline, found in leaves of *Peripterygia marginata*, is one of six closely related alkaloids in this plant. All of these alkaloids contain a 13-membered ring system derived from di-cinnamoylspermidine (102). Celacinnine was first reported in *Maytenus arbutifolia (M. serrata)* and *Tripterygium wilfordii* (48, 49). It has also been isolated from the bark of *Maytenus heterophylla* and the leaves of *Pleurostylia africana*, all members of the Celastraceae (99) and total synthesis has now been accomplished (59,104).

Periphylline

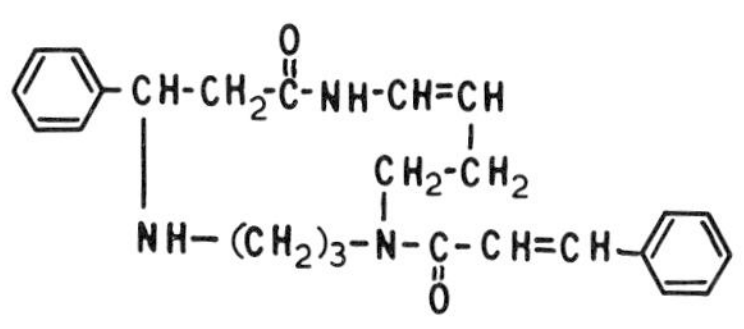

Celacinnine

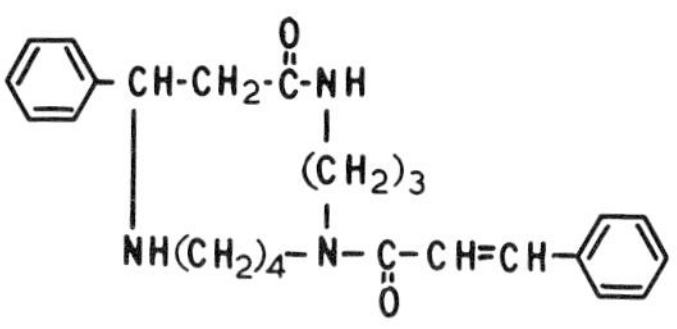

Another related alkaloid, pleurostyline, also occurs in the leaves of *Pleurostylia africana* (100). In *Maytenus mossambicensis* the alkaloid cyclocelabenzine occurs, together with an isomeric form, in which both cinnamic acid and benzoic acid are combined with spermidine (101). Mayfoline, isolated from aerial parts of *Maytenus buxifolia* which occurs in Cuba, is unusual in possessing an *N*-hydroxy moiety on the secondary nitrogen of the spermidine chains (76). Plants of this species found in one locality also contained mayfoline with the *N*-acetyl group in place of the *N*-hydroxy group (27).

Pleurostyline

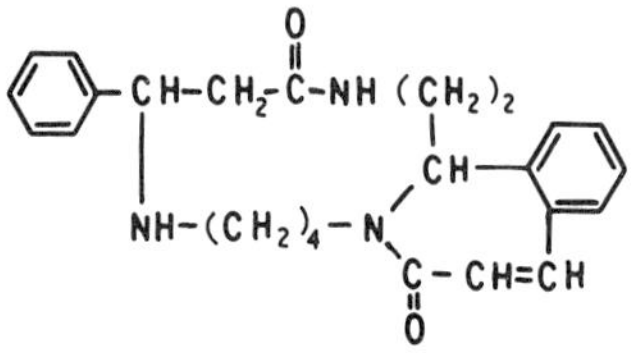

Cyclocelabenzine

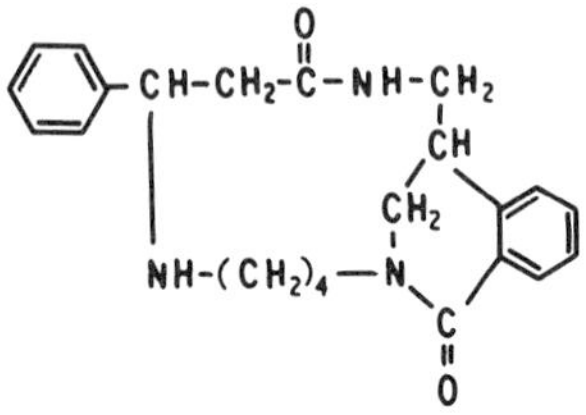

Mayfoline

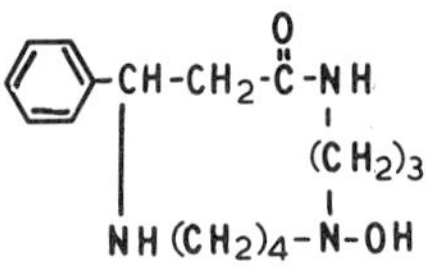

Lunarine, together with five closely related alkaloids, occurs in the seeds of *Lunaria biennis* and *L. rediviva*. One of these, lunaridine, differs from lunarine only in the inversion of the spermidine (65). Codonocarpine and *N*-methyl codonocarpine have been isolated from the bark of *Codonocarpus attenuatus* (Phytolaccaceae). The structure has been established by synthesis of tetrahydrocodonocarpine (74).

Lunarine

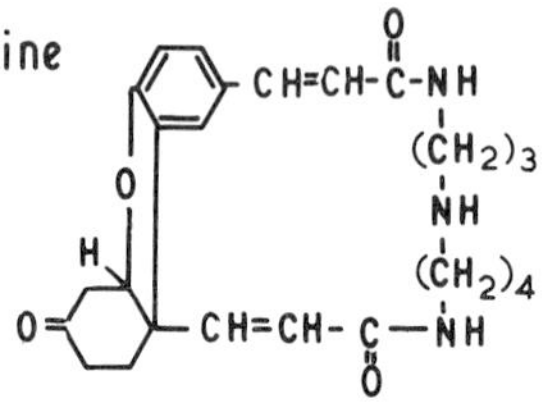

Codonocarpine

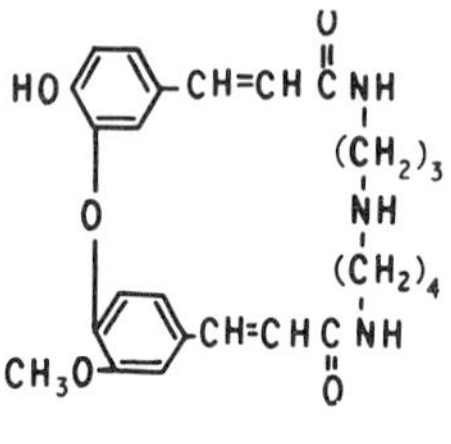

Aphelandrine

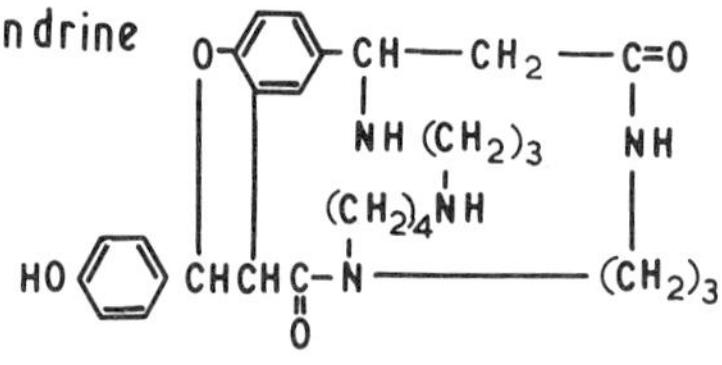

Aphelandrine, a spermine-based alkaloid, occurs in the roots of *Encephalosphaera lasiandra* and in various *Aphelandra* species including *A. squarrosa* (Acanthaceae), often grown as an ornamental house plant. This alkaloid, isolated from the roots of *A. squarrosa* with a yield of about 0.24% of the dry weight, possesses the phenylcoumaran moiety (24). This phenylcoumaran structure is also found in the hordatines (Fig. 2) which are considered separately in the next section.

Ephedradine, found in the roots of *Ephedra*, has hypotensive activity in rats (1.5 - 1.8 mg/kg i.v.). This alkaloid has a structure which is identical to that of aphelandrine, though of opposite absolute configuration (96). *O*-Methylorantine isolated from *Chaenorhinum minus* and *C. villosum* differs from ephedradine only in the possession of a methoxy group in place of the hydroxy group. The absolute configurations of ephedradine and methylorantine are identical (25).

Chaenorhine is also a spermine-based alkaloid obtained from the dried aerial parts of *Chaenorhinum origanifolium* (14). Homaline occurs in leaves of *Homalium pronyense* from New Caledonia (68,69).

Chaenorhine

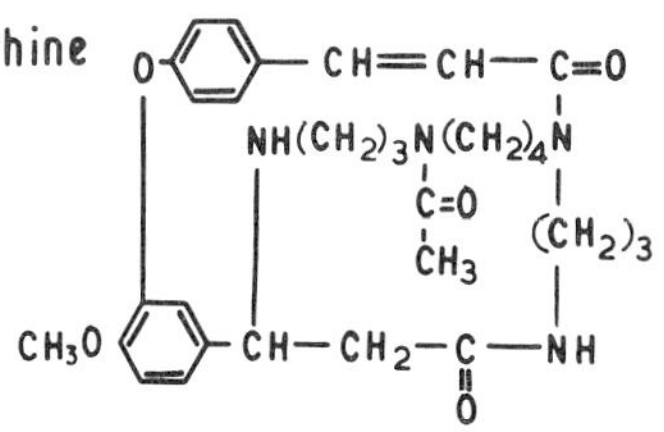

Homaline

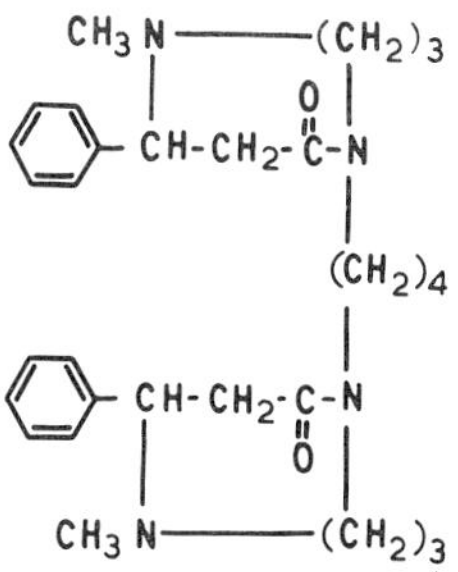

The occurrence of other alkaloids derived from the polyamines, combined with cinnamic acids or fatty acids has been reviewed by Smith (84), and Hesse and Schmid (38).

The function of the macrocyclic polyamine alkaloids is still obscure. They may be a part of a defence mechanism developed by certain plants in which the amine moiety occludes the nucleic acid of a pathogen, though this seems unlikely. One further possibility is suggested by the discovery of an antibiotic, Nocardamine, isolated from a *Nocardia* species. This has three

Nocardamine

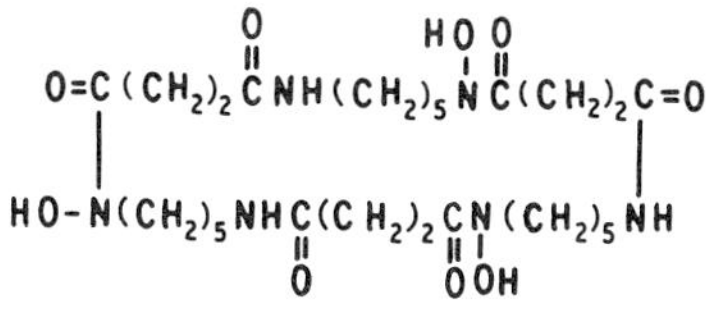

N-hydroxy groups (*cf.* Mayfoline) and has a 33 membered ring which consists of three cadaverine units in alternate amide linkage with three succinic acid units. This forms an iron III complex known

as ferrioxamine E (45,66). Desferrioxamine is now used as a treatment for iron toxicity in children. In *Nocardia* and in *Pseudomonas*, Nocardamine may be produced as an essential siderophore, and as an antibiotic, interfering with metal ion uptake in other competing organisms (60). Other siderophores have been found in *Agrobacterium* and *Paracoccus* which are based on phenolic amides of spermidine and which contain a threonine moiety. Agrobactin (73) from *A. tumefaciens* has the following structure:-

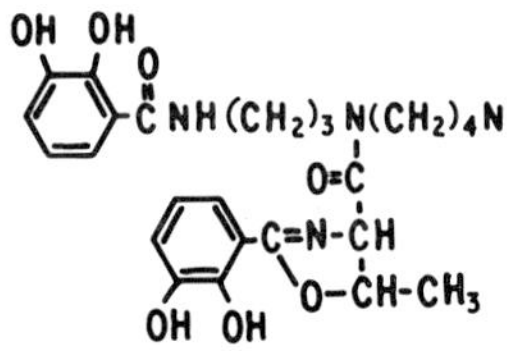

The possibility that the polyamine alkaloids in plants are concerned in ion transport may justify consideration.

THE HORDATINES

Structure and Fungitoxicity of the Hordatines

Resistance of very young barley seedlings to infection by *Helminthosporium sativum* is attributed to the occurrence of the antifungal dimers of 4-coumaroylagmatine known as the hordatines in the barley shoots (51,93). Purified hordatines inhibit the germination of a wide range of fungal spores; for example, 100% inhibition of *Monilinia* spores occurs at 7×10^{-6}M (93). The inhibitory properties are reduced *in vitro* by Mn^{2+} and Ca^{2+} and less so by Mg^{2+} and Fe^{2+}, while addition of the chelating agent EDTA restores the inhibition (51,85,93). A 2-fold increase in Ca^{2+} concentration occurs in seedlings after 5-6 days of growth and the antagonistic effect of Ca^{2+} probably explains the loss of resistance to fungal infection found after this stage despite a high concentration of hordatines at this time (85). Although high concentrations of calcium have been shown to increase the susceptibility of barley to infection by powdery mildew (*Erysiphe graminis* (39,95,98), this effect of calcium has not as yet been directly related to the antagonism of the antifungal effect of the hordatines.

The structures of the hordatines were established by Stoessl (90) who was able to isolate several active fractions from butanol extracts of barley shoots, none of which could be crystallized. Chemical studies of these fractions revealed the following optically active compounds (Fig. 2). Hordatine A ($R_1 = R_2 = H$) appears to be structurally a condensation product of two

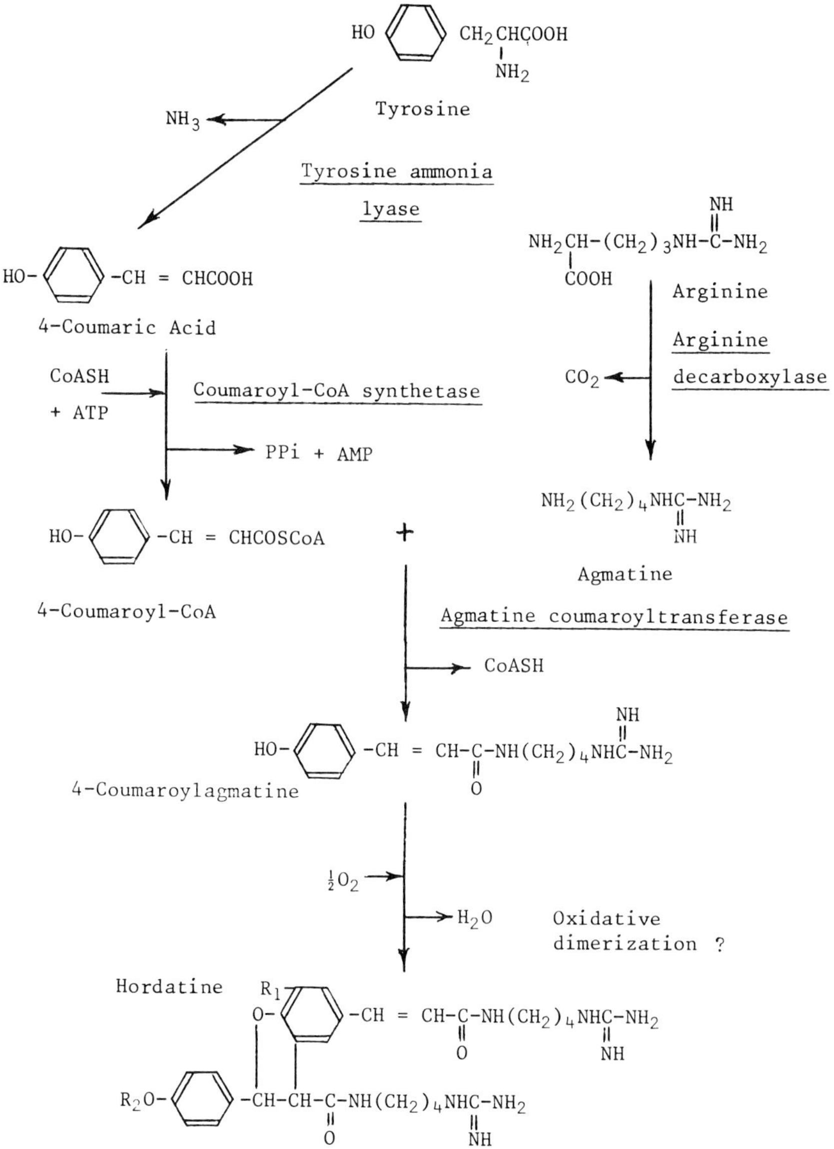

FIG. 2. Probable pathway of coumaroylagmatine metabolism in the shoots of barley seedlings. Dimerisation of coumaroylagmatine gives hordatine A (R_1 = R_2 = H). Barley seedlings also contain hordatine B (R_1 = OMe, R_2 = H) and hordatine M (R_1 = H or OMe, R_2 = D-glucopyranosyl). The biosynthesis of the hordatines from coumaroylagmatine is speculative at present (see text).

coumaroylagmatine molecules. The double bond is predominantly in the *trans* configuration, but exposure to UV light induces the formation of the *cis* isomer which shows a reduced antifungal activity. Hordatine B (R_1 = OMe, R_2 = H), which is similar to hordatine A, has a methoxy group on the coumaran skeleton and this dimer might be formed by condensation of coumaroylagmatine and feruloylagmatine. Hordatine M (R_1 = H or OMe, R_2 = D-glucopyranosyl-) is a mixture of the glucosides of hordatines A and B. An optically inactive but biologically active (±) hordatine A has been synthesized by oxidative coupling of coumaroylagmatine, using hydrogen peroxide in the presence of horseradish peroxidase (90). The stereochemistry on the coumaran ring has not been determined and it is difficult to distinguish between the *cis* and *trans* configuration; however it is possible that this isomerism could determine the degree of biological activity (91). It has not proved possible to define a simple structure-function relationship: a free phenolic hydroxyl is not required; agmatine and coumaroylagmatine alone are not active; there is no activity when the agmatine residues are removed from the hordatines; partial destruction of the coumaran ring and removal of one agmatine reduces antifungal activity by a factor of 8 (93).

The diguanidine streptomycin is known to inhibit protein synthesis and the effect of hordatines on protein synthesis was therefore studied (97). Hordatine A was found to inhibit protein synthesis by 50% in pea stems but by only 15% in barley. The effect of hordatines on protein synthesis in fungal spores is not known.

The synthetic fungicide guazatine, which is used as a cereal seed dressing has a structural similarity to the hordatines (Fig. 3) (85). The two guanidino groups are separated by 16 carbons and 2 nitrogens in hordatines and by 16 carbons and 1 nitrogen in guazatine. Substitution of an oxygen atom for the central secondary amino group appeared to cause loss of activity (2,4).

The antifungal antibiotic eulicin (Fig. 3) produced by a species of *Streptomyces* similar to *S. parvus* also has a structural similarity to guazatine and the hordatines (37). In eulicin the two guanidino groups are separated by 19 carbon atoms and one nitrogen atom. It is of interest that two unrelated species (barley and *Streptomyces*) should produce structurally related compounds with similar properties. It is anticipated that compounds like these also occur in other organisms.

The site of action of guazatine is probably the membrane of fungal spores, and it seems likely that the hordatines are also membrane toxins (87). Even so, hordatines are similar in structure to the antiviral guanidino-amidino antibiotic, congocidine, which is known to have a high affinity for double stranded nucleic acid (36).

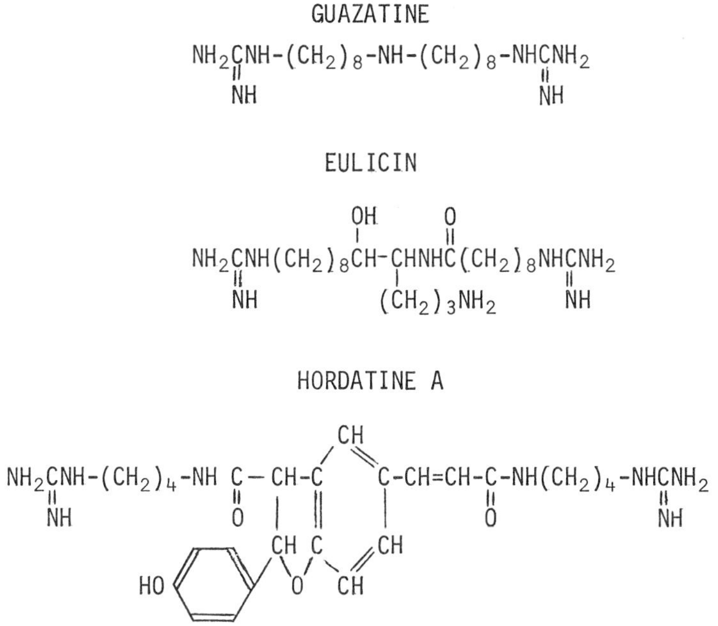

GUAZATINE

$$NH_2\underset{\underset{NH}{\|}}{C}NH-(CH_2)_8-NH-(CH_2)_8-NH\underset{\underset{NH}{\|}}{C}NH_2$$

EULICIN

$$NH_2\underset{\underset{NH}{\|}}{C}NH(CH_2)_8\overset{OH}{\overset{|}{C}}H-\underset{\underset{(CH_2)_3NH_2}{|}}{C}HNH\overset{O}{\overset{\|}{C}}(CH_2)_8NH\underset{\underset{NH}{\|}}{C}NH_2$$

HORDATINE A

$$NH_2\underset{\underset{NH}{\|}}{C}NH-(CH_2)_4-NH\underset{\underset{O}{\|}}{C}-CH-C \ldots C-CH=CH-\underset{\underset{O}{\|}}{C}-NH(CH_2)_4-NH\underset{\underset{NH}{\|}}{C}NH_2$$

FIG. 3. Structural similarity of the commercial fungicide, guazatine, the antifungal antibiotic, eulicin, and the naturally occurring antifungal compound, hordatine A.

Biosynthesis of Coumaroylagmatine

Free coumaroylagmatine can be isolated from young barley leaves, though in a lower concentration than the hordatines. Stoessl suggested that the likely precursors of this amide would be coumaric acid and agmatine (89), both of which are widely distributed in higher plants. A chemical synthesis from agmatine monohydrochloride heated with methyl-*trans*-4-coumarate yielded a product which was identical to the 4-coumaroylagmatine isolated from barley (89).

In our work, confirmation of the involvement of agmatine in the biosynthesis of coumaroylagmatine was obtained by feeding [U-^{14}C]agmatine to shoots cut from barley seedlings between 3 and 6 days after sowing (16). The agmatine, coumaroylagmatine and hordatines were extracted in acetic acid and separated by thin layer chromatography. Radioactivity could be detected in a spot which co-chromatographed with coumaroylagmatine. This showed maximum activity after about 3 hr, followed by a slow decline. Labelled hordatines could also be detected with a maximum after about 6 hr with little change for the next 20 hr.

Activation of the carbonyl group of the coumaric acid is necessary in the synthesis of *N*-coumaryl derivatives because the formation of the amide linkage is an endergonic process. Zenk (106) and Gross (35) have shown that cinnamic acids in their activated form as CoA thiol-esters play an important role in plant phenolic acid metabolism, and acyl-CoA derivatives are known to be involved in similar amide linkages in animals (82) and in micro-organisms (5). Conjugates of glycine with various aromatic acids found in human urine are also formed from the CoA-activated substrates (63). Coumaroyl-CoA was therefore a potential substrate for an enzyme synthesizing coumaroylagmatine in barley. 4-Coumaroyl-CoA synthetase (EC 6.2.1.12.) occurs widely in the plant kingdom and has been well characterized (50, 106).

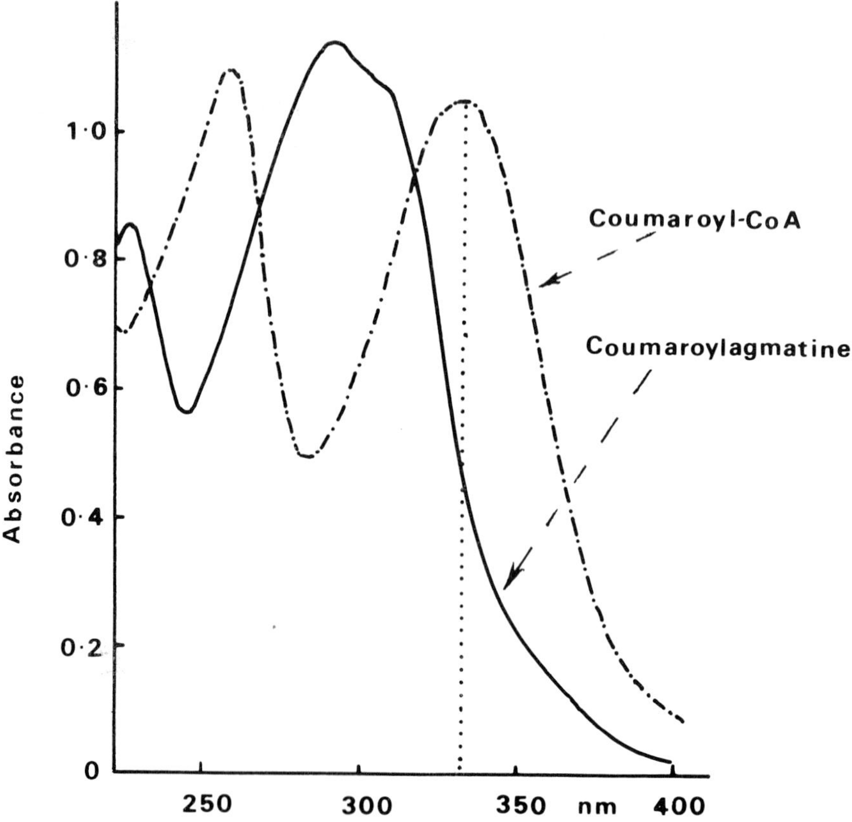

FIG. 4. Absorption spectra of coumaroyl CoA (dotted line) and coumaroylagmatine (continuous line). The transferase assay was conducted by recording the loss in absorbance at the peak of 333 nm for coumaroyl-CoA (vertical dotted line).

To investigate the involvement of coumaroyl-CoA, homogenates of 3-day-old barley seedlings were incubated with coumaroyl-CoA and [U-^{14}C]agmatine at pH 7.5. Radioactivity was incorporated into coumaroylagmatine, confirming that the formation of coumaroylagmatine is effected by a hitherto undescribed transferase (4-coumaroyl-CoA agmatine *N*-4-coumaroyltransferase; EC 2.3.1.-) as shown in Fig. 2 (16). The enzyme can be assayed spectrophotometrically by measuring the decline in absorbance at 333 nm due to cleavage of the thiol-ester bond of coumaroyl-CoA during the reaction, and this simple and rapid assay was used to characterize the enzyme (Fig. 4). A 75-fold purification has been achieved by a method involving affinity chromatography on an immobilized agmatine column as a final step. The molecular weight, estimated by gel filtration, is 40 000. The enzyme is specific for agmatine; no activity has been detected with the following similar amino compounds: *N*-carbamylputrescine, putrescine, spermidine, spermine, cadaverine, arginine, homoarginine and homoagmatine. The apparent *Km* for agmatine at 10 μM coumaroyl-CoA was 2.8 μM in crude extracts, rising to 12 μM in purified extracts, indicating some denaturation on purification. A wide range of substrates has been shown for the hydroxycinnamoyl CoA derivatives (Fig. 5). Activity decreases with

Carboxylic acid as coenzyme A derivative		Relative Activity nkat/g. fresh wt.
Cinnamic acid	-CH = CH-COOH	1.94
Coumaric acid	HO- -CH = CH-COOH	1.22
Caffeic acid	HO, HO- -CH = CH-COOH	0.58
Ferulic acid	CH_3O, HO- -CH = CH-COOH	0.53
Sinapic acid	CH_3O, HO, CH_3O -CH = CH-COOH	0.11

FIG. 5. Specificity of agmatine coumaroyl transferase

increasing hydroxylation and methoxylation of the aromatic ring. The apparent *Km* is 0.74 μM for coumaroyl-CoA, at 200 μM agmatine (15). This enzyme has hitherto only been studied in barley seedlings, and there has been no report of the mechanism of synthesis of other hydroxycinnamic acid amides which are known to occur in many plants (Fig. 1). However, it might be expected that an enzyme similar to that found in barley seedlings could be involved in their formation.

Chemical Synthesis of Hydroxycinnamic Acid Amides

Amides formed between hydroxycinnamic acids and amines may be prepared by reacting the acid chloride of the acetylated acid with the amine (29,62). An excess of the diamine is used to favour the formation of the monosubstituted form. In another method, used for the preparation of coumaroylagmatine, the methyl ester of coumaric acid is fused with half-neutralized agmatine (89). More recently amides have been synthesized by the use of activated esters. This method, which was originally developed for peptide synthesis may offer considerable advantages in selectivity. Activated esters of hydroxy-1-piperidine are suitable for selective amide formation with the primary amino groups of spermidine (41). Aminolysis of 3-acylthiazolidine-2-thione is also known to be an effective method for the synthesis of amides (33). Activated esters of *N*-hydroxysuccinimide can be obtained with dicyclohexylcarbodiimide (DCC) as a coupling agent and have been used as intermediates in the chemical synthesis of hydroxycinnamoyl-coenzyme A thiol esters (88). In this case chemical protection of phenolic hydroxy groups is difficult as the removal of any protecting group would in most cases lead to the hydrolysis of the thiol ester.

We found that it was possible to improve the method of acyl-CoA synthesis by using pyridine as solvent during the coupling of *N*-hydroxysuccinimide to the hydroxycinnamic acid. In a basic solvent, phenolic hydroxy groups show decreased reactivity as the activation of DCC depends on proton availability.

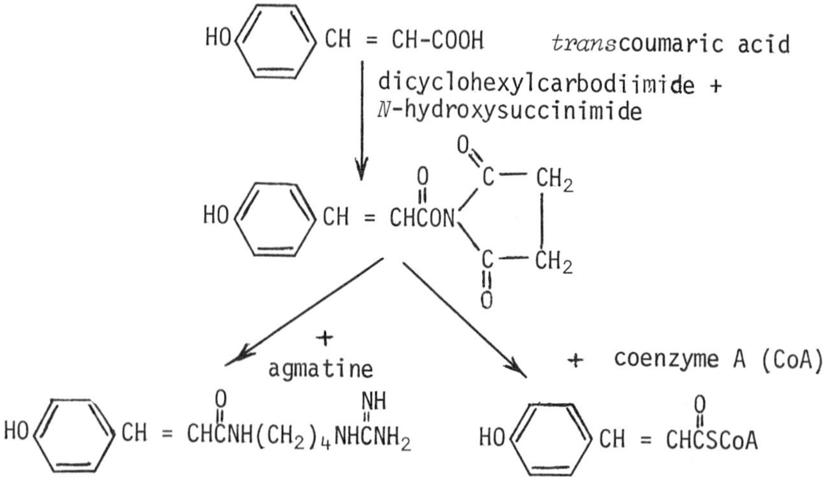

HO-C6H4-CH = CH-COOH *trans*coumaric acid

↓ dicyclohexylcarbodiimide + *N*-hydroxysuccinimide

HO-C6H4-CH = CHCON(CO-CH2)2 (succinimide ester)

+ agmatine → HO-C6H4-CH = CHC(=O)NH(CH2)4NHC(=NH)NH2

+ coenzyme A (CoA) → HO-C6H4-CH = CHC(=O)SCoA

FIG. 6. Chemical synthesis of coumaroylagmatine and coumaroyl-coenzyme A

The *N*-hydroxysuccinimide esters obtained by this method are sufficiently pure to be used in the transesterification reaction with coenzyme A, and also with agmatine to form 4-coumaroyl-agmatine without prior purification (Fig. 6).

However high yield monosubstitution of the symmetrical amines putrescine, cadaverine and spermine with phenolic acids remains difficult. No simple method of protection of only one amino group of these has been described. With spermidine, however, it should be possible to protect the aminopropane moiety using formaldehyde (59).

Biosynthesis of the hordatines

Although the biosynthesis of coumaroylagmatine has now been elucidated, the mechanism for the formation of the hordatines remains obscure. There are several possible theoretical mechanisms by which the hordatines may be synthesized.

The phenylcoumaran structure of the hordatines is analogous to that of dehydrodiconiferyl alcohol, a lignan identified during *in vitro* enzymatic studies on lignin biosynthesis using peroxidase (67). Indeed, it is possible to isolate hordatine A from the complex mixture formed on oxidation of 4-coumaroylagmatine in the presence of horse-radish peroxidase and hydrogen peroxide (90). Although a similar reaction may take place *in vivo*, the product of this reaction is optically inactive, unlike the natural hordatines.

As it is established that coumaroyl-CoA is involved in the formation of the amide linkage, and since coenzyme A esters of α, β-unsaturated acids are involved in the oxidation of the double bond (c.f. β-oxidation of fatty acids) a dimerization mechanism in which oxidation takes place prior to the addition of agmatine is also theoretically possible.

Distribution of the Hordatines

The distribution of the hordatines was studied by Smith and Best (85) using thin layer chromatography with a quantitative Sakaguchi assay. They showed that in both light- and dark-grown barley coumaroylagmatine and hordatine concentrations reached a peak in the first few days after germination and that this was followed by a progressive decline. No hordatines could be detected in the ungerminated seed and only very small amounts in the roots. These observations have been confirmed in a more detailed investigation of hordatine concentration and enzyme activity in dark grown barley (15). In this work an automated ion exchange chromatography system was used with two detection systems; *o*-phthaldialdehyde for the amines and the Sakaguchi reagent for guanidines (17). Coumaroylagmatine reached a peak after 3 days, whereas hordatines A + B and M had maximum concentration after 6 days. A similar pattern of change has been found in agmatine coumaroyltransferase activity, which is not detectable in the seed. This rises to a maximum after about 3-4 days, and undergoes a rapid decline to become again undetectable after 5-6 days. Coumaroyl-CoA synthetase activity appears to remain constant during this period. This apparent correlation between agmatine coumaroyltransferase activity and coumaroylagmatine concentration suggests that the enzyme regulates the duration of production of the hordatines. The arginine decarboxylase activity of barley shoots declines by 70-80% from 3-12 days (85). Potassium deficiency has been shown to increase the agmatine and putrescine content of barley (86). Arginine decarboxylase activity is also increased (83,85), and the concentration of the hordatines is considerably enhanced after 30 days growth in potassium deficient nutrient (Table 1) (85). This

TABLE 1. Coumaroylagmatine, hordatine and agmatine content of barley seedling shoots (nmole/g fresh weight) (85)

Age (days)	3	6	12	30	30
Potassium (K)	+ K	+ K	+ K	+ K	- K
Coumaroylagmatine	350	80	75	< 50	< 50
Hordatines A and B	680	430	350	< 50	315
Hordatine M	350	120	180	175	340
Agmatine	100	< 30	< 30	44	580

suggests that arginine decarboxylase may be a factor limiting the quantity of hordatine produced during the period of activity of the agmatine coumaroyltransferase.

In 3-day-old seedlings, hordatines and agmatine coumaroyltransferase were equally distributed between the shoot and the coleoptile (15). In 12-day-old light-grown shoots the coumaroylagmatine and hordatines were more abundant in the stems than the leaves (85). The hordatines (85) and agmatine coumaroyltransferase (15) have not been found in other members of the Gramineae and appear to be restricted to the genus *Hordeum*. Within this genus, hordatines have been found in several Eurasian species, although none were detected in *H. jubatum*, the only North American species tested (Table 2) (85). This species was grown for 27 days before sampling in order to obtain sufficient material for the assay. *Hordeum vulgare* 30 days old still contained 175 nmol/g of hordatine M (Table 1). Hordatines are not classified as phytoalexins because they are synthesized in uninfected barley seedlings. However infection of 11-day-old barley with powdery mildew (*Erysiphe graminis*) gave a 6-fold increase in hordatine A + B 13 days after infection (Table 3) (85).

TABLE 2. Coumaroylagmatine, hordatine and agmatine content of barley *(Hordeum)* species (nmol/g fresh weight) (85)

	H. bulbosum	*H. distichon*	*H. murinum*	*H. spontaneum*	*H. vulgare*	*H. jubatum*
Age (days)	18	15	18	14	12	27
Coumaroylagmatine	960	100	< 100	167	75	< 100
Hordatines A and B	1320	1100	1150	1260	350	< 10
Hordatine M	11	448	61	386	180	< 10
Agmatine	480	150	< 100	< 100	< 30	100
Geographical origin	Eurasian	Eurasian	Eurasian	Eurasian	Eurasian	N. American

TABLE 3. Effect of mildew *(Erysiphe)* on hordatines and related compounds in barley leaves (nmol/g fresh weight) (infected 11 days after germination; sampled 13 days after infection) (85)

	Non-infected	Infected
Coumaroylagmatine	100	100
Hordatines A and B	89	502
Hordatine M	43	105
Agmatine	68	119

ACKNOWLEDGEMENTS

J. Negrel is grateful to the British Council and French Government for financial support and C.R. Bird thanks the Agricultural Research Council for a Research Studentship.

REFERENCES

1. Aigner-Held, R., and Daves, G.D. (1980): Physiol. Chem. Phys. 12:389-400.
2. Anon (1981): Jpn Kokoi Tokkyo Koho JP. 81 95, 102 Chem. Abs. 96:29922.
3. Armstrong, M.D., Shaw., K.N.F., and Wall, P.E. (1956): J. Biol. Chem., 218:293-303.
4. Badcock, G.G., and Dyke, W.J.C. (1968): British Patent 1 114 155.
5. Barenholz, Y., Edelman, I., and Gatt, S. (1974): Biochim. Biophys. Acta, 358:262-274.
6. Bell, E.A. (1980): in Encyclopedia of Plant Physiology, edited by E.A. Bell and B.V. Charlwood Vol 8, Chapter 7, pp. 403-432. Springer Verlag, Berlin.
7. Berlin, J. (1980): Z. Pflanzenphysiol., 97:317-324.
8. Berlin, J. (1981): Phytochemistry, 20:53-55.
9. Berlin, J., and Forche, E. (1981): Z. Pflanzenphysiol., 101: 277-282.
10. Berlin, J., Knobloch, K.-H., Höfle, G., and Witte, L. (1982): J. Nat. Prod., 45:83-87.
11. Berlin, J., Kukoschke, K.G., and Knobloch, K.-H. (1981): Planta Med., 42:173-180.
12. Berlin, J., and Widholm, J.M. (1977): Plant Physiol., 59:550-553.
13. Berlin, J., and Witte, L. (1982): J. Nat. Prod., 45:88-93.
14. Bernhard, H.O., Kompiš, I., Johne, S., Gröger, D., Hesse, M., and Schmid, H. (1973): Helv. Chim. Acta, 56:1266-1303.
15. Bird, C.R., and Smith, T.A., (unpublished).
16. Bird, C.R., and Smith, T.A. (1981a): Phytochemistry, 20:2345-2346.

17. Bird, C.R., and Smith, T.A. (1981b): J. Chromatogr., 214: 263-266.
18. Booth, A.N., Masri, M.S., Robbins, D.J., Emerson, O.H., Jones, F.T., and Deeds, F. (1960): J. Biol. Chem., 235:2649-2652.
19. Buta, J.G., and Izac, R.R. (1972): Phytochemistry, 11:1188-1189.
20. Cabanne, F., Dalebroux, M.A., Martin-Tanguy, J., and Martin, C. (1981): Physiol. Plant., 53:399-404.
21. Cabanne, F., Martin-Tanguy, J., and Martin, C. (1977a): Physiol. Vég., 15:429-443.
22. Cabanne, F., Paynot, M., Javelle, F., Martin-Tanguy, J., and Martin, C. (1977b): Physiol. Vég., 15:445-451.
23. Chantrapromma, K. and Ganem, B. (1981): Tetrahedron Lett., 22:23-24.
24. Dätwyler, P., Bosshardt, H., Bernhard, H.O., Hesse, M., and Johne, S. (1978): Helv. Chim. Acta, 61:2646-2671.
25. Dätwyler, P., Bosshardt, H., Johne, S., and Hesse, M. (1979): Helv. Chim. Acta, 62:2712-2723.
26. Deletang, J. (1974): Ann. Tabac. (Sect. 2),:123-130.
27. Diaz, M., and Ripperger, H. (1982): Phytochemistry, 21:255-256.
28. Dumas, E., Perdrizet, E., and Vallée, J.-C. (1981): Physiol. Vég., 19:155-165.
29. Eagles, J., Self, R., and Synge, R.L.M. (1978): Organic Mass Spectrometry, 13:677-678.
30. Ellestad, G.A., Cosulich, D.B., Broschard, R.W., Martin, J.H., Kunstmann, M.P., Morton, G.O., Lancaster, J.E., Fulmor, W., and Lovell, F.M. (1978): J. Am. Chem. Soc., 100:2515-2524.
31. Englert, G., Klinga, K., Raymond-Hamet, Schlittler, E., and Vetter, W. (1973): Helv. Chim. Acta, 56:474-478.
32. Fischer, F.G., and Bohn, H. (1957): Liebigs Ann. Chem., 603: 232-250.
33. Fujita, E. (1981): Pure and Applied Chem., 53:1141-1154.
34. Gilbo, C.M., and Coles, N.W. (1964): Aust. J. Biol. Sci. 17: 758-763.
35. Gross, G.G. (1981): In: The Biochemistry of Plants, edited by E.E. Conn, Vol. 7, Chapter 11, pp. 301-316. Academic Press New York.
36. Grunicke, H., Puschendorf, B., and Werchav, H. (1976): Rev. Physiol. Biochem. Pharmacol., 75:69-96.
37. Harman, R.E., Ham, E.A., Bolhofer, W.A., and Brink, N.G. (1958): J. Am. Chem. Soc. 80:5173-5178.
38. Hesse, M., and Schmid, H. (1977): Organic Chemistry Series 2. International Review of Science (ed. K. Wiesner) Butterworths London, Boston 9:265-307.
39. Hirata, K. (1958): Ann. Phytopathol. Soc. Jpn. 23:139-144.
40. Hitchcock, M., and Smith, J.N. (1966): Biochem. J. 98:736-741.
41. Husson, H.-P., Poupat, C., and Potier, P. (1973): C.R. Acad. Sci. Paris (Sér C), 276:1039-1040.
42. Jaggy, H. (1978): Planta Med., 33:285.

43. Jordan, T.W., McNaught, R.W., and Smith, J.N. (1970): Biochem. J., 118:1-8.
44. Kato, K., Matsumoto, T., Koiwai, A., Misuzaki, S., Nishida, K., Noguchi, M., and Tamaki, E. (1972): Proc. IV IFS: Ferment. Technol. Today, 689-695.
45. Keller-Schierlein, W., and Prelog, V. (1961): Helv. Chim. Acta, 44:1981-1985.
46. Knobloch, K.-H., and Berlin, J. (1981): Planta Med., 42: 167-172.
47. Knobloch, K.-H., Beutnagel, G., and Berlin, J. (1981): Planta, 153:582-585.
48. Kupchan, S.M., Hintz, H.P.J., Smith, R.M., Karim, A., Cass, M.W., Court, W.A., and Yatagai, M. (1974): J. Chem. Soc. Chem. Communs., 329-330.
49. Kupchan, S.M., Hintz, H.P.J., Smith, R.M., Karim, A., Cass, M.W., Court, W.A., and Yatagai, M. (1977): J. Org. Chem., 42: 3660-3664.
50. Luderitz, T., Schatz, G., and Grisebach, H. (1982): Eur. J. Biochem., 123:583-586.
51. Ludwig, R.A., Spencer, E.Y., and Unwin, C.H. (1960): Can. J. Bot. 38:21-29.
52. Martin, C., and Martin-Tanguy, J. (1981): C.R. Acad. Sci. Paris (Sér III) 293:249-251.
53. Martin-Tanguy, J., Cabanne, F., Perdrizet, E., and Martin, C. (1978): Phytochemistry, 17:1927-1928.
54. Martin-Tanguy, J., Deshayes, A., Perdrizet, E., and Martin, C. (1979): FEBS Lett., 108:176-178.
55. Martin-Tanguy, J., Martin. C., and Gallet, M. (1973): C.R. Acad. Sci. Paris (Sér D), 276:1433-1435.
56. Martin-Tanguy, J., Martin, C., Gallet, M., and Vernoy, R. (1976): C.R. Acad. Sci. Paris (Sér D), 282:2231-2234.
57. Mbadiwe, E.I., (1973): Phytochemistry, 12:2546.
58. Mbadiwe, E.I., (1978): Qual. Plant., 28:261-269.
59. McManis, J.S., and Ganem, B., (1980): J. Org. Chem., 45:2041.
60. Meyer, J.-M., and Abdallah, M.A., (1980): J. Gen. Microbiol., 118:125-129.
61. Mizusaki, S., Tanabe, Y., and Noguchi, M., (1970): Agric. Biol. Chem., 34:972-973.
62. Mizusaki, S., Tanabe, Y., Noguchi, M., and Tamaki, E. (1971): Phytochemistry, 10:1347-1350.
63. Moldave, K., and Meister, A. (1957): J. Biol. Chem., 229:463-476.
64. Naarmann, H., (1959): Ph.D. Thesis, University of Würzburg, Germany, (cited by N. Seiler in Polyamines in Biology and Medicine edited by D.R. Morris and L.J. Marton) Marcel Dekker, New York, chapter 7.
65. Nagao, Y., Takao, S., Miyasaka, T., and Fujita, E. (1981): J. Chem. Soc. Chem. Communs., pp. 286-287.
66. Neilands, J.B., (1981): Ann. Rev. Biochem., 50:715-731.
67. Nimz, H., Naya, K., and Freudenberg, K., (1963): Chem. Ber., 96:2086-2089.

68. Païs, M., Sarfati, R., and Jarreau, F.-X. (1973a): Bull. Soc. Chim. France, 331-334.
69. Païs, M., Sarfati, R., Jarreau, F.-X., and Goutarel, R. (1973b). Tetrahedron, 29:1001-1010.
70. Panashchenko, A.D., and Ryabinin, A.A. (1955): Pharmacol. Toxicol., 18:9-17.
71. Panashchenko, A.D., and Ryabinin, A.A. (1961): Trudy Botan. Int. im V.L. Komarova Akad. Nauk SSSR, Ser. 5 (8):49-65. Chem. Abs., 56:12274c.
72. Perdrizet, E., and Prevost, J. (1981): Phytochemistry, 20: 2131-2134.
73. Peterson, T., Falk, K.-E., Leong, S.A., Klein, M.P., and Neilands, J.B. (1980): J. Am. Chem. Soc., 102:7715-7718.
74. Poupat, C. (1976): Tetrahedron Lett., (20):1669-1672.
75. Ranganathan, S., and Ramasarma, T. (1971): Biochem. J., 122: 487-493.
76. Ripperger, H. (1980): Phytochemistry, 19:162-163.
77. Ripperger, H., Budzikiewicz, H., and Schreiber, K. (1972): Biochem. Physiol. Alkaloide, p. 129 4th Int. Symp. Akademie Verlag, Berlin.
78. Ryabinin, A.A., and Il'ina, E.M. (1949): Dokl. Akad. Nauk. SSSR, 67:513-516.
79. Samborski, D.J., and Rohringer, R. (1970): Phytochemistry, 9:1939-1945.
80. Schlittler, E., Spitaler, U., and Weber, N. (1973): Helv. Chim. Acta, 56:1097-1099.
81. Seiler, N. (1981): in Polyamines in Biology and Medicine (edited by D.R. Morris and L.J. Marton) Marcel Dekker, New York, chapter 7.
82. Seiler, N., and Al-Therib, M.J. (1974): Biochim. Biophys. Acta, 354:206-212.
83. Smith, T.A. (1963): Phytochemistry, 2:241-252.
84. Smith, T.A. (1977): Progress in Phytochemistry, (edited by L. Reinhold, J.B. Harborne and T. Swain) 4:27-81.
85. Smith, T.A., and Best, G.R. (1978): Phytochemistry, 17:1093-1098.
86. Smith, T.A., and Richards, F.J. (1962): Biochem. J., 84:292-294.
87. Srivastava, S.K., and Smith, T.A. (1982): Phytochemistry, 21:997-1008.
88. Stöckigt, J., and Zenk, M.H. (1975): Z. Naturforsch., 30c: 352-358.
89. Stoessl, A. (1965): Phytochemistry, 4:973-976.
90. Stoessl, A. (1967): Can. J. Chem., 45:1745-1760
91. Stoessl, A. (1970): Recent Adv. Phytochemistry, 3:143-180
92. Stoessl, A., Rohringer, R., and Samborski, D.J. (1969): Tetrahedron Lett., (33):2807-2810.
93. Stoessl, A., and Unwin, C.H. (1970): Can. J. Bot., 48:465-470.
94. Takahashi, T. and Yoshida, D. (1960): Nippon dojo-hiryogaku zasshi, 31:39-41.
95. Takamatsu, S., Ishizaki, H., and Kunoh, H. (1978): Can. J. Bot., 56:2544-2549.

96. Tamada, M., Endo, K., Hikino, H., and Kabuto, C. (1979): Tetrahedron Lett., (10):873-876.
97. Venis, M.A. (1969): Phytochemistry, 8:1193-1197.
98. Vlamis, J., and Yarwood, C.E. (1962): Plant Disease Reporter, 46:886-887.
99. Wagner, H., and Burghart, J. (1977): Planta Med., 32A:9-14.
100. Wagner, H., and Burghart, J. (1981): Helv. Chim. Acta, 64: 283-296.
101. Wagner, H., Burghart, J., and Hull, W.E. (1978): Tetrahedron Lett., (41): 3893-3896.
102. Wasserman, H.H., and Matsuyama, H. (1981): J. Am. Chem. Soc., 103:461-462.
103. Wheaton, T.A., and Stewart, I. (1965): Nature, 206:620-621.
104. Yamamoto, H., and Maruoka, K. (1981): J. Am. Chem. Soc., 103: 6133-6136.
105. Yoshida, D. (1969): Plant Cell Physiol., 10:393-397.
106. Zenk, M.H. (1979): Recent Adv. Phytochemistry, 12:139-176.

Advances in Polyamine Research, Vol. 4, edited by U. Bachrach, A. Kaye, and R. Chayen. Raven Press, New York © 1983.

Some Properties of the Spermidine Synthase of Chinese Cabbage

Ram K. Sindhu and Seymour S. Cohen

Department of Pharmacological Sciences, State University of New York at Stony Brook, Stony Brook, New York 11794

As reported in previous Symposia (4,7), we have been studying polyamine biosynthesis in protoplasts of healthy and virus-infected Chinese cabbage (6). Turnip yellow mosaic virus (TYMV) is an icosahedral RNA virus, containing several hundred molecules of non-exchangeable spermidine. The polyamines appear to have become bound to the RNA before the packaging of the nucleate (5). The virus multiplies in chloroplast aggregates and we have asked if the chloroplasts synthesize the viral spermidine.

It is not yet proven that leaves contain an S-adenosylmethionine (SAM) decarboxylase similar to the bacterial, yeast or mammalian enzyme. An apparent decarboxylation of [1-^{14}C]-SAM by leaf extracts of Brassica, or of other plants is not stimulated by Mg^{++} as is the E.coli enzyme, or by putrescine, as is the yeast or mammalian enzyme. However an apparent decarboxylation is stimulated by pyridoxal phosphate, which is not known to be a co-factor of the other SAM decarboxylases. Nevertheless methionine does provide the propylamine moiety of spermidine in Cyanobacteria (8) and plants, including Brassica (6). For this reason we have sought spermidine synthase in Chinese cabbage.

Spermidine synthase, which catalyses the reaction, Decarboxylated S-adenosylmethionine + putrescine $\longrightarrow$ spermidine + methylthioadenosine + H^+, has been found in crude extracts of the viral host, Chinese cabbage and of leaf protoplasts (6). A significant portion of the enzyme has also been found associated with the chloroplast fraction (6). We have undertaken a purification of the enzyme from the cabbage leaves to enable us to improve the conditions of assay, to permit the characterization of structured and kinetic properties of the catalytic entity, and to facilitate the study of possible inhibitors. Relatively little work has been done on spermidine synthase because of the difficulty of obtaining decarboxylated S-adenosylmethionine (dSAM). A plant extract presents some complexities that limit the choice of a possible assay system. One product, methylthioadenosine, is rapidly degraded in the extract. Coupled enzymatic systems of assay for the other product, i.e. spermidine, are affected adversely by the presence of high levels of spermidine in the extract, as well as by enzymatic degradation of the intermediates.

Bypassing these difficulties, we have used labelled putrescine as one of the substrates and have isolated a purified labelled product, spermidine. This method introduces some different problems. An excess of the substrate, i.e. putrescine, must be separated from the small amount of labelled product, spermidine, and the latter is converted to the dansyl derivative which is purified and its radioactivity estimated.

The putrescine present in the plant extract markedly dilutes the radioactivity of the substrate and a true assay of the amount of product depends on a measure of the dilution. This problem has rarely been considered in estimating the activity of this enzyme in crude extracts. The ion-exchange separation of putrescine and spermidine provides a stage at which this dilution is readily determined. It will be seen that the determination of the specific radioactivity of the substrate present in the reaction mixture is essential to an estimation of the spermidine synthase present in an extract.

Following this stage of the work, the soluble enzyme has been purified to a stage at which a recently devised coupled enzyme assay (12) has become useful. This assay employs a plant amine oxidase which is inactive with putrescine but reacts with spermidine and spermine to produce H_2O_2. In the presence of peroxidase, the H_2O_2 generated converts a phenolic compound to a fluorescent derivative. At the stage of enzyme purification employed, the system is sensitive, specific for spermidine, and rapid, facilitating the determination of many kinetic properties of the enzyme. The use of the coupled system is also facilitated by the relatively low pH optimum of the plant spermidine synthase.

Results

Assay of spermidine synthase in Leaf and protoplast extracts: The growth of Chinese cabbage (*Brassica pekinensis*, Var. Pak Choy) from seed in a controlled environment and the preparation of protoplasts has been described (6). Leaves were homogenized at 4° with 10mM glycine-NaOH, pH 8.8 to a total volume of 2 ml per gm and the homogenate was filtered through cheesecloth. An aliquot of this crude extract was centrifuged at 10,000 x g for 20 minutes to obtain a supernatant fraction. Protoplasts were disrupted by passage of the cells, suspended in 10mM glycine-NaOH, pH 8.8 through a 20μ nylon mesh.

The reaction mixture for spermidine synthase contained 150mM glycine-NaOH, pH 8.8; 37μM [^{14}C]-putrescine·2 HCl (102mCi per mmol; 25μM dSAM; an appropriate amount of enzyme (0.03-8.0 units) and H_2O in a total volume of 0.325 ml. The K_m values for the two substrates have been determined with a purified enzyme. In the assay of crude extracts, dSAM is used at approximately four times the K_m, whereas putrescine is added at approximately the K_m in the linear range of enzyme activity. Control mixtures were set up lacking either dSAM or enzyme. The tubes were incubated at 37° for 1 hr and the reaction was stopped by adding 1ml of cold

5% PCA. After standing in ice, the precipitates were removed and washed, and the supernatant fluids were combined. The assay mixture for spermine synthase is similar, except that [^{14}C]-spermidine is substituted for [^{14}C]-putrescine.

As described earlier (6), it is necessary to separate the labelled substrate from the labelled product before conversion of the product to the dansyl derivative, thin-layer chromatographic analysis of the dansyl derivatives and estimation of the radioactivity in dansyl spermidine or dansyl spermine. In the assay for spermidine synthase, this is effected by addition of the acidic supernatant solution to a column of Dowex-50-H^+, and removal of the 93% of the residual putrescine by elution with 2.3M HCl. About 90% of the spermidine is retained and is then eluted in 6M HCl. In the assay for spermine synthase, 99% of the substrate spermidine is removed by 3.3M HCl, leaving 93% of the spermine on the column, to be eluted subsequently by 6M HCl.

These eluates containing the partially purified products are concentrated to dryness _in vacuo_, redissolved in 0.03M HCl, and dansylated. The dansyl derivatives are extracted into benzene and applied to preactivated (110°, 1 hr) silica gel plates. The plates were developed once in chloroform: triethylamine (80: 16,v/v). In this system the R_f values of dansyl spermine > dansyl spermidine > dansyl putrescine. In solvents in which this sequence is reversed, trailing from dansyl [^{14}C]-putrescine may leave 2% of the radioactivity in dansyl spermidine, whereas this cross contamination is reduced to a third or less in the chloroform: triethylamine solvent. The recoveries of spermidine and spermine from the plate were 82% and 86% respectively.

The results of such initial assays on crude extracts of rapidly growing cabbage leaves are presented in Table 1. Both spermidine synthase and spermine synthase are present at easily detectable levels but spermidine synthase appears to be present in about 20 times the activity of spermine synthase. Furthermore, the activities of both these enzymes appear significantly higher in the supernatant fraction. Comparable activities are present in extracts of protoplasts. As reported earlier (6), up to a quarter of the activity of a protoplast extract can be detected in the chloroplast fraction. It may be mentioned that an extract of mature spinach leaves contains about half the activity shown for the young cabbage leaves.

TABLE 1. Propylamine transferases in Chinese cabbage leaves[1]

	Spermidine synthase[2]	Spermine synthase[2]	Apparent spermidine synthase / spermine synthase activity
Crude extract	22.3	0.90	24.8
Supernatant fraction	25.4	1.20	21.2

[1]determined at 37°; [2]units/g fresh tissue

The effect of extract concentration on the apparent activity of spermidine synthase: The molar yield of radioactive spermidine is calculated from the specific activity of [^{14}C]-putrescine. It is seen in Figure 1 however that in this experiment the yield is proportional to the volume of extract for only low concentrations of enzyme, i.e. producing up to 25% of the yield of product. The leveling off and inhibition at higher concentrations were in fact so severe that it appeared possible that the effect might arise not from an inhibitor but from dilution of the radioactive putrescine by the putrescine in the extract. Specific activities were determined of the reisolated putrescine obtained from the 2.3M HCl eluates in the separation of substrate putrescine from product spermidine. The extensive dilution is presented in the Figure, as is the corrected yield of spermidine. The latter values reveal a maximal 30% inhibition of the apparent rate of synthesis at higher concentrations, and a total production equal to the input of active dSAM (8nmol of active synthetic stereoisomer).

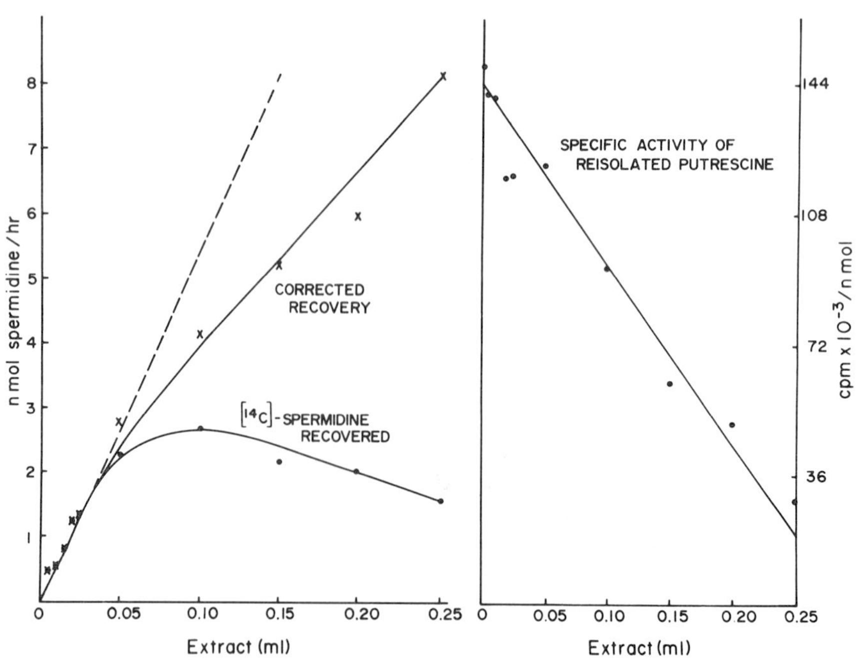

FIG. 1 Extract concentration and apparent spermidine synthase activity.
Left side - correction of apparent spermidine synthase activity by dilution of the substrate.
Right side - dilution of [^{14}C]-putrescine by putrescine in the extract.

<u>A partial purification and some properties of the cabbage enzyme:</u> The fractionation steps were carried out at 4°. A crude extract filtered through cheesecloth was centrifuged at 12,000 x g for 30 min, eliminating most of the chlorophyll at this step. The apparent activity of the supernatant fraction increased 11%. The fraction was adjusted to pH 5.1 with acetic acid and 3.5mg per ml of streptomycin sulfate was added slowly with stirring to precipitate nucleic acids and ribosomes. After standing for 20 min and centrifugation at 12,000 x g for 10 min, the supernatant was adjusted to pH 8.8 with NaOH. Ammonium sulfate (0.20 gm per ml) precipitated an inactive fraction, and an additional amount of $(NH_4)_2SO_4$ (0.21 gm per ml) to the supernatant precipitated the enzyme. This was collected by centrifugation at 12,000 x g for 30 min, and dissolved in a tenth volume of 25mM sodium phosphate. The enzyme loses most of its activity when dialysed at this stage. The presence of 1mM dithiothreitol and/or 1mM EDTA does not improve the recovery. The supernatant of a heat-inactivated enzyme does not restore the activity.

An equal volume of acetone (stored at -20°C) was added slowly with stirring and the initial heavy precipitate was discarded. To the clear supernatant fluid was added three additional volumes of cold acetone, precipitate was collected by centrifugation at 10,000 x g for 10 min, drained and dissolved in 25mM sodium phosphate, pH 7.2 containing 0.1 M KCl. Insoluble material was removed by centrifugation. At this stage the preparation showed an apparently 1.8 fold increase in total activity and an 82-fold increase in specific activity; i.e. units per mg protein. The material was now essentially free of nucleic acid and of contaminating putrescine. In contrast to the marked lability of the activity in protoplast extracts, the acetone-precipitated enzyme can be stored frozen for at least a month at -20°C.

Enzyme precipitated by acetone does not adhere to spermidine- or spermine-sepharose (11). Although the activity is retained on S-adenosyl (5')-3-thiopropylamine-Sepharose (10) only a small fraction can be eluted by high concentrations (3.5mM) of dSAM. The latter had been removed prior to assay.

A further purification of the enzyme, doubling the specific activity, was effected by fractionation on a Sephadex G-100 column, equilibrated with 25mM sodium phosphate buffer, pH 7.2 containing 0.1M KCl and eluted with the same buffer. All of the activity was recovered in a single sharp peak. Using a calibrated column with appropriate protein standards (1), the molecular weight of the enzyme was estimated to be 81,000.

The optimal pH for the 160-fold purified cabbage enzyme is 8.8 in glycine-NaOH buffer. No activity can be detected at pH 6.0 in phosphate or at 10.0 in glycine-NaOH. By contrast, the bacterial (2) and mammalian (10) spermidine synthases have been reported to have pH optima at 10.4 and 10 respectively. The activities at various pHs are slightly less in Tris buffer than in glycine-NaOH or in phosphate.

Some additional properties of the partially purified enzyme: Further studies were carried out with a coupled enzyme system to be discussed below. When the enzyme solution is heated at various temperatures between 40 and 70° for 10 minutes, cooled and assayed at 37°, it was found that the enzyme is fairly sensitive to temperature. Decreases of 35%, 65%, 86% and 95% were obtained at 45°, 50°, 55° and 60° respectively. Putrescine at 0.2mM did not stabilize the enzyme activity.

The time course of spermidine synthesis with the purified spermidine synthase was linear for at least 80% of the reaction, attained the theoretical limit, determined by the content of active dSAM. This system is presumably adaptable to a very sensitive analysis of this component. Increasing enzyme in the assay results in a proportional increase in the rate of synthesis of spermidine to about 50% of the maximal rate, as seen in Fig. 2. There was a slow fall off in this linear increase in rate until a rate was achieved limited once again by the amount of dSAM in the system.

The coupled enzyme assay: Suzuki et al (12) recently found that the oat seedling polyamine oxidase is quite active toward spermidine but devoid of activity to putrescine. The enzyme is also active on spermine. This method has been adapted by the Japanese workers for the assay of spermidine synthase in a dialysed supernatant fluid derived from rat prostate and other rat tissues (12). Their method is readily adapted to the assay of the partially purified cabbage spermidine synthase.

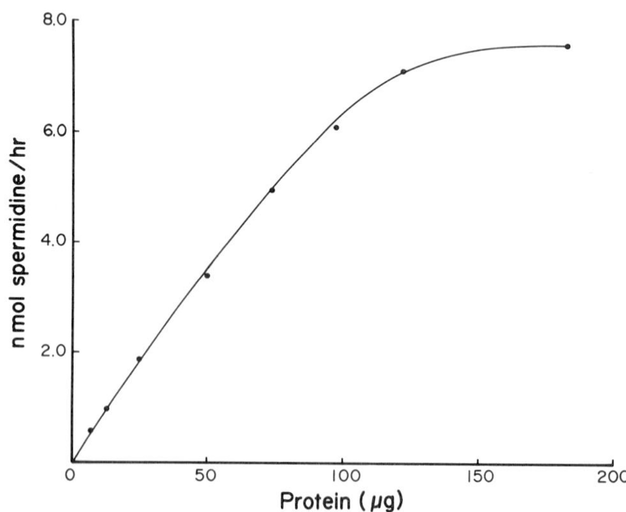

FIG. 2 The rate of spermidine synthesis as a function of concentration of spermidine synthase added to the coupled system.

We have grown oat seedlings (Avena sativa) and prepared the stable enzyme as they have described (12). Spermidine at nanomolar levels is incubated with the oat amine oxidase at 37° and coupled simultaneously with a peroxidase-homovanillic acid system as given in Fig. 3. The reaction is stopped by alkali, which also increases the fluorescence (excitation at 323nm, emission at 426nm) by several fold. The fluorescence is proportional to the spermidine present from 1 to 10nmoles.

It should be noted that the coupled system operates at pH 7.4, which is far from the reported pH optimum of the bacterial and mammalian systems. With the plant system in the recommended phosphate buffer, which we have adopted in our study, the activity of the enzyme was about 55% of that at the pH optimum. Nevertheless the ease with which many samples are analysed, as well as the possibly more physiological pH, i.e. 7.4 rather than 8.8, has led us to the use of this method in obtaining kinetic data. It may be mentioned that the oat seedling amine oxidase is devoid of spermidine synthase and our purified spermidine synthase is free of elements of the coupled enzyme system.

Coupled Enzymatic Analysis of Spermidine

1) $H_2N(CH_2)_3NH(CH_2)_4NH_2 + O_2 \xrightarrow{PAO}$ Δ^1-Pyrroline + $H_2N(CH_2)_3NH_2 + H_2O_2$

Δ^1-Pyrroline 1,3-Diaminopropane

2) 2 Homovanillic acid + $H_2O_2 \xrightarrow{Peroxidase}$ 2,2'-Dihydroxy-3,3'-dimethoxybiphenyl-5,5'-diacetic acid $+ 2H_2O$

Homovanillic acid 2,2'-Dihydroxy-3,3'-dimethoxybiphenyl-5,5'-diacetic acid

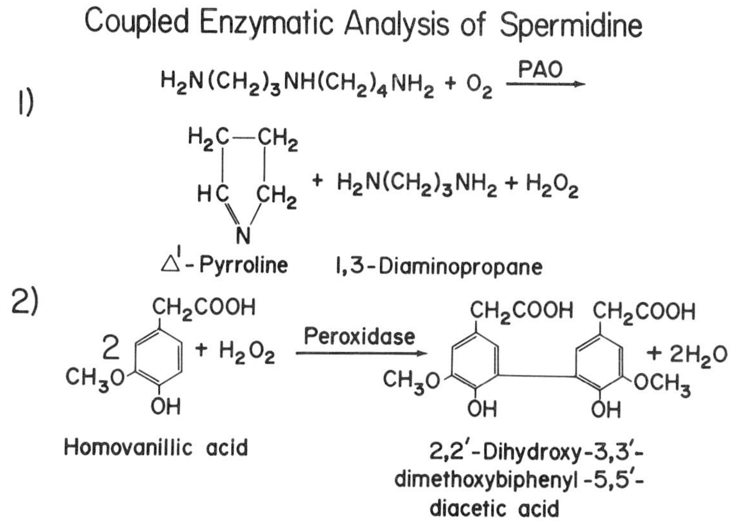

FIG. 3

Inhibition of spermidine synthase: The inhibition of the 160-fold purified enzyme over a range of concentrations was determined with the coupled enzyme system. It was demonstrated that the inhibitors tested did not affect the assay for spermidine, i.e. did not affect the activities of the oat seedling amine oxidase or the horseradish peroxidase. In Fig. 4 are presented the significant inhibitory activities of dicyclohexylamine (9), cyclohexylamine (9), and the transition state analogue,

S-adenosyl-3-thio-1,8-diaminooctane (13). Thus 50% inhibition was found at 2.4 x 10^{-7}M, 5.4 x 10^{-7}M and 1.3 x 10^{-6}M respectively. The linearity of the coupled assay in the absence of the inhibitor is also given in the Figure.

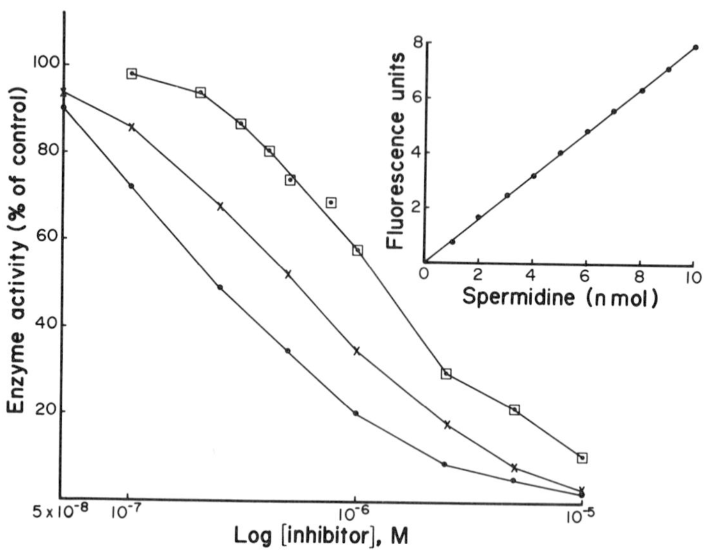

FIG. 4 Inhibition of spermidine synthase by dicyclohexylamine (·——·), cyclohexylamine (x——x) and S-adenosyl-3-thio-1,8-diaminooctane (□——□).

Discussion

The disruption of cells in buffers of relatively low osmolarity probably creates a single pool of readily dissociable putrescine, which permits the estimation of the specific radioactivity of the substrate, putrescine, in the assay of spermidine synthase. It is not at all clear if more tightly bound cell spermidine will also serve as a substrate similarly for spermine synthase in such circumstances. For this reason, it is uncertain that an estimate of the specific radioactivity of spermidine in the reaction mixture of spermine synthase will assist in a more reliable determination of this enzyme.

However the improvement in the assay for spermidine synthase in plant extracts will permit more careful comparisons of the content of this enzyme and its regulation as a function of the stages of growth and senescence of the plant, as well as of the effects of virus infection. The use of the coupled assay will obviously facilitate later stages of purification of the enzyme, e.g. column fractions, gel separations, as it already has been very helpful in the determination of molecular and kinetic properties, as well as the activities of inhibitors.

The fractionation has revealed the presence of only a single peak of enzyme activity in the acetone precipitate derived from the supernatant fraction. The assays were conducted to detect very small amounts of a possible second enzymatic component. Since this initial fraction, even from gently disrupted protoplasts, contains much of the chloroplast enzyme ribulose bisphosphate carboxylase (unpublished data), it is assumed that the chloroplasts have either leaked much of their spermidine synthase or have adsorbed some of the soluble enzyme. In either case it appears possible that there may be only a single form of this enzyme. It is therefore still possible that this enzyme may be derived entirely from the chloroplasts, the major site of synthesis of amino acids of the aspartate pathway, including methionine (3).

This work was supported by grants from the National Science Foundation (PCM78-0434) and the National Institutes of Health (1R01GM25522).

References

1. Andrews, P. (1964): Biochem. J., 91:222-233.

2. Bowman, W.H., Tabor, C.W., and Tabor, H. (1973): J. Biol. Chem., 248:2480-2486.

3. Burdge, E.L., and Wilson, K.G. (1981): FEBS Lett, 136: 322-324.

4. Cohen, S.S. (1981): In Advances in Polyamine Research, Vol. 3, edited by C.M. Caldarera, V. Zappia, and U. Bachrach, pp 249-257. Raven Press, New York.

5. Cohen, S.S., and Greenberg, M.L. (1981): Proc. Natl. Acad. Sci. U.S.A., 78:5470-5474.

6. Cohen, S.S., Balint, R., and Sindhu, R.K. (1981): Plant Physiol., 68:1150-1155.

7. Cohen, S.S., Balint, R., Sindhu, R.K., and Marcu, D. (1981): Med. Biol., 59:394-402.

8. Guarino, L.A., and Cohen, S.S. (1979): Anal. Biochem., 95:73-76.

9. Hibasami, H., Tanaka, M., Nagai, J., and Ikeda, T. (1980): FEBS Lett., 116:99-101.

10. Samejima, K., and Yamanoha, B. (1982): Arch. Biochem. Biophys. (in press).

11. Sindhu, R.K., and Cohen, S.S. (1982): In Methods in Enzymology, edited by H. Tabor and C.W. Tabor, Academic Press, New York (in press).

12. Suzuki, O., Matsumoto, T., Oya, M., Katsumata, Y., and Samejima, K. (1981): Anal. Biochem., 115:72-77.

13. Tang, K-C., Pegg, A.E., and Coward, J.K. (1980): Biochem. Biophys. Res. Commun., 96:1371-1377.

Advances in Polyamine Research, Vol. 4, edited by U. Bachrach, A. Kaye, and R. Chayen. Raven Press, New York © 1983.

The Control of Arginine Decarboxylase Activity in Higher Plants

A. W. Galston, *Y. -r. Dai, H. E. Flores, and N. D. Young

*Department of Biology, Yale University, New Haven, Connecticut 06511; *Department of Biology, Peking University, Beijing, People's Republic of China*

While polyamine biosynthesis in animal and microbial cells is initiated by ornithine decarboxylase (ODC; EC 4.1.1.17) (4,7) considerable evidence indicates a similar role in some higher plants for the enzyme arginine decarboxylase (ADC; EC 4.1.1.19)(8,17, 21,26). In both the stem and leaves of pea seedlings and the young green leaves of oats, we have found a variety of physiological stimuli and stress reactions to affect the activity of ADC. Since the changes in ADC activity are often very rapid, require protein synthesis, and are reflected in changes in polyamine titer, ADC synthesis may be central to numerous rapid growth and adaptive reactions in plants. We shall describe examples of control of ADC by light (absorbed by the pigment phytochrome), by a hormone (gibberellin), by osmotic stress and by pH stress.

MATERIALS AND METHODS

A. For Phytochrome and Gibberellin Experiments.

Plant material.
Seeds of *Pisum sativum* L. var Alaska were imbibed for 10 h in tap water in the dark and sown in prewashed vermiculite. After 6 days in the dark at 27° C and about 70% relative humidity, seedlings were selected for uniformity of third internode length (30-50 mm) and sharply recurved apical hooks.

Light sources and irradiation.
The red light (R) source (600-900 nm) consisted of four 15-w red fluorescent tubes (Sylvania) wrapped with two layers of Dupont red cellophane. Energy at the seedling level was 2.0 kiloergs cm^{-2} s^{-1}. The far-red (FR) source (710-760 nm) consisted of five 300-w internal reflector incandescent flood lamps

filtered through 16 cm water and a Westlake FR Plexiglas FRF 700 filter, emitting 19 kiloergs cm^{-2} s^{-1} energy at the seedling level. Both R and FR irradiations lasted for 5 min. After light treatment, the seedlings were kept in the dark for different time periods until harvest and assay. In the experiments involving photoreversibility, R or FR irradiation was followed immediately by FR or R, respectively.

Enzyme assay.

Buds or subapical 5-mm long epicotyl sections from the third internode were harvested as in (8), and ground in a chilled mortar with 100 mM phosphate buffer (pH 7.0). After centrifugation at 30,000 g for 15 min, the clear supernatant fraction was used for enzyme assay. ADC activity was determined by measuring labeled $^{14}CO_2$ release from L-[U-^{14}C]-arginine (ICN, 300 mCi/mmol) according to a method modified from Fong et al. (12), and described in (8). Enzyme activity is expressed as cpm/mg fresh weight or cpm/mg protein. Protein was determined in the crude extract according to Lowry et al. (19) using BSA as a standard.

Polyamine extraction and analysis.

Buds or 5 mm long epicotyl sections (3rd internode) were collected, weighed under photomorphogenetically inactive light, and extracted in cold 5% PCA with a usual ratio of about 100 mg fresh weight/ml PCA (9). After centrifugation at 27,000 g for 20 min, the supernatant fraction was used for polyamine analysis. Samples stored at -20° C remained stable for at least 2 months.

PCA extracts were analyzed by TLC according to a dansylation procedure modified from Seiler and Wiechmann (23). Dansyl-polyamines were separated on Whatman LK6D silica gel plates. Identification was accomplished by comparing the Rf values of unknowns with those of authentic dansyl-polyamine standards in two solvent systems: (a) cyclohexane:ethylacetate (5:4, v/v); and (b) chloroform:triethylamine (25:2, v/v). The spots were eluted in ethylacetate and quantified by spectrofluorimetry (excitation 350 nm, emission at 500 nm).

Extracts were also analyzed for free polyamines by HPLC according to a benzoylation procedure modified for higher plants (10). Separation, identification and quantification were done with a Beckman 320 HPLC model, equipped with a C-18 reverse-phase column (Altex, 5 μm) and Hewlett-Packard 3990A integrator. Benzoyl-polyamines were eluted with 64% methanol, at a flow rate of 1 ml/min, and detected at 254 nm. Results obtained by TLC and HPLC were reproducible and showed good correlation.

B. For Osmotic and pH Stress Experiments.

Victory oat seeds (Svalof, Sweden) were sown in vermiculite in control condition plant growth rooms. Seedlings were grown under 18-hour days (ca 12,000 lux) at 23° C, and subirrigated twice daily with a commercial mineral salt solution (Hyponex, Copley Chemical Co., Copley, Ohio, 1.2 g/l). The primary leaves of 7-9 day old leaves were harvested, the lower epidermis removed by stripping with a fine forceps, and the leaves floated in the light in the presence or absence of osmoticum or particular pH buffer.

RESULTS

A. Control By Phytochrome

In plants, the photoreversible chromoprotein phytochrome controls many aspects of plant growth and development, including seed germination, leaf and stem growth, flowering and the onset of senescence (16). Red light (optimum near 660 nm) converts the inactive red-absorbing form of the pigment (P_r) to the active far-red-absorbing form (P_{fr}); far-red light (optimum near 730 nm) causes a reversion of P_{fr} to P_r and annuls the effect of red light. Since P_{fr} promotes leaf growth but inhibits stem elongation (21), it was of great interest to see whether the activity of ADC, which we had already shown to be influenced by light (17) would be affected in reverse fashion in the two organs.

Six day old etiolated pea plants were given a 5 min exposure to red light and the terminal buds (mainly leaf tissue) and subapical epicotyl (stem) tissue harvested and compared with dark controls. In buds, where the specific activity of ADC rose gradually in the controls over the eight hour test period, red light caused a highly significant doubling of ADC specific activity within two hours, and a continued rise until the end of the experiment. (Fig. 1). This effect was mediated by phytochrome, as shown by mutual red/far-red reversibility (Fig. 2). In epicotyls, on the other hand, where the dark controls show a fairly steep rise in ADC activity over the eight hour test period, red light completely prevents this rise (Fig. 3). This effect, too, is completely red/far-red reversible (Fig. 4). Cycloheximide (5 µg/ml) strongly inhibits the increase of ADC activity in both rapidly-growing organs (darkened epicotyls and red-irradiated buds) but is without much effect on the already inhibited red-irradiated epicotyls and darkened buds. It

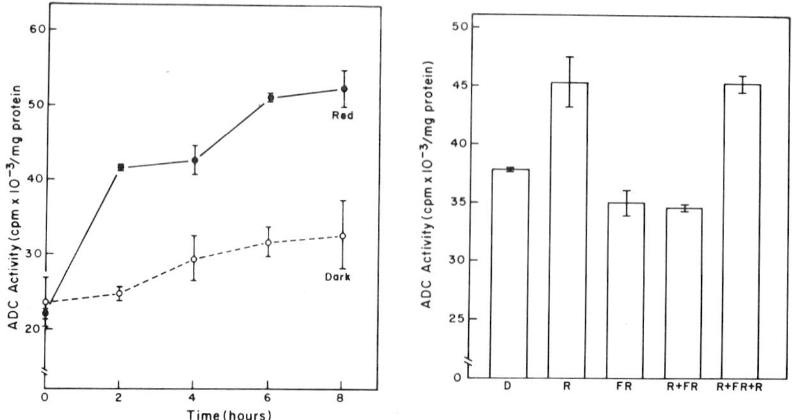

Fig. 1. (left) The kinetics of change of ADC specific activity in buds of etiolated pea seedlings. Bars indicate standard error of the mean.
Fig. 2. (right) FR reversibility of R-induced promotion of ADC activity in buds of etiolated pea seedlings. Measurements made 6 h after light treatment. Bars indicate standard error of the mean. D, dark; R, red: FR, far-red.

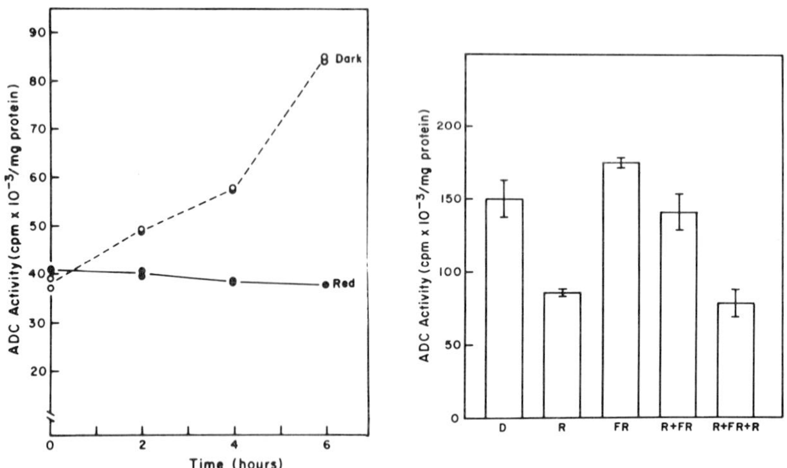

Fig. 3.(left) The kinetics of change in ADC specific activity in epicotyls of etiolated pea seedlings.
Fig. 4. (right) FR reversibility of the R-induced inhibition of ADC activity in epicotyls of etiolated pea seedlings.

would thus appear that phytochrome controls the de novo synthesis of ADC in both organs, but in opposite ways. ADC is the only enzyme yet known whose activity is controlled in this organ-specific, opposite manner. This correlation lends support to the view that polyamine titer may control growth processes in plants.

This conclusion is supported by our subsequent demonstration (14) that polyamine titer does change in etiolated pea seedlings following red light treatment. In buds, red light increased the titers of agmatine, putrescine and spermidine (Fig. 4), while in epicotyls red light decreased the titers of each of these polyamines (Fig. 5). In both organs, agmatine, the direct product of arginine decarboxylation, was the polyamine most affected.

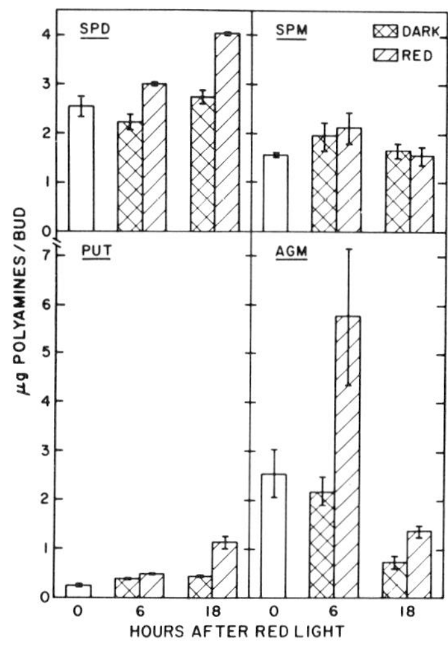

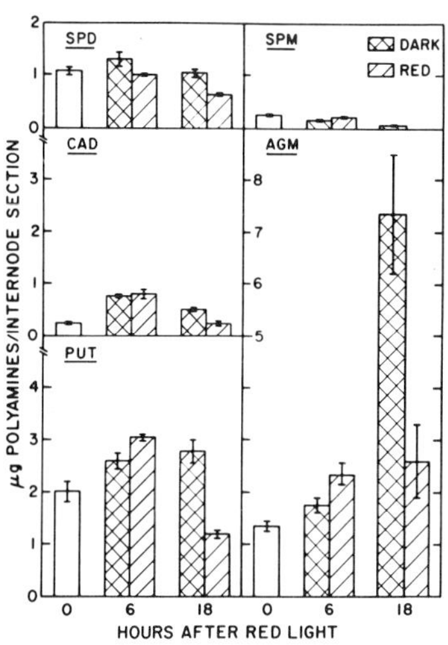

Fig. 5. Polyamine titer, determined by HPLC, in buds (left) and internodes (right) of 6-day-old etiolated seedlings: dark control and red light treated. Spd, spermidine; Spm, spermine; Put, putrescine; Agm, agmatine.

Is the rise or fall of ADC activity merely a secondary consequence of the rise or fall of growth rate? We used a peculiarity of the etiolated pea system to show that changes in ADC are in fact independent of growth. If the terminal bud is excised together with a long piece of epicotyl consisting of several internodes, then red light promotes the growth of the terminal bud, although to a lesser extent than in the intact plant (15). If the bud is excised with only a short piece of attached epicotyl, red light cannot promote bud growth. Yet in both systems, red light promotes ADC activity in the buds. This shows that red light has its effect on ADC activity irrespective of its linkage to growth.

B. Control By Gibberellin

The isoprenoid plant hormone gibberellin, which promotes stem elongation, especially in dwarf and rosette plants (6), is also known to reverse some effects of red light in etiolated plants (18) and to mimic phytochrome effects in other systems. We therefore asked whether gibberellin could induce greater ADC activity in systems wherein cell division and growth are stimulated. We chose the light-grown Progress dwarf pea seedling as the experimental material. Gibberellic acid [GA; Sigma; 5-20 µg/ml in 0.03 M pH 6.4 phosphate buffer containing .05% pluronic L101 (Wyandotte) to aid penetration] was sprayed onto nine day old plants. Pluronic in buffer was sprayed as a control. GA treatment caused a great increase in internode elongation; thus two days after spraying, fourth internode length had increased from 9.78 mm in controls to 16.03 mm in treated plants and fifth internodes from 6.68 mm in controls to 21.50 mm in treated plants. ADC specific activity in the fourth internodes was increased threefold by 5 µg/ml gibberellic acid spray; at the same time ODC activity was decreased threefold. Putrescine and spermidine contents per internode were increased by about 150% in internode 5, while spermine increased only 20-30% (Table I).

Table I. Effect of GA on polyamine content of fifth internodes of light-grown Progress pea seedlings. Measurements made 2 days after GA treatments and expressed as nmol per internode.

Treatment	Putrescine	Spermidine	Spermine
Control	5.39 ± 0.18	21.74 ± 0.32	13.68 ± 1.06
GA (20 µg/ml)	15.33 ± 2.87	39.37 ± 5.19	17.99 ± 1.69

A kinetic experiment showed that ADC activity was increased within 3 hours after application (Fig. 6) and peaked at about 9 hrs. ADC activity then declined and rose to a second (circadian?) peak at about 30 hours, accentuating a peak already apparent in the controls. We surmise that gibberellin has induced

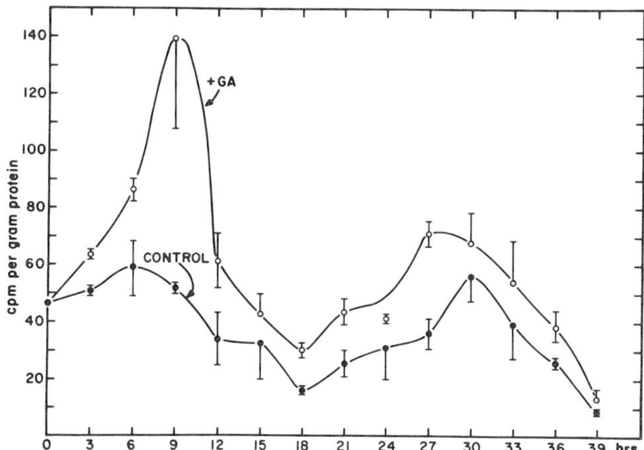

Fig. 6. Kinetics of GA-induced alteration of ADC activity in fifth internodes of light grown dwarf peas.

cell division, as it does in other systems (22), and that the necessary DNA synthesis is accompanied by polyamine synthesis. Similar linkages between polyamine biosynthesis and cell division in synchronized plant cell cultures have already been demonstrated (1). We are currently attempting to quantify mitotic figures in control and gibberellin-treated internodes of the dwarf pea, and also testing whether the specific inhibitor DFMA will prevent the gibberellin response.

C. Control By Osmotic Stress

For several years, we have been concerned with the isolation, cultivation and attempted regeneration of protoplasts derived from the leaves of seedling oats (5). We noted that protoplasts, isolated from cells by a 2 hour cellulolytic digestion of walls in the presence of 0.6 M mannitol or sorbitol as an osmotic protectant, had a markedly altered polyamine profile. Putrescine increased 5-10 fold in 2 hrs and up to 60 fold within six hours while spermidine and spermine were unaltered or slightly lower, and theii oxidation product, 1,3-diaminopropane, rose slightly. Control leaf segments floated on 1 mM phosphate buffer pH 5.8, showed no significant change in putrescine titer (Fig. 7) or in pH (see below). Wild oat, barley, corn and wheat leaves showed similar changes (Fig. 8); it is noteworthy that all were peeled except wheat, which showed distinct, albeit lower increases in putrescine without the complication of peeling injury.

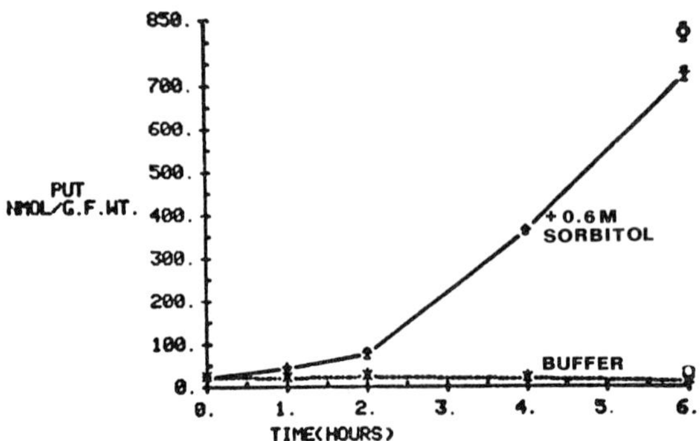

Fig. 7. Time course of putrescine accumulation in oat leaf segments under osmotic stress. * Buffer control. + 0.6 M sorbitol. □ difluoromethylarginine. ○ difluoromethylornithine. Values represent mean ± S.E.M.

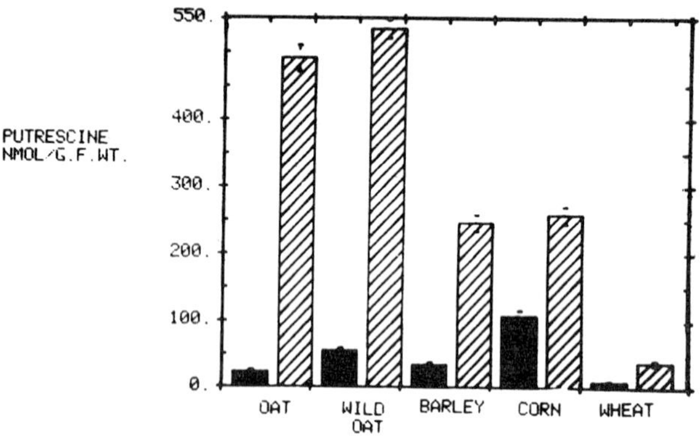

Fig. 8. Effect of osmotic stress on putrescine titer of cereal leaf segments. Solid bars, buffer control (4 hr). Hatched bars, 0.4 M sorbitol (4 hr). Bars represent the mean ± S.E.M.

The specific inhibitor of arginine decarboxylase DL-α-difluoromethyl arginine (DFMA) (0.1 to 1.0 mM) completely prevented the rise in putrescine level, while not affecting spermidine and spermine levels (Table II). DFMA also prevented the rise in ADC specific activity, but did not affect ODC. Conversely, DL-α-difluoromethylornithine (DFMO), a specific inhibitor of ODC in mammalian cells, had little effect on ODC activity, but significantly raised ADC activity activity and putrescine content. We conclude that the effect of sorbitol in increasing putrescine titer is mediated by an increase in ADC activity. Since other

Table II. Polyamine titer and biosynthetic enzyme activity in oat leaf segments under osmotic stress. Seven day old peeled leaf segments were floated in buffer (control), or buffer plus 0.6 M sorbitol in the presence or absence of polyamine synthesis inhibitors. After 4 hr incubation at 25-26° in darkness, triplicate samples were taken for polyamine analysis and enzyme assay (11). *Significantly different from the 4 hr buffer control ($P < .05$). **Significantly different at $P < .01$. Numbers for polyamine levels represent mean ± S.E.M. Numbers in parenthesis express enzyme activity as % of 4 hr buffer control. @ Protein was determined by the Coomassie Blue G-250 assay (28), using gamma globulin as the standard.

Treatment	Polyamine nmol/g.f.wt.			Enzyme activity (nmol CO_2/hr/mg protein) @	
	Put	Spd	Spm	ADC	ODC
0 hr. Control	16±1	161±18	41±4	2.27(115)	4.24(93)
4 hr. Control	13±2	152±10	38±3	1.98(100)	4.55(100)
0.6 M Sorbitol	102±5**	120±13*	21±2**	3.81(192)**	4.65(102)
" + 10^{-3}M DFMO	129±7**	186±10*	32±3	5.31(268)**	4.68(103)
" + 10^{-4}M DFMA	20±2*	123±7*	36±2	1.24(63)*	4.64(102)
" + 10^{-3}M DFMA	10±1	128±8	32±2	0.41(21)**	4.55(100)

osmotica such as mannitol, proline, betaine and, to a lesser extent sucrose, produce the same effect, we conclude that water stress provided by the osmotica serves to induce the appearance of ADC.

D. Control By pH

Low pH has also been reported to cause elevation of putrescine titer in cereal leaves (24). Accordingly we floated peeled oat leaves and excised pea leaf discs for 8 hours in the light on buffered mineral nutrient solutions containing KNO_3, 3.0 mM; NH_4 NO_3, 3.0 mM; $CaCl_2$, 2.0 mM; Mg SO_4, 0.5 mM and NH_4 H_2PO_4m 0.5 mM. Succinic acid (5.0 mM) coupled with variable concentrations of Tris-base provided a range of pH's from 8.0 to 3.5. Additional buffer systems were checked, and gave similar results, showing that the effects obtained were due to pH and not to choice of buffer.

In both oat and pea leaves, putrescine titer was low and constant at pH's greater than 6; putrescine rose slightly between pH 6 and 5, and steeply between pH 5 and 3.5 (Fig. 9). In oat, the effect of lowered pH was

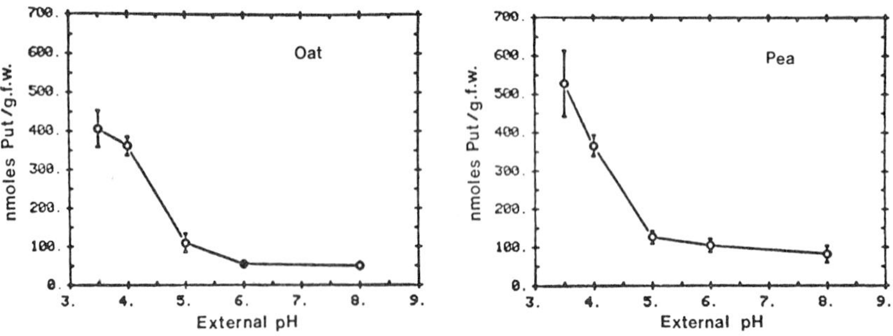

Fig. 9. Putrescine titer in excised and peeled oat leaf segments and pea leaf discs after 8 h incubation on nutrient solutions buffered to various external pH values. (Error bars represent ± 1 S. D.).

seen within 3 hours and peaked at about 9 hours (Fig. 10), and was sensitive to cycloheximide and DFMA but not to DFMO. It appears that the pH effect, like the osmotic effect, is due to activation of arginine decarboxylase.

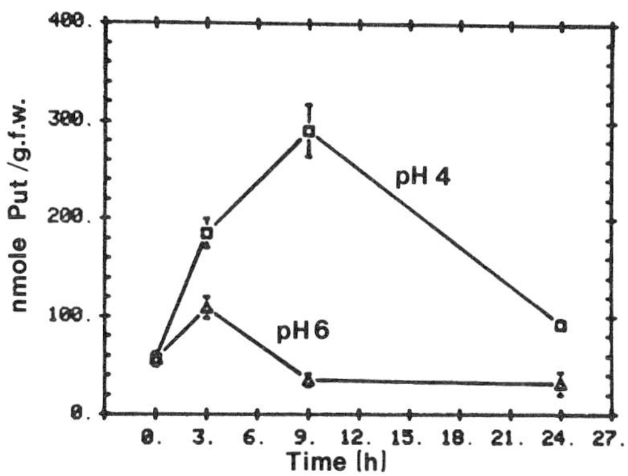

Fig. 10. Time course of changes in putrescine levels in excised and peeled oat leaf segments on solutions of two different external pH values, pH 4 (—— ——) and pH 6 (——Δ——). (Error bars represent ± 1 S. D.).

DISCUSSION

Sharp and rapid changes in polyamine content of plant cells have now been documented for two kinds of situations: 1. Where application of plant growth hormones gibberellin (8), auxin (2) or cytokinin (26) cause a promotion of growth through increase in cell division activity or cell elongation, the activity of polyamine-producing enzymes ADC or ODC are increased in specific plants. Following this, an increase is seen in the titers of putrescine, spermidine and sometimes spermine. Similarly, where P_{fr} phytochrome promotes growth, red light promotes both ADC activity and polyamine accumulation, and where P_{fr} phytochrome inhibits growth, ADC activity and polyamine titers are decreased. 2. Where stress is applied to particular plant tissues not undergoing cell division, the activity of ADC is sharply increased, resulting in elevated putrescine levels. In these instances, rapid transformation of the additional putrescine to spermidine and spermine is not observed. This is true whether the stress is osmotic (11), pH (24, 27), salinity (26) or K^+ deprivation (25).

The apparent inducibility and repressibility of ADC by a variety of stimuli points to its possible central role in mediating the translation of growth controlling agents into cellular growth mechanisms. We conjecture that the bimodal induction of ADC by gibberellin (Fig. 6) is linked to the induction of DNA synthesis and mitosis. This would imply that there is some control linking DNA and polyamine biosynthesis, not illogical in view of the apparently mandatory electrostatic linkage of these two classes of compounds in the cell.

Does the stress-induced rise in putrescine titer have any adaptive value for the stressed cell? It could, of course, itself serve as a buffer to resist changes in cellular pH, or to serve as a facilitator of the transport of H^+ or other cations across the plasma membrane. Putrescine is also known to inhibit the activity of ribonuclease (13) and acid protease (17), two enzyme systems associated with injury and senescence in plant cells (17). Putrescine can also protect membranes of fragile protoplasts against spontaneous or injury-induced lysis (3). To date, we have not been able to find any relation between polyamines and temperature stress, but this subject requires further investigation.

There are many productive experimental pathways for the plant physiologist interested in polyamines. We aim to continue our enjoyable excursion into this domain.

ACKNOWLEDGMENTS

We are grateful for grants from the National Science Foundation, National Aeronautics and Space Administration and Binational Agricultural Research and Development agency to A.W.G. Earlier collaboration with R. Kaur-Sawhney and L-m. Shih was essential to the development of ideas and techniques used in this work.

REFERENCES

1. Adlakha, R.C., Villanueva, R., Calvayrac, R. and Edmunds, L.N. (1980): Arch. Bioch. and Biophys. 201: 660-668.
2. Altman, A. and Bachrach, U. (1981): Adv. in Polyamine Res. 3: 365-375.
3. Altman, A., Kaur-Sawhney, R. and Galston, A.W. (1977): Pl. Physiol. 60: 570-574.
4. Bachrach, U. (1973): Function of Naturally-Occurring Polyamines. Academic Press, New York.
5. Brenneman, F. and Galston, A.W. (1975): Biochem. Physiol. Pflanzen. 168: 453-471.
6. Brian, P.W. and Hemming, H.G. (1955): Physiol. Plant. 8: 669-681.
7. Cohen, S.S. (1971): Introduction to the Polyamines. Prentice-Hall, Englewood Cliffs, N.J.
8. Dai, Y-r. and Galston, A.W. (1981): Pl. Physiol. 67: 266-269.
9. Dai, Y-r., Kaur-Sawhney, R. and Galston, A.W. (1982): Pl. Physiol. 69: 103-105.

10. Flores, H.E. and Galston, A.W. (1982): Pl. Physiol. 69: 701-706.
11. Flores, H.E. and Galston, A.W.: (submitted for publication).
12. Fong, W.F., Heller, J.S. and Canellakis, E.S. (1976): Biochem. Biophys. Acta. 428: 456-465.
13. Galston, A.W., Altman, A. and Kaur-Sawhney, R. (1978): Pl. Sci. Lett. 11: 69-79.
14. Goren, R., Palavan, N., Flores, H.E. and Galston, A.W. (1982): Pl. and Cell Physiol. 23: 19-26.
15. Goren, R. and Galston, A.W. (1966): Pl. Physiol. 41: 1055-1064.
16. Hendricks, S.B. and Borthwick, H.A. (1967): Proc. Natl. Acad. Sci. (U.S.A.) 58: 2125-2130.
17. Kaur-Sawhney, R., Shih, L-m., Flores, H.E. and Galston, A.W. (1982) : Pl. Physiol. 69: 405-410.
18. Lockhart, J.A.(1956) : Proc. Natl. Acad Sci. (U.S.A.) 42: 841-848.
19. Lowry, O.H., Rosebrough, N.J., Farr, A.L. and Randall, R.G. (1951) : J. Biol. Chem. 193: 265-275.
20. Montague, M.J., Armstrong, T.A. and Jaworski, E.G. (1979): Pl. Physiol. 63: 341-345.
21. Parker, M.W., Hendricks, S.B., Borthwick, H.A. and Went, F.W. (1949): Amer. J. Bot. 36: 194-209.
22. Sachs, R.M., Bretz, J.P. and Lang, A. (1959): Amer. J. Bot. 46: 376-384.
23. Seiler, N. and Wiechmann, M. (1967): Ztschr. Physiol. Chem. 348: 1285-1290.
24. Smith, T.A. and Sinclair, C. (1967): Ann. Bot. n.s. 31: 103-111.
25. Smith, T.A. (1977): Progr. in Phytochem. 4: 27-81.
26. Suresh, M.R., Ramakrishna, S. and Adiga, P.R. (1978): Phytochem. 17: 57-63.
27. Young, N.D. and Galston, A.W.: (in preparation).
28. Bradford, M.M. (1975): Anal Biochem. 248: 248-250.

Advances in Polyamine Research, Vol. 4, edited by U. Bachrach, A. Kaye, and R. Chayen. Raven Press, New York © 1983.

Alternative Metabolic Pathways for Polyamine Biosynthesis in Plant Development

Arie Altman, Ra'anan Friedman, and Nitsa Levin

Department of Horticulture, Hebrew University of Jerusalem, Rehovot, Israel

While the involvement of polyamines (PA) in the regulation of plant growth and physiology was limited in the past to few studies (7, 37, 38), accumulating evidence during the last 5-6 years indicates that PA can be regarded as a new class of plant growth substances (3, 5, 18). Thus, in recent years PA have been shown to participate in various aspects of plant growth processes and in the response to specific stress conditions and exogenous stimuli (3, 5).

The above mentioned conclusions have been inferred from various types of experiments, using several methodologies:

1. Application of exogenous polyamines and their immediate precursors (e.g. 1, 2, 3, 5, 23, 30).
2. Treatment of whole plants or excised tissues with structural analogues of PA and their immediate precursors, and with inhibitors of PA biosynthesis, such as L-canavanine, L-canaline, difluoromethyl ornithine (DFMO), difluoromethyl arginine (DFMA), and methylglyoxal bis (guanyl hydrazone)(MGBG) (e.g. 4, 8, 10, 12, 17).
3. Changes in endogenous PA levels in various plants and tissues (e.g. 3, 7, 24, 37, 38, 44).
4. Studies of PA biosynthesis, including the activity of several enzymes such as L-arginine decarboxylase (ADC), L-ornithine decarboxylase (ODC), spermidine and spermine synthases, S-adenosylmethionine decarboxylase, as well as the incorporation of labeled L-arginine, L-ornithine, L-citrulline, putrescine, L-methionine and related compounds (e.g. 3, 4, 13, 16, 20, 21, 24, 28, 29).
5. Studies of PA degradation and conjugation, and especially the activity of amine oxidases (e.g. 9, 37, 38).

If polyamines do have a role in the control of plant growth and senescence, and are not merely correlated with it, then meaningful metabolic pathways of PA biosynthesis should be elucidated. Unlike the classical concept of PA biosynthesis in mammalian tissues, i.e., a single step putrescine formation from ornithine via ODC, plants (and bacteria) seem to possess alternative routes for putrescine biosynthesis. Some of the recent data related to this issue will be reviewed, followed by a short discussion of PA levels, and effects of analogues and metabolic inhibitors.

POLYAMINE BIOSYNTHESIS

1. Pathways and Precursors

At least two alternative routes for putrescine biosynthesis exist side by side in plant tissues, i.e., one via the activity of ADC and the other via ODC. The interconversions of these two amino acids, via the arginine-ornithine-citrulline cycle, further stress the importance of the issue. This is outlined in Fig. 1,

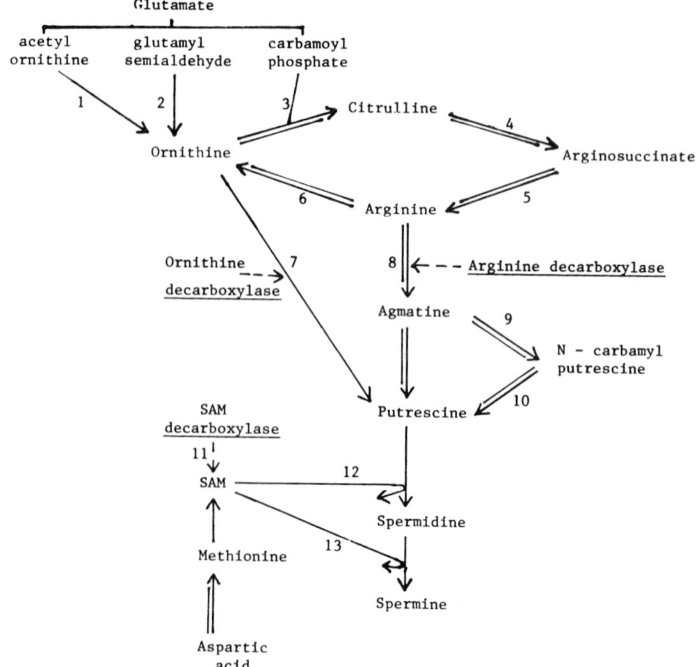

FIG. 1. Pathways of putrescine and polyamine biosynthesis in plants. Pathways which seem to be specific, or predominant in plants are marked with double lines. (1) N-acetyl ornithinase, (2) ornithine transaminase, (3) ornithine transcarbamoylase, (4) arginosuccinate synthetase, (5) arginosuccinate lyase, (6) ?, (7) ornithine decarboxylase, (8) arginine decarboxylase, (9) agmatine imino-hydrolase, (10) N-carbamylputrescine amidohydrolase, (11) S-adenosylmethionine decarboxylase, (12) spermidine synthase, (13) spermine synthase.

where routes which seem to be specific, or more predominant, in plants are indicated. The incorporation of nitrogen from glutamate and glutamine into arginine and ornithine involves 3 routes: (1) the 2-amino group of glutamate yields the 2-amino group of ornithine via acetylornithine and the activity of N-acetyl ornithinase; (2) ornithine is formed from glutamyl semialdehyde by ornithine transaminase, and ornithine then yields arginine; aminase; and (3) nitrogen is incorporated from glutamine via carbamoyl phosphate and the activity of ornithine transcarbamoylase. In addition, the interconversions of citrulline-arginine-ornithine are mediated by the activity of arginosuccinate synthetase, arginosuccinate lyase, and arginase (26).

With respect to PA, it may be significant that in addition to asparagine, arginine is the prevalent amino acid in storage proteins and in transport, especially in legumes and in trees. Arginine appears to have a specific role in long-distance transport and as an overwintering compound, both as a free amino acid and in polypeptide chains (26, 43). This unique property of arginine and asparagine is probably related to their specific position in N metabolism, having a high N/C ratio (4N:6C in arginine) and thus are economic in their use of carbon. It is possible therefore, that certain correlations of arginine with growth and development (e.g. shoot growth and flower formation in apple, 19, and personal communication) may be attributed to PA which derive from arginine.

Putrescine may arise either from L-ornithine, via the activity of L-ornithine decarboxylase (ODC), or from L-arginine, via the activity of L-arginine decarboxylase (ADC) (involving, a second step, the formation of N-carbamoylputrescine. Thus, ADC and ODC are the 2 key enzymes in polyamine biosynthesis.

2. Plant ADC and ODC

In contrast with the central role of ODC in mammalian cells (33), this enzyme seems to be of less importance in plants, and it has been claimed that putrescine is synthesized mainly from L-arginine by ADC via the intermediate agmatine (37, 38). This is supported by the relative high levels of ADC activity found in plant tissues in response to several growth stimuli. Thus, ADC activity was increased in embryonic cells of *Daucus carota* (28), in potassium-deficient barley (36), and in soybean seedlings grown on ammonium (25). Similarly, N^6-benzyladenine enhanced ADC activity in cucumber cotyledons (42) and in *Citrus* buds (Adamov and Altman, unpublished data), and pea ADC was found to be controlled by phytochrome and GA_3 (15). Other studies indicate that this is not always the case, and that both ADC and ODC are active and may contribute to PA biosynthesis (9, 34, 39, 46). It should be noted, however, that tomato fruit development was accompanied by a dramatic increase in ODC activity up to the 3rd day after pollination, while ADC activity was very low (21). In addition, it was found that auxin induces tomato fruit setting as well as a 3-fold increase in ODC activity (12). In view of the contradictory results regarding the activity and relative importance of ADC and ODC and their contribution to putrescine biosynthesis, a comparative characterization of these two enzymes in various tissues was undertaken.

Activity of ADC and ODC from mung bean tissues was found to increase linearly with increasing concentrations of crude enzymes, but the increase in ADC activity was considerably greater (4). The Km of crude ADC and ODC of mung beans was recently found to be 5.86×10^{-4}M and 2.53×10^{-4}M, respectively (Fig. 2). This is

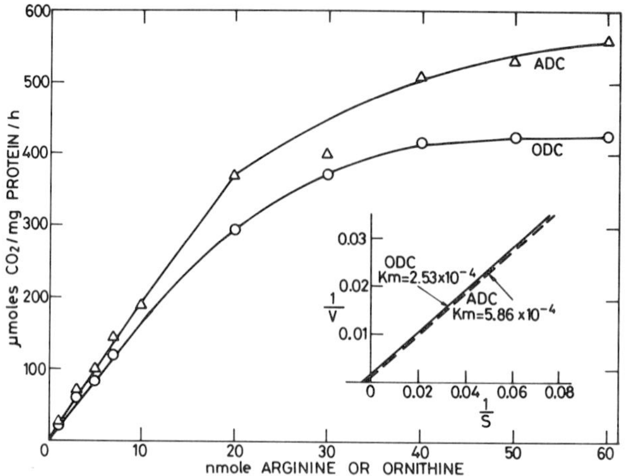

FIG. 2. Arginine decarboxylase (ADC) and ornithine decarboxylase (ODC) activity with increasing substrate concentrations (L-arginine and L-ornithine, respectively). Assayed according to Altman *et al*. (4).

close to the 1.4×10^{-4}M Km of partially purified ODC from tobacco cells (20). Recent data, from several laboratories, including ours, indicate that ADC and ODC activity may be similar with respect to some characteristics, while they may differ in others. Thus, both enzymes have a pH optimum of 7.2, but ADC evinces a greater decline in activity at higher and lower pH values (Fig. 3). Of special importance to the problem of alternative pathways of putrescine biosynthesis is the observation that treatment of excised roots with 1 mM α-DFMO resulted in a 55% inhibition of ODC activity and a concomitant 59% increase in ADC activity (4). This may represent a "compensation" mechanism for keeping a constant level of putrescine biosynthesis, due to the presence of two enzymes with a similar function. In an attempt to further characterize plant ADC and ODC we investigated the possibility that inhibitory proteins may interfere with the enzyme assay in crude extracts. When extracts were subjected to acid precipitation at pH 4.6 (and in other studies down to pH 4.1),

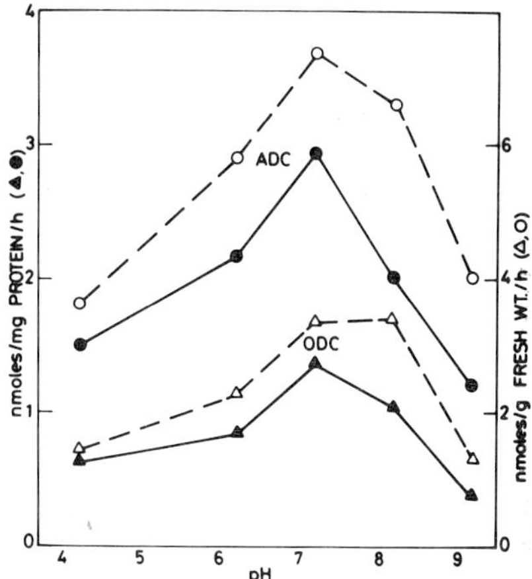

FIG. 3. Effect of pH on arginine (ADC) and ornithine (ODC) decorboxylase activity of mung bean hypocotyls.

most of the ADC and ODC activity was retained in the supernatant, although a considerable amount of protein was precipitated. The presence of phenylmethylsulfonyl fluoride (PMSF) in the extraction and assay media resulted in a 2-3 fold increase in the specific activity of both ADC and ODC (Table 1).

TABLE 1. Partial purification and "activation" of L-arginine and L-ornithine decarboxylases of mung bean[a]

	pmole CO_2 · mg $protein^{-1} h^{-1}$		% of crude control	
	ADC	ODC	ADC	ODC
Crude, control	221	66	100	100
Crude + PMSF	688	251	310	380
Acid prec., supernatant	278	75	125	114
Acid prec., supernatant + PMSF	657	186	297	281
Acid prec., pellet	27	6	12	9
Acid prec., pellet + PMSF	68	15	13	23

[a]ADC and ODC were extracted from leaves of 11-day old mung bean plants as described previously (4), either with or without phenylmethylsulfonyl fluoride (PMSF) at final concentration of 3×10^{-3} M. The 12000 x g supernatant was subjected to acid precipitation at pH 4.6 (phosphate buffer), and the resulting supernatant and pellet were assayed separately.

This effect was saturated at a concentration of 1.6×10^{-3} M PMSF. It is thus possible that plant ODC and ADC may have inhibitory factors similar to mammalian cell ODC inactivation factor (ODIF) (22), and that some data concerning ODC and ADC activity in various plant tissues and growth processes should be re-evaluated. In this respect, it should be mentioned that although most studies indicate that the effect of plant hormones on growth and development is accompanied by an increase in ADC and/or ODC activity (12, 15, 42, Adamov and Altman, unpublished), some exceptions do occur. Thus, treatment of mung bean seedlings of different age with gibberellic acid (GA_3) resulted in a marked inhibition of ADC activity, especially in older plants, while ODC activity was very little affected (Fig. 4). Levels of putrescine and spermidine, however, were considerably increased in these experiments due to the GA_3 treatment (Fig. 5). It is thus likely that ADC activity was inhibited by accumulation of its product, putrescine, or that putrescine conjugation to ADC modifies its activity, similar to the suggested transglutaminase-mediated incorporation of putrescine to ODC in mammalian cells (32).

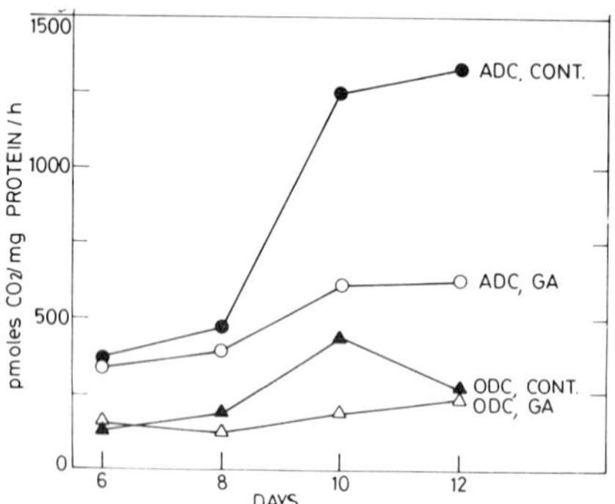

FIG. 4. Effect of gibberellic acid (GA) treatment on arginine (ADC) and ornithine (ODC) activity in the sub-apical region of mung bean seedlings of different age. Seedlings, germinated on different dates, were sprayed with water (control) or 10^{-4} M GA_3 2 days prior to harvesting. Seedlings were 6, 8, 10 or 12 days old when harvested. The 10 mm sub-apical region of the epicotyl was extracted and enzyme activities were determined according to Altman et al. (4). (Adapted from Galston, Altman and Levin, unpublished data).

Different ADC-to-ODC activity ratios and different patterns of activity are found in various tissues of a given plant, and in various plant species (4, 9). Recently it was found that the different patterns of ADC and ODC activity in the peel and pulp of developing Citrus fruit follow the growth pattern and other physiological activity of these two tissues (Nathan, Altman and

Monselise, unpublished data). The range of activities found by us in various plants and tissues is 0.3-750 nmoles CO_2/mg protein for ADC and 0.03-250 nmoles CO_2/mg protein for ODC. These values are in the range of ADC and ODC activity found in bacteria and ODC activity found in mammalian tissues, and in some cases considerably higher.

In addition, various exogenous stimuli, which result in growth responses and stress-related processes, induce differential activation of ADC and ODC. Thus, NaCl stress induces a 2-8 fold increase in ODC activity of mung bean roots, while ADC activity is either inhibited or not affected; on the other hand, this increase is not observed in roots of Pulicaria, a halophyte (Friedman, Altman and Bachrach, unpublished data). The possible differential activity of ADC and ODC in plant tissues is further supported by the observations that ADC and ODC activity in tobacco cell lines which accumulate high amounts of putrescine was increased 3- to 6 fold, as compared with wild type cells (9).

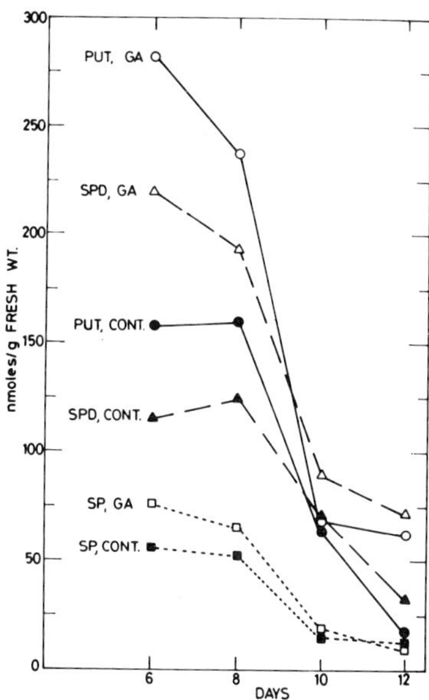

FIG. 5. Effect of gibberellic acid (GA) treatment on polyamine levels in the sub-apical region of mung bean seedlings of different age. Plants were treated as described in the legend for Fig. 4. Free polyamines were extracted and assayed according to Friedman et al. (16).

The above-mentioned studies, as well as additional data (37, 38), suggest that both ADC and ODC are active in plant tissues and that their relative contribution to putrescine (and polyamine) biosynthesis is dependent upon the type of tissue and the growth process. The presence of two enzymes with a similar function, resulting in two alternative pathways of putrescine biosynthesis, may have an important bearing on the control of polyamine levels and effects in plants. This will be discussed later.

3. Incorporation of Precursors

Very little is known on the interrelationships of plant growth and the control of precursor incorporation into various polyamines. Such studies should, ideally, look at the rates of conversion of immediate precursors (such as arginine, ornithine, and, possibly, citrulline) to putrescine, and at the incorporation of label from putrescine and/or methionine into spermidine and spermine. In addition, a correlation between the efficiency of incorporation of various precursors, the activity of corresponding enzymes (e.g. ADC, ODC, spermidine synthase), and growth processes, should be established. This has been done in very few cases only. Growing and differentiating carrot cell cultures have been shown to incorporate radioactive arginine into putrescine, spermidine and spermine; embryogenic cells incorporated arginine into putrescine at two times the rate of control cells, thus accounting for the higher level of putrescine in these cells (29). The formation of putrescine and its conjugates was also studied in two cell lines of tobacco, and it was found that labelled arginine and ornithine were equally well incorporated into the main conjugates caffeoyl and feruloyl putrescine, whereas arginine was preferentially incorporated into putrescine (9). Although arginine and ornithine seem to be the immediate precursors for putrescine biosynthesis, it was found that ^{14}C-carbamoyl citrulline is converted, in part, to N-carbamoyl putrescine and arginine (40), thus supporting the significance of the possible alternative routes for PA biosynthesis (Fig. 1). In a comparative study of polyamine synthesis from labelled methionine in healthy and virus-infected Chinese cabbage (13) it was found that protoplast preparations incorporate methionine into protein, spermidine and spermine more rapidly than do fresh leaf discs, and that disrupted protoplasts contain spermidine synthase which can account for the synthesis of cell and viral spermidine.

Recently, we have initiated a series of studies on arginine and ornithine incorporation into PA in several plants and tissues (*Vigna radiata*, *Phaseolus vulgaris*, *Zea mays*, *Nicotiana tabacum*, with respect to root initiation and growth, leaf maturation and senescence, response to salinity). In these studies it was noted that both precursors were available for putrescine, spermidine, and spermine biosynthesis, but their relative contribution to PA formation was dependent on the type of tissue and response. Thus, NaCl stress resulted in greater incorporation of arginine and ornithine into putrescine of mung bean roots, while considerably inhibiting the incorporation into putrescine in leaves

(Friedman, Altman and Bachrach, unpublished data). In senescing French bean leaves, the incorporation of U-^{14}C ornithine into putrescine was higher than from U-^{14}C arginine, and resembled more the pattern of PA change during the entire senescence period (Amir and Altman, unpublished data). A comparison of arginine, ornithine and putrescine incorporation (Table 2) indicates that in mung bean cuttings, the synthesis of putrescine and spermidine from arginine is several fold higher than from ornithine. IBA-induced root formation is accompanied by a considerable increase in the putrescine/spermidine incorporation ratio, while changes in accumulation of labelled spermine are very small. The data of incorporation of label from putrescine into the various measured polyamines may indicate a turn-over of putrescine and formation of compounds other than the assayed PA.

TABLE 2. Incorporation of U-^{14}C L-arginine, L-ornithine and putrescine into polyamines in root-forming mung bean cuttings [a]

	% of total radioactivity			
	put	spd	spm	put/spd
Arginine, control	16.2	23.8	2.3	0.75
IBA	20.1	12.1	0	1.66
Ornithine, control	1.3	7.8	5.0	0.18
IBA	12.1	11.0	3.5	1.10
Putrescine, control	8.3	26.5	4.9	0.28
IBA	13.1	18.2	4.3	0.73

[a]Mung bean hypocotyl cuttings were prepared as described previously (16) and incorporation of labeled precursors was carried out for 24h(17). Data are expressed as % of the total radioactivity (in the dansylated compounds applied to the plates) in putrescine, spermidine, and spermine. Putrescine/spermidine ratio was calculated from data of dpm/mg protein.

ANALOGUES AND METABOLIC INHIBITORS

Although the use of structural analogues and metabolic inhibitors was much criticized on grounds that their effect may not be specific, several recent studies with PA analogues and inhibitors (Fig. 6) indicate that they can be used to elucidate some effects of PA in plants. Thus, adventitious root formation in mung bean cuttings was inhibited by L-canavanine and L-canaline, and this inhibition was reversed by the corresponding amino acids L-arginine and L-ornithine (17). MGBG, an inhibitor of S-adenosylmethionine decarboxylase activity and PA biosynthesis was also found to inhibit rooting. All compounds at concentrations of $>10^{-4}$ M completely inhibited natural root formation, while at $<10^{-5}$ M only the IBA-induced root formation was inhibited. This lends support to our conclusion that the hormonal (IBA) effect can be mediated by polyamines. That a causal relationship between inhibitor treatments, polyamines, and root formation may indeed exist, is indicated by recent findings that canavanine inhibits putrescine accumulation in IBA-treated cuttings (Table 3). Similarly, the formation of lateral roots in decapitated primary roots of corn was also inhibited by canavanine, and the inhibition was relieved by addition of L-arginine or L-ornithine (Schwartz, Arzee and Altman, unpublished data). Recent work with *Helianthus tuberosus* explants (8) shows that canavanine and canaline inhibited explant growth as well as partially inhibiting putrescine and spermidine synthesis and accumulation, MGBG; methyl ornithine (MO) and other compounds did not affect growth. In addition, canavanine, canaline, difluoromethyl ornithine (DFMO), difluoromethyl arginine (DFMA), methyl ornithine (MO), as well as MGBG and other inhibitors have been shown previously to interfere with polyamine biosynthesis and/or effects (4, 20, 42).

TABLE 3. The effect of L-canavanine on polyamine levels and on root formation in mung bean cuttings[a]

	$nmole \cdot mg^{-1}$ protein			
	put	spd	put/spd	No. roots/cutting
Fresh cuttings	293	281	1.05	0
72-h, control	208	353	0.59	5.8
Control + canavanine	256	318	0.77	5.1
72-h, IBA	288	388	0.74	29.1
IBA + canavanine	178	419	0.43	4.1

[a] Hypocotyl cuttings from 8 day old mung bean seedlings were prepared as described previously (46), and treated for 72 h without or with 5×10^{-1} M indole-3-butyric acid (IBA), in the absence or presence of 10^{-5}M L-canavanine. Putrescine, spermidine and spermine were extracted and assayed according to Friedman *et al*. (16).

Both canavanine and canaline, the structural analogues of arginine and ornithine, respectively, may affect putrescine formation from these two amino acids (31). Similarly, MGBG can block spermidine and spermine biosynthesis by inhibiting S-adenosylmethionine decarboxylase. It is interesting to note that the first report of canavanine effect on plants indicated that it inhibited IAA-dependent elongation of *Avena* coleoptiles,

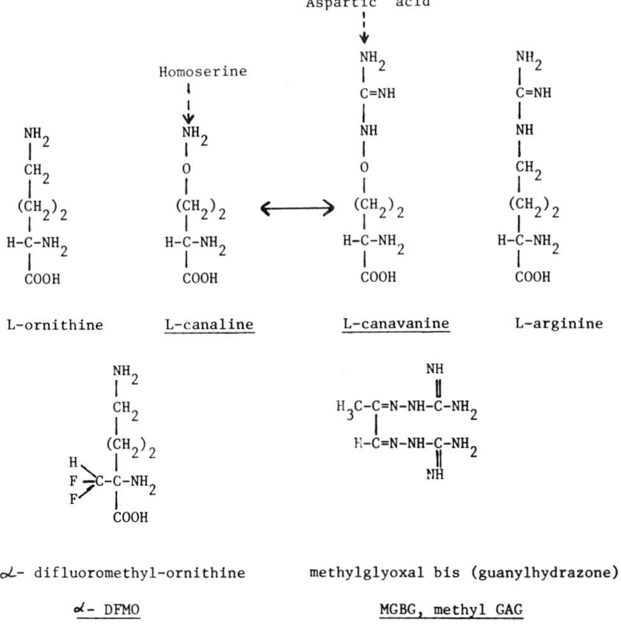

FIG. 6. Some structural analogues and inhibitors of polyamine biosynthesis.

and a specific role for arginine was suggested (11). Later, it was found that canavanine antagonized arginine effects on leaf senescence (35). It is thus possible that previously reported cases of inhibition by canavanine — e.g., growth of Phaseolus roots (45) and of soybean cells in suspension (14) — may be related also to its effect on polyamine metabolism.

POLYAMINE LEVELS

Levels of endogenous polyamines vary considerably in different plant species and organs, and this issue has been reviewed extensively by Smith (37, 38). Very little is known on distribution of polyamines in plant cell organelles and constituents, but they have been found at different ratios in the vacuoles, in ribosomes, mitochondria, chloroplasts and bound to several RNA species (7, 37). The range of putrescine concentrations in various plants and organs was found to be 0.5-5.0 µmoles/g fresh weight, and the mean ratio (µmoles/g fresh weight) of putrescine: spermidine:spermine is in the order of 10:5:0.5. However, polyamine levels are sometimes considerably higher (e.g., 6 µmole spermidine/g dry weight in seeds, or a 15-fold increase of putrescine content in potassium-deficient Hordeum leaves) (37). In our various experiments we found levels of 0.7 - 10 μmole putrescine, 0.005 - 5 μmoles spermidine, and 0.001 - 0.05 μmoles spermine (all per g fresh weight), depending on the type of tissue and response (3, 7).

In general, free putrescine, spermidine and spermine were the only amines measured, and only little attention has been given to bound or conjugated polyamines, which seem to be of great importance (9, 37, 38). Even less is known on the accumulation of polyamine metabolic products of oxidation, acetylation, etc. Thus, while the endogenous content, and especially the putrescine/spermidine ratio, may prove to be a good criterion of tissue growth and physiological activity, the dynamics of polyamine participation in the various growth processes can be better assessed by measuring the magnitude of the diversion of putrescine, spermidine and spermine to other metabolically-related compounds.

CONCLUDING REMARKS

It is now clear that polyamines fulfil at least several criteria to be regarded as growth regulators (5). In order to elucidate their role in the control of plant growth and physiology, several relevant questions ought to be asked: (1) what are the dose-response relationships with applied PA and their immediate precursors? (2) in what way are PA related to known hormonal effects? (3) how do the levels of putrescine, spermidine, spermine and related compounds change in various plant species, in different tissues, and in response to various growth and development processes?, and (4) what are the metabolic steps involved in PA biosynthesis and degradation?

From the above it is evident that no single route for PA biosynthesis in plants should be considered a priori. Moreover, while in few isolated cases either ADC or ODC activity was found, in the majority of the experiments simultaneous activity of ADC and ODC was detected, albeit at various rates. Because both ADC and ODC can be the rate limiting enzymes in putrescine formation, it seems that the regulation of their activity is crucial to the "selection" of either pathways. Thus, alternations in ADC and ODC activity can be regulated by: (a) concentration and availability of arginine and ornithine, (b) interconversion of arginine and ornithine and their metabolism other than to putrescine (Fig. 1), (c) activation, half-life, and the existence of specific inactivating factors, (d) separate compartmentation of ADC and ODC, and (e) specific de novo enzyme synthesis, or induction, in response to hormonal and physiological stimuli. So far, available data support (a), (b) and (e), while the other possibilities await further elucidation.

ACKNOWLEDGMENTS

We wish to acknowledge the collaboration with Mrs T. Adamov, Ms. D. Amir, Prof. T. Arzee, Prof. U. Bachrach, Prof. A.W. Galston, Prof. S.P. Monselise, Mr R. Nathan and Mrs M. Schwartz.

This research was supported by a grant from the United States-Israel Binational Agricultural Research and Development Fund (BARD).

REFERENCES

1. Altman, A. (1982 a): Physiol. Plant., 54:189-193.
2. Altman, A. (1982 b): Physiol. Plant., 54:194-198.
3. Altman, A., and Bachrach, U. (1981): In: Advances in Polyamine Research, edited by C.M. Caldarera, V. Zappia and U. Bachrach, Vol. 3, pp. 365-375. Raven Press, New York.
4. Altman, A., Friedman, R., and Levin, N. (1982). Plant Physiol. 69:876-879.
5. Altman, A., Friedman, R., and Levin, N. (1983): In: Proc. 11th Inter. Conf. Plant Growth Substances, in press.
6. Altman, A., Kaur-Sawhney, R., and Galston, A.W. (1977): Plant Physiol., 60:570-574.
7. Bagni, N., and Serafini-Fracassini, D. (1974): In: Plant Growth Substances 1973, Proc. 8th Inter. Conf. Plant Growth Substances, pp. 1205-1217, Hirokawa Publishing Co., Tokyo.
8. Bagni, N.P.Torrigiani , and Barbieri, P.(1981): Medical Biology, 59:403-409.
9. Berlin, J. (1981): Phytochemistry, 20:53-55.
10. Berlin, J., and Forche, E. (1981): Z. Pflanzenphysiol., 101: 277-282.
11. Bonner, J. (1949): Am. J. Bot., 36:323-332.
12. Cohen, E., Malis-Arad, S., Heimer, Y.M., and Mizrahi, Y. (1982): Plant Physiol., in press.
13. Cohen, S.S., Balint, R., and Sindhu, R.K. (1981): Plant Physiol., 68:1150-1155.
14. Constable, E., Kirkpatrick, J.W., Kao, K.N., and Kartha, K.K. (1975): Biochem. Physiol. Pflanzen., 168:319-325.
15. Dai, Y., Kaur-Sawhney, R., and Galston, A.W. (1982): Plant Physiol., 69:103-105.
16. Friedman, R., Altman, A., and Bachrach, U. (1982): Plant Physiol., in press.
17. Friedman, R., Altman, A., and Bachrach, U. (1983): Plant Physiol., in press.
18. Galston, A.W., and Kaur-Sawhney, R. (1980): What's New in Plant Physiology, 11:5-8.
19. Grasmanis, and Edwards, (1974): Austr. J. Plant Physiol., 1:99-105.
20. Heimer, Y.M., and Mizrahi, Y. (1982): Biochem. J. 201:373-376.
21. Heimer, Y.M., Mizrahi, Y., and Bachrach, U. (1979): FEBS Letters, 104:146-148.

22. Icekson, I., and Kaye, A.M. (1976); FEBS Letters, 61:54-58.
23. Kaur-Sawhney, R., and Galston, A.W. (1979): Plant Cell Environ., 2:189-196.
24. Kaur-Sawhney, R., Shih, L.M., Flores, H.E., and Galston, A.W. (1982): Plant Physiol., 69:405-410.
25. LeRudulier, D., and Goas, G. (1975): Phytochemistry, 14: 1723-1725.
26. Miflin, B.J., and Lea, P.J. (1977): Ann. Rev. Plant Physiology, 28:299-329.
27. Mizrahi, Y., and Heimer, Y.M. (1982): Physiol. Plant., 54: 367-368.
28. Montague, M.J., Armstrong, T.A., and Jaworski, E.G. (1979): Plant Physiol., 63:341-345.
29. Montague, M.J., Koppelbrink, J.W., and Jaworski, E.G. (1978): Plant Physiol., 62, 430-433.
30. Naik, B.I., and Srivastava, S.K. (1978): Phytochemistry, 17:1885-1887.
31. Rosenthal, G.A. (1977): The Quart. Rev. of Biology, 152: 155-178.
32. Russell, D.H. (1981): Biochem. Biophys. Res. Commun., 99: 1167-1172.
33. Russell, D.H., and Snyder, S.H. (1968): Proc. Natl. Acad. Sci. USA, 60:1420-1427.
34. Schuber, F., and Lambert, C. (1974): Physiol. Vég., 12: 571-584.
35. Shibaoka, H., and Thimann, K.V. (1970): Plant Physiol. 46: 212-220.
36. Smith, T.A. (1963): Phytochemistry, 2:241-252.
37. Smith, T.A. (1971): Biol. Rev., 46:201-241.
38. Smith, T.A. (1977): Progress in Phytochemistry, 4:27-81.
39. Speranza, A., and Bagni, N. (1977): Z. Pflanzenphysiol., 81: 226-233.
40. Speranza, A., and Bagni, N. (1978): Z. Pflanzenphysiol., 88: 163-168.
41. Suresh, M.R., and Adiga, P.R. (1977): Eur. J. Biochem., 79: 511-518.
42. Suresh, M.R., Ramakrishna, S., and Adiga, P.R. (1978): Phytochemistry, 17:57-63.
43. Tromp, P.J., and Ovaa, J.C. (1973): Physiol. Plant., 29:1-5.
44. Villanueva, V.R., Adlakha, R.C., and Cantera-Soler, A.M. (1978): Phytochemistry, 17:1245-1249.
45. Weaks, Jr., T.E., and Hunt, G.E. (1974): Bot. Gaz., 135: 45-49.
46. Yoshida, D. (1969): Plant Cell Physiol., 10: 393-397.

Advances in Polyamine Research, Vol. 4, edited by U. Bachrach, A. Kaye, and R. Chayen. Raven Press, New York © 1983.

In Vitro and *In Vivo* Effect of Ornithine and Arginine Decarboxylase Inhibitors in Plant Tissue Culture

N. Bagni, P. Torrigiani, and P. Barbieri

Istituto Botanico, Università di Bologna, 40126 Bologna, Italy

The activity of polyamine biosynthetic enzymes increases rapidly when growth is induced either in animal tissue (11) or in plant tissue (15,20). This enhanced activity leads to a rapid accumulation of polyamines in the tissue, usually preceding increases in nucleic acids and protein during the cell cycle (4,21), indicating that polyamines are especially involved in cell preparation for chromosomal DNA replication. In animal tissue putrescine is synthesized by direct decarboxylation of ornithine; in bacteria (29) and higher plants there is an additional pathway for the synthesis of putrescine by arginine decarboxylase (ADC) (28).

The inhibition of ornithine decarboxylase (ODC) by various substrate or product analogs and the consequent depletion of polyamines have been shown to significantly inhibit DNA synthesis and block replication in mammalian cells (19). A study, using E. coli, on the inhibition of ADC by the arginine analogs α-difluoromethylarginine (DFMA), a specific irreversible inhibitor recently synthesized, and α-methylarginine, a competitive inhibitor, has been reported; these compounds effectively block the enzyme activities in vivo and in vitro (16).

In previous works (4,6) the growth effect of various inhibitors of polyamine biosynthetic enzymes was investigated in dormant tuber explants of Helianthus tuberosus. This in vivo study represents one of the first experimental approaches to the problem. The presence in Helianthus tuber of both the biosynthetic enzymes ODC and ADC in an in vivo system (5,26) might represent an interesting model. In fact the above-mentioned synthesis of specific inhibitors of ADC, such as DFMA, gives us the possibility to block both the enzymes of putrescine synthesis and thus to study the effect of such inhibition on polyamine synthesis and growth in plant tissues.

In this paper the effect of ODC and ADC inhibitors on growth, alone or in combination *in vivo*, and on enzymes activity *in vitro* during the first cell cycle of *Helianthus* tuber is investigated.

MATERIAL AND METHODS

Plant Material

Helianthus tuberosus L. (Jerusalem artichoke) cv. OB1 was grown by vegetative reproduction in the Botanical Garden of Bologna University for several years. The tubers were harvested at the beginning of dormancy in November and stored in moist sand at 4°C. They were utilized during dormancy till sprouting at the end of March.

Activation and Tissue Culture

Sterilized cylindrical slices of homogeneous medullary parechyma (9 mm in diameter, 1 mm thick) from dormant tuber were aseptically excised transversely to the axis. About 10 g of slices were activated in 100 ml conical flasks containing 30 ml of Bonner and Addicott liquid medium (9) to which 4% sucrose plus 10 µM 2,4-dichlorophenoxyacetic acid (2,4-D) was added with or without the different inhibitors to be tested. The pH was adjusted to 5.5 with 1 N NaOH. Activation was carried out on rotatory shaker (75 rpm) for aeration in the dark.

Explants of tuber parenchyma tissue were grown in sterile culture on agarized medium containing 10 µM 2,4-D in all cases except one (O). The different inhibitors tested were added to the treatments containing 2,4-D. The explants were grown for 20 days at 24°C in light-dark cycles (12:12 hr, 5500 erg/cm^2/ /sec.). The data are a mean of at least 20 explants/treatment.

Assay of ADC and ODC Activities

Dormant and activated tuber tissue was homogenized in an Omni Mixer at 180 rpm in the assay buffer containing 50 µM ethylenediamine-tetracetic acid; 25 µM pyridoxal phosphate; 2.5 mM Tris-HCl, pH 7.1, and then centrifuged at 10,000xg for 20 min. ODC activity was determined by incubating 0.5 ml of the supernatant, previously filtered through filter paper with 14.8 kBq in 20 µl of L-[1-^{14}C]ornithine (1.93 GBq); the ADC activity was determined in the same manner except for the labelled compound which in this case was 14.8 kBq in 20 µl L-[U-^{14}C] arginine (12.43 GBq). The test tubes containing the reaction mixture, were capped with special rubber stoppers fitted with center wells containing each 0.2 ml of Protosol (NEN). The reaction was allowed to proceed for 2 hr at 37°C in a water bath and was termined by the injection of 0.2 ml of 6% perchloric acid. After an additional 60 min of shaking, the Protosol was removed from center well and placed in scintillation vials containing 4.5 ml of Lumagel additioned with 10% 0.5 N HCl and counted in a liquid scintillation counter (Isocap/300, Nuclear Chicago). In preliminary experiments the incubation time was controlled, and higher efficiency in capturing CO_2 was obtained with Protosol in respect to Hyamine. Each sample repeated 5 times.

Protein in the extracts was determined by the method of Lowry et al. (18) after precipitation with trichloroacetic acid and solubilization in 1 N NaOH.

Polyamine analysis was done as previously described (6).

RESULTS AND DISCUSSION

Effect of α-difluoromethylornithine and α-difluoromethylarginine on Tissue Growth in Culture

The growth inhibitory effect of DFMO and DFMA was assayed on Helianthus tuber explants after 20 days of culture. DFMO, a potent irreversible inhibitor of ODC in animal tissue (8) caused strong but not complete growth inhibition (63%) at 5 mM, while the lower concentrations 0.1 and 1 mM had no effect (Fig. 1); in mung bean roots a treatment with 1 mM DFMO (24 hr, 25°C, dark) inhibited ODC (69%) and enhanced ADC activity (1). On the other hand in Ophryotroca polyamine synthesis was blocked during oogenesis by 10 mM DFMO in sea water culture medium (13). In the same way DFMA, an irreversible inhibitor of ADC (16), had no effect on the growth at 0.1 and 1 mM concentrations (Fig. 1); in this case it was not possible to try the 5 mM concentration because of the lack of the compound. The use of 1 mM DFMO and 1 mM DFMA in combination reduced the growth by 66%, comparably to 5 mM DFMO.

Already in previous works Bagni et al. (4,6) found that various well-known inhibitors of polyamine biosynthesis in animal tissue, like methylglyoxal bis-(guanylhydrazone), 1,3-diaminopropane (DAP), 1,3-diaminopropan-2-ol, α-methylornithine, had

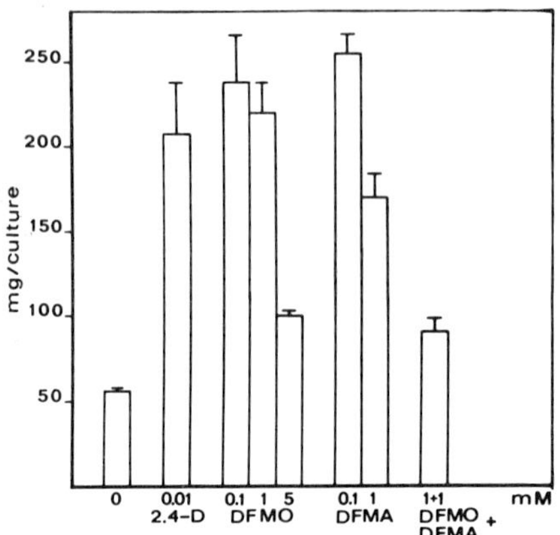

FIG. 1. Effect of α-difluoromethylornithine (DFMO) and α-difluoromethylarginine (DFMA), alone or in combination, on the growth of the explants from Helianthus tuberosus dormant tuber cultivated in vitro for 20 days on agarized medium. 2,4-dichlorophenoxyacetic acid (2,4-D) was added in all samples except in one case (0) in which the growth was zero. The growth of samples treated with 5 mM DFMO and 1 mM DFMO plus 1 mM DFMA differed significantly (t-test) from that of samples with 2,4-D alone.

no effect, at the same concentration used in animal tissue, on the growth of Helianthus tuber; on the contrary DAP had a stimulatory effect. The lack of growth inhibition for many of these compounds can be explained since, contrary to the previous literature in plant tissue (28), both biosynthetic enzymes of putrescine are active in the Helianthus tuber tissue (see later section); it is possible that the inhibition of one activates the other. In effect this hypothesis is supported by the fact that supplying 1 mM DFMO and 1 mM DFMA in combination and therefore blocking both biosynthetic pathways of putrescine, at the concentration which is inactive when supplied alone, there was strong inhibition of growth. Another feature of the Helianthus tuber is the presence of a great amount of endogenous arginine (3 mM) and glutamine (9 mM), precursors of putrescine, at the break of dormancy (26), which might strongly dilute the inhibitors and make them ineffective. The fact that 5 mM DFMO caused growth inhibition might confirm the dilution, due not only to endogenous precursors of polyamines, but also to the polyamines themselves, rapidly synthesized in proliferating tuber tissue during the first cell cycle (3,26). Moreover in tuber tissue there is a very high arginase activity (5,30) at the break of dormancy, responsible for supplying ornithine to the

ODC pathway.

An experiment was carried out supplying 1 mM DAP during the first cell cycle in activated sliced tissue because of the noticeable growth effect, with respect to the control, previously observed in tuber explants cultures treated with 1 mM DAP (6). No difference was revealed with regard to the polyamine content in the treated tissue; moreover the level of DAP in this tissue, after an initial rapid absorption, remained of the same magnitude (about 0.3 mM) during the whole cycle.

Effect of Analogs of Ornithine and Arginine in vitro on the ODC and ADC Activity and their Level during the first Cell Cycle

Fig. 2 shows a dramatic increase in ODC activity (30-fold from 0 hr to maximum) during the first cell cycle of Helianthus tuber; a peak of activity is reached at 12 hr of activation in late G_1 phase. In this period the level of polyamines increased in parallel to the ODC activity (26). This fact is well known as a general phenomenon for animal tissue (10,12,14) in which is one of the first events taking place after the cell is stimulated to proliferate; in fact ODC has a very short half-life (10 min about) and is not only degraded but also synthesized very rapidly. It is suggested that the late-G_1 stimulation of polyamine synthesis is involved in the cell's preparation for DNA chromosomal synthesis (11). At 0 hr, when the tissue is dor-

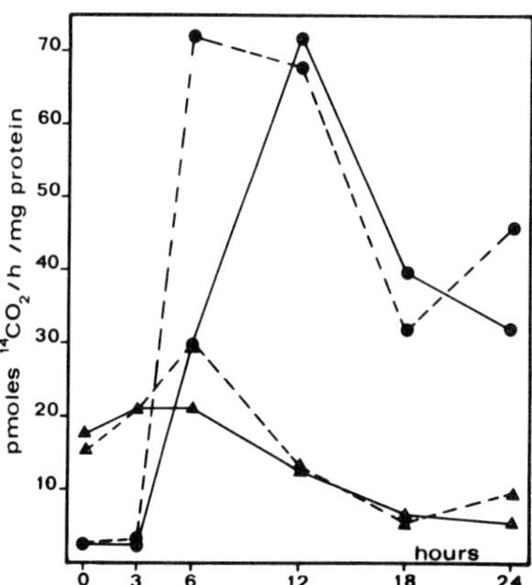

FIG. 2. Arginine decarboxylase (▲) and ornithine decarboxylase (●) activity during the first cell cycle in Helianthus tuberosus tuber slices activated on liquid medium containing 10 μM 2,4-D alone (continuous line) or in combination with 1,3-diaminopropane (DAP) (dotted line).

mant, ADC activity was 7-8 fold higher than ODC; after 6 hr of activation the situation was reverted, ODC becoming higher.

ADC activity showed no relevant changes during the first cell cycle with respect to that of ODC. A recent investigation (17) shows in sprouting potato tuber an analogous pattern: in fact breaking of dormancy and bud sprouting are accompanied by an increase in ODC and S-adenosylmethionine decarboxylase but not in ADC activity. An higher ODC activity than ADC was also found by Heimer et al. (15) in tobacco cells in suspension and in tomato ovaries after pollination, where a peak of activity was reached (about 10-fold increase) after 4-6 days of culture; in this case ADC was one tenth to one fourth of ODC. In other plants, however, ADC was predominant (28) during cell proliferation such as in embryogenic cells of Daucus carota (20) and in mung bean seedlings (1).

In plants polyamine biosynthesis via ADC has thus far been considered the more probable pathway, perhaps because more work has been done with ADC than with ODC (2,27).

The supply of 1 mM DAP in activating medium did not affect either ODC nor ADC activity (Fig. 2) during the first cell cycle except for an enhancement at 6 hr revealed both for ODC and ADC.

1 mM DFMO did not affect ODC in vitro, while at 5 mM it caused an inhibition of 50% (Fig. 3). On the contrary, neither 1 mM nor 5 mM DFMA inhibited ADC activity (Fig. 4). 1 mM canaline (CAL), an analogous of ornithine, strongly inhibited ODC activity during the whole cell cycle (Fig. 3) reaching a maximum of 78% at the 6th hr. Similarly, 1 mM canavanine

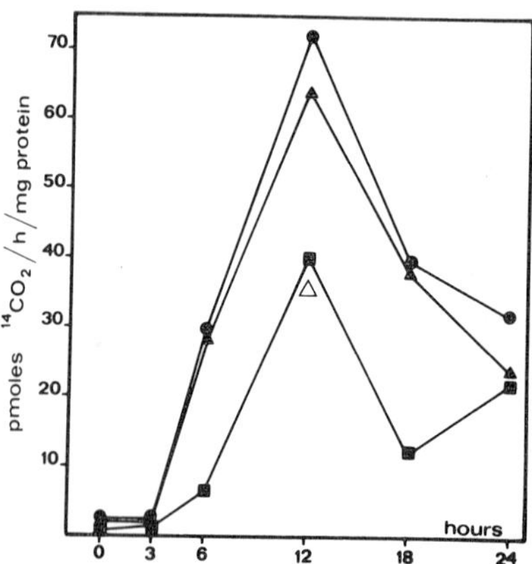

FIG. 3. In vitro effect of 1 mM DFMO (▲), 5 mM DFMO (△) and 1 mM canaline (■) in respect to the control (●) on ornithine decarboxylase activity during the first cell cycle in Helianthus tuberosus tuber slices activated on liquid medium containing 10 µM 2,4-D.

(CAV) inhibited ADC activity especially during the first phases of the cycle (Fig. 4). Only Ramakrishna and Adiga (23) found in *Lathyrus sativus* seedlings an *in vitro* inhibitory effect (50%) on highly purified ADC by 15 mM CAV.

The inhibitory effect of CAV on *Helianthus* tuber growth and on polyamine and RNA synthesis during the first cell cycle was known (6); also in other plant tissues and in animals its effects on RNA and protein synthesis are reported (24).

The strong inhibitory effect of CAL on growth in higher plants (6,25) is reported to be due to the stoichiometric, non-enzymic, irreversible binding with pyridoxal-phosphate, depending on the dissociation constant of the pyridoxal-phosphate-enzyme complex (22).

The lack of inhibition on ADC *in vitro* by 5 mM DFMA could be due to the presence in the supernatant of a great amount of small interfering molecules, for example endogenous arginine, and an accumulation of polyamines.

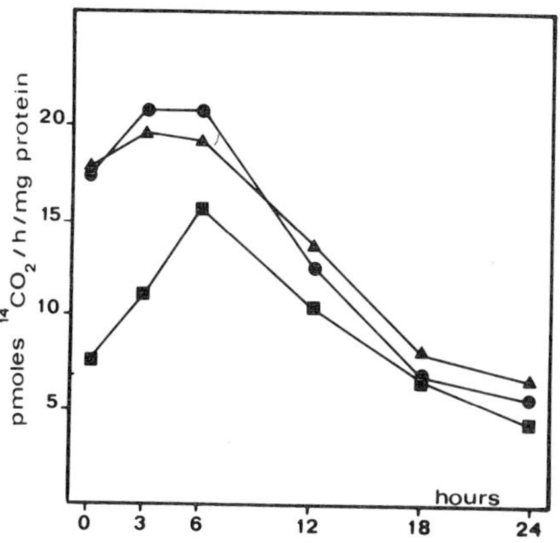

FIG. 4. *In vitro* effect of 1 mM DFMA (▲) and 1 mM canavanine (■) in respect to the control (●) on arginine decarboxylase activity during the first cell cycle in *Helianthus tuberosus* tuber slices activated on liquid medium containing 10 μM 2,4-D.

CONCLUSIVE REMARKS

In many plant systems an increase in growth occurs in parallel to the enhancement of ODC and/or ADC activity, similarly to animal systems where growth is accompanied by a marked ODC activity.

In spite of the high endogenous level of arginine and the high ADC/ODC ratio in dormant tuber, ODC appears to be very rapidly induced when the tissue is stimulated to grow and the

ADC/ODC ratio is reverted at 6 hr of activation.

The apparent correlation in our system of elevated ODC rather than ADC activity with cell division is in agreement with that observed in mammalian and bacterial cells.

The presence of both enzymes of putrescine biosynthesis makes the complete block of growth and of their activity by inhibitors difficult, because of the possible modulation of ODC and ADC as recently demonstrated by Altman *et al*. (1).

ACKNOWLEDGEMENTS

The generous supply of α-difluoromethylornithine and α-difluoromethylarginine by Merrel Research Center, Cincinnati, Ohio, is gratefully acknowledged. The work was supported by funds for scientific research from Ministero della Pubblica Istruzione, Italy.

REFERENCES

1. Altman, A., Friedman, R., and Levin, N. (1982): *Plant Physiol*., 69: in press.
2. Bachrach, U., (1973): In: *Function of Naturally Occurring Polyamines*. Academic Press, New York.
3. Bagni, N., Malucelli, B., and Torrigiani, P., (1980): *Physiol. Plant*., 49: 341-345.
4. Bagni, N., Serafini Fracassini, D., and Torrigiani, P. (1981): In: *Advances in Polyamine Research*, Vol. 3, edited by C. A. Caldarera, V. Zappia, and U. Bachrach, pp. 377 - 388. Raven Press, New York.
5. Bagni, N., and Speranza, A. (1977): In:*Plant Growth Regulators*, edited by T. Kudrev, I. Ivanova, and E. Karanov, pp. 75-78. Publ. House Bulgarian Acad. Sci., Sofia.
6. Bagni, N., Torrigiani, P., and Barbieri, P., (1981): *Med. Biol*., 59: 403-409.
7. Bertossi, F. (1959): *N. Giorn. Bot. Ital*., 66: 497-505.
8. Bey, P. (1978): In: *Enzyme-Activated Irreversible Inhibitors*, edited by N. Seiler, M. J. Jung, and J. Koch-Weser, pp. 27-41. Elsevier/North-Holland Biomedical Press, Amsterdam.
9. Bonner, J., and Addicott, F. (1937): *Bot. Gaz*., 99: 144-170.
10. Cress, A. E., and Gerner, E. W. (1980): *Biochem. J*., 188: 375-379.
11. Heby, O. (1981): *Differentiation*, 19: 1-20.
12. Heby, O., and Andersson, G. (1980): In: *Polyamines in Biomedical Research*, edited by J. M. Gaugas, pp. 17-34. John Wiley & Sons, Chichester, Sussex.

13. Heby, O., and Emanuelsson, H. (1981): Med. Biol., 59: 417-422.
14. Heby, O. and Jänne, J. (1981): In: Polyamines in Biology and Medicine, edited by D.R. Morris, and L. J. Marton, pp. 243-310. Marcel Dekker, Inc., New York.
15. Heimer, Y. M., Mizrahi, Y., and Bachrach, U. (1979): FEBS Letters, 104: 146-148.
16. Kallio, A., Mc Cann, P. P., and Bey, P. (1981): Biochemistry, 20: 3163-3166.
17. Kaur-Sawhney, R., Shih, L. M., and Galston, A. W. (1982): Plant Physiol., 69: 411-415.
18. Lowry, O. H., Rosebrough, N. J., Farr, A. L., and Randall, R. J. (1951): J. Biol. Chem., 193: 265-275.
19. Mamont, P. S., Duchesne, M. C., Joder-Ohlenbusch, A. M., Grove, J. (1978): In: Enzyme-Activated Irreversible Inhibitors, edited by N. Seiler, M. J. Jung, and Koch-Weser, pp. 43-54. Elsevier/North-Holland Biomedical Press, Amsterdam.
20. Montague, M. J., Armstrong, T.A., and Jaworski, E.G. (1979): Plant Physiol., 63: 341-345.
21. Morris, D. R. (1978): In: Advances in Polyamine Research, Vol. 1, edited by R.A. Campbell, D.R. Morris, D. Bartos, G. D. Daves, and F. Bartos, pp. 105-115. Raven Press, New York.
22. Rahiala, E. L., Kekomäki, M., Jänne, J., Raina, A., and Räihä, C. R. (1971): Biochim. Biophys. Acta, 227: 327-336.
23. Ramakrishna, S., and Adiga, R. (1975): Eur. J. Biochem., 59: 377-386.
24. Rosenthal, G.A. (1977): Q. Rev. Biol., 52: 155-178.
25. Rosenthal, G.A., Gulati, D.K., and Sabharwal, P.S. (1975): Plant Physiol., 56: 420-424.
26. Serafini Fracassini, D., Bagni, N., Cionini, P.G., and Bennici, A. (1980): Planta, 148: 332-337.
27. Smith, T.A. (1977): In: Progress in Phytochemistry, Vol. 4, edited by L. Reinhold, J.B. Harborne, and T. Swain, pp. 27-81. Pergamon Press, Oxford.
28. Smith, T.A. (1981): In: Polyamines in Biology and Medicine, edited by D.R. Morris, and L.J. Marton, pp. 77-82. Marcel Dekker, Inc., New York.
29. Tabor, C.W., and Tabor, H. (1976): Annu. Rev. Biochem., 45: 285-306.
30. Wright, L.C., Brady, C.J., and Hinde, R.W. (1981): Phytochemistry, 20: 2641-2645.

Advances in Polyamine Research, Vol. 4, edited by
U. Bachrach, A. Kaye, and R. Chayen. Raven Press,
New York © 1983.

Polyamines and Morphogenesis in *Helianthus Tuberosus* Explants

Donatella Serafini Fracassini and Milena Alessandri

Istituto Botanico dell'Università di Bologna, 40126 Bologna, Italy

Polyamines are known to sustain growth, as shown in polyamine deficient organisms, and their content seems to be proportional to the growth rate. In plant dormant tissues, such as tubers of Helianthus tuberosus, the amount of polyamines and other hormones is very low and the tissue is therefore responsive when treated with polyamines or IAA, 2,4-D etc. (5). The parenchyma cells are homogeneous and all in G_1 phase; by activation with hormones they synthesize polyamines, RNA, DNA and divide rather synchronously (25).

The first part of our work deals with the transition from dormancy to activation of the tuber. We checked if, during the G_1 (0-12 h) and S phases (12-21 h) of the first cell cycle, the observed biosynthesis of polyamines, RNA and DNA (probably partly nucleolar DNA), all induced by 2,4-dichlorophenoxyacetic acid (2,4-D) or indole-3-acetic acid (IAA) (3; 4; 25; 26; 27), might be correlated to an increase in volume of the nucleoli. In fact this enlargement does take place when an active metabolism of ribonucleoproteins occurs and moreover, in isolated nuclei of rat liver, polyamines induce the enlargement of nucleoli and the appearance of their vacuoles (13). In addition Heby and Emanuelsson (14) suggested that polyamine synthesis is essential for nucleolar formation and ribosomal gene expression in invertebrate oogenesis and embryogenesis.

We also examined cell division phase (21-30 h), as determined in previous years (25), to control if the lengh of the two cell cycles are sufficiently similar to make the results of the two studies comparable.

In the second part of our work we examined in this in vitro cultured tissue, the differentiation, at an anatomical level, caused by spermine and spermidine, compared with that induced by IAA and already well documented by Gautheret (12).

It is Known that polyamines similarly to IAA are able to promote an increase in fresh and especially dry weight of the explants (6).

Contrary to the data obtained in animal differentiation studies (14), at present the only direct evidence of polyamine action on plant morphogenesis is offered by the enhancement of flower bud differentiation (8) and by the inhibition of regeneration in fragments of Acetabularia by α-methylornithine (7). There is also a certain amount of other data supporting these observations: potato tuber sprouting (17), carrot embryogenesis (20; 21), mung-bean root formation (1) and tomato fruit development (15) are accompanied by high contents of polyamines or by increased ornithine decarboxylase (ODC) or arginine decarboxylase (ADC) activities.

MATERIALS AND METHODS

The activation of dormant tubers with 10μM 2,4-D for cytological observations during the transition from dormancy to activation, was performed on discs (9 mm. diam., 1 mm. high) as previously reported (25). The samples were collected after 0; 3; 4; 5; 18; 20; 21; 24; 27 and 30 h. 0.2% colchicine was added for 1 - 3h in the liquid medium at the 18th and 21st h.

The cultures for anatomical observations, were done on agarized medium (6) ± 0,1mM or 10μM spermine or spermidine or 0,35 μM IAA. The explants were collected after 2; 2,5; 3; 4; 4,5; 5; 7; 10; 15; 17; 25; 27; 32; 34; 47 and 60 days of growth at 24°C in light-dark cycles (12:12 h, 5500 ergs/cm^2/sec.).

The best fixative was found to be picroformol of Bouin. The explants were embedded, sectioned and stained according to several staining procedures; the best was safranin and fast-green.

RESULTS AND DISCUSSION

Cytological Aspect of the Explants during the Transition from Dormancy to Activation (first Cell Cycle: 0-30h).

During the G_1 and S phases, after a period of 3h, when no morphological modifications were observed, nuclei gradually become larger, were no longer adherent to the walls and contained less numerous (6→1), more stainable, larger (about 500% after 18-20h), and sometimes vacuolated nucleoli (Fig.1, 2, 3).

The enlargement of the nucleoli coincides with the synthesis of rRNA observed at various hours of activation (2; 3; 10; 15; 18h) whereas it is not directly correlated to the levels of total RNA, as discussed by Williams and Jordan (28). Total RNA in fact, simultaneously undergoes a synthesis but also a marked decrease probably due to a degradation of stored RNA (3; 4; 25; 26).

Nucleoli extrusion was occasionally observed (Fig.4) as reported in other critical phases (22; 23), when ribosomal gene amplification may also occur (2). Therefore the previously observed thymidine incorporation during G_1 phase (27) might probably be localized at the nucleolar level.

Around the 20th h, some nuclei are suspended in the center of cell (Fig. 5), cell division begins and continues until about the 30th. There is an increasing number of cells containing two nuclei separated by a thin wall that seems to extend from one lateral wall of the cell to the other. Nuclei are less densely stained and nucleoli more numerous and smaller (Fig. 6 and 9).

The nucleolar volume can be correlated,in agreement with the above mentioned reports (13; 14), with the amount of polyamines (25): both increase during G_1 and S phases, both decrease after cell division.

In the undivided cells nuclei sometimes appear less densely stained; the lose their compactness and have poorly defined margins (Fig. 7). They are thus similar to the chromocentric nuclei during the Z phase, interpreted as a very early prophase (16). Sometimes persistent nucleoli were observed, but nor discrete condensed chromosomes per se, mitotic figures, mitotic spindles, nor phragmoplasts and this even in the colchicine treated explants (about 2.000.000 nuclei were checked). As a consequence of the stress of explantation, the cells might react by undergoing non-mitotic divisions as also observed by Martin (cited by Gautheret (12)) in Helianthus tuberosus. Recently apparent amitotic nuclear divisions were reported by Miller (18). During the first cell cycle of the same type of parenchyma cells however, true mitoses have been previously observed (19; 24; 25; 29).

Nuti-Ronchi et al. (22) showed the coexistence of mitotic and amitotic divisions in Nicotiana explants, where nuclear fragmentation occurred simultaneausly with nucleolar extrusion (23); afterwards cells multiplied normally.

FIG. 1.,2.,3., Nuclei with decreasing number of nucleoli during the progression of G_1 phase. Nucleolar vacuols are visible in Fig. 3.

FIG. 4., Nucleolar extrusion at 18 h of culture.

FIG. 5., Nuclei at 18-20 h.

FIG. 6., Binucleate cells around 20 h.

FIG. 7., Poorly stained nucleus with undefined margins in an undivided cell at 20 h.

FIG. 8., Outermost part of the explant treated with 0.1 mM spermine after 48 h.

FIG. 9., Magnified detail of the same explant.

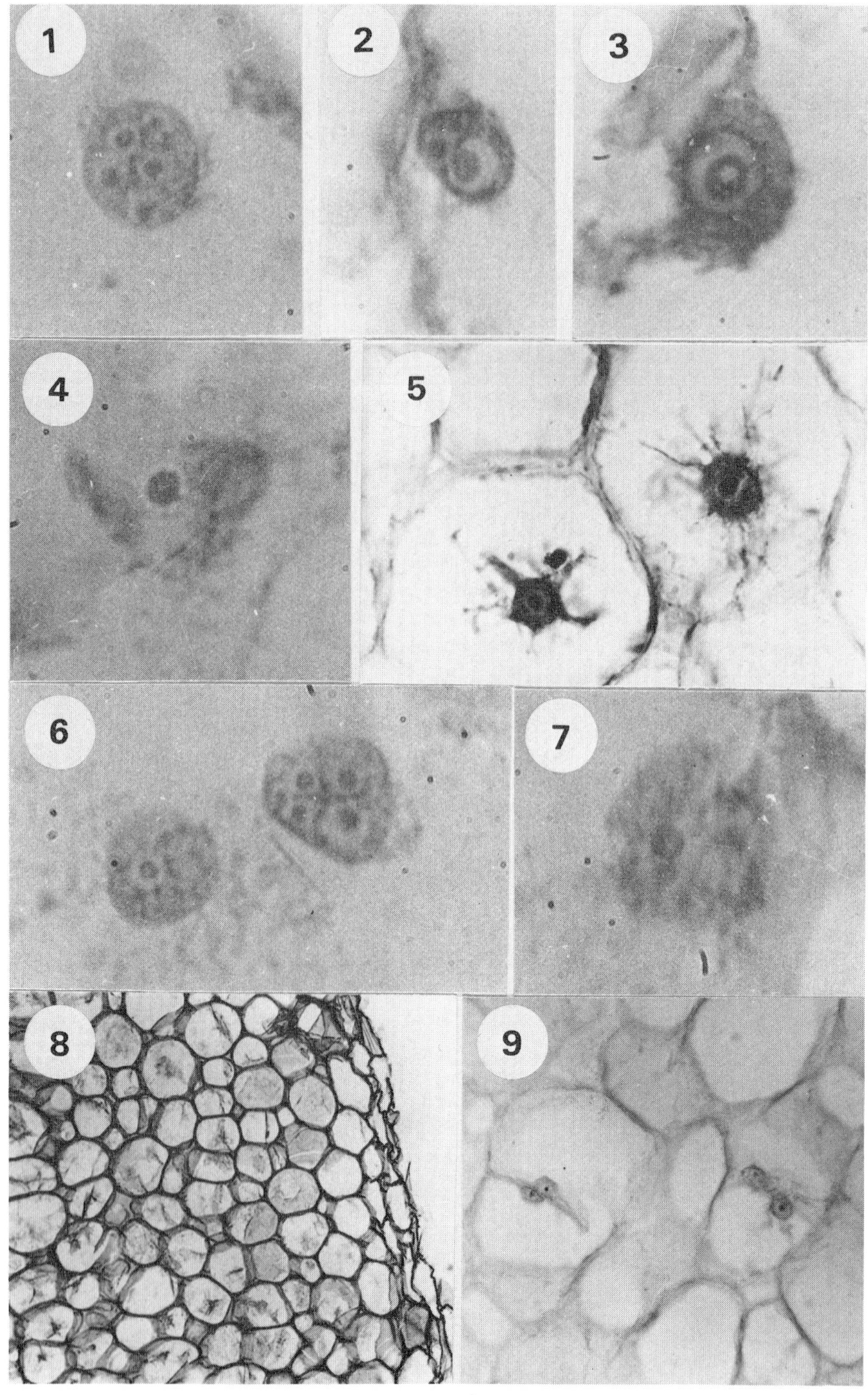
1
2
3
4
5
6
7
8
9

Anatomical Aspects of the Explants in long Term Cultures.

Both in 0,35 μM IAA- or 0,1mM polyamine-treated explants, a ring of dividing cells become evident 4-5 layers of cells beneath surface of the explant after two days of culture (Fig. 8 and 9). Mitoses become detectable, but very rarely. After 5 days the external cells suberized.

In the following 5 days the same cells repeteadly formed septa thus becoming smaller and smaller (Fig. 10 and 11). The multiplication rate was higher in the explants treated with IAA than in the polyamine-treated ones. The thickening of irregular reticulate tracheids was seen after 4 days in IAA-treated explants and after 7 days in 0,1mM polyamine-treated explants; in the same tissue but under different growth conditions, the first tracheids differentiated in 3 days (9; 24).

A ring of collateral vascular bundles differentiated around the 10th day; the were larger in IAA-treated, (Fig. 13) smaller in 0,1mM polyamines (Fig. 12) and undetectable in 0,01mM polyamine-treated tissue. Controls in basal medium showed only rare divisions.

Spiral tracheids, occasionally pre-existent among the parenchyma cells, and widely scattered secretory ducts, formed centers of induction for an early and rapid differentiation of new

---

FIG. 10., A parenchyma cell having divided twice after 2 days of treatment with 0.35 μM IAA.
FIG. 11., A parenchyma cell with numerous septa after 5 days of treatment with 0.1 mM spermidine.
FIG. 12., Differentiation of the first vascular bundles and suberization of the outermost cells of an explant after 10 days of treatment with 0.1 mM spermidine. Note the irregular shape and direction of the tracheids.
FIG. 13., A collateral vascular bundle of an IAA-treated explant after 10 days. (1) tracheids, (2) sieve tubes, (3) secretory ducts.
FIG. 14. and 15., Differentiation of new reticulate tracheids (2) induced by:
- a preexisting spiral tracheid (1) in explants treated for 7 days with 0.35 μM IAA (Fig. 14)
- a secretory duct (1) in 0.1 mM spermine-treated explants after 25 days (Fig. 15).

FIG. 16. and 18., Differentiation induced after 47 days in explants treated with:
- 0.1 mM spermidine (Fig. 16)
- 0.35 μM IAA (Fig. 18).

FIG. 17., Clusters of vascular elements in the pith of explants treated with 0.1 mM spermine for 25 days.

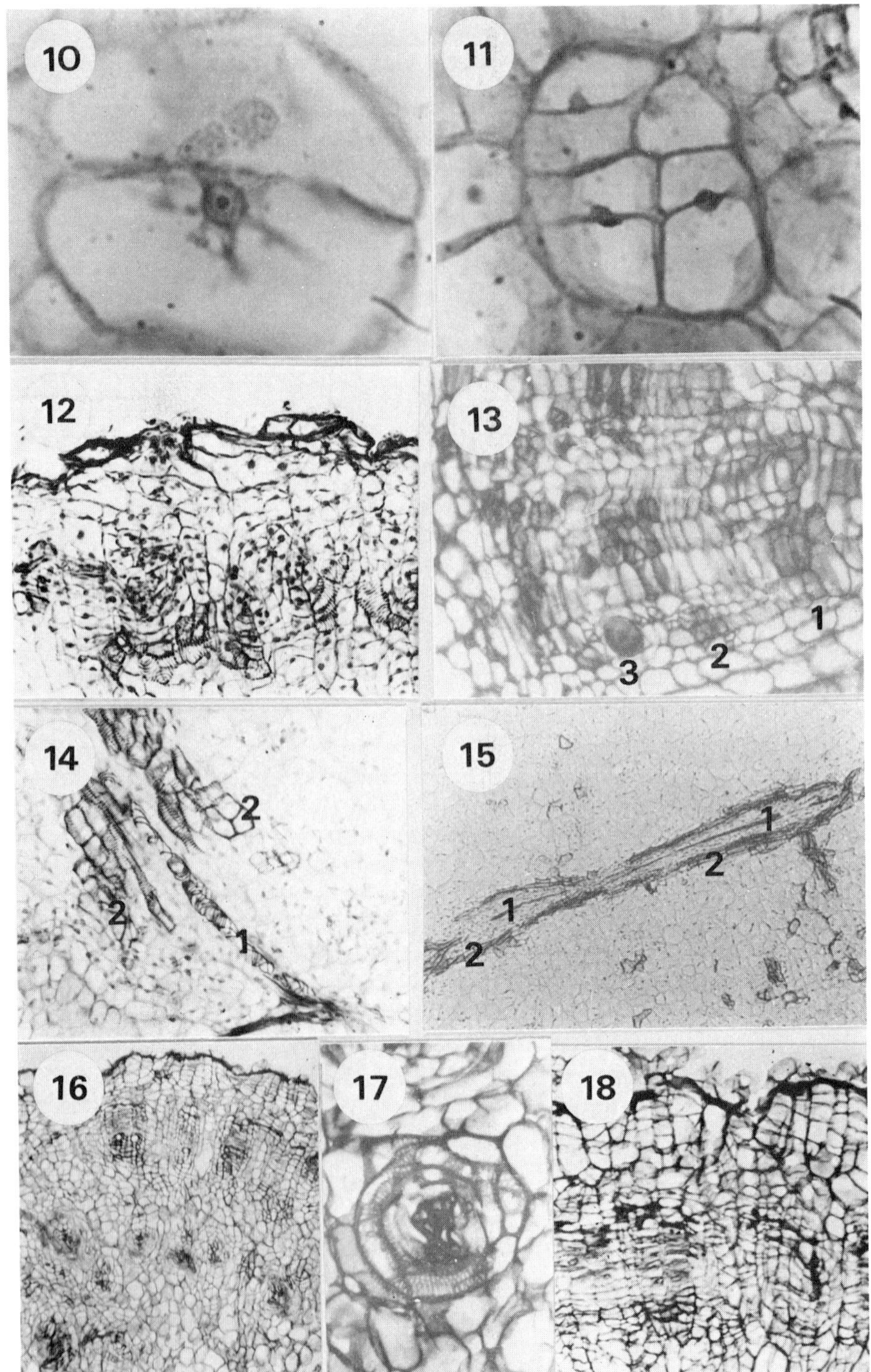
10
11
12
13
1
2
3
14
2
2
1
15
1
2
1
2
16
17
18

xylem elements independently from the type of hormonal treatment (Fig. 14 and 15). Similar induction centers were seen by Gautheret (12).

Around the 15-17th days, explants turned greenish, and a phellogen was formed.

In the pith of the explants, clusters of vascular elements, arranged in a perixylematic or perifloematic manner, differentiated and frequently surronded neoformed secretory ducts (Fig. 15 and 17).

Rarely a rhizogenetic process was observed beginning on the 25th day at the basal part of the explants treated with IAA or polyamines and, contrary to the observations of the group of Gautheret, independently from the concentration (10; 11).

Maximum differentiation and enlargement of the explants was reached on the 27-30th day and remained approximately constant till the 60th day. Along the explant a marked gradient of differentiation was detectable.

0,1mM spermine and spermidine exerted the same effect and were more effective at that concentration than at 10µM concentration, but less active than 0,35µM IAA which induced anatomical structures very similar to those observed by Gautheret (11) (Fig. 16 and 18).

CONCLUSION

Many of the morphogenetic effects induced by polyamines (both spermine and spermidine) such as the first cell divisions, which are apparently amitotic, the arrangement of the cambial zones, the differentiation of the vascular bundles together with the previously observed physiological effects, are very similar to those induced by other plant hormones, supporting the hypothesis that polyamines may also be considered growth substances or that they have the same effectors, or, finally, that they induce the synthesis of the other hormones.

Acknowledgements

This work was supported by funds for scientific research from Ministero della Pubblica Istruzione. One of us (D.S.F.) acknowledges the accurate work of Mrs. Adriana Schiavina, done during the period of her thesis.

REFERENCES

1. Altman, A., and Bachrach, U.(1981):In: Advances in Polyamine Research, vol. 3, edited by C.M. Caldarera, V. Zappia and U. Bachrach, pp. 365-375. Raven Press, New York.
2. Avanzi, S., Cionini P.G., and D'Amato, F. (1970): Caryologia, 23: 605-638.

3. Bagni, N., Donini, A., and Serafini-Fracassini, D. (1972): Physiol. Plant., 27: 370-375.
4. Bagni, N., Corsini, E., and Serafini-Fracassini, D. (1971): Physiol. Plant., 24: 112-117.
5. Bagni, N., Serafini-Fracassini, D., and Torrigiani, P. (1981): In: Advances in Polyamine Research, vol. 3, edited by C.M. Caldarera, V. Zappia, and U. Bachrach, pp. 377-388. Raven Press, New York.
6. Bertossi, F., Bagni, N., Moruzzi, G., and Caldarera C.M. (1965): Experientia, 21: 80-81.
7. Brachet, J., Mamont, P., Boloukère, M., Baltus, E., and Hanocq-Quertier, J. (1978): C.R. Acad. Sc., Paris, Série D, 287: 1282-1292.
8. Costa, D., and Bagni, N. (1981): Acta Hort.,Symposium on Growth Regulators in Fruit Production, 120: 239.
9. D'Alessandro, G. (1973): Plant Cell Physiol., 14: 1167-1176
10. Garcia-Rodriguez, M.-J. (1976): C.R. Acad. Sc., Paris, Série D, 283: 765-768.
11. Gautheret, R.-J. (1953): Rev. Gén. Bot., 60: 129-173.
12. Gautheret, R.-J. editor (1959): La culture des Tissues Végétaux, Masson et Cie, Paris
13. Gfeller, E., Stern, D.N., Russell; D.H., Levy, C.C., and Taylor, R.L. (1972): Z. Zellforsch., 129: 447-454.
14. Heby, O., and Emanuelsson, H. (1981): Med. Biol., 59: 417-422.
15. Heimer, Y. M., Mizrahi, Y., and Bachrach, U. (1979) FEBS Letters, 104: 146-148.
16. Heitz, E. (1929): Ber. dtsch. bot. Ges., 47: 274-341.
17. Kaur-Sawhney, R., Shih, L.-M., and Galston, A.W. (1982) Plant Physiol. 69: 411-415.
18. Miller, R.H. (1980): Ann Bot., 46: 567-575.
19. Mitchell, J.P. (1967): Ann. Bot., 31: 427-435.
20. Montague, M.J., Armstrong, T.A., and Jaworski, E.G. (1979): Plant Physiol., 63: 341-345.
21. Montague, M.J., Koppenbrink, J.W., and Jaworski, E.G. (1978): Plant Physiol., 62: 430-433.
22. Nuti-Ronchi, V., Bennici, A., and Martini, G. (1973): Cell Diff., 2: 77-85.
23. Nuti-Ronchi, V., and Martini, G. (1977): Protoplasma, 91: 409-415.
24. Phillips, R., and Dodds, J.H. (1977): Planta, 135: 207-212.
25. Serafini-Fracassini, D., Bagni, N., Cionini, P.G., and Bennici, A. (1980), Planta, 148: 332-337.
26. Serafini-Fracassini, D. and Filiti, N. (1982): C.R. Acad. Sc., Paris, Série D (in press).
27. Torrigiani, P., and Serafini-Fracassini, D. (1980): Z. Pflanzenphysiol., 97: 353-359.
28. Williams, L.M., and Jordan, E.G. (1980): J. exp. Bot., 31: 1613-1619.
29. Yeoman, M.M., and Evans, P.K. (1967): Ann. Bot., 31:323-332.

Advances in Polyamine Research, Vol. 4, edited by U. Bachrach, A. Kaye, and R. Chayen. Raven Press, New York © 1983.

Macromolecular Effectors of Ornithine Decarboxylase Activity in Germinating Barley Seeds

Dimitrios A. Kyriakidis

Laboratory of Biochemistry, Aristotelian University of Thessaloniki, School of Science, Thessaloniki, Greece

The synthesis of polyamines in plants has been studied by many investigators (3,20,23,24). Putrescine is formed either indirectly through L-arginine by L-arginine decarboxylase (ED 4.11.19, ADC) via agmatine, or directly from L-ornithine by L-ornithine decarboxylase (EC 4.11.17, ODC) (1,9,18,22,25).

The activity of ornithine decarboxylase can be inhibited by a specific protein inhibitor, named antizyme of ODC. The term antizyme was suggested for this type of non-competitive inhibitor, whose synthesis is induced by the product of the reaction it inhibits (10). Antizyme has been found in normal rat liver (11), in different cell lines (10,12) as well as in *Esherichia coli* (14). In addition to the antizyme, an activator of ODC has also been extracted from *Esherichia coli* (14). The complexity of the interaction of the ODC activator with ODC and the antizyme has been recently reported (5-7).

This paper describes the effect of polyamines on ODC activity during the germination of barley seeds. In addition it provides evidence for the first time, for the existence of a macromolecular inhibitor and a macromolecular activator of ODC.

EXPERIMENTAL

Sterilization of barley seeds: All solutions used were sterilized by autoclaving. Seeds of *Hordeum vulgare* var Beca were washed with water in a Buchner funnel and sterilized with 0.1% $HgCl_2$ for three minutes (19). The seeds were washed with sterile distilled water followed by 3 mM

EDTA, 0.1 M NaCl and finally with a lot of sterile water.
Germination of barley seeds: Seeds were germinated in 9 cm Petri dishes in the dark at 26°C, on a filter paper which was kept moist by underlying cotton soaked in distilled water. The barley seedlings were collected at various times following germinations and homogenized either with glass beads or in a Waring Blender with four volumes of 50 mM Tris:HCl pH 8.3, 0.3 mM EDTA, 50 μM pyridoxal phosphate and 5 mM dithiothreitol (buffer A).
ODC reaction: Enzyme reactions were performed in 50 μl volume as previously described (12). The optimum conditions for assaying ODC activity from germinating barley seeds extracts were found to be that of buffer A. The stoichiometry of the reaction of ODC was examined by measuring the formation of putrescine and CO_2. Uniformly labeled L-ornithine was used and the radioactive putrescine was measured as described by Heimer et al. (9) on thin layer chromatography plate. One unit of ODC activity was defined as 1 nmol of $^{14}CO_2$ released per hour. The assay of antizyme or the activator was performed as previously described (12,14), using partially purified cytosolic ODC (460 fold).
Protein determination: Protein was determined by the method of Lowry et al. (15) as modified by Ross and Scharz using bovine serum albumin as a standard (21).

RESULTS

Stimulation of ornithine decarboxylase activity by gibberellic acid.

The basal ODC activity increases with time during the germination of barley seeds. The presence of 10^{-5} M gibberellic acid in the medium raises the basal ODC activity by more than four fold (figure 1). The stimulation achieved plateau values at 100 hr and remained at this level for another 50 hr and then slowly declined.

Partial purification of ornithne decarboxylase.

Ornithine decarboxylase was partially purified from seeds which hab been germinated 120 hr in the presence of 10^{-5} M gibberellic acid. Germinated seeds (220 gr) were homogenized in 400 ml of buffer A in a Waring Blender, filtered through a cheese cloth and centrifuged at 30.000xg for 10 min. The supernatant was applied on a DEAE Sephadex A25 column (50 cmx1.6 cm) equilibrated before in buffer A. The column was eluted stepwise

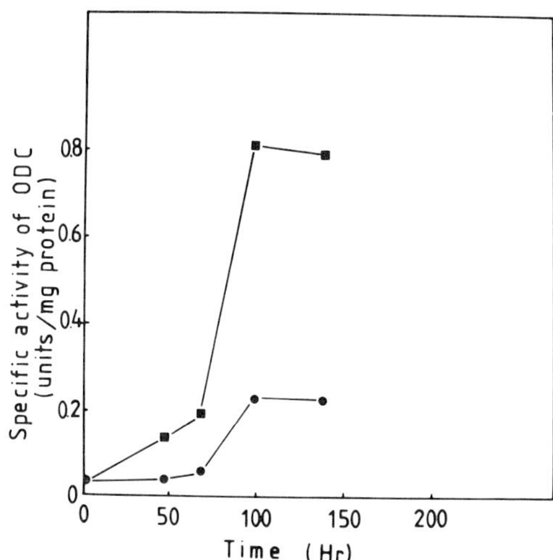

Figure 1. Effect of gibberellic acid on the stimulation of ODC activity during the germination of barley seeds. ●——●, control ; ■——■, 10^{-5} M gibberellic acid.

with 100 ml of buffer A and later with the same volume of 0.1 M NaCl or 0.2 M NaCl in buffer A. The active fractions of ODC eluted with 0.2 M NaCl were pooled, concentrated by ultrafiltration (Amicon PM10 filter) and applied on a CM-cellulose column which was equilibrated

TABLE I

Purification of barley seed ornithine decarboxylase

Step	Total protein (mg)	Total Units	Specific activity (units/mg protein)	Relative purification	Yield
30.000 x g supernatant	2.400	2.040	0.85	1	100
DEAE-Sephadex A_{25}	110	1.980	18.0	21	97.0
CM-Cellulose-1,8-diaminooctane	9	1.450	161.1	189	71.0
Sephadex G-200	0.3	1.242	4140	4870	60.9

with 1,8-diaminooctane. ODC activity was eluted with 0.2 M NaCl and the active fractions were concentrated by ultrafiltration. The purification at that stage was 190 fold with 70% recovery (table I). Furthermore ODC was purified by a Sephadex G200 column which was equilibrated with buffer A. The final purification of ODC was 4870 fold and it was really satisfactory for testing antizyme or activator activity.

Induction of antizyme during the germination of barley seeds.

When both polyamines putrescine and spermidine were added during germination of barley seeds at a concentration of 10^{-3} M each, the activity of ODC was remarkably inhibited (figure 2).

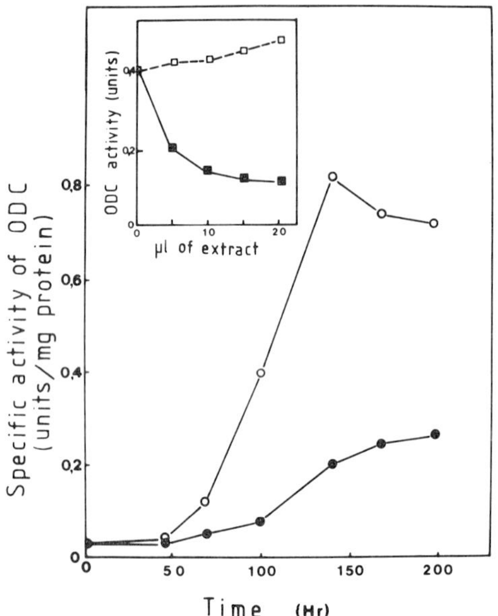

Figure 2. Effect of putrescine and spermidine on ODC activity during the germination of barley seeds. o——o, control; ●——●, 10^{-3} M putrescine and 10^{-3} M spermidine. Inset: Mixing experiment of the partially purified ODC (0.2 units) with inreasing amounts of dialysed extracts of germinated seeds for 72 hr in the presence of 10^{-3} M putrescine and 10^{-3} M spermidine (■——■). □----□ theoretical expected values.

In the control culture ODC activity achieved plateau at 120 hr and remained at this level for another

50 hr and then slowly declined. In the culture which were grown in the presence of polyamines, ODC activity was stayed at zero level for the first 100 hr and then gradually was increased. Mixing experiments of partially purified ODC (4870 fold) with the dialyzed extracts from the germinated for 72 hr barley seeds in the presence of polyamines, showed the presence of the antizyme in the latent extracts (inset of figure 2).

Partial purification of antizyme from germinated barley seeds.

Germinated for 72 hr barley seeds in the presence of 10^{-3} M putrescine and 10^{-3} M spermidine (200 gr), were homogenized in a Waring Blender in 400 ml of 50 mM Tris·HCl pH 8.2 and 0.3 mM EDTA (buffer B). The homogenate was centrifuged at 10.000xg for 10 min. In the supernatant solid ammonium sulfate was added until 40% saturation and the pellet after centrifugation at 10.000 xg for 10 min was dissolved in 50 ml of buffer B, dialyzed overnight against buffer B and applied on a DEAE-

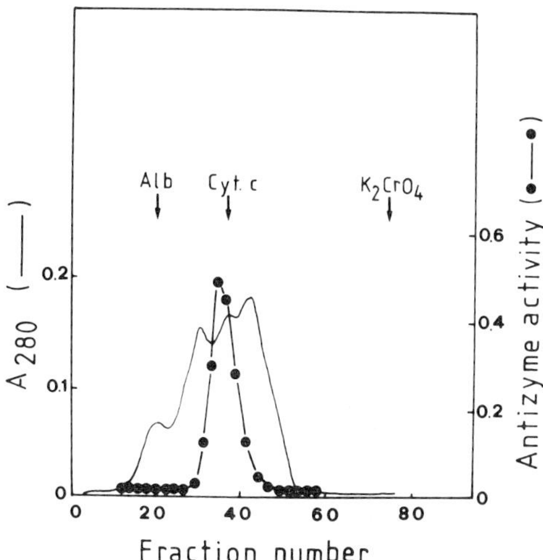

Figure 3. Chromatography of antizyme on a Sephadex G75 column. Antizyme purified by DEAE-Sephadex A25 column was applied on a Sephadex G75 column (70 cm x 1,6 cm) equilibrated with buffer A. To each assay 10 µl of partially purified ODC (0.4 unit) was added and 25 µl of each fraction. Bovine serum albumin (MW 67.000), cytochrome C (MW 12.300) and K_2CrO_4 were used as standards for calibration of the G75 column.

Sephadex A25 column, equilibrated with buffer B. The inhibitory fractions which eluted from the column with 0.1 M NaCl were concentrated by ultrafiltration (Amicon UM2 filter) and the concentrated fraction was applied on a G25 column equilibrated with buffer B (figure 3). The molecular weight of antizyme as analyzed by this Sephadex G75 column, revealed a componed of 16.000 ± 2.000 daltons. The final purification of antizyme after this step was about 3200 fold (table II).

TABLE II

Purification of barley seed antizyme

Step	Total protein (mg)	Total Units	Specific activity (units/mg protein)	Relative purification	Yield
30.000 x g supernatant	2204	708	0.3	1	100
0-40% Ammonium sulfate	205	1.720	1.9	5.9	242.9
DEAE-Sephadex A25	12	1.540	128.3	401.0	217.5
Sephadex G_{75}	1.4	1.440	1028.5	3214.3	203.3

Extraction of macromolecular activator of ODC from germinated barley seeds.

Barley seeds germinated for 94 hr in the presence of 10^{-3}M putrescine and 10^{-3} M spermidine (60 hr) were homogenized in a Waring Blender in 150 ml of buffer B. The homogenate was centrifuged at 10.000xg for 10 min. Concentrated HCl was added to the supernatant until final drop of pH at 1.8, followed by addition of concentrated $HClO_4$ to final concentration of 0.2 M. The mixture was stirred at 0°C for 15 min and the 10.000xg pellet suspended in 10 ml of buffer B. After stirring at 0°C for 2 hr the 10.000xg supernatant was applied on a Sephadex G75 (76 cm x 1.6 cm) and then on a Bio-gel P6 (90 cm x 1.6 cm) column (figure 4). After this step of purification (120 fold) the activation of ODC activity was 400-500% over the control value. The apparent molecular weight of this activator is 6.000 ± 1.000 daltons.

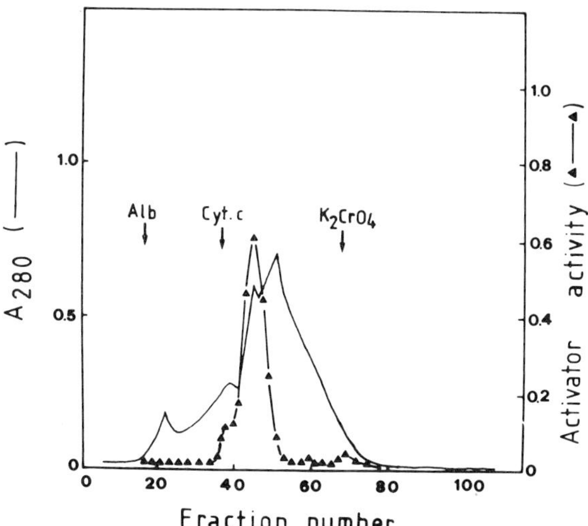

Figure 4. Chromatography of activator of ODC on a Bio-gel P6 column. Purified activator by Sephadex G75 column was applied on a Bio-gel P6 (90 cm x 1.6 cm) column. Assay of activator and calibration conditions of the column were the same as described at the legend of figure 3.

In vitro interaction of ODC with antizyme and activator from germinated barley seeds.

Figure 5a shows the change of ODC activity when ODC is titrated with increasing amounts of the activator. A continuous activation of ODC occurs up to 7.5 µg. By adding increasing amounts of antizyme to such activated ODC, a slight inhibition is obtained, which inhibition accounts only for the 10% from that expected.

The results of figure 5b show the reverse experiment. When ODC is titrated with increasing amounts of antizyme, decrease in ODC activity is obtained. However, addition of activator to this already inhibited ODC, increases the activity of ODC as if it has not been inhibited by antizyme.

DISCUSSION

Evidences have been accumulated suggesting that regulation of ODC activity occurs mainly at the post-transcriptional level. E.S. Canellakis and his group suggested that ODC activity is regulated by the antizyme and activator of ODC (5-7, 14). Other investigators have also reported the presence of ODC-inhibitor in different cell types (4,8,13,16). Interesting enough are the reported similarities between antizyme and a polyamine dependent protein kinase in the *Physarum polycephalum*. This kinase catalyses phosphorylation of a unique acidic nucleolar protein which prooved to be ODC (2).

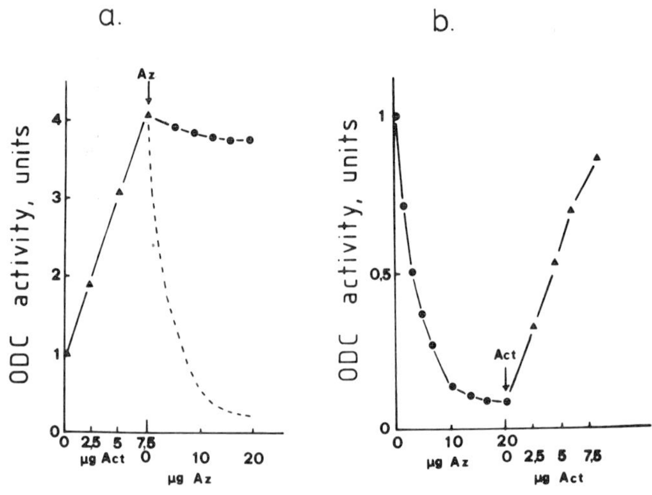

Figure 5. Effect of antizyme and activator on ODC activity.
a) ODC (1.0 unit) was titrated with increasing amounts of activator (▲—▲) and then titrated with increasing amounts of antizyme (●—●). --- Theoretical expected activity.
b) ODC (1.0 unit) was titrated with increasing amounts of antizyme (●—●) and then titrated with increasing amounts of activator (▲—▲).

Another post-transcriptional control mechanism of ODC is that provided by Mitchell and his colleagues (17), suggesting the alteration of ODC in two forms with different catalytic activities which depends on the concentration of pyridoxal phosphate.

The present study establishes the existence of positive and negative macromolecular effectors of ODC in another biological system. Data on the purification of cytosolic ODC (4870 fold), antizyme (3200 fold) and activator (120 fold) are presented as well. Further purification of these three molecules is still in progress and more detailed information will be published shortly. The interaction between these three molecules is also studied in this paper. The results show that this interaction is more complex and varies with the order of each addition. Addition of activator to ODC or to ODC-antizyme complex, activates ODC activity in an equivalent fashion, while addition of antizyme to already activated ODC in the reverse experiment does not counteracts the effect of the added antizyme. The limitation of assayable antizyme activity can be explained by this experiment, since the action of antizyme depends on the fact whether the enzyme is free or the activator is bound to it.

REFERENCES

1. Altman, A. and Bachrach, U. (1981) Advances in Polyamine Research, ed. by C.M. Caldarera et al. Vol. 3, pp. 365-375, Raven Press, New York
2. Atmar, V.J. and Kuehn G.D. (1981) Proc. Natl. Acad. Sci. USA 78, 5518-5522
3. Bagni, N. (1970) New Phytol. 69, 159-164
4. Branca, A.A. and Herbst E.J. (1980) Biochem J. 186, 925-931
5. Canellakis, E.S., Heller, J.S. and Kyriakidis, D.A. (1981) Advances in Polyamine Research, ed. by C.M. Caldarea et al. Vol. 3, pp. 1-13, Raven Press, New York
6. Canellakis, E.S., Heller, J.S., Kyriakidis, D.A. and Viceps-Madore D. (1981) Polyamines in Biology and Medicine, ed. by D.R. Morris and L.S. Marton pp 83-107, Marcel Dekker, New York
7. Canellakis, E.S., Kyriakidis, D.A., Heller, J.S. and Pawlak, J.W. (1982) Medical Biology, In press
8. Friedman, Y., Park, S., Levasseur, S. and Burke, G. (1977) Biochim. Biophys. Acta 500, 291-303
9. Heimer, Y.M., Mizrahi, Y. and Bachrach, U. (1979) FEBS Lett., 104, 146-148
10. Heller., J.S., Fong, W.F. and Canellakis, E.S. (1976) Proc. Natl. Acad. Sci., USA 73, 1858-1862
11. Heller, J.S., Kyriakidis, D.A., Fong, W.F. and Canellakis, E.S. (1977) Eur. J. Biochem, 81, 545-550
12. Heller, J.S., Chen, K.Y., Kyriakidis, D.A., Fong, W. F. and Canellakis, E.S. (1978) J. Cell Physiol., 96, 225-234
13. Kallio, A., Löfman, M., Pösö, H. and Jänne, J. (1979) Biochem. J. 177, 63-69
14. Kyriakidis, D.A., Heller J.S. and Canellakis, E.S. (1978) Proc. Natl. Acad. Sci. USA, 75, 4699-4703

15. Lowry, O.H., Rosebrough, N.J., Farr, A.L. and Ran dall, R.G. (1951) J. Biol. Chem., 193, 265-275
16. McCann, P.P., Tardif, C. and Mamont, P.S. (1977) Biochem Biophys Res. Comm. 75, 948-945
17. Mitchell, J.L.A., Augustine, T.A. and Wilson, J.M. (1981) Biochim. Biophys. Acta. 657, 257-267
18. Montague, M.J., Koppenbring, J.W. and Jaworski, E. G. (1978) Plant Cell Physiol., 62, 430-433
19. Papageorgakopoulou, N. and Georgatsos, J.G. (1977) Int. J. Biochem 8,611-618
20. Ramakrishna, S. and Adiga, P.R. (1975) Phytochemistry 14, 63-68
21. Ross, E. and Schatz, G. (1973) Anal. Biochem., 54, 304-306
22. Smith, T.A. (1975) Phytochemistry 14, 865-890
23. Smith, T.A. and Wilshire, G. (1975) Phytochemistry 2341-2346
24. Smith, T.A. and Best, G.R. (1977) Phytochemistry 16, 841-846
25. Yoshida, D. (1969) Plant Cell Physiol., 10, 393-397

Advances in Polyamine Research, Vol. 4, edited by U. Bachrach, A. Kaye, and R. Chayen. Raven Press, New York © 1983.

Applied Polyamines Inhibit Macromolecular Synthesis in Plant Tissue

Akiva Apelbaum and Isaac Icekson

Division of Fruit and Vegetable Storage, ARO, The Volcani Center, Bet Dagan 50250, Israel

Exogenously applied polyamines and some of their precursors have been shown to stabilize protoplasts against lysis, inhibit the dark-induced rise in RNase and protease activity, and reduce the rate of chlorophyll loss in leaves and protoplasts (2,3,8,9, 10). Supportive evidence for the protective role of polyamines at the membrane level was presented by Popovic et al. (14), who showed that spermine prevented the loss of chlorophyll from thylakoid membranes during dark-induced senescence of detached leaves. Altman and Bachrach (2) showed that applied polyamines also inhibited wound-induced RNase activity in potato tubers. All these effects provoked by polyamines oppose those produced by ethylene, a hormone known to be associated with plant senescence. Furthermore, Apelbaum et al. (4) reported that applied polyamines at concentrations similar to those that retard senescence (2,10), inhibited ethylene production in several plant tissues. The pathway for ethylene biosynthesis has recently (1) been described in more detail: Methionine → SAM → ACC → C_2H_4.

Apelbaum et al. (5) and Yu et al. (16) showed that the ACC level in the tissue is the rate-limiting factor in the pathway of ethylene production in mung bean stems and apple fruit. Apelbaum et al. (5,6) also showed that the integrity of membrane structure is required for the continuous conversion of ACC to ethylene. Thus it appears that the two crucial reactions in this pathway were the formation of ACC from SAM and the degradation of ACC to ethylene. The present work was undertaken in an attempt to understand the nature of the inhibition of ethylene biosynthesis by polyamines and to determine at which step in ethylene biosynthesis it occurs.

MATERIALS AND METHODS

Plant material.

Apple fruits (*Malus domestica* Borkh., cv. Golden Delicious) were harvested at the preclimacteric stage and stored at 0^oC until used. Fifteen to 18 hours before the beginning of the experiment apples were transferred to 22°C. The fruit was enclosed in glass jars for 1 h and ethylene production by the whole fruit was determined as described below. Discs, 9 mm in diameter and 2 mm thick, were cut and incubated for various periods of time at 25°C.

Incubation conditions.

Six discs (approximately 1.7 g) were incubated in a 25-ml Erlenmeyer flask containing 2 ml of 0.6 M sorbitol in 10 mM Na-phosphate buffer, pH 7.0, with or without the indicated concentration of the diamines or polyamines. Chloramphenicol was added to give a final concentration of 10 μg per ml. L-[^{14}C]-leucine (0.5 μCi/ml), [^{14}C]-uridine (0.5 μCi/ml) or 5[^{3}H]-uridine (4 μCi/ml) was added 2 h before the end of the incubation.

Ethylene determination.

Flasks were flushed with air before being closed with a rubber cap. A 1-ml sample of the gas contained in the flasks was analyzed by gas chromatography after a 2-h accumulation period.

Rate of macromolecular synthesis.

A 2-h pulse of radioactive precursor was used throughout, at the end of which the flasks were rapidly cooled to 0°C, and the radioactive incubation medium removed. The apple discs were washed three times with cold (0°C) incubation medium, transferred to a glass homogenizer, and homogenized at 0°C in 2.5 ml of distilled water. The homogenate was centrifuged for 5 min at 1500 x *g*, the pellet was discarded, and the supernatant was used for further analysis.

Aliquots (0.1 ml) of the supernatant were spotted on Whatman No. 3 paper strips, dried, and counted with Instagel II fluid in a liquid scintillation spectrometer to determine the total uptake of precursor. In parallel, aliquots, 0.5 ml, were brought to 10% v/v with trichloroacetic acid (TCA) and allowed to precipitate for 20 min at 2°C. The precipitates were collected on Whatman GF/C filters by vacuum filtration, washed twice with TCA 10%, twice with TCA 5%, and once with absolute ethanol. Filters were dried under an infrared lamp and counted with Instagel II liquid scintillation fluid.

ACC level determination.

ACC was determined in the same supernatant used for the incorporation studies by the chemical degradation method of Lizada and Yang (11). The efficiency of ACC conversion to ethylene was 70-80% under our conditions.

RESULTS AND DISCUSSION

Diamines and polyamines inhibited ethylene production and incorporation of radioactively labelled leucine or uridine into TCA-insoluble material in apple fruit discs, while having little or no effect on the ACC content of the tissue (Table I). The inhibitory effect was in general more pronounced with higher molecular weight amines.

The diamines had no effect on the ACC content of the tissue, while the polyamines spermidine and spermine slightly reduced the

TABLE I. Effect of diamines and polyamines on ethylene production ACC content, and macromolecular synthesis

Treatment	C_2H_4	ACC	Incorporation into TCA insoluble material	
			$[^{14}C]$Leucine	$5[^3H]$Uridine
	% C	% C	% C	% C
Control	100	100	100	100
1-3-diaminopropane	60	90	44	97
Putrescine	54	109	63	56
Cadaverine	56	-	34	72
Spermidine	28	84	19	45
Spermine	31	87	17	42

ACC level in the tissue (Table I). The inhibitory effect of spermine (Fig. 1) on ethylene production and similarly on labeled uridine incorporation progressed rapidly until 6 h and at much lower rate until 24 h. The effect of spermine on leucine incorporation was more pronounced than that on uridine incorporation, after 6 h of incubation; leucine incorporation was inhibited by 90% and uridine by 64%.

Fig. 1. Time course of inhibition of ethylene production and macromolecular synthesis by 5 mM spermine.

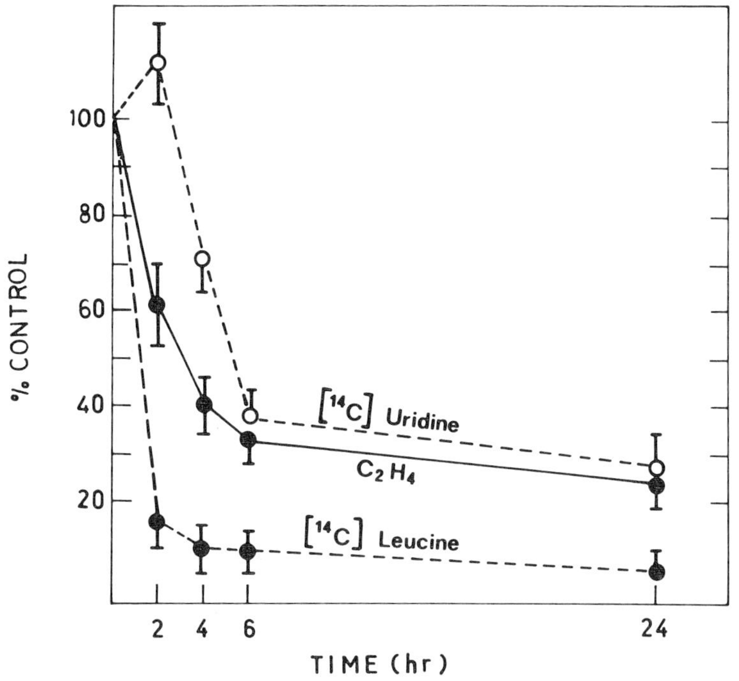

When apple discs were incubated with different concentrations of spermine (Fig. 2), ethylene production and leucine incorporation were progressively inhibited with the increase in concentrations of the polyamine. These inhibitory effects were observed within a relatively narrow range of concentrations. Spermine at 0.1 mM inhibited ethylene production by 14% and leucine incorporation by 55%, while lower concentrations of the polyamines were ineffective. An almost maximal effect was obtained with 1 mM spermine, where ethylene production was inhibited by 60% and leucine incorporation by 90%. In all cases, spermidine displayed effects similar to those of spermine (data not shown).

Fig. 2. Effects of different concentrations of spermine on C_2H_4 production and L-[^{14}C]leucine incorporation.

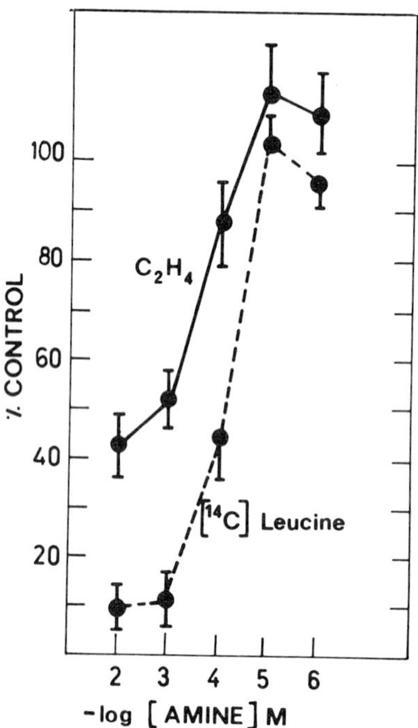

It has been well established that the biosynthesis of ethylene is inhibited by protein synthesis inhibitors (1). More recently, both the formation of ACC and the conversion of ACC to ethylene have been shown to require continuous de novo protein synthesis (16,17).

Our results show that polyamines inhibit macromolecular synthesis when applied exogenously at concentrations between 0.1 and 10 mM, and it is conceivable that this inhibition of macromolecule synthesis may be responsible, at least in part, for the observed inhibition of ethylene production reported by Apelbaum et al. (4) and Suttle (15). The conversion of ACC to ethylene has been shown to be inhibited rapidly by 10 mM spermidine or spermine in tobacco leaves (4) and in soybean hypocotyls (15). The fact that the ACC level was not observed to increase in apple discs, even though ethylene production was inhibited by 80%, would indicate that the polyamines must also inhibit the formation of ACC, probably by interference with the continuous de novo synthesis of ACC synthase which was shown to be required in order to maintain the level of ACC in the tissue (16).

The inhibition of macromolecular synthesis produced by polyamines, as reported here, might have, in addition to the effects on ethylene biosynthesis, effects of a more general nature. It has been shown that applied 0.1-10 mM polyamines retarded the dark-induced senescence phenomena, like the decrease in chlorophyll degradation, or the reduction in the increase of RNase (2). However, similar effects were reported for cycloheximide, a well-known protein synthesis inhibitor; it inhibited protease synthesis in senescing oat leaves (13) and the loss of chlorophyll from leaflets of Anacharis canadensis ("Elodea") in the dark (12). It is suggested therefore that the above reported effects produced by 0.1-10 mM polyamines are due to a pharmacological effect resulting from a lack of continuous de novo protein synthesis in the tissue.

REFERENCES

1. Adams, D.O., Yang, S.F. (1979): Proc. Natl Acad. Sci. USA 76:170-174.
2. Altman, A., Bachrach, U. (1981): In: Advances in Polyamine Research, edited by C.M. Caldarere et al., pp. 365-375. Raven Press, New York.
3. Altman, A., Kaur-Sawhney, R., and Galston, A.W. (1977): Plant Physiol. 60:570-574.
4. Apelbaum, A., Burgoon, A.C., Anderson, J.D., Lieberman, M., Ben-Arie, R. and Mattoo, A.K. (1981):Plant Physiol. 68:453-56.
5. Apelbaum, A., Burgoon, A.C., Anderson, J.D., Solomos, T., and Lieberman, M. (1981) Plant Physiol. 67:80-84.
6. Apelbaum, A., Wang, S.Y., Burgoon, A.C., Baker, J.E., and Lieberman, M. (1981):Plant Physiol. 67:74-79.
7. Cohen, A.S., Popovic, R.B., Zalik, S. (1979): Plant Physiol. 64:717-720.
8. Galston, A.W., Altman, A., Kaur-Sawhney, R. (1978): Plant Sci. Letters 11:69-79.
9. Kaur-Sawhney, R., Adams, W.R.Jr., Tsang, J., Galston, A.W. (1977):Plant Cell Physiol. 18:1309-1317.
10. Kaur-Sawhney, R., Galston, A.W. (1979): Plant Cell, Environment 2:189-196.
11. Lizada, M.C.C., Yang, S.F. (1979): Anal. Biochem. 100:140-145.
12. Makovetski, S., Goldschmidt, E.E. (1976): Plant Cell Physiol. 17:859-862.
13. Martin, C., Thimann, K.V. (1972): Plant Physiol. 49:64-71.
14. Popovic, R.B., Kyle, D.J., Cohen, A.S., Zalik, S. (1979):

Plant Physiol. 64:721-726.
15. Suttle, J.C. (1981): Phytochemistry 20:1477-1480.
16. Yu, Y.Y., Adams, D.O., Yang, S.F. (1979) Plant Physiol. 63: 589-590.
17. Yu, Y.Y., Yang, S.F. (1980): Plant Physiol. 66:281-285.

Advances in Polyamine Research, Vol. 4, edited by U. Bachrach, A. Kaye, and R. Chayen. Raven Press, New York © 1983.

Involvement of Polyamines in Cell Division of Various Plant Tissues

*E. Cohen, †Y. M. Heimer, §S. Malis-Arad, and *§Y. Mizrahi

**Department of Biology and §Applied Research Institute, Ben-Gurion University of the Negev, Beer-Sheva 84110, Israel; †Nuclear Research Center—Negev, Beer-Sheva 84190, Israel*

The hypothesis underlying this study is that polyamines play a role in the regulation of cell division in plants as they do in other biological systems. To support this hypothesis, we are studying plant systems in which we compare dividing and non-dividing tissues with respect to ODC and ADC activities, characteristics and levels, as well as the concentration of endogenous putrescine, spermidine and spermine.

ODC and ADC Activities

During tomato fruit development, there is a distinct period of intensive cell division, which occurs 10 days after anthesis, after which fruit growth is mainly due to cell expansion. During these first 10 days, ODC activity rises around 40-50 fold. The same results are obtained when fruit is induced to set and develop by auxin. ADC, although present in the fruits, does not show the same pattern as ODC.

Meristematic tissues (characterized by intensive cell division) are located at the tip of roots and shoot buds. In tomato and potato, ODC, and not ADC, was found to be correlated with cell division, i.e. maximal ODC activity occurred at the tip of roots and shoot buds, decreasing with the distance from the tip.

Tomato fruit growth could be inhibited by ODC-specific inhibitors when applied at the period of cell division only, putrescine counteracting this inhibition. ADC-specific inhibitor did not interfere with the normal fruit growth. (For details see Cohen *et al.* on ODC and ADC inhibitors in poster session.)

S-adenosyle methionine decarboxylase was also found in tomato ovaries.

Polyamine Concentration

The content of the three major polyamines, putrescine, spermidine, and spermine, was determined in the above-mentioned plant tissues.

In developing tomato fruits the concentration of the endogenous polyamines was found to be about three times higher during cell division than in the following period characterized by cell expansion. Among the three polyamines studied, only spermidine qualitatively correlated with ODC activity. Nevertheless, its quantity was only 2.5 as much as in non-dividing tissues as compared with the 40-50-fold difference in ODC activity.

In the tip of tomato shoot buds, the polyamine concentration was 50% to 100% higher than in the other parts of the organ, on fresh weight and protein basis, respectively. In tomato root tips, the polyamine concentration was much higher than in the rest of the organ.

Five-day-old tomato fruits pretreated with ODC and ADC inhibitors and/or with putrescine did not show any meaningful differences in polyamine content when compared with the nontreated control. (For details see Cohen *et al.* on ODC and ADC inhibitors in poster session.)

Labeled ornithine and arginine incorporated similarly into the three polyamines, putrescine, spermidine, and spermine. Spermidine had twice as much counts than putrescine and spermine. Of the total labeled substrate uptake, only about 10% were found in these polyamines, indicating that the label was incorporated into other metabolites.

CONCLUSIONS

1. Polyamines are essential for the regulation of cell division in plants as in other biological systems.

2. In the systems described above, ODC is the main enzyme catalyzing the biosynthesis of polyamines required for cell division.

3. Since ADC does not correlate with cell division, it is assumed that it is not involved in this process. Hence, compartmentalization should occur in the cells in order to distinguish between the two biochemical pathways leading to the formation of the same products.

4. The three common polyamines, putrescine, spermidine, and spermine, cannot account for all the metabolites which result from ODC activity. We should look for other polyamines and metabolites in order to get a better understanding concerning their role in the regulation of cell division in plants.

Advances in Polyamine Research, Vol. 4, edited by U. Bachrach, A. Kaye, and R. Chayen. Raven Press, New York © 1983.

Effect of ODC and ADC Inhibitors on Tomato Fruit Development

*E. Cohen, †S. Malis-Arad, §Y. M. Heimer, and *†Y. Mizrahi

**Department of Biology and †Applied Research Institute, Ben-Gurion University of the Negev, Beer-Sheva 84110, Israel; §Nuclear Research Center—Negev, Beer-Sheva 84190, Israel*

Following pollination of tomato fruit, there is a period of intensive cell division (10 days) after which fruit growth is a result of cell enlargement and not cell division. We have previously shown that during the 10 days of cell division, ODC activity increases 40-50-fold over that of the control, while ADC activity does not change.

In order to investigate the essentiality of ODC and its product - putrescine - for cell division in tomato fruits, we used specific inhibitors of ODC - α-DFMO and α-MO - and exogenous putrescine. α-DFMA, a specific inhibitor of ADC, was also tested in this system. The inhibitors were supplied to the ovary through the stem immediately, 24 h, and 10 days after pollination. Five days after the application of the chemicals, fruits were analyzed for fresh weight, cell number, and ODC and ADC activities.

The specific inhibitors of ODC caused a decrease in the abovementioned parameters when applied at the beginning of cell division (immediately and 24 h after pollination), while being uneffective after cell division ceased. The treatment with α-DFMO (an irreversible inhibitor of ODC) led to a decrease in fruit's fresh weight and cell number, to 31% and 36% of control fruits, respectively. ODC activity was reduced to 2-3% that of the control fruits on both a fresh weight and protein basis. The fresh weight of fruits treated with α-MO (a reversible inhibitor of ODC) was 55%, cell number 60%, and ODC activity 86% those of control fruits, respectively. The inhibition of both fresh weight and cell number was completely counteracted by putrescine given together with the inhibitors despite the fact that ODC activity was reduced to 13% and 40% by α-DFMO and α-MO, respectively. The ODC inhibitors had no effect on ADC activity.

Treatment with α-DFMA did not change fruit fresh weight and caused reduction in cell number per fruit of only 25% that of the control fruit. Under this treatment, the activity of ADC was reduced to 50% of the control, while ODC activity was not changed.

The level of the endogenous polyamines putrescine, spermidine, and spermine on a fresh weight or protein basis did not change following ODC or ADC inhibitors treatment.

CONCLUSIONS

1. In developing tomato ovaries an increased activity of ODC (and not ADC) is important for polyamine biosynthesis during the process of cell division.
2. Putrescine, which is the product of the ODC-catalyzed pathway, is essential for cell division.
3. The endogenous level of the polyamines putrescine, spermidine and spermine does not correlate with ODC activity. Hence, other metabolites are, apparently, involved in the regulation of cell division.

Advances in Polyamine Research, Vol. 4, edited by U. Bachrach, A. Kaye, and R. Chayen. Raven Press, New York © 1983.

Inhibition of Tomato Fruit Growth *In Vitro*: Possible Involvement of Polyamines

*D. C. Teitel, †Shoshana Malis-Arad, †E. Birnbaum, and *†Y. Mizrahi

**Department of Biology and †Applied Research Institute, Ben-Gurion University of the Negev, Beer-Sheva 84100, Israel*

Tomato fruits grown in *in vitro* culture were compared to *in vivo* fruits throughout their development.

In vitro fruits, transferred to culture two days after pollination, exhibited a ripening pattern similar in all parameters except taste to *in vivo* fruits at the termination of growth. However, no seeds were produced, and the maximal size of the fruits was greatly reduced. During 58 days of culture, *in vitro* fruits attained a final weight of 0.86 g, which represents a 74-fold increase over the initial weight but only 0.66% of the weight of equivalent fruits grown *in vivo*.

Cell division was also affected. The final number of cells in the *in vitro* fruits was 1/5 (4.6×10^6) that of *in vivo* fruits.

During *in vivo* fruit development, the DNA content per cell increased 20-fold after the cessation of cell division. The *in vitro* fruits displayed only a five-fold increase in DNA per cell throughout the entire period of their development.

The activity of ODC was greatly reduced in the *in vitro* fruits relatively to that of *in vivo* fruits five days after anthesis, i.e. at its highest level. This activity remained reduced throughout the entire development of the *in vitro* fruits.

In *in vitro* fruits, the levels (on a fresh weight basis) of the three polyamines putrescine, spermidine and spermine remained constant throughout fruit development, being the same as in *in vivo* fruits at 5 days after pollination. All three polyamines exhibited similar levels.

It is suggested that the inhibition of cell division and DNA replication in *in vitro* fruits is correlated with the reduced activity of ODC. The lack of correlation between cell division and the level of these polyamines may indicate that other metabolites are involved in the regulation of this process.

*This work was supported by a grant from the National Council for Research and Development and the Ministry of Energy and Infrastructure.

Advances in Polyamine Research, Vol. 4, edited by U. Bachrach, A. Kaye, and R. Chayen. Raven Press, New York © 1983.

The Involvement of Polyamines in the Cell Cycle of *Chlorella*

*E. Cohen, †Shoshana Malis-Arad, §Y. M. Heimer, and *†Y. Mizrahi

**Department of Biology and †Applied Research Institute, Ben-Gurion University of the Negev, Beer-Sheva 84110, Israel; §Nuclear Research Center-Negev, Beer-Sheva 84190, Israel*

The involvement of polyamines and their biosynthetic enzymes in various dividing plant systems has recently been shown. Their exact role in this process has not, however, been clarified yet.

The photosynthetic unicellular alga *Chlorella* can be readily synchronized by alternating dark-light cycles (DLD), thus providing an optimal model system of an eukaryotic plant cell in which the role of polyamines during the life cycle can be studied.

For achieving that purpose,we first characterized the enzymes catalyzing the first step of polyamine biosynthesis in *Chlorella* - ornithine decarboxylase (ODC) and arginine decarboxylase (ADC), and the optimal conditions for their activity. Some of the characteristics, which differ from those described in other plant systems, are presented in Table 1.

TABLE 1. Some characteristics of *Chlorella* ODC and ADC

Characteristic	ODC	ADC
K_m	$0.6x10^{-4}$	$1.0x10^{-4}$
Optimal pH	7.5	8.5
Buffer type and ionic strength	phosphate 0.025 M	Tris-HCl 0.25 M
Enzyme requirement		
DTT	+	+
pyridoxal phosphate	+	+
Inhibition by L-canaline	+	-
Heat inactivation		
45°C	82%	28%
55°C	100%	100%

In an asynchronous culture of *Chlorella vulgaris* grown under continuous illumination, active polyamine biosynthesis occurs during the logarithmic phase of growth. Polyamine biosynthesis catalyzed by ODC and by ADC increases during this period 6 and 2.5-fold, respectively, both peaks decreasing thereafter prior to the entrance of the cells to the stationary phase of growth.

In a *Chlorella* culture growing synchronously under an 8 h light 16 h dark regime, four daughter cells (autospores) are released from each mother cell. DNA replication occurs between the 4th and the 6th hour of the cycle, the main protein synthesis occurring between the 2nd and the 8th hour. The maximum of ODC activity was reached after the 2nd hour of the cycle, thus preceding DNA peak. The activity of ADC did not change significantly throughout the course of the cell cycle. The level of endogenous polyamines putrescine and spermine did not correlate with ODC activity nor with DNA replication. Spermidine content peaked at the 4th hour of the cycle, that is about 2 hours before and after the peak in DNA synthesis and ODC activity, respectively.

It is suggested that the sequence of events concerning polyamines and cell division during the life cycle of *Chlorella* is: increase in ODC activity, spermidine formation, and DNA replication. This pattern leads to cell division at the end of the cycle. Probably, polyamines other than those analyzed are involved in this process.

Advances in Polyamine Research, Vol. 4, edited by U. Bachrach, A. Kaye, and R. Chayen. Raven Press, New York © 1983.

Polyamines in Roots of *Zea Mays* Seedlings

M. Schwartz, T. Arzee, Y. Cohen, and *A. Altman

*Department of Botany, The George S. Wise Faculty of Life Sciences, Tel Aviv University, Israel; *Department of Horticulture, The Hebrew University of Jerusalem, Rehovot, Israel*

The root axis represents an open system of growth which undergoes gradual and continuous differentiation. At the tip of the corn root (Zea mays L.) a zone of meristematic tissue exists, whose derivatives gradually differentiate along the root axis. At about 2-3 cm from the tip other centers of meristematic activity can be observed which will eventually give rise to lateral roots. These root primordia initiate in an area where differentiation of root tissue is almost complete. The aim of this study was to investigate whether there is a correlation between root morphogenesis and polyamine content and metabolism.

The endogenous level of polyamines was determined for 1 cm segments along the root axis, using the dansylation T.L.C. method (Seiler and Wiechmann, 1967). It was found that the level of putrescine was low in the apical segment and increased along the axis, while spermidine was highest at the apex and lower in the other segments. Spermine content was generally low throughout, being only slightly higher in the apical segment (Table 1). In the first 6 cm of root the putrescine/spermidine ratio increased with increasing distance from the root apex. Where meristematic activity was highest the putrescine/spermidine ratio was lowest and rose as tissue differentiation progressed.

TABLE 1. Endogenous polyamine levels along the root axis

Distance from	nmole/20 segments ± SE			
Apex (cm)	Put	Spd	Spm	Put/Spd
0-1	83.0±27.8	135.0±2.8	17.2±3.8	0.61±0.19
1-2	194.2±44.6	45.7±9.8	5.2±0.7	4.22±0.06
2-3	237.7±53.7	28.5±7.5	4.5±0.0	8.43±0.35
3-4	220.5±45.4	21.7±2.3	3.0±0.0	10.03±1.03
4-5	271.5±51.4	20.2±0.7	3.0±0.0	13.32±2.04
5-6	298.5±59.0	15.0±3.0	3.0±0.0	19.91±0.08

The pattern of UC^{14} arginine and UC^{14} ornithine incorporation into the various polyamines resembled the pattern obtained for endogenous levels of these same polyamines (Table 2).

TABLE 2. Incorporation of $U^{14}C$-Arginine and $U^{14}C$-Ornithine into Polyamines along the Root Axis (± SE).

	Distance from Apex (cm)	% of total radioactivity ± SE			
		Put	Spd	Spm	Put/Spd
	0-1	21.5±1.0	21.3±0.8	1.0±0.1	1.00±0.06
Arginine	1-2	42.3±1.8	6.4±1.0	0	7.10±1.10
	4-5	57.0±5.5	4.2±0.5	0	13.50±0.70
	0-1	27.7±2.5	25.1±3.0	1.0±0.07	1.10±0.03
Ornithine	1-2	26.0±1.8	26.0±0.6	1.0±0.28	1.00±0.07
	4-5	59.6±1.9	2.7±0.3	0.2±0.03	22.00±2.70

Decapitation of 0.5 cm root tip prevents elongation of roots above the cut surface. However, the remaining axis continues to differentiate, thus providing us with a well defined zone of lateral root initiation and dvelopment in the area just above the cut surfaces. For the 72 hrs period following decapitation, during which time primordia appear and lateral roots are formed, endogenous levels of polyamines were determined for the 0.5 cm segment above the plane of decapitation (Table 3). The putriscine/spermidine ratio decreased with time as meristematic activity increased and approached the level found in the 0.5 cm apical segment which is highly meristematic.

TABLE 3. Time-course analysis of endogenous polyamine level (nmoles/20 root segments, ± SE) in 0.5 cm segment above the cut surface.

Time (hrs)	Put	Spd	Put/Spd
0	115.5±15.1	54.0±2.8	2.13±0.15
24	59.2± 9.8	25.5±1.4	2.35±0.82
48	53.2±15.7	30.7±0.7	1.74±0.55
72	72.0±12.1	88.5±10.5	0.84±0.23
Apical segment	51.7±9.8	114.7±2.2	0.44±0.07

Lateral roots failed to develop in roots treated with canavanine 10^{-3}M for 3 hours following decapitation. Simultaneous application of arginine or ornithine cancelled the inhibitory effect of canavanine. A similar result was reported by Weaks and Hunt (1974) in *Zea mays* root elongation and by Altman and Bachrach (1981) in adventitious roots of mung bean.

The preliminary results obtained for our intact decapitated systems show that the levels of putrescine and its bio-synthesis increased gradually along the root axis from the meristematic root apex up to the area of lateral root initiation, while the level of spermidine decreased, thus the putrescine/spermidine ratio is low when the rate of cell division is high and as meristematic activity diminishes and cell differentiation occurs, this ratio is increased.

Acknowledgement

We wish to thank Mrs. L. Cohen, Mrs. N. Levin and R. Friedman for the useful conversations and assistance.

This study was partially supported by the Botany Research Fund dedicated to the memory of Tzvi Meriminski and his sister Sonia Meriminski.

References

1. Altman, A. and U. Bachrach, 1981. In : "Advances in polyamine research" (Eds. C.M. Caldarera et al.) Vol. 3, pp. 365-375. Raven Press, New York.
2. Seiler, N., M. Wiechmann, 1967. Z. Physiol. Chem. 348 : 1285-1290.
3. Weaks, T. E. and G. E. Hunt, 1974. Bot. Gaz. 135, 45-49.

Advances in Polyamine Research, Vol. 4, edited by U. Bachrach, A. Kaye, and R. Chayen. Raven Press, New York © 1983.

Polyamine Biosynthesis and Function in *Escherichia Coli*

Herbert Tabor and Celia White Tabor

Laboratory of Biochemical Pharmacology, National Institute of Arthritis, Diabetes, and Digestive and Kidney Diseases, National Institutes of Health, Bethesda, Maryland 20205

BIOSYNTHETIC PATHWAY FOR POLYAMINES IN ESCHERICHIA COLI

The following figure summarizes the biosynthetic pathway in *E. coli* and indicates the genes that are responsible for the various steps (13, 20, 25, 26, 36, 39).

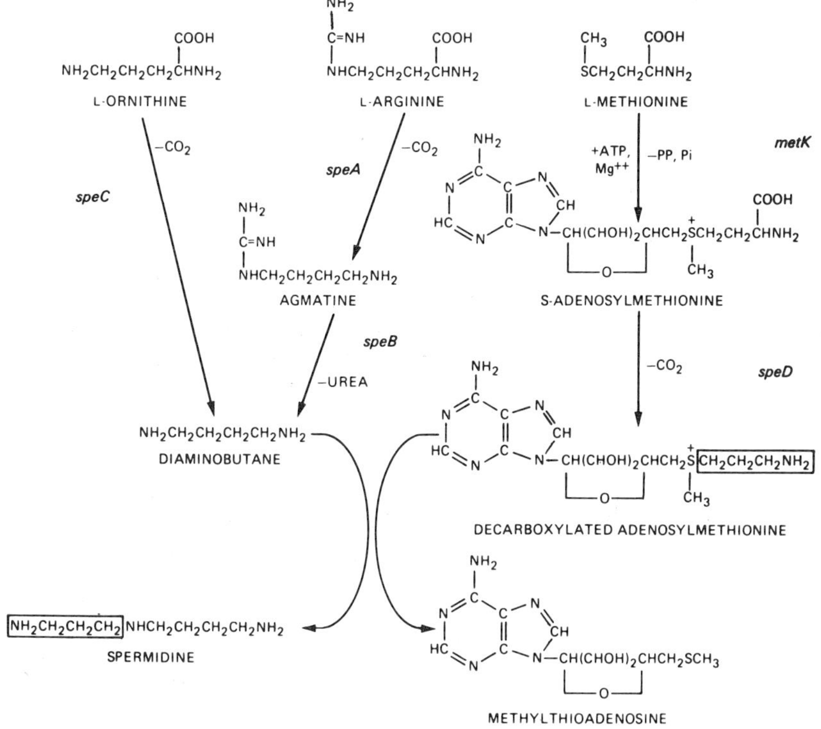

FIG. 1. Biosynthetic pathway for polyamines in E. coli (reprinted fron reference 32).

GENES AND ENZYMES CONCERNED WITH POLYAMINE BIOSYNTHESIS

metK is the gene for adenosylmethionine synthetase (methionine adenosyltransferase). It is located near 63 minutes on the E. coli chromosome (16); the gene order is glc speC metK speA speB serA (5, 11, 12, 20). The gene has been cloned into a multicopy plasmid which overproduces the enzyme. In a derepressed strain of metJ containing this plasmid, the amount of adenosylmethionine synthetase is increased 80-fold (22). Substantial amounts of homogeneous enzyme can be obtained easily after a three-step purification; 43 mg of enzyme are obtained from 50 g of packed cells. The enzyme has been crystallized (9, 21), and its kinetic properties have been studied. [For further details on our studies with this enzyme, see Markham et al. (22).] Since large quantities of this enzyme are obtained so easily, it is particularly useful for the preparation of isotopically-labeled adenosylmethionine for studies of adenosylmethionine decarboxylase, or for the synthesis of various analogs of adenosylmethionine, such as adenosylselenomethionine (34).

speD is the gene for adenosylmethionine decarboxylase and is located at 2.7 minutes on the chromosome (37). This gene has also been cloned into a multicopy plasmid (23). Strains containing this plasmid overproduce the enzyme, permitting the purification of substantial quantities of homogeneous enzyme (23). Adenosylmethionine decarboxylase is a particularly interesting enzyme since it contains covalently-bound pyruvate which is required for enzyme activity (23, 43). With George Markham we have recently found (23) that adenosylmethionine decarboxylase contains one pyruvate moiety on each of six subunits. The enzyme is inactivated by $NaCNBH_3$, but only in the presence of substrate or product and a divalent metal; inhibition is accompanied by incorporation of one mole of the product (decarboxylated adenosylmethionine) per mole of enzyme subunit, suggesting that the pyruvoyl group participates in catalysis by formation of a Schiff base with the substrate.

speA, speB, and speC are the genes for arginine decarboxylase, agmatine ureohydrolase, and ornithine decarboxylase, and are located near metK on the chromosome (see above) (12, 20). The separate genes have been cloned on multicopy plasmids (5, 35), and these experiments have helped in the elucidation of the gene order given (particularly the gene order of speA and speB, since these two genes are too close together to be separated by transduction techniques). Although strains containing these genes overproduce the respective enzymes, we have not purified these enzymes, since the two decarboxylases have been purified previously (3, 25).

The gene for putrescine aminopropyltransferase has not been mapped or cloned into a plasmid, although the enzyme has been purified to homogeneity (4).

cadA and cadR should be mentioned even though they are not listed in Figure 1. These two genes (46 minutes and 92 minutes on the chromosome, respectively) code for lysine decarboxylase (cadA) and the regulation (cadR) of this enzyme (38). Ordinarily there is no cadaverine in E. coli if it is grown on a purified medium; however, as shown by Dion and Cohen (7), the amounts of cadaverine and its aminopropyl derivative increase in putrescine deficiency.

MUTANTS DEFECTIVE IN POLYAMINE BIOSYNTHESIS

Mutants have been described for most of the above genes. The initial mutants were point mutants with an incomplete loss of enzyme activity (20, 26, 37). Although these mutants were useful, we decided to obtain strains with a complete loss of enzyme activities and of polyamines by developing deletion mutations in speA, speB, speC, and speD (13). Subsequently we added a point mutation in cadA, resulting in a strain with the genotype Δ(speA speB) ΔspeC ΔspeD cadA (38). This strain contains no detectable polyamines.

We constructed these strains in order to study the physiological role of the polyamines in vivo, i.e., what cellular functions are disturbed in the total absence of polyamines. Although polyamines have many effects in vitro, we still do not know which of these effects are significant in vivo.

As we reported previously with Edmund Hafner (13, 38) we were rather surprised at the relatively few defects shown by these mutants, even though they have no polyamines. Our previous observations may be summarized as follows:

(i) The amine-deficient cells grow indefinitely, although the growth rate is decreased by about 75%.

(ii) Good multiplication is observed with bacteriophages T4, T7, and f2. (However, see below for later studies with amber mutants of bacteriophages T4 and T7.)

(iii) There is a striking defect in the multiplication of bacteriophage λ, either after infection or after temperature-induction of a lysogenic strain. After induction, excision occurs, but no phage are produced, nor is the host killed.

(iv) There is a marked decrease in the number of recombinants in an Hfr cross if the donor organism is amine-deficient. There is also decreased adsorption of the male bacteriophages f2, Qβ, and f1. The decrease in adsorption is greater if the cultures were shaken vigorously during growth (unpublished data). These data indicate a defect in the male pili in the absence of amines.

EFFECT OF POLYAMINE-DEFICIENCY ON PROTEIN BIOSYNTHESIS IN VIVO

We have continued our studies with these and related polyamine-deficient strains in an effort to find which biochemical systems require polyamines in vivo. Our more recent studies have shown that optimum translation in protein biosynthesis in vivo requires polyamines, and we postulate that these effects are a result of the action of polyamines on the structure and conformation of the protein-synthesizing ribosomal complex.

Production of an Absolute Growth Requirement for Amines by Insertion of an *rpsL* Mutation into a Δ(*speA speB*) Δ*speC* Mutant

In the first set of experiments (41) we showed that introduction of certain specific rpsL (strA) mutations into amine-deficient E. coli causes a partial amine requirement for growth to become absolute. HT395, a $rpsL^+$ strain with deletions in speA, speB, and speC, grows indefinitely in the absence of polyamines at a rate approximating 25% of that obtained in the presence of amines. HT414 is a rpsL transductant of HT395; in contrast to the results with HT395, this strain will not grow in the absence of polyamines (Fig. 2). The cessation of growth of HT414 is seen within 10 hours after the cells are placed in amine-deficient medium, at which time there is only a little decrease in the growth rate of HT395. It should be noted that at this time the cells had not grown enough to completely dilute out their stored amines.

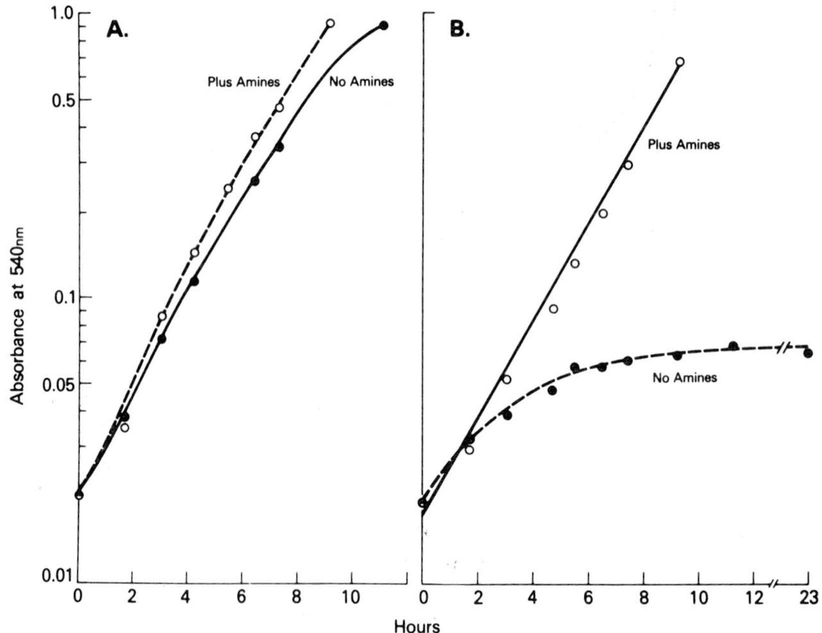

FIG. 2. Growth rate of HT395 (A) and of its rpsL transductant (HT414) (B) in amine-free and amine-supplemented medium (reprinted from reference 41).

It is well known from the work of Gorini, Nomura, and others (6, 27, 28) that rpsL mutations affect the structure of the S12 ribosomal protein, and that such changes in this ribosomal protein affect the fidelity of translation both in vivo and in vitro. Therefore, our results can best be explained by assuming that, in the absence of polyamines, the protein-synthesizing ribosomal complex is partially defective, accounting for the decreased growth rate. The rpsL mutation adds an additional defect to the complex, and the net result is complete cessation of growth. However, other less direct explanations are also possible, such as the biosynthesis of defective proteins by the rpsL ribosomes and their stabilization or activation by polyamines.

In very recent experiments (unpublished) we have studied this particular strain (HT414) further. If one observes the amine-free cultures for several days after growth has ceased, one obtains "bypass" organisms which grow at a rate comparable to the parent organism. Since deletion mutations were involved in the construction of the original strain, these bypasses are not the result of a reversion of the various mutations; also amine analyses showed that no amines are present in the bypass cultures. We have carried out preliminary mapping of the bypass mutation, and found that this mutation cotransduced 47% with rpsL, indicating that it probably represents a mutation in a ribosomal gene. This finding is reminiscent of the ram mutation described by Gorini [pages 791-803 in (28)], which causes a reversal of the rpsL effects, and which was shown to cause a change in the S4 ribosomal protein.

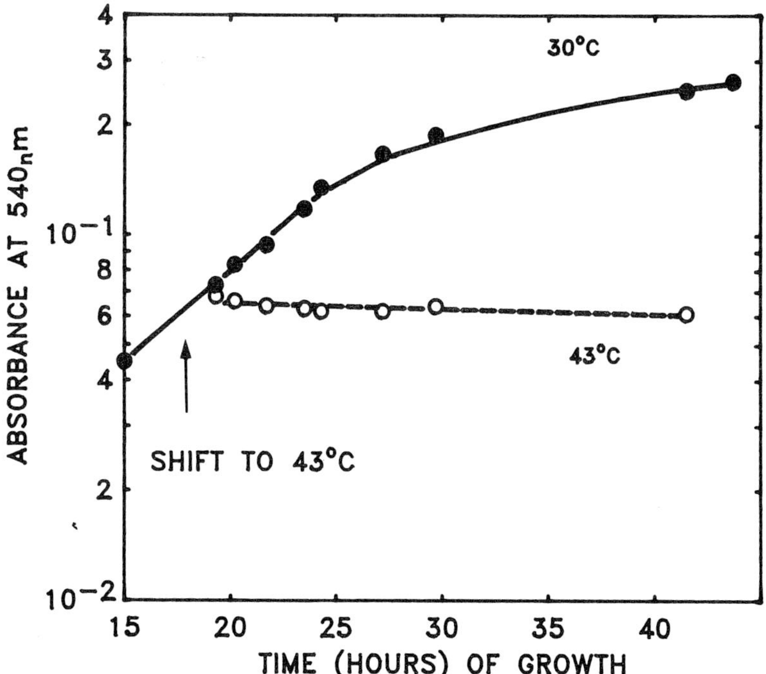

FIG. 3. Growth of HT411 in the absence of polyamines. Effect of temperature shift to 43°C. See text on next page.

A Temperature-Inducible Absolute Polyamine Requirement in Another *rpsL* Derivative of a Δ(*speA speB*) Δ*speC* Mutant

We have recently isolated another streptomycin-resistant derivative (HT411) of HT395, which only showed a marked requirement for polyamines for growth after the cultures had grown for several days at 30°C without polyamines. After only 20 hours at 30°C there was relatively little difference in growth rate between amine-deficient and amine-supplemented strains. However, the effect of the polyamine deficiency could be accentuated by a temperature shift; if the deficient cells were shifted to 43°C, they stopped growing within 1-2 hours (Fig. 3). Amine-supplemented cells continued growing indefinitely at both 30°C and at 43°C (data not shown).

Effect of Polyamine Depletion on the Multiplication of Amber Mutants of Bacteriophage T7

In our most recent studies (40) we have used another system to study the translation system *in vivo*. We have tested whether amber mutants of bacteriophage T7 multiply as well as wild type bacteriophage T7 in amine-deficient strains of *E. coli*. We have found that the multiplication of certain amber mutants of bacteriophage T7 is markedly decreased in amine-deficient cells, compared to wild type bacteriophage T7 growing in the same host. This marked decrease in the multiplication of these amber bacteriophage mutants is not seen if the host had been grown in the presence of polyamines.

We first compared the multiplication of three amber mutants of bacteriophage T7 with a wild type bacteriophage T7 in HT306

Table 1. Polyamines are necessary for multiplication of T7 amber bacteriophages (gene 3^- and gene 4^-)

Time (min)	Wild type	3^-*am*29	4^-*am*208
T7 Phage Production in Amine-Deficient Cells (plaque-forming units per ml)			
3	3,100	5,400	2,700
77	60,000	12,000	3,100
94	400,000	17,000	2,700
113	800,000	30,000	4,000
T7 Phage Production in Amine-Supplemented Cells (plaque-forming units per ml)			
4	3,850	6,900	4,200
25	124,000	95,000	71,200
40	190,000	165,000	112,000
52	340,000	270,000	185,000

Host: HT306 Δ(*speA speB*) Δ*speC* Δ*speD* *cadA* *rpsL* *supE44*

Δ(speA speB) Δ(speC glc) ΔspeD cadA supE rpsL and found that two of the amber phages showed no multiplication unless the cells had been grown with polyamines (Table 1). Similar results were also found in another polyamine-deficient strain (HT411) [Δ(speA speB) Δ(speC glc) rpsL sup and its rpsL+ parent (HT395)]. The results with the rpsL+ strain are of particular significance since they show that the effects seen in polyamine-deficient cells do not require the presence of an abnormal S12 ribosomal protein (which is present in rpsL strains).

In order to show more directly that the polyamines actually affected the translation of the amber mutation, we examined the protein products after infection with T7 gene 1⁻am342 bacteriophage, and with T7 wild type bacteriophage. As seen in Figure 4, lane 3, in the absence of amines there is a relative increase in the formation of the amber fragment (molecular weight 95,000) and a decrease in the full-sized protein (molecular weight 106,000), compared to the results with the wild type bacteriophage (lane

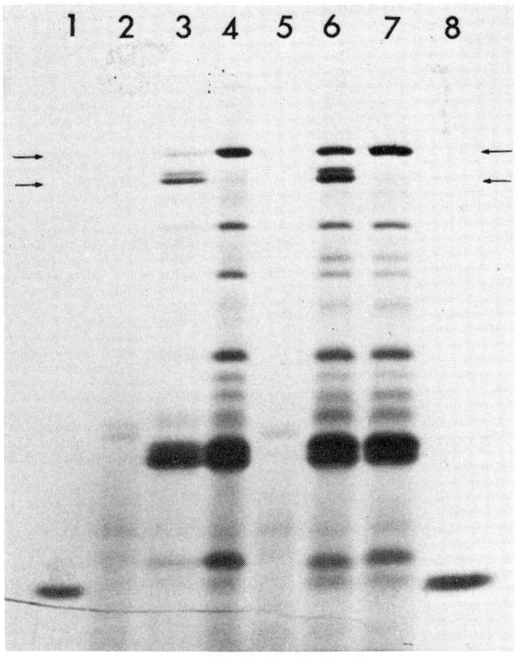

FIG. 4. Radioautograph of electrophoretic patterns of extracts of HT306, labeled with [^{35}S]methionine, after T7 bacteriophage infection (reprinted from reference 40). Lanes 2, 3, and 4 are cultures depleted of amines; lanes 5, 6, and 7 are cultures supplemented with amines. Lanes 2 and 5: uninfected cells; lanes 3 and 6: after infection with T7 1⁻am342 bacteriophage; lanes 4 and 7: after infection with wild type T7 bacteriophage. Lanes 1 and 8 are protein standards.

4), or to the results in cells grown with polyamines (lanes 6 and 7). Essentially similar results were found with either HT411 or with HT306 as the host.

Effect of Polyamine Depletion on the Multiplication of Amber Mutants of Bacteriophage T4

We had previously shown (13) that bacteriophage T4 multiplied well in amine-deficient bacteria, although the time of the burst was delayed. We have now shown, that in contrast to these results with wild type T4 bacteriophage, certain amber mutants of bacteriophage T4 do not multiply well in polyamine-deficient bacteria, even though these bacteria contain a suppressor mutation. All of the phage multiply well if the bacteria were grown with amines (Table 2).

Table 2.

T4 Phage Production in Amine-Deficient Cells (plaque-forming units per ml)

Time (min)	T4 Wild type	T4 am43⁻	T4 lysozyme⁻ (am)
10	6,000	2,500	4,000
120	57,500	6,000	11,500
140	154,000	5,600	11,200
160	220,000	10,800	18,900
181	264,000	9,000	21,200
205	262,000	4,000	29,400

Host: HT411 Δ(speA speB) ΔspeC rpsL sup

DISCUSSION

The effects of polyamines on protein synthesis in vitro have been described by a number of investigators and have included studies on initiation (2, 18), translation (2, 8), tRNA structure (29, 30, 31, 33), fidelity of translation (1, 10, 42), and fidelity of tRNA aminoacylation (19) (see reference 32 for other references). Of particular relevance to the present studies are the in vitro studies of Morch and Benicourt (24), who showed that the readthrough of amber termination codons on viral messenger RNA is stimulated by spermidine. Polyamines have also been reported to stimulate synthesis of the readthrough protein of Qβ (14) and of β-globin mRNA (15) in cell-free systems.

In in vivo studies, Young and Srinivasan (44) reported that, when amines were added to partially amine-deficient E. coli, protein synthesis increased before the synthesis of DNA and RNA. Jorstad and Morris (17), studying similar amine-deficient mutants, found that the rate of elongation of the peptide chain was decreased, and Goldemberg and Agranati (10) found that polyamine deficiency led to a relative decrease in the formation of proteins with a higher molecular weight. In these studies, the comparison between amine-supplemented and amine-deficient cultures was complicated by the differences in growth rate of these two cultures. In contrast, in the studies presented in this paper, we have compared the growth of wild type and amber bacteriophage in parallel amine-deficient cultures growing at the same growth rate.

In conclusion, we postulate that an important function of polyamines is concerned with translation and with the structure and function of the protein-synthesizing ribosomal complex. Even though our data are based on bacteriophage experiments, it is very possible that these findings have considerably wider implications for the action of polyamines in bacteria and other cells. It is important to emphasize, however, that this demonstration of one *in vivo* action of polyamine does not exclude other possible functions for polyamines *in vivo*.

REFERENCES

1. Abraham, A. K., Olsnes, S., and Pihl, A. (1979): *FEBS Lett.*, 101: 93-96.
2. Algranati, I. D., and Goldemberg, S. H. (1981): *Biochem. Biophys. Res. Commun.*, 103: 8-15.
3. Applebaum, D. M., Dunlap, J. C., and Morris, D. R. (1977): *Biochemistry*, 16: 1580-1584.
4. Bowman, W. H., Tabor, C. W., and Tabor, H. (1973): *J. Biol. Chem.*, 248: 2480-2486.
5. Boyle, S. M., Markham, G. D., Hafner, E. W., Wright, J. M., Tabor, H., and Tabor, C. W., manuscript in preparation.
6. Chambliss, G., Craven, G. R., Davies, J., Davis, K., Kahan, L., and Nomura, M., editors (1980): *Ribosomes. Structure, Function, and Genetics*. University Park Press, Baltimore.
7. Dion, A. S., and Cohen, S. S. (1972): *J. Virol.*, 9: 423-430.
8. Echandi, G., and Algranati, I. D. (1975): *Biochem. Biophys. Res. Commun.*, 67: 1185-1191.
9. Gilliland, G. L., Markham, G. D., and Davis, D. R. (1980): *American Crystallographic Association Program and Abstracts*, 8: 43.
10. Goldemberg, S. H., and Algranati, I. D. (1981): *Eur. J. Biochem.*, 117: 251-255.
11. Hafner, E. W., Tabor, C. W., and Tabor, H. (1976): *Fed. Proc.*, 35: 1547.
12. Hafner, E. W., Tabor, C. W., and Tabor, H. (1977): *J. Bacteriol.*, 132: 832-840.
13. Hafner, E. W., Tabor, C. W., and Tabor, H. (1979): *J. Biol. Chem.*, 254: 12419-12426.
14. Hirashima, A., Harigai, H., and Watunabe, I. (1979): *Biochem. Biophys. Res. Commun.*, 88: 1046-1051.
15. Hryniewicz, M., Chiou, L., and Vonder Haar, R. A. (1982): *Fed. Proc.*, 41: 1285.

16. Hunter, J. S. V., Greene, R. C., and Su, C.-H. (1975): J. Bacteriol., 122: 1144-1152.
17. Jorstad, C. M., and Morris, D. R. (1974): J. Bacteriol., 119: 857-860.
18. Konecki, D., Kramer, G., Pinphanichakarn, P., and Hardesty, B. (1975): Arch. Biochem. Biophys., 169: 192-198.
19. Lövgren, T. N. E., Pettersson, A., and Loftfield, R. B. (1978): J. Biol. Chem., 253: 6702-6710.
20. Maas, W. K. (1972): Mol. Gen. Genet., 119: 1-9.
21. Markham, G. D., and Gilliland, G. L. (1981): Fed. Proc., 40: 1685.
22. Markham, G. D., Hafner, E. W., Tabor, C. W., and Tabor, H. (1980): J. Biol. Chem., 255: 9082-9092.
23. Markham, G. D., Tabor, C. W., and Tabor, H. (1982): J. Biol. Chem., 257, in press.
24. Morch, M. D., and Benicourt, C. (1980): Eur. J. Biochem., 105: 445-451.
25. Morris, D. R., and Fillingame, R. H. (1974): Annu. Rev. Biochem., 43: 303-325.
26. Morris, D. R., and Jorstad, C. M. (1970): J. Bacteriol., 101: 731-737.
27. Nomura, M. (1970): Bacteriol. Rev., 34: 228-277.
28. Nomura, M., Tissières, A., and Lengyel, P., editors (1974): Ribosomes. Cold Spring Harbor Laboratory, Cold Spring Harbor, New York.
29. Nöthig-Laslo, V., Weygand-Durasevic, I., Zivkovic, T., and Kucan, Z. (1981): Eur. J. Biochem., 117: 263-267.
30. Pochon, F., and Cohen, S. S. (1972): Biochem. Biophys. Res. Commun., 47: 720-726.
31. Quigley, G. J., Teeter, M. M., and Rich, A. (1978): Proc. Natl. Acad. Sci. U.S.A., 75: 64-68.
32. Raina, A., and Jänne, J., editors (1981): Med. Biol., 59: 269-461.
33. Schrier, A. H., and Schimmel, P. R. (1975): J. Mol. Biol., 93: 323-329.
34. Stadtman, T. C. (1982): personal communication.
35. Tabor, C. W., Hafner, E. W., and Tabor, H. (1979): Proceedings of the 11th International Congress of Biochemistry, Toronto, Canada, July 8-13, 1979.
36. Tabor, C. W., and Tabor, H. (1976): Annu. Rev. Biochem., 45: 285-306.
37. Tabor, C. W., Tabor, H., and Hafner, E. W. (1978): J. Biol. Chem., 253: 3671-3676.
38. Tabor, H., Hafner, E. W., and Tabor, C. W. (1980): J. Bacteriol., 144: 952-956.
39. Tabor, H., and Tabor, C. W. (1972): Adv. Enzymol., 36: 203-268.
40. Tabor, H., and Tabor, C. W. (1982): Proc. Natl. Acad. Sci. U.S.A., 79, in press.
41. Tabor, H., Tabor, C. W., Cohn, M. S., and Hafner, E. W. (1981): J. Bacteriol., 147: 702-704.

42. Thompson, R. C., Dix, D. B., Gerson, R. B., and Karim, A. M. (1981): J. Biol. Chem., 256: 6676-6681.
43. Wickner, R. B., Tabor, C. W., and Tabor, H. (1970): J. Biol. Chem., 245: 2132-2139.
44. Young, D. V., and Srinivasan, P. R. (1972): J. Bacteriol., 112: 30-39.

Advances in Polyamine Research, Vol. 4, edited by U. Bachrach, A. Kaye, and R. Chayen. Raven Press, New York © 1983.

Biochemical and Genetic Studies of Polyamines in *Saccharomyces Cerevisiae*

Celia White Tabor, Herbert Tabor, and Anil K. Tyagi

National Institute of Arthritis, Diabetes, and Digestive and Kidney Diseases, National Institutes of Health, Bethesda, Maryland 20205

Several years ago we decided to study the regulation of polyamine biosynthesis in a eukaryotic system. We chose the organism Saccharomyces cerevisiae for three reasons. First it is a simple eukaryote that is easy to grow in both the haploid and the diploid state; second, it has a rapid doubling time, about 2-3 hours; and third, the genetics have been well-studied, mutants are available, and techniques for genetic manipulation have been described.

BIOSYNTHETIC PATHWAY IN YEAST

Table 1 shows the polyamine content in a wild type strain grown on a glucose-salts medium (4). In studies carried out with Murray Cohn, we have characterized the steps in biosynthetic pathway for polyamines in S. cerevisiae and some of the genes involved in these steps. This pathway is shown in Figure 1.

TABLE 1. Amine levels in wild type Saccharomyces cerevisiae

	µmol/gm Wet Weight
Putrescine	0.4
Spermidine	2.0
Spermine	0.3
Adenosylmethionine	Trace
Decarboxylated adenosylmethionine	Trace

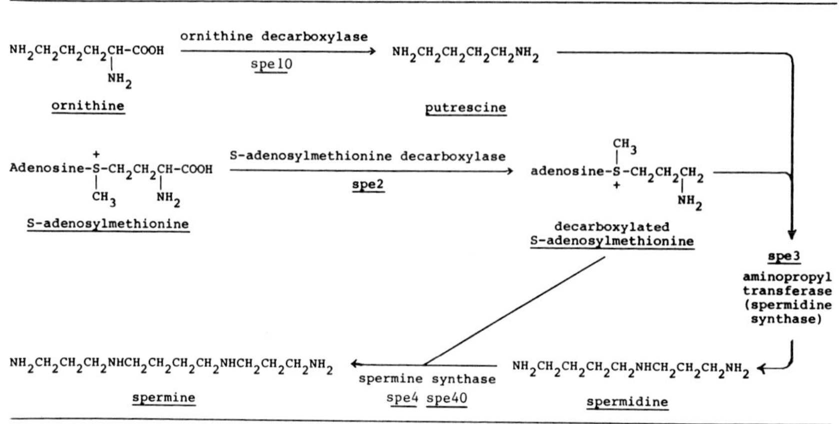

FIGURE 1. Biosynthetic pathway in Saccharomyces cerevisiae (reprinted from reference 9).

Yeast forms putrescine only by the decarboxylation of ornithine. There is no evidence for arginine decarboxylase, or agmatine ureohydrolase, which form an alternate pathway to putrescine in E. coli. We have defined one gene, spe10, which appears to be a regulatory gene for ornithine decarboxylase (4).

The gene which codes for adenosylmethionine decarboxylase is the spe2 gene. This enzyme, like the enzyme in eukaryotic systems, requires putrescine for activation (2). Mg^{++}, which is necessary for the E. coli enzyme, has no effect on the yeast enzyme. Yeast adenosylmethionine decarboxylase contains covalently linked pyruvate, which is essential for activity (1).

Two aminopropyltrasferases are active in yeast, as in other eukaryotic cells. One, putrescine aminopropyltransferase (or spermidine synthase), coded for by spe3 gene, uses putrescine as the substrate, and with decarboxylated adenosylmethionine forms only spermidine. The second enzyme, spermidine aminopropyltransferase (or spermine synthase), uses only spermidine as the substrate, and with decarboxylated adenosylmethionine, forms only spermine (4). We have found that two unlinked genes, spe4 and spe40, are important for the activity of spermine synthase. The spe4 gene is recessive, and therefore appears to be the gene for the structural protein. The spe40 gene is dominant, and presumably is a regulatory gene.

GENETIC STUDIES IN YEAST

Isolation of Mutants Defective in the Biosynthetic Enzymes for Polyamines

We utilized nitrosoguanidine mutagenesis in order to obtain strains with mutations in the biosynthesis of the amines. From our studies with *E. coli*, in which a long period of depletion was needed to dilute out intracellular amines and in which completely deficient strains continued to grow slowly even after depletion, we were not surprised that our first searches failed to yield any auxotrophs for polyamines. For this reason, we used a mass screening method in which we assayed isolated clones in a microtiter plate directly for the release of $^{14}CO_2$ from carboxy-labeled substrates (2, 4).

TABLE 2. Mutants unable to synthesize spermidine or spermine

Strain	Enzyme Defect	Amines Present*
spe10-3	Ornithine decarboxylase	0
spe2-2	Adenosylmethionine decarboxylase	Putrescine
spe3	Putrescine aminopropyltransferase	Putrescine

* After growth in amine-free medium for over 13 generations.

TABLE 3. Mutants unable to synthesize spermine

Strain	Enzyme Activity Absent	Amines Present*
spe10 spe40	Spermidine aminopropyltransferase	Putrescine Spermidine
spe10 spe4	Ornithine decarboxylase Spermidine aminopropyltransferase	None
spe4	Spermidine aminopropyltransferase	Putrescine Spermidine

* After growth in amine-free medium for over 13 generations.

We isolated a number of mutant strains. Table 2 shows examples of mutants which cannot make either spermidine or spermine, and indicates the enzymes and genes involved. In Table 3 we have listed examples of the strains which cannot make spermine, but which make spermidine and putrescine, and we have indicated the enzymes defective in these strains and the genes involved in their biosynthesis (9, 10).

Mutants in genes affecting ornithine decarboxylase activity. We have isolated 10 mutants which have no detectable ornithine decarboxylase (*spe10* mutants) activity (4). One of these does not grow at all in amine-free medium; the others have growth rates of 30 to 60 hours. Table 4 shows the phenotypic characteristics of these strains. These mutants all map at the same locus. This gene appears to be a regulatory gene, since phenotypic reversions of these strains occur with the reappearance of a high ornithine decarboxylase activity, increased growth rate, and the formation of putrescine. This phenotypic reversion is the result of a new mutation in the spermine synthase pathway. I will describe these in greater detail later in this paper.

TABLE 4. *Spe10 Mutants*

1. Lack ornithine decarboxylase.
2. Cannot make putrescine; consequently lack spermidine and spermine.
3. Cannot grow in amine-free medium unless supplemented with any of the three polyamines.
4. Have increased decarboxylated S-adenosylmethionine.
5. Cannot sporulate in the absence of polyamines.
6. Cannot maintain the killer plasmid unless supplemented with polyamines.
7. Occurrence of suppressor mutants indicates that *spe10* is a regulatory rather than a structural gene.

We found in collaborative studies with Reed Wickner and Murray Cohn that the *spe10* mutants were unable to maintain the double-stranded RNA killer plasmid of yeast, and were unable to sporulate (4). From studies with other mutant strains, we have found that these defects result from the inability of these strains to make spermidine or spermine.

We have also isolated 3 mutants with increased ornithine decarboxylase activity. These appear to be complex genetically, and we have not yet characterized the defects which lead to this phenotype. We have mutagenized these strains to obtain some of our *spe10* mutants, since the selection of deficient mutants from a parent with high ornithine decarboxylase activity is much easier than selection from a wild type parent with low enzymatic activity. Yeast mutants lacking ornithine decarboxylase have also been isolated by Whitney and Morris (12); Hosaka and Yamashita also isolated one mutant using the growth requirement of this strain for amines as the selection step (5). We do not yet know if the genes involved in these mutations are the same as the spe10 gene in our strains. We have tried to isolate other mutants lacking ornithine decarboxylase which might be in the gene for the structural protein and which map in a different site from our spe10 mutants, but we have not yet found any of this type.

Mutants in the gene coding for adenosylmethionine decarboxylase. We have isolated four mutants which have no detectable adenosylmethionine decarboxylase activity (spe2 mutants). Two of these have a 12 hour doubling time, while two have much longer doubling times. Table 5 shows the phenotypic characteristics of these strains (2). These strains have a 6-fold increase of ornithine decarboxylase activity over the wild type strains. We have used these derepressed strains for the purification of ornithine decarboxylase which I will describe later in this paper. These mutants also cannot maintain the double-stranded RNA "killer" plasmid of yeast and the diploids are unable to sporulate (3). These spe2 mutants have higher concentrations of putrescine than the wild type strain, but they lack spermidine and spermine. Thus these mutants show that putrescine alone is not sufficient for the maintenance of the killer plasmid or for sporulation. Spermidine and spermine are essential for these processes.

TABLE 5. spe2 Mutants

1. Lack S-adenosylmethionine decarboxylase and cannot make decarboxylated adenosylmethionine.
2. Consequently lack spermidine and spermine.
3. Overproduce ornithine decarboxylase and have increased putrescine when grown in amine-free medium.
4. Grow very slowly (13-hour doubling time) unless supplemented with spermidine or spermine (doubling time 2 hours).
5. Cannot sporulate in the absence of spermidine and spermine.
6. Cannot maintain the killer plasmid in the absence of spermidine and spermine.
7. Linked to arg1. arg1 is on chromosome XV (Hilger and Mortimer).

Mutants in the gene coding for putrescine aminopropyltransferase (spermidine synthase). Table 6 shows the phenotypic characteristics of a mutant lacking putrescine aminopropyltransferase (*spe3*). This strain, like the *spe2* strains, contains only putrescine, and therefore resembles the *spe2* strains in some of its characteristics (9).

TABLE 6. *spe3* Mutants

1. Lack putrescine aminopropyltransferase (spermidine synthase).
2. Cannot make spermidine and spermine *in vivo* from endogenous putrescine.
3. Can make spermine from added spermidine *in vivo* and *in vitro*, indicating that spermidine and spermine synthases are two distinct enzymes.
4. Have increased putrescine and decarboxylated S-adenosylmethionine.
5. Grow slowly (13-hour doubling time) unless supplemented with spermidine or spermine (2-hour doubling time).
6. Cannot sporulate in the absence of spermidine or spermine.
7. Cannot maintain the killer plasmid in the absence of speridine or spermine.

This strain demonstrates that yeast has two separate enzymes for making spermidine and spermine. The *spe3* strain cannot make spermidine either *in vivo*, or *in vitro* from putrescine and decarboxylated adenosylmethionine. However, it can make spermine both *in vivo* if grown with spermidine, and *in vitro* if spermidine and decarboxylated adenosylmethionine are present. Thus only the putrescine aminopropyltransferase has been lost.

Mutants lacking spermidine aminopropyltransferase (spermine synthase). Mutants of this type were found as bypasses in a lawn of amine-deficient *spe10* organisms, which grew very slowly in the absence of amines. These bypasses, which grew up more rapidly than the parent, fell into two classes, *spe10 spe4* and *spe10 spe40* mutants. Tables 7 and 8 show the phenotypic characteristics of these strains. The *spe10 spe40* strains, which grow at almost wild type rates, have recovered the ability to decarboxylate ornithine, have a high ornithine decarboxylase activity, and make significant amounts of putrescine and spermidine. However, they cannot make spermine. The *spe40* gene is tightly linked to the *spe10* gene, but is not linked to the *spe4* gene (4).

The *spe4* mutants are also found as bypasses to *spe10* mutations. The phenotypic characteristics are listed in Table 8. The *spe10 spe4* mutants, in contrast to the *spe10 spe40* mutants, grow slowly (with a 10-hour doubling time), and require amines to restore the growth rate to that of wild type strains. Consistent

TABLE 7. spe10 spe40 Mutants

1. Isolated as suppressors of spe10 mutants.
2. Have no spermidine aminopropyltransferase (spermine synthase).
3. Have no spermine.
4. Have high ornithine decarboxylase activity.
5. Have low putrescine and spermidine concentrations.
6. spe10 spe40 mutants have a doubling time of 2 hours in the absence of amines.
7. Tightly linked to spe10 (never found independent of spe10).
8. Not linked to spe4 gene.

TABLE 8. spe10 spe4 Mutants

1. spe10 spe4 strains have a doubling time of 10 hours in the absence of amines, but 2 hours in the presence of amines.
2. spe10 spe4 strains have no detectable ornithine decarboxylase, and no putrescine, spermidine or spermine.
3. spe10 spe4 strains have almost normal putrescine aminopropyltransferase.
4. spe10 spe4 strains have no spermidine aminopropyltransferase (spermine synthase).

with this is the finding that ornithine decarboxylase activity and putrescine are too low to detect. The spe4 gene is not linked to either the spe10 gene or the spe40 gene; therefore, we have been able to obtain the spe4 genotype in a $spe10^{+}$ background. This strain lacks only spermidine aminopropyltransferase and cannot make spermine, but can make putrescine and spermidine and has a normal growth rate (Table 9) (4). The spe4 gene is recessive, and probably codes for the structural protein, while the spe40 is dominant and is probably a regulatory gene. These studies with the spe4 and spe40 mutants indicate that intracellular spermine and spermine synthase are important in the control of ornithine decarboxylase.

TABLE 9. spe4 Mutants

1. Isolated as suppressors of spe10 mutants
2. Lack spermidine aminopropyltransferase (spermine synthase).
3. Lack spermine; putrescine and spermidine levels are normal.
4. spe4 mutants have a 2-hour doubling time in amine-free medium.
5. Not linked to spe10.

Yeast Ornithine Decarboxylase: Purification and Characterization of the Enzyme

We have previously reported the 1500-fold purification of homogeneous ornithine decarboxylase from a derepressed strain (spe2) by standard procedures, with a 40% yield of activity (11). The purified protein is stable on storage at -70°, contains a single subunit, and has a molecular weight of 68,000. The specific activity is 42 μmol/hr/mg at pH 8.0 (11), about 25-fold lower than the other eukaryotic enzymes which have been purified to homogeneity (6, 7). In view of the low specific activity of the purified enzyme, we have carried out the following experiments to show that the majority of the enzyme molecules are catalytically active, and hence that the low final specific activity is not related to inactivation of a large fraction of the pure protein. For these studies we used the technique of binding [^{14}C] α-difluoromethylornithine (kindly furnished by Dr. Peter McCann of Merrell Co.) to the enzyme as described by Pegg and his colleagues (8). About 75% of our pure protein molecules bind this irreversible inhibitor, and therefore are active. Thus the maximum specific activity of the homogeneous yeast enzyme is much lower than that found for the purified enzyme from rat liver by Kameji *et al* (6) or mouse kidney by Pegg (7).

Immunoprecipitation-electrophoresis studies of crude extracts. We have recently found by immunoprecipitation that the enzyme in crude extracts does not exist as the 68,000 molecular weight form, but as an 86,000 Mr protein. The latter form is converted to the 68,000 form during the purification procedures (10). Table 10 shows the method that we have used for immunoprecipitation of ^{35}S-labeled protein, and subsequent radioautographic studies. With this technique, taking particular care to work rapidly and to keep the extracts at 0°, we have found that the immunoprecipitable protein in crude extracts of the spe2 strain has a molecular weight of 86,000, with a small component of 81,000. None of the 68,000 form is found. If this extract is stored at -70° for 8 days, little change occurs. However, if it is stored at 4°, the 86,000 band decreases, and a band of 68,000

TABLE 10. Immunoprecipitation of Yeast Ornithine Decarboxylase

Reaction mix:

1. Extract of culture grown with [^{35}S]methionine (containing 0.6 mg of protein in 100 μl)
2. Rabbit antibody to pure yeast ornithine decarboxylase, purified by $(NH_4)_2SO_4$ precipitation (15 μl)
3. Phenylmethylsulfonyl fluoride 0.001 M
4. Tris buffer 0.05 M, pH 7.4, to bring volume to 500 μl

Incubation: 40 min at 4°

Add: protein A suspension in Tris buffer - 175 μl

Incubation: 45 min at 4° with gentle mixing

Centrifuge and wash

SDS electrophoresis and radioautography

molecular weight appears. We have further shown that this conversion of 86,000 to 68,000 is inhibited if the proteolytic inhibitor phenylmethylsulfonyl fluoride is present during the breaking of the cells and during the immunoprecipitation procedure. Thus the enzyme appears to exist in crude extracts predominantly as an 86,000 molecular weight protein, which undergoes partial proteolytic breakdown at 4°, to form a stable active 68,000 molecular weight protein. The relatively specific splitting of the 86,000 molecular weight form to the 68,000 form in such good yield implies that the proteolytic cleavage is either specific or that the native ornithine decarboxylase has certain domains that are more sensitive to such cleavage.

Purification on an antibody-Sepharose column. In order to prepare pure 86,000 molecular weight form of the enzyme, we have made an antibody column consisting of rabbit antibody (prepared against our purest 68,000 molecular weight enzyme preparations), linked to cyanogen bromide activated Sepharose 4B. Table 11 summarizes the purification procedure that we have developed. With this affinity column we have obtained in a single step a 70% purified enzyme, with almost complete recovery of activity from this column. Upon electrophoresis of the purified protein and staining with Coomassie Blue, the major band (about 50% of the total protein) appears to be 86,000 in molecular weight. The other prominent protein bands are 68,000 (20%) and 35,000 molecular weight (30%). Since the yield of activity units from the antibody column is about 90%, the above findings suggest that the 86,000 molecular weight form is about as active as the 68,000 form that we have previously purified. We are now trying to purify the 86,000 molecular weight form to homogeneity, in order to compare the kinetic characteristics of the 86,000 and the 68,000 forms.

TABLE 11. Purification of Yeast Ornithine Decarboxylase on an Antibody Column

Column: Rabbit antibody bound to Sepharose 4B cyanogen bromide

Column volume: 10 ml

Capacity: Crude extract (150 ml) from about 30 g of yeast

Equilibration buffer: Tris Cl buffer, 0.025 M, with EDTA, $MgCl_2$ and DTT

Wash column with 500 ml of 1 M NaCl in equilibration buffer.

Elute enzyme with 100 ml of 4.5 M $MgCl_2$. Concentrate by ultrafiltration.

Yield: 90%

Purification: 70%

Regulation of Yeast Ornithine Decarboxylase: Post-translational Modification of Ornithine Decarboxylase

In view of these new findings, we decided to restudy the effect of addition of spermine and spermidine to the growth medium on the amount of ornithine decarboxylase protein found in the yeast cells. We have previously shown that addition of amines to the medium resulted in a complete loss of ornithine decarboxylase activity within 6 hours; this inactivation required protein synthesis (Fig. 2) (11). In contrast to the loss of enzymatic activity, there was no significant loss of immunoreactive 68,000 protein. When this experiment was repeated with our improved immunoprecipitation procedure, we found complete retention of the 86,000 molecular weight protein, despite complete loss of enzyme activity. Thus we have evidence that a post-translational modification of the 86,000 form occurs following growth in amine-supplemented medium. This modification is unrelated to the proteolytic cleavage of the native enzyme.

We have also prepared immunoprecipitates from one of the *spe10* mutants which lack ornithine decarboxylase activity, to determine if these strains contain residual inactive protein. We found that these inactive extracts contained an amount of 86,000 molecular weight protein equal to that found in the very active extracts obtained from the derepressed *spe2* strain. Thus we again have evidence for regulation of the enzyme activity by a modification which is not related to the proteolytic changes which we have described. We are at present trying to characterize the post-translational changes in the protein. As we have reported previously, we had no evidence for the regulation of yeast ornithine decarboxylase by antizyme, by alterations in the binding of pyridoxal phosphate, by feedback inhibition by spermine or spermidine, by repression of synthesis of the enzyme

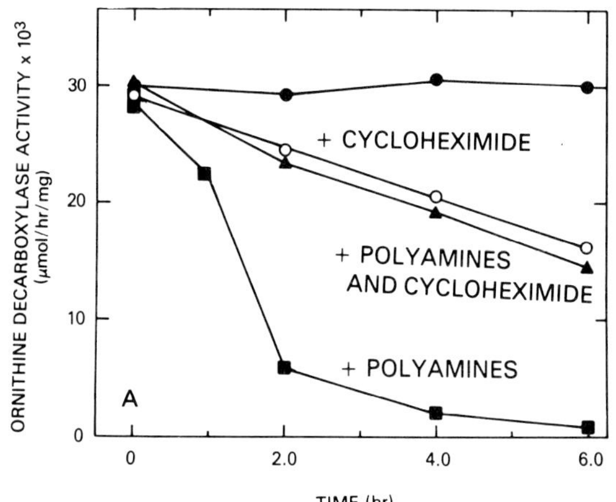

FIGURE 2. Effects of spermidine and spermine in the growth medium on ornithine decarboxylase activity (reprinted from reference 11).

protein (11), or by incorporation of polyamines into the enzyme protein (unpublished).

Summary

We have prepared mutants in each of the steps in the polyamine biosynthetic pathway in yeast. We have shown that S. cerevisiae has an absolute requirement for spermidine and spermine for growth, for maintenance of the killer plasmid, and for sporulation. We have purified ornithine decarboxylase to homogeneity by standard procedures, characterized the enzyme, and raised an antibody in rabbits. An affinity column, consisting of antibody linked to Sepharose has been prepared and we are developing a new rapid purification procedure. By careful immunoprecipitation of the enzyme in crude extracts, we have found evidence that the enzyme is present as an 86,000 molecular weight protein, and is converted by proteolysis to a stable 68,000 form. We have shown that the 86,000 protein form is retained when the activity is lost after growth in the presence of amines. Also the spe10 mutant, which has no enzymatic activity as a result of mutation in a regulatory gene, has the same amount of immunoprecipitable 86,000 molecular weight protein as the derepressed spe2 mutant. These findings indicate that post-translational modification of the enzyme results in loss of activity, and that the proteolytic cleavages which we have observed are not related to this enzyme modification.

REFERENCES

1. Cohn, M. S., Tabor, C. W., and Tabor, H. (1977): J. Biol. Chem., 252: 8812-8216.

2. Cohn, M. S., Tabor, C. W., and Tabor, H. (1978): *J. Bacteriol.*, 134: 208-213.
3. Cohn, M. S., Tabor, C. W., Tabor, H., and Wickner, R. B. (1978): *J. Biol. Chem.*, 253: 5225-5227.
4. Cohn, M. S., Tabor, C. W., and Tabor, H. (1980): *J. Bacteriol.*, 142: 791-799.
5. Hosaka, K., and Yamashita, S. (1981): *Eur. J. Biochem.*, 116: 1-6.
6. Kameji, T., Murakami, Y., Fujita, K., Noguchi, T., and Hayashi, S. (1981): *Med. Biol.*, 59: 296-299.
7. Pegg, A. E. (1982): personal communication.
8. Pritchard, M. L., Seely, J. E., Pösö, H., Jefferson, L. S., and Pegg, A. E. (1981): *Biochem. Biophys. Res. Commun.*, 100: 1597-1603.
9. Tabor, C. W. (1981): *Med. Biol.*, 59: 272-278.
10. Tabor, C. W., Tabor, H., Tyagi, A. K., and Cohn, M. S. (1982): *Fed. Proc.*, 41: in press.
11. Tyagi, A. K., Tabor, C. W., and Tabor, H. (1981): *J. Biol. Chem.*, 256: 12156-12163.
12. Whitney, P. A., and Morris, D. R. (1978): *J. Bacteriol.*, 134: 214-220.

Advances in Polyamine Research, Vol. 4, edited by
U. Bachrach, A. Kaye, and R. Chayen. Raven Press,
New York © 1983.

Unusual Polyamines in an Extreme Thermophile, *Thermus Thermophilus*

Tairo Oshima

Mitsubishi-Kasei Institute of Life Sciences, Machida, Tokyo 194, Japan

EXTREME THERMOPHILE AND *in vitro* PROTEIN SYNTHESIS

An extreme thermophile, *Thermus thermophilus* strain HB8(=ATCC 27634), isolated from a hot spring in Japan, is capable of growing in a temperature range of 47-85° (14). The cell constituents, such as enzymes, proteins, nucleic acids, ribosomes, membranes and a phage of the thermophile are heat resistant without exception (11,15).

When *in vitro* protein synthesis was investigated using a cell-free extract of the thermophile, no activity was found at the physiological temperatures such as 65° or 75°. Since the components involved in protein synthesis such as soluble enzymes, factors, tRNA, and ribosomes were stable to heat, this finding suggested the existence of some factor that plays a crucial role in the thermophile, but is missing in the cell-free system. It was found that the cofactor is polyamine (6). The polyamine requirement is specific and essential.

The protein synthesis at high temperatures was restored by the addition of spermine. The optimum concentration of spermine was 3 mM. The action mechanism of spermine was not to protect any single component of the protein synthesizing machinery from heat denaturation, but to maintain the active conformation of the ribosome-mRNA-aminoacyl tRNA ternary complex (6).

The polyamine requirement was rather specific. Magnesium ion, putrescine and other diamines were no effect on the reaction. Spermidine was less effective. These results urged the author to investigate polyamine composition of *T. thermophilus*.

ISOLATION AND IDENTIFICATION

Polyamines in the cells of *T. thermophilus* were extracted with 5% trichloroacetic acid, and separated with a cation exchange resin column chromatography. So called "long column" in an automatic amino acid analyzer was used for the chromatography. The chromatography was repeated until the pure compound was obtained. Polyamine hydrochloride salt can be crystallized by adding an excess amount of ethanol : methanol mixture (1 : 1) to the concentrated solution, or by cooling a hot 50% ethanol solution. For details of the isolation procedures, see ref. 10.

The chemical structure of a polyamine isolated from the thermophilic cells was easily determined by measuring its ^1H NMR, ^{13}C NMR and mass spectra. In ^1H NMR spectrum, CH_2 groups adjoining to a nitrogen atom give signals at around 3.2 ppm, and the internal CH_2 groups of propyl and butyl groups give signals at around 2.2 ppm and 1.8 ppm, respectively. In ^{13}C NMR spectrum, signal(s) at 37-39 ppm can be assigned to carbons adjoining to an amino group, and those at 45-48 ppm to carbons adjoining an aza group. The internal carbon atoms of propyl or butyl group(s) give signal(s) at 22-28 ppm. Thus the chemical structure can be speculated based on NMR spectra.

The proposed structure can be examined by the mass spectrum. Since N-C bond is generally more stable than C-C bond in mass fragmentation, a molecule containing an aminobutyl terminal gives three fragment peaks, m-30(NH_2CH_2), m-44($NHCH_2CH_2$), and m-58 ($NH_2CH_2CH_2CH_2$), as major peaks along with the molecular peak. In contrast, a molecule containing only aminopropyl terminals may not give a peak correspond to m-58.

An example is shown in Table 1 (12). An unknown polyamine was extracted from *T. thermophilus* cells grown at 80°, and was purified by using a cation exchange resin (CK-10S) column. The ^1H NMR, ^{13}C NMR and mass data of the isolated polyamine (hydrochloride salt) are summarized in Table 1. ^1H NMR data suggest that the carbon skeltons of the molecule consist of only propyl groups. ^{13}C NMR suggest that the ratio of carbon atoms adjacent to an amino nitrogen : carbons adjacent to an aza nitrogen : internal carbons is 1 : 3 : 2. Based on these observations, a structure, $NH_2(CH_2)_3NH(CH_2)_3NH(CH_2)_3NH(CH_2)_3NH_2$ can be speculated. Mass data in Table 1 strongly support this structure.

The chemical structure can finally be confirmed by comparison of the isolated polyamine with the authentic compound. For this purpose, IR spectrum would provide the simplest and most sensitive way.

Table 1. Spectral data for caldopentamine

	Signals observed	Relative intensity	Possible assignment[a]
^{1}H NMR	2.1 ppm	1	(b)+(e)
	3.2 ppm	2	(a)+(c)+(d)+(f)
^{13}C NMR	26.8 ppm	1	(e)
	28.0 ppm	1	(b)
	38.8 ppm	1	(a)
	46.4 ppm	1	(c)
	46.8 ppm	2	(d)+(f)
Mass[b]	245	100	Molecular peak
	215	30	m-(NH_2CH_2)
	201	42	m-($NH_2CH_2CH_2$)

[a] (a) (b) (c) (d) (e) (f)
NH_2 CH_2 CH_2 CH_2 NH CH_2 CH_2 CH_2
NH
NH_2 CH_2 CH_2 CH_2 NH CH_2 CH_2 CH_2

[b] Peaks above m/e170 are listed.

UNUSUAL POLYAMINES IN THE EXTREME THERMOPHILE

The extreme thermophile, *T. thermophilus*, contains several unusual polyamines. FIG. 1 shows the polyamine composition of cells grown at low temperature (65°), the optimum temperature (75°), and high temperature (80°) in a synthetic medium. Chemical structures of the polyamines in FIG. 1 are summarized in Table 2 (8-10,13). Spermine was not detected in the cells grown at 75° (10).

The polyamine composition of the extreme thermophile depends on the growth temperature as shown in FIG. 1. At low temperature, the cells contained more triamines. The cells at high temperature contained more tetraamines and pentaamines. Among tetraamines, thermine content was positively correlated to the growth temperature, whereas thermospermine content was rather negatively correlated to the environmental temperature. Likewise caldopentamine increased with the raise of the growth temperature.

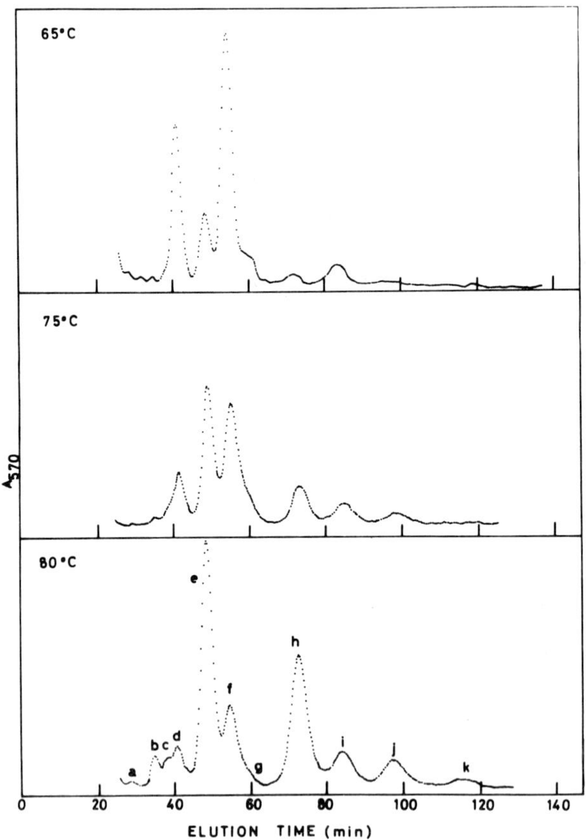

FIG. 1., Polyamine compositions of *T. thermophilus*. The cells were grown at 65°, 75° or 80°C. Names of polyamines are a = putrescine, b = norspermidine, c = spermidine, d = sym-homospermidine, e = thermine, f = thermospermine, g = unknown, h = caldopentamine, i = homocaldopentamine, j and k = unknown. Polyamines were analyzed using a CK-10S ion exchange column developed with potassium citrate-HCl buffer, pH 6, containing 2.01 M KCl and 5% ethanol (v/v).

Table 2. Polyamines found in *T. thermophilus*

Chemical formula	Trivial name (Systematic name)
$NH_2(CH_2)_3NH(CH_2)_3NH_2$	Norspermidine[a] (1,7-Diamino-4-azaheptane)
$NH_2(CH_2)_3NH(CH_2)_4NH_2$	Spermidine (1,8-Diamino-4-azaoctane)
$NH_2(CH_2)_4NH(CH_2)_4NH_2$	sym-Homospermidine (1,9-Diamino-5-azanonane)
$NH_2(CH_2)_3NH(CH_2)_3NH(CH_2)_3NH_2$	Thermine (1,11-Diamino-4,8-diazaundecane)
$NH_2(CH_2)_3NH(CH_2)_3NH(CH_2)_4NH_2$	Thermospermine (1,12-diamino-4,8-diaza-dodecane)
$NH_2(CH_2)_3NH(CH_2)_3NH(CH_2)_3NH(CH_2)_3NH_2$	Caldopentamine (1,15-diamino-4,8,12-triaza-pentadecane)
$NH_2(CH_2)_3NH(CH_2)_3NH(CH_2)_3NH(CH_2)_4NH_2$	Homocaldopentamine (1,16-Diamino-4,8,12-tri-azahexadecane)

[a]Also call "caldine".

Thermine and thermospermine supported *in vitro* protein synthesis at high temperature. Spermine and thermospermine showed the highest activity although spermine is absent in the thermophile. Protein synthesis at 65° in the presence of thermine was about two-thirds of that in the presence of thermospermine. Younger cells contained more polyamines, whereas the cells at stationary stage contained only a small amount of polyamines. Thus it was speculated that *in vivo* protein synthesis is regulated by the concentrations of these tetraamines (9).

CHEMICAL SYNTHESIS

Thermine and caldopentamine can be synthesized by treating acrylonitrile with norspermidine and thermine, respectively, in ethanol. The reaction product was reduced with $NaBH_4$ in the presence of $CoCl_2$ at pH 6 (17).

$$R\text{-}NH_2 + CH_2{=}CHCN \rightarrow R\text{-}NHCH_2CH_2CN \xrightarrow[CoCl_2]{NaBH_4} R\text{-}NHCH_2CH_2CH_2NH_2$$

sym-Homospermidine, thermospermine and homocaldopentamine were synthesized by treating N-(4-bromobutyl)phthalimide(Aldrich Chemical) with putrescine, norspermidine and thermine, respectively, in 2-propanol at 105° for 16 h (7). The product was hydrolyzed in 6N HCl at 105° for 8 h.

$$R\text{-}NH_2 + Br(CH_2)_4N(CO)_2C_6H_4 \rightarrow R\text{-}NH(CH_2)_4N(CO)_2C_6H_4$$

$$\xrightarrow{HCl} R\text{-}NH(CH_2)_4NH_2 + (HOOC)_2C_6H_4$$

In every case, the product was separated from the unreacted raw materials and the by-products using a cation exchange column chromatography. It was often necessary to repeat the chromatography many times to obtain a pure preparation. The overall yield was around 20-25%.

DISTRIBUTION IN OTHER ORGANISMS

Extreme thermophiles belonging to the genus *Thermus* such as *T. thermophilus* strain HB27 and *T. aquaticus*, produced norspermidine, spermidine, thermine, thermospermine, and pentaamines. Norspermidine and thermine were also found in cells of extreme acido-thermophiles, *Caldariella acidophila* (1) and *Sulfolobus acidocaldarius* (unpublished), but *sym*-homospermidine and pentaamine were not detected in these cells. Moderate thermophiles such as *Bacillus stearothermophilus* contained no norspermidine,

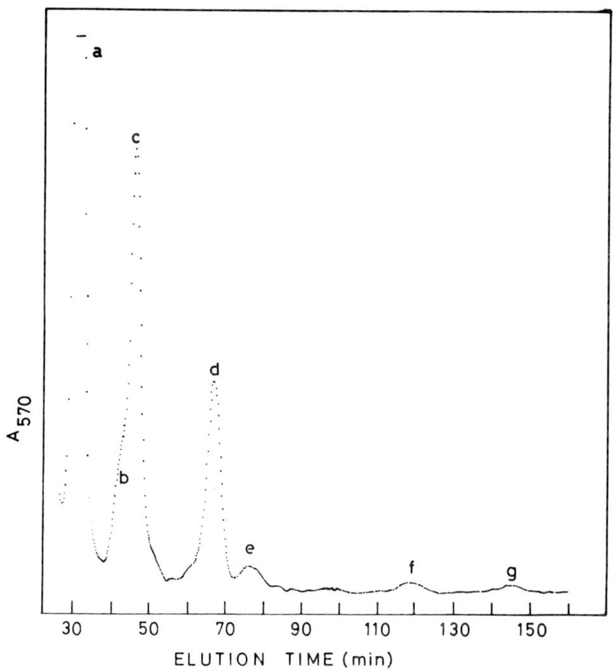

FIG. 2., Polyamine composition of *Periplaneta brunnea*. Larvae of the cockroach were homogenized in 5% trichloroacetic acid and the supernatant was analyzed using a CK-10S ion exchange column developed with sodium acetate buffer, pH 5, containing 2.5 M NaCl (final sodium concentration was 2.78 M) and 5% 2-propanol (v/v). Peaks were identified as a = putrescine, b = norspermidine, c = spermidine, d = thermine, e = spermine/thermospermine, f = caldopentamine, g = unknown.

thermine nor pentaamine.

Novel polyamines found in *T. thermophilus* were also identified in a variety of mesophilic organisms. Norspermidine and thermine are widely distributed in bacteria (22), algae (4,16,21) sea animals and anthropods (19,23). *sym*-Homospermidine is present in vertebrate animals (2), plants (5,20), algae (3,4,16,21), and bacteria (18,20).

Polyamine composition of a cockroach was qualitatively similar to that of the extreme thermophile as shown in FIG. 2. In the insect cells, norspermidine, thermine and caldopentamine were present in addition to putrescine, spermidine and (thermo)spermine. Recently the author succeeded in an attempt to establish an analytical method for the determination of spermine and thermospermine in mixed solutions. These tetraamines were separated on an analytical CK-10S ion exchange column chromatography developed with sodium borate buffer, pH 8 (unpublished data). The presence of thermospermine in the cockroach will be analyzed by this method.

It will be interesting to point out that some unidentified, strongly basic compounds were present in *T. thermophilus* cells as shown in FIG. 1 and in the cockroach as shown in FIG. 2.

REFERENCES

1. DeRosa, M., DeRosa, S., Gambacorta, A., Carteni-Farina, M., and Zappia, V. (1976): *Biochem. Biophys. Res. Commun.*, 69: 253-261.
2. Hamana, K., and Matsuzaki, S. (1979): *FEBS Lett.*, 99: 325-328.
3. Hamana, K., and Matsuzaki, S. (1982): *J. Biochem.* 91: 1321-1328.
4. Kneifel, H., Schuber, F., Aleksijevic, A., and Grove, J. (1978): *Biochem. Biophys. Res. Commun.*, 85: 42-46.
5. Kuttan, R., Radhakrishnan, A.N., Spande, T., and Witkop, B. (1971): *Biochemistry*, 10: 361-365.
6. Ohno-Iwashita, Y., Oshima, T., and Imahori, K. (1975): *Arch. Biochem. Biophys.*, 171: 490-499.
7. Okada, M., Kawashima, S., and Imahori, K. (1979): *J. Biochem.*, 85: 1235-1243.
8. Oshima, T. (1975): *Biochem. Biophys. Res. Commun.*, 63: 1093-1098.
9. Oshima, T. (1978): In: *Biochemistry of Thermophily*, edited by S.M. Friedman, pp.211-220, Academic Press, New York.
10. Oshima, T. (1979a): *J. Biol. Chem.*, 254: 8720-8722.
11. Oshima, T. (1979b): In: *Strategies of Microbial Life in Extreme Environments*, edited by M. Shilo, pp.455-469, Dahlem
12. Oshima, T. (1982): *J. Biol. Chem.*, in press.

13. Oshima, T., and Baba, M. (1981): *Biochem. Biophys. Res. Commun.*, 103: 156-160.
14. Oshima, T., and Imahori, K. (1974): *Intern. J. Syst. Bacteriol.*, 24: 102-112.
15. Oshima, T., Sakaki, Y., Wakayama, N., Watanabe, K., Ohashi, Z., and Nishimura, S. (1976): In: *Symposium on Enzymes and Proteins from Thermophilic Microorganisms*, edited by H. Zuber, pp.317-331, Birkhäuser-Verlag, Basel.
16. Rolle, I., Hobucher, H.-E., Kneifel, H., Paschold, B., Riepe, W., and Soeder, C.J. (1977): *Anal. Biochem.*, 77: 103-109.
17. Satoh, T., Suzuki, S., Suzuki, Y., Miyaji, Y., and Imai, Z. (1969): *Tetrahedron Lett.*, 4555-4558.
18. Smith, T.A. (1977): *Phytochem.*, 16: 278-279.
19. Stillway, L.W., and Walle, T. (1977): *Biochem. Biophys. Res. Commun.*, 77: 1103-1107.
20. Tait, G.H. (1979): *Biochem. Soc. Trans.*, 7: 199-201.
21. Villanueva, V.R., Adlakha, R.C., and Calvayrac, R. (1980): *Phytochem.*, 19: 787-790.
22. Yamamoto, S., Shinoda, S., and Makita, M. (1979) *Biochem. Biophys. Res. Commun.*, 87: 1102-1108.
23. Zappia, V., Porta, R., Carteni-Farina, M., DeRosa, M., and Gambacorta, A. (1978): *FEBS Lett.*, 94: 161-165.

Advances in Polyamine Research, Vol. 4, edited by U. Bachrach, A. Kaye, and R. Chayen. Raven Press, New York © 1983.

Ornithine Decarboxylase Activity from an Extremely Thermophilic Bacteria, *Clostridium Thermohydrosulfuricum*

Hannu Pösö and Lars Paulin

Research Laboratories of the State Alcohol Monopoly (Alko), SF-00101 Helsinki 10, Finland

The activity of ornithine decarboxylase (ODC) was detected first time in extracts of an thermophilic bacteria, Clostridium thermohydrosulfuricum. The temperature optimum of the thermoresistant ornithine decarboxylase was 55°C and the pH optimum was 7.5. It required pyridoxal phosphate and dithiothreitol for activity. The activity was stimulated by GTP and dGTP. Other analogues of GTP caused a smaller stimulation than GTP itself. Magic spot I and II inhibited powerfully the enzyme activity as did also ppGp and cppGp. The activity of ODC was highest during the logarithmic growth instead of arginine decarboxylase (ADC) which had very low activity during growth (1/10 from the corresponding ODC activity). ODC was inhibited irreversible by 50 mM difluoromethylornithine (DFMO) and ADC was inhibited irreversible by difluoroarginine (DFMA).

INTRODUCTION

Escherichia coli has two distinct ODC (biodegradative and biosynhetic and both have been purified to homogeneity (1,2). The other route to synthesize putrescine in E. coli is the decarboxylation of arginine (9). During the growth, however, E. coli derives most of its polyamines from putrescine, indicating that arginine pathway is only an alternative route when arginine is exogenously supplied (11).

Normal mesophilic bacteria contains only putrescine or spermidine (11). More recently it has been discovered that thermophilic bacteria (growing between 60 to 80°C) contain "odd" polyamines, such as thermine and thermospermine (10). Since there are so striking differences between the distribution of polyamines in bacteria depending on the growth temperature it is possible that odd polyamines are necessary for the thermostability of thermophiles. So, it was interesting to study the occurrence of ODC in an extremely thermophilic bacteria, Cl. thermohydrosulfuricum. It appears the ODC activity is somehow connected to the growth of the the thermophile and its activity is strongly regulated in vitro by GTP and its analogues.

MATERIALS AND METHODS

Cultivation of Clostridium thermohydrosulfuricum

Cl. thermohydrosulfuricum was grown under anaerobic conditions as described in (8). Cell extracts were prepared by using X-press.

Analytical methods.
ODC activity (4) and ADC activity (12) were measured as described earlier.
Chemicals.
L-(1-^{14}C)Ornithine (56 Ci/mol) and L-(U-^{14}C)ornithine (285 Ci/mol) were from Radiochemical Centre. DL-(1-^{14}C)Arginine (20 Ci/mol was purchased from C.E.A. Research International. DL-α-difluoromethylornithine and DL-α-difluoromethylarginine were a generous gift from Merrell Research Centre, Cincinnati. NTPs, dNTPs and all derivatives of nucleoside phosphates were from PL Biochemicals.

RESULTS AND DISCUSSION

pH and temperature dependence of ornithine decarboxylase

The enzyme had as pH-optimum at 7.5 with or without GTP (a stimulator of the enzyme activity in vitro; 4). When the activity of ODC was measured in crude dialyzed extracts at optimum pH it had highest activity at the temperature of 55°C (results not shown).

Pyridoxal phosphate and thiolreagent requirement. Pyridoxal-5′-phosphate was necessary for the full activity and 2 μM concentration of PLP yielded the maximum stimulation. L-canaline (0.5-1 mM) totally inhibited the enzyme activity, thus showing that PLP is the coenzyme (results not shown).
When dithiothreitol was omitted from the enzyme solution and assay mixture, ODC activity was 10 % of the control. The addition of dithiothreitol (2 mM) to either the assay mixture or the enzyme solution totally restored the activity. This indicates that the thermoresistant ODC resembles the corresponding enzyme from E. coli, which is also PLP and thioldependent enzyme (11).

The activity of ornithine decarboxylase during the growth of Cl. thermohydrosulfuricum. As shown in Fig. 1, the activity of ODC was highest at early periods of the growth, the peak being in the middle of the logarithmic growth. This is in contrast to E. coli (9) where the activity of ODC remains constant during the growth. The activity of ADC did not change during the growth and it was always less than 10 % of ODC activity (results not shown).

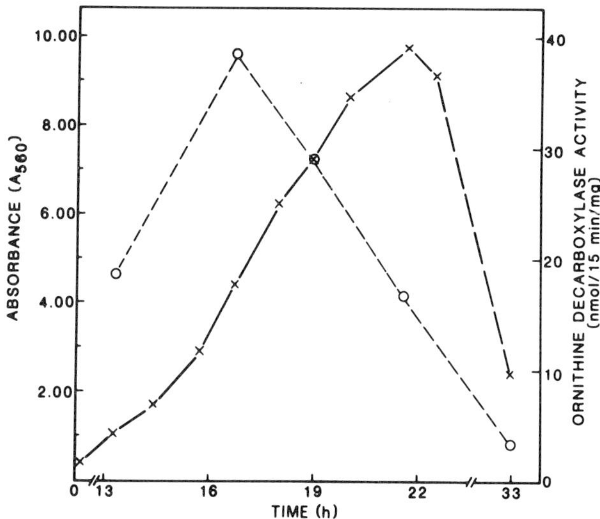

FIG. 1. The activity of ODC during the growth of the bacteria
x——x growth of the bacteria
o——o ornithine decarboxylase activity

So, it is possible that increased polyamine synthesis is needed in the thermophile during fast growth and that ODC is the main route to synthesize putrescine.

The effect of GTP, analogues of GTP, other NTPs or dNTPs on ODC activity. TABLE 1 lists the effect of different nucleotides on ODC activity. GTP and dGTP were the most potent activators of the enzyme. The change in the triphosphate part of GTP lead to a smaller activation of the enzyme than caused by GTP/dGTP. It appeared that when triphosphate part was lengthened with phosphate(s) $G(P)_4$, $G(P)_5$ or modified with a methylene group, the ability to activate the enzyme was decreased. The specificity of the base part of GTP seemed also to be very important in the stimulation of the enzyme. 7-met-GTP did not stimulate ODC at all. ITP, lacking the amino group on carbon 2 of GTP was much less effective than GTP and ATP was inactive (TABLE 1). The data presented here suggests that two or more sites are needed for the binding of GTP to the protein.

Inhibition of ODC by the analogues of GTP having a substitution at 3′-carbon

Fig. 2 shows the effect of the "magic spot" I and II (ppGpp and pppGpp; 3), ppGp and ppApp on ODC activity. In agreement with the previous report (5) it was found that ppGpp is a potent inhibitor of the enzyme *in vitro*. As a new finding it was noticed that the corresponding triphosphate derivative (pppGpp) was as good inhibitor as ppGpp. Since also ppGp was clearly inhibitory to the activity (Fig. 2) it appears that the 3′-carbon in the ribose molecule could be important for the binding on GTP to ODC.

To elucidate further the possibility to inhibit ODC by modification of the 3′-carbon, the effect of close analogues of magic spots were tested. It was found that guanosine-5′-diphosphate, 2′, 3′-cyclic phosphate (cppGp) was as good inhibitor as pppGpp and ppGpp (results not shown). The structural specificity of the ribose molecule (triphosphate at 5′-carbon and hydroxyl group at 3′-carbon) suggests that it may be possible to make compounds resembling triphosphate-ribose. That will inhibit the bacterial ODC. Such compounds could be used to inhibit the synthesis of polyamines in bacteria and to study the metabolic consequences of inhibition.

TABLE 1. The effect of GTP and its analogues on ODC activity

Compound	Conc. (μM)	ODC act. (%)	Compound	Conc. (μM)	ODC act. (%)
None	-	100	$G(P)_4$	10	245
GTP	10	390	$G(P)_5$	10	220
GDP	10	307	7-met-GTP	10	160
GMP	10	160	dGTP	10	450
α,β-GTP	10	163	ITP	10	168
			ATP	10	80

The compounds were tested by addition to the standard mixture at concentrations shown. The percentage change in the activity of ODC due the addition is shown. α,β-GTP, α,β-methyleneguanosine-5′-triphosphate, $G(P)_4$, guanosine-5′-tetraphosphate, $G(P)_5$, guanosine-5-pentaphosphate, 7-met-GTP, 7-methylguanosine-5′-triphosphate.

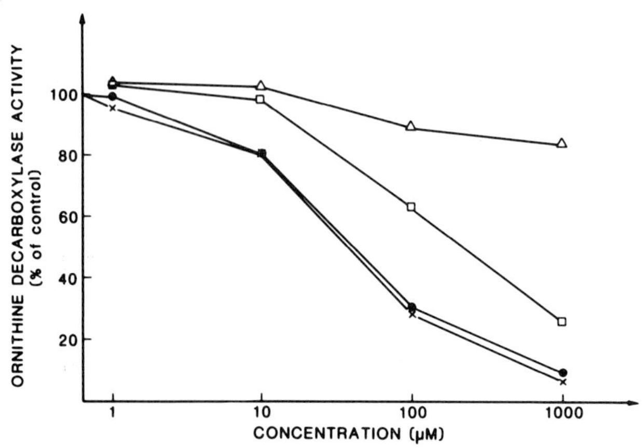

FIG. 2. The effect of the "magic" spots on ODC activity
x——x pppGpp, o——o ppGpp, —— ppGp, Δ——Δ ppApp

Effect of difluoromethylornithine (DFMO) on ODC activity. ODC was inhibited by 83 % when preincubated with 45 mM DFMO (results not shown. In further experiments (TABLE 2) it was shown that in case of thermoresitant ODC the inhibition caused by DFMO was irreversible since the lost of the activity was not recovered after an exhaustive dialysis (TABLE 2). This finding is in striking contrast to the report by Kallio and McCann where high concentration of DFMO was needed to inhibit ODC from *E. coli*, but the inhibition was reversible (6).

TABLE 2. The effect of DFMO on ODC activity

Treatment of the enzyme	ODC activity (nmol/mg/15 min)	Inhibition (%)
Control	58	-
+ DFMO	21	64
Dialyzed control	63	-
Dialyzed + DFMO	14	78

Enzyme was incubated with (+ DFMO) or without (Control) 50 mM DFMO 30 min at 55°C in standard reaction mixture and then assayed or dialyzed against 1000 volumes of incubation buffer for 48 h and then assayed.

The effect of difluoromethylarginine (DFMA) on ADC activity. TABLE 3 shows the effect of DFMA on the activity of ADC. The compound is an excellent inhibitor of the activity and the loss was irreversible since the activity was not recovered after dialysis. This finding confirms the results of Kallio and McCann (7).

TABLE 3. The effect of DFMA on ADC activity

Treatment of the enzyme	ADC activity (nmol/mg/15 min)	Inhibition (%)
Control	1.60	-
+ DFMA	0.32	80
Dialyzed control	1.65	-
Dialyzed + DFMA	0.24	85

Details as in TABLE 2.

The effect of polyamines on ODC activity. When the effect of putrescine, putreanine, spermidine and spermine was tested *in vitro*, there was no inhibition caused by these compounds up to 10 mM concentration. Thermine which is a new polyamine found in thermophilic bacteria (10) and also in *Cl. thermohydrosulfuricum* (Paulin and Pösö, unpublished) caused 35 % inhibition at 10 mM concentration (results not shown). In this connection it should be mentioned that when the formation of radiolabelled putrescine from (U-^{14}C)ornithine *in vitro* was followed, it was noticed that the formation of putrescine was stoichiometric to the amount of $^{14}CO_2$ released (results not shown). This indicates that the decarboxylation of ornithine at high temperature was a specific enzymatic process.

In conclusion it can be sead that that the thermoresistant ODC has common features with other bacterial ODCs. This means that the beginning of the synthesis of polyamines in thermophiles is "normal", indicating that the enzymes synthesizing higher polyamines in thermophiles would be very interesting to study.

REFERENCES

1. Applebaum, D., Dunlop, J.C. and Morris, D.R. (1977) *Biochemistry* 16: 1580-1584.
2. Applebaum, D., Sabo, D.L. and Morris, D.R. (1975) *Biochemistry* 14: 3675-3681.
3. Gashel, M. and Gallant, J. (1969) *Nature* 221: 838-841.
4. Hölttä, E., Jänne, J. and Pispa, J. (1972) *Biochem. Biophys. Res. Commun.* 47: 1165-1171.
5. Hölttä, E., Jänne, J. and Pispa, J. (1974) *Biochem. Biophys. Res. Commun.* 59: 1104-1114.
6. Kallio, A. and McCann, P.P. (1981) *Biochem. J.* 200: 69-75.
7. Kallio, A. and McCann, P.P. (1981) *Biochemistry* 20: 3163-3166.
8. Klaushofer, H. and Parkkinen, E. (1965) *Z. Zuckerind.* 15: 445-449.
9. Morris, D.R. and Fillingame, R.H. (1974) *Annu. Rev. Biochem.* 43: 303-325.
10. Oshima, T. and Baba, M. (1981) *Biochem. Biophys. Res. Commun.* 103: 156-160.
11. Pegg, A.E. and Williams-Ashman, H.G. (1981): In: *Polyamines in Biology and Medicine*, edited by D.R. Morris and L.J. Marton, pp. 3-42. Marcel Dekker, New York.
12. Wu, W.H. and Morris, D.R. (1973) *J. Biol. Chem.* 248: 1687-1695.

Advances in Polyamine Research, Vol. 4, edited by U. Bachrach, A. Kaye, and R. Chayen. Raven Press, New York © 1983.

Inhibition of Bacterial Polyamine Biosynthesis and Consequent Effects on Cell Proliferation

Alan J. Bitonti, *Arja Kallio, Peter P. McCann, and Albert Sjoerdsma

*Merrell Dow Research Center, Cincinnati, Ohio 45215; *Department of Biochemistry, University of Helsinki, 00170 Helsinki 17, Finland*

Bacteria are able to synthesize putrescine by at least two alternate routes (Fig. 1), unlike higher organisms which have a single synthetic pathway. Putrescine arises either directly from the decarboxylation of ornithine by ornithine decarboxylase (ODC) or by the decarboxylation of arginine by arginine decarboxylase (ADC), producing agmatine, which is then converted to putrescine by agmatinase or the sequential action of agmatine iminohydrolase and putrescine transcarbamoylase (20). Some bacteria contain two distinct ornithine decarboxylases and two distinct arginine decarboxylases which are classified as "biosynthetic" and "biodegradative". The biosynthetic enzymes' primary function is the formation of putrescine. Biosynthetic ODC is normally the major route for putrescine biosynthesis (20), but as will be shown below, biosynthetic ADC also functions to maintain intracellular putrescine concentrations. The biodegradative enzymes are induced by media of low pH and have as their primary function degradation of amino acids for control of intracellular acidity rather than maintenance of putrescine levels.

Bacteria synthesize spermidine but are unable to convert spermidine to spermine, unlike most eukaryotes. The enzyme spermidine synthase converts putrescine to spermidine by the transfer of a propylamine moiety from decarboxylated S-adenosylmethionine to one of the terminal amino groups of putrescine (25). Decarboxylated S-adenosylmethionine is produced from S-adenosylmethionine by the enzyme S-adenosylmethionine decarboxylase (18,25).

Many studies have linked increased polyamine biosynthesis with rapid cell growth (12,19,21). An exogenous requirement for polyamines has been demonstrated in the bacterium *Hemophilus influenzae* (4), and mutants of the fungi *Aspergillus nidulans* (22) and *Neurospora crassa* (2). Furthermore, mutants of *E. coli* which lack the polyamine biosynthetic enzymes and contain no polyamines, grow at rates considerably slower than wild-type strains (3,6,23,24). These same mutants grow at rates equal to those of wild-type strains if polyamines are supplied in the media.

Based on the above evidence we attempted to restrict the proliferation of a variety of cell types through inhibition of polyamine biosynthesis. It had been shown previously that inhibition of ornithine decarboxylase in cultured rat hepatoma cells by the addition of α-methylornithine to the medium resulted in slower cell growth (14). More recently, a potent enzyme-activated, irreversible inhibitor of ornithine decarboxylase, D,L-α-difluoromethylornithine (DFMO), was shown to more effectively limit the growth of cultured tumor cells (15) and tumors *in vivo* (12,21), as well as the proliferation of parasitic protozoa (16). That the inhibitory effects of DFMO are specific for polyamine biosynthesis was shown by the fact that the effects of DFMO are completely reversed by putrescine or spermidine (15,16).

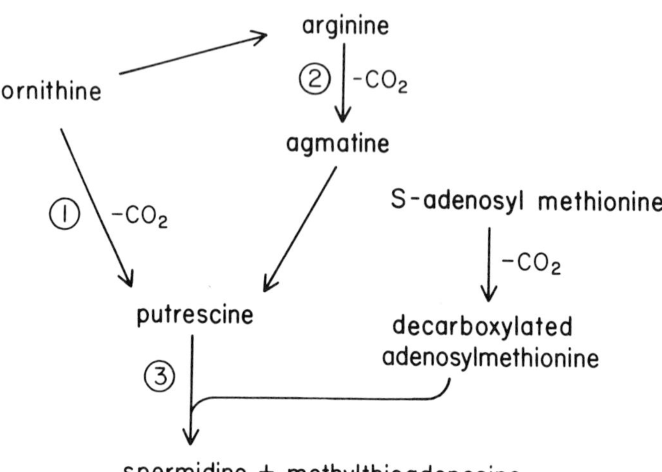

FIG. 1. Biosynthesis of putrescine and spermidine. The numbered reactions are catalyzed by the following enzymes: ① ornithine decarboxylase, ② arginine decarboxylase and ③ spermidine synthetase.

The successful use of DFMO to inhibit proliferation in eukaryotic cells encouraged the search for inhibitors of polyamine synthesis that would be effective in limiting bacterial growth. We can now describe the inhibition of what appear to be three critical steps (Fig. 1; 1,2,3) in bacterial polyamine biosynthesis, namely ornithine decarboxylase, arginine decarboxylase and spermidine synthase and the effects of combined inhibition on bacterial proliferation.

Inhibition of Bacterial Ornithine Decarboxylase

The decarboxylation of ornithine by ornithine decarboxylase (ODC) appears to be the major route for putrescine formation in bacteria (20) and therefore was a logical first target for inhibition. Difluoromethylornithine effectively inhibited the activities of both biosynthetic and biodegradative ODC extracted from *Pseudomonas aeruginosa* but had no effect on the same enzymes derived from *Escherichia coli* (Table 1; Ref. 8). ODC activities from *Klebsiella pneumoniae* were also unaffected by DFMO. Conversely, D,L-α-monofluoromethylputrescine, an analog of the enzyme product putrescine, proved to be an excellent inhibitor of biosynthetic ODC activity from *E. coli* and *K. pneumoniae* but had only a slight effect on ODC activity of *P. aeruginosa* (10). Recently D,L-α-monofluoromethylornithine (13) was shown to inhibit ODC activity derived from *E. coli*, *P. aeruginosa* and *K. pneumoniae* (1). The inhibition of ODC produced by difluoromethylornithine, monofluoromethylputrescine and monofluoromethylornithine in cell extracts was time-dependent, followed pseudo first-order kinetics and was slowed by the natural substrate L-ornithine, suggesting that the inhibitors were all active site-directed. Inactivation of ODC by the three compounds was apparently irreversible since extensive dialysis of the inactivated enzymes failed to restore catalytic activity. The kinetic parameters calculated for the three inhibitors of ODC activity can be found in references 1, 8 and 10.

It is noteworthy that there is marked species variation in sensitivity of bacterial ornithine decarboxylases to difluoromethylornithine and monofluoromethylputrescine whereas monofluoromethylornithine inhibits ODC activity from all three of the bacteria studied. The effectiveness of the compounds as inhibitors of bacterial ODC is apparently related to the number of fluorine atoms incorporated into the alpha-methyl group. At this time no generalizations can be made pertaining to the effectiveness of a given compound on various bacterial ornithine decarboxylases.

TABLE 1. Difluoromethylornithine inhibits biosynthetic ornithine decarboxylase activity and formation of putrescine in P. aeruginosa but not E. coli in vitro

Bacterial Source	Concn. of DFMO	Ornithine decarboxylase activity	Putrescine formation
	(mM)	(μmol/h/mg of protein)	(μmol/h/mg of protein)
E. coli 59	0	0.011	0.010
E. coli 59	2	0.010	0.011
P. aeruginosa	0	0.047	0.039
P. aeruginosa	2	<0.003	<0.003

Extracts of cells were preincubated without substrate in the absence or presence of 2 mM difluoromethylornithine (DFMO). After 20 min L-[U-^{14}C]ornithine was added, and the reaction was stopped with 1 M HCl after 20 min. The CO_2 released and putrescine formed were determined as described elsewhere (8).

Inhibition of Bacterial Arginine Decarboxylase

D,L-α-Difluoromethylarginine was shown to inhibit the E. coli and P. aeruginosa biosynthetic and biodegradative arginine decarboxylase (ADC) (9), the former being the key enzyme in the alternate route for putrescine biosynthesis. The inhibition was time-dependent, followed pseudo first-order kinetics and was slowed by the natural substrate L-arginine, indicating that the inhibitor was active site-directed (Fig. 2). The effect of difluoromethylarginine was apparently irreversible since extensive dialysis did not restore activity to the inactivated enzyme. Kinetic analysis of ADC inhibition by difluoromethylarginine was done by the method of Kitz and Wilson (11) as modified by Jung and Metcalf (7). From a plot of half-life of the enzyme activity vs. the inverse of difluoromethylarginine concentration (Fig. 3) it was possible to calculate dissociation constants (biosynthetic K_i=0.8 mM; biodegradative K_i=0.14 mM) and the half-life of each enzyme at an infinite concentration of difluoromethylarginine (biosynthetic $t_{1/2}$=1 min; biodegradative $t_{1/2}$=2.1 min).

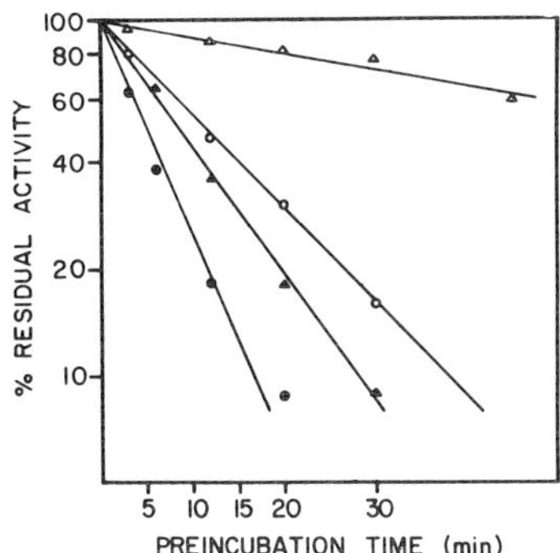

FIG. 2. Time-dependent loss of biosynthetic and biodegradative arginine decarboxylase activity upon incubation with DL-α-difluoromethylarginine and protection against enzyme inactivation by the substrate, L-arginine. Biosynthetic arginine decarboxylase obtained from E. coli was incubated with 0.1 mM DL-α-difluoromethylarginine in the absence (○) or presence of 1 mM arginine (△) as was the biodegradative enzyme (●; + 1 mM arginine, ▲). Aliquots of the incubates were taken at the indicated times, and the residual enzymatic activity was measured as described previously (9).

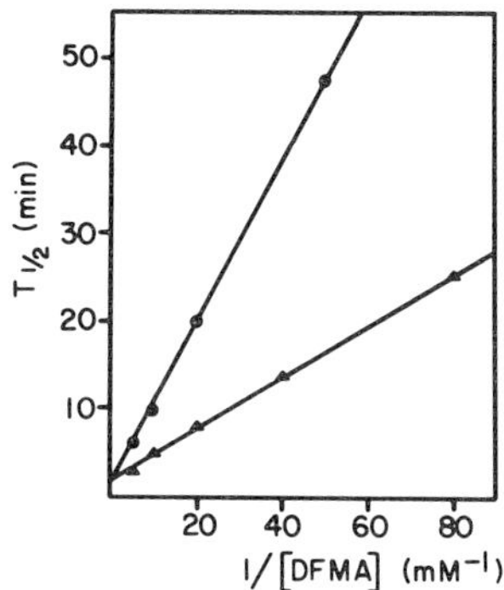

Fig. 3. Dependence of the half-life for inhibition of arginine decarboxylase on the inverse concentration of D,L-α-difluoromethylarginine. (●) E. coli biosynthetic arginine decarboxylase. (▲) E. coli biodegradative arginine decarboxylase. The rate of irreversible inhibition of both decarboxylases was followed by withdrawing samples for assays at different time points after mixing of enzyme and DL-α-difluoromethylarginine. The half-lives of the enzyme activity at different concentrations of DFMA (0.0125-0.2 mM) were determined according to Kitz & Wilson (11).

Inhibitor Effects in Intact Bacteria

Addition of difluoromethylornithine to growing P. aeruginosa cells resulted in a concentration-dependent decrease in ODC activity and a concomitant slight increase (35%) in ADC activity (Fig. 4). The increased ADC activity may be an attempt by the cell to maintain putrescine at a normal level by increasing an alternate route for synthesis of the diamine. Monofluoromethylputrescine added to E. coli caused a 95% reduction in ODC activity while having no effect on ADC activity (Table 2).

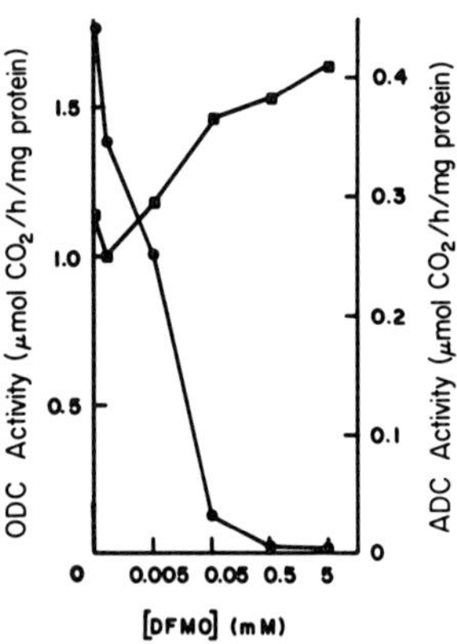

FIG. 4. Inhibition of ornithine decarboxylase and stimulation of arginine decarboxylase by several concentrations of DL-α-difluoromethylornithine in situ in P. aeruginosa cells. The cells were grown for 5h in minimal medium in the presence and in the absence of various concentrations of DL-α-difluoromethylornithine (DFMO), and thereafter assayed for enzymic activities as described (8). ●, Ornithine decarboxylase activity; ■, arginine decarboxylase activity.

When difluoromethylarginine was added to growing E. coli, ADC activity was decreased 85% while ODC activity was essentially unchanged (Table 2). Intracellular polyamine concentrations were relatively unaffected in the presence of either monofluoromethylputrescine or difluoromethylarginine (Table 3). When both monofluoromethylputrescine and difluoromethylarginine were added to growing E. coli, ODC and ADC activities were inhibited more than 90% (Table 2). As a consequence of simultaneous inhibition of both enzymes, the intracellular putrescine concentration of E. coli was decreased about 70% (Table 3) but spermidine concentrations actually increased 40%.

TABLE 2. In situ effects of difluoromethylarginine and monofluoromethylputrescine on ornithine and arginine decarboxylase activities of E. coli

Compound	Ornithine Decarboxylase (μmol/h/mg protein)	Arginine Decarboxylase (μmol/h/mg protein)
None	0.97	0.64
DFMA	1.28	0.03
MFMP	0.05	0.73
DFMA + MFMP	0.06	0.05

E. coli was grown in minimal media in the absence or presence of 1 mM difluoromethylarginine (DFMA), 1 mM monofluoromethylputrescine (MFMP) or a combination of these two inhibitors. After 4 hours cells were harvested and enzyme activities were measured as described (10).

TABLE 3. Effect of monofluoromethylputrescine (MFMP), difluoromethylarginine (DFMA) or their combination on polyamine levels in growing E. coli

Additions	Putrescine	Spermidine
	$nmol/10^8$ cells ± S.D.	
None	1.58 ± 0.15	0.34 ± 0.06
MFMP - 2.5 mM	1.29 ± 0.16	0.36 ± 0.06
DFMA - 2.5 mM	1.18 ± 0.15	0.42 ± 0.08
MFMP + DFMA - 2.5 mM	0.44 ± 0.01	0.48 ± 0.07

E. coli was grown as detailed in Table 2. Numbers of cells at initiation of experiment = $2.43 \pm .07 \times 10^8$ cells/ml; in controls and all experimental groups after 4 hours = $19.9 \pm .09 \times 10^8$ cells/ml. Results are mean values from 4 experiments.

Growth of E. coli or P. aeruginosa was unaffected by inhibition of either ODC or ADC alone or by simultaneous inhibition of both biosynthetic enzymes. Lack of effect on growth when only one of the decarboxylases is blocked is not unexpected since polyamine levels remain unchanged. When both ODC and ADC were inhibited it is assumed that growth was not restricted because there was a rise in spermidine concentration to offset the 70% decrease in putrescine. It was felt that if the increase in spermidine could be blocked in conjunction with ODC and ADC inhibition, then bacterial proliferation could be slowed. To achieve this we used dicyclohexylammonium sulfate whose inhibition of spermidine synthase in mammalian cell lines and tumors has recently been described (5). In collaboration with A.E. Pegg we have shown that dicyclohexylammonium sulfate is also a potent competitive inhibitor of spermidine synthase from E. coli and P. aeruginosa (unpublished observations).

To restrict bacterial growth we added a combination of three inhibitors to bacterial cultures: 1)monofluoromethylornithine to inhibit ODC; 2)difluoromethylarginine to inhibit ADC and 3)dicyclohexylammonium sulfate to inhibit spermidine synthase. The combination proved to be effective in reducing bacterial growth rates. The generation time (the time necessary for the number of cells to double) of E. coli increased from 41 min to 57 min and the generation time of P. aeruginosa increased from 35 min to 60 min (Figs. 5 and 6; Ref. 1). It was also evident that the three-drug combination caused a lengthening of the lag phase in P. aeruginosa (Fig. 6). The specificity of these effects for inhibition of polyamine biosynthesis was demonstrated by the fact that addition of 0.1 mM putrescine and 0.1 mM spermidine along with the three drug combination resulted in restoration of the normal growth rates in both E.coli and P.aeruginosa (Figs. 5 and 6).

Changes in intracellular polyamines, in particular spermidine, were shown to correlate well with changes in growth caused by monofluoromethylornithine, difluoromethylarginine and dicyclohexylammonium sulfate. Combination of MFMO and DFMA decreased the putrescine concentration in E. coli by 85% and in P. aeruginosa by 35% (Table 4), but spermidine levels were increased 42% and 20% in E. coli and P. aeruginosa, respectively. The reduction in putrescine caused by MFMO and DFMA was not sufficient to inhibit growth. Dicyclohexylammonium sulfate alone had no effect on putrescine and only a minimal effect on the spermidine level in E. coli, but decreased spermidine in P. aeruginosa >90% and apparently caused a slight decrease in putrescine (20%). It is noteworthy that DCHA had no effect on E. coli growth but markedly inhibited the growth of P. aeruginosa, the growth inhibition reflecting the greater reduction in polyamines in P. aeruginosa as compared to E. coli. The three-drug combination (MFMO, DFMA and DCHA) effectively reduced both putrescine (34% of control) and spermidine (50% of control) in E. coli and decreased the rate of proliferation of this bacterium by 40%. Similarly, MFMO, DFMA and DCHA reduced putrescine (20% of control) and spermidine (5% of control) in P. aeruginosa and decreased the rate of growth by 70%. When 0.1 mM putrescine and 0.1 mM spermidine were added to cultures of E. coli and P. aeruginosa, along with the three-drug combination, there was uptake of putrescine with intracellular putrescine returning to 40-50% of control, while enough spermidine was absorbed to return intracellular spermidine to control levels. The rates of growth of both bacteria were equal to those of control cultures when putrescine and spermidine were added to inhibited cultures.

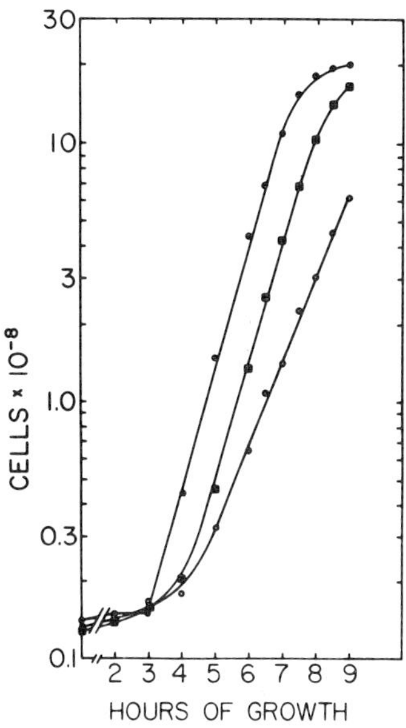

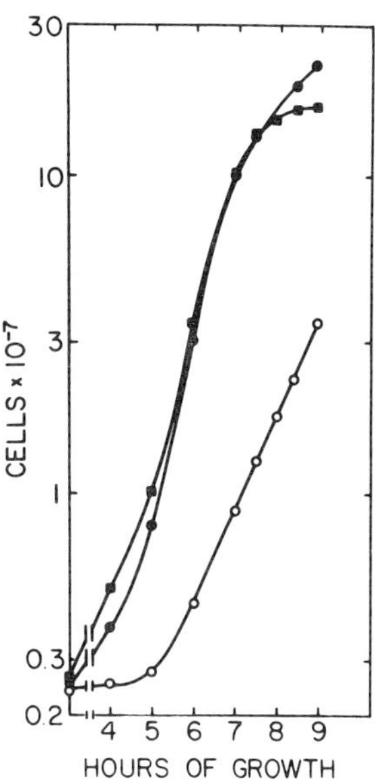

FIG. 5. Inhibition of E. coli growth by a combination of monofluoromethylornithine (MFMO), difluoromethylarginine (DFMA) and dicyclohexylammonium sulfate (DCHA) and reversal by polyamines. E. coli was grown without additions (●) or with 2 mM MFMO, 2.5 mM DFMA and 10 mM DCHA without (○) or with (■) 0.1 mM putrescine and 0.1 mM spermidine. Generation times during the exponential phase of growth were calculated to be: control (●), 41 min.; MFMO/DFMA/DCHA (○), 57 min.; MFMO/DFMA/DCHA plus putrescine and spermidine (■), 43 min. The experiment was repeated 5 times and results from the separate experiments were similar. Data are from a single representative experiment.

FIG. 6. Inhibition of P. aeruginosa growth by a bination of monofluoromethylornithine (MFMO), difluoromethylarginine (DFMA) and dicyclohexylammonium sulfate (DCHA) and reversal by polyamines. P. aeruginosa was grown without additions (●) or with 2 mM MFMO, 2.5 mM DFMA and 5 mM DCHA without (○) or with (■) 0.1 mM putrescine and 0.1 mM spermidine. Generation times during the exponential phase of growth were calculated to be: control (●), 35 min.; MFMO/DFMA/DCHA (○), 60 min.; MFMO/DFMA/DCHA plus putrescine and spermidine (■), 37 min. The experiment was repeated 3 times and results from the separate experiments were similar. Data shown are from a single representative experiment.

TABLE 4. Effect of monofluoromethylornithine (MFMO), difluoromethylarginine (DFMA) and dicyclohexylammonium sulfate (DCHA) on putrescine and spermidine levels in E. coli and P. aeruginosa

	E. coli		P. aeruginosa	
	putrescine	spermidine	putrescine	spermidine
	nmol/10^8 cells (% control)			
Control	1.70 (100)	0.26 (100)	3.11 (100)	0.64 (100)
MFMO+DFMA	0.24 (14)	0.37 (142)	2.04 (66)	0.76 (120)
DCHA	1.68 (99)	0.22 (85)	2.44 (78)	0.05 (8)
MFMO+DFMA+DCHA	0.57 (34)	0.13 (50)	0.63 (20)	0.03 (5)
MFMO+DFMA+DCHA + putrescine + spermidine	0.72 (42)	0.27 (104)	1.51 (49)	0.66 (103)

Drug concentrations were: MFMO, 2 mM; DFMA, 2.5 mM; DCHA, 10 mM for E. coli and 5 mM for P. aeruginosa. Cells were grown for 9 hours.

DISCUSSION

It is evident that inhibition of polyamine biosynthesis in bacteria is more complicated than in eukaryotic cells where there is only a single enzyme, ornithine decarboxylase, for putrescine synthesis. The alternate arginine decarboxylase pathway to putrescine maintains putrescine levels when almost 100% of ornithine decarboxylase activity is inhibited. Because of the two synthetic routes a combination of inhibitors of both decarboxylase activities must be used to decrease putrescine. When putrescine is markedly reduced, spermidine is increased, offsetting the effect on putrescine and necessitating the use of a spermidine synthase inhibitor to depress spermidine levels. Only when spermidine levels are reduced is growth restricted.

In previous studies it was shown that mutant E. coli cells, unable to synthesize polyamines due to the lack of the biosynthetic enzymes, grow slower than wild-type strains (3,6,23,24). In the presence of exogenous polyamines these mutants grow at the same rate as wild-type strains, thus establishing a clear-cut dependence on polyamines for optimal bacterial growth. Even in the E. coli mutants which completely lack putrescine and spermidine (23,24) growth is only reduced by about two thirds. Using a wild-type P. aeruginosa we have been able to restrict proliferation to approximately the same extent without complete depletion of polyamines. More complete depletion of the polyamines by chemical means in P. aeruginosa might result in even slower growth. There may also be other bacteria that have an absolute dependence on polyamines, requiring them not only for optimal growth but for any growth at all. Evidence for this possibility exists in the literature (4). We have also recently shown that the combined enzyme inhibition caused by MFMO, DFMA and DCHA may be bactericidal to Serratia marcescens (unpublished observations). Therefore it appears worthwhile to test a wide variety of bacteria looking for specific polyamine dependence. Further studies are currently in progress to test the hypothesis that bacterial polyamine synthesis is a target amenable to pharmacologic intervention as has been the case for polyamine biosynthesis in eukaryotic cells.

REFERENCES

1. Bitonti, A.J., McCann, P.P., and Sjoerdsma, A. (1982): Biochem. J., in press.
2. Deters, J., Miskimen, J. and McDongall, K.J. (1974): Genetics 77:S16-S17.
3. Hafner, E.W., Tabor, C.W. and Tabor, H. (1979): J. Biol. Chem. 254: 12419-12426.
4. Herbst, E.J. and Snell, E.E. (1948): J. Biol. Chem. 176:989-990.
5. Hibasami, H., Tanaka, M., Nagai, J. and Ikeda, T. (1980): FEBS Letters 166: 99-101.
6. Jorstad, C.M., Harada, J.J. and Morris, D.R., (1980): J. Bacteriol. 141: 456-463.
7. Jung, M.J. and Metcalf, B.W. (1975): Biochem. Biophys. Res. Comm. 67:301-306.
8. Kallio, A. and McCann, P.P. (1981): Biochem. J. 200:69-75.
9. Kallio, A., McCann, P.P. and Bey, P. (1981): Biochemistry 20:3163-3166.
10. Kallio, A., McCann, P.P. and Bey, P. (1982): Biochem. J. 204:771-775.
11. Kitz, R. and Wilson, I.B. (1962): J. Biol. Chem. 237:3245-3249.
12. Koch-Weser, J., Schechter, P.J., Bey, P., Danzin, C., Fozard, J.R., Jung, M.J., Mamont, P.S., Prakash, N.S., Seiler, N. and Sjoerdsma, A. (1981): In: Polyamines in Biology and Medicine, edited by D.R. Morris and L.J. Marton, pp. 437-453, Marcel Dekker, New York.
13. Kollonitsch, J., Patchett, A.A., Marburg, S., Maycock, A.L., Perkins, L.M., Doldouras, G.A., Duggan, D.E. and Aster, S.D. (1978): Nature 274:906-908.
14. Mamont, P.S., Böhlen, P., McCann, P.P., Bey, P., Schuber, F., and Tardif, C. (1976): Proc. Natl. Acad. Sci. USA 73:1626-1630.
15. Mamont, P.S., Duchesne, M.C., Grove, J., and Bey, P. (1978): Biochem. Biophys. Res. Commun. 81:58-66.

16. McCann, P.P., Bacchi, C.J., Hanson, W.L., Cain, G.D., Nathan, H.C., Hutner, S. and Sjoerdsma, A. (1981): In: Advances in Polyamine Research Vol. 3, edited by C.M. Caldarera, V. Zappia, and U. Bachrach, pp. 97-110, Raven Press, New York.
17. Metcalf, B.W., Bey, P., Danzin, C., Jung, M.J., Casara, P. and Vevert, J.P. (1978): J. Am. Chem. Soc. 100:2551-2553.
18. Morris, D.R. and Fillingame, R.H. (1974): Annu. Rev. Biochem., 43:303-325.
19. Pegg, A.E. and Coward, J.K. (1981): In: Advances in Polyamine Research Vol. 3, edited by C.M. Caldarera, V. Zappia, and U. Bachrach, pp. 153-162, Raven Press, New York.
20. Pegg, A.E. and Williams-Ashman, H.G. (1981): In: Polyamines in Biology and Medicine, edited by D.R. Morris and L.J. Martin, pp. 3-42, Marcel Dekker, New York.
21. Sjoerdsma, A. (1981): Clin. Pharmacol. Ther. 30:3-22.
22. Sneath, P.H.A. (1955): Nature 175:818.
23. Tabor, C.W., Tabor, H. and Hafner, E.W. (1978): J. Biol. Chem. 253:3671-3676.
24. Tabor, H. Tabor, C.W., Cohn, M. and Hafner, E.W. (1981): J. Bacteriol. 174:702-704.
25. Williams-Ashman, H.G. and Pegg, A.E. (1981): In: Polyamines in Biology and Medicine, edited by D.R. Morris and L.J. Marton, pp. 43-76, Marcel Dekker, New York.

Advances in Polyamine Research, Vol. 4, edited by U. Bachrach, A. Kaye, and R. Chayen. Raven Press, New York © 1983.

The Antiviral Effect of Inhibitors of Polyamine Metabolism

A. Stanley Tyms, Bandhana K. Rawal, Helen M. Naim, and John D. Williamson

Department of Medical Microbiology, St. Mary's Hospital Medical School, Paddington, London W2 1PG, United Kingdom

The availability of inhibitors of polyamine metabolism has led during the past decade to a recognition of the essential requirement for polyamines in the growth of animal cells. Earlier studies had shown that cell proliferation is accompanied by increased formation of putrescine due to a rapid elevation of ornithine decarboxylase activity followed in turn by an intracellular accumulation of the polyamines, spermidine and spermine (1). Although such observations demonstrated the concomitant nature of cell growth and polyamine biosynthesis, the physiological significance of this correlation remained unresolved. In 1972, however, the anti-tumour agent methylglyoxal bis(guanylhydrazone) (MGBG) was shown to be an efficient inhibitor of S-adenosylmethionine decarboxylase (19). This enzyme is required for the formation of spermidine and spermine in a complex series of biosynthetic events involving the successive transfer to putrescine of the aminopropyl group produced by the decarboxylation of S-adenosylmethionine. Fillingame & Morris (5) showed that DNA synthesis, which is normally increased in lymphocytes activated by concanavalin A, was markedly reduced in the presence of MGBG although both RNA and protein synthesis were unimpaired. More recently other inhibitors of polyamine metabolism have been developed, most notably -difluoromethylornithine (DFMO), a catalytically-activated, irreversible inhibitor with specific action against ornithine decarboxylase (16). Inhibition of this enzyme activity prevents the proliferation of cell cultures after endogenous polyamines have been depleted and the cell cycle is arrested before the initiation of DNA synthesis at the end of the G1 phase (12). The inhibitory effect can be reversed by the exogenous provision of putrescine, spermidine or spermine showing, therefore, that there is an absolute requirement for polyamines in cell proliferation.

Viruses are obligate, intracellular parasites and during their replication the infective genome modifies the metabolism of the host cell to produce virus-specific nucleic acid and proteins. Since polyamines are required for macromolecular synthesis during the multiplication of both prokaryotic and eukaryotic cells, it is reasonable to propose that these aliphatic bases should also be required for the replication of viruses. Polyamines have, in fact, been shown to be essential in the replication of different bacterial viruses but little is known of similar requirements in animal virus replication (14). Extrapolation from prokaryotic into eukaryotic systems is unwise, however, in view of the apparently unique association of spermine with animal cells. The replication of certain animal viruses in cell cultures is prevented by inhibitors of polyamine biosynthesis (4, 11, 20, 22, 25) but relatively few studies have been made of polyamine metabolism in virus-infected, animal cells. This is an unfortunate ommission since such models may answer fundamental questions of a general nature concerning the biological function(s) of polyamines. More specifically, altered polyamine metabolism which is consequent upon virus infection may be pertinent in relation to the pathogenesis of virus diseases. The anti-viral effect of polyamine inhibitors demonstrated *in vitro* also encourages the possibility of their use in a chemotherapeutic context.

This paper reports the results of our investigations of the anti-viral properties of polyamine inhibitors using cell cultures infected with different DNA-containing animal viruses. In these studies the dose-response curves have been obtained by the method of plaque or virus yield reduction. Certain characteristics which are typical of such studies are described using vaccinia virus-infected cultures as a model system. Further investigations have been made with four members of the Herpetoviridae: herpes simplex virus type 1 (HSV-1), herpes simplex virus type 2 (HSV-2), human and murine cytomegalovirus (CMV). Polyamine metabolism in mouse embryo fibroblasts infected with murine cytomegalovirus has been examined as a potential experimental model of the human disease.

MATERIALS AND METHODS

A continuous, human cell line, HeLa; a human diploid cell line, MRC-5; and primary cell cultures of mouse embryo fibroblasts (MEF) were prepared and maintained as described previously (21, 22, 25). The viruses used were: vaccinia virus (Lister strain); clinical isolates of HSV-1 and HSV-2; human CMV (AD169 and the 'West' strain); and murine CMV (Osborn strain). Virus stocks were prepared and infectivity titrations were carried out as described previously (21, 22, 25).

The effect of polyamine inhibitors on virus replication was assayed by the plaque reduction method (21) which has been used extensively for studies of other anti-viral compounds (2). In virus yield experiments the titres of infective, progeny virus were determined under one-step growth conditions.

Quantitative analysis of the benzoyl derivatives of polyamines recovered from uninfected and virus-infected cell cultures was made by high-pressure liquid chromatography (HPLC) based on the method of Redmond & Tseng (18).

MGBG was obtained from Aldrich Chemical Co. Inc., Milwaukee; α-methylornithine (α-MO) and DFMO were generous gifts from Centre de Recherche Merrell International, Strasbourg, France.

RESULTS

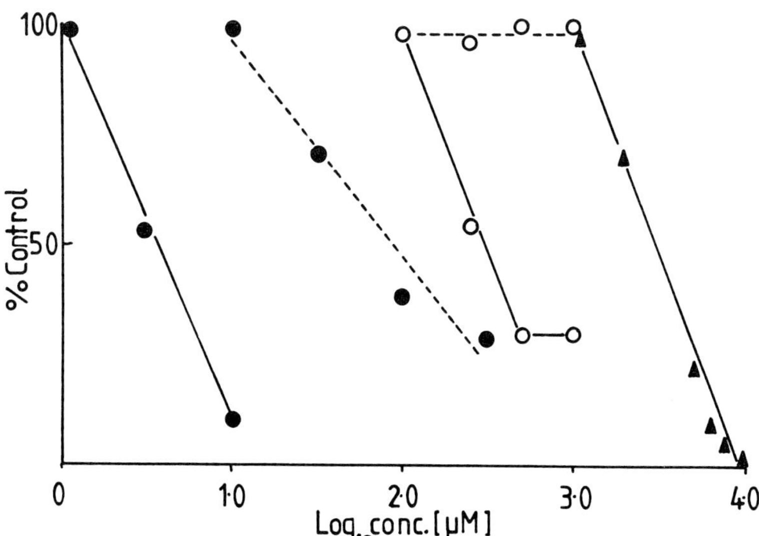

FIG.1. A comparison of the dose-response curves of the antiviral effect of MGBG and DFMO against vaccinia virus replication in HeLa cells (open symbols) and in MRC-5 cells (closed symbols). The broken lines refer to the virus sub-population which is resistant to MGBG in HeLa cells: triangular symbols refer to DFMO.

In an earlier investigation MGBG was shown to inhibit partially the production of infectious, progeny virus in vaccinia-infected HeLa cells (25). This effect is shown more clearly in the dose-response curve obtained (Fig.1). The virus yield decreased progressively until at 500 μM-MGBG a 70% reduction in virus yield had occurred but further decreases were not obtained with higher concentrations of the inhibitor. An ED_{50} value of 330 μM was calculated for the MGBG-sensitive virus population. However, repeated passage in HeLa cells maintained in the presence of 500 μM-MGBG selected a virus sub-population completely resistant to the compound up to the highest concentration without cytotoxic effect (Fig.1).

The anti-viral effect of MGBG against the vaccinia virus stock and the drug-resistant population generated in HeLa cells was also examined in MRC-5 cells, a human diploid cell line. In this system the ED_{50} value was 3.5 μM with unpassaged virus but a higher value of 79 μM was obtained with the MGBG-passaged virus population (Fig.1).

These results with vaccinia virus show that markedly different ED_{50} values can be obtained for MGBG against the same virus in different cell systems. In further studies, however, it has been found that both virus populations were sensitive to the effect of DFMO with similar ED_{50} values of 5.9-mM (results not shown) and 4.4 mM in HeLa or MRC-5 cells, respectively (Fig.1). There was no indication with this inhibitor of a drug-resistant population. The results obtained with MGBG show also the facility with which drug-resistant virus populations may be obtained. Similar observations have been made with diaminopropane, another inhibitor of polyamine metabolism (17), but α-MO has a complete inhibitory effect (our unpublished results).

Since one objective of these studies has been an assessment of the therapeutic potential of the anti-viral effect shown by polyamine inhibitors, further extensive investigations have been carried out with more important human pathogens. Earlier studies showed that the replication of human CMV in MRC-5 cells is inhibited by either MGBG or α-MO whereas HSV-1 and HSV-2 were unaffected (22). These different effects were consistent with the inhibition of polyamine biosynthesis in HSV-infected cells compared with the stimulation of such metabolism after human CMV infection. Thus, it appears that HSV replication may not require *de novo* polyamine synthesis but is dependant upon their availability within the host cell at the time of infection. Exhaustive depletion of polyamines, therefore, by exposure of cell cultures to appropriate inhibitors before infection could result in the subsequent inhibition of HSV replication. As shown in Table 1 the growth of HSV-2 was completely inhibited in MRC-5 cells treated with 100 μM-MGBG both prior to and after infection. When MGBG was present only from the time of infection, however, the inhibitory effect was markedly reduced and the lower virus yield compared with untreated controls may be due to an effect of MGBG on virus adsorption. MRC-5 cell cultures exposed to 100 μM-MGBG before infection but maintained subsequently in the absence of the compound produced a virus yield similar to that obtained with untreated, infected cultures. Unlike HSV-2, the replication of human CMV was completely inhibited by 10 μM-MGBG without treatment prior to infection (Table 1).

TABLE 1. The effect of MGBG on the replication of HSV-2 and human CMV in MRC-5 cells cultures

		Inhibitor treatment:		
	MGBG (μM)	Before infection*	After infection	Infectivity titre**
HSV-2	100	Yes	Yes	3.5×10^2
		Yes	No	2.0×10^7
		No	Yes	4.0×10^5
		No	No	2.4×10^7
Human CMV	10	No	Yes	$< 10^1$
		No	No	1.6×10^6

* MRC-5 cells were treated with 100 uM-MGBG for 18 hours
** Infectivity titrations (plaque-forming units/ml) were made at 25 hours p.i. (HSV-2) and 144 hours p.i. (human CMV).

These results are consistent with the hypothesis that the replication of HSV-2 is affected by the utilization of polyamines available within the host cell at the time of infection. In a separate study by Tuomi et al. (20) prolonged treatment of BHK-21 cell cultures with DFMO was shown to impair their subsequent ability to support the replication of HSV-2 in the continued presence of the inhibitor. In MRC-5 cell cultures, however, plaque formation by HSV-2 in dose-response experiments was unaffected in the presence of 10 mM-DFMO whereas this concentration completely inhibited the growth of human CMV (ED_{50}-3.2 mM).

The results presented above and elsewhere (23, 24), which have been obtained by in vitro studies, demonstrate the essential nature of polyamine biosynthesis in human CMV replication. In view of the species specificity exhibited by this virus, experimental studies in vivo are not possible. Murine CMV has biological properties similar to human CMV and the disease caused in mice may provide a suitable model of cytomegalovirus infections (9). Since murine CMV will not grow in either HeLa or MRC-5 cells, further studies have been made with primary mouse embryo fibroblast (MEF) cultures.

In a series of dose-response experiments plaque formation by murine CMV in MEF cultures was inhibited by MGBG and ED_{50} values

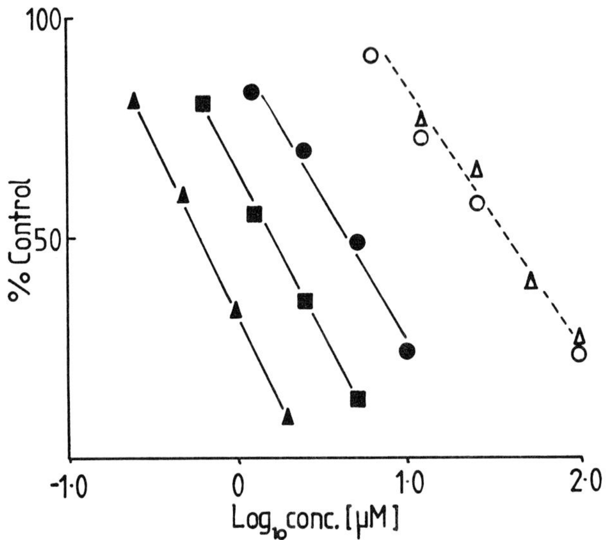

FIG. 2 Dose-response curves obtained by the plaque reduction method of the anti-viral effect of MGBG against murine CMV (closed symbols) and HSV-2 (open symbols) in MEF cultures.

between 0.6 and 5.6 μM were calculated. The results of typical experiments are shown in Fig. 2. By comparison, ED_{50} values between 32 and 40 μM were obtained with HSV-2 in this cell system. These results show that HSV-2 can grow in MEF cultures in the presence of MGBG at concentrations which inhibit murine CMV replication. In these studies the inhibitor was present from the time of infection only.

Unlike human CMV in MRC-5 cells, the replication of murine CMV in MEF cultures continued in the presence of 10 mM-DFMO. Although some inhibitory effect was obtained with higher concentrations of the compound accurate ED_{50} values could not be calculated. It has not been possible to demonstrate in this cell system any anti-viral effect of DFMO against HSV-2 replication.

Previous investigations using radiotracer techniques have shown that polyamine metabolism in MRC-5 cells is stimulated by human CMV infection (23). In order to make further comparisons of the polyamine requirement in the replication of human and murine CMV, quantitative analysis of spermidine and spermine recovered from infected cells has been carried out after their separation by HPLC as benzoylated derivatives (18). Appropriate cell cultures were infected with virus inocula calculated to achieve one-step growth conditions. The growth curve of human CMV in MRC-5 cells is characterized by an eclipse phase of 48 hours and from this time the production of infective, progeny virus increases to give maximum titres at about 96 hours post-infection (p.i.) which are maintained until at least 192 hours p.i. (23). The replication cycle of murine CMV in MEF cultures occupies a shorter time with an eclipse phase of about 12 hours duration, maximal production of progeny virus is attained at 48 hours p.i. and maintained until 96 hours p.i. (3).

In both human CMV and murine CMV-infected cultures polyamine biosynthesis has been examined at times which include the eclipse phase and the exponential phase of virus growth. It is apparent from Table 2 that spermidine and spermine levels in MRC-5 cells increased markedly after infection with human CMV. The accumulation of spermine results in a decrease in the spermidine: spermine ratio to reach a minimum value at 3 days p.i. which is the mid-point of the exponential phase of virus growth. After this time the spermidine: spermine ratio increased to values comparable with uninfected cells but the overall levels of polyamines remained higher in human CMV-infected cells. A similar pattern was obtained with MEF cultures infected with murine CMV (Table 3). The levels of spermine increased with a concomitant decrease in the spermidine: spermine ratio whereas the amount of both polyamines in uninfected MEF cultures remained relatively constant.

In contrast to the results obtained with the two cytomegaloviruses, polyamines in MEF cultures decreased markedly after infection with HSV-2. In typical experiments using 1.0×10^6 cells, 11.6 nmoles spermidine and 4.5 nmoles spermine were recovered from MEF cultures immediately after infection but these intracellular concentrations had fallen to 0.91 nmoles and 0.05 nmoles, respectively, at 18 hours p.i. At the end of the same period 10.8 nmoles spermidine and 5.3 nmoles spermine were measured in uninfected cultures. The later time is equivalent to the end of the exponential phase of the HSV-2 replication cycle in MEF cultures (3). A similar effect of HSV-2 infection on polyamine biosynthesis has been described in other host cell systems (14).

TABLE 2. Spermidine and spermine content (nmoles/10^6 cells) of uninfected and human CMV-infected MRC-5 cells at various times after infection

		Time (hours)				
		24	48	72	96	120
Uninfected	Spermidine (Sd)	3.2	10.0	9.8	8.0	8.0
	Spermine (Sp)	2.2	12.0	7.6	6.8	8.4
	Sd : Sp ratio	1.4	0.8	1.3	1.2	1.0
Infected	Spermidine	3.4	5.0	3.6	15.0	16.0
	Spermine	6.0	14.3	14.0	23.0	23.2
	Sd : Sp ratio	0.6	0.3	0.3	0.6	0.7

TABLE 3. Spermidine and spermine content (nmoles/10^6 cells) of uninfected and murine CMV-infected MEF cultures at various times after infection

		Time (hours)			
		0	15	24	36
Uninfected	Spermidine (Sd)	28.8	17.9	15.8	20.7
	Spermine (Sp)	5.1	5.7	4.8	7.9
	Sd : Sp ratio	5.7	3.1	3.3	2.6
Infected	Spermidine (Sd)	20.9	29.3	18.2	17.9
	Spermine (Sp)	5.3	10.5	7.9	10.5
	Sd : Sp ratio	4.0	2.8	2.3	1.7

DISCUSSION

The present study may help to resolve the confusion which has risen from earlier investigations into the anti-viral properties of polyamine inhibitors (20). It is clear that the nature of the host cell system can play an important part in both the quantitative and qualitative character of the results obtained. For example, it has been shown that there is a hundred-fold difference between the ED_{50} values for MGBG against vaccinia virus grown in HeLa or MRC-5 cells. The dose-response curves with vaccinia virus and human CMV in MRC-5 cells also demonstrate the dissimilar effects of MGBG against these viruses in the same cell system. Since such results show the ability of vaccinia virus to replicate in MRC-5 cells under conditions which completely inhibit human CMV replication, they provide important elements in the interpretation of the specificity of this anti-viral action. Further examples of the discriminatory activity of MGBG were seen with HSV-2 and both cytomegaloviruses in their appropriate host cells.

Although an anti-viral effect of DFMO has been demonstrated against vaccinia virus and human CMV in MRC-5 cells with ED_{50} values of about 5mM in each instance, the replication of HSV-2 was unaffected by 10mM-DFMO. Likewise, the same concentrations of the compound did not inhibit the replication of HSV-2 or murine CMV in MEF cultures. The growth of HSV-2 in BHK-21 cells has been shown to be affected by DFMO but only after exposure of the cell cultures to the inhibitor before infection (20). In the present study, such pre-treatment with DFMO as an integral part of the plaque reduction assay has not been effective. However, the anti-viral property of MGBG against HSV-2 could only be demonstrated if a similar regimen was used. These apparent discrepancies may be resolved by the observation that, unlike the results obtained in the human cell lines, 10mM-DFMO does not inhibit vaccinia virus replication in MEF cultures although some effect is manifest by higher concentrations of the compound (our unpublished results). It is conceivable that DFMO does not achieve a sufficient intra-cellular concentration in certain host cell systems to exert an inhibitory effect against HSV-2 replication. The generation in HeLa cells of a vaccinia virus population resistant to the effect of MGBG and the variation of the anti-viral effect of the same inhibitor against different strains of human CMV (24) suggests alternatively that virus-specific factors may determine sensitivity to polyamine inhibitors.

In addition to the innate, biological characteristics of either virus or host cell, it is obvious that the effect of specific inhibitors will depend upon the requirement for polyamines during virus replication. The anti-viral effects of DFMO can be related to increased or continued expression of ornithine decarboxylase activity in cells infected with vaccinia virus (8) or human CMV (10). In contrast, HSV-2 does not stimulate this enzyme activity (7) and in other HSV infections the synthesis of polyamines is inhibited (13, 22). The failure of DFMO to inhibit HSV-2 replication in the cell systems used in the present study also suggests that polyamine biosynthesis is not required for the replication of this virus. Other results show that spermidine and spermine are present in the virions (7) and cell-free synthesis of HSV DNA is influenced by the addition of polyamines at concentrations close to their cellular levels (6). Such requirements for HSV replication may be satisfied, however, by the utilization of polyamines extant at the time of infection. This conclusion is supported by the anti-viral effects of MGBG against HSV-2 which can be demonstrated after the presumed depletion of intracellular pools by prolonged exposure to polyamine inhibitors.

Direct examination of polyamine metabolism by the determination of spermidine and spermine levels has shown increased polyamine biosynthesis in MRC-5 cells infected with human CMV and in MEF cultures infected with murine CMV. These metabolic changes and the decreased levels of polyamines in HSV-2-infected MEF cultures are consistent with the different anti-viral effects of the polyamine inhibitors. Inhibition by MGBG of the replication of either cytomegalovirus can be prevented by the addition of spermidine but it has not been possible to demonstrate a similar reversal by putrescine of the anti-viral effect of DFMO against human CMV (our unpublished results). Although such equivocal results do not confirm the essential nature of de novo polyamine biosynthesis for the replication of human and murine CMV, the anti-viral properties of DFMO and MGBG indicate their potential as chemotherapeutic agents. A recent study has shown that MGBG has a more potent anti-viral effect in vitro than other drugs used clinically in the treatment of human CMV infections (24). The comparisons made in this study suggest that murine CMV serves as a suitable in vivo model to investigate the chemotherapy of cytomegalovirus infections by polyamine inhibitors.

ACKNOWLEDGEMENTS

We are grateful to Dr. G.H. Tait for help and advice with HPLC and to Mrs. J. Beh for preparation of this typescript. This work was supported by grants from the Medical Research Council and an anonymous donor.

REFERENCES

1. Bachrach, U. (1973): Function of Naturally Occurring Polyamines. Academic Press, New York.
2. Collins, P. & Bauer, D.J. (1977): Ann. N.Y. Acad. Sci., 284: 49-59.
3. Eizuru, Y., Minamishima, Y., Hirano, A. & Kitimura, T. (1978): Microbiol. Immunol., 22: 755-764.
4. Ferrari, W., Loddo, B. & Gessa, G.L. (1965): Ann. N.Y. Acad. Sci., 130: 404-411.
5. Fillingame, R.H., & Morris, D.R. (1973): Biochemistry, 4479-4487.
6. Francke, B. (1978): Biochemistry, 17: 5494-5499.
7. Gibson, W. & Roizman, B. (1971): Proc. Nat. Acad. Sci. U.S.A., 68: 2818-2821.
8. Hodgson, J. & Williamson, J.D. (1975): Biochem. Biophys. Res. Commun., 63 308-312.
9. Hudson, J.B. (1979): Archiv. Virol., 62: 1-29.
10. Isom, H.C. (1979): J. gen. Virol., 42: 265-278.
11. Küchler, C., Küchler, W. & Schulze, W. (1968): Acta Virol., 12: 441-445.
12. Mamont, P.S., Duchesne, M.C., Grove, J. & Bey, P. (1978): Biochem. Biophys. Res. Commun., 81: 58-66.
13. McCormick, F. (1978). Virology, 91: 496-503.
14. McCormick, F. & Cohen, S.S. (1979): Ad. Virus Res., 24: 331-387.
15. McCormick, F. & Newton, A.A. (1975): J. gen. Virol. 27: 25-33.
16. Metcalf, B.W., Bey, P., Danzin, C., Jung, M.J., Casara, P. & Vevert, J.P. (1978): J. Am. Chem. Soc., 100: 2551-2553.
17. Pösö, H. & Jänne, J. (1976): Biochem. J., 158: 485-488.
18. Redmond, J.W. & Tseng, A. (1979): J. Chromatog., 170: 479-481.
19. Schenone, A. & Williams-Ashman, H.G. (1972): Biochem. Biophys. Res. Commun., 46: 288-295.
20. Tuomi, T., Mäntyjärvi, R. & Raina, A. (1980): FEBS Lett., 121: 292-294.
21. Tyms, A.S., Scamans, E. & Naim, H.M. (1981): J. Antimicrob. Chemother., 8: 65-72.
22. Tyms, A.S., Scamans, E. & Williamson, J.D. (1979): Biochem. Biophys. Res. Commun., 86: 312-318.
23. Tyms, A.S. & Williamson, J.D. (1980): J. gen. Virol., 48: 183-191.
24. Tyms, A.S. & Williamson, J.D. (1982): Nature, 297: 690-691.
25. Williamson, J.D. (1976): Biochem. Biophys. Res. Commun., 73: 120-126.

Advances in Polyamine Research, Vol. 4, edited by U. Bachrach, A. Kaye, and R. Chayen. Raven Press, New York © 1983.

Level of Polyamines in *Escherichia Coli* Carrying the *MetA* Gene on a Multicopy Plasmid

Shulamit Michaeli, Sonia Rozenhak, and Eliora Z. Ron

Department of Microbiology, Tel-Aviv University, Tel-Aviv, Israel

The purpose of this work was to study the effect of intracellular methionine concentrations on the cellular level of polyamines. Since spermidine biosynthesis is dependent on the availability of methionine,we were interested in finding out whether an elevated level of cellular methionine would affect the regulation of the biosynthesis of spermidine.

Strains with an elevated level of intracellular methionine were obtained by the introduction of multicopy plasmids containing the metA gene which codes for the first enzyme in the methionine pathway - homoserine transsuccinylase (HTS). The plasmids were obtained by cloning the metA gene from a λmetA bacteriophage into the multicopy resistance plasmid pBR322(Michaeli et al, 1981). One of the plasmids which were obtained and contained the metA gene is pMA-3. Strains carrying plasmid pMA-3 are overproducers of methionine , as can be seen from their increased resistance to methionine analogues and from measurements of methionine levels. The effect of pMA-3 on the level of polyamines was measured in wild type strains , as well as in strains which are derepressed for methionine biosynthesis as a result of a mutation in the regulatory gene metJ. The results are presented in Table 1 and indicate that in the presence of elevated intracellular methionine concentrations there is an increase in the level of spermidine. This decrease is correlated with a decrease in the level of putrescine, and brings about a significant change in the ratio of putrescine to spermidine.

The intracellular level of spermidine appears to depend on the cellular level of methionine, and there is no indication for the existance of a regulatory mechanism which represses the biosynthesis of spermidine when the substrate for its synthesis -methionine - or the end products are present in excess.

References

Michaeli, S, Ron, E.Z. and Cohen, G. 1981. Construction and physical mapping of plasmids containing the metA gene of Escherichia coli K-12. Mol. Gen. Genet. 182:349-354.

Cohen, S.S., Morgan, S. and Stribel, E. 1969. The polyamine content of the tRNA of E. coli. Proc. Nat. Acad. Sci. U.S.A. 64:669-676.

Table 1:

Concentration of Polyamines

μg/ml10^7 bacteria

	Putrescine	Spermidine	Spd/Pu
EH80	0.152	0.027	0.18
EH80 (pMA-3)	0.150	0.031	0.21
EH47 *metJ*	0.138	0.030	0.22
EH47 *metJ* (pMA-3)	0.065	0.048	0.73

Polyamine analysis was performed as described by Cohen et al, 1969.

Advances in Polyamine Research, Vol. 4, edited by
U. Bachrach, A. Kaye, and R. Chayen.s36Raven Press,
New York © 1983.

A Mutant of *Escherichia Coli* Deficient in *S*-Adenosylmethione

Bracha Kimchi and Eliora Z. Ron

Department of Microbiology, The George S. Wise Faculty of Life Sciences, Tel Aviv University, Ramat Aviv 69978, Israel

S-adenosylmethionine synthetase (SAM synthetase) of E. coli is the enzyme which catalyses SAM synthesis in-vitro and is coded for by the metK gene. Mutants in the metK gene have been isolated and are characterized by their ability to grow on analogues (ethionine and norleucine) and by having a reduced in-vitro activity of SAM synthetase. However, all the metK mutants grow in the absence of exogenously added SAM and have a relatively high intracellular level of SAM. This finding might indicate that a complete block of SAM mutants is lethal. Yet, mutants were found which have a temperature sensitive metK mutation, but even they grow well at the temperature in which the enzyme is inactive. These findings raised the possibility that SAM-synthetase is not the only enzyme involved in the biosynthesis of SAM in E. coli.

We are interested in studying the role of SAM in the growth processes of E. coli, and for this purpose it was essential to study cells under conditions of low intracellular SAM concentrations. In this communication we show that it is possible to get cells that have a very low intracellular SAM concentration and that under such conditions growth is totally inhibited. Moreover, since the mutation responsible for the reduction in SAM level is in the metK gene, our findings also indicate that SAM synthetase is the enzyme which catalyses SAM biosynthesis in E. coli cells.

In order to get cells deficient in SAM we constructed a double mutant which had a metA mutation in addition of a metK mutation. Mutants in the metA gene are capable of synthesizng methionine when given cystathionine in the growth medium. However, since cystathionine is transported very poorly into E. coli cells, its intracellular concentration is the limiting factor for methionine synthesis. We assumed that the inefficiency of a mutated SAM-synthetase would be better expressed under conditions of limiting methionine concentration and therefore we expected that a metKmetA mutant will be deficient in SAM when grown on cystathionine.

The double mutant was constructed by transducing the metK mutation from strain E. coliK-12 EWH193 metJmetK (Hafner, 1977) into strain NO 1820 which has a metA deletion but is able to grow without methionine as it is lysogenic for λmetA$_2$ (Yamamoto, 1976). Transduction was with P1 vir and transductants were selected for growth on 10mM ethionine at 42°C. Transductants were then replica plated and screened for their ability to grow at 32°C in the presence of 42mM ethionine. This step was introduced to distinguish between metJ and metK transductants, as the latter mutation is temperature sensitive and therefore metK transductants (contrary to metJ transductants) would not be ethionine resistant at 32°C. The following step was to cure metK transductants of the λmetA in order to express the metA mutation to methionine auxotrophy.

The mutant RK-6 was obtained by this procedure and expresses the temperature sensitive metK mutation. When grown in 15 µg/ml cystathionine its growth rate at 32°C was identical to that of the metA mutant but at 43°C growth was completely inhibited (Table 1). Under these conditions we could show a substantial decrease in level of intracellular SAM and of spermidine as compared to metA or metK mutants.

It is important to note that although the level of methionine is reduced in metAmetK strains growing on cystathionine, the methionine pool is higher at 43°C then it is at 32°C. This finding rules out the possibility that inhibition of growth at 43°C is due to the limitation of methionine, and suggests that the limiting factor is either SAM or spermidine. Since the addition of spermidine at 43°C did not improve growth, it appears that the inhibtion of growth is due to the reduced level of SAM.

The results presented here indicate that depletion of intracellular SAM prevents growth of E. coli. The results also indicate that SAM depletion can be obtained in SAM synthetase mutants, suggesting that under our experimental conditions there is no alternative pathway for the synthesis of SAM.

REFERENCES

1. Hafner, W.H., C.W. Tabor and H. Tabor. 1977. Isolation of a metK mutant with a temperature sensitive SAM synthetase. J. Of Bact. 132:832-840.

2. Yamaoto, M. and Nomura, H. 1976. Isolation of λ transducing phages carrying mRNA genes at the metA-purD region of the E. coli chromosome. FEBS Lett. 72:256-261.

Strain	GENERATION TIME (min)			
	Minimal + Methionine 50μg/ml		Minimal + Cysththionine 15μg/ml	
	32°C	43°C	32°C	43°C
Ab2569 *metA*	60	60	180	160
RKG *metK* t.s.	60	75	180	NO GROWTH

Table 1: Growth rate of *metA* and *metA metK* mutants

Advances in Polyamine Research, Vol. 4, edited by U. Bachrach, A. Kaye, and R. Chayen. Raven Press, New York © 1983.

Evidence for Sulfhydrylic Involvement in the Inactivation of Rat Heart Ornithine Decarboxylase by Oxygen Radicals and Hyperoxia

C. M. Caldarera, F. Flamigni, C. Muscari, C. Clô, and C. Guarnieri

Istituto di Chimica Biologica e Centro Studi e Ricerche sul Metabolismo Cardiaco, Facoltà di Medicina e Chirurgia, Università di Bologna, 40126 Bologna, Italy

Ornithine decarboxylase (ODC;E.C. 4.1.1.17.), the apparent rate limiting enzyme in polyamine biosynthesis, has been extensively studied since an increase in its activity is well correlated with an increased rate of cell division and growth (37,19,40). Many studies suggest that ODC activity is regulated at the level of transcription or translation (8,31,5); nevertheless, interest in the mechanisms involved in the control of the enzymatic activity at post-translational levels has recently focused much attention. Several studies have shown that the conversion of multiple physically and kinetically distinct forms of ODC (6,35,39,26) as well as the modification of the enzyme by phosphorylation(2) or by conjugation with putrescine (38) may represent some events able to modulate ODC activity. In addition, the activity of the enzyme may be also changed by the macromolecular inhibitor antizyme (15), by a non-dialyzable activator (25) and by several chemical species including polyanions (22), phospholipids (23), and detergents (24).

In the present research, we have studied the possibility that the alteration of ODC thiol groups represents and additional post-translational process capable of modifying ODC activity. Other Authors in early investigations (17,18) and more recently Mitchell (32) and ourselves (12) have indicated that ODC requires the presence of thiol compounds or thiol reducing agents in order to reach its maximum activity. Since there are abundant evidences that reactive species of oxygen, such as superoxide radicals ($O_2^{\cdot}$) are toxic for enzymes containing SH groups (13), we have studied the effects of these radicals or hyperoxic conditions on heart ODC activity.

MATERIALS AND METHODS

DL-(1-^{14}C) ornithine ·HCl (51.3 mCi/mmol) was obtained from New England Nuclear; DL isoproterenol·HCl dibutyryl-cyclic AMP, pyridoxal-5' phosphate, dithiothreitol (DTT), diethyldithiocarbamate and superoxide dismutase (SOD) were obtained from Sigma Chemical,Co. Xanthine, xanthine oxidase and glutathione were purchased from Boehringer Mannheim. Male albino Sprague-Dawley rats, body weight 150 g were used. Exposure of rats to hyperoxic conditions was done in 40 liter plastic chambers fluxed with a continuous flow of 100% O_2. Chamber atmosphere was monitored with an oxygen analyzer (Beckman). The animals were sacrificed 4 hours after the intraperitoneally injection of isoproterenol (3 mg/Kg). The hearts were homogenized in 3 vol of 20 mM Tris-HCl pH 7.2, 0.1 mM EDTA (buffer A) and the supernatant was obtained after centrifugation at 17,600 g for 40 min. Heart ODC was partially purified from the supernatant by ammonium sulfate fractionation (20%-50% saturation). The final pellet was dissolved in buffer A containing 2.5 mM DTT and then dialyzed overnight against the same buffer. The dialyzed sample was desalted on Sephadex G-25 with buffer A; the sample (2 ml) containing 7.5 mg of protein was loaded on activated-thiol Sepharose 4B column (0.8 x 6 cm) equilibrated with 100 mM NaCl, 5 mM phosphate buffer, pH7.8 (buffer B). The column was washed with buffer B (flow rate 5 ml/h) and the bound ODC was removed by buffer B additionated with 5 mM DTT.

ODC activity was measured by trapping $^{14}CO_2$ into paper disk saturated with Protosol (New England Nuclear). The assay mixture was 19.5 μM DL-(1-^{14}C) ornithine in 100 mM glycylglycine buffer pH 7.2, 5 μM pyridoxal-5' phosphate. DTT was added when expected by the experiments. Exogenous $O_2^{\cdot}$ radicals were generated by incubating at 37°C the cardiac supernatant or the purified ODC in presence of 3 mM xanthine and 0.05 μM xanthine oxidase dissolved in buffer A, pH 7.6 (12). Primary beating heart cells, obtained from 10 day old chick embryos, were grown as described elsewhere (7). When complete confluency was reached, the plates were transferred into suitable sealed chambers with the desired oxygen tension. Gas mixture prepared with calibrated flowmeters (Brooks Instrument Co.) consisted of N_2 and O_2. The oxygen content was controlled as above described. For ODC assay, the cells were treated according to Bachrach et al. (3). The released $^{14}CO_2$ was measured by incubating 0.5 ml of enzyme supernatant in presence of 50 μM ornithine and of 0.4 μCi DL-(1-^{14}C) ornithine in a final volume of 0.6 ml. Polyamines were quantified by fluorescence determination of their dansylated derivates separated by high pressure liquid chromatography (Waters Associates) according to Newton et al. (33). Proteins were determined by the method of Bradford (26) using serum albumin as standard.

RESULTS

Table 1 shows that in isoproterenol-treated rats the cardiac ODC activity is significant decreased after 4 hours of exposure to hyperoxia (100% O_2) in comparison with normoxic condition (air).

TABLE 1. Effect of hyperoxia on cardiac ODC activity and polyamine contents in isoproterenol-treated rats[a]

Determinations	Control	Hyperoxia
ODC activity (pmoles $^{14}CO_2$/mg prot·h)	102.0 ± 10.4	61.5 ± 12.6
Putrescine (nmoles/mg prot)	1.18 ± 0.5	0.8 ± 0.2
Spermidine (nmoles/mg prot)	8.99 ± 1.1	2.6 ± 0.8
Spermine (nmoles/mg prot)	7.12 ± 0.8	2.3 ± 0.6

[a]The rats injected with isoproterenol (3mg/kg) were exposed for 4 hours to air (control) or to 100% oxygen (hyperoxia) as described in Materials and Methods. Data are the means ± S.D. of four separate experiments.

Also the cardiac polyamine contents were strongly decreased, particularly that of spermidine and spermine (Table 1).

The effects of different oxygen tensions on ODC activity of cultured chick embryo heart cells are shown in Fig.1. In comparison to 20% O_2, a lower ODC activity was measured at high oxygen tension (80%), while an opposite behaviour was observed when the cultures were exposed to 5% O_2, which leads to a significant increase of the activity, with a peak after 6 hours.

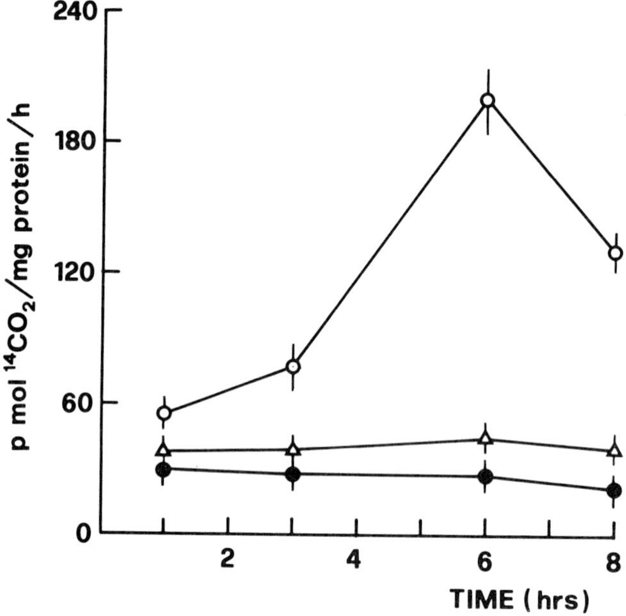

FIG. 1. Effect of different oxygen tensions on ODC activity in chick embryo heart cell cultures. After 20 hours serum-restriction (0.5%), confluent cultures were fed with fresh serum-free medium and exposed to different O_2 tensions. Groups of 4 plates were used for each experimental condition. Data are the means ± S.D. of four separate experiments. (○) 5%O_2; (△) 20%O_2; (●) 80%O_2.

The dependence of ODC activity on the environmental oxygen tension is well documented also in Fig.2. The data indicate that in confluent heart cells maintained at 20% O_2, the addition of isoproterenol caused within 3 hours a 15-fold increase in ODC activity. On the other hand, 80% O_2, while decreasing ODC activity, strongly reduced the capability of the cells to respond to the β-adrenergic agonist. In this condition the value of ODC activity remained under the control value (20%O_2) during the whole period of incubation.

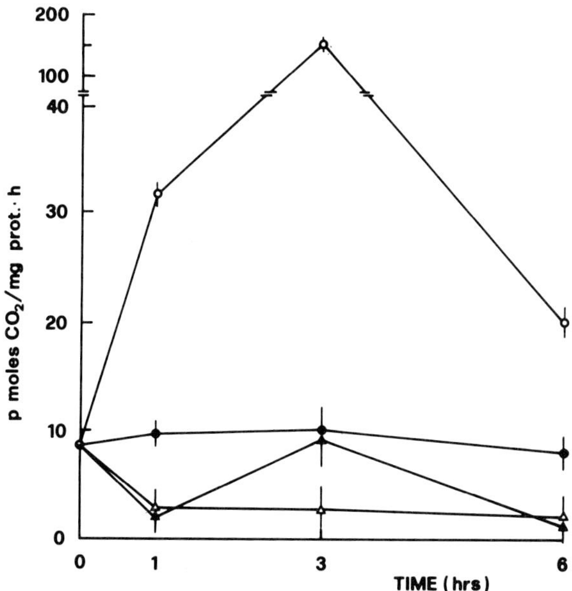

FIG. 2. Changes in ODC activity in isoproterenol-treated heart cells exposed to different oxygen tensions. Experimental conditions were as given in Fig.1. After 1 hour of exposure to the different O_2 tensions, isoproterenol was added to the cells at the final concentration of 10 μM (time zero) and the cells were further exposed to the different O_2 tensions. Data are the means ± S.D. of four separate experiments. (●) 20%O_2; (○) 20%O_2 plus Isoproterenol; (△) 80%O_2; (▲) 80% plus Isoproterenol.

Table 2 shows that dibutyryl cyclic AMP caused a net increase in ODC activity in 20%O_2 treated cells. Hyperoxia completely prevented ODC stimulation by the cyclic AMP congener.

TABLE 2. Effect of dibutyryl-cyclic AMP on ODC activity in heart cell cultures exposed to different oxygen tensions[a]

Oxygen tension (%)	Time (hour)	pmoles $^{14}CO_2$/mg prot·h
20	0	10.2 ± 0.1
	2	15.6 ± 0.2
	4	38.9 ± 0.2
80	0	11.1 ± 0.1
	2	8.8 ± 0.2
	4	8.6 ± 0.3

[a]Experimental conditions were as given in Fig. 2, except that 1 mM dibutyryl-cyclic AMP was added instead of isoproterenol. Data are the means ± S.D. of four separate experiments.

The effect of exogenous superoxide radicals, generated by the xanthine/xanthine oxidase reaction, on ODC activity of cardiac supernatants from isoproterenol-treated rats, is showed in Table 3.

TABLE 3. Effect of $O_2^{\cdot}$ radicals on heart ODC activity[a]

Treatment	pmoles $^{14}CO_2$/mg prot·h	(%)
Control($O_2^{\cdot}$ absent)	102.0 ± 10.7	100
Incubated($O_2^{\cdot}$ generated)	37.8 ± 19.4	37.0
Incubated plus diethyldithiocarbamate	14.0 ± 8.4	13.7
Incubated plus superoxide dismutase(2µM)	86.3 ± 10.4	84.6
Incubated plus glutathione(2.5mM)	89.9 ± 9.8	88.2
Incubated plus dithiothreitol(2.5mM)	87.1 ± 8.0	85.4

[a]The cardiac supernatants were incubated for 10min in the presence of xanthine and xanthine oxidase as described in Materials and Methods. The incubation further continued for 50min in presence of the incubation medium for ODC assay. The other compounds were added 5min before the xanthine/xanthine oxidase incubation. Data are the means ± S.D. of four separate experiments.

ODC activity was strongly inhibited by the xanthine/xanthine oxidase incubation and this negative effect was further enhanced when diethyldithiocarbamate, a powerful inhibitor of SOD (30), was added to the supernatant. On the other hand, the presence of exogenous SOD or of thiol protective agents, such as glutathione or DTT, prevented the decrease in ODC activity.

Fig. 3 shows that the incubation of ODC with exogenous $O_2^{\cdot}$ radicals modified the elution profile of the enzyme from the thiol-activated Sepharose 4B column.

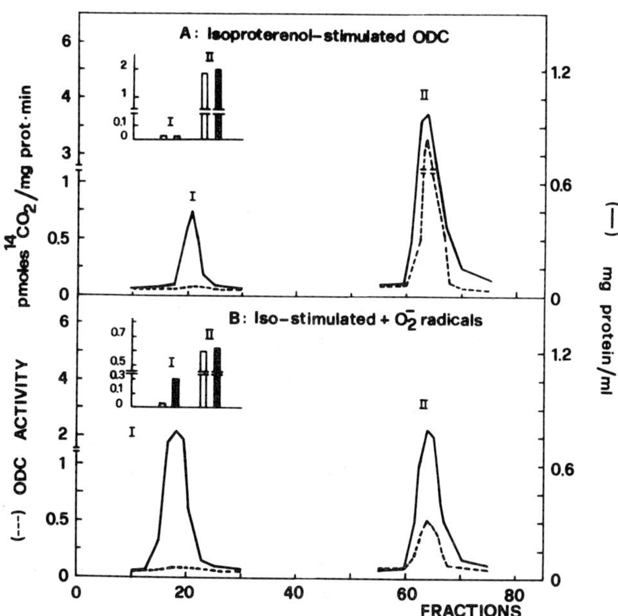

FIG. 3. Fractionation of heart ODC by thiol-covalent chromatography. The ODC stimulated by isoproterenol was purified as described in Materials and Methods. The enzyme was incubated with $O_2^{\cdot}$ radicals and then was loaded to an activated thiol-Sepharose 4B column. Fractions of 0.8 ml were collected and monitored for protein content (26) and for ODC activity (without dithiothreitol). The (□) and (■) columns indicate ODC activity measured in the pooled fractions (I and II) in the absence or presence of dithiothreitol (2.5mM), respectively.

In fact, $O_2^{\cdot}$-exposed ODC collected in the second peak (panel B), was significantly reduced in comparison to the profile of the untreated enzyme reported in panel A. The inserts report the ODC activity of the pooled fractions measured in the presence or absence of DTT. In the control experiment, DTT did not modify ODC activity of both fractions. On the contrary, when DTT was added to the pooled fractions obtained from $O_2^{\cdot}$-inactivated ODC chromatography, there was in fraction I a significant recovery of ODC activity.

DISCUSSION

The present paper provides evidence that experimental conditions that cause in the heart cell oxygen-mediated damages are not favourable conditions in increasing ODC activity. Of course, in the "in vivo" experiments, it is quite difficult to establish whether hyperoxia directly is able to reduce cardiac ODC activity or its effect is mediated by other events, because of the many variables that can be affected by hyperoxia. Nevertheless, the capability of isoproterenol to stimulate cardiac ODC was strongly reduced by hyperoxic exposure. The contents of heart polyamines, particularly of spermidine and spermine,were also considerable reduced suggesting that an increased polyamine catabolism, might occur under this environmental condition. It is known that heart muscle is particularly affected by oxygen-mediated damages (14,10,11) as a consequence of the formation of reactive species of oxygen such as superoxide radicals ($O_2^{\cdot}$) or $OH^{\cdot}$ radicals (34). These radicals are particularly toxic for cell integrity since they cause lipid peroxidation breakdown (42), macromolecular denaturation (16), myoblast growth arrest (29) and free thiol oxidation (1). The cells are protected against $O_2^{\cdot}$ toxicity by the presence of SOD which catalyzes the inactivation of $O_2^{\cdot}$ radicals through their dismutation to O_2 and H_2O_2 (28). Hyperoxia enhances $O_2^{\cdot}$ production and, with time, induces the synthesis of SOD (9). This enzyme is therefore important against superoxide-induced damages elicited by different conditions, including hyperoxia (20) and radiation exposure (27). Since ODC is an enzyme that requires SH groups for its optimal catalytic activity (18), it is reasonable that the inactivation of cardiac ODC by hyperoxia may be due to the perturbation of the redox state of the enzyme thiol-group. This effect could be consequent to a direct oxidation of the ODC SH-groups, and/or seconday to the reduction of the cellular thiol protective compound, such as glutathione. These possibilities may be also considered in the experiments carried out by exposing heart cells cultures to a hyperoxic environment. In fact, an high pO_2 level caused a reduction of the basal ODC activity and prevented the stimulatory effect of isoproterenol on the enzyme. Conversely, at low oxygen tension (5%) the cells exhibited higher ODC activity than normoxic treated cells. Moreover, the fact that hyperoxia blocked the activation of heart ODC by cyclic AMP suggests that high pO_2 levels did not affect the β-receptor system, but very likely reduced the functional status of the enzyme. An additional evidence supporting the idea that ODC may be inactivated by oxygen-mediated reactions, comes from the results obtained by incubating the cardiac enzyme with the reaction xanthine/xanthine oxidase, which is known to produce $O_2^{\cdot}$ radicals (36). In this condition, there was a remarkable reduction of ODC activity, particularly evident when cytosolic SOD was inhibited by diethyldithiocarbamate. The protection provided by thiol-reducing agents, such as glutathione or dithiothreitol on $O_2^{\cdot}$-inactivated ODC, clearly indicates that $O_2^{\cdot}$ radicals impaired ODC activity by oxidizing SH groups essential for enzyme activity. This conclusion is also supported by the experiments showing that the enzyme exposed to $O_2^{\cdot}$ radicals was less immobilized by the thiol-affinity chromatography than was the control enzyme. Moreover, the fact that the inactive enzyme collected during the eluition from the column partially recovered its activity when incubated with dithiothreitol, clearly indicates that the $O_2^{\cdot}$-induced damages are reversible and can be removed by thiol reducing treatment.

From our data we cannot establish whether the oxidation of ODC-thiol groups leads to ODC polymeration, as demonstrated by Jänne and Williams-Ashman (18) in other experimental conditions. In fact, we cannot exclude that $O_2^{\cdot}$ radicals may oxidize essential thiol group(s) needed for ODC activity, as it has been reported for

other enzymes (1).
The meaning of the ODC inactivation by $O_2^{\cdot}$-mediated reactions is not clear. However, in view of the reported protective role played by polyamines against peroxidative processes (21,43) or in membrane stabilization (41), we can speculate that a polyamine deficiency, secondary to a reduced ODC activity, could contribute to an increased cell damage by hyperoxic conditions.
In conclusion, we suggest that rat heart ODC may be inactivated by reactions which accumulate $O_2^{\cdot}$ radicals in the cells. Therefore those conditions that cause an increase of $O_2^{\cdot}$ production or a decrease of the protective mechanism(s) against these radicals, are unfavourable conditions for ODC activity. These results also indicate that a possible mechanism of ODC inactivation is represented by the oxidation of the thiol group(s) of the enzyme, thus suggesting that the alteration of the SH groups might be an important event capable of modifying ODC activity at the post-translational level.

ACKNOWLEDGMENTS

We would like to acknowledge Dr. C. Rossoni, Dr. B. Tantini and Dr. C. Pignatti for their excellent technical assistance. This work was supported by grants from Consiglio Nazionale delle Ricerche and Ministero Pubblica Istruzione, Rome, Italy.

REFERENCES

1. Armstrong, D.A., and Buchanan, J.D. (1978): Photochem. Photobiol., 28: 743-755.
2. Atmar, V.J., and Kuehn, G.D. (1981): Proc. Natl. Acad. Sci. U.S.A., 78: 5518-5522.
3. Bachrach, U. (1976): Biochem. Biophys. Res. Commun., 72: 1008-1013.
4. Bradford, M.M. (1976): Anal. Biochem., 72: 246-254.
5. Clark, J.L. (1974): Biochem., 13: 4668-4674.
6. Clark, J.L., and Fuller, J.L. (1976): Eur. J. Biochem., 67: 303-314.
7. Clô, C., Coccolini, M.N., Tantini, B., and Caldarera, C.M. (1980): Life Sci., 27: 67-73.
8. Fausto, N. (1971): Biochim. Biophys. Acta, 238. 116-128.

9. Fridovich, I. (1974): Adv. Enzymol., 41: 82-84.
10. Guarnieri, C., Flamigni, F., and Caldarera, C.M. (1980). J. Mol. Cell. Cardiol., 12: 797-808.
11. Guarnieri, C., Flamigni, F., Ventura, C., and Rossoni Caldarera, C. (1981): Biochem. Pharmacol., 30: 2174-2176.
12. Guarnieri, C., Flamigni, F., Davalli, P., Clô, C., and Caldarera, C.M. (1982): Ital. J. Biochem., 31: 63-64.
13. Halliwell, B. (1978): Cell. Biol. Intern. Rep., 2: 113-128.
14. Hughson, M., Balentine, J.D., and Daniell, H.B. (1977): Lab. Invest., 37: 516-521.
15. Heller, J.S., and Canellakis, E.S. (1981): J. Cell. Physiol., 107: 209-217.
16. Ishida, R., and Takahashi, T. (1975). Biochem. Biophys. Res. Commun., 66: 1432-1438.
17. Jänne, J., and Williams-Ashman, H.G. (1970): Biochem. J., 119: 595-597.
18. Jänne, J., and Williams-Ashman, H.G. (1971): J. Biol. Chem., 246: 1725-1732.
19. Jänne, J., Poso, H., and Raina, A. (1978): Biochim. Biophys. Acta, 473: 241-293.
20. Kinball, R.E., Reddy, K., Peirce, J.H., Schwartz, L.W., Mustafa, M.G., and Cross, C.E. (1976): Am. J. Physiol., 230: 1425-1432.
21. Kitada, M., Igarashi, K., Hirase, S., and Kitagawa, H. (1979): Biochem. Biophys. Res. Commun., 87: 388-394.
22. Kitani, T., and Fujisawa, H. (1981): FEBS Lett., 125: 83-85.
23. Kitani, T., and Fujisawa, H. (1981): FEBS Lett., 132: 296-298.
24. Kitani, T., and Fujisawa, H. (1981): Eur. J. Biochem., 119: 177-181.
25. Kyriakidis, D.A., Heller, J.S., and Canellakis, E.S. (1978): Proc. Natl. Acad. Sci. U.S.A., 75: 4699-4703.
26. Lau, C., and Slotkin, T.A. (1982): Eur. J. Pharmacol., 78: 99-105.
27. Lin, P.S., Kwock, L., and Butterfield, C.E. (1979) Radiat. Res., 77: 501-511.
28. McCord, J.M., and Fridovich, I. (1969): J. Biol. Chem., 244. 6049-6055.

29. Michelson, A.M., and Buckingham, M.E. (1974): Biochem. Biophys. Res. Commun., 58. 1079-1086.
30. Misra, H.P. (1979): J. Biol. Chem., 254: 11623-11628.
31. Mitchell, J.L.A., and Rusch, H.P. (1973): Biochim. Biophys. Acta, 297: 503-516.
32. Mitchell, J.L.A. (1981): In: Advances in Polyamine Research, vol.2, edited by C.M. Caldarera, et al., pp. 15-26.Raven Press, New York.
33. Newton, N.E., Ohno, K., and Abdel-Monem, M. (1976): J. Chromat., 124: 277-285.
34. Nohl, H., Hegner, D., and Summer, K.H. (1981): Biochem. Pharmacol., 30: 1753-1757.
35. Obenrader, M.F., and Prouty, W.F. (1977): J. Biol Chem., 252: 2860-2865.
36. Porras, A.G., Olson, J.S., and Palmer, G. (1981): J. Biol. Chem., 256. 9096-9103.
37. Russell, D.H., and Snyder, S.H. (1968): Proc. Natl Acad. Sci. U.S.A., 60: 1420-1427.
38. Russell, D.H. (1981): Biochem. Biophys. Res. Comm. 99: 1167-1172.
39. Sedary, M.J., and Mitchell, J.L.A. (1977): Exp. Cell Res., 107: 105-110.
40. Stoscheck, C.M., Florini, J.R., and Richman, R.A. (1980): J. Cell Physiol., 102: 11-18.
41. Tabor, H., and Tabor, C.W. (1964): Pharmacol.Rev. 16: 245-300.
42. Tyler, D.D. (1975): FEBS Lett., 79: 180-183.
43. Vanella, A., Pinturo, R., Geremia, E., Tiriolo,P., D'Urso, G., Rizza, V., Di Silvestro, I., Grimaldi, R., and Brai, M. (1980): IRCS Med. Sci., 8: 940.

Advances in Polyamine Research, Vol. 4, edited by
U. Bachrach, A. Kaye, and R. Chayen. Raven Press,
New York © 1983.

ODC Activity Evaluated in Perfused Rat Heart in Response to Catecholamine

C. Guarnieri, F. Flamigni, C. Muscari, and C. M. Caldarera

Istituto di Chimica Biologica e Centro Studi e Ricerche sul Metabolismo Cardiaco, Facoltà di Medicina e Chirurgia, Università di Bologna, 40126 Bologna, Italy

In heart muscle, ornithine decarboxylase (ODC;E.C. 4.1.1.17.) plays a crucial role because its activity is increased by conditions which are correlated to an elevated functional activity of the muscle. In fact, heart ODC activity increases rapidly during exercise (28), during hypoxia or hyperbaric conditions (12) or following increased cardiac work produced by aortic constriction (20) or by pulmonary arterial constriction (18). Numereous hormones are also able to stimulate cardiac ODC activity. In particular, the activation of ODC has been reported to be promoted by treatment with adrenaline (6), noradrenaline (5,8), triiodothyronine (15), growth hormone (24) and by an increased noradrenergic input via β-receptor stimulation (28,8, 6).

Since it has been proposed that noradrenaline (13), isoproterenol (25) or the circulating adrenaline (31) can activate the cardiac noradrenergic system and lead to the formation of heart hypertrophy, it is evident that by catalyzing the production of polyamines, ODC may play a relevant role in translating the stimuli from the adrenergic system to the biochemical processes leading to the macromolecular synthesis. On the other hand, until now, the mechanisms by which the catecholamines lead to cardiac ODC activation have not been exactly known. Some researches have shown that an increase in ODC activity is controlled at the transcriptional level by cyclic AMP through the activation of a specific cyclic AMP dependent protein kinase (21)

More recently, Lau and Slotkin (16) described that the activation of cardiac ODC in response to isoproterenol might result from a shift of the enzyme from its low-affinity state for ornithine to a high-affinity form.

The experiments performed in the present study were designed to give further information on the biochemical events in heart muscle involved in the activation of ODC secondary to catecholamine stimulation. We have determined the ODC activity by a new method that monitors the production of $^{14}CO_2$ from infused (1-^{14}C) ornithine rat heart perfused with the catecholamines.

MATERIALS AND METHODS

Isoproterenol, noradrenaline, adrenaline, propranolol, cycloheximide were purchased from Sigma Chemical Co., St. Louis, Mo., U.S.A.; A23187 was obtained from Calbiochem-Behring, San Diego, Ca., U.S.A. Verapamil from Knoll AG, Liestal, Switzerland; Trifluoperazine (TFP) was a gift from Maggioni Laboratory, Milan, Italy; and D,L-α-difluoromethylornithine (DFMO) was a gift from Merrell Research, Strasbourg, France.

Female Wistar rats (180-200 g) which were fed "ad libitum" were slightly anesthetized with diethylether. The hearts were rapidly excised and placed in an ice cold perfusion medium. They were then mounted on the non-recirculating perfusion apparatus as described elsewhere (10) and perfused by using a modified Krebs-Henseleit buffer solution containing in mM: NaCl, 115.0; $NaHCO_3$, 25; KCl, 4.0; KH_2PO_4, 0.9; $MgSO_4$, 1.1; $CaCl_2$, 1.5 and glucose, 11. The perfusion medium was saturated with 95% O_2 - 5% CO_2 (pO_2 >600 mm Hg) at 37°C and was delivered to the aortic inflow canula with a constant flux of 12 ml/min. The heart was initially perfused for a 20 min stabilization period. (1-^{14}C) ornithine (42.7 mCi/mmol; New England Nuclear, Dreieich, W. Germany) was dissolved in the perfusion medium and added to the heart by a cannula adapted above aorta ligation.

The isotope flux was 1 ml/min and the heart received 0.4 μCi/min for 30 min. Ornithine decarboxylation in the perfused heart was determined by monitoring the production of $^{14}CO_2$ from the infusion of

(1-^{14}C) ornithine. Samples of the effluent perfusate collected at time intervals for 1 min were placed in flasks which contained 0.5 ml of 0.5 M NaOH, 0.125 M EDTA. The perfusion fluid in the closed flasks was acidified by injection of 0.4 ml of 40% trichloroacetic acid to liberate the $^{14}CO_2$. The evolved $^{14}CO_2$ was trapped by 20 µl of Protosol (New England Nuclear) on a filter paper suspended above the reaction mixture in a plastic well (Kontes). The flasks were agitated at room temperature for 30 min and the filter paper was then placed in toluene-PPO.The radioactivity was determined in a liquid scintillation spectrometer (Prias PLD Tricarb Packard, U.S.A.).

A sample of the perfusate collected at zero time before aorta incannulation was collected for 1 min and acidified in order to evaluate blank $^{14}CO_2$.

The catecholamines or the drugs were added to the perfused heart separately from the isotope at a flux of 1 ml/min.

Heart samples were stored at -20°C until assayed. ODC activity was determined by the method of Jänne and Williams-Ashman (11). Frozen tissue (approximately 400 mg) was homogenized in 20 mM Tris-HCl, pH 7.2, 0.1 mM EDTA, 2.5 mM dithiothreitol with Ultra-Turrax homogenizer. The homogenates were centrifuged at 15,000 g x 30 min and three 100 µl aliquots of the resulting supernatant were assayed for ODC activity. The assay mixture contained in a total volume of 0.2 ml: 20 mM Tris HCl, pH 7.2, 2.5 µM pyridoxal phosphate, 0.2 µCi (1-^{14}C) ornithine. Incubations were conducted at 37°C for 60 min. The reaction was terminated by acidification and the $^{14}CO_2$ released was counted as described above. Proteins were determined by the Bradford method (2) with bovine serum albumin as standard.

RESULTS

Figure 1 shows the kinetics of $^{14}CO_2$ production from (1-^{14}C) ornithine in the perfused rat heart. In the control experiments, the ornithine decarboxylation increased slightly after 15 min until 1 pmoles $^{14}CO_2$/min·g and remained constant during the remaining 15 min.

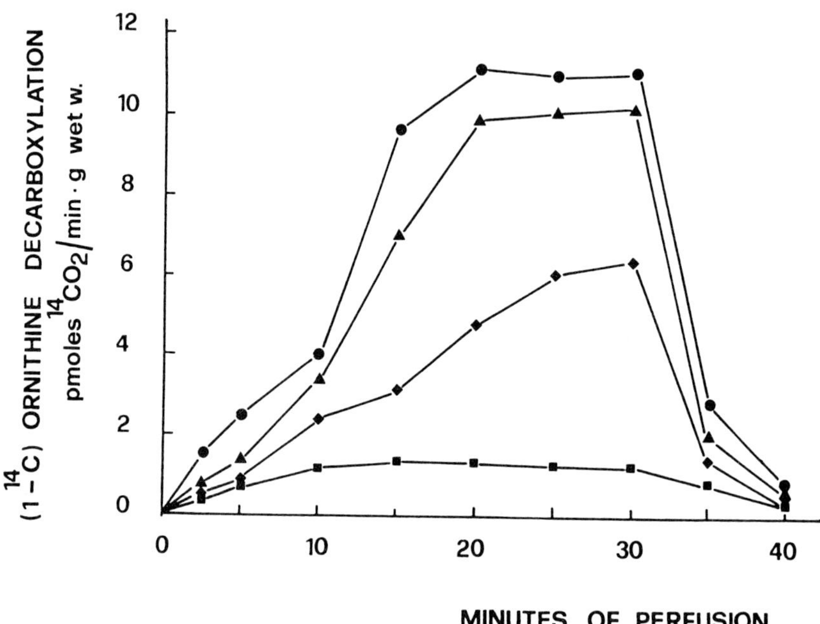

FIG. 1. Effect of catecholamines on the decarboxylation of (1-^{14}C) ornithine in the isolated perfused rat heart. Each catecholamine was injected for 30 min at 1 ml/min; final concentrations 2 μM; Control hearts received only the isotope. (●) Isoproterenol; (▲) Noradrenaline; (◆) adrenaline; (■) control. Each point represents the mean of four observations.

Isoproterenol and noradrenaline produced a marked stimulation of $^{14}CO_2$ production by elevating its value to 10 pmoles/min·g after 20 min of perfusion. Less evident was the effect of adrenaline that produced a maximum release of 6 pmoles $^{14}CO_2$/min·g after 15 min of (1-^{14}C)-ornithine infusion. Upon terminating the infusion of (1-^{14}C) ornithine, the decrease in the rate of production or the washout of the $^{14}CO_2$ from the heart was rapid. ODC activity evaluated in the homogenates prepared after the catecholamine infusions is shown in Table 1.

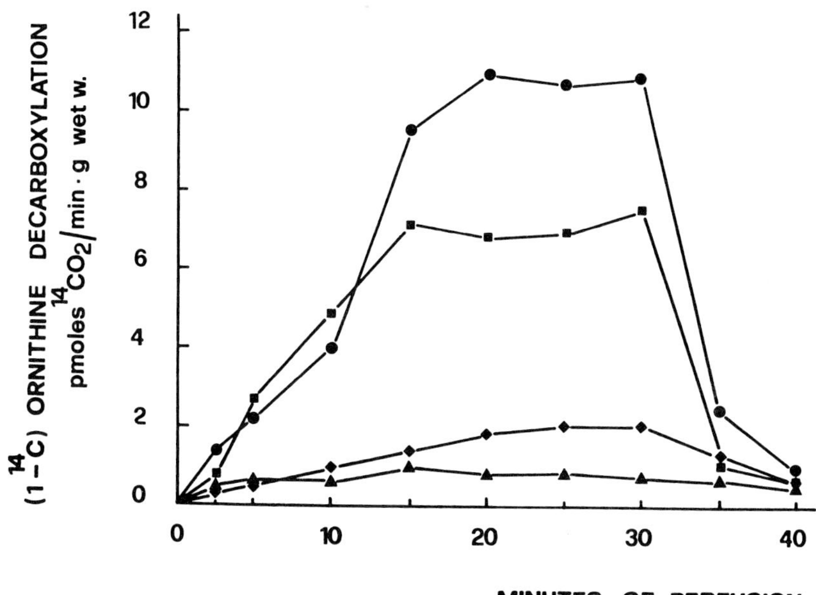

FIG. 2. Effect of various agents on the decarboxylation of (1-^{14}C) ornithine stimulated by isoproterenol The hearts were perfused for 30 min in the presence of isoproterenol. The drugs were added separately 15 min before the introduction of isoproterenol and were continued for the following 30 min. (●) Isoproterenol, 2μM; (■) Cycloheximide, 30μM; (◆) Propranolol, 50μM; (▲) DFMO, 2mM.
Each point represents the mean of four observations.

The experiments described in Figure 3 indicate that calcium is involved in the activation of ornithine decarboxylation promoted by isoproterenol. In fact, verapamil, a potent inhibitor of the uptake of Ca^{++} by plasma membranes (14), prevented the isoproterenol-induced $^{14}CO_2$ production. Also the omission of Ca^{+2} from the perfusion medium decreased the decarboxylation of the labeled ornithine stimulated by the catecholamine.

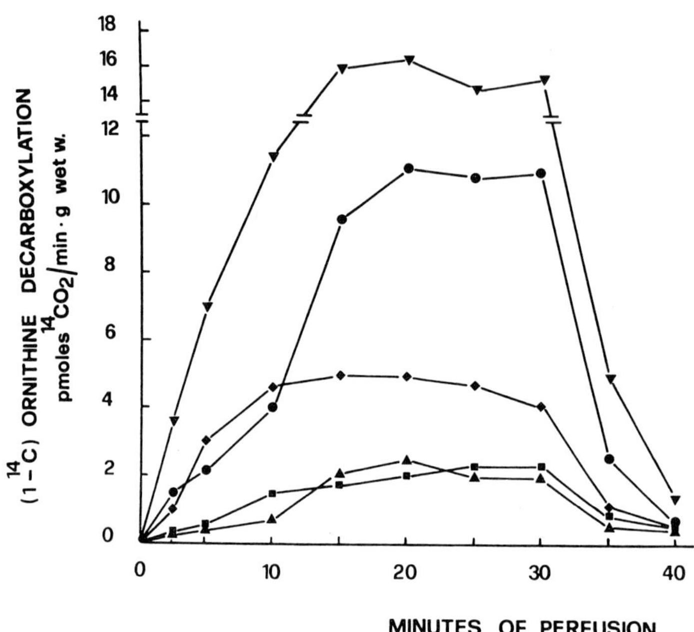

FIG. 3. Involvement of Ca^{+2} in the decarboxylation of (1-^{14}C) ornithine stimulated by isoproterenol. The hearts were perfused for 30 min in the presence of isoproterenol. The drugs were injected 15 min before the introduction of isoproterenol and were continued for the following 30 min. Ca^{+2}-free perfusions started at 0 time. (●) Isoproterenol, 2μM; (▼) A23187, 10μM; (◆) Ca^{+2}-free; (■) Verapamil, 10μM; (▲) Trifluoperazine, 10μM.
Each point represents the mean of four observations.

The same figure shows that the presence of trifluoperazine, a phenothiazine compound used as an inhibitor of the Ca^{+2}-calmodulin complex (29) produced an inhibition of the $^{14}CO_2$ production, that was comparable to the effect of verapamil.
Conversely, the calcium-ionophore A23187 strongly increased ornithine decarboxylation stimulated by isoproterenol by elevating CO_2 production to 16-17 pmoles/min·g after 15 min of perfusion.

TABLE 1. Effect of catecholamines on ODC activity in the perfused rat heart[a]

Treatment	Minutes of perfusion	
	15	30
Control	6.78 ± 0.8	5.95 ± 0.9
Isoproterenol	16.50 ± 1.2	17.66 ± 1.0
Noradrenaline	10.75 ± 0.6	19.50 ± 1.1
Adrenaline	9.00 ± 0.8	12.50 ± 0.8

[a] The hearts were perfused with or without catecholamines (final concentration 2 μM) for 15 and 30 min in absence of the labeled ornithine as described in Materials and Methods. ODC activity is expressed as pmoles $^{14}CO_2$/mg prot·h·±SEM of four experiments.

After 15 min, ODC activity was clearly higher in the heart given isoproterenol, reaching a value almost three times the control value, while the stimulation produced by noradrenaline and adrenaline did not exceed 60% of the control value. After 30 min of perfusion, high values of ODC activity were observed in the hearts perfused with noradrenaline and isoproterenol, while a less activated enzyme was produced by adrenaline.

Figure 2 shows that the decarboxylation of (1-^{14}C) ornithine decarboxylation stimulated by isoproterenol was prevented by propranolol, a β-antagonist compound. Also the addition of DFMO, a known irreversible inhibitor of ODC (1) strongly reduced the rate of (1-^{14}C)-ornithine decarboxylation.

The addition of cycloheximide reduced, but did not block, the flux of $^{14}CO_2$ increased by isoproterenol and during the last 15 min of perfusion ornithine decarboxylation remained close to 6-7 pmoles $^{14}CO_2$/min·g^{-1}.

TABLE 2. Change in heart ODC activity by substances which modify cellular calcium availability[a]

Treatment	pmoles $^{14}CO_2$ / mg prot·h
Control	5.95 ± 0.9
Isoproterenol	17.66 ± 1.0
Verapamil	7.78 ± 1.8
Trifluoperazine	3.15 ± 1.6
A23187	24.08 ± 0.8

[a]The hearts were perfused for 30 min as described in Materials and Methods. The drugs were injected as reported in figure 3.
Values are expressed ± SEM of four experiments.

Table 2 shows that ODC activity measured in the heart homogenates was differently stimulated by isoproterenol according to the experimental conditions which modify cell calcium availability.
In fact, ODC was activated by the Ca^{+2} ionophore A23187, while a strong inhibitory effect was evident when verapamil and trifluoperazine were injected in the perfused rat hearts.

DISCUSSION

The present experiment demonstrates that the measurement of $^{14}CO_2$ production from infused (1-^{14}C) ornithine can provide an accurate monitor of the ornithine decarboxylase activity in the perfused rat heart. In fact, when the catecholamines were infused in the isolated organs, the (1-^{14}C) ornithine decarboxylation was stimulated in a manner and magnitude similar to that measured by assaying tissue ODC activity. Moreover, the inhibition that DFMO produced on the labeled ornithine decarboxylation confirms that $^{14}CO_2$ collected from the perfusate derived mainly from the cardiac ODC activity.

The possibility of monitoring ODC activity "in situ" during the infusions with catecholamines has given important information in understanding the mechanisms involved in the activation of heart ODC.

First of all, in accordance with previous research (6, 26) it appears that myocardial ODC is differently stimulated by catecholamines with isoproterenol and noradrenaline more active than adrenaline. This effect could be the consequence of the high β-receptor stimulating activity possessed by isoproterenol and noradrenaline, while adrenaline can interact also with the α-receptors (23).

The direct involvement of the β-receptor activation in the stimulation of cardiac ODC is confirmed also by the experiments with propranolol, a pure β-antagonist agent, which prevented the ODC stimulation. This result is consistent with the recent paper of Copeland et al. (6) which described that β_2-adrenoceptors are selectively coupled to the regulation of cardiac ODC following catecholamine stimulation.

In addition, the results of this study provide evidence for the partecipation of Ca^{+2} in the activation of ODC. In fact, the omission of Ca^{+2} in the perfusion medium or the reduction of the cellular uptake of Ca^{+2} by the presumptive antagonist of transmembrane Ca^{+2} influx, specifically verapamil (14) were conditions capable of reducing the isoproterenol-stimulated ODC activity.

Conversely, treatment with a dicarboxylic inophore, A23187, which is capable of enhancing intracellular accumulation of Ca^{+2} in many tissues (19,30), evoked a significant increase in the stimulatory effect of isoproterenol. Therefore, it is possible that an enhanced Ca^{+2} flux caused by β-receptor stimulation (22) represents an important step in the activation of heart ODC by isoproterenol.

The involvement of Ca^{+2} in the activation of ODC in other experimental conditions has been described recently. Costa and Nye (7) affirm that Ca^{+2} is essential for the asparagine and dibutyryl cyclic AMP mediated stimulation of ODC in Chinese Hamster ovary cells while Gibbs et al. (9) suggest that Ca^{+2} is essential for ODC activation stimulated by β-adrenergic agonists in rat astracytoma cells. Again, Canellakis et al. (4)

implicate Ca^{+2} in the activation of ODC in the rat hepatoma cells, while Veldhuis and Hammond (27) report a regulatory effect of Ca^{+2} in the modulation of ODC in isolated pig granulosa cells.

The mechanisms by which Ca^{+2} is implicated in the ODC activation is not clear. However, the fact that trifluoperazine, an antipsychotic drug which is able to combine to Ca^{+2}-calmodulin (17) and bloch the metabolic effects of this complex (29,3), was able to prevent ODC activation, suggests that the active Ca^{+2}-calmodulin complex is in some way implicated with the stimulation of cardiac ODC by isoproterenol.

Other evidence that suggests that the activation of ODC cannot be completely due to de novo synthesis of the enzyme results from the experiments with cycloheximide which showed that the isoproterenol-induced ODC activation was not blocked by the inhibition of the protein synthesis.

In conclusion, the present study supports the evidence that the activation of heart ODC by isoproterenol mediated by β-receptor stimulation occurs with the participation of Ca^{+2} ions. The molecular mechanism/s by which such regulatory effects is/are exerted by Ca^{+2} remain unexplained, nevertheless on the basis of the inhibitory effect showed by trifluoperazine, we can suggest an involvement of the Ca^{+2}-calmodulin complex.

ACKNOWLEDGMENTS

This work was supported by grants from Consiglio Nazionale delle Ricerche and Ministero Pubblica Istruzione Rome (Italy).

REFERENCES

1. Bey, P. (1978): In: Enzyme Activated Irreversible Inhibitors, edited by N. Seiler, M.J. Jung, and J. Koch-Waser, pp. 27-41. Elsevier North Holland, New York.
2. Bradford, M.M. (1976): Anal. Biochem. 72: 246-254.
3. Brostrom, C.O., and Wolff, D.J. (1981): Biochem. Pharmacol. 30. 1395-1405.
4. Canellakis, Z.N., Theoharides, T.C., Bondy, P.K., and Triarhos, E.T. (1981): Life Sci. 29: 707-710.

5. Casti, A., Corti, A., Reali, N., Mezzetti, G., Orlandini, G., and Caldarera, C.M. (1977): Biochem. J. 168: 333-340.
6. Copeland, J.G., Larson, D.F., Roeske, W.R.,Russell D.H., and Womble, J.R. (1982): Br. J. Pharmacol. 75: 479-483.
7. Costa, M., and Nye, J.S. (1978): Biochem. Biophys. Res. Commun. 85: 1156-1164.
8. Fuller, R.W., and Hemrick, S.K. (1978): J. Mol. Cell. Cardiol. 10: 1031-1036.
9. Gibbs, J.B., Hsu, C., Terasaki, W.L., and Brooker, G. (1980): Proc. Natl.Acad. Sci. U.S.A. 77: 995-999.
10. Guarnieri, C., Ferrari, R., Visioli, O., Caldarera, C.M., and Nayler, W.G. (1978): J. Mol. Cell. Cardiol. 10: 893-906.
11. Jänne, J., and Williams-Ashman, H.G. (1971): J. Biol. Chem. 246: 1725-1732.
12. Krelhaus, W., Gibson, K., and Harris, P. (1975): J. Mol. Cell. Cardiol. 7: 63-69.
13. Laks, M.M. (1976): Am. Heart J. 91: 674-675.
14. Langer, G.A., Serena, S.D., and Nudd, L.M. (1975): Am. J. Physiol. 229: 1003-1007.
15. Lau, C., and Slotkin, T.A. (1979): J. Pharmacol. Exp. Ther. 208: 485-490.
16. Lau, C., and Slotkin, T.A. (1979): Mol. Pharmacol. 16: 504-512.
17. Levin, R.M., and Weiss, B. (1979): J. Pharmacol. Exp. Ther. 208: 454-459.
18. Matsushita, S., Sogani, R.K., and Raben, M.S. (1972): Circ. Res. 31: 699-709.
19. Reed, P.W., and Lardy, H.A. (1972): J. Biol. Chem. 247: 6970-6977.
20. Russell, D.H., Shiverick, K.T., Hamrell, B.B., and Alpert, N.R. (1971): Am. J. Physiol. 221: 1287-1291.
21. Russell, D.H., Byus, C.V., and Manen, C.A. (1976): Life Sci. 19: 1297-1306.
22. Schneider, J.A., and Sperelakis, M. (1975): J.Mol. Cell. Cardiol. 7: 249-273.
23. Schüman, M.J. (1980): Trends Pharmacol. Sci. 4: 195-197.
24. Sogani, R.K., Matsushita, S., Mueller, J.F., and Raben, M.S. (1972): Biochim. Biophys. Acta 279:

377-386.

25. Stanton, H.C., Brenner, G., and Mayfield, E. (1969) Am. Heart J. 77: 72-80.
26. Veldhuis, J.D., Harrison, T.S., and Hammond, J.M. (1980): Biochim. Biophys. Acta 627: 123-130.
27. Veldhuis, J.D., and Hammond, M. (1981): Biochem. J. 196: 795-801.
28. Warnica, J.W., Antony, P., Gibson, K., and Harris, P. (1975): Cardiovasc. Res. 9: 793-796.
29. Weiss, B., Prozialeck, W., and Cimino, M. (1980): In: Advances in Cyclic Nucleotide Research, edited by P. Hamet, and H. Sands, pp. 213-225. Raven Press, New York.
30. Weissmann, G., Anderson, P., Serhan, C., Samuelsson, E., and Goodman, E. (1980) : Proc. Natl.Acad. Sci. U.S.A. 77: 1506-1510.
31. Womble, J.R., Larson, D.F., Copeland, J.G., Brown B.R., Haddax, M.K., Russell, D.H. (1980): Life Sci. 27: 2417-2420.

Advances in Polyamine Research, Vol. 4, edited by U. Bachrach, A. Kaye, and R. Chayen. Raven Press, New York

Catecholamine-Stimulated β_2-Receptors Coupled to Ornithine Decarboxylase Induction and to Cellular Hypertrophy and Proliferation

J. R. Womble and Diane Haddock Russell

Departments of Pharmacology and Surgery, University of Arizona College of Medicine, Tucson, Arizona 85724

Catecholamines, particularly epinephrine, can stimulate growth and/or proliferation of the myocardium, salivary glands, and selective other tissues and cell lines. In the fetal murine heart, ornithine decarboxylase activity has been used to characterize the nature of the receptor coupled to a catecholamine-stimulated trophic response. The murine fetal heart ornithine decarboxylase response is coupled through a β_2-adrenoceptor. Based on this observation and other studies of β_2-receptor linkage to ornithine decarboxylase induction by catecholamines, it is tempting to speculate that catecholamines may be involved in the regulation of growth and proliferation primarily through β_2-receptors whereas the general metabolic effects of catecholamines appear to be related to β_1-receptor coupling.

Catecholamines can initiate the biochemical events known to occur in a trophic cascade in rodent salivary glands and in the hearts of a wide variety of animals (22,32,36,41,42,59,67). In the adult heart, it appears that catecholamine-stimulated hypertrophy of myocytes occurs without significant stimulation of DNA synthesis and proliferation (61,62). However, in rodent salivary glands, different analogs of catecholamines can stimulate hypertrophy and/or proliferation (22). In most instances, with certain exceptions, the growth cycle always accompanies the division cycle, although hypertrophy can occur without proliferation. The tight coupling of cellular proliferation and hypertrophy is explained by the necessity to double cell mass prior to division. If this were not true, cells would get smaller and smaller as a function of the number of division cycles. In general, cell cycle progression involves a temporal cascade of biochemical events remarkably similar in various cell lines and tissues.

Although catecholamines have been implicated in the regulation of hypertrophy and proliferation, the majority of studies of catecholamine action have focused on alterations in secretion from exocrine organs or on muscle contraction-relaxation (23,51,58). It has remained puzzling how catecholamines can exert such widely variable responses in various target organs. We will summarize the evidence that certain of these apparently disparate actions may be related to the presence of specific β_2-receptor coupling to trophic and proliferative responses.

CATECHOLAMINE-STIMULATED HYPERTROPHY AND PROLIFERATION

One of the earliest reports of the proliferative potential of catecholamines was by Selye and coworkers in 1961 (59). They found that high doses of isoproterenol induced adult rat and mouse salivary gland acinar cells to initiate the necessary biochemical cascade of events culminating in DNA synthesis within 24 hours. The mechanism of this growth stimulation later was demonstrated to be the action of circulating isoproterenol binding to acinar cell β-receptors linked to the cyclic AMP generating enzyme, adenylate cyclase (49,63). Catecholamines, acting through the β-receptor/adenylate cyclase system, are now known to increase cyclic AMP in many tissues resulting in altered metabolism and, in some cases, in hypertrophy and/or proliferation. The generality of growth stimulation by catecholamines is demonstrated in Table 1. These references, with one exception, include only those studies in which catecholamine administration resulted in increased DNA synthesis, RNA synthesis, or tissue mass. The study of Sen et al. (60) was included because they demonstrated that cardiac hypertrophy was prevented by elimination of the major source of plasma epinephrine via bilateral adrenalectomy. Also, they found elevated catecholamine concentrations in the ventricles and suggested an important role for catecholamines in modulating cardiac hypertrophy.

To date, epinephrine has been implicated as the physiological circulating amine trophic hormone which parallels cardiac hypertrophy in the dog after aortic constriction (Table 2) (70,71). Further, stress-induced cardiac hypertrophy in the dog can be blocked by denervation of the adrenal medullae, a procedure which results in reduction of the circulating epinephrine concentration to near zero (71). Cardiac hypertrophy also was blocked by the administration of propranolol, a nonspecific β-adrenoceptor antagonist (71). By the use of terbutaline, a specific β_2-agonist, we have implicated β_2-adrenoceptors as selectively coupled to the regulation of murine cardiac ornithine decarboxylase following catecholamine stimulation (18).

THE IDENTIFICATION OF β_1- AND β_2-RECEPTOR SUBTYPES

The gradual definition of β-adrenoceptor subtype has an extensive history. In 1948, Ahlquist (1) described the existence of two types of adrenergic receptors which he referred to as α and β for the purpose of accommodating the variable responses which occurred following the administration of sympathomimetic agents.

TABLE 1. Generality of β-receptor mediated hypertrophy and/or hyperplasia in response to catecholamines

Catecholamine	Target Tissue	Reference
Isoproterenol	Salivary glands	Selye et al. (59)
		Barka (3) (4)
		Malamud & Baserga (41) (42)
		Guidotti et al. (30)
		Chang & Barka (17)
		Durham et al. (22)
		Inoue et al. (34)
		Burke & Barka (9)
		Tsang et al. (64)
Isoproterenol	Heart	Stanton et al. (61)
		Wood et al. (72)
		Warnica et al. (66)
		Byus et al. (13)
		Bareis & Slotkin (2)
		Bartolomé et al. (6)
		Haddox et al. (32)
Epinephrine	Heart	Womble et al. (70) (71)
		Sen et al. (60)
		Fell et al. (24)
		Copeland et al. (18)
Norepinephrine	Heart	Gans & Cater (29)
		Laks (36)
		Laks et al. (37) (38)
		Corti et al. (19)
		Casti et al. (15)
		Bareis & Slotkin (2)
		Fuller & Hemrick (26)
Epinephrine Norepinephrine Isoproterenol	Granulosa cells	Veldhuis et al. (65)
Isoproterenol	Hepatocytes	Brønstad & Christoffersen (8)
Epinephrine	Diaphragm	Hopkins & Manchester (33)
Epinephrine	Stomach epithelium	Frankfurt (25)
Isoproterenol	Kidney	Malamud & Malt (43)
		Burns et al. (12)
		Burns (10)
		Catanzaro & Marzi (16)
Isoproterenol	Corneal epithelium	Burns (10)
	Duodenum	Burns and Scheving (11)
		Burns et al. (12)
Isoproterenol	Prostate	Winter (69)
Epinephrine	Lymphocytes	MacManus et al. (40)
		Whitfield et al. (67)

TABLE 2. Plasma epinephrine and norepinephrine levels after aortic constriction and adrenal medulla denervation

	pg/ml plasma			
	24 h	48 h	72 h	96 h
Sham coarctation (N=6)				
Epinephrine	99± 4	116± 13	100± 4	110± 7
Norepinephrine	415±105	225± 21	320±86	325±52
Aortic coarctation (N=6)				
Epinephrine	228± 64[a]	317± 38[a]	246±18[a]	336±51[a]
Norepinephrine	427±121	230±115	384±70	343±84
Aortic coarctation plus denervation (N=6)				
Epinephrine	9± 9[b]	10± 6[b]	16± 7[b]	29± 9[b]
Norepinephrine	397± 97	210± 24	208±31	324±99

Data are expressed as the mean ± S.E.M.
[a]Data differ from controls ($p < 0.05$).
[b]Data differ from controls and coarcted alone ($p < 0.001$).
Reprinted by permission from Life Sciences (71).

Twenty years later, Lands et al. (39) presented data supportive of two types of β-receptors, with those mediating cardiac rate and contractility, lipolysis and small bowel motility inhibition, referred to as β_1, and those responsible for bronchodilation and vascular relaxation referred to as subtype β_2. Studies by Lands and coworkers (39) revealed that norepinephrine was the most selective β_1 agonist whereas epinephrine was a more potent β_2 agonist. One of the first reports of heterogeneous populations of β-receptor subtypes in the same organ apparently subserving the same biological response was in 1972 by Carlsson et al. (14). They studied heart rate response *in vivo* and *in vitro* in the cat and found that, based on agonist-antagonist studies, both β_1- and β_2-receptor subtypes affected the same response, heart rate, although to a different degree. The evidence presented was that practolol, a β_1 antagonist, blocked the heart rate response to norepinephrine in the order of isoproterenol > epinephrine > salbutamol whereas propranolol, a β_1 and β_2 antagonist, blocked in the order of epinephrine > isopropterenol = norepinephrine. H35/25, a specific β_2 blocker, resulted in blockade with the order of salbutamol > epinephrine > isoproterenol > norepinephrine. If the cat heart contained only β_1-receptors, then each blocker should have inhibited heart rate response to all agonists to about the same degree. Therefore, the activation of both β_1- and β_2-receptors was implicated in physiological regulation. Furchgott (27) concurred with the coexistence of both β_1- and β_2-receptors in studies on dog heart and guinea pig trachea. Barnett et al. (5) supplied direct evidence for two types of β-receptor binding sites in lung tissue. Minneman et al. (48)

performed simultaneous determinations of β_1 and β_2 adrenergic receptors in tissues containing both subtypes and calculated ratios of β_1/β_2 to be: 83/17 in the heart, 15/85 in the lungs, 81/19 in the cerebral cortex, 76/24 in the caudate, 15/85 in the cerebellum, 81/19 in the hippocampus, 71/29 in the diencephalon.

With increasing evidence that both β_1- and β_2-receptors are present in a variety of tissues, the hypothesis that these receptor subtypes subserve different biological functions was tested in our laboratory. Since β_2-receptors have a high affinity for epinephrine and a much lower affinity for norepinephrine, epinephrine again appears to be the specific physiological agonist (hormone) for β_2-receptors.

ORNITHINE DECARBOXYLASE AS A MARKER OF CATECHOLAMINE-STIMULATED GROWTH

We selected ornithine decarboxylase as a marker of catecholamine-mediated trophic action because of its remarkably constant pattern of expression. An induction of ornithine decarboxylase has been reported as a general, rapid event (within hours) in response to all trophic hormones studied to date and suggests that target tissues for trophic hormones could be ascertained by using ornithine decarboxylase as a biological marker (57). It has been found that the extent of ornithine decarboxylase induction is hormone dose-dependent in a variety of growth-stimulated tissues (54). In Figure 1 we have shown the major sequential steps in a catecholamine-generated trophic response. Although the initiation of DNA synthesis is shown in this schema, the trophic response is not always coupled to DNA synthesis and proliferation as previously discussed. The evidence for type I cyclic AMP-dependent protein kinase activation resulting in the transcriptional induction of ornithine decarboxylase is reviewed elsewhere (31,53,56). Type II cyclic AMP-dependent protein kinase involvement in the triggering of DNA synthesis has been reported (7,20). Russell and coworkers have implicated a post-translational modification of ornithine decarboxylase in the direct regulation of ribosomal RNA synthesis (44-46,55).

CATECHOLAMINE STIMULATION OF ORNITHINE DECARBOXYLASE ACTIVITY IN THE FETAL MURINE HEART

Initially, we established the developmental pattern of ornithine decarboxylase in the fetal murine heart as a function of days of embryogenesis and compared the activity to the growth rate, which was determined by the percent increment in heart protein (32) (Fig. 2). At 13 days of gestation, the earliest time that hearts could be dissected, ornithine decarboxylase activity was maximal and then declined progressively throughout the remainder of gestation. The change in enzymatic activity closely paralleled the change in growth rate of the heart. The mothers were then injected with varying concentrations of a β-adrenergic agonist, isoproterenol, and 4 hours later ornithine decarboxylase activity was determined in the fetal heart. The maximal ornithine decarboxylase response occurred with 10 mg/kg

Figure 1

MODEL OF MAJOR SEQUENTIAL STEPS IN A CATECHOLAMINE-GENERATED TROPHIC RESPONSE

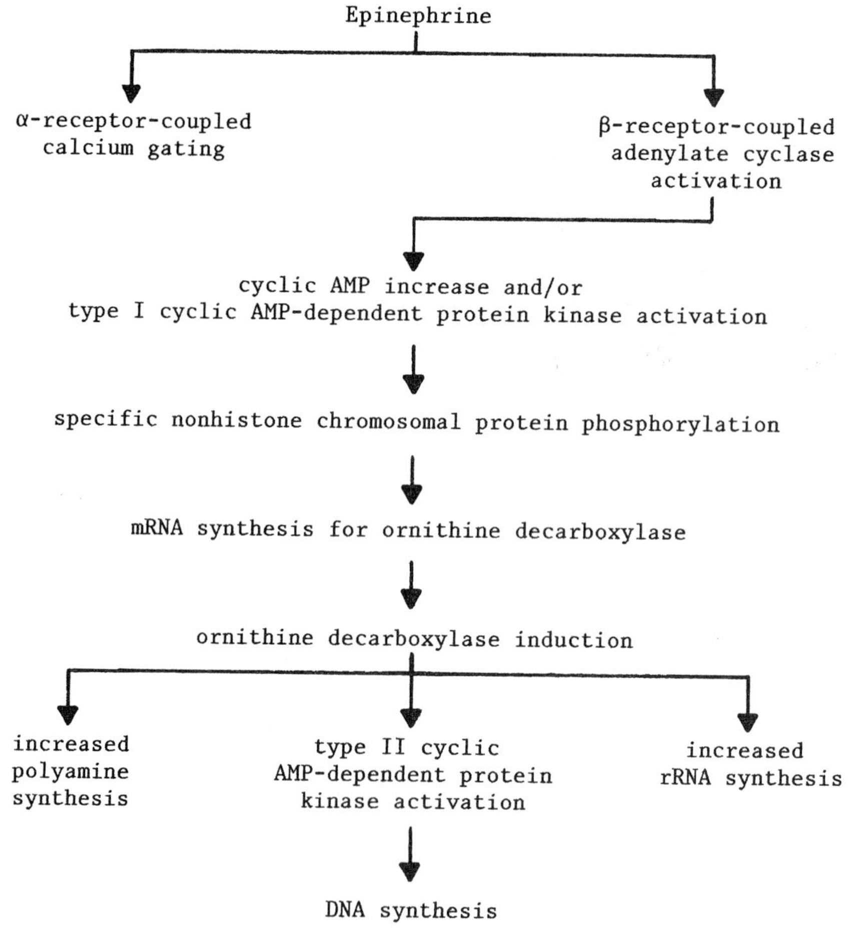

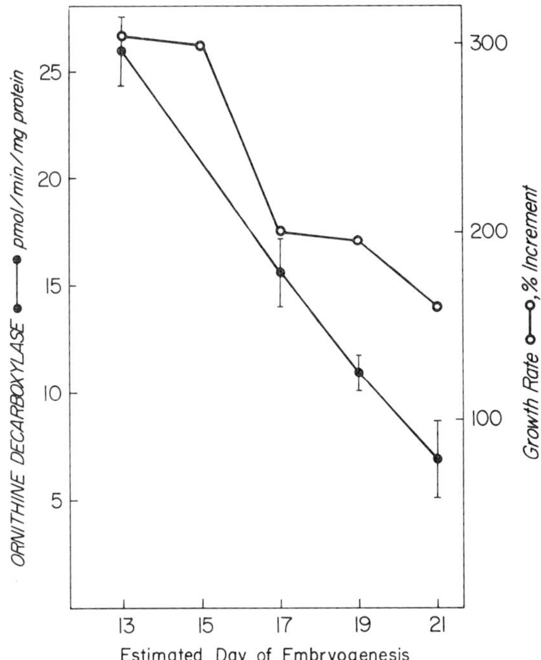

FIG. 2. Changes in Ornithine Decarboxylase Activity in the Fetal Mouse Heart during Embryogenesis. The activity of ornithine decarboxylase (O) in mouse hearts obtained at the indicated day of gestation was assayed. Each data point represents the mean ± standard deviation of measurements of 3 or 4 heart pools. The rate of growth of the heart (O) was calculated as the percentage increment in heart protein during each 2-day period. Reprinted by permission from Molecular Pharmacology (32).

isoproterenol. Response to the drug was biphasic since higher concentrations of isoproterenol were less effective. To insure that maternal injection was comparable to fetal administration, IM injections of 2 mg/kg and 10 mg/kg were given to fetuses at 17 days of gestation. All of those receiving 10 mg/kg died, but ornithine decarboxylase activity was stimulated in hearts of fetuses given 2 mg/kg (Table 3). Stimulation after direct fetal injection did not differ from that of injected mothers. These experiments were added in proof of stimulation of fetal heart β-receptors after maternal injection of β-agonists. Propranolol administration blocked the increase in ornithine decarboxylase activity in response to isoproterenol after either fetal or maternal injection.

TABLE 3. Fetal heart ornithine decarboxylase activity after fetal or maternal injection

	Ornithine Decarboxylase Activity ($pmol\ ^{14}CO_2/min/mg$ protein)	
Isoproterenol	49.5 ± 6.3	(n = 8)
0.9% NaCl solution	12.1 ± 2.5	(n = 6)
Isoproterenol + propranolol	6.2 ± 0.9	(n = 6)
Mother		
Isoproterenol	54.0 ± 4.7	(n = 5)
0.9% NaCl solution	8.2 ± 0.9	(n = 5)
Isoproterenol + propranolol	5.8 1 0.06	(n = 5)

Fetuses were injected IM in the hip area (2 mg/kg). The average weight of fetuses at 17 days was 500 mg, so they received 5 µg/25 µl. Mothers received 10 mg/kg s.c. of isoproterenol. Propranolol (10 mg/kg s.c. maternal or IM fetal) was given 10 min prior to isoproterenol in the mothers, but simultaneously in fetuses. Reprinted by permission from Molecular Pharmacology (32).

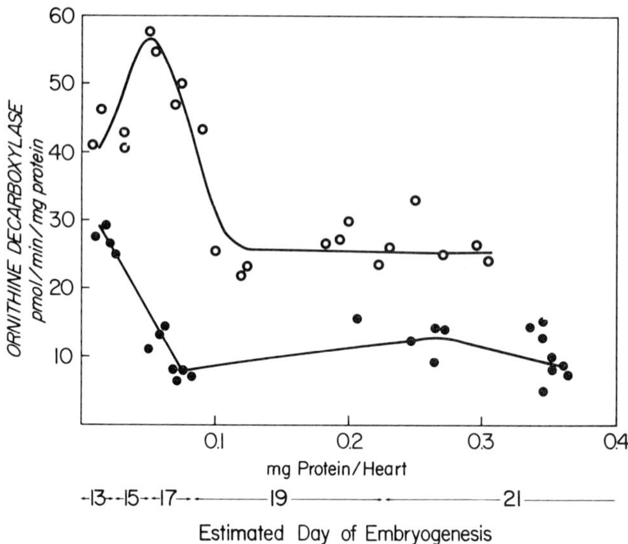

FIG. 3. Isoproterenol Stimulation of Ornithine Decarboxylase Induction in the Fetal Mouse Heart at Different Times of Gestation. Isoproterenol (10 mg/kg; 0) or 0.9% NaCl solution (control; 0) was administered i.p. to mothers at the indicated days of gestation. The activity of ornithine decarboxylase in the fetal heart was measured 4 hours after drug administration. Reprinted by permission from Molecular Pharmacology (32).

In order to study the maturation of the heart trophic response to β-receptor stimulation, the ability of maternal injections of isoproterenol to elevate ornithine decarboxylase activity in the fetal heart was examined throughout embryogenesis. As shown in Figure 3, the β-agonist, isoproterenol, promoted an increase in cardiac ornithine decarboxylase activity at all times assayed during fetal development. The highest drug-stimulated activity, 57 pmol/min/mg protein, occurred at 16-17 days of gestation compared to the unstimulated activity of 12 pmol/min/mg protein. After this, elevated ornithine decarboxylase activity detected in response to isoproterenol administration declined, although the magnitude of the stimulation remained at a level approximately 2.5- to 3-fold above control until birth. Therefore, we found that trophic responsiveness of the fetal heart was functionally coupled to the β-receptor from 13 days of gestation through birth. However, the same relationship was not apparent when we compared the effects of isoproterenol administration on heart rate. The heart rate response was not statistically elevated over the basal rate at 13 days after isoproterenol administration (Fig. 4). However, by 17 days, heart rate was increased to 131% above basal activity and was increased further by days 19 and 21 to 170% and greater than 190% of basal activity, respectively. These data compare favorably with those of Wildenthal (68), who demonstrated a 110% increase in heart rate at 13 days with either isoproterenol or norepinephrine. Therefore, development of the cardiac trophic response to β-adrenergic stimulation preceded by several days the expression of the chronotropic response. The data were in harmony with the ability to detect β-receptors during murine heart development prior to the development of significant responsiveness to β-agonists as measured by a detectable heart rate response (68). These data also suggested there might be a different subclass of β-receptors coupled to ornithine decarboxylase induction and hypertrophy compared to those coupled to chronotropic responses such as heart rate.

β_2-ADRENOCEPTORS REGULATE THE INDUCTION OF MYOCARDIAL ORNITHINE DECARBOXYLASE ACTIVITY IN MICE IN VIVO

In order to characterize the nature of the receptor coupled to the isoproterenol-stimulated ornithine decarboxylase response in the murine fetal heart, we chose specific agonists and antagonists for α_1-, α_2-, β_1- and β_2-receptors (18). Epinephrine, which has α_1, α_2, β_1 and β_2 activity, isoproterenol, which has β_1 and β_2 activity, and terbutaline, which has only β_2 activity, significantly elevated ornithine decarboxylase activity in the fetal mouse heart at 18 days of gestation (Table 4). Susphrine, a sustained release suspension of epinephrine, produced a 2-fold elevation of ornithine decarboxylase activity in the fetal mouse heart at 18 days of gestation. The adrenoceptor antagonists, prazosin (α_1), yohimbine (α_2), metoprolol (β_1) and propranolol (β_1 and β_2) failed to reduce control enzyme activity and therefore were not cytotoxic. Further, the stimulatory effects of

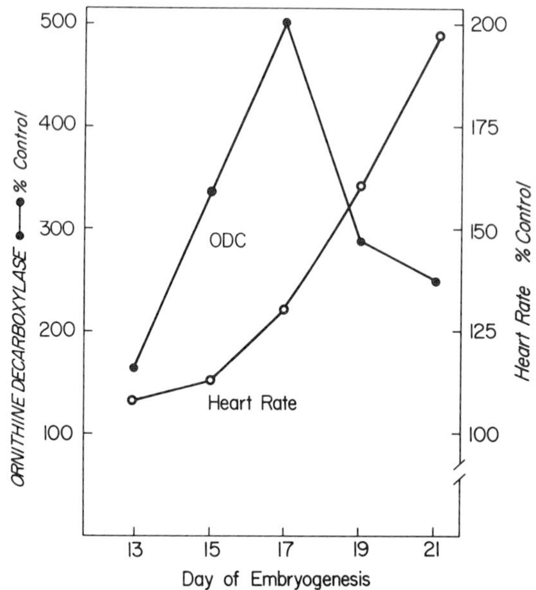

FIG. 4. Changes in the Trophic and Chronotropic Response of Fetal Mouse Hearts to Isoproterenol during Embryogenesis. The percentage increase stimulated by isoproterenol above basal ornithine decarboxylase activity in fetal heart at the indicated days of embryogenesis was calculated from the data in Figure 3. The data showing the change in heart rate of the fetal mouse heart in response to a maximally stimulating dose (10^{-3} M) of isoproterenol are taken from Roeske et al. (52). Reprinted by permission from Molecular Pharmacology (32).

isoproterenol on ornithine decarboxylase activity were not inhibited by prazosin or yohimbine, the α_1- and α_2-receptor blockers, respectively. However, the isoproterenol-induced increase in ornithine decarboxylase activity was blocked by propranolol, a nonspecific β-adrenoceptor blocker. Metoprolol, a relatively specific β_1 blocker, failed to inhibit the increase in ornithine decarboxylase activity induced by either isoproterenol or epinephrine. The specific involvement of the β_2-receptor was demonstrated by the ability of terbutaline, a selective β_2-receptor agonist, to increase ornithine decarboxylase activity over 3-fold. This effect could be blocked by propranolol but not by metoprolol.

This study suggests that physiological β_2-receptors in mouse heart are coupled to the induction of ornithine decarboxylase. A β_2 response to catecholamine stimulation is in agreement with the results of Veldhuis et al. (65) who examined catecholamine modulation of ornithine decarboxylase in isolated granulosa cells from immature pig ovarian follicles. Catecholamines produced

TABLE 4. Effects of adrenoceptor agonists and antagonists on ornithine decarboxylase activity in the fetal murine heart

Drug(s)	Ornithine Decarboxylase Activity (pmol/min/mg protein)	% of Control
Control (saline)	12.1 ± 2.5	100
Yohimbine (10 mg/kg)	12.7 ± 1.2	105
Prazosin (1 mg/kg)	13.6 ± 0.9	119
Metoprolol (10 mg/kg)	13.8 ± 1.6	114
Propranolol (10 mg/kg)	12.5 ± 1.4	103
Isoproterenol (10 mg/kg)	49.5 ± 6.3[a]	409
Epinephrine (1 mg/kg)	37.5 ± 2.8[a]	301
Susphrine (1 mg/kg)	26.4 ± 3.0[a]	218
Terbutaline (10 μg/fetus)	37.2 ± 3.2[a]	308
Isoproterenol + propranolol	6.2 ± 0.9	51
Isoproterenol + metoprolol	43.5 ± 6.0[a]	360
Isoproterenol + prazosin	39.9 ± 2.6[a]	330
Isoproterenol + yohimbine	37.6 ± 5.4[a]	311
Epinephrine + metoprolol	38.1 ± 4.0[a]	315
Terbutaline + propranolol	9.8 ± 0.8[a]	82
Terbutaline + metoprolol	49.7 ± 3.3[a]	411

[a]Values differ significantly from controls ($p < 0.001$).

All experiments were conducted on fetal hearts at 18 days of gestation. Each value represents duplicate determinations of pools of at least 5 fetal hearts. Each value contains data from 5 to 10 separate samples. Drugs administered in combination were at the same doses as for single agents. Values are expressed as the mean ± S.E.M.
Reprinted by permission from British Journal of Pharmacology (18).

concentration-dependent increases in ornithine decarboxylase activity which were not altered by α-receptor antagonists or by a β_1-antagonist, metoprolol, but were blocked by the antagonists propranolol (β_1, β_2), butoxamine (β_2) and timolol (β_1, β_2). In addition, terbutaline, a preferential β_2 agonist, stimulated ornithine decarboxylase in a dose-dependent manner. Other authors have reported β_2 coupling to adenylate cyclase activation in human neutrophils (28) and fat cells (35) as well as β_2-mediated cyclic AMP increases in mouse epidermis (21) and Ehrlich ascites tumor cells (50). Further, alkaline phosphatase induction in rat liver occurs in response to β_2 catecholamine administration (47). It is, therefore, intriguing to speculate that catecholamines may regulate specific target tissue hypertrophy and/or proliferation selectively through β_2-receptors, whereas catecholamines may exert their general effects on metabolism through β_1-receptor coupling.

REFERENCES

1. Ahlquist, R.P. (1948): Am. J. Physiol., 153: 586-600.
2. Bareis, D.L., and Slotkin, T.A. (1978): J. Pharmacol. Exp. Ther., 205: 164-174.
3. Barka, T. (1965): Exp. Cell Res., 37: 662-679.
4. Barka, T. (1967): Exp. Cell Res., 48: 53-60.
5. Barnett, D.B., Rugg, E.L., and Hanorski, S.R. (1978): Nature, 273: 166-168.
6. Bartolomé, J., Huguenard, J., and Slotkin, T.A. (1980): Science, 210: 793-794.
7. Boynton, A.L., and Whitfield, J.F. (1980): Exp. Cell Res., 126: 477-481.
8. Brønstad, G., and Christoffersen, T. (1980): FEBS Lett., 120: 89-93.
9. Burke, G.T., and Barka, T. (1978): Biochim. Biophys. Acta, 539: 54-61.
10. Burns, E.R. (1975): Pavlov. J. Biol. Sci., 10: 161-185.
11. Burns, E.R., and Scheving, L.E. (1973): J. Cell. Biol., 56: 605-608.
12. Burns, E.R., Scheving, L.E., and Tsai, T.-H. (1972): Science, 175: 71-73.
13. Byus, C.V., Chubb, J.M., Huxtable, R.J., and Russell, D.H. (1976): Biochem. Biophys. Res. Commun., 73: 694-702.
14. Carlsson, E., Ablad, B., Brandstrom, A., and Carlsson, B. (1972) Life Sci., 11: 953-958.
15. Casti, A., Corti, A., Reali, N., Mezzetti, G., Orlandini, G., and Caldarera, C.M. (1977): Biochem. J., 168: 333-340.
16. Catanzaro, O.L., and Marzi, A. (1974): Experientia, 30: 1334-1335.
17. Chang, W.W.L., and Barka, T. (1974): Anat. Rec., 178: 203-210.
18. Copeland, J.G., Larson, D.F., Roeske, W.R., Russell, D.H., and Womble, J.R. (1982): Br. J. Pharmacol., 75: 479-483.
19. Corti, A., Casti, A., Reali, N., and Caldarera, C.M. (1976): Biochem. Biophys. Res. Commun., 71: 1125-1130.
20. Costa, M., Gerner, E.W., and Russell, D.H. (1976): J. Biol. Chem., 251: 3313-3319.
21. Duell, E.A. (1980): Biochem. Pharmacol., 29: 97-101.
22. Durham, J.P., Baserga, R., and Butcher, F.R. (1974): Biochim. Biophys. Acta, 372: 196-217.
23. Exton, J.H. (1981): Mol. Cell. Endocrin., 23: 233-264.
24. Fell, R.D., Terblanche, S.E., Winder, W.W., and Holloszy, J.O. (1981): Am. J. Physiol., 241: C55-C58.
25. Frankfurt, O.S. (1968): Exp. Cell Res., 52: 220-232.
26. Fuller, R.W., and Hemrick, S.K. (1978): J. Mol. Cell. Cardiol., 10: 1031-1036.
27. Furchgott, R.F., Wakade, T.D., Sorace, R.A., and Stollak, J.S. (1976): Fed. Proc., 34: 794.
28. Galant, S.P., and Allred, S.J. (1980): J. Lab. Clin. Med., 96: 15-23.
29. Gans, J.H., and Cater, M.R. (1970): Life Sci., 9: 731-740.

30. Guidotti, A., Weiss, B., and Costa, E. (1972): Mol. Pharmacol., 8: 521-530.
31. Haddox, M.K., Roeske, W.R., and Russell, D.H. (1979): Biochim. Biophys. Acta, 585: 527-534.
32. Haddox, M.K., Womble, J.R., Larson, D.F., Roeske, W.R., and Russell, D.H. (1981): Mol. Pharmacol., 20: 382-386.
33. Hopkins, D., and Manchester, K.L. (1981): Biochim. Biophys. Acta, 678: 388-394.
34. Inoue, H., Tanioka, H., Shiba, K., Asada, A., Kato, Y., and Takeda, Y. (1974): J. Biochem., 75: 679-687.
35. Kather, H., and Simon, B. (1980): Horm. Metab. Res., 12: 695-697.
36. Laks, M.M. (1976): Am. Heart J., 91: 674-675.
37. Laks, M.M., Garner, D., and Wong, V. (1979): Am. Heart J., 98: 732-735.
38. Laks, M.M., Morady, F., and Swan, H.J.C. (1973): Chest, 64: 75-78.
39. Lands, A.M., Arnold, A., McAuliff, J.P., Luduena, F.P., and Brown, T.G., Jr. (1967): Nature, 214: 597-598.
40. MacManus, J.P., Whitfield, J.F., and Youdale, T. (1971): J. Cell. Physiol., 77: 103-116.
41. Malamud, D., and Baserga, R. (1967): Life Sci., 6: 1765-1769.
42. Malamud, D., and Baserga, R. (1968): Exp. Cell Res., 50: 581-588.
43. Malamud, D., and Malt, R.A. (1971): Lab. Invest., 24: 140-143.
44. Manen, C.A., and Russell, D.H. (1975): Life Sci., 17: 1769-1776.
45. Manen, C.A., and Russell, D.H. (1977): Science, 195: 505-506.
46. Manen, C.A., and Russell, D.H. (1977): Biochem. Pharmacol., 26: 2379-2384.
47. Mary, P.L., and Rao, J.P. (1981): Br. J. Pharmacol., 72: 8-9.
48. Minneman, K.P., Hegstrand, L.R., and Molinoff, P.B. (1979): Mol. Pharmacol., 16: 34-46.
49. Murad, F., Chi, Y.M., Rall, T.W., and Sutherland, E.W. (1962): J. Biol. Chem., 237: 1233-1238.
50. Onaya, T., Akasu, F., Takazawa, K., and Hashizume, K. (1978): Endocrinology, 103: 1122-1127.
51. Reuter, H. (1974): J. Physiol. (Lond.), 242: 429-451.
52. Roeske, W.R., Chen, F.M.C., and Yamamura, H. (1979): In: Catecholamines: Basic and Clinical Frontiers, Vol. 1, edited by E. Usdin, J. Kopin, and J. Barchas, pp. 779-781. Pergamon Press, New York.
53. Russell, D.H. (1978): In: Advances in Cyclic Nucleotide Research, Vol. 9, edited by W.J. George and L.J. Ignarro, pp. 493-506. Raven Press, New York.
54. Russell, D.H. (1980): Pharmacology, 20: 117-129.
55. Russell, D.H. (1981): Med. Biol., 59: 286-295.
56. Russell, D.H. (1981): In: Polyamines in Biology and Medicine, edited by D.R. Morris and L.J. Marton, pp. 109-125. Marcel Dekker, Inc., New York.
57. Russell, D.H., and Haddox, M.K. (1979): In: Advances in Enzyme Regulation, edited by G. Weber, pp. 61-87. Pergamon Press, Oxford-New York.

58. Schümann, H.J., Endoh, M., and Brodde, O.E. (1975): Naunyn Schmeidebergs Arch. Pharmacol., 289: 291-302.
59. Selye, H., Veilleux, R., and Cantin, M. (1961): Science, 44-45.
60. Sen, S., Tarazi, R.C., and Bumpus, F.M. (1979): Clin. Sci., 56: 439-443.
61. Stanton, G.C., Brenner, G., and Mayfield, E.D. (1969): Am. Heart J., 77: 72-80.
62. Stanton, H.C., and Schwartz, A. (1967): J. Pharmacol. Exp. Ther., 157: 649-658.
63. Sutherland, E.W., and Rall, T.W. (1960): Pharmacol. Rev., 12: 265-299.
64. Tsang, B.K., Rixon, R.H., and Whitfield, J.F. (1980): J. Cell. Physiol., 102: 19-26.
65. Veldhuis, J.D., Harrison, T.S., and Hammond, J.M. (1980): Biochim. Biophys. Acta, 627: 123-130.
66. Warnica, J.W., Antony, P., Gibson, K., and Harris, P. (1975): Cardiovasc. Res., 9: 793-796.
67. Whitfield, J.F. (1980): In: Adrenergic Activators and Inhibitors, edited by L. Szekeres, pp. 267-317. Springer-Verlag, Berlin, Heidelberg, New York.
68. Wildenthal, K. (1973): J. Clin. Invest., 52: 2250-2258.
69. Winter, W.A. (1974): Beitr. Pathol., 153: 73-79.
70. Womble, J.R., Haddox, M.K., and Russell, D.H. (1978): Life Sci., 23: 1951-1958.
71. Womble, J.R., Larson, D.F., Copeland, J.G., Brown, B.R., Haddox, M.K., and Russell, D.H. (1980): Life Sci., 27: 2417-2420.
72. Wood, W.G., Lindenmayer, G.E., and Schwartz, A. (1971): J. Mol. Cell. Cardiol., 3: 127-138.

Advances in Polyamine Research, Vol. 4, edited by U. Bachrach, A. Kaye, and R. Chayen. Raven Press, New York © 1983.

Effect of Spermidine Depletion on the Response of Hepatic Pyruvate Kinase and Lipogenesis to Refeeding

Margaret E. Brosnan and Yu-Wan Hu

Department of Biochemistry, Memorial University of Newfoundland, St John's, Newfoundland, Canada A1B 3X9

During starvation, liver metabolism shifts from glycolysis and lipogenesis to gluconeogenesis. The key regulatory enzymes in the glycolytic pathway, glucokinase, phosphofructokinase and pyruvate kinase, are depressed in activity, while those of the gluconeogenic pathway, pyruvate carboxylase, phosphoenolpyruvate carboxykinase, fructose bisphosphatase and glucose-6-phosphatase, are increased. This pattern of enzyme activities is rapidly changed upon refeeding to give increased glycolysis and lipogenesis and decreased gluconeogenesis, without the need for concomitant protein synthesis (7). It has recently been shown that starvation of rats (17) or glucagon treatment of perfused liver (2) or hepatocyte suspensions (17) causes cyclic AMP-dependent phosphorylation of pyruvate kinase. This covalent modification has been shown to cause a change in the kinetic properties of the enzyme so that its affinity for its substrate, phosphoenolpyruvate is decreased, although maximal velocity obtained at saturating substrate concentration is unaffected (2). This change results in a rapid fall in enzyme activity under physiological conditions. Refeeding of rats, or treatment of perfused liver or hepatocyte suspensions with insulin, results in reactivation of pyruvate kinase (2, 17) by activation of a phosphoprotein phosphatase which dephosphorylates the enzyme (15).

In 1973, Domschke and Söling (5) reported that the activity of ornithine decarboxylase in liver of starved rats was very low, and that this activity could be increased rapidly by feeding the animals. S-adenosylmethionine decarboxylase activity and the concentrations of putrescine and spermidine were also depressed by starvation and increased on refeeding. It was proposed at the time that these changes in polyamine synthesis were coupled to changes in RNA synthesis and concentration (5). However, it has since been shown that polyamines could influence the activity of several phosphoprotein phosphatases from different tissues (10,16). We have recently reported (9) that depletion of hepatic polyamines by diaminopropanol pretreatment could inhibit the rapid lipogenic response of diabetic liver to insulin. This increase in lipogenesis in response to insulin is thought to involve dephosphorylation and activation of several enzymes (8). Thus we have designed the present study to determine the effect of refeeding on the activity of polyamine-synthetic decarboxylases and on polyamine concentrations relative to the effect on pyruvate kinase and lipogenesis. We have also used diaminopropanol to deplete the hepatic spermidine pool in order to study the effect of refeeding on pyruvate kinase and lipogenesis in the absence of an elevation in hepatic spermidine concentration.

METHODS

Animals

Adult male rats of the Sprague-Dawley strain were purchased from Canadian Breeding Laboratories, St. Constant, Quebec, Canada. Rats received no food for 64 hours before the experiments, but had free access to water at all times. On the day of the experiment 1,3-diaminopropan-2-ol (100 ug/100 g body weight) was given in three subcutaneous injections (09:00, 11:00, 13:00 h). Refed rats were each given 4 grams Purina rat chow at 11:00 h. For decarboxylase assays, rats were killed three hours after feeding (13:00 h) in order to determine the peak response. When the rate of lipogenesis was measured, ^3H_2O (0.5 mCi/rat) was injected intraperitoneally at 14:00 h. Rats were killed by cervical dislocation at 15:00 h.

Lipogenesis

Rates of lipogenesis *in vivo* were measured by intraperitoneal injection of ^3H_2O one hour before death (as above). A sample of blood was collected into a heparinized tube for determination of specific radioactivity of plasma water. Weighed samples of liver were added to 5M NaOH (10 ml/g), and lipid was saponified and extracted by the method of Lowenstein (11).

Polyamines, amino acids and metabolites

One lobe of liver was rapidly frozen between aluminum clamps precooled in liquid nitrogen (18). Frozen liver was ground to a fine powder in a mortar under liquid nitrogen. One sample of powdered liver was extracted with cold sulfosalicylic acid (5%, w/v) and the supernatant obtained by centrifugation at 20000g for 10 min was used for analysis of free amino acids and polyamines (3). A second sample of frozen liver was extracted with perchloric acid (6%, w/v) and used for assay of the following metabolites: lactate (12), pyruvate (4), phosphoenolpyruvate (4), and fructose-1,6-bisphosphate (13).

Decarboxylase assays

A portion of fresh liver was weighed and homogenized in sucrose medium (0.25 M sucrose/2 mM EDTA/5 mM dithiothreitol/2 mM Hepes, pH 7.4; 10 ml/g liver), as described (3). To prepare cytosol, homogenate was centrifuged at 105000g for 70 min. Ornithine decarboxylase and S-adenosylmethionine decarboxylase were assayed as described previously (3).

Pyruvate kinase assay

A weighed portion of liver was homogenized in Krebs-Ringer bicarbonate buffer (100 ml/g liver), 1 mM mercaptoethanol was added and the liver extract was centrifuged at 20000g for 10 min (17). Pyruvate kinase activity was assayed as described by Blair and coworkers (2). Activity is expressed as the ratio of the activity at 1.3 mM phosphoenolpyruvate to that found at 6.6 mM substrate (maximal velocity). There was no difference in maximal velocity in any of the experimental groups.

RESULTS AND DISCUSSION

On starvation, rats in all groups lost approximately 15% of their body weight. Average blood glucose was 3.3 mM. Diaminopropanol-treated rats tended to eat less on refeeding than did their saline-treated controls, so it was necessary to give all

TABLE 1. Polyamine synthetic decarboxylases and polyamine concentrations in liver of starved rats

	Starved control	Starved refed	Starved refed + diaminopropanol
Ornithine decarboxylase	5±1 (8)	1690±1385(6)	8±3 (9)
S-adenosylmethionine decarboxylase	122±42(8)	630±374 (6)	124±15(9)
Putrescine	13±5 (12)	41±20 (12)	63±27(10)
Spermidine	663±82(12)	772±142(12)	418±47(10)
Spermine	859±99(12)	734±119(12)	728±93(10)

Enzyme activities are given as pmol CO_2 formed per mg cytosol protein per hour. Concentrations of polyamines are expressed as nmol polyamine per gram fresh liver. Values represent Means±SD for (n) rats.

rats an amount of food (4 grams) which experience showed would be eaten by the inhibitor-treated group. This represented about one-half to one-third the quantity of food which would have been eaten by rats refeeding ad libitum. There was no effect of diaminopropanol treatment on the concentration of glucose (refed 6.7 ± 0.8 mM, n = 5; refed + diaminopropanol 6.2 ± 0.8 mM, n = 6) or insulin (refed 39 ± 9 μU/ml plasma, n = 10; refed + diaminopropanol 41 ± 15 μU/ml, n = 8) measured in the portal vein of these rats one hour after refeeding.

Polyamine synthesis and concentrations

Ornithine decarboxylase activity, shown in Table 1, was barely detectable in livers of starved rats. On refeeding, there was a rapid increase in activity to reach a peak approximately 350-fold higher than control by 3 hours after refeeding. This pattern of enzyme activity is similar to that observed by Domschke and Söling (5) in starved, refed rats. The increased activity of ornithine decarboxylase was completely inhibited by injection of diaminopropanol before and during refeeding. In our experience, the effect of the inhibitor on ornithine decarboxylase wears off after 2 to 3 hours, and thus it is necessary to give frequent injections.

We have also observed a rapid increase in the activity of

Table 2. Lipogenesis, pyruvate kinase activity and concentration of several metabolites in liver of starved rats.

	Starved control	Starved refed	Starved refed + diaminopropanol
Lipogenesis	3.06±0.74 (12)	7.96±2.48 (12)	5.09±1.72 (12)
Pyruvate kinase	0.14±0.02 (9)	0.23±0.06 (9)	0.15±0.05 (7)
Pyruvate	0.024±0.001(3)	0.019±0.003(3)	0.020±0.005(3)
Lactate	0.203±0.078(3)	0.343±0.055(3)	0.173±0.032(3)
Alanine	0.127±0.018(3)	0.167±0.029(3)	0.149±0.025(3)
Phosphoenol-pyruvate	0.087±0.015(3)	0.167±0.029(3)	0.113±0.023(3)
Fructose-1,6	0.026±0.002(3)	0.038±0.001(3)	0.021±0.003(3)

Lipogenesis is given as mol ^{3}H incorporated into lipid in vivo per gram liver per hour. Pyruvate kinase is given as the ratio of the activity at 1.3 mM phosphoenolpyruvate to that found at 6.6 mM substrate. Metabolite concentrations are expressed as μmol metabolite per gram fresh liver. Values represent Means SD for (n) rats.

S-adenosylmethionine decarboxylase after refeeding (Table 1), and this increase is also inhibited by pretreatment of rats with diaminopropanol. The cause for this effect of diaminopropanol is not known. We have, however, observed a similar depression of S-adenosylmethionine decarboxylase in liver of diabetic rats treated with insulin (9), and Grillo and coworkers (6) have made a similar observation in liver of chickens treated with insulin. The inhibition was not due to formation of macromolecular inhibitors (6,9) or to diaminopropanol interaction with the enzyme in vitro (9).

The concentration of putrescine and spermidine increased on refeeding, as might be expected from the increased activity of the decarboxylases responsible for their synthesis. Spermine concentration decreased somewhat on refeeding, to give a marked shift in the relative proportion of spermidine to spermine as previously observed by Domschke and Söling (5). It appears possible that there is conversion of spermine to spermidine by the acetylase/oxidase pathway (14); this would increase the spermidine concentration more rapidly than by depending on de novo synthesis from ornithine alone.

The increase in putrescine concentration was exaggerated in the diaminopropanol-treated refed rats, and even more so in diaminopropanol-treated starved rats (84 ± 28 nmol/gram, n = 12). This putrescine is probably generated from the degradation of spermidine by the acetylase/oxidase pathway. It appears likely that diaminopropanol activates this pathway, since the rapid marked fall in hepatic spermidine concentration observed in this study, and in diabetic rats in a previous study (9), would be unlikely otherwise. Spermidine concentration fell by one-third in the five hours following diaminopropanol treatment, whereas the half-life of spermidine is usually quoted in days (1).

Metabolic effects

The rate of lipogenesis in vivo (Table 2) increased approximately 2.5-fold in refed rats. In rats receiving diaminopropanol before refeeding, the increase was only 25% of that expected. Thus, in starved-refed rats (present study) and in diabetic rats (9), diaminopropanol pretreatment is able to depress the insulin-induced increase in the total lipogenic pathway in liver.

It is now thought that at least three of the key enzymes in the conversion of glucose to lipid in liver (pyruvate kinase, pyruvate dehydrogenase, acetyl CoA carboxylase) are activated by an insulin-induced dephosphorylation (8). The activity of one of these enzymes, pyruvate kinase, measured at low substrate concentration, doubled in response to refeeding, as previously shown by Blair and coworkers (2). This data is given in Table 2. Pretreatment of rats with diaminopropanol prevented the change in kinetic properties of pyruvate kinase, presumably by preventing dephosphorylation, without affecting the activity measurable at saturating phosphoenolpyruvate concentration. This effect on pyruvate kinase appeared to be dependent upon depletion of spermidine. In two rats which had received diaminopropanol before refeeding, the spermidine concentration inexplicably did not fall (0.629 and 0.606 umol/gram liver), although the ornithine decarboxylase activity was completely depressed (6 and 2 pmol/mg/h). In these two rats, there was an increase (to 0.20 in the pyruvate kinase activity ratio.

The concentrations of several metabolites in the glycolytic pathway are also given in Table 2. Lactate and alanine, which may be derived from glycolysis, are excellent gluconeogenic substrates. They are normally low in the liver of starved animals because of their rapid deployment for glucose synthesis (7). When starved rats are refed, liver metabolism switches from gluconeogenesis to glycolysis, and we observe a resultant increase in concentration of lactate and alanine. There is also a slight increase in the glycolytic intermediates, phosphoenolpyruvate and fructose-1,6-bisphosphate. All of these changes in metabolite concentrations, indicative of a shift from gluconeogenesis to glycolysis, are depressed by pretreatment of rats with diaminopropanol.

In conclusion, the rapid insulin-induced conversion of liver from gluconeogenesis to glycolysis and lipogenesis can be depressed by depletion of hepatic spermidine by pretreatment of rats with diaminopropanol. The data encourage us to speculate on a possible role of spermidine in insulin action. We would like to propose that spermidine is required for insulin to activate the phosphoprotein phosphatase which dephosphorylates pyruvate kinase and possibly other enzymes activated by insulin, such as acetyl CoA carboxylase and glycogen synthase. In support of this hypothesis, Huang and Chang (10) have provided evidence that muscle glycogen synthase phosphatase, another phosphoprotein phosphatase known to be activated by insulin, can be activated by polyamines. Should the hepatic phosphatase have a similar requirement for spermidine, then the depletion of spermidine in the diaminopropanol treated refed animals could account for the inhibition of insulin action in the livers of these rats.

ACKNOWLEDGEMENT

This research was supported by a grant from the Medical Research Council of Canada (MA-6635).

REFERENCES

1. Antrup, H. and Seiler, N. (1980): Neurochem. Res., 5:123-143.
2. Blair, J.B., Cimbala, M.A., Foster, J.L., and Morgan, R.A. (1976): J. Biol. Chem., 251:3756-3762.
3. Brosnan, M.E., Roebothan, B.V., and Hall, D.E. (1980): Biochem. J., 190:395-403.
4. Czok, R. and Lamprecht, W. (1974): In: Methods of Enzymatic Analysis, edited by H.U. Bergmeyer, pp. 1446-1451. Academic Press, New York.

5. Domschke, S. and Söling, H.D. (1973): Horm. Metab. Res., 5:97-101.
6. Grillo, M.A., Bedino, S. and Testore, G. (1980): Int. J. Biochem. 11:37-42.
7. Guynn, R.W., Veloso, D., and Veech. R.L. (1972): J. Biol. Chem., 247:7325-7331.
8. Hardie, G. (1981): Trends Biochem. Sci., 6:75-77.
9. Hu, Y.-W., Hall, D.E., and Brosnan, M.E. (1982): Can. J. Physiol. Pharmacol., 60: in press.
10. Huang, L.C. and Chang, L.Y. (1980): Biochim. Biophys. Acta, 613:106-115.
11. Lowenstein, J.M. (1971): J. Biol. Chem. 245:629-632.
12. Lowry, O.H. and Passonneau, J.V. (1972): A Flexible System of Enzymatic Analysis. Academic Press, New York.
13. Michal, G. and Beutler, H.-O. (1974): In: Methods of Enzymatic Analysis, edited by H.U. Bergmeyer, pp. 1314-1319. Academic Press, New York.
14. Pegg, A.E., Matsui, I., Seely, J.E., Pritchard, M.L., and Pösö, H. (1981): Med. Biol., 59:327-333.
15. Titanji, V.P.K., Zetterquist, O., and Engström, L. (1976): Biochim. Biophys. Acta, 422:98-108.
16. Usui, H., Imazu, M., Imaoka, T., and Takeda, M. (1978): Biochim. Biophys. Acta, 526:163-173.
17. Van Berkel, T.J.C., Kruijt, J.K., and Koster, J.F. (1977): Eur. J. Biochem., 81:423-432.
18. Wollenberger, A., Ristau, O., and Schoffa, G. (1960): Pflugers Archiv., 270:399-412.

Advances in Polyamine Research, Vol. 4, edited by
U. Bachrach, A. Kaye, and R. Chayen. Raven Press,
New York © 1983.

Immunocytochemical Demonstration of Ornithine Decarboxylase

Lo Persson, Elsa Rosengren, *Frank Sundler, and *Rolf Uddman

*Department of Physiology, University of Lund, S-223 62 Lund, Sweden; *Department of Histology, University of Lund, S-223 62 Lund, Sweden*

SUMMARY

Antibodies against purified mouse kidney ornithine decarboxylase (ODC) were used for the cellular localization of ODC by immunocytochemistry. The specimens examined were tissues known to be rich in ODC activity, i.e. the kidneys of mice subjected to testosterone administration, rat ovary after human chorionic gonadotropin (HCG) administration and rat placenta. In the kidneys intensely immunoreactive cells were found in the cortical tubules. In the rat ovary an occurrence of ODC-immunoreactivity was observed in the interstitial gland tissue as well as in the thecal layer. Placental ODC was confined to trophoblast cells of the labyrinth in the fetal placenta. In the present report we also show results demonstrating ODC-immunoreactivity in a neuronal tissue, namely in the external nerve fibres of the guinea-pig cochlear spiral and in the cell bodies in the spiral ganglion. Hence, our results show that ODC can be localized cellularly by immunocytochemistry. Knowledge of the cellular localization of the enzyme seems of general importance in the search for the biological roles of the polyamines.

INTRODUCTION

Although the exact biological function(s) of the polyamines putrescine, spermidine and spermine is unknown it is a well-known fact that the biosynthesis of the polyamines is closely related to cellular growth and proliferation (8,9,13,30). Ornithine decarboxylase (L-ornithine carboxy-lyase, EC 4.1.1.17) (ODC), which catalyzes the formation of putrescine, is considered a key enzyme in the biosynthetic pathway of polyamines. Induction of ODC has been reported to occur in various kinds of tissues after numerous different growth-promoting stimuli (13,30). Most studies concerning ODC are based on measurements of enzyme activity. Since the induction of ODC in several instances may be restricted to a few cells in the target tissue, a technique for the cytochemical localization of ODC appears highly desirable. Hence, the localization of ODC at the cellular level could provide new information in the search for the biological roles of the polyamines.

Immunocytochemical techniques have frequently been used as a tool for the localization of small concentrations of peptides and proteins, including enzymes. Although antibodies to mammalian ODC have been prepared in several laboratories (12,14-16,20,35,36) there is until recently (26,27) no reports on their use in immunocytochemistry. This may be due to earlier lack of antibodies of sufficient affinity and specificity to ODC. However, ODC was recently purified to apparent homogeneity from the kidneys of testosterone-treated mice (22,23). Using the purified enzyme it was possible to generate a high-titre antiserum containing seemingly monospecific antibodies against ODC (24). These antibodies have been used for the immunocytochemical localization of ODC in kidneys of mice subjected to testosterone administration (26) and in ovaries of prepubertal rats treated with human chorionic gonadotropin (HCG) (27).

In the present report these recent results are summarized together with the presentation of some new results on the immunocytochemical localization of ODC.

METHODS

Animals

The following tissues were studied: mouse kidney, the ovary and the placenta of rats and the cochlea of guinea-pigs. Male mice of the NMRI strain were gonadectomized and used for experiments 3-4 weeks later. Testosterone propionate, suspended in 50 μl of Arachis oil, was administered subcutaneously in a dose of 200 μg per day for 7 days. In other experiments prepubertal (28 days of age) and adult female Sprague-Dawley rats as well as adult guinea-pigs were used. HCG was given to the prepubertal rats subcutaneously in a dose of 50 IU in 0.1 ml saline 5 h before sacrifice. Pregnant rats were obtained by allowing mating in the proestrus or early estrus phase. The next day was considered as day 1 of pregnancy.

Tissue preparation

The animals were anesthetized and perfused via the heart with cold formaldehyde solution (4 % formaldehyde in 0.1 M phosphate buffer, pH 7.2, containing 5 mM dithiothreitol). Specimens from the kidneys (mouse), ovaries (rat) and placentas (rat) were immersed overnight in the same solution as used for perfusion. The temporal bones from the guinea-pigs were removed and the cochlea was exposed. The spiral ganglia were dissected out and immersed overnight in the perfusion solution. After thorough rinsing in phosphate buffer containing 5-25 % sucrose specimens were frozen on dry ice and sectioned on a cryostat at 10-15 μm.

Alternatively, spread preparations of the cochlear spiral were fixed in a buffered solution of formaldehyde and picric acid overnight. After dehydration they were cleared in xylene and hydrated in a series of ethanol solutions.

Immunocytochemistry

The sections and spread preparations were processed for the immunohistochemical demonstration of ODC using a rabbit antiserum (code no. 8111) obtained against purified ODC from kidneys of testosterone-treated mice (24). The site of the antigen-antibody reaction was revealed by the indirect immunofluorescence technique (IF) of Coons et al. (2) or by the peroxidase-antiperoxidase (PAP) method (32).

In the immunofluorescence procedure, sections and spread preparations were incubated with ODC antiserum, diluted 1:640 (kidney) or 1:80 (ovary, placenta and cochlea), for 3 h at room temperature. After thorough rinsing in phosphate buffered saline (PBS) containing 0.25 % Triton X-100 the immunoreaction was visualized by incubation for 30 min with fluorescein-conjugated goat anti-rabbit IgG diluted 1:20.

In the PAP procedure, sections and spread preparations were incubated with the ODC antiserum, diluted 1:2560 (kidney), 1:320 (ovary) or 1:80 (cochlea), for 18 h at 4 °C. After washing in PBS containing 0.25 % Triton X-100 the sections and spread preparations were incubated for 30 min with sheep anti-rabbit IgG diluted 1:30. After further washing the sections and spread preparations were incubated for 30 min with rabbit peroxidase-antiperoxidase complex diluted 1:160. After rinsing in Tris buffer, pH 7.6, the peroxidase activity was visualized by incubation with 3,3'-

diaminobenzidine tetrahydrochloride, 60 mg/100 ml, and 0.01 % hydrogen peroxide in 0.05 M Tris buffer, pH 7.6, for 1 h.

After staining, sections and spread preparations were rinsed in PBS containing 0.25 % Triton X-100 and mounted in phosphate-buffered glycerine (IF) or in Permount after dehydration (PAP).

Controls were run as recommended by Sternberger (32) and included sections exposed to ODC antiserum inactivated by the addition of the pure enzyme (100 µg/ml diluted antiserum).

RESULTS

Immunohistochemical localization of ODC in kidneys of testosterone-treated mice

The kidney of the male mouse contains high ODC activity. Orchidectomy results in a total disappearance of the enzyme activity within a few weeks, concomitant with a cellular atrophy in the kidney (10,11). Administration of testosterone reverses these effects and after testosterone treatment the renal activity of ODC reaches very high levels (11).

Using the antibodies generated against purified mouse kidney ODC attempts were made to localize ODC in the mouse kidney by immunocytochemistry (26). As seen in Fig. 1 numerous ODC-immunoreactive cells were found in the kidney of testosterone-treated mice. The cells were restricted to the cortical part of the kidney. Most, if not all, of the ODC-immunoreactivity was located in the cytoplasmic part of the cell. It was not established if the immunoreactive cells were located in proximal and/or distal tubules. However, the distribution pattern of these cells closely resembled that of proximal tubules. The glomeruli did not contain any immunoreactive material (Fig 2). The cells of the Bowman's capsule, however, displayed a weak immunoreactivity.

In agreement with biochemical results (11) the kidneys of gonadectomized male mice given the vehicle alone were found to completely lack immunoreactive ODC (Fig. 1).

Immunohistochemical localization of ODC in ovaries of rats treated with HCG

ODC activity is markedly stimulated in the rat ovary at proestrus, i.e. between the surge of luteinizing hormone and ovulation, indicating a role for polyamines in the ovulatory process (17). Likewise,

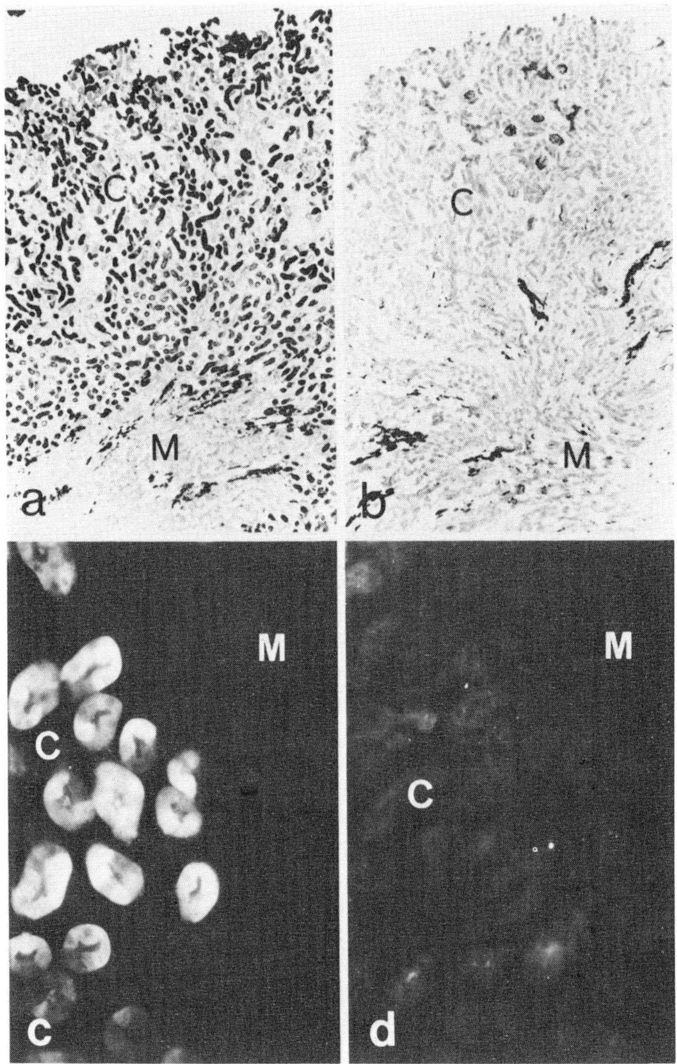

Fig. 1 a - d. Immunocytochemical localization of ODC in mouse kidney. (a) Section of kidney from mouse treated with testosterone, exhibiting immunoreactivity (PAP) in the cortex. x 40. (b) Section of kidney from a castrated control, demonstrating the absence of immunoreactivity (PAP). x 40. (c) Higher magnification of a section from a testosterone-stimulated mouse kidney, showing immunofluorescent cells in the cortex. x 250. (d) Higher magnification of a section of kidney from castrated control, showing absence of ODC-immunoreactive (IF) cells. x 250. C, Cortex; M, Medulla.

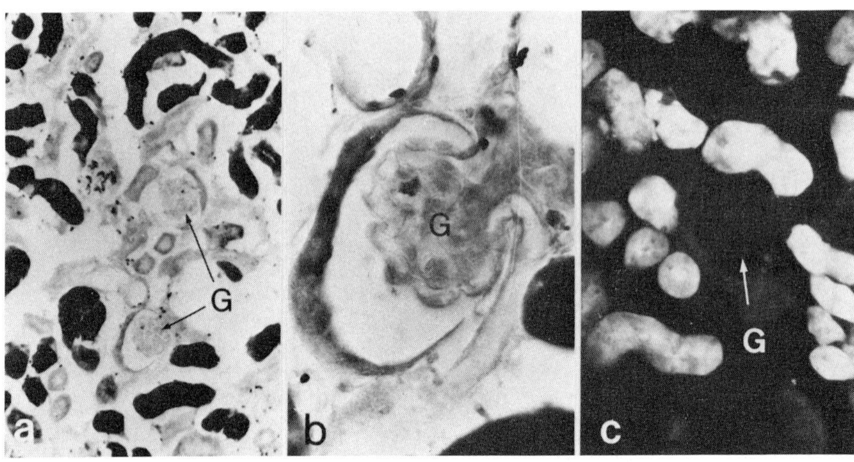

Fig. 2 a - c. Immunocytochemical localization of ODC in mouse kidney. (a) Section of kidney from mouse treated with testosterone, showing absence of immunoreactivity (PAP) in the glomerulus. x 180. (b) Higher magnification of (a). x 750. (c) Same as (a) but stained with the immunofluorescence technique. x 250. G, Glomerulus.

ovarian ODC is greatly increased after the administration of either luteinizing hormone or HCG (6,17). Hence, it appeared of interest to use immunocytochemical techniques to define the cellular distribution of ODC in the rat ovary after treatment with gonadotropins (27).

Administration of HCG to prepubertal female rats resulted in the appearance of numerous immunoreactive cells in the ovary (Fig. 3). These cells were restricted to the thecal layer surrounding the individual follicles and to the ovarian interstitial gland tissue. In the thecal layer, the immunoreactivity seemed to be confined to the cells of the theca interna, whereas the theca externa cells appeared to be devoid of immunoreactive ODC. No immunoreactivity was observed in the granulosa cells, the ovum or in the ovarian stroma.

A few weakly immunoreactive cells were observed in the thecal layer and in the interstitial gland tissue of ovaries from prepubertal rats treated with saline alone (Fig. 3).

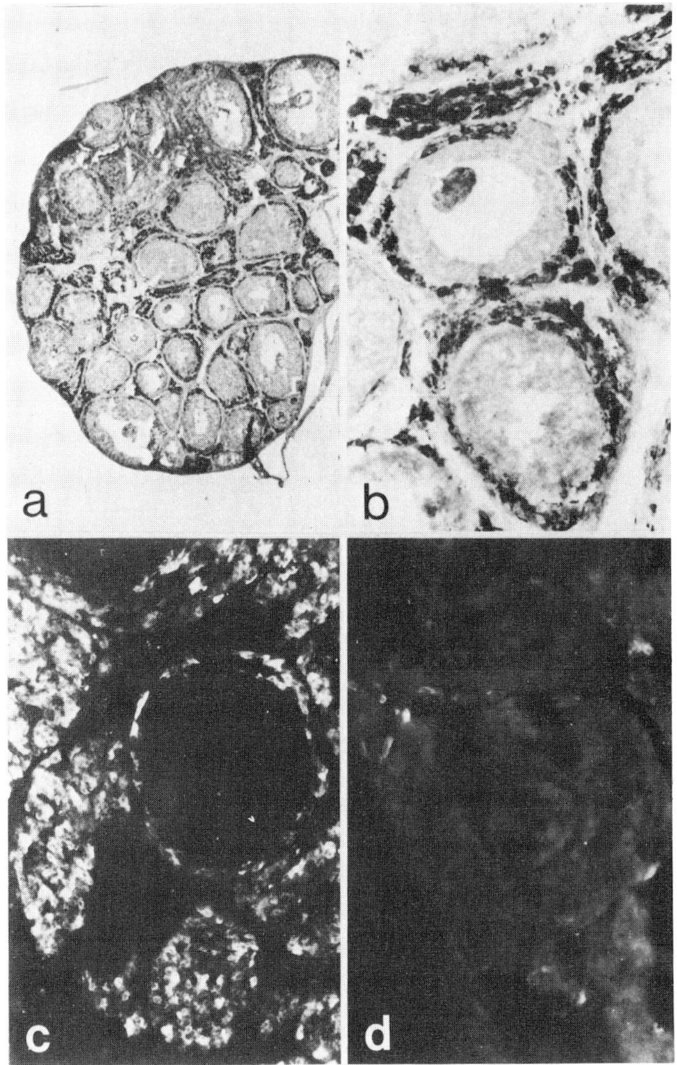

Fig. 3. a – d. Immunocytochemical localization of ODC in rat ovary. (a) Section of ovary from rat treated with HCG, showing immunoreactivity (PAP) in the thecal cells and in the interstitial gland tissue. x 40. (b) Higher magnification of (a). x 180. (c) Section of ovary from rat treated with HCG, demonstrating intensely immunofluorescent cells in the interstitial gland tissue as well as in the thecal layer. x 250. (d) Section of ovary from rat treated with vehicle alone, showing a few weakly immunofluorescent cells around a follicle. x 250.

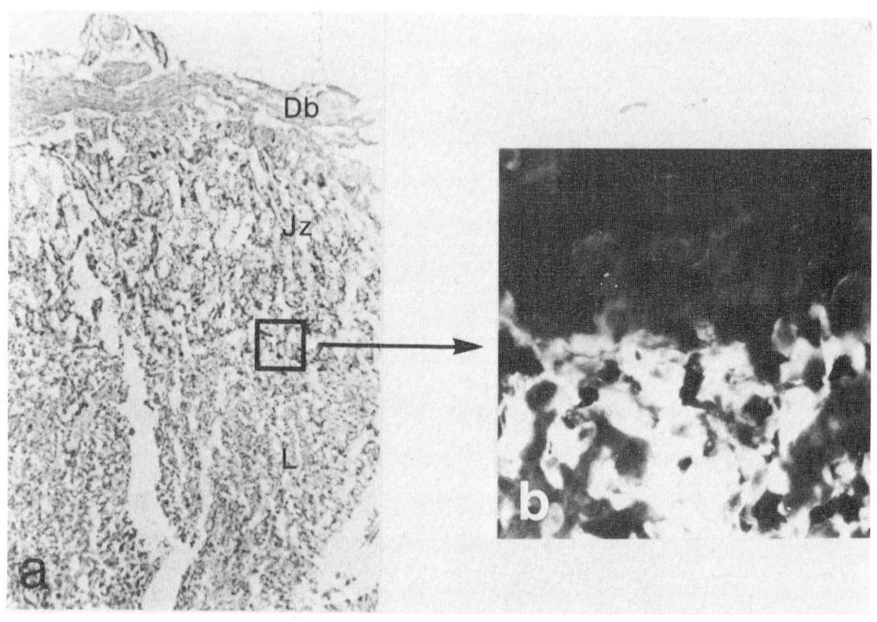

Fig. 4 a - b. Immunocytochemical localization of ODC in rat placenta at day 17 of pregnancy. (a) Section stained with hematoxylin and eosin. x 40. (b) Higher magnification of a section stained for ODC-immunoreactivity (IF). x 250. Db, Decidua basalis; Jz, Junctional zone; L, Labyrinth.

Immunohistochemical localization of ODC in the rat placenta

During pregnancy pronounced changes occur in the metabolism of polyamines in the placental tissue. The activity of diamine oxidase, which catalyzes the degradation of putrescine as well as other diamines, is remarkably high in the maternal part of the placenta, whereas the fetal part contains lower enzyme activity (5,18,33). The developing rat placenta also exhibits high activities of ODC. Placental ODC activity increases steadily up to day 17 of pregnancy, whereafter there is a slight decline until parturition (5,18). In contrast to the distribution pattern of diamine oxidase activity, ODC activity is mainly confined to the fetal part of the placenta.

The peak activity of placental ODC is comparable to

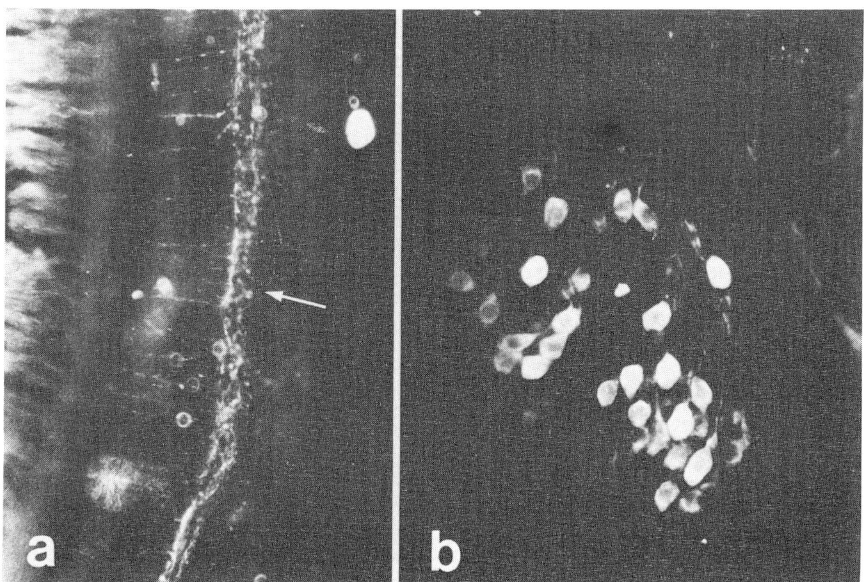

Fig. 5 a - b. Demonstration of ODC-immunoreactivity (IF) in (a) the external nerve fibres of a cochlear spiral and (b) in the cell bodies in the spiral ganglion. x 325.

that found in the ovaries of rats treated with HCG. Hence, it appeared possible to examine the cellular distribution of ODC in the rat placenta using immunocytochemistry. As shown in Fig. 4 immunocytochemistry gave positive results and in agreement with earlier biochemical results the ODC-immunoreactivity was shown to be restricted to cells in the fetal part of the placenta. Intense immunofluorescence was observed in the trophoblast cells of the labyrinth, whereas no ODC-immunoreactivity was found in the junctional zone or in the decidua basalis.

Immunohistochemical demonstration of ODC in a neuronal tissue

So far only the inner ear of guinea-pig has been examined in some detail. Spread preparations of the cochlear spiral were examined for ODC-immunoreactivity. As shown in Fig. 5 these preparations revealed intense ODC-immunoreactivity in the external spiral nerve fibres which innervate the external hair cells of the cochlea.

The cell bodies of these nerve fibres are located in the spiral ganglion. Hence, sections from this ganglion were processed for ODC immunocytochemistry. As seen in Fig. 5 intensely immunofluorescent cells were revealed in the spiral ganglion.

DISCUSSION

The localization of ODC by immunocytochemistry has been hampered for two main reasons. One is the extremely small quantities of enzyme protein present in the tissues (29). The other has been due to the lack of a specific and high-titre antiserum against ODC. Recently, the latter obstacle was overcome by the generation of apparently monospecific antibodies using ODC purified from kidneys of testosterone-treated mice (24). As shown in the present study as well as in previous reports (26,27) this antiserum was useful for the immunohistochemical localization of ODC in tissues known to exhibit high activities of the enzyme (6,11, 18). The specificity of the immunoreaction was supported by the close parallelism observed between the immunocytochemical staining and the presence of high enzyme activity. In addition the administration of 1,3-diaminopropane, a non-physiological "repressor" of ODC (25,28), resulted in a marked reduction in immunoreactivity as studied in the mouse kidney (26). Furthermore, in accordance with the well-known rapid turnover of ODC (31) administration of cycloheximide gave rise to a fast reduction in immunoreactive material in the kidney (26). Using another antiserum (kindly provided by Dr. S. Hayashi, Tokyo) raised against partly purified rat liver ODC (15) a similar distribution of immunoreactivity was observed in the kidneys of testosterone-treated mice, whereas no immunoreactive material was found in the kidneys of castrated controls (unpublished observations). Though this latter antiserum may contain antibodies against other proteins the finding strongly supports the specificity of the immunoreaction. Unfortunately, the titre of this antiserum was too low for defining the cellular distribution of immunoreactive material in the other tissues examined.

Recently, Gilad and Gilad (3,4) presented a technique for the histochemical localization of ODC utilizing α-difluoromethylornithine (DFMO), an enzyme-activated irreversible inhibitor of ODC (19), conjugated to the fluorescent molecule rhodamine. This technique was used to localize ODC in the developing rat cerebellum. The validity of this technique has, however, been questioned (21) due to the fact that the enzyme appears not to accept substrates with bulky groups on the terminal amino group (1). Another approach is the autoradiographical localization of ODC after administration of labelled DFMO. However, the possibility of a spurious distribution of radioactivity due to the inhibitor being metabolized cannot be excluded. In kidneys of mice treated with testosterone Pegg, Seely and Zagon (21) have recently employed ^{14}C-DFMO in the autoradiographical localization of ODC. In agreement with our results (26) they found that the enzyme was predominantly located in the cells of the proximal tubules.

The observation that the distribution of ODC in the mouse kidney after testosterone administration closely resembled that of proximal tubules is in accordance with the close connection generally observed between polyamine biosynthesis and cellular growth. The renal hypertrophy after testosterone treatment seems to be restricted to the cells of the proximal tubules (34). The induction of ODC in the cells of the theca interna and the interstitial gland tissue of the ovary by HCG is also in line with this concept. Gonadotropins have been shown to cause hypertrophic growth of these cells in the ovary (7).

The finding of ODC-immunoreactivity in the external spiral nerve fibres and in their cell bodies in the spiral ganglion is of great interest. The observation must, however, be interpreted with caution until it has been confirmed, using for example ^{3}H-DFMO. Such experiments are in progress in our laboratory. Nevertheless, immunocytochemical techniques may enable a closer examination of the distribution of ODC in neuronal tissues.

ACKNOWLEDGEMENTS

This study was supported by grants from the Swedish Medical Research Council (04X-02212 and 14X-14499) and the Faculty of Medicine, University of Lund.

REFERENCES

1. Bey, P., Danzin, C., van Dorsselaer, V., Mamont, P., Jung, M. and Tardiff, C. (1978): J. Med. Chem., 21: 50-55.
2. Coons, A.H., Leduc, E.H. and Connolly, J.M. (1955): J. Exp. Med., 102: 49-60.
3. Gilad, G.M. and Gilad, V.H. (1980): Biochem. Biophys. Res. Commun., 96: 1312-1316.

4. Gilad, G.M. and Gilad, V.H. (1981): In: Advances in Polyamine Research, edited by C.M. Caldarera, V. Zappia and U. Bachrach, vol. 3, pp. 111-121. Raven Press, New York.
5. Guha, S.K. and Jänne, J. (1976): Biochim. Biophys. Acta, 437: 244-252.
6. Guha, S.K. and Jänne, J. (1977): Biochem. Biophys. Res. Commun., 75, 136-142.
7. Guraya, S.S. (1978): In: International Review of Cytology, edited by G.H. Bourne and J.F. Danielli, pp. 171-245. Raven Press, New York.
8. Heby, O. (1981): Differentiation, 19: 1-20.
9. Heby, O. and Jänne, J. (1981): In: Polyamines in Biology and Medicine, edited by D.R. Morris and L.J. Marton, pp. 243-310. Marcel Dekker, Inc., New York.
10. Henningsson, S. and Rosengren, E. (1975): J. Physiol., 245: 467-479.
11. Henningsson, S., Persson, L. and Rosengren, E. (1978): Acta Physiol. Scand., 102: 385-393.
12. Hölttä, E. (1975): Biochim. Biophys. Acta, 399: 420-427.
13. Jänne, J., Pösö, H. and Raina, A. (1978): Biochim. Biophys. Acta, 473: 241-293.
14. Kallio, A., Löfman, M., Pösö, H. and Jänne, J. (1977): FEBS Lett., 79: 195-199.
15. Kameji, T., Murakami, Y., Fujita, K., Noguchi, T. and Hayashi, S. (1981): Med. Biol., 59: 296-299.
16. Kitani, T. and Fujisawa, H. (1981): Eur. J. Biochem., 119: 177-181.
17. Kobayashi, Y., Kupelian, J. and Maudsley, D.V. (1971): Science, 172: 379-380.
18. Maudsley, D.V. and Kobayashi, Y. (1977): Biochem. Pharmac., 26: 121-124.
19. Metcalf, B.W., Bey, P., Danzin, C., Jung, M.J., Casara, P. and Vevert, J.P. (1978): J. Amer. Chem. Soc., 100: 2551-2553.
20. Obenrader, M.F. and Prouty, W.F. (1977): J. Biol. Chem., 252: 2866-2872.
21. Pegg, A.E., Seely, J. and Zagon, I.S. (1982): Science, in press.
22. Persson, L. (1981): Acta Chem. Scand., B 35: 451-459.
23. Persson, L. (1981): Acta Chem. Scand., B 35: 737-738.
24. Persson, L. (1982): Acta Chem. Scand., in press.
25. Persson, L. and Rosengren, E. (1979): Acta Chem. Scand., B 33: 537-540.
26. Persson, L., Rosengren, E. and Sundler, F. (1982): Biochem. Biophys. Res. Commun., 104: 1196-1201.

27. Persson, L., Rosengren, E. and Sundler, F. (1982): Histochemistry, in press.
28. Pösö, H. and Jänne, J. (1976): Biochem. Biophys. Res. Commun., 69: 885-892.
29. Pritchard, M.L., Seely, J.E., Pösö, H., Jefferson, L.S. and Pegg, A.E. (1981): Biochem. Biophys. Res. Commun., 100, 1597-1603.
30. Russell, D.H. (1980): Pharmacology, 20, 117-129.
31. Russell, D.H. and Snyder, S.H. (1969): Mol. Pharmacol., 5: 253-262.
32. Sternberger, L.A. (1979): Immunocytochemistry. 2nd edn. John Wiley, New York.
33. Swanberg, H. (1950): Acta Physiol. Scand., 23: suppl. 79.
34. Tessmann, D. (1968): Z. Zellforsch., 86: 61-73.
35. Theoharides, T.C. and Canellakis, Z.N. (1976): J. Biol. Chem., 251: 1781-1784.
36. Weiss, J.M., Lembach, K.J. and Boucek, R.J. (1981): Biochem. J., 194: 229-239.

Advances in Polyamine Research, Vol. 4, edited by U. Bachrach, A. Kaye, and R. Chayen. Raven Press, New York © 1983.

Visualization of Rhodamine-Labeled α-Difluoromethylornithine in Viable Neuroblasts: Towards *In Vivo* Localization of Ornithine Decarboxylase

Gad M. Gilad and Varda H. Gilad

Department of Isotope Research, The Weizmann Institute of Science, 76100 Rehovot, Israel

SUMMARY

The present work describes the use of rhodamine-labeled α-difluoromethylornithine as a fluorescent marker for ornithine decarboxylase within viable neuroblasts in tissue culture.

INTRODUCTION

Affinity labeling of identified protein constituents in viable cells with fluorescent ligands has been used extensively to study the dynamic changes in localization, quantity and regulation of these molecules in situ. These studies delt mostly with cell surface constituents (1). Their labeling has been achieved by the use of fluorescent analogues of the naturally occurring ligands and the distribution of the corresponding membrane receptors has been studied (2,3,4).

Studies of intracellular proteins are complicated by the existence of the plasma membrane barrier. To circumvent this problem intracellular microinjections of proteins, labeled with fluorescent ligands into living cells,have been used to study the dynamic behavior of these proteins (5,6,7,8). However, in addition to being tedious, this technique may be damaging to the cell. The availability of a penetrable fluorescently labeled ligand would clearly be of great advantage in studies of intracellular behavior of proteins. Indeed, a fluorescent derivative of methotrexate has been recently developed, which retained the inhibitory characteristics for the enzyme, dihydrofolate reductase, of its unsubstituted precursor (9). Furthermore, this labeled inhibitor can be taken into and accumulate specifically in viable cells which contain the enzyme. Using a similar approach we have recently synthesized a fluorescently (rhodamine)-labeled derivative of α-difluoromethylornithine (α-DFMO), a suicidal inhibitor of the enzyme ornithine decarboxylase (ODC). This new compound which retained most of the inhibitory characteristics of α-DFMO has been used for cytochemical localization of ODC in tissue sections (10,11). We describe here a series of preliminary experiments demonstrating that rhodamine-labeled α-DFMO can be taken up by viable neuroblasts grown in tissue culture and inhibit ODC activity in these cells. Furthermore, the fluorescent inhibitor can be visualized within these viable cells. After preincubation of the cultures for 1h with the labeled inhibitor, all cells fluoresce. However, while the label is rapidly lost from differentiated cells, it is selectively retained within undifferentiated proliferating cells.

MATERIALS AND METHODS

The suicidal inhibitor of ODC, α-difluoromethylornithine (RMI 71.782), was a generous gift from Centre de Recherche Merrell International Subsidiary of the Dow Chemical Company, Strasbourg (France).

Preparation of Labeled Inhibitor

Alfa-DFMO was conjugated with rhodamine as described before (11). The inhibitor 0.03M was dissolved in H_2O and reacted with excess (150 mg/ml) cupric carbonate to mask the α-amino group from reacting with rhodamine, leaving the δ-amino group free. Rhodamine-B-isothiocyanate (Sigma) was added slowly using a 1:10 molar ratio, with the inhibitor-copper chelate in a 0.4M sodium carbonate-bicarbonate buffer, pH 8.4, and mixed overnight at 4°C. The mixture was then brought to pH 3 with 1M HCl and the precipitate formed was lyophylized. The lyophylized material was dissolved in 50 mM Tris-HCl buffer, pH 7.2, cleared from excess free rhodamine and inhibitor by high voltage paper electrophoresis and used for cytochemical studies.

Cell Culture

Primary cell cultures were prepared from cerebelli of 6-7d old Sprague-Dawley rats by a modification of the method of Yavin and Yavin (12). The cerebelli were rapidly excised and mechanically dissociated in Dulbeco's modified Eagle medium (GIBCO). After centrifugation at 200g for 5 min the pellet was resuspended in fresh medium containing 20% fetal calf serum and a mixture of antibiotic-antimycotic solution (GIBCO). After an additional washing the cells($1x10^6$ in 3 ml) were seeded in 5 cm^2 Falcon culture dishes precoated with poly (1-lysine). Cultures were maintained at 37°C in 5% CO_2, 95% air atmosphere. Experiments were carried out in 1, 2 and 3d old cultures. Figure 1 is a photomicrograph of a representative 2d old culture demonstrating differentiated neurons with extended neurites and undifferentiated proliferating neuroblasts in strings and clusters.

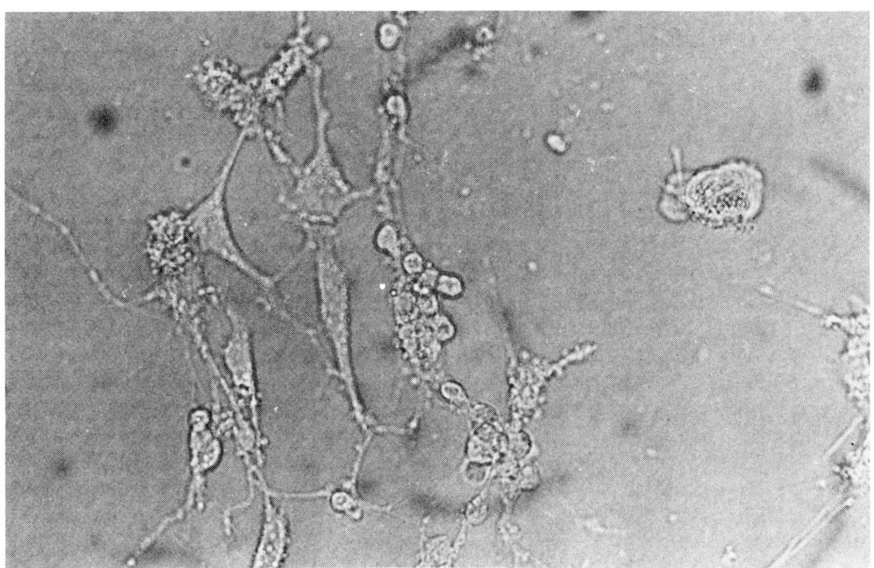

Fig. 1. A phase contrast micrograph of a representative field demonstrating neuroblasts growing in culture 48h after seeding. Note the differentiated cells with extended neurites and the smaller undifferentiated cells growing in clusters or strings. Original magnification x 160.

Biochemical Assays

Cells were scraped off the culture dish and homogenized in 100 mM Tris-HCl buffer (pH 7.4) containing 7.5 mM dithiothreitol, 6 mM ethylenediamine tetraacetic acid (EDTA) and 40 μM pyridoxal-5'-phosphate. ODC activity was assayed as previously described (11). Protein concentrations were determined according to the method of Lowry et al. (13).

Fluorescence Cytochemistry

In this series of experiments 1 or 2d old cultures were incubated for 1h in the presence of medium containing 50 μM rhodamine-labeled α-DFMO. Thereafter the cultures were washed with fresh medium and examined at different time intervals by fluorescence microscopy.

RESULTS

Inhibition of Ornithine Decarboxylase Activity

After seeding, ODC activity was greatly decreased as compared to activity in homologous cerebellar homogenates (Table 1). After 24h in culture however, the activity achieved levels comparable to those observed in homogenates. But, from then on ODC activity gradually declined to very low negligible levels (Table 1).

After treatment with rhodamine-labeled α-DFMO, ODC activity was inhibited by 82 and 88% in the 24 and 48h old cultures respectively (Table 1).

TABLE 1. Activity of ODC in cerebellar cultures and its inhibition after 1h incubation with rhodamine-labeled α-DFMO

Time in culture (h)	ODC activity (pmol $^{14}CO_2$/mg protein/h)	
	without labeled inhibitor	with labeled inhibitor
Original[(a)] homogenate	288	
Cell suspension	180	
24	305	95
48	220	48.5
72	37	37

[(a)]Enzyme activity was determined in homogenates of 6d old cerebelli, in cell suspensions prepared for culture, and in cultured cells at different times after seeding. Results are the mean values of two separate determinations done in duplicate.

Cytochemical Localization of Rhodamine-Labeled α-DFMO Within Viable Cells in Culture

Figure 2 demonstrates the localization of the labeled α-DFMO as observed by fluorescence microscopy in 2d old cultures. Fluorescence can be detected 1h after incubation with labeled α-DFMO within most cells in the culture (Fig. 2A and a). After 2h, fluorescence has been greatly reduced within differentiated cells which have extended neurites, but was selectively retained in undifferentiated smaller cells, which usually appeared in clusters (Fig. 1B and b). Later, 4h after preincubation, fluorescence was observed only in some of the undifferentiated cells (Fig. 1C and c).

DISCUSSION

The present work demonstrates that rhodamine-labeled α-DFMO can enter viable neuroblasts grown in tissue culture and inhibit their ODC activity. Furthermore, the labeled inhibitor can be visualized by fluorescence microscopy and is found to be retained selectively in undifferentiated proliferating neuroblasts.

As revealed by fluorescence microscopy the labeled-inhibitor initially enters all cells in the culture. However, it is not retained in all cells. Rather, over a period of 3 to 4h, it gradually disappears from differentiated cells in culture while in undifferentiated cells it is selectively retained.

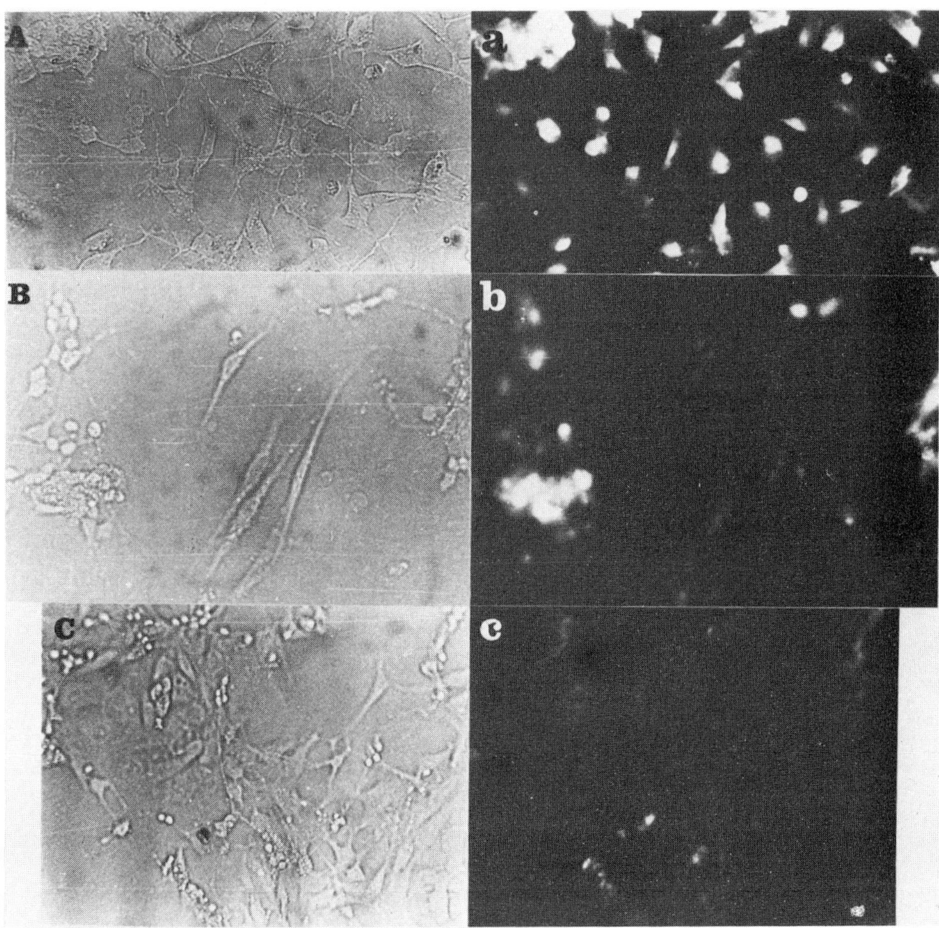

Fig. 2. Photomicrographs of 48h old cultures preincubated for 1h with rhodamine-labeled α-DFMO. Photomicrographs were taken immediately after preincubation (A and a), 2h later (B and b) and 4h later (C and c). On the left column (capital letters) are phase contrast micrographs of the corresponding fields on the right (small letters) photographed by fluorescence microscopy. Note that immediately after incubation fluorescence is observed in most of the cells (a), but 2h later fluorescence is observed selectively in undifferentiated cells in clusters (b), and after 4h only some of the smaller undifferentiated cells still fluoresce (c). Original magnifications: A and a, x 160; B and b, x 250; C and c, x 160.

These preliminary data indicate that using this fluorescently labeled inhibitor,it will be possible to study the dynamics of intracellular regulation of ODC in undisturbed viable cells.

REFERENCES

1. Schlessinger, J., and Elson, E.L. (1981): in Membrane Receptors. Methods for Purification and Characterization, Receptors and Recognition series B, Vol. 11, Edited by Jacobs, S. and Cuatrecasas, P., Chapman and Hall, London, pp. 155-170.
2. Axelrod, D., Ravdin, P., Koppel, D.E., Schlessinger, J., Webb, W.W., Elson, E.L., and Podleski, T.R. (1976): Proc. Natn. Acad. Sci. U.S.A., 73:4594-4598.
3. Schlessinger, J., Shechter, Y., Cuatrecasas, P., Willinghem, M.C., and Pastan, I. (1978): Proc. Natn. Acad. Sci. U.S.A., 75:5353-5357.
4. Shechter, Y., Hernaez, L., Schlessinger, J., and Cuatrecasas, P. (1979): Nature, 278:835-838.
5. Diacumakus, E.G. (1973): Methods Cell Biol., 7:287-311.
6. Feramisco, J.R. (1979): Proc. Natn. Acad. Sci. U.S.A., 76: 3967-3971.
7. Kreis, T.E., Winterhalter, K.H., and Birchmeier, W. (1979): Proc. Natn. Acad. Sci. U.S.A., 76:3814-3818.
8. Taylor, D.L., and Wang, Y.L. (1978): Proc. Natn. Acad. Sci. U.S.A., 75:857-861.
9. Henderson, G.B., Russell, A., and Whiteley, J.M. (1980): Arch. Biochem. Biophys., 202:29-34.
10. Gilad, G.M., and Gilad, V.H. (1980): Biochem. Biophys. Res. Comm., 96:1312-1316.
11. Gilad, G.M., and Gilad, V.H. (1981): J. Histochem. Cytochem., 29:687-692.
12. Yavin, Z., and Yavin, E. (1980): Develop. Biol., 75:454-459.
13. Lowry, O.H., Rosebrough, M.J., Farr, A.L., and Randall, R.S. (1951): J. Biol. Chem., 193:265-275.

Advances in Polyamine Research, Vol. 4, edited by U. Bachrach, A. Kaye, and R. Chayen. Raven Press, New York © 1983.

Mammalian Cell Mutants with Altered Levels of Ornithine Decarboxylase Activity

Carolyn Steglich, Jung Choi, and Immo E. Scheffler

Department of Biology, University of California, San Diego, La Jolla, California 92093

The enzyme ornithine decarboxylase has attracted much attention not only because it is the required, rate limiting enzyme in the synthesis of essential polyamines, but also because of its highly regulated pattern of activity during the cell cycle in quiescent and proliferating cells and tissue (for reviews see ref. 1 and 12). Evidence exists for control by both transcriptional (9,10) and posttranscriptional mechanisms (5,13), and perhaps by enzyme modification. The unusually short half life of the enzyme has been documented, and possibly is correlated with the induction of a specific antizyme which appears to inhibit the enzyme stoichiometrically (7,8). However, definitive studies have not yet been done because of the very low abundance of ODC protein even in induced cells, the difficulty of obtaining adequate amounts of highly purified enzyme, and the general lack of good, specific antisera. Determinations of ODC mRNA levels during induction have not been feasible because of low message levels and lack of a probe. This makes all evidence for transcriptional or posttranscriptional control indirect and subject to controversy.

In our laboratory we began some years ago with attempts to select and exploit mutants of mammalian cells in culture with alterations in ODC expression. A priori a large variety of mutations can be expected to contribute to our understanding of ODC and its regulation: 1) mutations in the structural gene for ODC could be responsible for a total lack of activity, alterations in specific activity, alterations in the half life of the active enzyme, changes in the interaction with the antizyme, and possibly changes in the modification of the enzyme by posttranslational mechanisms; 2) mutations in postulated regulatory genes could also have a variety of consequences, ranging from total suppression of activity to enzyme overproduction; 3) based on precedents in the somatic cell genetics literature (2,14,16) gene amplification is another possible genetic change affecting ODC activity. While some of these mutations might help to shed light on physiologically important steps in ODC regulation, others were expected to be particularly useful for the purification of large amounts of enzyme, for the purification of mRNA and the preparation of cDNA clones, or for gene transfer experiments also aimed at cloning the ODC gene and its flanking sequences.

We report below on progress towards these goals. We have been successful in isolation of clones which overproduce the enzyme, and clones which exhibit a total lack of ODC activity. Many of these studies have been published, and we refer to these original publications for details on experimental procedures (4,15).

RESULTS

A) Mutants which overproduce the enzyme

In our earlier attempts Chinese hamster ovary cells (CHO) were mutagenized and after a few days of growth in the absence of any drug resistant clones were selected in the presence of 1.7 mg/ml α-methylornithine. Subsequently cultures of the original resistant cells (C51) were subjected to further selection in the presence of 200 μg/ml α-difluoromethylornithine, and several clones were isolated, of which C51.18 is the best overproducer of ODC characterized so far. It should be noted that the phenotype is stable in the absence of selection. In the case of intraspecific hybrids between C51 cells and wild type CHO cells the drug resistance phenotype is codominant.

A comparison of the plating efficiencies of the wild type CHO cells and the variants C51 and C51.18 as a function of the inhibitor concentration is shown in Figure 1. Wild type cells start to be affected at ~1 μg/ml of αDFMO, and rarely some survive above 5 μg/ml; C51 cells are resistant to an almost tenfold higher concentration, and C51.18 cells survive and plate at concentrations up to 200 μg/ml.

In all the cases examined so far the drug resistance is associated with significantly elevated levels of ODC activity in the cells. When ODC activity is monitored in cells synchronized by serum starvation and refeeding, the kinetics of induction and then disappearance of ODC activity during the G1 phase of the cell cycle are identical in the wild type and overproducers, except that the overproducers reach a higher peak. In the overproducers the maxima are strikingly dependent on the history of the cells prior to the synchronization: cells which had been exposed to the inhibitor (α-MO) show four- to five-fold higher levels of induction compared to cells which had been maintained in the absence of drug for weeks (Fig. 2). To a lesser extent this observation can also be made with wild type cells which

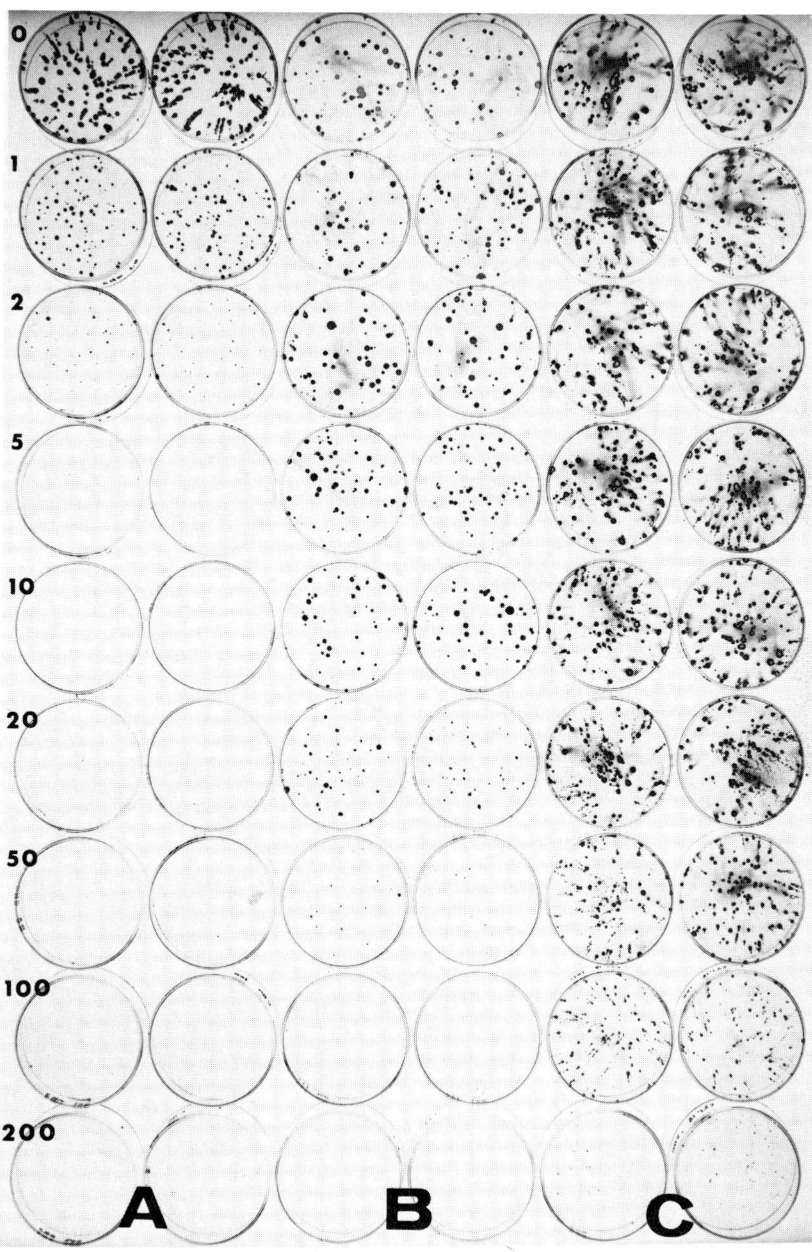

FIG. 1. Plating efficiencies of wild type (A), C51 (B), and C51.18 (C) Chinese hamster ovary cells in DME medium with different concentrations of αDFMO, given in μg/ml on the left side of the panel.

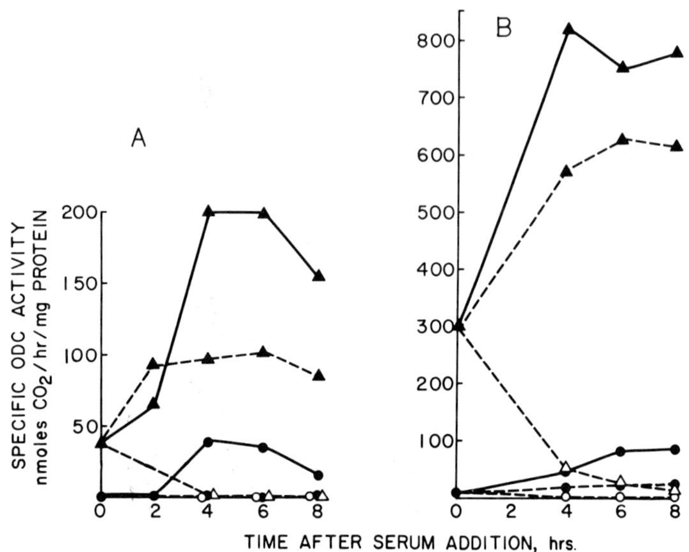

FIG. 2. Induction of ornithine decarboxylase in wild-type and mutant cells in the presence of various drugs. (A) without αMO treatment, (B) with 48-h pulse of αMO (2 mg/ml) prior to serum starvation. Cells were grown to near-confluence in medium containing 5% fetal calf serum, then switched to medium containing 0.1% serum after washing with TD buffer. After 24 h of serum starvation, cells were again washed and stimulated to grow by adding growth medium with 5% fetal calf serum, or with the same medium with 5 mM DRB, or with 100 μg/ml emetine. ●—●, wild type control; ●---●, wild type with DRB; ○ wild type with emetine; ▲—▲, C51 control; △---△, C51 with DRB; △, C51 with emetine.

have been exposed to the inhibitor for a non-lethal period of time. From these data we conclude that in the fully induced state C51 has about 5 to 10 times more activity than wild type, and C51.18 has 10 to 20 times more activity than wild type.

The issue has been raised whether the induction of ODC in cells after serum stimulation is due to transcriptional or post-transcriptional activation (5,9,10,13). Some of our studies with C51 cells, shown in Figure 2, address this problem. First of all, we note that the C51 cells have considerable activity at t=0 hours, compared to the wild type cells. When protein synthesis is inhibited at the time of stimulation, this activity decreases immediately with a half life of about 2 hours, but inhibition of mRNA synthesis by dichlororibofuranosyl-benzimidazole (DRB) prevents the increase in ODC activity only partially. We believe that this high level of enzyme activity in C51 cells at t=0 reflects the presence of a considerable amount of mRNA, and that the subsequent increase in ODC activity is therefore due to an increased rate of translation (in the presence of DRB) as well as additional synthesis of ODC message (in the absence of DRB). These mechanisms are obscured in the wild type cells because the message level at t=0 is very low.

The question can be raised whether the increase in activity of ODC in C51 and C51.18 is due to an increase in the amount of ODC protein, or due to some alteration in the enzyme itself. Comparative studies of the half lives, the K_m values for the substrate ornithine, and the inhibition by α-methylornithine had shown the enzyme in C51 cells to be identical to the wild type enzyme (4). More recently the tritiated form of α-difluoromethylornithine has become commercially available (New England Nuclear), and this has made it possible to titrate the amount of active ODC protein in an extract and to compare this value with the activity measured in a standard ODC assay. Figure 3 shows a correlation of our data obtained with a highly purified preparation of ODC from wild type cells, and crude extracts from C51 and C51.18 cells. Clearly, all the points fall on a straight line which passes through the origin, indicating that the enzymes in all the extracts in all the cell lines have identical turnover rates; i.e., they are physically indistinguishable. We conclude that overproduction of ODC activity is due to the overproduction of the ODC enzyme, which is likely, but not yet proved, to be the result of excess mRNA. The mechanism responsible for this elevation of ODC mRNA levels remains to be explored. It appears to be a relatively specific phenomenon, since S-adenosylmethionine decarboxylase is not affected.

We have also examined some of the physiological consequences of excess ODC activity. Our studies suggest that though intracellular levels of polyamines (specifically putrescine) can increase, substantial quantities of polyamines are secreted into the medium.

B) Mutants which lack ODC activity

A procedure for selecting mutants with little or no ODC activity has been published by us (15). It is summarized in outline in Figure 4. The basis for enriching a population of cells in ODC^- mutants is that they survive an exposure to [^{3}H] ornithine of high specific activity, while wild type cells convert it to polyamines which accumulate in the cells, leading to radiation suicide. From reconstitution experiments we estimate the enrichment to be of the order of 1000 to 10,000 fold. Two independent experiments have yielded mutant clones from Chinese hamster ovary cells which a) have an absolute requirement for

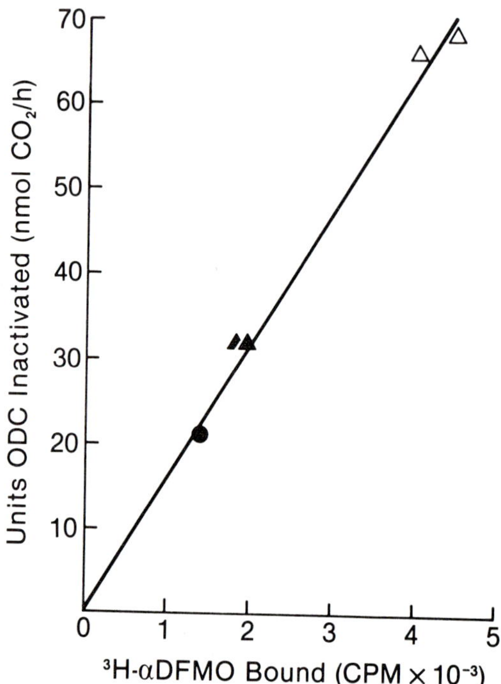

FIG. 3. Correlation of [^{3}H]- DFMO binding with ornithine decarboxylase activity inactivated by the analogue. 1) ● highly purified enzyme from CHO wild type cells; 2) ▲ crude extract from C51 cells; 3) △ crude extract from C51.18 cells.

putrescine in the medium (Fig. 5) and b) have very little or no ODC activity. Whole cells do not incorporate significant amounts of radioactivity derived from [^{3}H] ornithine (Fig. 6), and assays of ODC activity in synchronized, stimulated cells indicate that some mutants have 2-3% activity at the maximum when compared to wild type cells (Fig. 7), while others have no detectable activity.

While the enrichment procedure is quite efficient, the frequency at which these recessive mutants have been isolated from CHO cells is quite high from the standpoint of somatic cell genetics. It would be consistent with the existence of a single, intact or active gene for ODC in these cells. Even more striking is our observation that in the absence of putrescine in the medium a substantial number of small "revertant" colonies can be observed from populations of some mutants. These "revertants" have about 25% of the ODC activity of wild type cells.

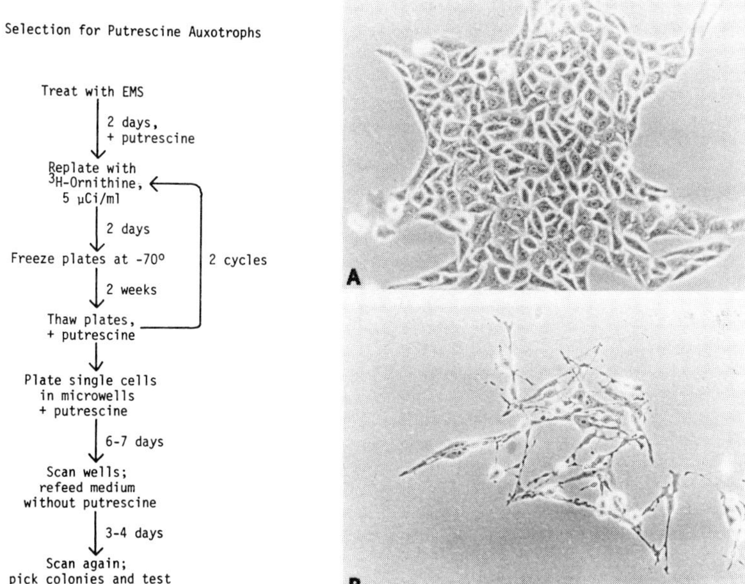

FIG. 4. Selection for putrescine auxotrophs. Details have been published elsewhere (15). EMS, ethylmethane sulfonate.

FIG. 5. 6-day-old colonies of the mutant C54 grown with (A) and without (B) putrescine supplementation of the medium.

Other ODC-deficient mutants yield no revertants. Further examination reveals that so far "revertants" have been observed only from mutants which still possess some residual activity, while mutants with zero activity have yielded no revertants even after an active search. One possible explanation for which we are still seeking proof is that a mechanism similar to the one giving rise to α-methylornithine-resistant cells leads to overproduction of a partially defective enzyme. Such a mechanism will obviously not help overcome the dependence on putrescine when the enzyme is completely inactive.

A major challenge of the future is to establish whether all of these ODC^- mutants are altered in the structural gene of the enzyme or in some other gene with a regulatory function. Complementation tests in somatic cell hybrids will help to establish whether we are dealing with more than one complementation group. The results of these experiments will determine the feasibility of using these cells in hybridization experiments aimed at mapping the ODC gene on a human (or mammalian)

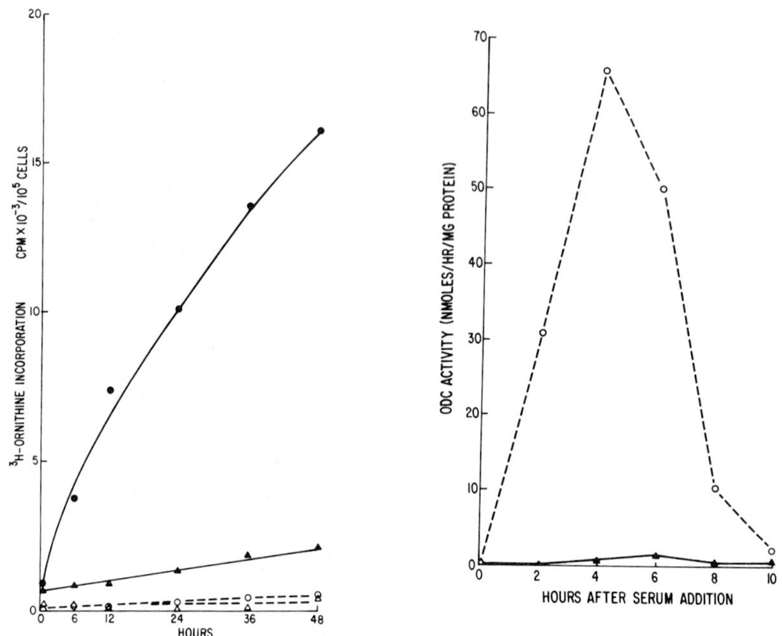

FIG. 6. Incorporation of [^{3}H] ornithine by wildtype CHO (●,○) and C54 (▲,△) into acid precipitable (open symbols) and acid-soluble (closed symbols) fractions.

FIG. 7. Cell cycle dependent ornithine decarboxylase activity of wildtype CHO (○) and mutant C54 (▲). Cells that had been growing in medium containing 5 x 10^{-4} M putrescine were replated in medium without putrescine. After 36 h the cells were washed and fed DME without serum for 12 h. At zero time the cells were refed serum containing medium. Plates were harvested at 2-h intervals for ODC assays.

chromosome, or of employing these cells as recipients in gene transfer experiments with the goal of cloning the ODC gene.

C) Selection of Mutants altered in ODC activity and the choice of tissue culture medium

In the course of our selections of α-methylornithine-resistant mutants and during testing of ODC-deficient mutants we noted a striking difference between two commonly employed tissue culture media: Dulbecco's modified Eagle's medium (DME) and Ham's F-10 medium. Neither medium contains putrescine, and the fetal calf serum required was the same in both. It appeared that cells had a significantly lower requirement for polyamines when growing in F-10 medium, as illustrated by the following series of experiments.

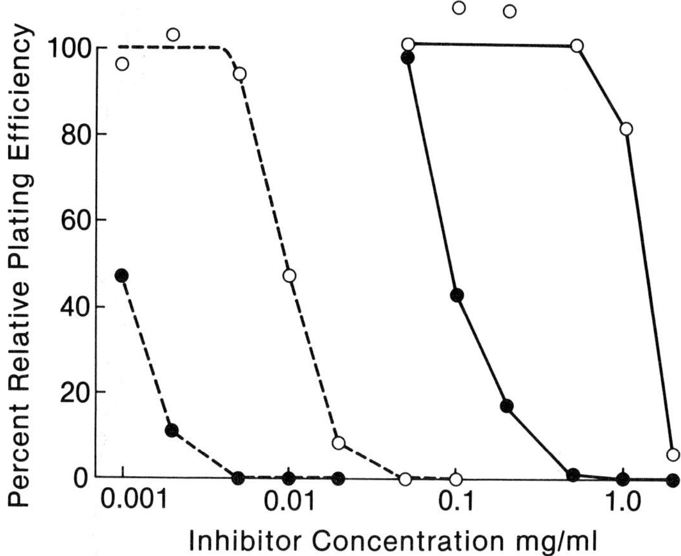

FIG. 8. Sensitivity of wildtype cells to ornithine analogues in DME and Ham's F-10 medium. Solid lines: αMO; dotted lines: αDFMO; ●, DME; O, F-10.

Figure 8 shows the plating efficiency of CHO wild type cells as a function of the concentration of α-methyl- and α-difluromethylornithine in either DME or F-10 medium. The cells are much less sensitive to growth inhibition by the analogues when grown in F-10. A similar effect is noted when ODC-deficient cells are plated in either DME or F-10 with or without added putrescine (Table 1). In DME, putrescine is an absolute requirement for growth of C54 cells, while F-10 without added putrescine supported the cells quite adequately. Growth curves shown in Figure 9 support this conclusion. When C54 cells are switched from DME with a putrescine supplement to DME alone, they stop dividing after three to four days, while no such arrest is seen when a similar shift is made in F-10.

This phenomenon was explored further with two clones (C55.7 and C55.15) which had no detectable ODC activity and so far had not yielded any revertants. In contrast to C54, these nonleaky mutants had a requirement for putrescine in F-10 medium. We tentatively conclude that the level of ODC activity required for cell proliferation in polyamine-free media can differ widely in DME and F-10, and this fact should be taken into account in selecting mammalian cell variants with altered ODC levels and in

Table 1
Plating efficiencies of ODC$^-$ mutants in different media

	DME		F-10	
	+ptr	-ptr	+ptr	-ptr
C54	84.7	<0.3	54.3	24.3
C55.7	59.5	<0.5	53.0	<0.5
C55.15	30.5	<0.5	22.0	<0.5

C54 is a leaky, revertible mutant with 2-3% residual ODC activity. C55.7 and C55.15 are nonreverting mutants with no detectable enzyme activity. Plating efficiencies are given as percent of cells plated.

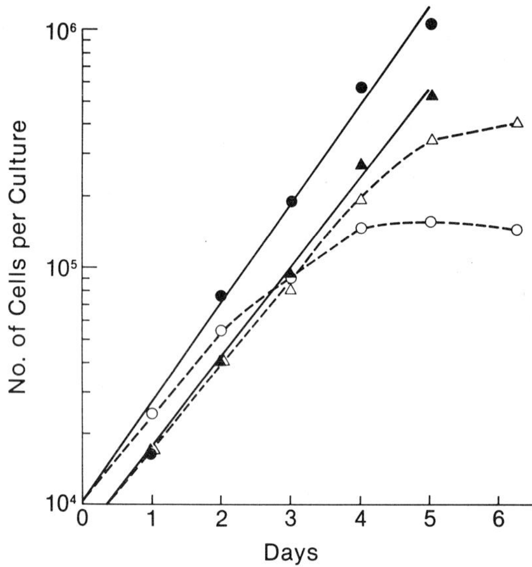

FIG. 9. Growth curves of ODC-deficient C54 cells in DME and F-10 media with (closed symbols) and without (open symbols) putrescine supplementation. ●, ○: DME; ▲, △: Ham's F-10.

designing and interpreting experiments dealing with polyamine depletion in cultured cells.

CONCLUSIONS

Our selections so far have yielded two new types of somatic cell mutants altered in ODC activity: overproducers can be isolated as cells able to proliferate in the presence of the analogues αMO and αDFMO, and mutants completely deficient in ODC activity can be selected by a suicide enrichment procedure with [^{3}H] ornithine.

While the overproducers are immediately useful as a source of substantial quantities of enzyme, they also present the challenge of having to establish the mechanisms by which increased amounts of enzyme are produced and maintained. Mamont and colleagues (11) have isolated rat hepatoma cells resistant to the antiproliferative action of α-MO. It appears that in this case elevated enzyme levels are the result of an increased stability of the enzyme. The mutants we describe are quite different. Indirect experiments suggest elevated levels of ODC mRNAs in these cells, but it will require a cloned probe to establish whether the cause is gene amplification, an alteration in the promoter region of the gene, or possibly a relatively stable epigenetic change. Whatever the answer, interesting insights into the regulation of expression of ODC can be expected.

The ODC-deficient cells raise a number of different questions. Of immediate concern is whether we are dealing with regulatory gene or structural gene alterations. A structural gene mutation could be exploited for a variety of goals, including the mapping of the gene on a mammalian chromosome and the cloning of the gene. Specific regulatory genes in mammalian cells are the subjects of frequent speculations, but few, if any well studied examples exist. Some ODC deficient yeast mutants described by Cohn _et al_. (6) have been tentatively characterized to carry regulatory gene mutations, but the mechanism of this regulation and the nature of the defective gene remain to be elucidated. If the induction of a specific inhibitor were to be postulated to account for the low ODC activity in C54 cells, two other observations would require explanation: a) in somatic cell hybrids the ODC$^-$ phenotype is recessive; b) in mixing experiments with mutant and wild type extracts excess inhibitor cannot be demonstrated (15).

It is our hope that after years of strictly biochemical approaches to the study of ODC and its regulation in mammalian cells the isolation of mutants such as those we described will open up a successful new approach to the study of this fascinating enzyme.

REFERENCES

1. Bachrach, U., (1981): In: Polyamines in Biomedical Research, edited by J.M. Gaugas, pp. 81-107, John Wiley and Sons, New York.

2. Beach, L.R., and R.D. Palmiter (1981): Proc. Nat. Acad. Sci. USA, 78: 2110-2114.
3. Canellakis, E.S., D. Viceps-Madore, D.A. Kyriakidis, and J.S. Heller (1979): Curr. Topics Cell. Regul., 15: 155-202.
4. Choi, J., and Scheffler, I.E. (1981): Somat. Cell Genet., 7: 219-233.
5. Clark, J.L., and S. Greenspan (1979): Exp. Cell Res., 118: 253-260.
6. Cohn, M.S., C. W. Tabor, and H. Tabor (1980): J. Bact., 142: 791-799.
7. Heller, J.S., and E.S. Canellakis (1981): In: Polyamines in Biomedical Research, edited by J.M. Gaugas, pp. 135-146, John Wiley and Sons, New York.
8. Heller, J.S., and E.S. Canellakis (1981): J. Cell. Physiol., 107: 209-217.
9. Landy-Otsuka, F., and I.E. Scheffler (1978): Proc. Nat. Acad. Sci. USA, 75: 5001-5005.
10. Landy-Otsuka, F., and I.E. Scheffler (1980): J. Cell. Physiol., 105: 209-220.
11. Mamont, P.S., M.-C. Duchesne, J. Grove, and C. Tardif, (1978): Exp. Cell Res., 115: 387-393.
12. McCann, P.P. (1981): In: Polyamines in Biomedical Research, edited by J.M. Gaugas, pp. 109-124, John Wiley and Sons, New York.
13. McCormick, F. (1977): J. Cell Physiol. 93: 285-292.
14. Schimke, R.T., R.J. Kaufman, F.W. Alt, and R.F. Kellems (1978): Science, 202: 1051-1055.
15. Steglich, C., and Scheffler, I.E. (1982): J. Biol. Chem., 257: 4603-4609.
16. Wahl, G.M., R.A. Padgett, and G.R. Stark (1974): J. Biol. Chem., 254: 8679-8689.

Advances in Polyamine Research, Vol. 4, edited by
U. Bachrach, A. Kaye, and R. Chayen. Raven Press,
New York © 1983.

The Use of α-Difluoromethylornithine to Study Eukaryotic Ornithine Decarboxylases

Hannu Pösö, *James E. Seely, **Ian S. Zagon, and
*Anthony E. Pegg

*Research Laboratories of the State Alcohol Monopoly (Alko), SF-00107, Helsinki 10, Finland;
Department of *Physiology, and **Anatomy, Milton S. Hershey Medical Center,
Pennsylvania State University, Hershey, Pennsylvania 17033*

α-Difluoromethylornithine (DFMO) is a potent, enzyme-activated irreversible inhibitor of ornithine decarboxylases (ODC) from mammalian tissues (1,12,20,23,24) and lower eukaryotes (4,14,19). DFMO is a highly specific inhibitor which appears not to have significant effects on any other metabolic process and as such has been widely used to investigate the role of polyamines in cellular physiology. Its specificity is confirmed by numerous observations that the effects of DFMO on many different cells are reversed by addition of putrescine, the product of the ODC reaction (see reviews in 5,9,10,11,25). DFMO has, therefore, been an important tool in furthering the understanding of the function of polyamines and is likely to continue to be of great value in such research. These aspects of the biochemistry and pharmacology of DFMO are not considered further in the present article which describes the use of DFMO in another aspect of polyamine biochemistry, namely the study of ODC. Since DFMO forms a stable covalent bond with ODC and can be synthesized in a radiolabeled form, it can be used in a number of ways to obtain information concerning this enzyme. Some examples of these which have already been demonstrated are discussed below. Additional potential uses of DFMO in this field which seem probable, but have not yet been fully described are also indicated. There has been intense research interest in ODC for many reasons. It provides the only metabolic route to putrescine in mammalian cells and many other eukaryotes. ODC is highly inducible and can change many fold within a few hours of exposure to growth-promoting stimuli. The enzyme protein appears to have a very rapid rate of turnover and may also be subject to regulation by post-translational modifications. These and other properties of ODC are covered more extensively in several recent reviews (2,3,17,21).

TITRATION OF ODC WITH DFMO

The rationale for the synthesis of DFMO as a potential inhibitor of ODC was that it would act as a substrate for the enzyme. Such decarboxylation would generate an intermediate carbanionic species which would lose a fluorine producing a highly reactive intermediate able to alkylate a nucleophilic site on the enzyme (2,12). This brilliant concept of Bey, Metcalf and colleagues was strikingly justified by their finding that DFMO did lead to irreversible inactivation of ODC when incubated with the enzyme under conditions permitting the expression of enzyme activity. Kinetic analysis was consistent with the postulated mechanism since the inactivation proceeded in a manner which was compatible with an enzyme-activated "suicide" mechanism having a Ki of 39 µM and a $t_{1/2}$ of 3.1 min for loss of activity at infinite inhibitor concentration (2,12). As expected from this mechanism, the rate of inactivation of ODC was much reduced by the presence of L-ornithine, putrescine, or α-methylornithine which compete with DFMO for binding at the active site (2,12).

Further support for this scheme was provided by studies in our laboratory in which [5-^{14}C]DFMO was incubated with rat liver ODC (15,20). There was an exact parallel between the incorporation of radioactivity into the protein and the loss of enzyme activity (Table 1). Such incorporation did not occur when the incubation conditions were not appropriate for ODC activity to be manifest or when no ODC activity was present. The rate of incorporation and inactivation was greatly reduced when excess L-ornithine was present. In the absence of L-ornithine, the incorporation of [5-^{14}C]DFMO was complete within 60 min of incubation at 37° in the presence of 4 µM [5-^{14}C]DFMO (20,23). The labeled protein was quite stable to extensive dialysis, precipitation with acid, alcohol or ammonium sulfate or to polyacrylamide gel electrophoresis under denaturing conditions. These results confirm that a stable covalent bond

TABLE 1. Effect of DFMO concentration on binding to rat ODC and inactivation of the enzyme

[5-^{14}C]DFMO concentration	% Ornithine decarboxylase activity remaining after 60 min	[5-^{14}C]DFMO bound (cpm) in 60 min
0	100	--
0.2	74	19
0.5	40	44
1.0	25	74
2.0	9	95
3.0	6	118
5.0	4	120

Further details are given in the text and in reference 20.

between the drug and the ODC protein was made as a consequence of the inactivation.

When highly purified or crude rodent ODC preparations were inactivated by incubation with [5-^{14}C]DFMO, only a single protein became labeled and this corresponded exactly to ODC on analysis by gel filtration or polyacrylamide gel electrophoresis (20,23,24). This protein had a molecular weight of about 100,000 under native conditions and 50,000 under denaturing conditions which is consistent with the mammalian enzyme being comprised of two identical subunits (see below). Assuming that only one molecule of DFMO is needed for the inactivation of each subunit, the number of molecules of ODC present in tissue extracts from animals after various physiological manipulations could be calculated (15,20,22-24). These results indicated that the changes in rat liver ODC brought about by partial hepatectomy, thioacetamide, carbon tetrachloride or cycloheximide were due to changes in the number of molecules of the enzyme rather than to changes in the catalytic center activity of the enzyme (23). However, even after maximal induction of ODC in rat liver by thioacetamide, the enzyme represented only about 1 part in 850,000 of the total soluble protein which corresponds to less than 15,000 molecules per cell (assuming that the enzyme is present predominantly in the hepatocytes). Without induction, ODC is of the order of 1 part in 60 million of the hepatic soluble protein corresponding to only 100-200 molecules per cell.

Similar findings that changes in ODC activity appeared to be due to changes in the amount of enzyme protein were made when extracts from mouse kidneys exposed to androgens or to cycloheximide were studied (22,23). ODC is highly androgen sensitive in the mouse and after stimulation with excess testosterone, it was found by titration with DFMO that ODC represented about 1 part in 10,000 of the soluble protein in the mouse kidney. The titration wth [5-^{14}C]DFMO has also been used to estimate the amount of ODC present in the yeasts, <u>Saccharomyces cerevisiae</u> and <u>Saccharomyces uvarum</u> (19).

This technique for estimating the amount of ODC protein present in a cell extract has several advantages over immunochemical determinations previously described for ODC, particularly since the specificity of these antibodies was not firmly established and the low amount of ODC necessitated indirect methods for quantitating the immune complex (see 3,17,22). The DFMO titration method is, therefore, likely to find wider applications. However, it suffers from several problems which must be taken into account before its use or in interpretation of the results. Firstly, the minute quantities of ODC present in most tissues cannot be studied without substantial purification and concentration if the [5-^{14}C]DFMO which has a specific activity of only 60 mCi/mmol is used. Since this is almost completely replaced with ^{14}C at the 5 position and there are only 4 other suitable carbons available for substitution with ^{14}C, there is little potential to increase this specific activity without changing the isotope. Recently [5-^{3}H]DFMO of specific activity 15,000-30,000 mCi/mmol has been synthesized by New England Nuclear Corp., Boston, MA and is commercially available from this source. This may prove suitable for titration experiments where low ODC activities are involved, but the use of ^{3}H which could exchange during storage or work-up of the samples will require careful experimental justification before results with it can be accepted. An even more troublesome problem is that the rate of inactivation of ODC by DFMO is concentration dependent and it may be difficult to achieve complete inactivation with the low concentration of the drug commensurate with a high specific activity.

Finally, the method can only be used for estimating the number of ODC molecules which are enzymatically active during the assay procedure since DFMO has to be activated by decarboxylation at the enzymes active site. Cryptic forms of ODC which are totally inactive (2,3) under the assay conditions will not become labeled as noted by Mitchell and colleagues for the form of ODC from Physarum polycephalum which has an unphysiologically high Km for pyridoxal phosphate (13,14).

MONITORING OF PURIFICATIONS OF ORNITHINE DECARBOXYLASE

The titration of crude cell extracts with labeled DFMO can be used as described above to compare various sources for the purification of ODC. Our finding that more than 800,000-fold purification would be needed to obtain pure ODC from thioacetamide-induced rat liver and that one rat liver contains less than 1 μg of the enzyme protein (15,20) was a powerful consideration in our decision to look for another mammalian source from which to purify this enzyme. The DFMO titration data suggested that in the androgen induced mouse kidney only a 10,000-fold purification was needed and we, therefore, decided to use this as a source. The mouse kidney ODC was purified 10,350-fold and the resulting enzyme appeared to be homogeneous by a variety of criteria (24). The specific activity of the final material (about 50 μmol CO_2 produced per mg protein) was in excellent agreement with that predicted from the DFMO binding data and suggests that the enzyme preparation does not contain significant amounts of inactive protein (24). Recently, this proof of homogeneity has also been applied to the enzyme from yeast (28) and rat liver (6). Hayashi and his colleagues (7,8) have carried out the impressive feat of purifying rat liver ODC almost 400,000-fold to homogeneity starting from livers induced not only with thioacetamide, but also with δ-amino-α-hydrazinovaleric acid (which doubles the starting specific activity over that obtained with thioacetamide alone). Thus, their purifications and final specific activity of 20 μmol CO_2 per min per mg are in excellent agreement with our predictions based on the titration with DFMO. Furthermore, their titrations of the purified rat enzyme with [5-^{14}C]DFMO also confirm that the material binds the expected amount of 1 molecule of drug per subunit (6).

In addition to the prediction of final specific activity, the labeling of ODC with radioactive DFMO can be used to confirm (or deny) the identity of a protein fraction or band on a gel. Comigration of the labeled and native protein would be expected except under very narrow range isoelectric focusing conditions where any change difference due to the bound drug could lead to a separation. Another possible approach which has not yet been exploited is to inactivate the enzyme with labeled DFMO and purify the radioactive protein which could then be used for structural or any other studies not requiring the active enzyme. Whilst the purification of a radioactively labeled protein has obvious advantages, this approach does have the difficulty that affinity chromatography which gives an excellent purification for native ODC (6,7,19,24) is unlikely to work when the active site is blocked by DFMO.

COMPARISONS OF ORNITHINE DECARBOXYLASES FROM VARIOUS SEPCIES

After labeling with radioactive DFMO, which can be carried out even in crude extracts, the ODC protein (which is the only one which becomes labeled) can be separated by standard techniques of protein fractionation including analysis on polyacrylamide gel electrophoresis in the presence of sodium dodecylsulfate. In this way, ODC molecules isolated from different species can be compared. Although rat liver and mouse kindey ODC have quite similar size and properties, they can be separated in this way and the rat liver enzyme consists of two subunits of M.W. about 55,000 whereas the mouse kidney enzyme is a dimer with two subunits of M.W. about 53,000 (24). In contrast, ODC from S. cerevisiae or S. uvarum consists of a monomer of M.W. about 70,000 (19,27).

The stoichiometry of binding of DFMO to ODC can also be used to compare the catalytic center activity (enzyme turnover number) of the enzyme from different species. For example, the mouse kidney enzyme was inactivated by the binding of 430 fmol of drug per unit whereas the rat liver enzyme requires 840 fmol/unit [1 unit is 1 nmol of CO_2/min]. Since the size of the enzymes is quite similar (see above) this indicates that the mouse enzyme has a higher catalytic center activity (23) and the Vmax values for the purified enzymes given above are in agreement with this (7,24). The yeast ODC bound about 1410 fmol/unit suggesting that it has a substantially lower catalytic center activity than the mammalian enzyme (19), but the predicted specific activity from this study is still about 10 μmol CO_2/min/mg which is twenty times that of the homogenous enzyme preparation described by Tyagi et al. (27) from a mutant S. cerevisiae which overproduces the enzyme. Whether the differences in strain is responsible for this discrepancy remains to be determined.

AUTORADIOGRAPHIC LOCALIZATIONS OF ORNITHINE DECARBOXYLASE

When [5-^{14}C]DFMO was administered to mice and extracts of the kidneys prepared 0.5 to 10 h later and analyzed by polyacrylamide gel electrophoresis only one protein band became labeled and this corresponded to ODC (22). This remarkable specificity of in vivo incorporation of the label into only one protein was confirmed by observations that incorporation was much higher in the kidney than in liver or brain which have much lower contents of ODC and that incorporation into all tissues was greatly decreased when the mice were pretreated with cycloheximide for 6 h to reduce ODC activities (16). Therefore, the incorporation of radioactivity from [5-^{14}C]DFMO into macromolecular form can be used for autoradiographic localization of ODC. Such studies indicate that the majority of the enzyme in the androgen-induced mouse kidney is in the cells of the proximal tubules (16). This is in agreement with a recent study of the localization of this enzyme by immunofluorescence techniques (18) and strengthens the evidence linking ODC to cell growth. Hypertrophy of the proximal tubules is known to occur after androgen treatment in the mouse and other enzymes such as β-glucuronidase induced by testosterone are known to be present in these cells (26). However, unlike β-glucuronidase which is located in the lysosomes and microsomes (26) the ODC appeared to be mainly cytoplasmic (16). Our results with [^{14}C]DFMO show that most and possibly all of the ODC is in the cytoplasm, but the resolution was not sufficient to rule out the nucleus as a minor site. Studies with [^{3}H]DFMO which should give a better resolution are in progress to provide more evidence on this point.

The use of [^{3}H]DFMO for autoradiographic localization of ODC should permit the visualization of this enzyme in cells and tissues which could not be studied with [^{14}C]DFMO because they contained much lower amounts of enzyme than the mouse kidney. The use of the drug with a much higher specific activity should obviate this problem, but it will be necessary to devise an in vitro labeling technique in which tissue samples are incubated with the drug under conditions permitting a high enough concentration to enter the cell for a sufficient time to bring about substantial inactivation. These experiments are in progress using the mouse kidney where the distribution is known as a control.

In our experiments with the mouse kidney system, the dose of [5-^{14}C]DFMO given in vivo was sufficient to bring about the loss of at least 85% of the renal ODC activity. Therefore, this percentage of the enzyme became labeled and there can be no question that the localization suggested by the incorporation represents the distribution of the enzyme. If only a small fraction of the enzyme activity is lost when [5-^{3}H]DFMO is used in studies of other tissues some proof that the distribution of label incorporated is a function of enzyme distribution rather than the selective uptake of drug into these cells must be obtained.

TURNOVER OF ORNITHINE DECARBOXYLASE PROTEIN

It has been known for some time that ODC activity is lost rapidly after inhibition of protein synthesis. Indeed ODC has one of the shortest half lives of mammalian enzymes (17,21). However, some doubts have been expressed that the loss of activity in such experiments is actually due to degradation of the enzyme protein (2,3,13). Confirmation with antibodies to quantitate the protein has been reported, but convincing evidence of the monospecificity of these antibodies was not given (17,22). The specific labeling of ODC in vivo by DFMO provides a system in which the turnover of the protein can be measured directly. A single dose of [5-^{14}C]DFMO was given to mice and 30 min later a large excess dose of unlabeled DFMO was administered to prevent further incorporation of the radioactive drug. The rate of decay of the labeled protein was then measured (Table 2). It was found that the protein was degraded with a half life of 80 min in androgen-treated mice and 40 min in control male CD1 mice. As can be seen from Table 2, these values were exactly the same as those obtained by following the loss of enzyme activity after treatment with cycloheximide (22). It might be argued that the DFMO-protein complex is regarded as an abnormal protein by the cell and, therefore, degraded rapidly, but the exact correspondence between the loss of enzyme activity when protein synthesis is blocked and the loss of this labeled protein is good evidence against this particularly since the androgenic status changes the half life. These results, therefore, confirm directly that ODC does indeed turn over very rapidly in the mouse kidney.

Further use of the DFMO-labeled protein may be possible in following the degradation of ODC. For example, the possible transfer of the protein into lysosomes and the appearance of protein fragments might be detectable. However, in the mouse kidney extracts the only labeled bands on the gels corresponded to the intact protein. This could indicate that after an initial cleavage further degradation was very rapid, but another possibility is that the cleavage produces a small peptide containing the labeled bound DFMO and a larger piece remains but is unlabeled.

DEVELOPMENT OF A RADIOIMMUNOASSAY FOR ORNITHINE DECARBOXYLASE

The DFMO labeled ODC can be used as a convenient antigen to test for antibody production in rabbits immunized with ODC or to screen potential hybridomas making antibodies to ODC. By incubation of 28 μg of ODC from mouse kidney with 100 μCi of [5-^{3}H]DFMO (15,000 μCi/μmol), 40 μM pyridoxal phosphate, 2.5 mM dithiothreitol, 50 mM Tris HCl, pH 7.5 for 1 h at 37°, approximately 9.5 μCi of labeled enzyme was obtained. Aliquots of about 5-7 nCi were used to test putative antisera. This mater-

TABLE 2. Half-life of mouse kidney ODC activity and DFMO-labeled ODC protein

Treatment of mice	Time after unlabeled DFMO chase	cpm DFMO bound/ mg protein	% Zero-time	Time after cycloheximide (min)	Units enzyme activity/ mg protein	% Zero-time
Androgen	0	426 ± 58	100	0	247 ± 25	100
Androgen				30	198 ± 30	80
Androgen	45	237 ± 40	56			
Androgen				60	178 ± 25	72
Androgen	90	213 ± 34	50			
Androgen				120	69 ± 20	28
Androgen	180	112 ± 20	26	180	51 ± 12	21
Control	0	77 ± 19	100	0	30 ± 13	100
Control				30	15 ± 4	50
Control	60	29 ± 9	37	60	10 ± 6	33

Details for this experiment are described in reference 22 and in text. 1 unit of enzyme released 1 nmol of CO_2 per 30 min.

ial was all precipitable by addition of serum from a rabbit immunized with ODC followed by addition of protein A. The binding could be competed against by unreacted ODC. This system can, therefore, be used as a competitive RIA which permits measurement of the amount of ODC protein. Preliminary results with such an RIA are shown in Table 3. It can be seen that very small amounts of ODC protein can be quantitated in this system. The comparison of results obtained with the RIA which should detect all forms of ODC and those with titration with labeled DFMO which as described above detects only the active form should provide a sensitive method for the search for cryptic or inactive forms.

TABLE 3. Competition of native ODC with [^{3}H]DFMO-labeled ODC for binding with anti-ODC antibodies

Units native ODC added	Bound [^{3}H]DFMO labeled ODC/Free [^{3}H]DFMO-labeled ODC
0	1.36
5.5	0.65
11	0.52
22	0.37
55	0.22
110	0.13

Rabbits were immunized with 40 µg purified ODC and boosted with 25 µg 6 weeks later and bled 2 weeks after the boost. A 1:300 dilution of serum was used in the experiment. Native ODC (1 unit releases 1 nmol CO_2/30 min and is equal to about 1.5 ng protein), [^{3}H]DFMO-labeled-ODC (~4000 cpm), and serum were incubated for 90 minutes at room temperature. Twenty-five µl of Staphylococcus Aureus ghosts were added and incubated for an additional 2 hrs at room temperature. The samples were then centrifuged to pellet the ghosts and the supernatant counted for radioactivity. The ghosts were washed twice to remove any unbound radioactivity and counted.

MECHANISM OF INACTIVATION BY DFMO AND COMPARISONS OF THE ACTIVE SITE OF ORNITHINE DECARBOXYLASE

As described above, the general outline of the mechanism by which DFMO leads to the loss of ODC activity are clear, but at present the exact nature of the alkylating speces, the site of attack on the protein and the ratio of decarboxylations carried out to inactivation events are not known. Studies of all of these parameters are in progress using the labeled drug. By preparing protease digests of the labeled protein, a peptide containing the bound DFMO can be identified and sequenced. This peptide is likely to contain at least part of the active site. The most likely amino acid acceptor of the activated species from DFMO is lysine and such addition should prevent cleavage by trypsin at this site and alter the tryptic peptide map. In any case, complete digestion with a mixture of non-specific proteases should lead to the release of the DFMO-amino acid adduct.

An important property of an enzyme-activated irreversible inhibitor is the ratio of turnover of the suicide substrate to inactivation of the enzyme. This ratio is not yet known for DFMO, but can easily be investigated by comparing the binding of [5-^{14}C]DFMO with the production of $^{14}CO_2$ from [1-^{14}C]DFMO and this experiment is in progress. However, it appears likely that this ratio will be quite close to 1. In our experiments with [5-^{3}H]DFMO and ODC to prepare labeled antigen described above, there was only a 10-fold molar excess of DL-[5-^{3}H]DFMO over the amount of enzyme protein. Since only the L-isomer of DFMO is active and 20% of this was incorporated into the enzyme, the ratio must be less than 5.

ACKNOWLEDGEMENTS

We thank Dr. P. P. McCann for his advice and encouragement in the development of these studies. This work was supported by grants CA-18138 and training grant HL-07223 from NIH and by a research fellowship from the Research Council for Natural Sciences of the Academy of Finland and of a travel grant from the League of Finnish American Societies Scholarship Foundation (thanks to Scandinavia grant) to H. P. We also thank Mrs. Bonnie Merlino for manuscript preparation.

REFERENCES

1. Bey, P. (1978): In: Enzyme-Activated Irreversible Inhibitors, edited by N. Seiler, M. J. Jung, and J. Koch-Weser, pp. 27-41. Elsevier/North Holland, Amsterdam.
2. Canellakis, E. S., Kyriakidis, D. A., Heller, J. S., and Parolak, J. W. (1981): Med. Biol. 59:279-285.
3. Canellakis, E. S., Viceps-Madore, D., Kyriakidis, D. A., and Heller, J. S. (1979): In: Current Topics in Cellular Regulation, edited by B. L. Horecker and E. R. Stadtman, Vol. 5, pp. 155-202. Academic Press, New York.
4. Garofalo, J., Bacchi, C. J., McLaughlin, S. P, Mockenhaupt, D., Trueba, G., and Hutner, S. H. (1982): J. Protozoology, in press.
5. Heby, O. (1981): Differentiation 14:1-20.
6. Kameji, T., and Hayashi, S. (1982): personal communication.
7. Kameji, T., Murakami, Y., Fujita, K., and Hayashi, S. (1982): Biochim. Biophys. Acta, in press.
8. Kameji, T., Murakami, Y., Fujita, K., Noguchi, T., and Hayashi, S. (1981): Med. Biol. 59:296-299.
9. Mamont, P. S., Duchesne, M.-C., Joder-Ohlenbusch, A.-M., and Grove, J. (1978): In: Enzyme-Activated Irreversible Inhibitors, edited by N. Seiler, M. J. Jung and J. Koch-Weser, pp. 43-54. Elsevier/North Holland Biomedical

Press, New York.

10. McCann, P. P., Bacchi, C. J., Clarkson, A. B., Jr., Seed, J. R., Nathan, H. C., Arnele, B. O., Hutner, S. H. and Sjoerdsma. A. (1981): Med. Biol. 59:434-440.
11. McCann, P. P., Bacchi, C. J., Hanson, C. D., Cain, G. D., Nathan, H. C., Hutner, S. H., and Sjoerdsma, A. (1981): In: Advances in Polyamine Research, Vol. 3, edited by C. M. Caldarera, V. Zappia, and U. Bachrach, pp. 97-110. Raven Press, New York.
12. Metcalf, B. W., Bey, P., Danzin, C., Jung, M. J., Casara, P., and Vevert, J. (1978): J. Am. Chem. Soc. 100:2551-2553.
13. Mitchell, J. L. A. (1981): In: Advances in Polyamine Research, Vol. 3, edited by C. M. Calderera, V. Zappia, and U. Bachrach, pp. 15-26. Raven Press, New York.
14. Mitchell, J. L. A., Yingling, R. A., and Mitchell, G. K. (1981): FEBS Lett. 131:305-309.
15. Pegg, A. E., Matsui, I., Seely, J. E., Pritchard, M. L., and Pösö, H. (1981): Med. Biol. 59:327-333.
16. Pegg, A. E., Seely, J. E., and Zagon, I. S. (1982): Science, in press.
17. Pegg, A. E., and Williams-Ashman, H. G. (1981): In: Polyamines in Biology and Medicine, edited by D. R. Morris, and L. J. Marton, pp. 3-42. Marcel Dekker, New York.
18. Persson, L., Rosengren, E. and Sundler, F. (1982): Biochem. Biophys. Res. Commun. 104:1196-1201.
19. Pösö, H., and Pegg, A. E. (1982) FEBS Lett., submitted.
20. Prichard, M. L., J. E. Seely, H. Pösö, L. S. Jefferson, and Pegg, A. E. (1981): Biochem. Biophys. Res. Commun. 100:1597-1603.
21. Russell, D. H. (1980): Pharmacology 20:117-129.
22. Seely, J. E., Pösö, H., and Pegg, A. E. (1982): J. Biol. Chem., in press.
23. Seely, J. E., Pösö, H., and Pegg, A. E. (1982): Biochem. J., in press.
24. Seely, J. E., Pösö, H., and Pegg, A. E. (1982): Biochemistry, in press.
25. Seiler, N., Danzin, C., Prakash, N. J., and Koch-Weser, J. (1978): In: Enzyme-Activated Irreversible Inhibitors, edited by N. Seiler, M. J. Jung, and J. Koch-Weser, pp. 27-41. Elsevier/North Holland, New York.
26. Swank, R. T., Paigen, K., Davey, R., Chapman, V., Labarca, C., Watson, G., Ganshow, R., Brandt, e. J., and Novak, E. (1978): Rec. Prog. Horm. Res. 34:401-436.
27. Tyagi, A. K., Tabor, C. W., and Tabor, H. (1981): J. Biol. Chem. 256:12156-12163.
28. Tyagi, A. K., Tabor, C. W., and Tabor, H. (1982): Fed. Proc. 41:Abst. 6591.

Advances in Polyamine Research, Vol. 4, edited by
U. Bachrach, A. Kaye, and R. Chayen. Raven Press,
New York © 1983.

New Perspectives on Polyamine-Dependent Protein Kinase and the Regulation of Ornithine Decarboxylase by Reversible Phosphorylation

Glenn D. Kuehn and Valerie J. Atmar

Department of Chemistry, New Mexico State University, Las Cruces, New Mexico 88003

INTRODUCTION

This chapter presents two disparate topics relating to the enzyme which we have termed, polyamine-dependent protein kinase. First, we shall present results which show that interferon induces the enzymic activity of this protein kinase two orders of magnitude in Ehrlich ascites tumor cells. Induced polyamine-dependent protein kinase activity and double-stranded RNA-dependent protein kinase activity, previously reported by other investigators, have several common properties. Second, we shall discuss our most recent findings that calcium ion and calmodulin can modulate the capacity of polyamine-dependent protein kinase, from Physarum polycephalum, to phosphorylatively inactivate ornithine decarboxylase. This observation suggests a plausible mechanism for calcium ion in the regulation of ornithine decarboxylase activity.

REGULATION OF ORNITHINE DECARBOXYLASE BY REVERSIBLE PHOSPHORYLATION

The discovery of the regulation of ornithine decarboxylase (OrnDCase) by reversible phosphorylation in the slime mold, Physarum polycephalum, originated in our earlier observation of the actual phosphorylation reaction which yielded phospho-OrnDCase. In 1975, we discovered a phosphorylation reaction in nuclei and nucleoli of P. polycephalum that was absolutely dependent on the polyamines, spermidine and spermine (22). The major phosphorylated product of this reaction was an acidic nonhistone, phosphoprotein of molecular weight (M_r) 70,000 (1). Spermidine and spermine synergistically activated the phosphorylation reaction. Putrescine antagonized the activation by spermidine and spermine. The phosphoprotein was subsequently purified to homogeneity (20,21). It was shown to demonstrate specific associative properties toward palindromic, ribosomal DNA from the nucleolus and it also demonstrated properties suggestive of a regulatory role in in vitro rRNA gene transcription by RNA polymerase I (2). These properties were dependent on the phosphorylation state of the M_r 70,000 phosphoprotein.

A protein kinase of M_r 26,000 was discovered and subsequently purified from nuclei or nucleoli which catalyzed phosphorylation of the M_r 70,000 protein in a spermidine/spermine-dependent reaction (10). Moreover, a portion of the protein kinase isolable from phosphocellulose chromatography copurified as a complex tightly associated with its natural phosphate-acceptor protein, the M_r 70,000 peptide. The finding that salt, such as 160 mM NaCl, could dissociate the complex formed between the two proteins (3,21), coupled with additional characterization of the M_r 70,000 phosphoprotein, led us to recognize that many properties of the protein kinase-acceptor protein complex resembled those described earlier for the interaction of OrnDCase with its antizyme (17,27). Indeed, we subsequently showed by several rigorous criteria including amino acid composition, molecular weight, enzymatic activity, and immunological properties, that the M_r 70,000 phosphate-acceptor protein and authentic OrnDCase were the same protein (3). Stoichiometric phosphorylation of the OrnDCase subunit by one phosphate group was sufficient to inhibit the capacity of the enzyme to catalyze decarboxylation of L-ornithine. A functional role for phosphorylation of OrnDCase was thus established.

Many mechanisms have been proposed to account for apparent, rapid turnover (appearance and disappearance) of OrnDCase activity. These have been discussed in earlier volumes of this series (26). They include: (i) control at the level of mRNA synthesis; (ii) translational control by polyamines; (iii) antizyme induction or release by polyamines; and (iv) unspecified transitions between active and less active forms. Thus, our original report on control of OrnDCase by reversible phosphorylation was treated with skepticism. Critics were reluctant to accept that the phosphate-acceptor protein for polyamine-dependent protein kinase was OrnDCase. Our value for the monomer subunit M_r of 70,000 for OrnDCase from P. polycephalum was higher than the 55,000 value for OrnDCase from rat liver (32) or from calf liver (15). The subunit M_r of OrnDCase prepared from rat liver has also been reported to have values that range from 50,000 to 90,000 for alledged, homogeneous preparations (6,29,30). Recently, Tyagi et al. (40) have found that the M_r for the OrnDCase subunit from another fungus, Saccharomyces cerevisiae, is indeed, 68,000. Moreover, the yeast-enzyme has been isolated as a phosphorylated peptide by immunological precipitation (39).

Polyamine-dependent protein kinase activity has been demonstrated in nuclei from rat liver (unpublished results), Ehrlich ascites tumor cells (36), P. polycephalum (3), and in bovine spermatozoa (4). The fractionated enzyme from all of these sources has the capacity to phosphorylate purifed OrnDCase from

P. polycephalum. These observations suggest that the regulation of OrnDCase by reversible phosphorylation may be broadly distributed among eukaryotes. It should not be surprising that reversible phosphorylation may be an important element for control of this enzyme. Biochemists have speculated for years that likely candidates for discovery of new examples of enzymes controlled by this mechanism would be those known to be subject to hormonal control. In this regard, OrnDCase was a prime candidate. A marked increase in OrnDCase activity, on the order of a thousand-fold in some cases, has been found to occur during the early phase of cellular response to a host of hormones. Virtually every hormone tested, which includes at least 26 reported examples, has been shown to increase OrnDCase in appropriate target tissues (5,28). Some hormones decrease its activity. Shortly after the discovery of OrnDCase in animal tissues, the enzyme was reported to have the shortest known activity half-life of any known eukaryotic enzyme. To a biochemist, these exceptional properties taken collectively hinted of an enzymatic process involving reversible, post-translational modification of the enzyme.

An appraisal of the evidence for early proposals which attempted to explain rapid control of OrnDCase activity supports a strong case for some type of post-transcriptional control. Moreover, whatever the mechanism(s) for rapid regulation of OrnDCase, increasing evidence indicates that the polyamines themselves, namely spermidine and spermine, are involved. These polyamines participate in some type of negative feedback process which is antagonized by putrescine (25,31). Recent investigations by Paulus and Davis (31), working with Neurospora, have elegantly identified spermidine, and perhaps spermine, as the active agents which exert a strict negative control on OrnDCase. Perceived inhibition of OrnDCase exerted by exogenous putrescine was found to be due to efficient intracellular conversion to spermidine and spermine. Restoration of the normal intracellular putrescine content, by addition of putrescine to the growth medium of polyamine-deficient cells, transiently increased OrnDCase activity. Thereafter, intracellular conversion of putrescine to spermidine was accompanied by inactivation of the enzyme at a rate that was similar to that found on addition of spermidine alone. Thus, putrescine antagonized the inhibition of OrnDCase by spermidine. Mamont et al. (25), working with cultured rat hepatoma cells, have reached similar conclusions regarding polyamine regulation of S-adenosylmethionine decarboxylase. These conclusions, derived from intact cell studies, are in accord with our findings on the influence of the polyamines on the polyamine-dependent protein kinase reaction. Spermidine and spermine activate the polyamine-dependent protein kinase reaction which phosphorylatively inactivates OrnDCase. Activation of the protein kinase by spermidine and spermine is reversed by putrescine. Interestingly, the observation that S-adenosylmethionine decarboxylase is controlled in intact cells (25) by mechanisms similar to those that regulate OrnDCase (31), suggests that reversible phosphorylation may be found to control the former enzyme as well.

INTERFERON INDUCTION OF POLYAMINE-DEPENDENT PROTEIN KINASE IN EHRLICH ASCITES TUMOR CELLS

Recognition that the polyamine-dependent protein kinase inactivated OrnDCase by phosphorylation drew our attention to a previously recognized consequence of interferon treatment of mammalian cells (12,13,23,37). Exposure of sensitive cells to interferon induces an antiviral state. This state results in part from an inhibition of viral mRNA translation. The state is established by several biochemical processes requiring ATP hydrolysis, among them being: phosphorylation of certain ribosome-associated proteins, synthesis of a small 2',5'-oligo(adenylic acid)nucleotide inhibitor of protein synthesis, activation of a nuclease, and induction of a new protein kinase activity. The protein kinase requires double-stranded RNA (dsRNA) for maximum catalytic activity. This enzyme phosphorylates two endogenous polypeptides of apparent M_r 35,000 and M_r 67,000-70,000 (18). The M_r 35,000 polypeptide is proposed to be the smallest of the subunits of one of the initiation factors, eIF-2, involved in protein synthesis (34). The identity or function of the M_r 70,000 phosphopeptide is unknown. The phosphorylation of the M_r 70,000 peptide is the most apparent difference between interferon-treated and control cell fractions. It can serve as a marker for the action of interferon (33). It appears only under conditions where the antiviral state is induced.

The dsRNA-dependency of interferon-induced protein kinase has been exploited for purification of the enzyme. Affinity chromatography on the agarose-linked synthetic dsRNA polymer, poly-(I:C), comprised of the nucleosides inosine (I) and cytidine (C), has been successful for purification of the protein kinase. Interestingly, purification trials on poly(I:C)-Sepharose of cell extracts from cultured mouse L cells revealed that the protein kinase and the M_r 70,000 phosphate-acceptor protein copurified (18). The similarities of M_r of the phosphate-acceptor proteins for the polyamine-dependent and the dsRNA-dependent protein kinases, coupled with the recent findings that interferon treatment of mammalian cells results in a rapid loss of OrnDCase activity (12,13,23,37), prompted us to investigate the relationship of interferon treatment to the expression of these two protein kinase activities.

Phosphocellulose Chromatography

Polyamine-dependent protein kinase has been purified from four eukaryotic sources by phosphocellulose chromatography (4,10,36). A portion of the protein kinase and its M_r 70,000 phosphate-acceptor substrate protein copurify as a complex by this procedure. Thus, the protein kinase can be detected by measuring its capacity to phosphorylate the endogenous substrate protein. Table 1 (column iii) shows that a similar protein kinase activity could also be fractionated from the nuclear and cytoplasmic compart-

TABLE 1. Effect of polyamines and poly(I:C) on protein kinase, fractionated by phosphocellulose chromatography and by affinity chromatography on poly(I:C)-agarose, from control and interferon-treated ascites tumor cells

Protein kinase preparation (i)	Modifier addition[a] (ii)	Specific activity of protein kinase: from phosphocellulose chromatography (iii)	Specific activity of protein kinase: from poly(I:C) affinity chromatography (iv)
		(pmoles/min/mg enzyme)[b]	
Control cytoplasmic[c]	None	85	32
	Polyamines	374	109
	dsRNA	359	115
	Polyamines + dsRNA	371	152
Control nuclear[c]	None	28	19
	Polyamines	29	27
	dsRNA	43	38
	Polyamines + dsRNA	44	43
Interferon-treated cytoplasmic[d]	None	490	475
	Polyamines	1401	2418
	dsRNA	1436	2030
	Polyamines + dsRNA	1997	2556
Interferon-treated nuclear[d]	None	355	809
	Polyamines	977	3429
	dsRNA	1015	3909
	Polyamines + dsRNA	1376	4008

[a]Polyamines were a mixture of spermidine plus spermine, each 0.5 mM. Double-stranded RNA (dsRNA) was poly(I:C), potassium salt, and was added to a final concentration of 1 μg/ml.

[b]The specific radioactivity of [γ-^{32}P]ATP was 3.49 x 10^8 cpm/μmol for assays given in column (iii) and 2.85 x 10^8 cpm/μmol for assays given in column (iv). Each assay contained 34 μg of enzyme preparation. Other details were as in (3).

[c]Control cells which had not been exposed to interferon were fractionated into separate cytoplasmic and nuclear protein preparations. Each control preparation was subsequently fractionated further by chromatography on Bio-Rex 70 and then on phosphocellulose (10) (column iii) or on poly(I:C)-agarose (18) (column iv).

[d]Interferon-treated cells were exposed to 400 units/ml of mouse fibroblast interferon for 21 hr before isolation of cytoplasmic and nuclear protein fractions (36). Each interferon-treated preparation was also fractionated further by chromatography on Bio-Rex 70 and then on phosphocellulose (10) (column iii) or on poly(I:C)-agarose (18) (column iv).

ments of Ehrlich ascites tumor cells. The control cytoplasmic fraction (column iii) contained a protein kinase activity that was stimulated over 4-fold by a combination of 0.5 mM each of spermidine and spermine. Stimulation by dsRNA was similar. A comparable enhancement was produced by simultaneous additions of polyamines and dsRNA. Little protein kinase activity was demonstrated by the control nuclear fraction (column iii) which was essentially unaffected by polyamines, dsRNA, or their combination.

Table 1 (column iii) also shows that the interferon-treated, cytoplasmic fraction demonstrated elevated levels of protein kinase activity compared to the control cytoplasmic preparation. The activity was enhanced 2.8-fold by the polyamines or poly-(I:C). Polyamines and dsRNA supplied simultaneously produced a 4-fold stimulation compared to reactions that were conducted without the addition of a modifier. Markedly enhanced protein kinase activity was demonstrated by the interferon-treated nuclear fraction compared to the control nuclear preparation. The same levels of enhanced protein kinase activity elicited by the polyamines (2.7-fold), dsRNA (2.80-fold) or a combination of the polyamines plus dsRNA (3.9-fold) were observed in the interferon treated cytoplasmic preparation.

The most striking comparison of protein kinase specific activities (column iii) were observed between the control nuclear and the interferon-treated nuclear preparations. An apparent 33-fold induction of polyamine-dependent protein kinase activity (cf. 43 vs 1015) by interferon was found in the nucleus. A similar 31-fold increase was evoked by combined additions of the polyamines and of dsRNA (cf. 44 vs 1376). These results suggested that a single protein kinase enzyme may be activated both by a spermidine/spermine mixture and by dsRNA, since the combined additions evoked an increase which was not significantly greater than that observed with single additions of the polyamines or dsRNA. Moreover, this proposal was supported further by exhaustive phosphorylation experiments (data not shown). Protein kinase reactions were activated by polyamines and were allowed to continue for 60 min. After 60 min, the phosphate-acceptor substrate protein available to the polyamine-dependent reaction was depleted. No additional phosphorylation was observed on addition of dsRNA to these reaction mixtures. Similar reactions in which the order of additions of the polyamines and dsRNA were reversed, also yielded no additional phosphorylation. These experiments indicated that the dsRNA-dependent protein kinase and the polyamine-dependent protein kinase may utilize the same endogenous substrate protein of M_r 70,000 (see below).

Poly(I:C)-Agarose Chromatography

dsRNA-dependent protein kinase that has been induced by interferon can be fractionated by affinity chromatography on poly(I:C)-agarose (18). Control and interferon-treated cytoplasmic and nuclear preparations from cultured Ehrlich ascites tumor cells were therefore fractionated on poly(I:C)-agarose. The objective was to compare the responses of interferon-induced, dsRNA-dependent protein kinase to various combinations of polyamines and dsRNA, as were determined for polyamine-dependent protein kinase isolated by phosphocellulse chromatography. Table 1 (column iv) shows the results. Low levels of protein kinase activity were found in the control cytoplasmic and control nuclear preparations. The nuclear fraction demonstrated little protein kinase activity that could be activated by the single or combined modifiers. In contrast, markedly elevated protein kinase activities which were strongly responsive to stimulation by either the polyamines or dsRNA were found in the corresponding interferon-treated fractions. In the interferon-treated nuclear preparation, the protein kinase specific activity was stimulated 127-fold over the control (cf. 38 vs 3909). The interferon-treated nuclear protein kinase was activated 93-fold over the control (cf. 43 vs 4008) by a combination of the polyamines and dsRNA. That one protein kinase may have been responsible for all of these phosphorylation reactions was again suggested by the fact that the combined modifiers enhanced the protein kinase activity comparable to that of each individual modifier (cf. 4008 vs 3909).

Endogenous Phosphate-Acceptor Proteins

The proposal in the previous sentence above gained support from analyses of the labeled peptide products of the phosphorylation reactions. Electrophoretic separations of labeled peptides in sodium dodecylsulfate ($NaDodSO_4$) polyacrylamide gels revealed that the phosphorylated product in all reactions containing samples from the interferon-treated preparations plus polyamines and/or dsRNA, was a [^{32}P]phosphopeptide of M_r 68,000-70,000 (36). Putrescine antagonized the activations by spermidine/spermine. In contrast, no labeled phosphopeptide was observed in reactions conducted in the absence of the polyamines and/or dsRNA. Also, none was produced by control cytoplasmic or control nuclear enzyme preparations in the absence or presence of the polyamines and/or dsRNA. The small amount of phosphorylation observed by the control cytoplasmic or the control nuclear preparations in the absence of the modifiers were phosphorylated products which could not be detected on the polyacrylamide gel system used. These results showed that the phosphorylation reactions which yielded the M_r 70,000 phosphopeptide were critically dependent on the modifiers, polyamines or dsRNA, for activation. It is still unknown whether the polyamine-dependent protein kinase and the dsRNA-dependent protein kinase catalyze phosphorylation of the same identical polypeptide. The only valid conclusion that can be made is that the reaction product in either case was a phosphoprotein of M_r 68,000-70,000.

Remarks

Several findings reported here suggest that a single protein kinase enzyme may be activated by the strikingly dissimilar modifiers, spermidine/spermine or dsRNA. First, the polyamine-dependent protein kinase and the dsRNA-dependent protein kinase activities cofractionated on phosphocellulose or on poly(I:C)-agarose. Second, both activities yielded endogenous, phosphorylated products of the same M_r. Lastly, the two enzyme activites could not be differentiated in a common preparation on the bases of their respective activations by the polyamines or dsRNA.

The M_r 68,000-70,000 phosphate-acceptor protein for both polyamine- and dsRNA-dependent protein kinase in Ehrlich ascites tumor cells has not yet been rigorously identified. However, nuclear and cytoplasmic preparations which contain the phosphate-acceptor protein, also demonstrate OrnDCase activity when assayed in high salt solutions (280 mM Na^+). Phosphorylation of the phosphate-acceptor protein correlates with the loss of the OrnDCase activity. There are strong parallels between the properties reported here for phosphorylation of the M_r 68,000-70,000 polypeptide in Ehrlich ascites tumor cells and those that we have discovered in four other eukaryotic sources. Phosphorylation of OrnDCase with concomitant suppression of polyamine synthesis may be another target for interferon induction of protein kinase.

POSSIBLE ROLE OF Ca^{2+}-CALMODULIN-STIMULATED AUTOPHOSPHORYLATION OF POLYAMINE-DEPENDENT PROTEIN KINASE IN MODULATING OrnDCase

The polyamines probably serve many functions in metabolism, that are alternatively satisfied by inorganic cations. It should not be surprising therefore that polyamine syntheses may be integrated, in some as yet unknown manner, with general cation availability to cells. The ubiquitous Ca^{2+} ion receptor and effector protein, calmodulin, has been found in all eukaryotic cells. It is now regarded as a multifunctional, intracellular receptor which mediates many regulatory actions of calcium in the cell (7). Moreover, a general principle seems to be emerging that all enzymes that are regulated by low levels of Ca^{2+} have calmodulin, or a calmodulin-like protein, as one of their components. Calcium ion-calmodulin complex activates several enzyme systems such as protein kinases, phosphodiesterase, adenylate cyclase, thymidylate synthetase, transglutaminase, and phospholipase (43). Indeed, the Ca^{2+}-calmodulin complex has been demonstrated to modulate protein phosphorylation by numerous types of protein kinases in several tissues (35, references therein). Calmodulin is increasingly recognized as an important modulator of cyclic nucleotide-independent protein kinases in eukaryotes. Inasmuch as several Ca^{2+}-calmodulin-modulated protein kinases have been demonstrated, we examined whether the polyamine-dependent protein kinase which phosphorylated OrnDCase responded in some capacity to this bivalent metal complex. This was of special interest since it has been shown that there is a specific dependence on intracellular Ca^{2+} ion of both basal and hormone-stimulated OrnDCase in cultured animal cells (9,11,14, 41). Intracellular Ca^{2+} ion is also involved in the control of OrnDCase induction by tumor promoters (42). These important regulatory effects appear to operate without the participation of cyclic AMP. The mechanisms by which intracellular Ca^{2+} ion concentrations might regulate OrnDCase activity are unknown.

As described earlier, NaCl-gradient elution chromatography on phosphocellulose separated the polyamine-dependent protein kinase catalytic enzyme from a complex formed by the protein kinase and OrnDCase, the natural phosphate-acceptor protein. Fig. 1 illustrates this separation (3). The protein kinase-OrnDCase complex copurified at 0.7 M NaCl (fraction _b_). A second fraction of less protein kinase activity (fraction _a_) preceded elution of the polyamine-dependent protein kinase-OrnDCase complex. Fraction _a_ contained only the protein kinase of subunit M_r 26,000. As illustrated in Fig. 1, the protein kinase activity in fraction _a_ was inhibited by the polyamines, spermidine/spermine. By contrast, the protein kinase activity in fraction _b_ was markedly activated by spermidine/spermine. Analyses of the labeled peptides produced by these two protein kinase activities using $NaDodSO_4$-polyacrylamide gel electrophoresis explained these observations (10). Fraction _b_ produced [^{32}P]OrnDCase in a reaction that was spermidine/spermine-dependent. Fraction _a_ produced [^{32}P]protein kinase in the absence of spermidine/spermine and OrnDCase. Thus, autophosphorylation of polyamine-dependent protein kinase occurred in the absence of its natural phosphate-acceptor protein. Autophosphorylation was inhibited by spermidine/spermine as shown in Fig. 1.

The influences of Ca^{2+} and calmodulin were studied on polyamine-dependent phosphorylation of OrnDCase and on the autophosphorylation of polyamine-dependent protein kinase (Table 2). Calcium ion alone inhibited the phosphorylation of OrnDCase by polyamine-dependent protein kinase but only at very high concentrations (exp. #1). Inhibition of 82% occurred at 500 μM Ca^{2+}, the highest concentration tested. Calmodulin in concentration up to 20 μM did not affect polyamine-dependent phosphorylation of OrnDCase (exp. #2). Simultaneous addition of Ca^{2+} and calmodulin also inhibited phosphorylation of OrnDCase in a manner similar to that of free Ca^{2+} (exp. #3). Thus, calmodulin did not appear to exert a role in the direct phosphorylation reaction of OrnDCase. Interestingly, however, the autophosphorylation of polyamine-dependent protein kinase, which occurs in the absence of its natural substrate, OrnDCase, was markedly stimulated by Ca^{2+}-calmodulin (exp. #4). Calmodulin or Ca^{2+} independently did not stimulate autophosphorylation. Autophosphorylation is a common property of nuclear protein kinases. In most cases, autophosphorylation of protein kinases has been found to be without kinetic significance (16). In the case of polyamine-dependent protein kinase, however, autophosphorylation correlated with inhibition of the capacity of the enzyme to phosphorylate

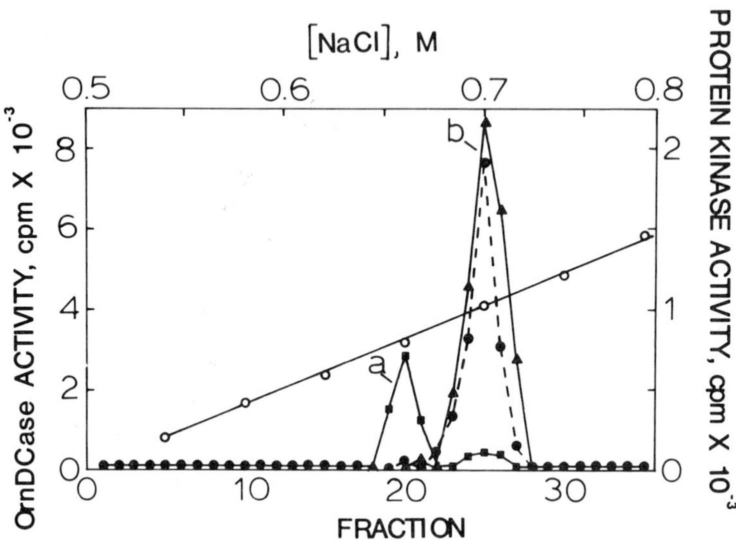

FIG. 1. Cofractionation of polyamine-dependent protein kinase (● and ■) and OrnDCase (▲) enzymatic activities from nuclear preparations by phosphocellulose chromatography. Crude nuclear extracts from P. polycephalum were treated batchwise with Bio-Rex 70 (Na^+ form, 400 mesh, Bio-Rad) before column chromatography. The partially purified nonhistone protein (120 mg) in 42 ml of 50 mM Tris•HCl, pH 7.5, was applied to a phosphocellulose column (3.5 x 10 cm) equilibrated with 50 mM Tris•HCl, pH 7.5. The column was eluted with this buffer until the absorbance of the effluent was <0.02. Next, a 200-ml gradient of NaCl, 0.5 to 1.0 M (○), in 50 mM Tris•HCl, pH 7.5, was applied. Collection of 3-ml fractions was begun. Protein kinase activity was measured in each fraction in the presence (●) and absence (■) of 0.5 mM spermidine/0.5 mM spermine.

OrnDCase (unpublished data). Inhibitory phosphorylation of OrnDCase by polyamine-dependent protein kinase could thus seemingly be prevented by autophosphorylation of the kinase.

Calcium-ion-calmodulin involvement in modulating the catalytic capacity of polyamine-dependent protein kinase through autophosphorylation suggests a biochemical mechanism for observed Ca^{2+} induction of OrnDCase activity in various cell types. This proposal is diagrammed in Fig. 2 along with known facts regarding the phosphorylation of OrnDCase. Prevention of inhibitory phosphorylation of OrnDCase by autophosphorylation of the protein kinase would not in itself result in apparent induction of OrnDCase. However, inactivation of the protein kinase would be

TABLE 2. Calcium ion and calmodulin activation of autophosphorylation by polyamine-dependent protein kinase

Enzyme preparation used[a]	Additions to complete assay[b]			Sp. act. (nmol/min/mg protein)
	polyamines,[c] mM	$CaCl_2$, μM	calmodulin, μM	
1. protein kinase + OrnDCase (fraction b, Fig. 1)	1	-0-	-0-	83.7
	1	100	-0-	31.1
	1	250	-0-	28.2
	1	500	-0-	25.8
2. protein kinase + OrnDCase (fraction b, Fig. 1)	1	-0-	-0-	67.8
	1	-0-	10	81.6
	1	-0-	15	78.9
	1	-0-	20	82.2
3. protein kinase + OrnDCase (fraction b, Fig. 1)	1	-0-	-0-	81.6
	1	10	10	40.7
	1	15	15	44.5
4. protein kinase (fraction a, Fig. 1)	-0-	-0-	-0-	19.0
	0.5	-0-	-0-	6.2
	-0-	10	-0-	17.4
	-0-	50	-0-	10.0
	-0-	-0-	10	18.5
	-0-	-0-	50	17.3
	-0-	10	10	48.0
	-0-	50	50	51.5
	0.5	50	50	44.1

[a]Assays contained 0.02 μg of protein from fractions containing either protein kinase (fraction a, Fig. 1) or protein kinase + OrnDCase (fraction b, Fig. 1).

[b]The complete assay contained: 100 mM disodium β-glycerolphosphate, pH 6.8/20 mM NaF/1 mM Na_2EDTA/10 mM magnesium acetate/2 mM [γ-^{32}P]ATP (specific radioactivity: 8.40 x 10^8 cpm/μmol)/60 mM NaCl.

[c]Polyamines were 0.5 mM each of spermidine and spermine.

mandatory to prevent further conversion of active OrnDCase to a phosphorylated, inactive form. Immediate induction of OrnDCase would necessitate the simultaneous action of a phosphoprotein phosphatase. The phosphatase should have the capacity to dephosphorylate phospho-OrnDCase in order to convert inactive phospho-OrnDCase back to its catalytically active, dephosphorylated form. Intuitively, one would predict that such a phosphoprotein phosphatase should be dependent on or stimulated by Ca^{2+}-calmodulin. We have previously reported on the occurrence of a potent phosphoprotein phosphatase activity in nuclei of *P. polycephalum* which dephosphorylates OrnDCase (10). Recently we have found an alkaline phosphatase activity that fractionates from phosphocellulose column chromatography at about 0.4 M NaCl.

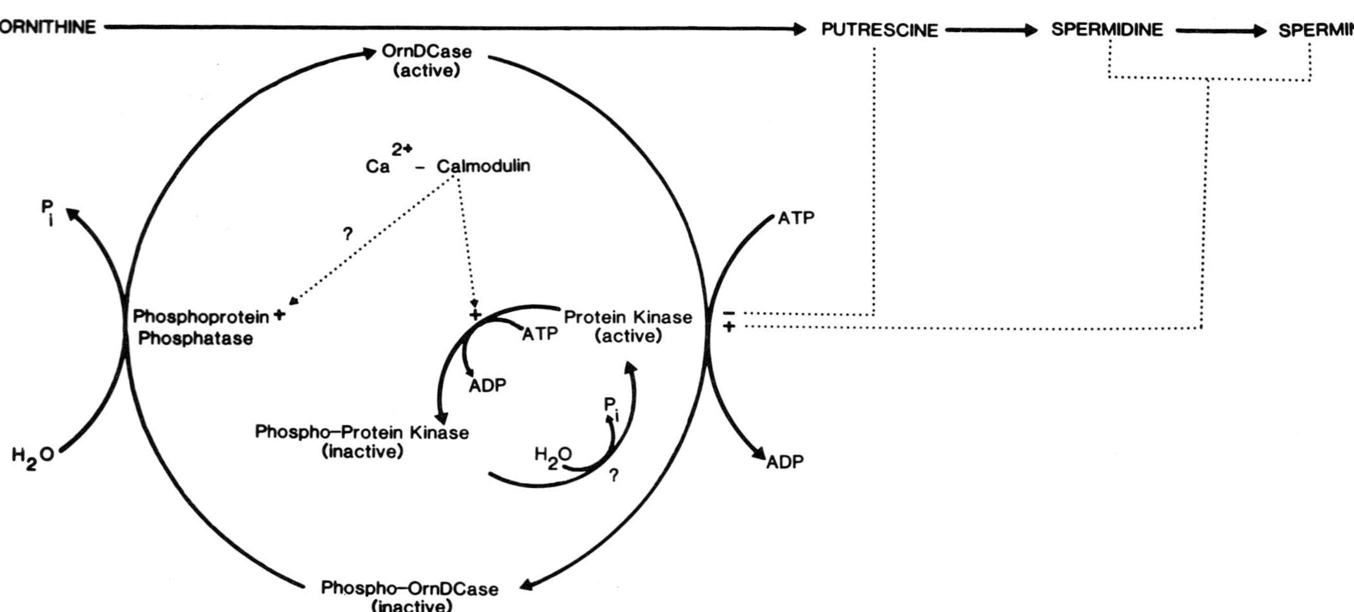

FIG. 2. Regulation of OrnDCase from _Physarum polycephalum_ by a polyamine-mediated protein kinase reaction and by Ca^{2+}-calmodulin. The stipled lines indicate the capacities of spermidine and spermine to activate (+) the protein kinase reaction that phosphorylates OrnDCase, the capacity of putrescine to antagonize (-) the phosphorylation, and the capacity of Ca^{2+}-calmodulin to stimulate (+) inhibitory autophosphorylation of polyamine-dependent protein kinase. Transformations, indicated by arrows, over which question marks are written (?) indicate speculation for which there is yet no supporting evidence.

The same column simultaneously yields the protein kinase and the protein kinase-OrnDCase complex which elute near 0.7 M NaCl (Fig. 1). We believe that this alkaline phosphatase is a prime candidate to function as phospho-OrnDCase phosphatase. It may be activated by Ca^{2+}-calmodulin. This has not yet been tested because purification of the enzyme is not yet completed. Few investigations have been conducted to study the involvement of calmodulin in protein dephosphorylation. However, there are precedences for activation (38) and inhibition (19) of specific phosphoprotein phosphatases by Ca^{2+}-calmodulin. Additional research should reveal many interesting details for the role of Ca^{2+}-calmodulin in the regulation of OrnDCase by the mechanisms implicated in the findings reported here.

METABOLIC END PRODUCT-MEDIATED ENZYME PHOSPHORYLATION

A final comment of significance pertains to implications of these findings for the regulation of cellular metabolism in general. Inhibition of the catalytic function of biosynthetic enzymes through phosphorylation by protein kinases has been well documented (8). However, the discovery of the property of activation of a protein kinase reaction by the end products (spermidine and spermine) of a biosynthetic pathway which phosphorylatively inactivates the first enzyme of that pathway (OrnDCase) was unprecedented. This represented a newly recognized dimension of feedback control by end products of a biosynthetic pathway. It also implicates a likely new role for many protein kinases in general metabolism through what can be termed, end-product mediated enzyme phosphorylation. There is no reason to expect that this type of control will be restricted to OrnDCase and the polyamines. Indeed, another example of this type of control was reported recently (24). AcetylCoA carboxylase, the first enzyme in long chain fatty acid synthesis, is inhibited after phosphorylation by a coenzyme A-dependent protein kinase reaction.

REFERENCES

1. Atmar, V. J., Daniels, G. R., and Kuehn, G. D. (1978): Europ. J. Biochem., 90:29-37.
2. Atmar, V. J., Daniels, G. R., and Kuehn, G. D. (1980): FEBS Lett., 114:205-208.
3. Atmar, V. J. and Kuehn, G. D. (1981): Proc. Natl. Acad. Sci. USA, 78:5518-5522.
4. Atmar, V. J., Kuehn, G. D., and Casillas E. R. (1981): J. Biol. Chem., 256:8275-8278.
5. Bachrach, U. (1980) In: Polyamines in Biomedial Research, edited by J. M. Guagas, pp. 81-107. John Wiley, New York.
6. Canellakis, Z. N. and Theoharides, T. C. (1976): J. Biol. Chem., 251:1781-1784.

7. Cheung, W. Y. (1980): Science, 207:19-27.
8. Cohen, P., editor (1981): Recently Discovered Systems of Enzyme Regulation By Reversible Phosphorylation. Molecular Aspects of Cellular Regulation, vol. 1. Elsevier-North Holland, Amsterdam.
9. Costa, M. and Nye, J. S. (1978): Biochem. Biophys. Res. Commun., 85:1156-1164.
10. Daniels, G. R., Atmar, V. J., and Kuehn, G. D. (1981): Biochemistry, 20:2525-2532.
11. D'Amore, P. A. and Shepro, D. (1978): Life Sci., 22:571-576.
12. Gazitt, Y. (1981): Cancer Res., 41:2959-2961.
13. Gazitt, Y. and Friend, C. (1980): Cancer Res., 40:1727-1732.
14. Gibbs, J. B., Hsu, C-Y., Terasaki, W. L. and Brooker, G. (1980): Proc. Natl. Acad. Sci. USA, 77:995-999.
15. Haddox, M. K. and Russell, D. H. (1981): Biochemistry, 20:6721-6729.
16. Hathaway, G. M. and Traugh, J. A. (1979): J. Biol. Chem., 254:762-768.
17. Heller, J., Fong, W. F., and Canellakis, E. S. (1976): Proc.Natl. Acad. Sci. USA, 73:1858-1862.
18. Hovanessian, A. G. and Kerr, I. M. (1979): Eur. J. Biochem., 93:515-526.
19. Khandelwal, R. L., Kotello, M. C., and Wang, J. H. (1980): Arch. Biochem. Biophys., 203:244-250.
20. Kuehn, G. D., Affolter, H. U., Atmar, V. J., Seebeck, T., Gubler, U., and Braun, R. (1979): Proc. Natl. Acad. Sci. USA, 76:2541-2545.
21. Kuehn, G. D. and Atmar, V. J. (1982): Fed. Proc., 41:in press.
22. Kuehn, G. D., Atmar, V. J., Garcia, G., and Westland, J. (1976): Tenth International Congress of Biochemistry Abstracts, Hamburg, Germany, p. 397.
23. Lee, E. J., Larkin, P. C., and Sreevalsan, T. (1980): Biochem. Biophys. Res. Commun., 97:301-308.
24. Lent, B. and Kim, K-H. (1982): J. Biol. Chem., 257:1897-1901.
25. Mamont, P. S., Joder-Ohlenbusch, M. N., and Grove, J. (1981): Biochem. J., 196:411-422.
26. McCann, P. P. (1980): In: Polyamines in Biomedical Research, edited by J. M. Gaugas, pp. 109-123. John Wiley, New York.
27. McCann, P. P., Tardif, C., and Mamont, P. S. (1977): Biochem. Biophys. Res. Commun., 75:948-954.
28. Morris, D. R. and Fillingame, R. H. (1974): Annu. Rev. Biochem., 43:303-325.
29. Obenrader, M. F. and Prouty, W. F. (1977): J. Biol. Chem., 252:2866-2872.
30. Ono, M., Inoue, H., Suzuki, F., and Takeda, Y. (1972): Biochim. Biophys. Acta., 284:285-297.

31. Paulus, T. J. and Davis, R. H. (1981): J. Bacteriol., 145: 14-20.
32. Pritchard, M. L., Seely, J. E., Pösö, H., Jefferson, L. S., and Pegg, A. E. (1981): Biochem. Biophys. Res. Commun., 100:1597-1603.
33. Revel, M. and Groner, Y. (1978): Annu. Rev. Biochem., 47: 1079-1126.
34. Roberts, W. K., Hovanessian, A. G., Brown, R. E., Clemens, M. J., and Kerr, I. M. (1976): Nature (Lond.), 264:477-480.
35. Schubart, U. K., Erlichman, J., and Fleischer, N. (1982): Fed. Proc., 41:2278-2282.
36. Sekar, V., Atmar, V. J., Krim, M., and Kuehn, G. D. (1982): Biochem. Biophys. Res. Commun., 106:305-311.
37. Sreevalsan, T., Taylor-Papadimitriou, J., and Rozengurt, E. (1979): Biochem. Biophys. Res. Commun., 87:679-685.
38. Stewart, A. A., Ingebritsen, T. S., Manalan, A., Klee, C. B., and Cohen, P. (1982): FEBS Lett., 137:80-84.
39. Tyagi, A. K., Tabor, H., and Tabor, C. W. (1982): Fed. Proc., 41 Abs.:1398.
40. Tyagi, A. K., Tabor, C. W., and Tabor, H. (1981): J. Biol. Chem., 256:12156-12163.
41. Veldhuis, J. D. and Hammond, J. M. (1981): Biochem. J., 196:795-801.
42. Verma, A. K. and Boutwell, R. K. (1981): Biochem. Biophys. Res. Commun., 101:375-383.
43. West, W. L. (1982): Fed. Proc., 41:2251-2252.

ACKNOWLEDGEMENTS

This research was supported in part by the following institutions: the National Institutues of Health through research grants GM18538 and RR08136; New Mexico State University Research Center through a travel grant to partially defray the expenses of G. D. K. to attend this conference.

Advances in Polyamine Research, Vol. 4, edited by U. Bachrach, A. Kaye, and R. Chayen. Raven Press, New York © 1983.

Effects of Spermidine and Its Monoacetylated Derivatives on Phosphorylation by Nuclear Protein Kinase NII

Samson T. Jacob, Kathleen M. Rose, and *Zoe N. Canellakis

*Department of Pharmacology, Milton S. Hershey Medical Center, Pennsylvania State University College of Medicine, Hershey, Pennsylvania 17033; *Departments of Pharmacology and Medicine, Yale University and West Haven Veterans Administration Medical Center, New Haven, Connecticut 06510*

The suggestion that phosphorylation of nuclear nonhistone proteins may have a regulatory role in gene expression (2,15) has stimulated considerable interest in the identification of unique phosphorylation reactions in the cell nucleus. These studies have led to the observation that polyamines can stimulate nuclear protein kinase activities (1,3,10,11,16). Recently, we have purified essentially to homogeneity a nuclear protein kinase (NII) from a rat hepatoma (18). Phosphorylation of casein (18) or RNA polymerase I (12,21) by this kinase was markedly stimulated by polyamines and completely inhibited by very low concentrations of heparin (18). The ability of RNA polymerase I to serve as a substrate for the NII kinase and the activation of the polymerase by phosphorylation (9) have raised the possibility that protein kinase NII may control ribosomal gene expression. To this end, we have investigated the mechanism by which polyamines influence protein phosphorylation catalyzed by protein kinase NII.

RESULTS

Effect of Polyamines on the NII Protein Kinase Using Casein as the Substrate

Previous studies in our laboratory have shown that spermine can stimulate phosphorylation of casein about 2-fold (18). Relatively high concentrations of spermine were used in these experiments to achieve maximal stimulation of the NII kinase. Using a more purified preparation of casein, we have now been able to demonstrate virtual dependence of the kinase on polyamines. The kinase used in these studies was purified essentially to homogeneity from Morris hepatoma 3924A. Isolation of nuclei, enzyme extraction and purification protocol were essentially as described previously (18). As shown in Fig. 1, as much as a 15-fold increase in the phosphorylation was noted at 0.5 mM spermine and a remarkable 27-fold augmentation in the kinase activity was observed at 2.5 mM spermine. At relatively low concentrations, spermidine also could stimulate phosphorylation of casein. For example, at 2.5 mM, spermidine enhanced the phosphate transfer to casein nearly 12-fold. The NII kinase activity increased in a near linear manner from 0.1 mM to 0.5 mM spermine and 0.5 mM to 2 mM spermidine.

Using 7.5 mM spermidine, the time course of the phosphorylation reaction was measured at time intervals ranging from 10 min to 60 min. Biphasic curves were obtained both for the control reaction (evident when plotted on an expanded scale) and for the reaction in the presence of polyamine (Fig. 2). At the 60 min time point, there was almost a 25-fold increase in the phosphorylation in the presence of spermidine.

Effect of Spermidine on V_{max} and the Apparent K_m^{ATP} of the Protein Kinase

The protein phosphorylation reaction was conducted in the presence or absence of spermidine (7.5 mM) using ATP concentrations ranging from 0.27 µM to 27 µM. The double reciprocal plot of velocity versus ATP concentration was constructed using linear regression analysis (Fig. 3). Addition of spermidine clearly altered both the maximum reaction velocity and apparent K_m with respect to ATP. Specifically, the V_{max} increased from 3 pmol phosphate/30 min to 780 pmol phosphate/30 min and the apparent K_m^{ATP} from 7 µM to 164 µM upon addition of spermidine.

Effect of N^1- and N^8-acetyl Spermidine on the NII Kinase Activity

Both spermidine and spermine are known to be monoacetylated by chromatin-associated enzymes (23). Since the acetylated compounds exist *in vivo*, it was of interest to know what effects they exert on protein phosphorylation *in vitro*. Accordingly, the effect of N^1- and N^8-acetyl spermidine on casein phosphorylation by the NII kinase was determined. Varying concentrations of the derivatives ranging from 2.5 to 10 mM were used (Fig. 4). Even at the highest concentration (10 mM), there was no significant stimulation of phosphorylation. At the same concentration, spermidine produced a substantial increase in phosphate transfer. These data show that both N^1 and N^8-NH_2 groups of spermidine must be present to enhance the NII kinase activity significantly and suggest that the intracellular concentrations of spermidine and acetyl spermidine may regulate the activity of the NII protein kinase *in vivo*.

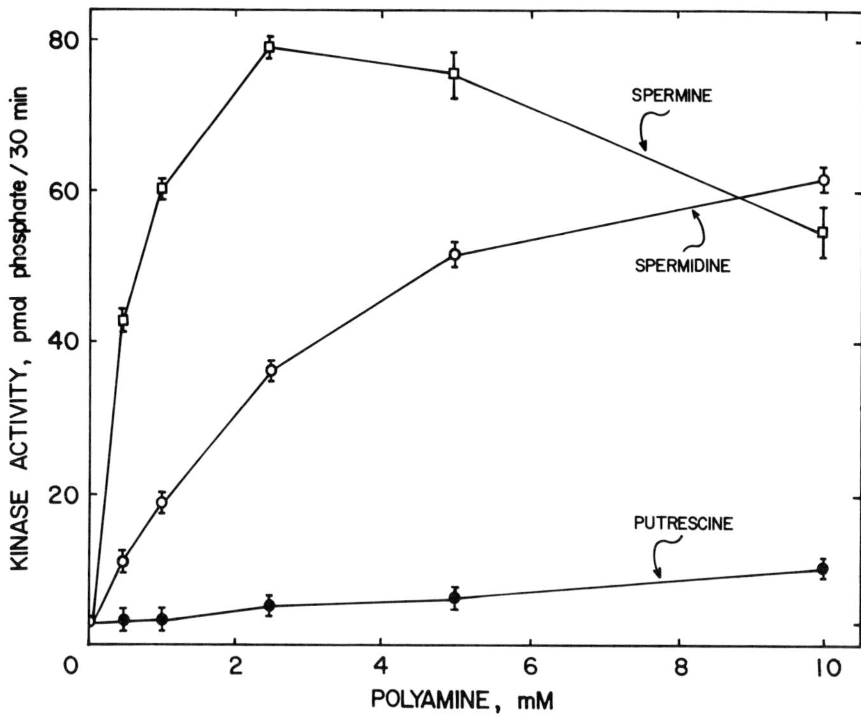

FIG. 1. Effect of varying concentrations of spermine, spermidine and putrescine on protein kinase NII purified from Morris hepatoma 3924A. The polyamines spermine (□), spermidine (O), and putrescine (●) were used at concentrations as indicated. Casein concentration was 2 mg/ml. The kinase activity was determined as described previously (18).

Specificity of Polyamines for NII Kinase-Catalyzed Phosphorylation

Since both of the two major cyclic nucleotide independent protein kinases, NI and NII, are capable of phosphorylating casein, it was of interest to know whether polyamines also exert a pronounced effect on the NI kinase. NI protein kinase was obtained from Morris hepatoma 3924A in the course of purification of nuclear poly(A) polymerase; this kinase, which can be dissociated from the majority of poly(A) polymerase during phosphocellulose chromatography (19) or hydroxylapatite chromatography, was further purified by affinity chromatography on a casein sepharose column. This kinase was cyclic nucleotide-independent, was active in the presence of $MgCl_2$ or $MnCl_2$ and had specific activities ranging from 13,000 to 70,000 pmol phosphate transferred to casein/mg of protein in 15 min using [γ-^{32}P]ATP as phosphate donor. Casein was used as the substrate

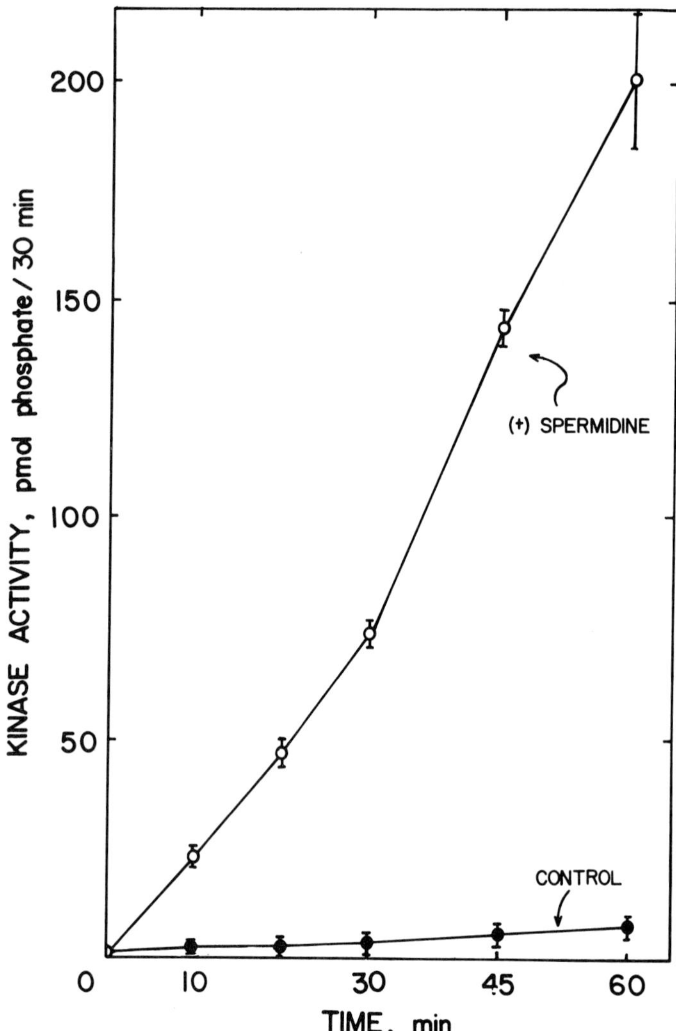

FIG. 2. Time course of the phosphorylation reaction catalyzed by the NII kinase. Phosphorylation of casein by the NII kinase was conducted for various times as indicated in the presence (O) or absence (●) of spermidine (7.5 mM). The enzyme assay was essentially as described previously (18).

for NI and NII kinases, and spermidine was added at varying concentrations ranging from 0.25 mM to 10 mM. Even at a concentration as high as 10 mM, spermidine had only a minimal effect on the phosphorylation of casein by NI kinase (Fig. 5). Since threonine was preferentially phosphorylated by spermidine (see the subsequent section), the lack of stimulation of phosphorylation by NI kinase in the presence of the polyamine might be due to its relative inability to phosphorylate this amino acid.

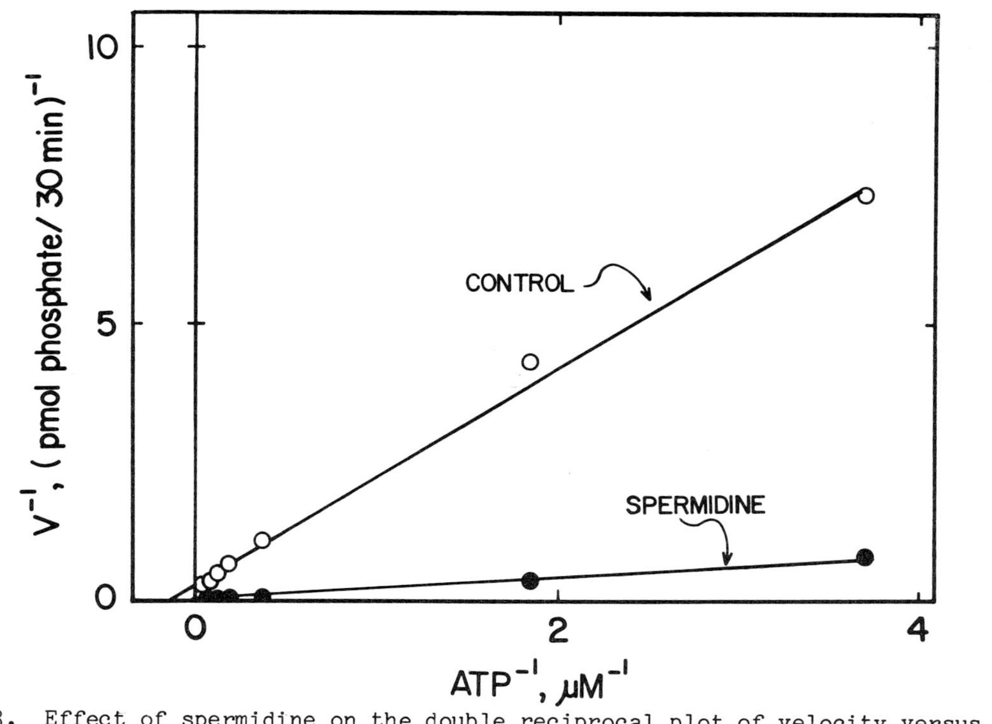

FIG. 3. Effect of spermidine on the double reciprocal plot of velocity versus ATP concentration. The activity of protein kinase NII at varying ATP concentrations (0.27 μM to 27 μM) was measured in the absence (O) and presence (●) of spermidine (7.5 mM). The reaction velocity ranged from 0.14 ± 0.01 to 3.49 ± 0.47 and 1.27 ± 0.08 to 57.9 ± 2.9 pmol phosphate/30 min in the absence and presence of the polyamine, respectively. Assays were performed in triplicate and data analyzed by linear regression analysis.

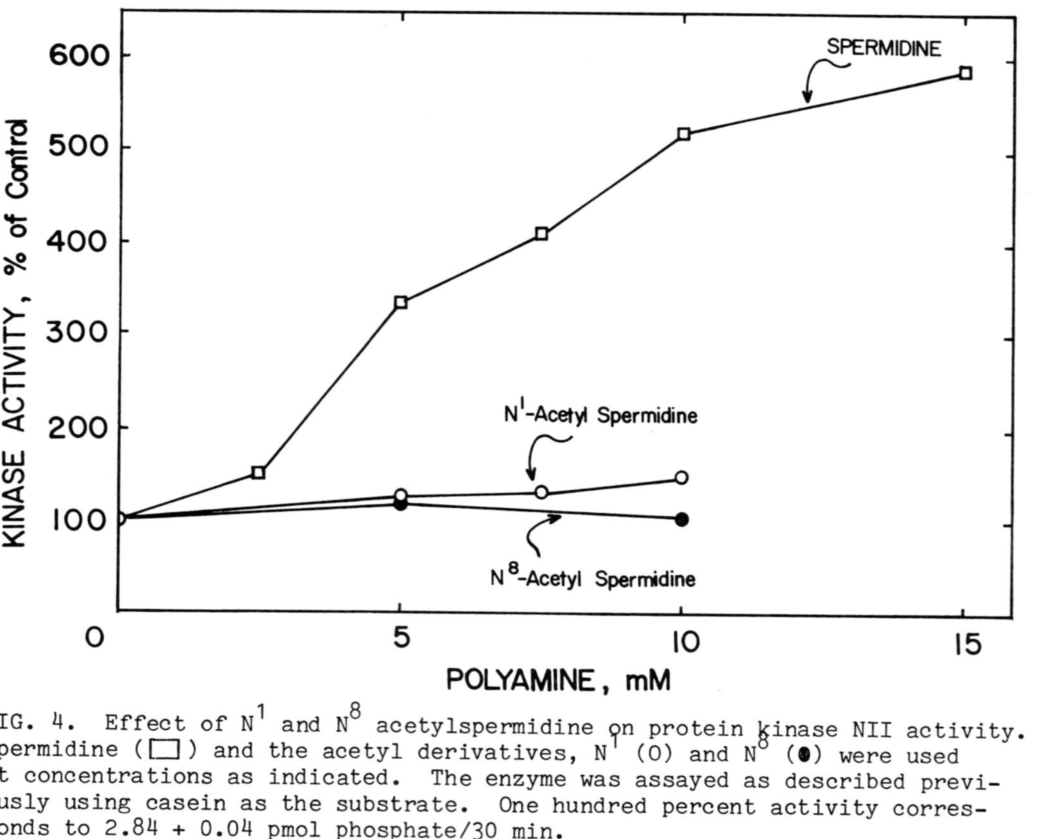

FIG. 4. Effect of N^1 and N^8 acetylspermidine on protein kinase NII activity. Spermidine (□) and the acetyl derivatives, N^1 (O) and N^8 (●) were used at concentrations as indicated. The enzyme was assayed as described previously using casein as the substrate. One hundred percent activity corresponds to 2.84 ± 0.04 pmol phosphate/30 min.

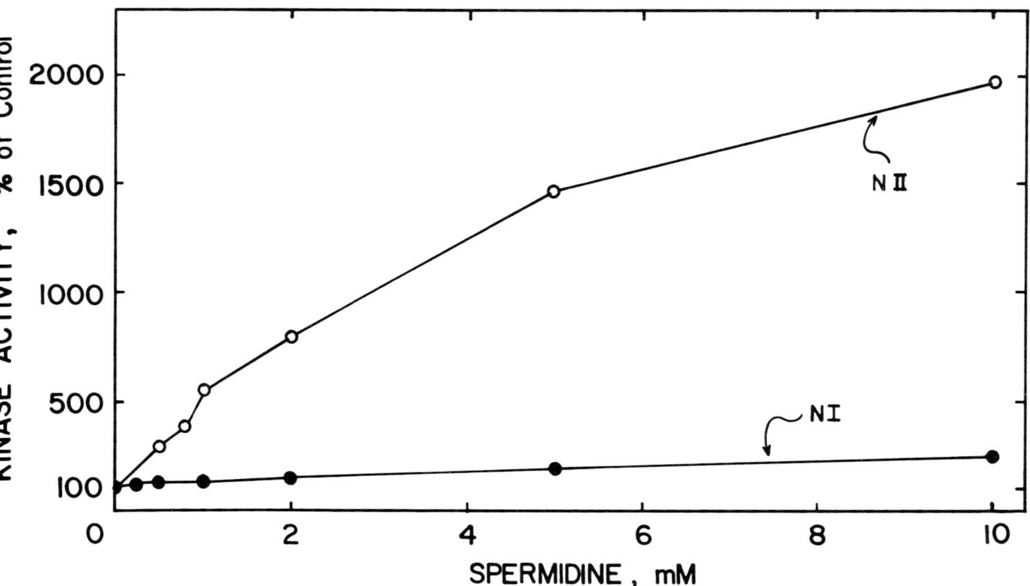

FIG. 5. Effect of spermidine on NI and NII protein kinases purified from Morris hepatoma 3924A. NI and NII kinases were purified as described previously (19,18). Both enzymes were assayed using casein as the protein substrate and at different concentrations of spermidine as indicated. One hundred percent activity was equivalent to 2.92 ± 0.64 and 4.77 ± 0.40 pmol phosphate/30 min for protein kinases NII (O) and NI (●), respectively.

Effect of Spermidine on Prephosphorylated Casein

Although polyamines exert a dramatic effect on protein phosphorylation catalyzed by NII kinase, it has not been established whether they could enhance phosphorylation after the reaction in the absence of polyamine was nearly complete. To test this possibility, the enzyme assay was carried out using limiting amounts of casein and excess NII kinase. Under these conditions, the reaction reached a plateau after 60 min of incubation. When spermidine (5 mM) was added at this time point and the reaction continued, a dramatic stimulation of phosphate transfer to casein was observed (Fig. 6); after 2 hr, the phosphorylation was enhanced nearly 9-fold. These data indicate that polyamines phosphorylate sites on the substrate that were not available to the kinase in their absence. It is possible that at least part of this additional phosphorylation by the polyamine is due to the preferential transfer of phosphate to threonine moieties (see the following section).

Preferential Phosphorylation of Threonine in the Presence of Spermidine and its Acetyl Derivatives

Previous studies in this laboratory have shown that spermine can preferentially increase the phosphorylation of threonine residues (12,18). It was of interest to see whether spermidine has a similar action and, if so, whether acetyl spermidine can prevent such an effect. The ratio, phosphate transfer to threonine/phosphate transfer to serine (thr/ser), was elevated from 2.4 in the control sample to 3.2 and 4.3 in the presence of 5 mM and 10 mM spermidine, respectively (Fig. 7). Under the same conditions, spermine increased this ratio to 5.4 which is consistent with our earlier data (18). Interestingly, although the acetyl compounds did not significantly stimulate NII kinase activity, both N^1 and N^8 acetyl spermidine shifted the phosphorylation pattern in favor of threonine residues. In fact, the ratio, thr/ser, was elevated to an even greater extent in the presence of either N^1 (6.7) or N^8 (6.6) acetyl spermidine than in the presence of spermidine itself. The altered p-thr/p-ser ratio in response to the acetyl derivatives of spermidine was essentially due to stimulation of threonine phosphorylation and a proportional decrease in serine phosphorylation. It is likely that both spermidine and its analogs interact with the protein substrate in such a way that new sites containing threonine residues are exposed to the NII kinase. However, both terminal amino groups of the polyamine are required for the marked stimulatory effect on phosphorylation. It is not known whether the preferential phosphorylation of threonine residues by polyamines occurs *in vivo*. Further studies are needed to explore this possibility.

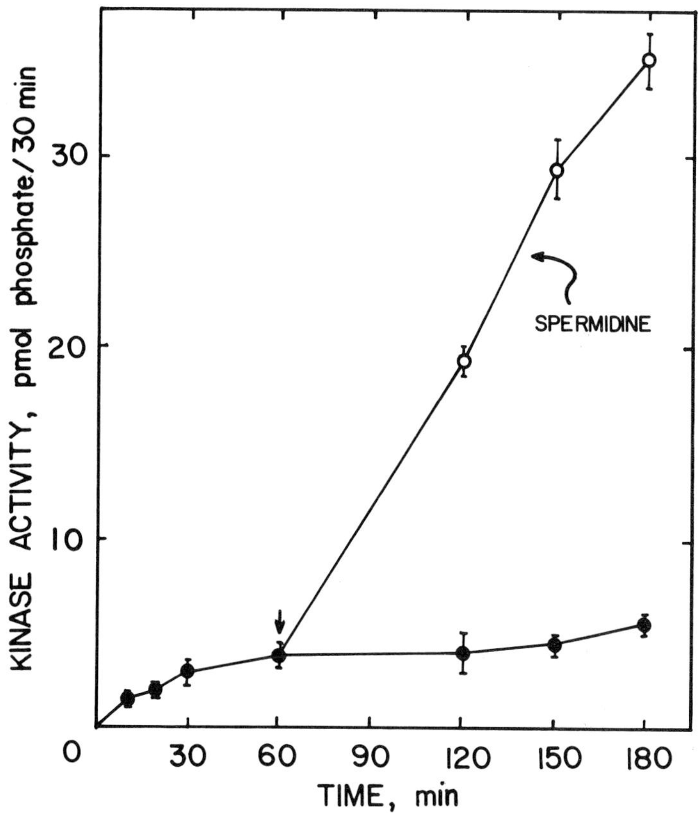

FIG. 6. Effect of spermidine on prephosphorylated casein. Phosphorylation of casein was measured at different time points using limiting amounts of casein (0.4 mg/ml) and excess NII kinase. After 60 min, spermidine (5 mM) was added to one set of reaction mixtures and H_2O to another set. Incubations were continued and the kinase activity was determined at different time points as described previously (18).

Effect of Polyamines on Transcription: Relationship to Their Effect on Phosphorylation

Polyamines are known to have a pronounced effect on *in vitro* transcription (5,13,14,17,20). All three classes of RNA polymerases are stimulated by these compounds (13). The acetyl derivatives of spermidine did not exhibit a stimulatory effect on RNA synthesis *in vitro* (unpublished data). More recently, we have demonstrated that NII protein kinase activity is closely associated with highly purified RNA polymerase I

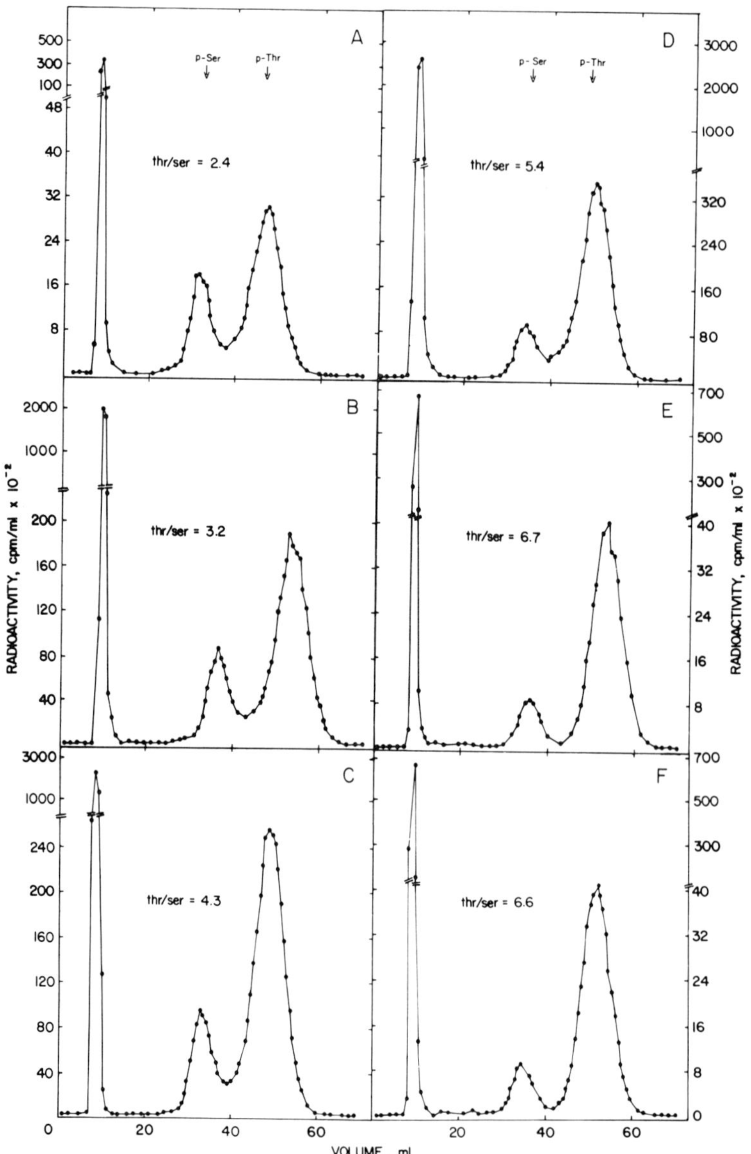

FIG. 7. Effect of polyamines on phosphorylation of serine and threonine. Protein kinase NII-catalyzed phosphorylation of casein was conducted as described in the legend to Fig. 1. Serine and threonine were separated on Dowex 50W - χ8 as described previously (18). Arrows indicate position of elution of P-ser and p-thr standards. Note differ-ences in scales in each panel. The ratio of radioactivity in threonine to that in serine (thr/ser) was determined from the area under the peak for each sample. Panel A, no polyamine; Panel B, 5 mM spermidine; Panel C, 10 mM spermidine; Panel D, 5 mM spermine; Panel E, 5 mM N^1-acetylspermidine and Panel F, 5 mM N^8-acetylspermidine.

(21). Three out of eight polypeptides (S2, M_r 120,000; S3, M_r 65,000 and S5, M_r 24,600) of RNA polymerase I were phosphorylated by the endogenous kinase (21). Since this kinase was not present in excess, addition of exogenous kinase could result in further phosphorylation of RNA polymerase I. Spermine at a relatively low concentration (0.5 mM) could enhance phosphorylation of RNA polymerase I polypeptides either by endogenous (12) or by exogenous NII kinase (9). Concomitant with phosphorylation of RNA polymerase I the activity of the enzyme was elevated as a result of an increase in elongation of RNA chains (9). Since the NII kinase is closely associated with highly purified RNA polymerase I, one can raise the question whether the polyamine-mediated stimulation of transcription by polymerase I is due to enhanced phosphorylation of the enzyme. Because polyamines stimulate transcription conducted in the presence of Mn^{2+} and this divalent cation is not utilized by the NII kinase (18), it is probable that polyamines do not always alter transcription via the kinase. In order to distinguish the effect of the aliphatic amines on transcription due to RNA polymerase I phosphorylation from that due to other effects, two purine analogues, 5'-adenylyl imidodiphosphate (AMP-PNP) and 5'guanylylimidodiphosphate (GMP-PNP) were substituted for ATP. Due to the presence of nonhydrolyzable phosphate in the γ position the kinase is unable to transfer the terminal phosphate of these compounds. In contrast, RNA polymerases, which require hydrolyzable phosphate at only the α position, can utilize these derivatives in RNA synthesis (25). When Mg^{2+} was used as divalent cation, RNA polymerase I-catalyzed RNA synthesis was enhanced 2-to 4-fold by addition of spermine (Table 1). However, when the β, γ imido derivatives replaced ATP and GTP, a lesser stimulation (1.4-to 3-fold) was observed. Since phosphorylation is manifest under normal Mg^{2+} conditions, but only factors other than phosphorylation prevail in the presence of the derivatives, the contribution of protein phosphorylation to the overall reaction will be the difference between the extent of activation under the two conditions. Computing this difference then, it is clear that phosphorylation accounts for approximately 35% of the total polyamine activation of RNA polymerase I (Table 1). It is noteworthy that when Mn^{2+} replaced Mg^{2+}, the polyamine enhanced RNA synthesis to a similar extent in the presence or absence of the derivative. This latter observation not only confirms the ability of the ATP/GTP derivatives to be utilized as substrates in RNA synthesis but also implicates the Mg^{2+}-requiring protein kinase NII in polymerase I phosphorylation.

In contrast to RNA polymerase I, protein phosphorylation contributed very little (15%) to RNA synthesis catalyzed by RNA polymerase II. Further, the effect was not restricted to Mg^{2+}, but also occurred in the presence of Mn^{2+}.

TABLE 1. Effect of ATP and GTP Analogues on Spermine-Activated RNA Synthesis In Vitro

Assay Conditions	RNA Polymerase Activity (+ Spermine/-Spermine x 100) ATP + GTP (X)	AMP PNP + GMP PNP (Y)	$(\frac{X-Y}{X}) \times 100$
RNA Polymerase I			
Mg^{2+}	208[a]	139	33
	340[b]	188	45
	405[c]	299	26
Mn^{2+}	656[d]	708	-8
RNA Polymerase II			
Mg^{2+}	265	224	15
Mn^{2+}	533	448	16

Partially purified RNA polymerases I and II were used for these experiments. Superscripts refer to separate experiments. One hundred percent activity (-spermine) ranged from 12 $\pm$ 1.3 to 107 $\pm$ 10 pmol UMP/20 min under normal assay conditions (Mn^{2+} or Mg^{2+}; ATP + GTP). Activity in the presence of the derivatives was 11% to 50% of that observed under normal conditions.

DISCUSSION

The noteworthy findings of the present studies are as follows: (a) both spermine and spermidine at physiological concentrations (0.5 - 1 mM) can markedly enhance the activity of the NII protein kinases, but not of the NI kinase; (b) the terminal NH_2 groups on the spermidine molecule are required for mediating the stimulation of the phosphorylation reaction; and (c) polyamines can expose new sites for phosphorylation.

To our knowledge, there has been no report on the virtual dependence of a highly purified nuclear protein kinase from higher organisms on spermine or spermidine. It is possible that other purified nuclear casein kinases (6,22) might be related to the hepatoma NII kinase although the effect of polyamines has not been tested on these purified kinases. A protein kinase which can be stimulated by spermine has been partially purified from bovine epididymal spermatozoa (4), the cytosolic (soluble) fractions of Morris hepatoma 3924A (7) and rat liver (24). It is likely that the cytosolic kinases were derived from the nuclei as a result of tissue homogenization in isotonic buffers. However, the spermine-induced stimulation of casein phosphorylation by the cytosolic kinase does not exceed 2-to

3-fold (24). Although a polyamine-activated protein kinase has been purified from nuclei of the slime mold Physarum polycephalum (8), the specific activity of this enzyme in the presence or absence of polyamines is at least 200-fold less than that reported for the tumor enzyme. The exact relationship of the polyamine-activated kinases from the hepatoma and from Physarum is not evident. For example, the kinase purified from the rat tumor consists of two subunits of M_r 42,000 and M_r 25,000 (18) whereas the kinase from the slime mold appears to consist of a single polypeptide of M_r 26,000. Unlike the tumor kinase, the slime mold protein kinase is closely associated with a single protein acceptor of M_r 70,000 which appears to be similar to ornithine decarboxylase (ODC) with respect to amino acid composition and immunological crossreactivity (8).

The hepatoma NII kinase has been found associated with RNA polymerase I purified to homogeneity from the same tumor (21). The kinase subunits correspond to two of the eight polypeptides of RNA polymerase I with respect to molecular weights, immunological cross-reactivity and ATP-binding site (21). Antibodies raised against purified RNA polymerase I interact with six subunits of the enzyme with molecular weights ranging from 25,000 to 190,000 and with the two subunits of the purified NII kinase (21). These antibodies do not cross-react with [^{3}H]DFMO-labelled ODC purified from rat liver (unpublished data) which suggests that six of the polypeptides of RNA polymerase I with molecular weights above 25,000 are not antigenically related to ODC. We have shown that three polypeptides of RNA polymerase I [M_r 120,000, M_r 65,000, and M_r 25,000 (21)] are phosphorylated by the NII kinase *in vitro* or *in vivo* (unpublished data). Although the subunit of M_r 65,000 is a likely candidate for being a protein analogous to the M_r 70,000 phosphate acceptor of Physarum, structural analysis, such as peptide mapping, is required to establish whether these proteins are similar.

Finally, the close association of the polyamine-dependent NII protein kinase with RNA polymerase I, phosphorylation of three of the eight subunits of the polymerase by the endogenous kinase and the preferential modification of threonine moieties by physiological concentrations of spermine or spermidine suggest that phosphorylation of RNA polymerase I may have a regulatory role in ribosomal gene expression. It is feasible that the phosphate groups on threonine might enhance interaction between RNA polymerase I and the promoters or regulatory sequences in rDNA which, in turn, might facilitate accurate initiation of rRNA synthesis. Studies to test this hypothesis are currently underway.

SUMMARY

The nuclear protein kinase NII purified to homogeneity from the Morris hepatoma 3924A required polyamines for optimal activity. At a concentration of 1 mM, spermine and spermidine stimulated phosphorylation of casein by the NII kinase 20-fold and 6-fold, respectively. The kinase activity was stimulated by 27-fold and 17-fold at 2.5 mM and 5 mM spermine and spermidine, respectively. The phosphorylation reaction in the presence of spermidine was essentially linear up to 60 min. V_{max} (pmoles phosphate/30 min) and apparent K_m (μM) with respect to ATP were increased from 3 to 780 and from 7 to 164, respectively, upon addition of spermidine. The requirement for the free NH_2 groups at N^1 and N^8 positions of spermidine was shown by the lack of stimulation of phosphorylation by either N^1-acetyl or N^8-acetylspermidine.

Unlike the NII kinase, the NI kinase activity was not affected by spermidine. Spermidine had a marked stimulatory effect when added after the phosphorylation reaction reached a plateau in the absence of the polyamine. The ratio of phosphorylation of threonine to that of serine (thr/ser) was significantly higher in the presence of spermine or spermidine than in the absence of polyamines. Even though the acetyl derivatives had no marked effect on the phosphorylation reaction, a shift in the thr/ser ratio was observed in their presence, apparently as a result of inhibition in serine phosphorylation and concurrent stimulation in threonine phosphorylation. Using the purine analogues, 5'-adenylyl and 5'-guanylyl imidodiphosphates which contain non-hydrolyzable γ-phosphate, it was shown that the polyamine-mediated increase in the in vitro transcription by RNA polymerase I was, at least in part, a result of phosphorylation of the polymerase.

ACKNOWLEDGEMENTS

The expert technical assistance of Quang Bui is greatly appreciated. This work was supported by grants by the United States Public Health Service.

REFERENCES

1. Ahmed, K., Wilson, M.J., and Gouelhi, S.A. (1978): Biochem. J., 176:739-750.

2. Allfrey, V.G., Inoue, A., Karn, J., Johnson, E.M., and Vidali, G.(1973): Cold Spring Harbor Symp. Quant. Biol., 38:785-801.

3. Atmar, V.J., Daniels, G.R., and Kuehn,G.D. (1978): Eur. J. Biochem., 90:29-37.

4. Atmar, V.J., Kuehn, G.D., and Casillas, E.R. (1981): J. Biol. Chem., 256:8275-8278.

5. Barbiroli, B., Masotti, L., Moruzzi, M.S., Monti, M.G., and Moruzzi, G. (1978): Advances in Polyamine Research, 1:217-229.

6. Baydoun, H., Hoppe, J., Jacob, G., and Wagner, K.G. (1980): FEBS Lett., 122:231-233.

7. Criss, W.E., Yamamato, M., Takai, Y., Nishizuka, Y., and Morris, H.P. (1978): Cancer Res., 38:3532-3539.

8. Daniels, G.R., Atmar, V.J., and Kuehn, G.D. (1981): Biochemistry, 20:2525-2532.

9. Duceman, B.W., Rose, K.M., and Jacob, S.T. (1981): J. Biol. Chem., 256:10755-10758.

10. Farron-Furstenthal, F., and Lightholden, J.R. (1978): Biochem. Biophys. Res. Commun., 83:94-100.

11. Imai, H., Shimoyama, M., Yamamato, S., Tanigawa, Y., and Ueda, I. (1975): Biochem. Biophys. Res. Commun., 66:856-862.

12. Jacob, S.T., Duceman, B.W., and Rose, K.M. (1981): Medical Biology, 59:381-388.

13. Jacob, S.T., and Rose, K.M. (1976): Biochim. Biophys. Acta, 425:125-128.

14. Jänne, O., Bardin, C.W., and Jacob, S.T. (1975). Biochemistry, 14:3589-3597.

15. Kleinsmith, L.J. (1975): In: The Regulation of Gene Expression, edited by G.S. Stein, and L.J. Kleinsmith, pp. 45-57. Academic Press, New York.

16. Kuroda, Y., Hashimoto, E., and Nishizuka, Y. (1977): J. Biochem., 82:1167-1172.

17. Mandel, J.L., and Chambon, P. (1974): Eur. J. Biochem., 41:367-378.

18. Rose, K.M., Bell, L.E., Siefken, D.A., and Jacob, S.T. (1981): J. Biol. Chem., 256:7468-7477.

19. Rose, K.M., and Jacob, S.T. (1979): J. Biol. Chem., 254: 10256-10261.

20. Rose, K.M., Ruch, P.A., Morris, H.P., and Jacob, S.T. (1976): Biochim. Biophys. Acta, 432:60-72.

21. Rose, K.M., Stetler, D.A., and Jacob, S.T. (1981): Proc. Natl. Acad. Sci., U.S.A., 78:2833-2837.

22. Thornburg, W., and Lindell, T.J. (1977): J. Biol. Chem., 252:6660-6665.

23. Williams-Ashman, H.G., and Canellakis, Z.N. (1979): Persp. Biol. Med., 22:421-439.

24. Yamamato, M., Criss, W.E., Takai, Y., Yamamura, H., and Nishizuka, Y. (1979): J. Biol. Chem., 254:5049-5052.

25. Yount, R.G., Babcock, D., Ballantyne, W., and Ojala, D. (1971): Biochemistry, 10:2684-2489.

Advances in Polyamine Research, Vol. 4, edited by U. Bachrach, A. Kaye, and R. Chayen. Raven Press, New York © 1983.

Multiple-Protein Complex with [Calmodulin]-Polyamine Responsive Protein Kinase Activity

*†W. E. Criss, *†Y. Morishita, **Q. Watanabe, **C. Akogyeram, †A. Sahai, †B. Deu, and ††T. Oka

*Departments of *Human Nutrition and Food, **Pharmacology, †Zoology, School of Human Ecology, Cancer Center, Howard University, Washington, D.C. 20059; ††Section on Intermediary Metabolism, Laboratory of Biochemistry and Metabolism, National Institute of Arthritis, Metabolism, and Digestive Diseases, National Institutes of Health, Bethesda, Maryland 20205*

Polyamines are ubiquous in living organisms and are considered to be directly involved in the regulation of processes associated with cellular growth (1,2). They increase markedly during the rapid growth of tumors (3,4). A relationship is currently being established between polyamines and protein kinase activity in several different laboratories (5-14). For example, polyamines inhibit cAMP dependent protein kinase (5-7), while they stimulate two cyclic nucleotide independent protein kinases. One of these latter protein kinases is nuclear (8-11); the other is cytoplasmic (12-17). The current manuscript will review the findings on the cytoplasmic form of the polyamine (calmodulin) responsive protein kinase. The physiological roles of either the nuclear or the cytoplasmic polyamine responsive kinases are as yet unexplored as compared to the cAMP and cGMP dependent protein kinases.

METHODOLOGIES

In order to measure polyamine responsive kinase activity in rat tissues, it is necessary to obscure or minimize the activities of cAMP and cGMP dependent protein kinases and also to minimize protein phosphatase activities. Therefore, partially purified components of this polyamine responsive protein kinase system must be prepared and employed such that verification of the enzymatic activity of specifically this enzyme system is indeed evaluated. For this reason, a detailed section of methods

This work was supported in part by grants and contracts from the USPHS-CA 26102 and CA-14345-39.

and techniques will follow.

Materials

ACI rats from Simonsen Laboratories, California, are inoculated with Morris hepatoma 3924A. The tumors are grown subcutaneously and transplanted to other ACI rats as necessary. All rat tissues to be evaluated, including the hepatoma tissues, are removed, cooled to 4°C, cleaned of necrotic portions, minced, and homogenized with five volumes of buffer containing 20 mM Tris-HCl (pH 7.5) 50 mM 2-mercaptoethanol, 2 mM EDTA and 0.25 sucrose in a teflon glass homogenizer at 4°C. The homogenate is filtered through 4 layers of gauze and centrifuged for 90 minutes at 100,000 x g at 4°C. The supernatant is used to prepare enzyme.

Enzyme Assay

Polyamine responsive protein kinase activity is assayed in a reaction mixture (0.25 ml) which contains 5 μmol of Tris-Cl (pH 7.5) 1.25 μmol of magnesium acetate, 7.5 μmol of 2-mercapto-ethanol, 2.5 nmol of (γ-32P) ATP (10^5 cpm/nmol), 50 μg of protein (enzymatic components) and without or with 10 μg poly-lysine. It does not require addition of any exogenous protein substrates. The incubation is carried out for 5 or 10 min. at 37°C. The reactions are stopped by the addition of 10% trichloroacetic acid. Acid-precipitable materials are collected on a polycarbonate membrane filter (pore size 0.8 μm). The radioactivity is determined by counting in a Beckman LS-9000 Scintillation Counter.

Catalytic Component Assay

The catalytic components of the enzyme complex are separated from all other components, therefore catalysis requires addition of new substrates. The enzyme-substrate complex is purified by DEAE-cellulose and Sephacryl-300 chromatographies (12-13) and is boiled for 10 min to destroy all enzymatic activity. 50 μl of the heat stable substrate, from such a preparation, is added to each assay as endogenous phosphate accepting proteins.

Substrate Component Assay

The substrate components of the enzyme complex are separated from all other components, therefore catalysis requires addition of new catalytic components. 100 μl of a purified catalytic fraction is added to each assay as a catalytic unit (14-16).

Inhibitor and Activator Assay

Evaluations of both inhibitory and activating activities are performed by adding approximately 5,000 dpm of enzyme activity to each assay. The enzyme activity is a catalytic-substrate complex which is free of all activators and/or inhibitors (12-17). Such a complex is purified through DEAE cellulose and Sepharose column chromatographies.

Purification of Enzymatic Complex

The 100,000 x g supernatant fraction from a 20% homogenate of adult rat liver and/or Morris hepatoma 3924A tissue is applied to a DEAE-cellulose (DE-52) column which is equilibrated with 20 mM EDTA, 50 mM 2-mercaptoethanol, 70 mM NaCl and 10 uM cAMP (14-16). The column is extensively washed with the same buffer. Elution is then accomplished with the application of a linear concentration gradient of NaCl (70 to 500 mM). Fractions are collected, and assayed for enzymatic activity and inhibitor/activator activity. Under these conditions, an enzymatic activity peak appears which is attributed to polyamine responsive protein kinase enzyme activity (14-16). The exact same fractions also contain measurable enzyme inhibitor activity. Such an enzyme preparation contains no responsiveness to cyclic AMP or cyclic GMP, and does not phosphorylate histone (12-13,16).

Separation of Enzyme into Several Components

25 mg protein of the enzymatic activity peak from the DEAE cellulose column is mounted onto a 2 x 150 cm column of either Sephacryl 300 or onto a 2 x 3.5 column of phosphocellulose equilibrated with the Tris-HCl buffer. Proteins are fractionated on the Sephacryl column by continued buffer flow or on the phosphocellulose column by step wise gradient elution with 30 ml increments of 0.2 M NaCl. Employing the above assay procedures, the polyamine responsive protein kinase complex can be readily identified in all rat tissues, in subcellular compartments, and can be separated into several components. Component composition varies from tissue to tissue.

RESULTS

Over 75% of the polyamine responsive protein kinase activity is located in the cytosols of adult rat liver and Morris hepatoma 3924A. Therefore all subsequent analysis of this enzymatic system has been routinely performed on liver and tumor cytosol (100,000 x g for 90 min) preparations.

TABLE 1. Comparison of the liver and hepatoma polyamine responsive protein kinase systems

Normal adult rat liver	Morris fast growing hepatoma
low cellular polyamines	high cellular polyamines
low cytoplasmic PPK activity	high cytoplasmic PPK activity
3 catalytic components	3 catalytic components
4 substrate components	4 substrate components
protein inhibitors I, II, III	protein inhibitors I and II
very low protein activator activity	high protein activator activity
activated by polyamines (> 500Å)	activated by polyamines (> 500Å)
activated by polyamines + calmodulin	activated by polyamines + calmodulin
phosphorylation of serine & threonine	phosphorylation of serine & threonine

Component Comparisons of Liver and Hepatoma Enzymatic Complexes

Table 1 summarizes a comparison of the component compositions of polyamine (calmodulin) responsive, self-phosphorylating, protein kinase complexes from normal adult rat liver and a fast growing (Morris) hepatoma (3924A). Most of the measurable (and verifiable) enzymatic activity is in the 100,000 x g supernatant fractions of these tissue homogenates. The hepatoma tissue has higher levels of the three "natural" polyamines (putrescine, spermidine, and spermine) and higher measurable levels of polyamine responsive protein kinase activity. Purification and separation of the tumor enzymatic complex reveals: 3 catalytic components, 4 substrate components (phosphate accepting proteins), 2 inhibitor proteins, and an activator protein. This enzyme system is activated by polyamines and calmodulin. The liver enzyme complex separated into: 3 catalytic components, 4 substrate components (phosphate accepting proteins), and 2 inhibitor proteins. The latter enzyme system is activated by polyamines and calmodulin also.

Protein Inhibitors

Three protein inhibitors of polyamine responsive protein kinase can be isolated from the enzymatic complex. Inhibitors I and II are found in the hepatoma tissue. Inhibitors I, II, III are isolated from liver tissue. Each inhibitor prevents the polyamine activation of the purified enzyme by decreasing the Vmax. They do not affect the Km for ATP (Table 2). The molecular weights are 1,800, 800, and near 35,000 daltons for protein inhibitors I, II, and III, respectively. It is not yet known if these inhibitors affect other enzymes (17).

TABLE 2. Protein inhibitors of polyamine responsive protein kinase activity

---

Protein inhibitor I
- molecular weight = 1,800 Å
- ID_{50} = 20 μg of protein (hepatoma)
- no affect on Km (ATP)
- decreases Vmax
- heat stable

Protein inhibitor II
- molecular weight = 800 Å
- ID_{50} = 45 μg of protein (hepatoma)
- no affect on Km (ATP)
- decreases Vmax
- heat stable

Protein inhibitor III
- molecular weight = 30,000 - 50,000 Å
- decreases Vmax

Protein Activator

The protein activator can be readily identified and purified from the hepatoma enzymatic complex (Table 3). It is a polypeptide with a broad molecular weight region identified on sephadex as 20,000 to 50,000 daltons. This protein activator has calmodulin-like activity. It will activate the polyamine responsive protein kinase only in the presence of polyamines. The calcium ion and/or EGTA have no affect.

TABLE 3. Protein activator and polyamine responsive protein kinase activity

---

Activator Characteristics
- polypeptide composition
- 20,000 to 40,000 Å molecular weight range
- calmodulin-like activity (stimulates PDE)
- does not require calcium for activation
- activation not affected by EGTA
- protein activator requires polyamine for activation
- protein activator increases polyamine stimulation

- purified calmodulin + polyamine activates PPK
- purified calmodulin (only) will not activate PPK
- purified calmodulin + polyamine activation independent of Ca^{2+}
- purified calmodulin + polyamine activation independent of EGTA

Purified calmodulin will also activate the polyamine responsive protein kinase activity in the presence of polyamines. Similar to the purified protein activator, the calcium ion and/or EGTA do not affect this calmodulin response. Because the cytoplasmic levels of calmodulin increase in these tumors (15), it is postulated that the endogenous protein activator of the cytoplasmic polyamine responsive protein kinase system is calmodulin, and that calmodulin increases the polyamine stimulation of this protein kinase system by direct protein-protein interaction.

Activation by Polyamines

Ornithine and putrescine do not activate the polyamine responsive protein kinase system (Table 4). However, spermidine, spermine, and a wide range of polyamino-compounds do activate this system. The Ka for each varies inversely with the molecular weight (for polylysine and/or polyornithine). Activation requires closely spaced amino groups covalently attached to 3-4 carbon chain moieties (side groups). Bulky side chains interfer, and imines or imidazole groups have no affect.

TABLE 4. Polyamine activation of polyamine responsive protein kinase

molecular weight		Ka (M)
ornithine	(132Å)	no activation
putrescine	(84Å)	no activation
spermidine	(144Å)	8×10^{-4} (weak activation)
spermine	(202Å)	4×10^{-4} (weak activation)
polylysine	(350,000Å)	1×10^{-8}
polylysine	(60,000Å)	2×10^{-8}
polylysine	(12,000Å)	2×10^{-7}
polylysine	(4,000Å)	3×10^{-6}
polylysine	(539Å)	4×10^{-5}
polylysine	(274Å)	no activation
lysine	(182Å)	no activation
polyornithine	(110,000Å)	4×10^{-8}
polyornithine	(30,000Å)	6×10^{-8}
polyornithine	(10,000Å)	3×10^{-7}
ornithine	(280Å)	no activation
polyarginine	(100,000Å)	very weak activation*
polyarginine	(30,000Å)	very weak activation*
polymethionine	(40,000Å)	no activation
polyhistidine	(10,000Å)	no activation
spermidine + spermine	(½ x ½)#	2×10^{-4} (weak activation)
lysine - alanine peptide	(50% each)	excellent activation(10^{-7})
lysine - tyrosine peptide	(50% each)	no activation

*The very weak activation is probably due to 1-5% polyornithine contamination in the polyarginine preparations.

#Activators were ½ spermine + ½ spermidine.

CONCLUSIONS AND SUMMARY

A multiple protein complex with calmodulin-polyamine responsive protein kinase activity has been identified and characterized from several rat tissues. It is measurable in the cytoplasm of all tissues, and the component composition varies from tissue to tissue. Comparison of this enzymatic system in normal adult rat liver and in fast growing Morris hepatoma 3924A reveals the following: liver and hepatoma have 4 protein substrates, 3 catalytic units and protein inhibitor I. Liver tissue also has protein inhibitor III; while tumor tissue also has protein inhibitor II and a protein activator. The latter activator is probably calmodulin. Activation by polyamines and calmodulin do not require Ca^{2+}, are independent of EGTA, and increase the Vmax while not affecting the Km (ATP) of the enzymatic system.

Comparison of this cytoplasmic polyamine responsive protein kinase system with the reported nuclear polyamine responsive protein kinases indicate several important differences. These include: NaCl requirement for activity, molecular weights of the catalytic units, component compositions, inhibition by putrescine, phosphorylation of tyrosine residues, etc. It is probable that two polyamine responsive protein kinase systems exist in cells, nuclear and cytoplasmic. They each may contribute to the polyamine regulation of cell function in different ways.

ACKNOWLEDGEMENTS

The authors greatly acknowledge the expert technical assistance of Ms. Charity Jackson, Mr. Calvin Small, and Dr. Julius Fakunle, and Mr. Cornelius Diya.

REFERENCES

1. Russell, D.H., editor (1973): Polyamines in Normal and Neoplastic Growth, Raven Press, New York.
2. Bachrach, U., editor (1973): Function of Naturally Occurring Polyamines, Academic Press, New York.
3. Andersson, G. and Heby, O. (1972): J. Natl. Cancer Inst., 48: 165-172.
4. Williams-Ashman, H.G. Coppoc, G.L., and Weber, G. (1972): Cancer Res., 32:1924-1932.
5. Takai, Y., Nakaya, S., Inoue, M., Kishimoto, A., Nishiyama, K., Yamamura, H., and Nishizuka, Y. (1976): J. Biol. Chem., 251:1481-1487.
6. Murray, A.W., Froscio, M., and Rogers, A. (1976): Biochem. Biophys. Res. Commun., 71:1175-1181.
7. Hochman, J., Katz, A., and Backrach, U. (1978): Life Sci., 22:1481-1484.
8. Atmar, V.J., Kuehn, G.D., and Casillas, E.R. (1981): J. Biol. Chem., 256:8275-8278.

9. Daniels, G.R., Atmar, V.J. and Kuehn, G.D. (1981): Biochem., 20:2525-2532.
10. Rose, K.M., Bell, L.E., Siefken, D.A., and Jacob, S.T. (1981) J. Biol. Chem., 256:7468-7477.
11. Duceman, B.W., Rose, K.M., and Jacob, S.T. (1981): J. Biol. Chem., 256:10755-10758.
12. Criss, W.E., Yamamoto, M., Takai, Y., Nishizuka, Y., and Morris, H.P. (1978): Cancer Res., 38:3532-3539.
13. Criss, W.E., Yamamoto, M., Takai, Y., Nishizuka, Y., and Morris, H.P. (1978): Cancer Res., 38:3540-3545.
14. Morishita, Y., Akogyeram, C., Liederman, L., Oka, T., and Criss, W.E. (1982): Med. & Ped. Oncology, 10:102.
15. Criss, W.E., and Kakiuchi, S. (1982): Federation Proceedings, 41:67-69.
16. Sahai, A., Morishita, Y., Akogyeram, C., Oka, T., and Criss, W.E. (1982): J. Protein Chemistry, submitted.
17. Morishita, Y., Sahai, A., Akogyeram, C., Hollis, V., Oka, T., and Criss, W.E. (1982): in press in J. Cyclic Nucleotide Res.

Advances in Polyamine Research, Vol. 4, edited by U. Bachrach, A. Kaye, and R. Chayen. Raven Press, New York © 1983.

Hormonal Regulation and Function of Polyamine Responsive Protein Kinase Activity in the Mouse Mammary Gland

*†Lisa J. Leiderman, **††Wayne E. Criss, **††Yorihiko Morishita, and *Takami Oka

**Laboratory of Biochemistry and Metabolism, National Institute of Arthritis, Diabetes, Digestive Diseases and Kidney Diseases, National Institutes of Health, Bethesda, Maryland 20205; Departments of †Pharmacology, **Human Nutrition and Food and Human Ecology, ††Cancer Center, Howard University, Washington. D.C. 20059*

The enzymatic phosphorylation of a protein was first described by Burnett and Kennedy (2) in 1954. This protein kinase (EC 2.7. 1.37) derived from rat liver mitochondria, was shown to phosphorylate the seryl and threonyl residues of casein and phosvitin. It has since been shown that this type of protein kinase activity exists in a variety of mammalian tissues (18). These enzymes have been found to function independently of cyclic nucleotides. They are able to utilize ATP and/or GTP as the phosphate donor and prefer acidic proteins, such as casein and phosvitin, as exogenous substrates. In contrast to these protein kinases, cyclic nucleotide dependent kinases have been shown to be active only with ATP as the phosphate donor and preferentially phosphorylate basic proteins, such as histone and protamine, in the assay (8,22). Recently, the cyclic nucleotide independent protein kinase activity has been shown to be enhanced by polyamines (1,3,4,5,6, 9,12).

The work presented in this paper provides evidence for the hormonal regulation of the cyclic nucleotide independent, polyamine responsive protein kinase activity in the mouse mammary gland. Attempts have also been made to assess their role in the development of mammary tissue. The mammary gland, which can serve as a model system for the study of cell growth and differentiation, undergoes extensive morphological and biochemical changes during pregnancy and lactation. It has previously been shown (20) that these changes can be induced *in vitro* by culturing the mammary explants in a chemically defined medium containing the appropriate combinations of hormones. The addition of insulin, prolactin and cortisol has been shown to stimulate the synthesis and phosphorylation of the milk protein, casein, and the accumulation of polyamines in cultured tissue (13-16,20). Furthermore, polyamines have been implicated as key regulatory substances in the process of lactogenesis *in vitro* (13,14). These collective observations led us to utilize this organ culture system for the study of the polyamine responsive protein kinase activity.

MATERIALS AND METHODS

Organ culture

The method of the mouse mammary gland culture has been described previously (21). The hormone concentrations and dibutyryl cyclic adenosine monophosphate (dbcAMP) concentration are indicated in the results. C_3H/HeN mice were obtained from the NIH Animal Breeding Facility. Other materials were obtained from the following sources: Medium 199 (Hanks' salts) from GIBCO, cortisol(F) from Calbiochem, crystalline porcine zinc insulin (I) from Lilly, bovine prolactin(P) from the Hormone Distribution Program, NIADDK, and progesterone and dbcAMP from Sigma.

Protein kinase assay

Protein kinase activity was determined by measuring the incorporation of (γ-^{32}P)ATP into acid-precipitable materials. This method is similar to that of Criss *et al*. (4). The enzyme preparation was made by homogenizing the mammary tissue with a glass homogenizer and a teflon pestle (25 strokes) at 4°C in a solution containing 0.25 M sucrose, 15 mM 2-mercaptoethanol and 10 mM tris(hydroxymethyl)-aminomethane HCl (Tris-HCl, pH 7.5). The homogenate was centrifuged at 105,000 x g for one hour at 4°C. The supernatent (i.e., the cytosol fraction) was used for the enzyme solution (see below). The assay cocktail (200 μl) contained 5.2 μmol of Tris-HCl (pH 7.5), 7.8 μmol of 2-mercaptoethanol, 8.75 μmol-17.5 μmol of magnesium acetate, 2.4 nmol of ATP(Calbiochem) and 10-30 x 10^4 cpm of (γ-^{32}P)ATP (Amersham). The enzyme solution (50 μl) contained the cytosol fraction derived from 0.7 mg-0.8 mg of mammary tissue, 0.75 μmol of 2-mercaptoethanol, 0.5 μmol of Tris-HCl (pH 7.5) and 2.5 μmol of sucrose. Unless otherwise indicated, 10 μg of polylysine (M.W. 350 K, Sigma) was included in the solution to stimulate the enzyme activity. Bovine casein (Schwartz/Mann) was added, where indicated, to serve as the phosphate acceptor protein. Spermine tetrahydrochloride (Sigma), spermidine trihydrochloride (Calbiochem), and putrescine dihydrochloride (Sigma) were included in the solution in the amounts as specified in Figure 1.

The reaction was initiated by the addition of the assay cocktail to the enzyme solution. The reaction mixture was then incubated for 5 minutes at 37°C in a shaking water bath and the reaction was terminated by the addition of 2 ml of 10% trichloroacetic acid (TCA). The acid-precipitable materials were collected on polycarbonate membrane filters (Nucleopore, pore size 0.8 μm, diameter 25 mm) and washed once with 3 ml of 10% TCA. The filters were then air-dried and each was placed in a vial containing 10 ml of scintillation cocktail. The radioactivity of the samples was determined with a liquid scintillation spectrometer. The enzyme reaction was linear with respect to time and enzyme concentration. The values for the polyamine responsive protein kinase activity presented in this study were corrected for the basal activity seen in the absence of polyamines. The net polyamine responsive kinase activity is expressed as pmol/min/mg tissue which is equivalent to one unit. Unless otherwise stated, enzyme activity represents the phosphorylation of endogenous substrates in the mammary gland.

Slab gel electrophoresis and autoradiographic analysis

The electrophoretic procedure was a modification of the method of Laemmli (7). The separation gel solution contained the following: 12.5% acrylamide, 0.23% N,N'-(bis)-methylene acrylamide, 0.2% sodium dodecyl sulfate(SDS), 0.375 M Tris-HCl (pH 8.8), 0.025% ammonium persulfate and 0.1% tetramethylethylenediamine (TEMED). The gel was prepared in a dual vertical slab electrophoresis cell (Biorad). The stacking gel solution contained the following: 3.0% acrylamide, 0.08% N,N'-(bis)-methylene acrylamide, 0.2% SDS, 0.125 M Tris-HCl (pH 6.8), 0.025% ammonium persulfate and 0.1% TEMED. The electrode buffer contained 0.025 M Tris (pH 8.3), 0.19 M glycine and 0.1% SDS.

All sample preparations (30 µl-60 µl) were made in the following manner: the acid-precipitable materials, obtained from the protein kinase assay after the addition of TCA, was centrifuged at 200 x g for 15 minutes at 4°C. The pellet was washed with 250 µl of ethyl ether and centrifuged another 15 minutes at the same speed and temperature. A buffer was then added to the pellet. Each sample preparation contained the equivalent of approximately 1 mg-8 mg tissue, 1.0% SDS, 7.5% glycerol, 0.5% β-mercaptoethanol and 0.25% bromophenol blue as the dye. Mouse casein controls and molecular weight marker proteins were run in each experiment. The proteins were completely dissociated by immersing the samples for 2 minutes in boiling water. The samples were then added to the lanes of the stacking gel. Electrophoresis was carried out with 50 volts until the bromophenol blue marker reached the bottom of the stacking gel. The voltage was then increased to 100 volts as the dye approached the separation gel. Electrophoresis was continued until the dye reached the bottom of the separation gel (about 3.5 hours total). The gel was removed from the electrophoresis cell and placed for 30 minutes in a protein staining solution which contained 50% methanol, 10% acetic acid and 0.1% Coomassie brilliant blue (CBB). The gel remained for an additional 10 minutes in a solution which contained 12.5% methanol, 10% acetic acid and 0.025% CBB. It was then destained for about

16 hours in 10% acetic acid, during which time the 10% acetic acid bath was changed once. After destaining, the gel was placed on filter paper and covered with plastic wrap. It was then placed on a slab gel dryer (Biorad) and allowed to dry for about 1 hour at 100°C.

Autoradiograms were prepared by using x-ray film (Kodak X-Omat AR film). The cassette, which contained the film and the gel, was placed in complete darkness at -70°C for 3-4 days.

RESULTS AND DISCUSSION

Protein kinase activity, which was able to function independently of cyclic nucleotides and was stimulated by polyamines, was observed in mammary glands from virgin, pregnant and lactating mice. This activity was dependent on ATP as the phosphate donor and magnesium as the cofactor. The activity of the crude cytosolic enzyme(s) remained stable for up to one month when stored at 0°C in a buffer containing 0.25 M sucrose.

Figure 1 shows the stimulatory effects of putrescine, spermidine and spermine on protein kinase activity in the presence of 100 μg of bovine casein. It can be seen that spermine, a tetra-amine, stimulated the optimal activity when added to the assay in concentrations between 0.1 mM and 1 mM. Putrescine, a di-amine, stimulated optimal activity when the concentrations were between 10 mM and 40 mM. For spermidine, a tri-amine, the concentrations that produced optimal activity were 1 mM-5 mM which was intermediate between those of the other amines. The effective concentrations of spermine and spermidine were within the physiological range (14). These data seem to indicate that the protein kinase activity varied as a function of the relative amounts of cationic charges that were present. Polylysine, which contains the greatest number of cationic charges, stimulated the phosphorylation of endogenous substrates. The other polyamines, which stimulated activity in the presence of bovine casein, had little effect on endogenous phosphorylation in contrast to polylysine (data not shown). It was for this reason that polylysine was utilized in this study.

Table 1 shows that the polylysine responsive protein kinase activity in the mammary gland varied as a function of the reproductive stage of the mice. The enzyme activity was quite low in virgin mice and markedly increased during the periods of pregnancy and lactation.

It can be seen in Table 2 that polylysine responsive protein kinase activity in the cultured mammary gland changed in response to different hormone combinations. Maximal enzyme activity was attained with insulin (I), prolactin (P) and cortisol (F), the combination of hormones which has been previously shown to induce casein synthesis at optimal levels (15,16). The IP combination gave a somewhat less than optimal response, the magnitude of which varied depending upon the stage of pregnancy. During early pregnancy, the endogenous cortisol levels are lower than they are

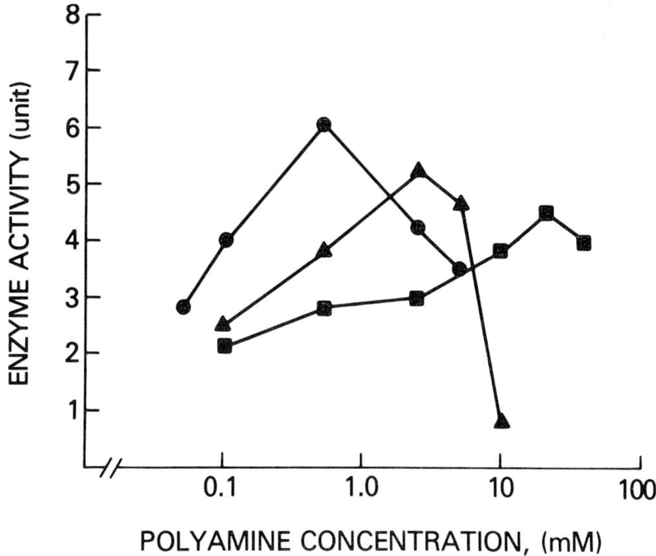

FIG. 1. Effects of spermine, spermidine and putrescine on protein kinase activity in cultured mammary gland. Spermine (●—●), spermidine (▲—▲), and putrescine (■—■) were added to the enzyme solution at the indicated concentrations along with 100 μg of bovine casein. The magnesium concentration in the assay was 2 mM. The enzyme preparation was made from midpregnant mouse mammary explants cultured for 72 hours in the presence of I(5 μg/ml), P(5 μg/ml), and F(1 μg/ml).

during advanced pregnancy (10). Mammary tissue from early pregnant mice therefore produced a greater kinase activity when cortisol was added along with I and P. A less pronounced increase in enzyme activity was seen in mammary tissue from advanced pregnant mice when this steroid was added. This was probably due to the fact that advanced pregnant mice contained greater levels of circulating cortisol (10). This phenomenon has been observed in regard to casein synthesis (16). The IF combination as well as I alone in the organ culture system induced enzyme activity which was always less than the activity induced by the IPF and IP combinations. Only a slight amount of enzyme activity was present without hormones or by P alone, F alone, or the combination of PF.

TABLE 1. Polylysine responsive protein kinase activity in the mammary glands of virgin, midpregnant and midlactating mice[a]

Stage of reproduction	Enzyme activity (unit)
Virgin	0.02
Midpregnancy	1.50
Midlactation	2.20

[a]Virgin mice were three months of age, midpregnancy correlates to 11-12 days into the term, and midlactation correlates to 10 days after the birth of pups.

TABLE 2. Effects of various hormones on the polylysine responsive protein kinase activity in cultured mammary gland[a]

Culture condition	Enzyme activity (unit)
IPF	2.32
IP	1.89
IF	0.45
PF	$\leq$ 0.15
I	0.35
P	$\leq$ 0.15
F	$\leq$ 0.15
NH	$\leq$ 0.15

[a]Mammary explants from 12 day pregnant mice were cultured for 48 hours in medium containing the indicated combination of I (1 μg/ml), P(1 μg/ml), and F(1 μg/ml). NH refers to no hormone addition.

Figure 2 shows the effects of the various concentrations of I, P and F on the polylysine responsive protein kinase activity in cultured mammary gland. The enzyme activity increased maximally as the prolactin concentration increased to 0.5 μg/ml-5 μg/ml in the presence of optimal concentrations of I and F (panel A). Insulin also produced maximal enhancement of enzyme activity at concentrations of 0.5 μg/ml-5 μg/ml (panel B). The effect of cortisol was maximal at approximately 1 μg/

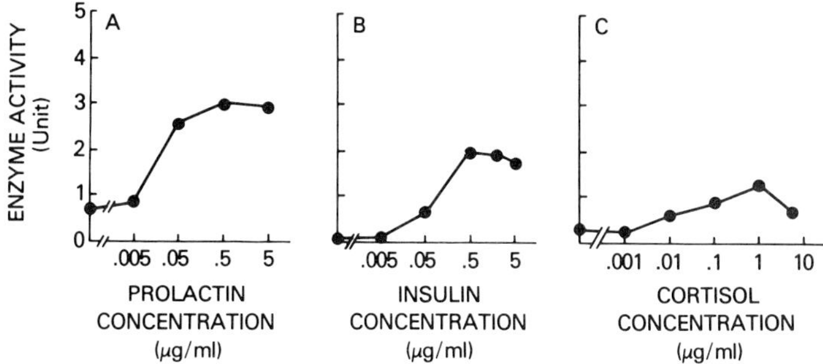

FIG. 2. The effects of various concentrations of I, P and F on the polylysine responsive protein kinase activity in cultured mammary gland from pregnant mice. The concentrations of hormones in the culture medium were as follows: (A) I(5 μg/ml), F(1 μg/ml), P, as indicated; (B) P(1 μg/ml), F(1 μg/ml), I, as indicated; (C) I(5 μg/ml), P(5 μg/ml), F, as indicated. The mammary explants, which were cultured for 48 hours, were taken from (A) 13 day pregnant mice, (B) 11 day pregnant mice and (C) 10 day pregnant mice.

ml) (panel C). These data directly correlate with those of Ono and Oka (15,16) in relation to casein accumulation.

Table 3 shows the induction of polylysine responsive protein kinase activity as a function of time. A steady increase in the enzyme activity in cultured mammary gland was observed over a 72 hour period. The extent of the increase varied as a function of the stage of pregnancy. The tissue from early pregnant mice gave a greater increase in enzyme activity (data not shown) when compared to tissue from advanced pregnant mice.

It has been demonstrated that progesterone and dbcAMP inhibit milk protein synthesis in the cultured mammary gland (17,19). The data presented in Table 4 show that the increase in polylysine responsive protein kinase activity was also inhibited by these two agents. Enzyme activity decreased when progesterone (100 ng/

TABLE 3. Time course of the polylysine responsive protein kinase activity in cultured mammary gland[a].

Length of culture (hours)	Enzyme activity (unit)
0	2.04
24	3.77
48	4.46
72	6.16

[a]The mammary explants from 14 day pregnant mice were cultured for the indicated lengths of time in medium containing I(1 μg/ml), P(1 μg/mg) and F(1 μg/ml).

ml) was added to the culture medium along with I(1 μg/ml), P(1 μg/ml) and F(1 μg/ml). When the cortisol concentration was reduced to 0.01 μg/ml, the inhibitory effect of progesterone became greater than that which was produced in the presence of 1 μg/ml of cortisol. The antagonistic interaction of cortisol and progesterone has also been demonstrated with regard to casein production (19). In addition, the inhibition of enzyme activity by progesterone varied as a function of the stage of pregnancy (data not shown). As mentioned previously, there is a progressive increase in the serum level of cortisol during the course of pregnancy (10). Thus, the progesterone-induced inhibition of enzyme activity in tissue from early pregnant mice was of a greater magnitude than that from advanced pregnant mice because of lower levels of circulating cortisol. This was also observed with regard to casein production (15,16). During the course of pregnancy there is also a progressive increase in endogenous progesterone levels (11). Therefore, the inhibition of the increase in polylysine-protein kinase activity in tissue from early pregnant mice by exogenous progesterone was of a greater magnitude than that from advanced pregnant mice. This also corresponds to previously described data on casein production (19). dbcAMP, at a concentration of 0.5 mM, caused a maximal decrease in enzyme activity when present in the culture medium along with I(1 μg/ml), P(1 μg/ml) and F(1 μg/ml).

The autoradiographic analysis of the endogenous proteins in the mammary gland that were phosphorylated by the polylysine responsive protein kinase(s) is shown in Figure 3. In the non-cultured gland, there were three distinct proteins that were phosphorylated (lane A). The predominant phosphorylated protein had a molecular weight of approximately 96 K. The other two had molecular weights of approximately 80 K and 48 K.

TABLE 4. Inhibitory effects of progesterone and dbcAMP on the polylysine responsive protein kinase activity in cultured mammary gland[a]

Culture condition	Enzyme activity (unit)
IPF[b]	3.52
IPF[b] + progesterone	2.44
IPF[c]	4.22
IPF[c] + progesterone	1.52
IPF[b]	3.68
IPF[b] + dbcAMP	0.61

[a]Mammary explants from 11-12 day pregnant mice were cultured for 72 hours in the presence of the indicated combinations of I, P, F, progesterone (100 ng/ml) and 0.5 mM dbcAMP. The concentrations of I, P and F varied as follows: [b]I(1 μg/ml), P(1 μg/ml), F(1 μg/ml); [c]I(1 μg/ml), P(1 μg/ml), and F(0.01 μg/ml).

A less distinct band, corresponding to a protein with a molecular weight of 29 K, appeared also. In contrast, tissue cultured with I, P and F (lane B and also lane A in Fig. 4) had a predominant phosphorylated protein with a molecular weight of about 22 K. Other distinct labeled proteins had molecular weights of about 29 K and 42 K. A band corresponding to a molecular weight of 17 K appeared faintly. It was noted that the phosphorylation patterns of non-cultured tissue as well as cultured tissue varied somewhat depending on the amount of sample preparation applied as well as the gestational stage of the mice. Non-cultured tissue from advanced pregnant mice had a pattern which approached that of IPF (data not shown). Similarly, cultured tissue treated with dbcAMP and progesterone (see below) varied to some extent for the reasons stated above (data not shown).

The effects of dbcAMP and progesterone on the polylysine responsive protein kinase phosphorylation of the endogenous proteins in cultured mammary gland were also analyzed (see Figure 4). The cultured explants treated with dbcAMP had a different protein phosphorylation pattern (lane B) from that of the IPF control (lane A). Although the protein with a molecular weight of 22 K was the predominant species, as it was in the IPF control, the dbcAMP-treated tissue also had very distinguishable bands corresponding to phosphorylated proteins with molecular

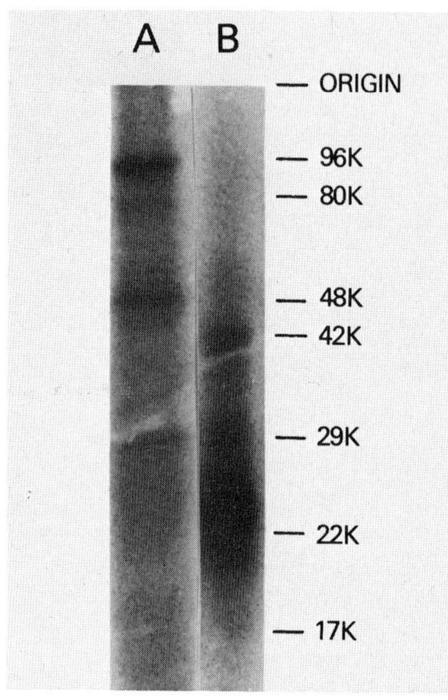

FIG. 3. Autoradiographic analysis of the polylysine responsive protein kinase phosphorylation of endogenous substrates in non-cultured and cultured mammary gland. Lane (A): non-cultured tissue from 10-11 day pregnant mice; the sample preparation applied to this lane contained the equivalent of approximately 4 mg tissue. Lane (B): explants were taken from 12-13 day pregnant mice and cultured for 72 hours in medium which contained I(5 μg/ml), P(5 μg/ml) and F(1 μg/ml). The sample preparation applied to this lane contained the equivalent of approximately 1.5 mg tissue.

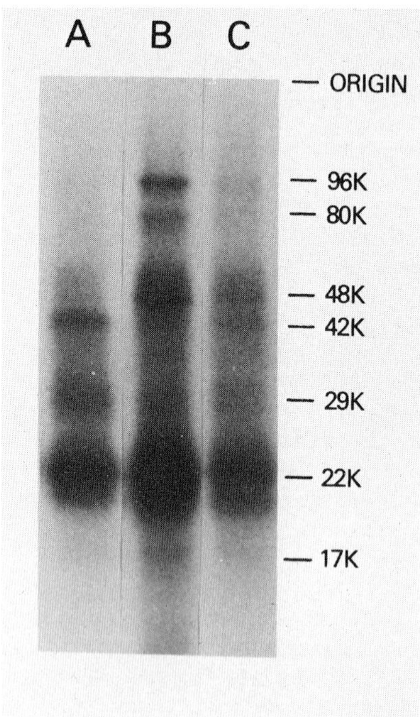

FIG. 4. Autoradiographic analysis of the polylysine responsive protein kinase phosphorylation of endogenous substrates in mammary tissue treated with dbcAMP and progesterone. Lane (A): explants were cultured for 72 hours in the presence of I(5 μg/ml), P(5 μg/ml), and F(1 μg/ml), lane (B): same conditions as in lane (A) except that the culture medium also contained 0.5 mM dbc-AMP. Lane (C): same conditions as in lane (A) except that the concentration of F was 0.01 μg/ml and the medium also contained progesterone (100 ng/ml). The sample preparations which were applied to each lane (A, B and C) contained approximately the equivalent of the following: 1.5 mg, 5 mg and 2 mg tissue, respectively. The explants were taken from 12-13 day pregnant mice (lane A) and 11 day pregnant mice (lanes B and C).

weights of 96 K, 80 K and 48 K. However, the labeled protein with a molecular weight of 42 K, which appeared in the IPF control, was absent in the dbcAMP-treated tissue. The labeled proteins with molecular weights of 29 K and 17 K appeared faintly. The cultured explants treated with progesterone (lane C) had a phosphorylation pattern which seemed to be intermediate between that of the IPF control and dbcAMP-treated tissue.

It is essential to point out that, in general, the majority of the labeled proteins in the IPF system as demonstrated by autoradiographic analysis, coelectrophoresed with authentic mouse caseins which had molecular weights of 42 K, 29 K and 22 K. This indicates that most of the phosphorylated endogenous substrates in IPF-treated tissue were caseins. In contrast, the indentity of the phosphorylated proteins of larger molecular weights in noncultured tissue is not yet known.

Both progesterone and dbcAMP were previously shown to inhibit casein synthesis in cultured mammary tissue (17,19). It is unknown, at the present time, whether the inhibition of enzyme activity produced by progesterone and dbcAMP resulted from a decrease in enzyme production or from a decrease in the relative amounts of endogenous acceptor proteins including casein. Autoradiographic analysis showed different patterns of protein phosphorylation in dbcAMP and progesterone treated tissue with respect to the IPF control. There was also a variation in the phosphorylation of proteins from non-cultured tissue in relation to the IPF control. These results clearly indicate that the patterns of the polylysine responsive protein kinase phosphorylation of casein and other endogenous substrates varied as a function of the developmental status of the mammary gland.

Further investigation must be carried out in order to understand the regulatory mechanism(s) of enzyme activity as well as the role of the unidentified phosphorylated substances in the growth and differentiation of the mammary gland. It will also be essential to isolate and purify this (these) kinase(s) in addition to investigating the possibility of the existence of endogenous activators and/or inhibitors.

SUMMARY

Protein kinase activity, which was stimulated by polyamines and was able to function independently of cyclic nucleotides, was identified in non-cultured mammary glands from virgin, pregnant and lactating mice. In cultured gland from midpregnant mice, optimal polylysine responsive enzyme activity was observed in explants incubated for 72 hours in the presence of insulin (0.5 μg/ml-5 μg/ml), prolactin (0.5 μg/ml-5 μg/ml) and cortisol (1 μg/ml). This kinase activity was shown to be inhibited by progesterone and dbcAMP. Autoradiographic analysis demonstrated that the endogenous substrates for this (these) enzymes(s), which include casein and unidentified proteins of high molecular weights, varied as a function of the developmental status of the mammary gland.

REFERENCES

1. Atmar, V.J., Kuehn, G.D. and Casillas, E.R. (1981): J. Biol. Chem., 256:8275-8278.
2. Burnett, H.G. and Kennedy, E.P. (1954): J. Biol. Chem., 211:969-980.
3. Cochet, C., Job, D., Pirollet, F. and Chambaz, E.M. (1980): Endocrinology, 106:750-757.
4. Criss, W.E., Yamamoto, M., Takai, Y., Nishizuka, Y. and Morris, H.P. (1978): Cancer Res., 38:3532-3539.
5. Daniels, G.R., Atmar, V.J. and Kuehn, G.D. (1981): Biochem., 20:2525-2532.
6. Imai, H. (1977): Bull. Osaka Med. Sc., 23:123-133.
7. Laemmli, U.K. (1970): Nature, 227:680-685.
8. Langan, T.A. (1973): In: Advances in Cyclic Nucleotide Research, edited by P. Greengard and G.A. Robison, Vol. 3, pp. 99-153, Raven Press, New York.
9. Maenpaa, P.H. (1977): Biochim. et Biophys. Acta, 498:294-305.
10. Malay, S., Giannopoulos, G. and Solomon, S. (1973): Endocrinology, 82:157.
11. McCormack, J.T. and Greenwald, G.S. (1974): J. Endocrinology, 62:101.
12. Morishita, Y., Akogyeram, C., Leiderman, L.J., Oka, T. and Criss, W.E. (1982): Med. and Ped. Oncology, 10:102.
13. Oka, T. and Perry, J.W. (1974): J. Biol. Chem., 249:7674-7652.
14. Oka, T., Perry, J.W., Takemoto, T., Sakai, T., Terada, N. and Inoue, H. (1981): Recent Prog. Polyamine Res., 3:309-320.
15. Ono, M. and Oka, T. (1980a): Science, 207:1367-1369.
16. Ono, M. and Oka, T. (1980b): Cell, 19:473-480.
17. Perry, J.W. and Oka, T. (1980): Proc. Natl. Acad. Sci. USA, 77:2093-2097.
18. Rubin, C.S. and Rosen, O.M. (1975): Annu. Rev. Biochem., 44:831-877.
19. Terada, N. and Oka, T. (1981): Fed. Proc., 40:1699.
20. Topper, Y.J. (1970): Recent Prog. Horm. Res., 26:287-308.
21. Topper, Y.J., Oka, T. and Vonderhaar, B.K. (1975): Methods in Enzymol., 39: 443-454.
22. Walsh, D.A. and Krebs, E.G. (1973): In: The Enzymes, edited by P.D. Boyer, 3rd edn., Vol. 8, pp. 555-581, Academic Press, New York.

Advances in Polyamine Research, Vol. 4, edited by U. Bachrach, A. Kaye, and R. Chayen. Raven Press, New York © 1983.

Regulation of Cyclic Nucleotide Metabolism by Polyamines in Heart Cell Cultures

C. Clô, B. Tantini, C. Pignatti, C. Guarnieri, and C. M. Caldarera

Istituto di Chimica Biologica, Facoltà di Medicina e Chirurgia, Università di Bologna, 40126 Bologna, Italy

The evidence accumulated to date favours an important role for the naturally occurring polyamines, spermine (SPM), spermidine (SPD) and putrescine (PTC) in the control of cell growth, division and differentiation (2,5,24,28,39,43). However, in spite of the great amount of literature dealing with the effects of polyamines on the processes leading to cell growth and division, there are still no specific biological functions that can be assigned to these compounds. In this regards, our attention has been focused on the relationship existing between polyamines and other intracellular factors, such as cyclic nucleotides, that experimental evidence indicates not only as mediators of many biological functions, but also as biochemical signals that can turn the cell towards its division or differentiation. In particular its has been suggested that cyclic GMP (cGMP) represents a positive signal for the induction of cell proliferation and cyclic AMP (cAMP) a negative signal, restricting proliferation(19,38,42). This comes from observations showing that in many types of cultured cells the addition of growth promoting factors causes an early and transient increase in intracellular cGMP and decrease in cAMP (36,40) and that exogenous cAMP or cGMP congeners reduce or stimulate cell growth respectively (20,43). However, according to MacManus et al. (33), it seems reasonable that the transitory changes in cyclic nucleotide contents, which occur usually within the first 30 min, cannot be solely responsable for the proliferative action of growth factors, such as serum, which has to be present as many as 5-7 hrs before full stimulation can be observed (17). Furthermore a concomitant change in the levels of the two nucleotides is not a general rule in cells undergoing division. It has therefore been suggested that a fall in cAMP/cGMP ratio is a more adeguate stimulus for the initiation of cell proliferation.

On the other hand an increase in the activity of ornithine decarboxylase (ODC), the rate limiting enzyme in polyamine biosynthesis, represents a constant event following a stimulus inducing cell growth. It is rather interesting that this increase is frequently maximal 6-8 hrs after stimulation (3,11,23) and that, in many experimental systems, its prevention by specific inhibitors abolishes the effects of growth promoting factors (25, 26,27,28,34), thus suggesting that polyamines are a requirement for cell proliferation.

These and other considerations prompted us to study the relationship existing between polyamines and cyclic nucleotides in order to obtain useful informations on their specific role in many cellular activities, including growth and division. Indeed our previous experiments have indicated that both in normal and neoplastic cell cultures exogenous SPM, SPD or PTC at the dose of 1 μM, 10 μM and 0,1 μM respectively, early and rapidly reduce cellular cAMP content and counteract the effect of different cAMP-mediated effectors (9). This report is dealing with the effect of polyamines on both cAMP and cGMP in primary chick embryo heart cell cultures. Heart cells in culture are an appropriate system, since their metabolic and functional activities are closely dependent on cAMP and cGMP and they retain their specialized characteristics, including beating ability and sensitivity to many hormones and chemical agents, in the same way as the intact organ of origin (11,13,18,35).
In our experiments confluent cultures were utilized as they constitute a population of non-dividing cells all synchronized in the G_1 phase of the cell cycle. Furthermore the use of stationary cultures reduces the possible interferences of pleiotypic effects with the specific effects under investigation.

METHODS

Primary cultures of heart cells from 10 day-old chick embryos were prepared according to DeHaan (16). The cells were grown in Eagle's minimum essential medium (MEM F-15, GIBCO) supplemented with 10% heat-inactivated foetal calf serum, 10% triptose phosphate broth, penicillin (100 units/ml), streptomycin (100 μg/ml) and buffered with 25 mM Hepes, pH 7.4 Confluent cultures were used routinely. Groups of 3-4 plates were pooled for each experimental condition.

Polyamines and Cyclic Nucleotide Analysis

At the end of the incubation periods, the monolayers were rinsed with cold 0.85% NaCl and then added with 0.6 M $HClO_4$. The cells were pooled by scraping, frozen and thawed twice and centrifuged at 15,000 g for 10 min. Aliquots of the supernatants were assayed for cyclic nucleotides, while the pellets were dissolved in 1 N NaOH and assayed for protein (32). Intracellular cAMP and cGMP contents were assayed in duplicate by using the kits from Radiochemical Centre, after purifying the cyclic nucleotides according to Chan and Lin (7). The recovery of cAMP and cGMP was estimated, by (^{3}H)-radiolabelled cyclic nucleotides, to be more than 90%.

Cyclic GMP-phosphodiesterase assay

Cyclic GMP phosphodiesterase was prepared by rinsing the cells twice with ice-cold phosphate-buffer saline (pH 7.4) and twice with homogenization buffer containing 40 mM Tris-HCl (pH 7.4) and 10% sucrose (w/v). The cells were harvested by scraping and homogenized at 4°C for 20 sec with a glass Dounce homogenizer. The homogenates were then centrifuged at 23,000 g for 30 min to yield particulate (pellets) and soluble (supernatants) fractions. Pellets were suspended in 40 mM Tris-HCl (pH 7.4) before assaying for activity, according to the two-step radioisotopic procedure of Thompson et al. (46). The reaction mixture contained 40 mM Tris-HCl (pH 7.4), 5 mM $MgCl_2$, 3.75 mM 2-mercaptoethanol, (^{3}H) cGMP (450,000 counts/min), 10 μM cGMP plus appropriate amounts of enzyme in a final volume of 200 μl. Proteins were determined according to Bradford (6).

Guanylate cyclase assay

At the end of the incubations, the monolayers were rinsed twice with cold 0.85% NaCl, pooled by scraping and homogenized at 4°C in 50 mM NaCl, 10 mM Hepes (pH 7.6), 0.1 mM dithiothreitol and 0.1 mM EGTA, by using a glass homogenizer with a Teflon pestle. Homogenates were centrifuged at 105,000 g for 60 min to separate soluble and particulate fractions. Pellets were suspended in a volume of buffer equal to that of the original homogenates. Solubilization of the homogenates and pellets was accomplished by incubation with 0,1% Triton X-100 for 30 min at 4°C. The homogenate and fraction activities were assayed as described by Coffey et al. (15). The reaction mixtures contained 30 mM Hepes (pH 7.6), 25 mM NaCl, 0.5 mM dithiothreitol, 0.05 mM EGTA, 2.5 mM phosphocreatine, 0.5 mM isobutylmethylxanthine, 0.5 mM GTP, 2 mM $MnCl_2$,

0.05 mg/ml creatine phosphokinase and 20 µl of appropriately diluted enzyme (20-40 µg protein) in a final volume of 200 µl. When kinetic analysis was performed, GTP concentration ranged from 5 µM to 1 mM.
Reactions were carried out at 37°C for 10 to 15 min following addition of enzyme, stopped by addition of 0.6 ml of 50 mM Tris-HCl containing 4 mM EDTA (pH 7.6), boiled and centrifuged at 2,000 g for 10 min. Aliquots of the supernatants were assayed for cGMP as described above. Boiled protein samples were used for estimating the blank values, which were then substracted from each experimental value. Proteins were determined according to Bradford (6).

RESULTS

Effect of Serum or Polyamines on Cyclic Nucleotide Contents

The cyclic nucleotide contents of cultured cells respond specifically to growth conditions. In particular the level of cAMP rises and that of cGMP declines in cells which approach confluency and stop dividing, while the opposite behaviour is observable with actively dividing cells (20,38,42). Furthermore cyclic nucleotide contents display a strict dependence on the presence and amount of serum in the culture medium (36,44). The data reported in Table 1 clearly indicate that serum-mediated changes in both cAMP and cGMP contents also occur in primary quiescent heart cells. The addition of serum (20%) to 0.5% serum-restricted cells

TABLE 1. Influence of serum on cyclic nucleotide contents of confluent heart cell cultures[a]

Addition	Time (min)	cAMP (pmoles/mg protein)	cGMP	cAMP/cGMP
Control	0	20.00	1.22	16.39
20% Serum	10	13.50	3.45	3.91
20% Serum	20	10.37	2.40	4.31
20% Serum	30	7.97	2.14	3.72

[a]Confluent cultures were maintained for 20 hrs in a medium containing 0.5% serum and then added with 20% foetal calf serum. Data are the means of six determinations from thee separate experiments. The S.E.M. was less than 10%.

causes an early and significant increase in cGMP and decrease in cAMP contents, with a maximum after 10 and 30 min respectively.

Because of the parallel but reciprocal changes in cGMP the cAMP/cGMP ratio reflects the changes more dramatically. It follows therefore that confluency associated with serum-restriction provides a useful model to study the effect of factors, such as polyamines, which in view of their positive correlation with cell growth, are expected to modulate cyclic nucleotide contents in the same manner elicited by the growth factors of serum.
Indeed the data reported in Fig. 1 clearly support our expectations. As in the case of serum readdition, the addition of PTC, SPD or SPM at relatively low doses, in the range of those found in physiological fluids (2,28), to confluent and serum-restricted heart cells causes an early and rapid decrease of cAMP and increase of cGMP levels. While cAMP progressively drops in time, cGMP is maximally elevated within 10 min. This is particularly

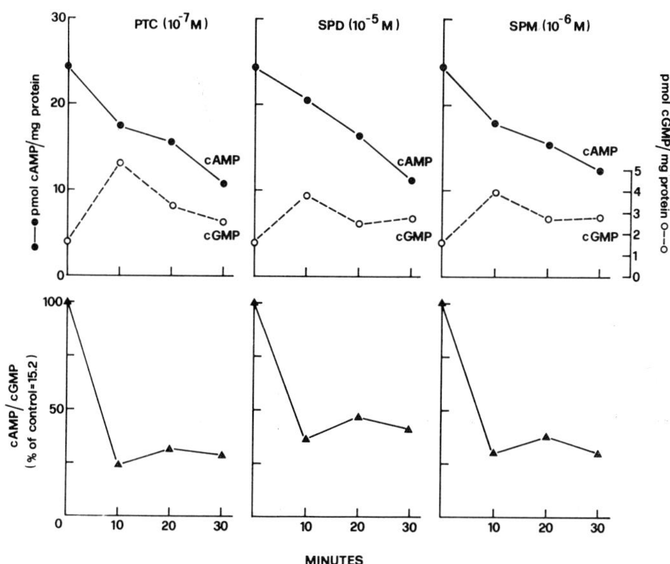

FIG. 1. Kinetics of cyclic nucleotide contents and of cAMP/cGMP ratio in polyamine treated heart cells. After a 20 hrs serum-restriction (0.5%), polyamines were added to confluent cells at the final doses indicated. Data are the means of six determinations from three separate experiments, differing by no more than 10%.

evident with PTC which causes a 3-fold increase of the cGMP value measured at time zero. Thereafter the cGMP content gradually declines and remains approximately constant, within a factor of more than 1.5 after 30 min of treatment. SPM and particularly PTC exhibit the highest efficiency in reducing the cAMP/cGMP ratio 10 min after their addition, while SPD causes the maximum fall at the 30th min. Since cGMP has been reported to increase the activity of cAMP-dependent phosphodiesterase (cAMP-PDE) (20,45), the possibility exists that the effects of polyamines on cGMP are mediated, or in part mediated, by cGMP itself. This hypothesis is indirectly supported by the fact that with each individual polyamine, the peak of cGMP preceeds the maximum fall of cAMP. However further experiments are needed which are courrently in progress in our laboratory.

Polyamines and the Enzymes of cAMP Metabolism

Data exist in literature supporting the capability of polyamines to early and rapidly inhibit adenylate cyclase activity of cultured cells when added at concentrations in the range of those we have used (1,48). Our previous reports have indicated that exogenous polyamines can regulate cAMP content of heart cell cultures also by affecting cAMP-PDE activity (11,12). A huge increase in the activity of high and particularly of low affinity forms of cAMP-PDE occurs shortly after exposure of the cells to each amine at the concentrations here used, possibly as consequence of an increase in the substrate affinity of the enzymes. Furthermore other data have indicated that in heart cell cultures polyamine accumulation is required in both cAMP-dependent and cAMP-independent mechanism of induction of cAMP-PDE (11,13).

Polyamines and cGMP-Phosphodiesterase Activity

A phosphodiesterase specific for cGMP is present in both soluble and particulate fractions of various cell types (20), and it has been proposed to be under separate genetic control from cAMP-PDE (41). The data reported in Fig. 2 indicate that similar to cAMP-PDE, cGMP-PDE is involved in the regulation of cGMP by exogenous polyamines. Most of the cGMP-PDE activity is present in the soluble fraction of confluent heart cells. The addition of each polyamine causes within 10 min a decrease of the particulate activity, particularly evident with 1 μM SPM. No significant differences are observable between 10 and 20 min of incubation with 0.1 μM PTC and 10 μM SPD, while in the case of SPM the enzymatic activity is partially restored after 20 min. The inhibitory effect of polyamines is more evident on the soluble cGMP-PDE. In this case SPD displays the highest efficiency within the first 10 min, while PTC and SPM reach their maximum effect after 20 min of incubation. These data therefore indicate that, differently from cAMP-PDE, whose particulate form appears to be a more

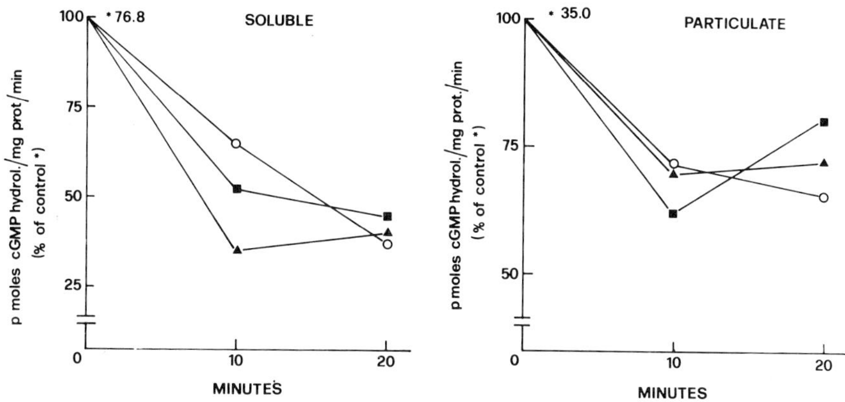

FIG. 2. Soluble and particulate cGMP-PDE of heart cells treated with polyamines. Experimental conditions were as given in Table 1. Data are the means of six determinations from 3 separate experiments, differing by no more than 10%. (○) 0.1 μM PTC; (▲) 10 μM SPD; (■) 1 μM SPM.

specific target for the action of polyamines and especially of PTC (11), the soluble component of cGMP-PDE is preferentially affected by each amine, with particular regard for SPD. The inhibitory effect of polyamines on cGMP-PDE is supported by the results of Kincaid et al. (30) showing that polyamines significantly decrease the activity of a specific soluble cGMP-PDE purified from bovine brain.

However, since PTC and SPM cause maximum accumulation of cGMP 10 min after their addition (see Fig.1), while cGMP-PDE activity is maximally decreased after 20 min, we have investigated whether guanylate cyclase enzyme could also be involved in the regulation of cGMP by exogenous polyamines. The data reported in the next figures clearly support this possibility.

Polyamines and Guanylate Cyclase Activity

Fig. 3 shows the changes in the total guanylate cyclase activity of confluent and serum-starved heart cells exposed to polyamines. A rapid increase in the enzymatic activity occurs shortly after the addition of each amine. PTC causes the maximum increase after 5 min (more than 3 fold the basal value), while SPM and particularly SPD maximally stimulate 10 min after their addition. In all the cases however the effect of each polyamine is still evident after 30 min.

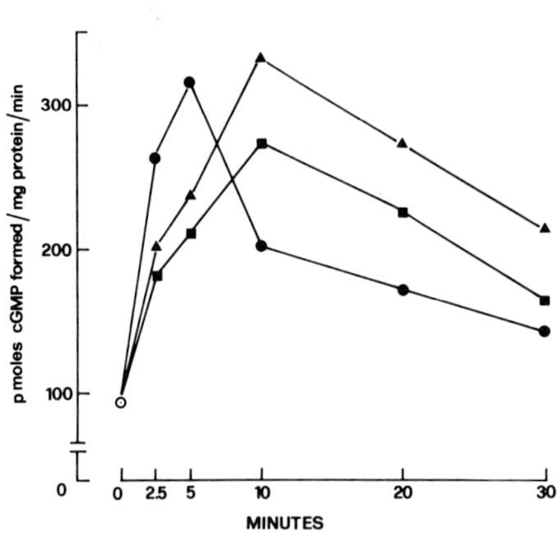

FIG. 3. Guanylate cyclase activity of heart cells treated with polyamines. Confluent cultures, maintained for 20 hrs in a serum-free medium, were incubated with 0.1μM PTC (●), 1μM SPM (■) or 10μM SPD (▲), directly added to the medium. Data are the means of six determinations from three different samples, differing by no more than 10%.

Guanylate cyclase activity is associated with both supernatant and particulate fractions of homogenates from a number of different cell types and the distribution of activity varies with the cell examined and with the state of growth (15,20,21,37). However at the present the significance of soluble and particulate enzymes is not clear. In spite of some common properties, such as similar pH optimum and requirement of Mn^{2+}, the more striking differences uncovered during comparative studies, including: molecular weight, kinetic properties, substrate requirement, sensitivity to different drugs and antigen reactivity, suggest that the two activities are different proteins or different forms of the same enzyme (20,29,36). In our cells, after treatment of homogenates with 0.1% Triton X-100 most of the guanylate cyclase activity is particulate, and as observed with whole heart (29), the sum of the activities of the soluble and particulate fractions exceeded that of the homogenate (not shown).

Fig. 4 compares the time course effects of the addition of each polyamine or of 10% foetal calf serum to confluent and serum-starved heart cells on the particulate guanylate cyclase activity. It may be seen that the behaviour of the activity exactly reflects that observed with the homogenate during the first 10 min of treatment (see Fig.3). PTC is the most efficient in increasing the activity within 5 min, while, similarly to serum factors, SPM and SPD maximally stimulate after 10 min. The activity of the soluble enzyme is also enhanced shortly after the addition of polyamines to the cells. However some differences are observable with respect to the particulate form. PTC stimulates progressively with time and after 10 min each polyamine displays a greater efficiency than with particulate enzyme. Furthermore it is quite inter-

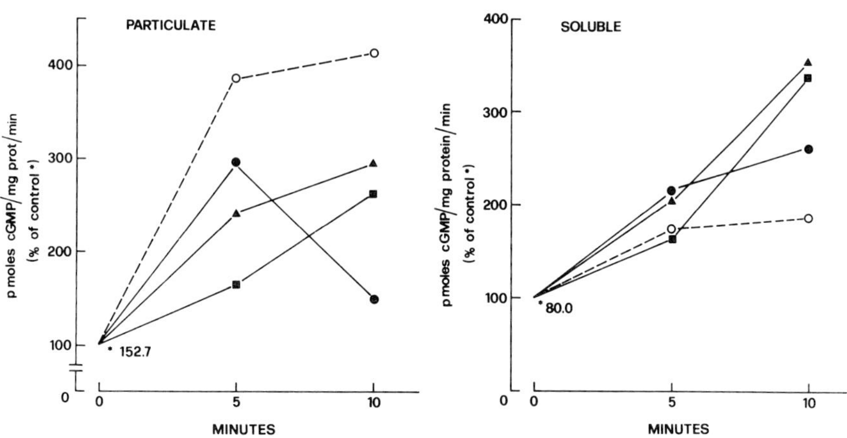

FIG. 4. Particulate and soluble guanylate cyclase activities of heart cells treated with polyamines or 10% foetal calf serum. Experimental conditions are as given in Fig. 3. Serum (○),PTC (●), SPM (■),SPD (▲).

esting that this form of guanylate cyclase is poorly affected by serum, whose effect after 10 min is significantly lower than that of each amine. All together these data therefore indicate that both soluble and particulate guanylate cyclases mediate the action of polyamines in the control of cellular cGMP, and that the two enzymes exhibit different sensitivity with respect to the different amines.

In order to obtain same informations on the mechanism by which polyamines stimulate soluble and particulate enzymes we have analyzed the kinetic parameters of both the enzymes after treatment of the cells with each individual polyamine (Table 2). Studies with the soluble enzyme from various tissue demonstrated classical Michaelis-Menten kinetic behaviour with respect to GTP, the apparent Km varying by using MnGTP, MgGTP or CaGTP (4, 20,29,37). The apparent Km for the soluble enzyme from confluent heart cells is 67.5 μM using Mn^{2+} as cofactor. Both Km and Vm values are changed after 10 min of incubation of the cells with each amine. PTC, SPD and SPM appear to be progressively more efficient in reducing the Km value. While SPM does not affect the Vm, SPD and particularly PTC bring about an increase.

Crude preparations of the particulate enzyme usually exhibit a higher substrate requirement than soluble one and a positive co

TABLE 2. Changes in the kinetic parameters of guanylate cyclase of heart cells exposed to polyamines[a]

Addition	Soluble Enzyme		Particulate Enzyme			
			High Affinity		Low Affinity	
	Km	Vm	Km	Vm	Km	Vm
None	67.5	52.6	140.3	117.6	555.5	238.0
PTC (0.1 μM)	37.3	72.7	64.5	153.8	166.6	151.1
SPD (10 μM)	30.5	61.5	75.7	140.8	210.5	156.2
SPM (1 μM)	26.6	51.5	90.1	136.9	275.8	166.6

[a]After a 20 hrs serum starvation, confluent cultures were added for 10 min with polyamines. In the case of the particulate enzymes PTC was added for 5 min. The kinetic parameters were obtained according to Lineweaver and Burk. Km are expressed in μM and Vm as pmoles of cGMP formed/mg protein/min. Data are the means of four separate samples. The S.E.M. was less than 10%.

operativity as the MnGTP concentration is varied (20,37). Table 2 shows that the particulate activity from confluent heart cells displays biphasic property for GTP resulting in two apparent Michaelis constant of 140.3 μM and 555.5 μM, each having a different maximum velocity. Very similar values of Km for soluble and particulate enzymes from whole heart have been reported (29), providing additional evidence that heart cells in culture retain the biochemical characteristics of the intact organ.
Like soluble enzyme, both the Km and Vm values of the high affinity component of the particulate enzyme are changed by exposing the cells to PTC for 5 min or to SPD or SPM for 10 min. In this case however PTC exhibits the highest efficiency, followed by SPD and SPM. The capability of each amine to reduce the Km value is more pronounced on the low affinity component of the enzyme. Again PTC is the most effective if compared to SPD and SPM. Differently from the other two enzymes, each amine also promotes a decrease in the Vm value.

All together these data indicate that exogenous polyamines, at low concentrations, stimulate guanylate cyclase by increasing its affinity for the substrate and, with the exception of the low affinity component of the particulate enzyme, its catalytic property.

Of course from these data it is quite difficult to make some speculation on the mechanism of action of polyamines, also at the light of the complex picture of possible interactions observed between the enzyme, substrate and cofactors (20,31,37,47). Both the soluble and particulate enzymes from rat heart (29) contain two or more interactive sites for Mn^{2+}. Furthermore, interactions of more than one cation cofactor with guanylate cyclase have been demonstrated (9,37).

Fig. 5 compares the effect of varying $MnCl_2$ or SPD concentrations at fixed GTP concentration (0.5 mM) on soluble activity from heart cells. Without added $MnCl_2$ or SPD there was no detectable activity. It may be seen that the activity increases by increasing the concentrations of $MnCl_2$ or SPD. At low concentrations (1 μM and 10 μM) SPD appears to be a more effective cofactor than Mn^{2+}, but at higher concentrations SPD either (100 μM) displays the same efficiency of Mn^{2+} or (2 mM) causes half of the activity observed with Mn^{2+}. Similarly to SPD, Mg^{2+} or Ca^{2+} has been reported to support only a fraction of the activity observed with optimal concentration of Mn^{2+} (20,29,37). These preliminary data therefore suggest that SPD, probably by interacting with the cation site of the enzyme, can substitute for Mn^{2+}, expecially when its concentration is below that of GTP. Additional studies to define the molecular mechanism by which polyamines might activate guanylate cyclase in intact cells are courrently in progress in our laboratory.

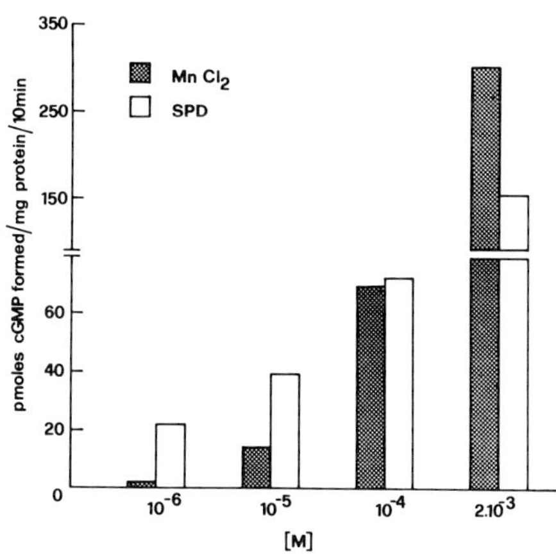

FIG. 5. Effect of $MnCl_2$ or SPD on soluble guanylate cyclase activity from heart cells. Data are the means of three determinations in duplicate, differing by no more than 10%.

CONCLUSION

The data presented in this study indicate that similarly to the growth promoting factors of serum, the addition of SPM, SPD or PTC at very low concentrations to confluent heart cells, while lowering cAMP content, significantly increases cGMP content. Since cAMP and cGMP represent respectively a negative and a positive signal for the induction of cell proliferation, their concomitant modulation may constitute a mechanism by which exogenous polyamines can act as growth promoters when used at relatively low doses (8,22).

The possibility that endogenous polyamines could also be involved in the control of the steady-state levels of cAMP and cGMP is supported by our recent results (14) obtained by treating heart cells with different concentrations of serum in the presence or absence of α-difluoromethylornithine, an irreversible inhibitor of ODC (34). It appears that progressively higher amounts of cGMP and lower amounts of cAMP are present in cells with progressively higher contents of polyamines and that polyamine-deficient cells display a significant reduced sensitivity to agents such as morphine or serum, notably able to increase cGMP and decrease cAMP contents. Conversely, polyamine-deficient cells exhibited higher sensitivity to norepinephrine than polyamine rich ones (10).

The data reported by Others on adenylate cyclase and our data on cyclic nucleotide phosphodiesterases and on guanylate cyclase demonstrate that alterations in the activities of all these enzymes could mediate the changes in cyclic nucleotide levels by polyamines. The scheme below summarizes the major aspects of the mechanism of action of polyamines in regulating cellular cyclic nucleotides:

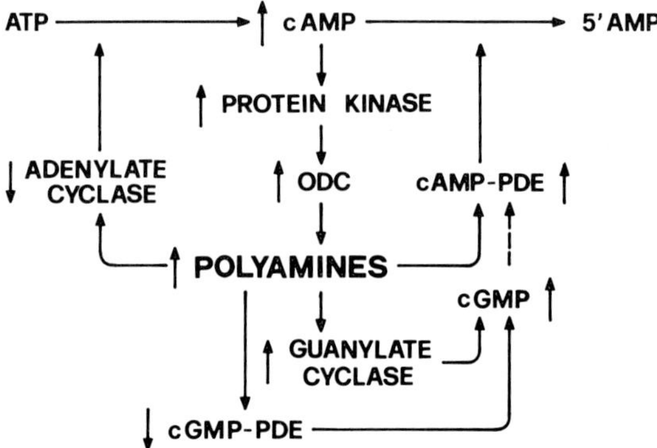

By increasing guanylate cyclase and by reducing cGMP-PDE activities polyamines promote cellular cGMP accumulation. On the other hand, polyamines reduce cAMP content by inhibiting adenylate cyclase and increasing cAMP-PDE activities. This last effect might be in part mediated by cGMP itself. The scheme also underlines that polyamine accumulation is a requirement for the cAMP-mediated induction of cAMP-PDE, as previously observed in heart cell cultures (11,13).

The capability of polyamines to differently affect cellular cyclic nucleotide contents by modulating their specific metabolic enzymes, stresses the importance of these polycations in the regulation of many cellular activities, and is particularly relevant for heart cells, whose metabolic and functional activities are closely dependent on cAMP and cGMP. Furthermore if a drop of cAMP/cGMP ratio is an adequate signal for the cell to enter the cell cycle again and to proliferate, than our findings provide additional evidence for a role of polyamines as growth regulators.

ACKNOWLEDGMENTS

The skilful technical assistance of Ms A. Marmiroli and Mr. S. Manfroni is gratefully acknowledged. This work was supported by the National Research Council project "Control of Neoplastic Growth" (grant n° 80.01495.96).

REFERENCES

1. Atmar, V.J., Wesland, J.A., Garcia, G., and Kuehan, G.D. (1976): Biochem. Biophys. Res. Commun., 68:561-568.
2. Bachrach, U. (1973): The Function of Naturally Occurring Polyamines, Academic Press, New York.
3. Bachrach, U. (1980): Biochem. J., 188:387-392.
4. Bartfai, T., Breakefield, X.O., and Greengard, P. (1978): Biochem. J., 176:119-127.
5. Bethell, D.R., and Pegg, A.E. (1981): Biochem. Biophys. Res. Commun., 102:272-278.
6. Bradford, M.H. (1976): Anal. Biochem., 72:248-254.
7. Chan, P.S., and Lin, M.C. (1974): Methods in Enzymology, 38: 38-41.
8. Clô, C., Orlandini, G.C., Casti, A., and Guarnieri, C. (1976). Ital. J. Biochem., 25:94-114.
9. Clô, C., Caldarera, C.M., Tantini, B., Benalal, D., and Bachrach, U. (1979): Biochem. J., 182:641-649.
10. Clô, C., Coccolini, M.N., Tantini, B., and Caldarera, C.M. (1980): Life Sci., 27:67-73.

11. Clô, C., Tantini, B., Coccolini, M.N., and Caldarera, C.M. (1981): Adv. Polyamine Res., 3:333-345.
12. Clô, C., Tantini, B., Coccolini, M.N., and Caldarera, C.M. (1981): J. Mol. Cell. Cardiol., 13:773-776.
13. Clô, C., Tantini, B., Pignatti, C., and Rossoni Caldarera,C. (1982): In: Heart Metabolism, edited by P. Harris and C.M. Caldarera, in press.Clueb, Bologna.
14. Clô, C., Pignatti, C., Tantini, B., and Caldarera, C.M. (1982): in preparation.
15. Coffey, R.G., Hadden, E.M., Lopez, C., and Hadden, J.W. (1978): Adv. Cyclic Nucleotide Res., 9:661-676.
16. DeHaan, R.L. (1967): Dev. Biol., 16:216-249.
17. Froehlich, J.E., and Rachmeler, M.J. (1974): J. Cell. Biol., 60:249-257.
18. Ghanbari, H., and McCarl, R.L. (1976): J. Mol. Cell.Cardiol., 8:481-488.
19. Goldberg, N.D., Haddox, M.K., Dunham, E., Lopez, C., and Hadden, J.W. (1974): In: Control of Proliferation of Animal Cells, edited by B. Clarkson and R. Baserga, pp. 571-579. Cold Spring Harbor, New York.
20. Goldberg, N.D., and Haddox, M.K. (1977): Ann. Rev. Biochem., 46:823-896.
21. Guiraud-Simplot, A., and Colobert, L. (1977): Biochem. Biophys. Res.Commun.,76:963-970.
22. Ham, R.G. (1964): Biochem. Biophys. Res. Commun., 14:34-38.
23. Heby, O., Gray, J.W., Lindl, P.A., Marton, L.J., and Wilson, C.B. (1976): Biochem. Biophys. Res. Commun., 71:99-105.
24. Heby, O. (1981): Differentiation, 19:1-20.
25. Heby, O., and Jänne, J. (1981): In: Polyamines in Biology and Medicine, edited by D.R. Morris and L.J. Marton, pp. 243-310. Marcel and Dekker, New York.
26. Herbs, E.J., and Elliot, Q.D. (1981): Medical Biology, 59: 410-416.
27. Höltta, E., Jänne, J., and Hovi, T. (1979): Biochem. J.,178: 109-117.
28. Jänne, J., Pösö, H., and Raina, A. (1978): Biochim. Biophys. Acta, 433:241-293.
29. Kimura, H., and Murad, F. (1974): J. Biol. Chem., 249:6910-6916.
30. Kindaid, R.L., Manganiello, V.C., and Vaughan, M. (1979): J. Biol. Chem., 254:4970-4973.
31. Limbird, L.E., and Lefkowitz, R.J. (1975): Biochim. Biophys. Acta, 377:186-196.

32. Lowry, O.H., Rosebrough, N.J., Farr, A.L., and Randall, R.J. (1951): J. Biol. Chem., 193:265-275.
33. MacManus, J.P., Boynton, A.L., and Whitfield, J.F. (1978): Adv. Cyclic Nucleotide Res., 9:485-491.
34. Mamont, P.S., Duchesne, M.C., Grove, J., and Bey, P. (1978): Biochem. Biophys. Res. Commun., 81:58-66.
35. Mark, G.E., and Strasser, F.F. (1966): Exp. Cell. Res., 44: 217-233.
36. Moens, W., Vokaer, A., and Kram, R. (1975): Proc. Natl. Acad Sci. U.S.A., 72:1063-1067.
37. Murad, F., Arnold, W.P., Mittal, C.K., and Branghler, J.M. (1979): Adv. Cyclic Nucleotide Res., 11:175-204.
38. Pastan, I.H., Johnson, G.S., and Anderson, W.B. (1975): Ann. Rev. Biochem., 44:491-522.
39. Raina, A., and Jänne, J. (1975): Medical Biology, 53:121-147.
40. Rudland, P.S., Seely, M., and Seifert, W.E. (1974): Nature, 251:417-419.
41. Russell, T.R., and Pastan, I.H. (1974): J. Biol. Chem., 249: 7764-7769.
42. Schönöfer, P.S., and Peters, H.D. (1977): In: Cyclic 3'-5' Nucleotides: Mechanism of Action, edited by C. Cramer and J. Schultz, pp. 107-131. John Wiley and Sons, London.
43. Seifert, W.E., and Rudland, P.S. (1974): Nature, 248:138-140.
44. Seifert, W.E., and Rudland, P.S. (1974): Proc. Natl. Acad. Sci. U.S.A., 71:4920-4924.
45. Takemoto, D.J., Kaplan, S.A., and Appleman, M.M. (1979): Biochem. Biophys. Res. Commun., 90: 491-497.
46. Thompson, W.J., Brooker, G., and Appleman, M.M. (1974): Methods in Enzymology, 38:205-212.
47. Wallach, D., and Pastan, I. (1976): Biochem. Biophys. Res. Commun., 72:859-865.
48. Wright, R., Buchler, B.A., and Rennert, O.M. (1976): Pediatr. Res., 10:373.

Advances in Polyamine Research, Vol. 4, edited by U. Bachrach, A. Kaye, and R. Chayen. Raven Press, New York © 1983.

Regulation of Hepatic Ornithine Decarboxylase by Antizyme and Antizyme Inhibitor

K. Fujita, Y. Murakami, T. Kameji, S. Matsufuji, K. Utsunomiya, R. Kanamoto, and S. Hayashi

Department of Nutrition, Jikei University School of Medicine, Tokyo 105, Japan

One of the most interesting aspects of the regulation of ornithine decarboxylase (ODC) is the role of ODC antizyme, a unique regulatory protein found by Canellakis and his coworkers (1, 2, 4, 5, 10). The antizyme specifically inactivates ODC, apparently by forming a complex with it (2, 5). It is induced by polyamines or their analogs and appears to participate, at least in part, in the process of rapid ODC decay caused by administration of polyamines (11). However, little is known about the nature, mode of action, or physiological role of antizyme.

Recently, we purified ODC from rat liver to near homogeneity and prepared a monospecific antiserum to the enzyme (7-9). To understand the regulatory mechanism of ODC at a molecular level, we also started studies on the antizyme and encountered another novel regulatory factor "antizyme inhibitor", which specifically inhibits antizyme and reactivates ODC after its inactivation by antizyme. In this paper we present the results of our study on the two regulatory factors. Part of this work has been published elsewhere (3).

Partial purification and characterization of antizyme

We first studied the antizyme to confirm the results reported by Canellakis's group. Antizyme activity was induced in rat liver by repeated injections of 1,3-diaminopropane, an analog of putrescine (6). Partial purification of antizyme was performed by DEAE-cellulose chromatography and Sephadex G-75 gel-filtration, as shown in TABLE 1, and the partially purified preparation was used throughout the present study.

On Sephadex G-75 gel-filtration analysis, antizyme was eluted after most of the other protein, and its molecular weight was estimated as 24,500 (FIG. 1), in close agreement with the reported value of 26,500 (2, 5).

TABLE 1. Partial purification of antizyme from rat liver[a]

Step	Total protein (mg)	Total activity (units)	Specific activity (units/mg)	Purif. (-fold)	Yield (%)
1. Crude extract	4610	3802	0.82	1	100
2. DEAE-cellulose	392	3446	8.78	11	90
3. Sephadex G-75	34.3	2046	59.6	72	54

[a]Eight Sprague-Dawley rats weighing about 300 g were injected twice intraperitoneally with 1,3-diaminopropane (1 mmol/kg b.w.) at 09:00h and 11:00h, and killed at 12:00h. Liver extract was prepared and subjected to purification procedures essentially as described in the previous paper (3). Antizyme activity was assayed by measuring inhibition of ODC activity. One unit of antizyme activity was defined as the amount inhibiting one unit of ODC activity, which in turn was defined as the amount forming 1 nmol of CO_2 per h.

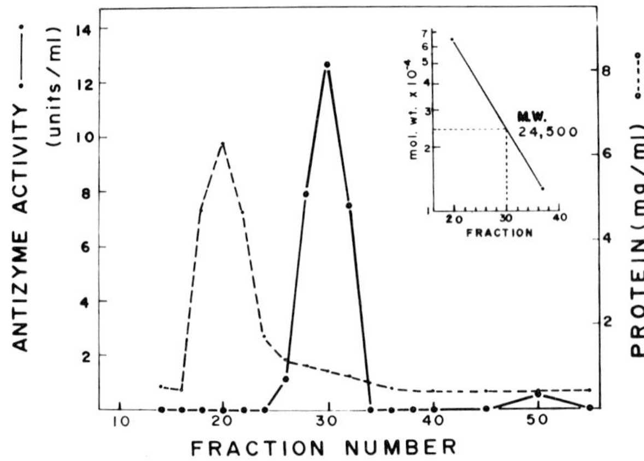

FIG. 1. Gel-filtration analysis of antizyme. Concentrated DEAE eluate containing 705 units of antizyme activity in 9.9 ml was filtered through a Sephadex G-75 column (2.8 x 64.5 cm). The elution buffer consisted of 50 mM Tris-HCl, pH 7.4, containing 1 mM dithiothreitol, 0.1 mM EDTA, 50 µM pyridoxal phosphate, and 50 mM NaCl. The fraction volume was 6.7 ml.

FIG. 2 shows the result of a titration experiment, in which increasing amounts of ODC were assayed in the presence and absence of a fixed amount of antizyme. The titration pattern indicates that antizyme inhibits ODC stoichiometrically, and thus supports the proposal that antizyme forms a complex with ODC.

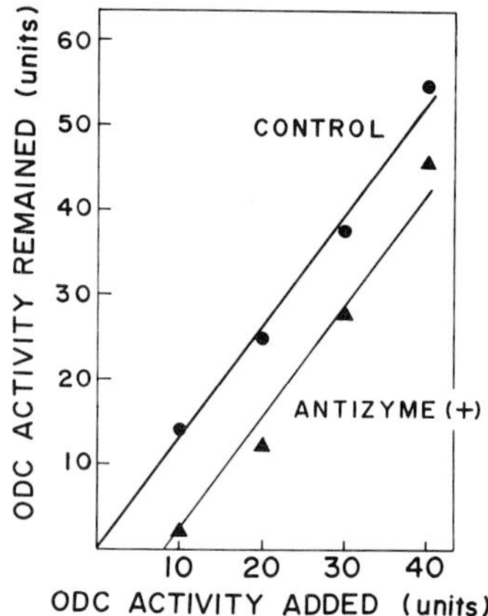

FIG. 2. Titration of antizyme with ODC. Partial purification of ODC was performed as described previously (3).

Presence of antizyme inhibitor in rat liver extracts

During the early stage of work, we sometimes had difficulty in assaying antizyme activity, finding that some ODC preparations were insensitive to antizyme. Results of mixing experiments suggested the presence of an inhibitor of antizyme in these ODC preparations.

To confirm its existence, we subjected a crude liver extract containing induced ODC activity to DEAE-cellulose chromatography. The liver extract was prepared from rats that had been treated with thioacetamide (150 mg/kg b.w., 20 h) and DL-α-hydrazino-δ-aminovaleric acid (HAVA, 75 mg/kg b.w., 6 h) as described before (3). As shown in FIG. 3, inhibitor activity was eluted as a single peak with 0.30-0.37 M NaCl and was separated from most of

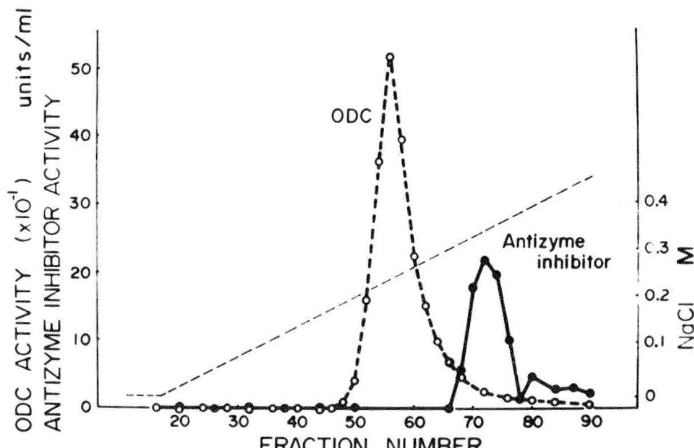

FIG. 3. Separation of ODC and antizyme inhibitor by DEAE-cellulose column chromatography (3). The broken line indicates NaCl concentration.

the ODC activity. The inhibitor activity was assayed by measuring reversal of ODC inhibition caused by antizyme. The total amount of inhibitor corresponded to about 5% of that of ODC. Since the inhibitor had no influence on free ODC activity in the absence of antizyme, we named it antizyme inhibitor.

Antizyme inhibitor was also found in a DEAE-cellulose eluate of liver extracts from rats in which hepatic ODC had been induced by feeding a high-protein diet (12). However, its activity was not detected in liver extracts from starved rats. Therefore, there seemed to be some correlation between the activities of ODC and antizyme inhibitor in rat liver extracts.

Properties of antizyme inhibitor

Antizyme inhibitor was a protein, because it was inactivated by heat treatment at 70°C and also by treatment with insolubilized trypsin. It was not diffusable and did not pass through a PM-10 membrane in an Amicon ultrafiltration apparatus. On Sephadex G-100 gel-filtration, antizyme inhibitor was eluted at about the same position as ODC (FIG. 4). This indicated that its molecular weight was about the same with that of ODC, which is 105,000 (9). The activity of antizyme inhibitor was not affected by treatment with rabbit antiserum monospecific to ODC. This absence of immuno-crossreactivity eliminated the possibility that antizyme inhibitor was derived from ODC either *in vivo* or *in vitro*.

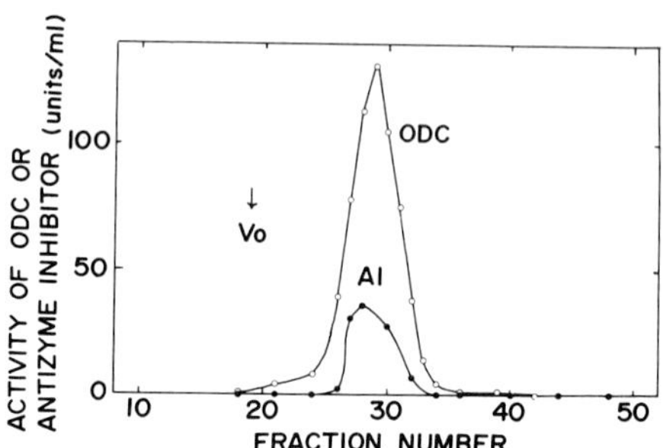

FIG. 4. Gel-filtration analysis of antizyme inhibitor (AI) and ODC. A DEAE eluate containing both antizyme inhibitor (1272 units) and ODC (3252 units) was concentrated by ultrafiltration and then applied to a column (1 x 78 cm) of Sephadex G-100. The elution buffer consisted of 50 mM Tris-HCl, pH 7.4, 0.1 mM EDTA, and 50 mM NaCl. The fraction volume was 4.2 ml.

Mode of action of antizyme inhibitor

The action of antizyme inhibitor was not time-dependent. Thus the ODC reaction proceeded linearly for at least 1 h when the enzyme was incubated at 37°C either alone, or with antizyme, or with antizyme and antizyme inhibitor (3). It is unlikely, therefore, that antizyme inhibitor is a protease specifically attacking antizyme. When ODC was assayed with a fixed amount of antizyme and increasing amounts of antizyme inhibitor, the inhibition by antizyme was stoichiometrically reversed by antizyme inhibitor (FIG. 5A). This suggested that antizyme inhibitor inactivated antizyme by forming a complex with it, just like antizyme inactivates ODC by forming a complex with it.

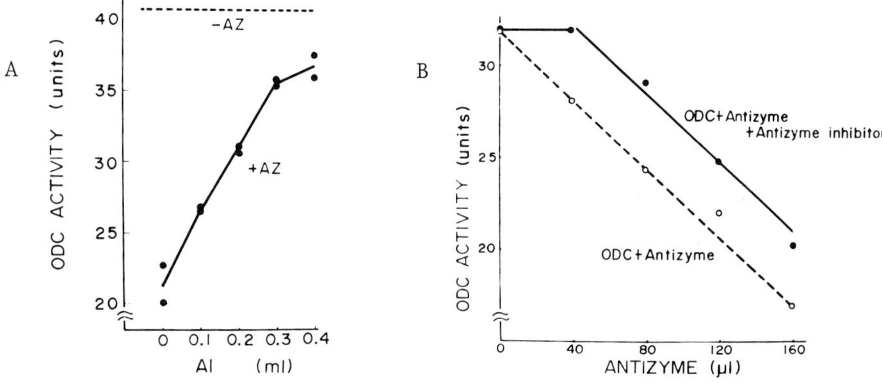

FIG. 5. (A) Effect of dose of antizyme inhibitor (AI) on ODC activity in the presence of a fixed amount of antizyme (AZ). The broken line indicates ODC activity assayed alone (3).
(B) Titration of ODC with antizyme in the presence and absence of a fixed amount of antizyme inhibitor (3).

When ODC was assayed with increasing amounts of antizyme in the presence of a fixed amount of antizyme inhibitor, ODC activity was inhibited only when the activity of antizyme added exceeded that of antizyme inhibitor present; then a linear dose-effect relationship was observed with a slope identical with that observed in the absence of antizyme inhibitor (FIG. 5B). This suggests that antizyme inhibitor has a much higher affinity than ODC to antizyme. Furthermore, antizyme inhibitor showed the same activity whether it was added to ODC before or after addition of antizyme. This means that antizyme inhibitor not only inhibited antizyme but also reactivated the inactive ODC-antizyme complex, apparently by replacing ODC in the complex.

For examination of complex formation between antizyme and antizyme inhibitor, the two factors were mixed and then subjected to Sephadex G-75 gel-filtration. As shown in FIG. 6, out of 65 units of antizyme inhibitor added 50 units were not recovered. At the same time, out of 103 units of antizyme added 57 units were not recovered. Thus, antizyme inhibitor was consumed as it inactivated an equivalent amount of antizyme. This strongly suggests that antizyme inhibitor acts by forming a complex with antizyme.

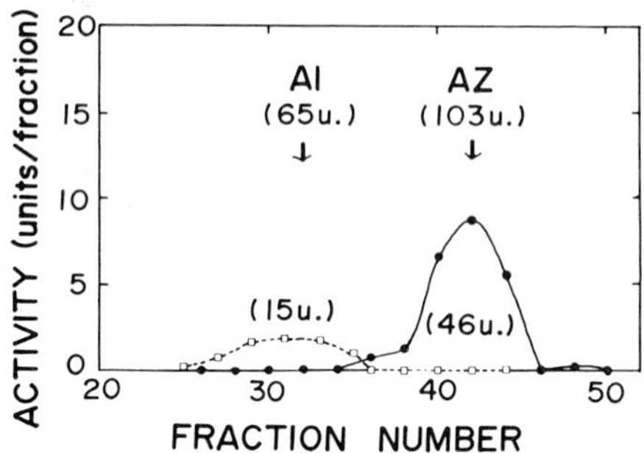

FIG. 6. Gel-filtration analysis of a mixture of antizyme (AZ) and antizyme inhibitor (AI). Partially purified preparations of antizyme (103 units in 0.9 ml) and antizyme inhibitor (65 units in 0.8 ml) were mixed and then passed through a column (1 x 73 cm) of Sephadex G-75. The elution buffer was 50 mM Tris-HCl, pH 7.4, containing 1 mM dithiothreitol, 0.1 mM EDTA, and 50 mM NaCl. The fraction volume was 1 ml. Arrows indicate the positions of elution of each factor, tested alone.

Assay of antizyme inhibitor in crude extracts

Activity of antizyme inhibitor is assayed by measuring reversal of ODC inhibition caused by antizyme. A correction has to be made for the ODC activity present in the test sample. Since antizyme inhibitor activity in crude liver extracts is usually less than 5% of their ODC activity, it is necessary to remove the ODC activity beforehand from liver extracts to measure the activity of the antizyme inhibitor accurately. We developed a simple method for this purpose: A monoclonal antibody against rat liver ODC was recently obtained in our laboratory by using the hybridoma technique, and ODC activity could be removed completely from liver extracts simply by filtering them through a small column of monoclonal anti-ODC-Sepharose 4B. Using this method, we plan to measure the level of antizyme inhibitor activity in rat liver under various *in vivo* conditions to obtain information on its physiological role.

Use of antizyme inhibitor for assay of ODC-antizyme complex

A serious difficulty in antizyme studies has been the lack of a simple method for assay of the ODC-antizyme complex (4). This difficulty can now be overcome since antizyme inhibitor can be used to reactivate the complex. We first applied this method to gel-filtration analysis of the complex formed *in vitro*. Partially purified preparations of ODC and antizyme were mixed and then passed through a Sephadex G-100 column. As shown in FIG. 7, the inactive ODC-antizyme complex was eluted slightly before ODC, indicating that it has a somewhat larger molecular size than ODC. This provided direct evidence for formation of the complex.

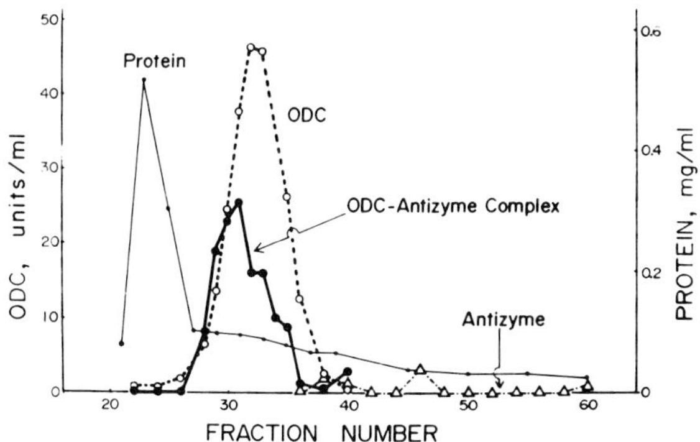

FIG. 7. Sephadex G-100 gel-filtration of the ODC-antizyme complex formed in vitro (3).

We then applied the assay method to measure the amount of ODC-antizyme complex formed in vivo. Rats were first treated with thioacetamide and DL-HAVA as before to induce hepatic ODC. They were then injected with 1,3-diaminopropane (1 mmol/kg b.w.) and killed 1 h later. This latter treatment caused almost complete loss of hepatic ODC activity. However, substantial increases in ODC activity were observed on addition of sufficient antizyme inhibitor to crude liver extracts (TABLE 2). This suggested that these liver extracts contained some ODC-antizyme complex. It should be noted that the amount of the complex corresponded to only 5 - 10% of the hepatic ODC present before the 1,3-diaminopropane treatment. However, this does not necessarily mean that antizyme played relatively small role in the process of ODC decay caused by diaminopropane. First, most of the complex that had been formed might have been degraded rapidly. Second, once formed to some level, the complex might have repressed the synthesis of ODC, and of antizyme itself. In any case, measurement of the ODC-antizyme complex should help in understanding the physiological role of the antizyme.

TABLE 2. Presence of ODC-antizyme complex in rat liver extract

Liver extract	ODC activity (cpm/0.5 ml extract)[a]		
	Antizyme inhibitor		
	minus	plus	difference
Rat 1	56	356	300
Rat 2	34	249	215
Rat 3	34	298	264

[a]Hepatic ODC at the time of 1,3-diaminopropane injection was was 2100-5400 cpm/0.5 ml of extract.

Preliminary studies in primary cultured rat hepatocytes

It would be useful to have a simple in vitro system for studying enzyme regulation. Recently, we made some preliminary studies on antizyme in primary cultures of rat hepatocytes. Hepatocytes were prepared from rat liver perfused with collagenase and were cultured as monolayers essentially by the method of Tanaka et al. (13). As shown in FIG. 8, addition of 10^{-2}M putrescine caused marked induction of antizyme activity in 2-4 h. The effect of putrescine concentration on ODC and antizyme activities was then studied 4 h after addition of the compound. As shown in FIG. 9, a concentration of 10^{-3}M was enough to repress ODC activity completely while 10^{-2}M was needed to induce antizyme activity to a level comparable with that of non-repressed ODC.

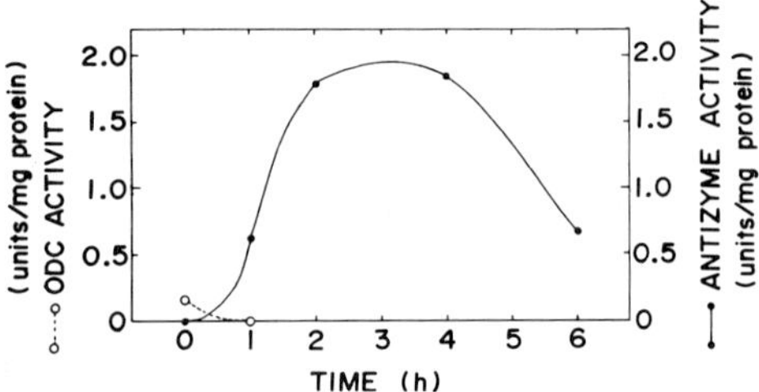

Fig. 8. Time course of the change in ODC and antizyme activities of primary cultured hepatocytes after addition of putrescine. Freshly prepared hepatocytes were first cultured in William's medium E (WE) containing 6% calf serum, 10^{-7}M insulin, and 10^{-6}M dexamethasone for 7 h and then cultured in WE alone for 10 h. The medium was then replaced with fresh WE containing 6% calf serum and 10^{-2}M putrescine and the culture was continued.

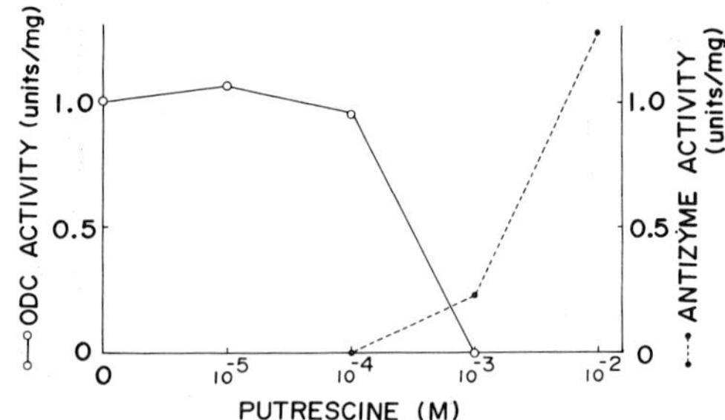

FIG. 9. Effect of varying concentration of putrescine on ODC and antizyme activities of primary cultured hepatocytes. The culture condition was the same as described under FIG. 8. Cells were harvested for assay 4 h after the transfer to new medium containing 6% calf serum and various concentration of putrescine.

Thus, primary cultures of hepatocytes seem to have the same pattern of regulation of ODC as whole rat liver and, therefore, should be useful in studies on ODC regulation. Based on the present results we now plan to study the physiological roles played by antizyme and antizyme inhibitor in both *in vivo* and *in vitro* systems.

References

1. Canellkis, E.S., Heller, J.S., and Kyriakidis, D.A. (1981): In: *Advances in Polyamine Research*, Vol. 3, edited by C.M. Caldarera, V. Zappia, and U. Bachrach, pp. 1-13. Raven Press, New York.
2. Fong, W.F., Heller, J.S., and Canellakis, E.S. (1976): *Biochim. Biophys. Acta*, 428: 456-465.
3. Fujita, K., Murakami, Y., and Hayashi, S. (1982): *Biochem. J.*, 204: 647-652.
4. Heller, J.S., and Canellakis, E.S. (1981): *J. Cell. Physiol.*, 107: 209-217.
5. Heller, J.S., Fong, W.F., and Canellakis, E.S. (1976): *Proc. Natl. Acad. Sci. USA*, 73: 1858-1862.
6. Kallio, A., Löfman, M., Pösö, H., and Jänne, J. (1979): *Biochem. J.*, 177: 63-69.
7. Kameji, T., and Hayashi, S. (1982): *Biochim. Biophys. Acta*, (in press).
8. Kameji, T., Murakami, Y., Fujita, K., and Hayashi, S. (1982): *Biochim. Biophys. Acta*, 717 (in press).
9. Kameji, T., Murakami, Y., Fujita, K., Noguchi, T., and Hayashi, S. (1981): *Med. Biol.*, 59: 296-299.
10. Kyriakidis, D.A., Heller, J.S., and Canellakis, E.S. (1978): *Proc. Natl. Acad. Sci. USA*, 75: 4699-4703.
11. McCann, P.P., Tardif, C., Hornsperger, J., and Böhlen, P. (1979): *J. Cell. Physiol.*, 99: 183-190.
12. Noguchi, T., Aramaki, Y., Kameji, T., and Hayashi, S. (1979): *J. Biochem.*, 85: 953-959.
13. Tanaka, K., Sato, M., Tomita, Y., and Ichihara, A. (1978): *J. Biochem.*, 84, 937-946.

Advances in Polyamine Research, Vol. 4, edited by
U. Bachrach, A. Kaye, and R. Chayen. Raven Press,
New York © 1983.

Ornithine Decarboxylase and Ornithine Decarboxylase Inactivating Factor in Rat Organs

A. M. Kaye and M. Yariv

Department of Hormone Research, The Weizmann Institute of Science, Rehovot 76100, Israel

This paper presents the recent progress in our attempts to answer the question - what is responsible for the 10-20 minute half life of ornithine decarboxylase (ODC) activity in mammalian organs?

These studies were initiated in collaboration with Dr. Isaac Icekson; The original report (10) of the existence of the ornithine decarboxylase inactivating factor (ODIF) in 1976, was a part of his doctoral thesis. With his departure from our laboratory to carry out post doctoral studies, this work was not extended further. The report of the ODC antizyme by Canellakis and his colleagues (4,7) overshadowed our announcement and only recently have we been able to resume work on this question.

PROVISION OF A SUITABLE SUBSTRATE FOR DEMONSTRATING ODIF

Our approach was the straightforward one of attempting to purify ODC to a point where it would be sufficiently stable to serve as a substrate for testing putative inactivating factors. As starting material, we chose the immature rat ovary which we had found to respond to luteinizing hormone by a characteristic increase in ODC activity after 4 h, parallel to the stimulation in a great variety of physiological and stress conditions (15,19). At that time, the specific activity obtained was among the highest reported in a mammalian system. We followed up the observation by asking the question - in which compartment of the ovary is the highest specific activity obtained? We measured ODC activity in ovaries from proestrus and from pregnant rats and from the corpora lutea and Graffian follicles of these ovaries. We found that the highest specific activity was obtained in luteinizing hormone (LH) treated Graffian follicles (9) and therefore we set out to obtain maximally folliculized ovaries which we could stimulate with (LH). We used the standard endocrinological procedure (20) of injecting immature rats (25 to 27 days old) with pregnant mare serum gonadotrophin (PMSG) to cause rapid and massive growth of the follicles and 52 hours later injecting luteinizing hormone to cause the maximal induction of ODC; 4 hours later, the rats were killed and the ovaries collected in liquid nitrogen. While PMSG alone caused a significant increase in the specific activity of ovarian ODC in both granulosa cells and the remainder of the ovary (Fig. 1), the combination of PMSG followed by LH provided us with our highest specific activity starting material plus an unexpected bonus - a stable enzyme.

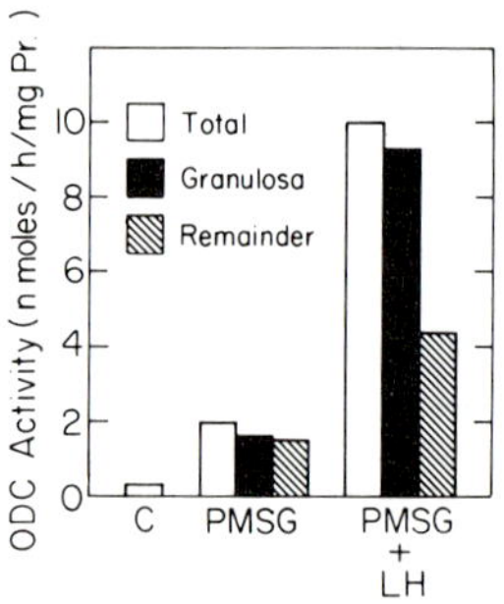

FIG. 1. ODC activity in immature rat ovaries and response to gonadotrophic hormones. Rats from the Hormone Research Dept. colony (Wistar derived) were injected subcutaneously with 15 i.u. pregnant mare serum gonadotrophin (PMSG, Gestyl, Organon, Oss, The Netherlands) at 8:30 A.M., and 52 h later, interperitoneally with 30 μg of ovine luteining hormone (NIAMDD-oLH-22) a gift of the NIAMDD, National Pituitary Agency, Baltimore, MD., U.S.A. Four hours after the LH injection, ovaries were extirpated and their large antral follicles punctured to release granulosa cells into homogenization buffer consisting of 0.25 M sucrose containing 10 mM Tris-HCl, pH 7.5, 5 mM dithiothreitol, (DTT) 0.1 mM disodium ethylene diamine tetraacetate (EDTA) and 0.1 mM pyridoxal phosphate. After washing by centrifugation in homogenization buffer, granulosa cells and the remainder of the ovary were homogenized separately in an Ultra-Turrax homogenizer and then ultracentrifuged at 4.5×10^6 g_{max} x min. The supernatant solutions, considered to be the cytosol fraction, were assayed for ODC activity (13) in 100 μl total volume in a mixture containing 50 mM Tris-HCl buffer, pH 7.5, 5 mM DTT, 1 mM pyridoxal phosphate, 0.1 mM EDTA, 0.25 mM L-ornithine, 0.02 mM $^{14}CO_2$-DL-ornithine (New England Nuclear Corp.) and from 0.1 to 0.9 mg of protein (Pr) as determined by the Bradford method (3).

The ODC activity in the cytosol fraction of gonadotrophic stimulated ovaries, which contained essentially all the ODC activity detected in the whole homogenate, was remarkably stable at 37° even in the absence of pyridoxal phosphate. This stability was in contrast to ODC instability in cytosol preparations from other organs such as the ventral prostate which showed a half life of ODC activity of only a few minutes (Fig. 2) equal to the half life seen in vitro in rat organs after treatment of rats with cycloheximide (6,17). The cytosol from hormone stimulated ovaries, which had a half life measured in hours rather than in minutes, thus provided, without the necessity for purification, a substrate for testing a putative inactivating factor.

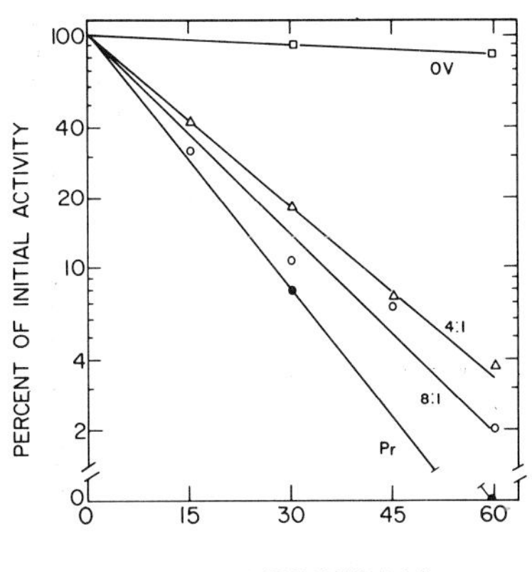

FIG. 2. Inactivation of ovarian ODC by a factor in ventral prostate. Cytosols were prepared (v. legend to Fig. 1) from ovaries of 26 day old gonadotrophin stimulated rats (□——□) and from ventral prostate glands of 4-month old rats (●——●), gel filtered through a Sephadex G-25 column and preincubated at 37° for 1 h. Mixtures of the two cytosols were made at a 4:1 (Δ——Δ) or 8:1 (O——O) ratio of prostatic to ovarian protein. After preincubation for the indicated times, ODC activity was measured as described in Fig. 1. Initial (100%) activity values were 1.79 nmol $^{14}CO_2$ liberated/h/mg protein for the prostatic and 32.3 nmoles/h/mg protein for the ovarian extract. (Reproduced with permission, Ref. 10).

DEMONSTRATION OF ODIF ACTIVITY IN RAT PROSTATE CYTOSOL

Cytosols of ovary and ventral prostate glands were passed through Sephadex G25 columns to rid them of pyridoxal phosphate and all other low molecular weight compounds. They were mixed at 2 different proportions, preincubated from 15 minutes to one hour and then assayed for ODC activity (13) by the addition of an assay mixture containing ^{14}C labelled ornithine; the $^{14}CO_2$ formed in the reaction was collected and measured (Fig. 2).

This mixing experiment showed that prostate cytosol contained a factor which caused the rapid but time dependent inactivation of ovarian ODC (10). The greater the degree of specificity of the inactivation, the more significant is this observation. Therefore, the activities of several other enzymes in the ventral prostate cytosol were tested for their stability at 37°, both in the presence and absence of 0.1 mM pyridoxal phosphate. Only ODC showed a significant loss of activity in 15 min (10) while S-adenosyl methionine decarboxylase, phosphoprotein phosphatase, glucose 6-phosphate dehydrogenase, and 6-phosphogluconate dehydrogenase retain full activities. In addition, we have tested creatine kinase in ventral prostate cytosol for its stability. This enzyme, with approximately the same isoelectric point (4.6) and similar amino acid composition to ODC and a slightly smaller molecular weight, also maintained its activity under conditions in which significant ODC activity was lost.

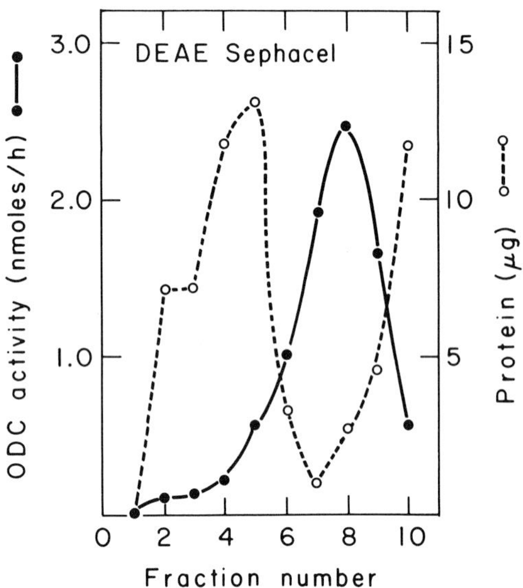

FIG. 3. DEAE Sephacel chromatography of ODC. A sample of the resuspended pellet from pH 4.6 precipitation (see Table 1) of cytosol from ovaries of PMSG + LH treated rats (see legend to Fig. 1) containing 3.9 mg protein was loaded on a DEAE Sephacel (Pharmacia) column, 6 ml volume, in 3 ml of TDE buffer (10 mM Tris-HCl, pH 7.5, 5 mM DTT and 0.1 mM Na_2EDTA). Fractions of 7 ml each were collected at a flow rate of 1.25 ml/min of a linear gradient of NaCl from 50 to 300 mM, in buffer containing 10 mM Tris-HCl pH 7.5, 5 mM DTT, 0.1 mM EDTA, 1 mM phenyl-methyl-sulfonyl fluoride and 60 mM pyridoxal phosphate. ODC activity and protein concentration were measured as described in the legend to Fig. 1, after desalting and concentration using a PM 10,000 Amicon membrane.

We therefore gave the factor in ventral prostate the non-prejudicial name of ornithine decarboxylase inactivating factor (ODIF) and sought to purify and characterize it. We defined a unit of activity (10) as that activity which results in a decrease of 1 unit of ODC during 1 h preincubation at 37°C. A unit of ODC was defined as that activity which catalyses the release of 1 nmol $^{14}CO_2$/h.

For provision of substrate we carried out two steps of purification of ODC. The use of isoelectric precipitation of cytosol at pH 4.6 as introduced by Ono et al. (16) proved a useful step in distinguishing ODC from ODIF in that it precipitated ODC while leaving ODIF in the supernatant solution. DEAE Sephacell chromatography was a useful next step for further ODC purification since ODC elutes from the column, when treated with a linear salt gradient (Fig. 3) at a position corresponding to a trough in protein elution. Thus, ovarian ODC could be purified to a specific activity of several hundred units (nmoles/h/mg protein) in two steps (Table 1).

TABLE 1. Partial Purification of ODC From Gonadotrophin-Treated Rat Ovaries[a]

Step	Total Protein (mg)	Total Units (nmol/h)	Specific Activity (units/mg	Recovery (%)	Purification (fold)
150,000 x g supernatant solution	360	7128	19.8	-	-
pH 4.6 precipitate	105	4515	43.0	63	2.2
DEAE Sephacell fraction	3.29	3064	932	43	47

a) Ovaries from immature rats treated with PMSG and LH were homogenized, centrifuged and assayed for ODC activity as described in the legend to Fig. 1, the supernatant solution pricipitated at pH 4.6 as previusly described (10) and the resuspended precipitate fractionated on a DEAE Sephacell column as described in the legend to Fig. 3.

In parallel, prostate cytosol was subjected to these same steps of purification. The ODIF activity in the pH 4.6 supernatant solution was not retained on DEAE Sephacell, permitting more than an order of magnitude purification at this step. Another order of magnitude purification was achieved by gel chromatography on Sephadex G150, resulting in an overall several hundred fold purification of ODIF (Table 2). During this purification, the ODC activity in prostate cytosol declines from a value of approximately 1 nmole/h/mg protein to a value indistinguishable from background. The gel filtration step (Fig. 4) permitted an estimate of the molecular weight of ODIF. The peak was found at a position corresponding to a molecular weight determined by appropriate markers to be slightly smaller than 20,000.

TABLE 2. Partial Purification of ODIF From Rat Prostate[a]

Step	Total Protein (mg)	Total Units (nmol/h)	Specific Activity (units/mg	Recovery (%)	Purification (fold)
150,000 x g supernatant solution	930	3116	3.4	-	-
pH 4.6 precipitate	496	3620	7.3	116	2.2
DEAE Sephacell effluent	23.2	3132	135	86	40
Sephadex G150 fraction	0.47	1239	2637	34	787

a) Ventral prostate glands from 3-4 month-old rats were homogenized, centrifuged, and assayed for ODIF activity as described in the legend to Fig. 2. The cytosol was precipitated at pH 4.6 as previously described (10), the resulting supernatant solution passed through columns of DEAE Sephacell packed in disposable plastic syringes and collected by centrifugation, and finally fractionated on a Sephadex G150 column as described in the legend to Fig. 4.

Another property of the ornithine decarboxylase inactivating factor is a strict temperature dependence (Fig. 5) with no activity at 0° and a graded response up to 37°. ODIF activity is destroyed by boiling but is stable overnight at room temperature (approximately 23°C) and to heating at 60° for 30 min. The activity is proportional to the protein concentration of the ODIF extract (Fig. 6), here illustrated for the first two steps in its purification.

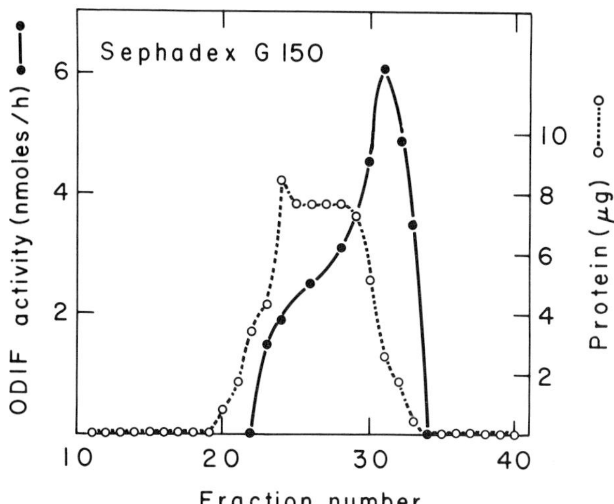

FIG. 4. Gel filtration of ODIF. A sample of the DEAE Sephacell effluent (see Table 2) of an ODIF preparation from rat prostate cytosol (1.4 ml, containing 1.9 mg protein) was subjected to gel filtration at room temperature on a column of Sephadex G150 (51 x 0.64 cm). The column was eluted with TDE buffer at a flow rate of 0.3 ml/min. Fractions of 1.3 ml were collected and 85 μl samples were assayed for protein concentration and for ODIF activity, using the pH 4.6 precipitate of ODC described in Table 1 as substrate and the assay described in the legend to Fig. 1.

COMPARISON OF ODIF WITH ANTIZYME, POLYAMINE DEPENDENT PROTEIN KINASE AND Ca^{2+} ACTIVATED TRANSGLUTAMINASE

ODIF is distinguishable from antizyme (4,7) in that ODIF shows a time dependent activity consistent with an enzymic reaction. Moreover, while crude prostate extracts show an immediate partial inhibition of ovarian ODC activity, purified fractions of prostatic ODIF (which no longer retain any ODC activity) no longer show any immediate inhibition of ODC activity. Speculation on the nature of the possible enzymic reaction catalyzed by ODIF is restricted by the fact that the extracts used to measure ODIF activity were subjected to procedures (gel filtration, dialysis) to rid them of small molecular weight components necessary in protein modification reactions. Thus it is unlikely that ODIF acts as a polyamine dependent protein kinase as described by Kuehn and his colleagues (1,5), since no polyamine nor ATP is required for the reaction, and ODIF preparations can be heated to 60° for 30 minutes, as mentioned above, without loss of activity. These same considerations apply to the possibility of identifying

ODIF as calcium dependent transglutaminase (18). Indeed, Isaac Icekson has recently tested our ODIF fractions for this activity and found that there is no increase in the specific activity of Ca^{2+} dependent transglutaminase during the several hundred fold purification of ODIF.

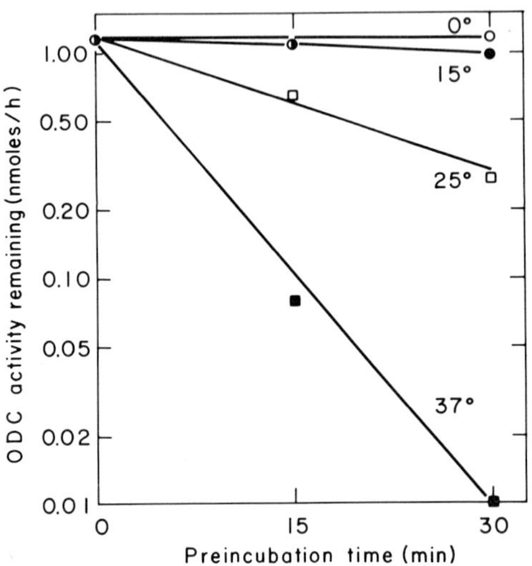

FIG. 5. Temperature dependence of ODIF activity. The cytosol fraction from prostate glands of 3-month old rats was preincubated at the indicated temperatures before being assayed for the remaining intrinsic ODC activity as described in the legend to Fig. 1. Each incubation mixture contained 1.16 mg of total prostatic soluble protein.

IS ODIF A PROTEASE?

Our original suggestion (10) that ODIF could be a proteolytic enzyme is consistent with the existence of proteases which do not require cofactors, which have a relatively low molecular weight, and have a very restricted substrate specificity. In fact, the group of Katunuma in Japan has described the so-called group specific proteases (12,14), serine proteases which cleave a group of enzymes which share a common coenzyme. In particular, enzymes which cleave pyridoxal phosphate apoenzymes have been characterized and crystallized. We therefore tested a variety of inhibitors of serine proteases (Table 3) and inhibitors which were effective against the three other main groups of proteases, the cystine, aspartic and metallo-proteases. We found no effective inhibitor of ODIF action among the antiproteolylytic agents tested. This raises the possibility that ODIF might be a new

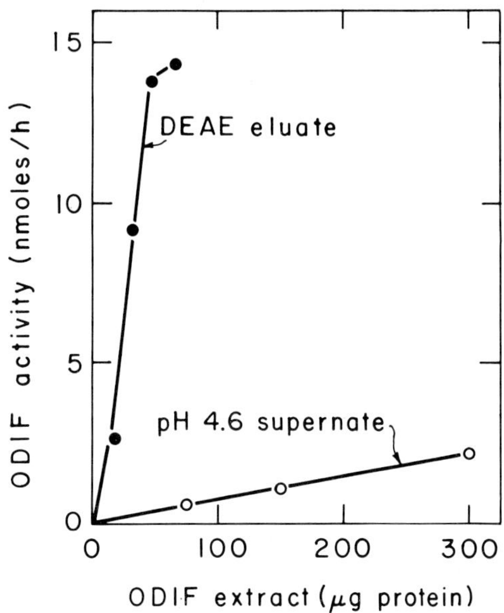

FIG. 6. Protein concentration dependence of ODIF activity. The supernatant solution from a pH 4.6 precipitation of prostatic cytosol was assayed for ODIF activity as described in Fig. 1, using 2.25 units of ovarian ODC as substrate. The DEAE Sephacell effluent was assayed for ODIF (using the standard 15 min preincubation time) and 6.5 units of ovarian ODC as substrate.

proteolytic enzyme which does not fit into any of the four major classes of proteases. The possibility that inactivation of ODC is indeed associated with proteolysis is consistent with reports that induction and subsequent reduction in ODC activity paralled the concentration of immunodetectable ODC (8).

The existence of exquisitely specific proteases for which no specific inhibitor has yet been discovered has precedents. One example is the signal peptidase within secretory cells (2,11) which cleaves the signal peptide of newly synthesized secreted proteins, during the process of their passage through the cell membrane. Another example is an enzyme described by Neurath and his colleages in Heidelberg (personal communication) from the hepato-pancreas of the crayfish which has a specificity for an amino acid sequence different from that of the 4 major groups of proteases. The hypothesis of the proteolytic nature of ODIF will be tested directly when a small sample of highly purified ODC becomes available.

TABLE 3. Antiproteolytic Agents Found Ineffective Against ODIF[a]

Agent	Maximum Conc. Tested	Class of Proteinase Inhibited
Diisopropyl fluorophosphate	1 mM	Serine
Phenyl-methyl-sulfonylchloride	1 mM	Serine
Trasylol	1 mg/ml	Serine
Benzamidine	8 mg/ml	Serine
Leupeptin	1 mM	Serine & cysteine
Tosyl-lysyl-chloromethylketone	2 mg/ml	Serine & cysteine
Pepstatin	1 mM	Aspartic -
1,10 Phenanthrolene	1 mM	Metallo -

a) Agents were tested in the ODIF assay system described in the legend to Fig. 2.

From a general viewpoint, if we now can number tens of ways (15,19) of inducing the synthesis of ODC, we should not be surprised if there may turn out to be several ways of controlling the concentration and utilization of ODC by means of enzyme inactivation.

ACKNOWLEDGEMENT

This research was supported by a grant from the United States-Israel Binational Science Foundation (BSF, Jerusalem, Israel). A.M.K. is the incumbent of the Joseph Moss Professorial Chair of Molecular Endocrinology.

REFERENCES

1. Atmar, V.J. and Kuehn, G.D. (1981): Proc. Nat'l. Acad. Sci. USA, 78:5518-5522.

2. Blobel, G., and Dobberstein, B. (1975): J. Cell Biol., 67:835-851.

3. Bradford, M.M. (1976): Anal. Biochem., 72:248-254.

4. Canellakis, E.S, Viceps-Madore, D., Kyriakidis, D.D., and Heller, J.S. (1979): Curr. Topics Cell Regulation, 15:155-202.

5. Daniels, G.R., Atmar, V.J., and Kuehn, G.D. (1981): Biochemistry, 20:2525-2532.

6. Harik, S.I., Hollenberg, M.D., and Snyder, S.H. (1974): Mol. Pharmacol., 10:41-47.

7. Heller, J.S., Fong, W.F., and Canellakis, E.S. (1976): Proc. Nat'l Acad. Sci. USA, 73:1858-1862.

8. Höltta, E. (1975): Biochim. Biophys. Acta, 399:420-427.

9. Icekson, I., Kaye, A.M., Lieberman, M.E., Lamprecht, S.A., Lahav, M., and Lindner, H.R. (1974): J. Endocr., 63:417-418.

10. Icekson, I., and Kaye, A.M. (1976): FEBS Letters, 61:54-58.

11. Jackson, R.G., and Blobel, G. (1980): Ann. N.Y. Acad. Sci., 343:391-404.

12. Katunuma, N., Kominami, E., Kobayashi, K., Banno, Y., Suzuki, K., Chichibu, K., Hamaguchi, Y, and Katsunuma, T., (1975): Eur. J. Biochem., 52:37-50.

13. Kaye, A.M., Icekson, I., Lamprecht, S.A., Gruss, R., Tsafriri, A., and Lindner, H.R. (1973): Biochemistry, 12:3072-3076.

14. Kominami,E., Kabayashi, K., Kominami, S., and Katunuma, N. (1972): J. Biol. Chem., 247:6848-6855.

15. Morris, D.R., and Fillingame, R.N.H. (1974): Ann. Rev. Biochem., 43:303-325.

16. Ono, M., Inoue, H., Suzuki, F., and Takeda, Y. (1972): Biochim. Biophys. Acta, 284:285-297.

17. Russell, D.H., and Snyder, S.H. (1969): Mol. Pharmacol., 5:253-262.

18. Russell, D.H. (1981): Biochem. Biophys. Res. Commun., 99:1167-1172.

19. Russell, D.H. (1981): In: Polyamines in Biology and Medicine, edited by D.A. Morris and L.J. Marton, pp.109-125. Marcel Dekker, Inc., New York.

20. Zarrow, M.X., and Wilson, E.D. (1961): Endocrinology, 69:851-855.

Advances in Polyamine Research, Vol. 4, edited by U. Bachrach, A. Kaye, and R. Chayen. Raven Press, New York © 1983.

Requirement for Polyamines in the Response of Sympathetic Neurons to Axonal Injury

Gad M. Gilad

Department of Isotope Research, The Weizmann Institute of Science, 76100 Rehovot, Israel

Recently, we have demonstrated that ornithine decarboxylase (ODC) activity is rapidly enhanced within nerve cell bodies in the superior cervical ganglion (SCG) of adult rats after their axon has been cut. This is one of the earliest biochemical changes so far detected during the response of the nerve cell body to injury of its axon (i.e. the axon reaction). In the present series of experiments we sought to determine: a) whether the ODC response is specifically associated with the axon reaction or, is it a nonspecific change, associated with other drastic metabolic alterations in the neuron, and b) whether the ODC response is a mandatory event for the progression of the axon reaction.

Studies were carried out on 3 months old female Sprague-Dawley rats. In one set of experiments ODC activity was determined in the SCG 6h and 7d after several noxious stimuli: 1) postganglionic nerve cut (axotomy); 2) preganglionic nerve cut (denervation); 3) end organ removal; and 4) 6-hydroxydopamine treatment. We have also measured choline kinase (CK) activity which is elevated in SCG after axotomy, during the period of axonal regeneration. After 6h the activity of ODC was increased after all treatments while CK activity remained unchanged. By 7d no changes were detected in ODC activity. In contrast, CK activity was increased, but only after axotomy; other treatments had no effect.

In another set of experiments polyamine biosynthesis was specifically inhibited prior to, during and for 6d after axotomy was performed. Inhibition was achieved by a combination treatment with a single intraperitoneal (i.p.) injection of 50 mg/kg methylglyoxal bis-(guanylhydrazone) and by repeated (every 12h) i.p. injections of 100 mg/kg α-difluoromethylornithine, starting at 2d before surgery. This treatment prevented the appearance of the chromatolytic response, and, furthermore, resulted in death of axonally injured neurons. In contrast, uninjured neurons remained intact.

We conclude that in adult rats: a) increased ODC activity is a common feature of drastic metabolic alterations in sympathetic neurons, and b) increased polyamine biosynthesis is obligatory for the progression of the axon reaction in these neurons.

This research was supported by a grant from the Muscular Dystrophy Association. -

Advances in Polyamine Research, Vol. 4, edited by U. Bachrach, A. Kaye, and R. Chayen. Raven Press, New York © 1983.

Cell Cycle Dependent Induction of Ornithine Decarboxylase and Its Antizyme in Ehrlich Ascites Tumor Cells Separated by Centrifugal Elutriation

S. Anehus, M. Linden, *Å. Borg, *E. Långström, and O. Heby

*Department of Zoophysiology, University of Lund, S-223 62 Lund, Sweden; *Department of Oncology, University of Lund, S-221 85 Lund, Sweden*

Mammalian ornithine decarboxylase (ODC) activity appears to be modulated by positive and negative macromolecular effectors (2, 8). An unusual mode of enzyme regulation was revealed for ODC by Fong et al. (4). They showed that putrescine, the product of the ODC-catalyzed reaction, elicited the appearance of an intracellular ODC inhibitory protein, which they named ODC-antizyme (4). This type of inhibitor was subsequently found to be inducible by putrescine treatment in many different cell lines. ODC-Antizyme was also isolated from Ehrlich ascites tumor cells that had been treated with putrescine while grown *in vivo* (10).

In an attempt to determine whether possible cell cycle-dependent changes in ODC activity are due to modulation by ODC-antizyme, the following experiment was performed. Ehrlich ascites tumor cells in plateau phase growth were transplanted into the peritoneal cavity of a new host in order to stimulate their ODC activity (1). They were then treated with putrescine for 4 hr, at a dose that caused complete eradication of the ODC activity, and release or synthesis of ODC antizyme (10). The putrescine-treated cells were then isolated and separated according to their size, using a centrifugal elutriation technique. ODC and ODC-antizyme activities were assayed for in the cell fractions collected.

CELLS

Hyperdiploid Ehrlich ascites tumor cells were grown in the peritoneal cavity of 6-week-old males of NMRI mice. For the ODC and ODC-antizyme experiments, mice were given an intraperitoneal

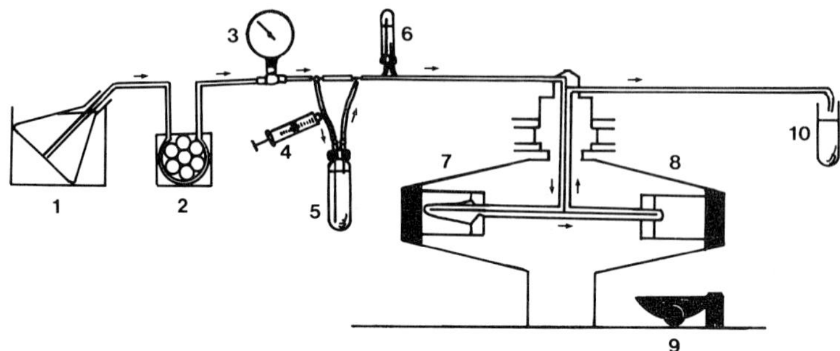

FIG. 1. Arrangement of the centrifugal elutriation system. (1) Reservoir with ice-cold medium; (2) 2115 Multiperpex peristaltic pump; (3) pressure gauge; (4) syringe for cell sample injection; (5) sample mixing chamber; (6) bubble trap; (7) separation chamber and (8) bypass chamber of the Beckman JE-6 elutriator rotor; (9) stroboscope; and (10) fraction collector.

inoculation of 0.5 x 10^8 and 2.8 x 10^8 cells, respectively. These cells were from mice that had been inoculated with 1.4 x 10^6 cells 12 days earlier.

INDUCTION OF ODC AND ODC-ANTIZYME

Increased ODC activity was induced by transplantation of tumor cells into a new host. The tumor cells were then collected in 10 ml of ice-cold phosphate-buffered saline (PBS), pH 7.2 (1).

The release or synthesis of ODC-antizyme was accomplished by treating tumor-bearing animals with putrescine as previously described (10). A dose of 25 μmoles of putrescine in 0.2 ml of PBS was injected intraperitoneally at 1-hr intervals. After 4 hr of treatment the tumor cells were collected in 10 ml of ice-cold PBS (1).

CELL SEPARATION BY CENTRIFUGAL ELUTRIATION

Centrifugal elutriation, or counterflow centrifugation, has been available as a technique for cell separation since 1948, when Lindahl (9) first described the principle of his "counter-streaming centrifuge". More recently, Beckman Instruments have developed an elutriator rotor (JE-6), which is commercially available. This elutriator rotor has been successfully used for separating cells according to their size. Thus, it has become possible to study individually various types of cells in a population (3, 5, 6), and cells in various phases of the cell cycle.

Centrifugal elutriation depends on the balancing of an outwardly directed centrifugal force acting on a cell, with inwardly directed fluid dynamic forces as well as buoyancy forces. A suspended cell reaches its equilibrium position in the separation chamber according to the sedimentation velocity, which is based on the size, density and shape of the cell. When either the rotor speed is decreased or the velocity of the fluid flow is increased, successive fractions of the cells in a population can be collected. The basic elements of the elutriator system used are shown in Fig. 1.

Tumor cells (putrescine-treated as well as untreated controls) were separated according to their size using a Beckman JE-6 Elutriator Rotor installed in a Beckman Model J-21C Centrifuge. Ice-cold PBS (pH 7.2) was pumped through the elutriation system using a 2115 Multiperpex peristaltic pump (LKB). Flow rates were calibrated by measuring the fluid volume collected per unit time at specific dial settings. The fact that the flow rates were highly reproducible, eliminated the need for a flow-meter. Both the fluid reservoir and the sample mixing chamber were kept on ice, and the temperature setting for the centrifuge well was 2 ^{o}C. Roughly 1.4 x 10^8 Ehrlich ascites tumor cells, suspended in 5 ml of PBS (single-cell suspension), were loaded into the sample mixing chamber at a rotor speed of 2700 r.p.m., and a fluid flow rate of 20 ml/min. After collecting two 250 ml fractions in Autoclear polycarbonate bottles, the rotor speed was lowered to 2600 r.p.m. and a new fraction was collected. Subsequent fractions of the original cell population were collected by step-wise reduction of the rotor speed.

The cell fractions collected were centrifuged at 1000 x g for 15 min at 4 ^{o}C in an International centrifuge. The cells were resuspended in 5-10 ml of PBS. About 95 % of the injected cells were recovered after elutriation. The cell fractions were processed for flow cytometric analysis and for assay of ODC and ODC-antizyme activities.

CELL CYCLE ANALYSIS BY FLOW CYTOMETRY

One to 2 x 10^6 cells from each fraction were fixed in absolute ethanol at -20 ^{o}C, treated with RNase and stained with propidium iodide. This fluorescent dye selectively and quantitatively intercalates into the double-stranded regions of nucleic acids. In RNase-treated cells its fluorescence intensity is proportional to the DNA content. The stained cells were analyzed with a Cytofluorograph (Ortho Instruments). The DNA histograms for a representative series of fractions, collected with the elutriation system, are shown in Fig. 2.

DETERMINATION OF ODC AND ODC-ANTIZYME ACTIVITIES

Cells from the various fractions were pelleted by centrifugation and sonicated in 200 μl of ice-cold homogenization medium

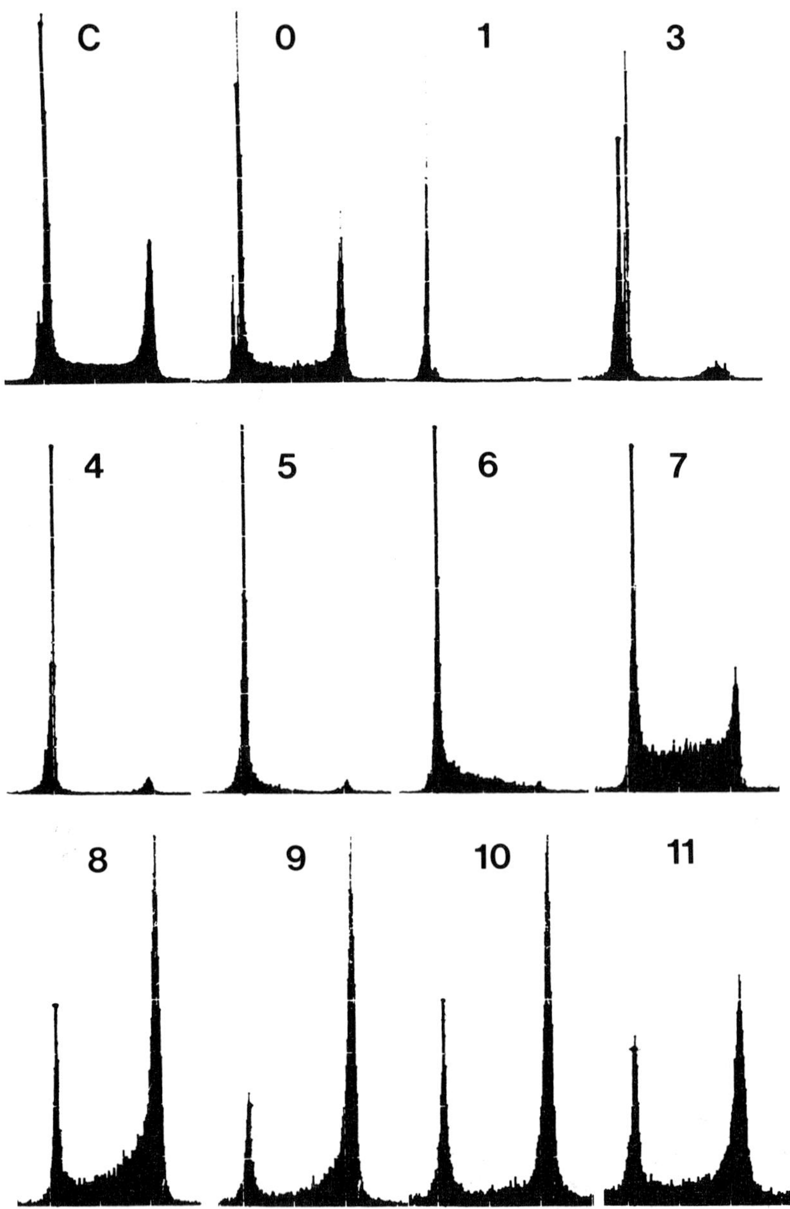

FIG. 2. DNA histograms representative of consecutive fractions obtained during centrifugal elutriation of Ehrlich ascites tumor cells. (C) Control 11-day-tumor; (0) putrescine-treated (4 hr) 11-day-tumor prior to elutriation; (1-11) cell fractions collected in sequence during elutriation.

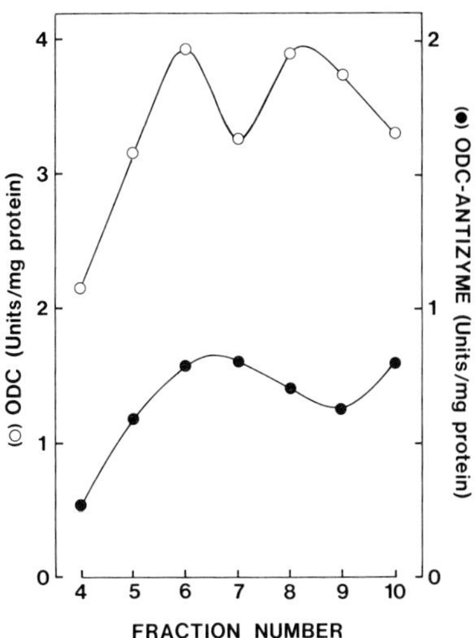

FIG. 3. ODC and ODC-antizyme activities in Ehrlich ascites tumor cells separated into fractions according to their size using a centrifugal elutriation technique. One unit of ODC activity is defined as 1 nmol of CO_2 released per hr. One unit of ODC-antizyme activity is defined as the amount of inhibitor that eradicates 1 unit of ODC activity. The distribution of cells in the cell cycle is shown for each fraction in Fig. 2.

(0.5 mM Na_2EDTA, 5 mM dithiothreitol, 0.05 mM pyridoxal 5'-phosphate in TRIS buffer, pH 7.2). Cell samples for ODC-antizyme assays were sonicated in homogenization medium that was supplemented with 250 mM NaCl. The assays for ODC (7) and ODC-antizyme (4, 10) have been previously described.

CELL CYCLE-RELATED CHANGES IN ODC AND ODC-ANTIZYME

Fig. 2 and 3 show that fractions containing a majority of cells in G1-S and in S-G2 exhibit maximum ODC activities. In the fraction that has the largest S phase population (fraction 7), the ODC activity shows a minimum. The fact that ODC-antizyme exhibits its highest activity in the corresponding fraction, indicates that this ODC-inhibitory protein may be involved in the regulation of ODC activity during the mammalian cell cycle.

ACKNOWLEDGEMENTS

This investigation was supported by grants from the Swedish Natural Science Research Council.

REFERENCES

1. Andersson, G., Österberg, S., and Heby, O. (1978): Int. J. Biochem., 9:263-267.

2. Canellakis, E.S., Kyriakidis, D.A., Heller, J.S., and Pawlak, J.W. (1981): Med. Biol., 59:279-285.

3. Flangas, A.L. (1974): Prep. Biochem., 4:165-177.

4. Fong, W.F., Heller, J.S., and Canellakis, E.S. (1976): Biochim. Biophys. Acta, 428:456-465.

5. Glick, D., von Redlich, D., Juhos, E.T., and McEwen, C.R. (1971): Exp. Cell Res., 65:23-26.

6. Grabske, R.J., Lake, S., Gledhill, B.L., and Meistrich, M.L. (1974): J. Cell. Physiol., 86:177-189.

7. Heby, O., Marton, L.J., Wilson, C.B., and Martinez, H.M. (1975): J. Cell. Physiol., 86:511-522.

8. Kyriakidis, D.A., Heller, J.S., and Canellakis, E.S. (1978): Proc. Natl. Acad. Sci. USA, 75:4699-4703.

9. Lindahl, P.E. (1948): Nature, 161:648-649.

10. Linden, M., Andersson, G., and Heby, O. (1980): Int. J. Biochem., 12:387-393.

Advances in Polyamine Research, Vol. 4, edited by U. Bachrach, A. Kaye, and R. Chayen. Raven Press, New York © 1983.

Ornithine Decarboxylase Activity in Cultured Bone Cells is Activated by Bone-Seeking Hormones and Physical Stimulation

D. Sömjen, *M. Yariv, *A. M. Kaye, **R. Korenstein, **H. Fischler, and I. Binderman

*Hard Tissue Unit, Ichilov Hospital, 64239 Tel-Aviv; *Departments of Hormone Research and **Membrane Research, The Weizmann Institute of Science, Rehovot 76100, Israel*

A variety of hormones and physical effectors are known to initiate a complex of cellular reactions such as changes in ion transport, in nutrient intake and in intracellular concentrations of cyclic nucleotides, which influences the proliferative activity of cells. Previously, we have demonstrated that bone cells in culture responded to physical forces (1) prostaglandins (PGE_2) (2), parathyroid hormone (PTH) (3,4) and to capacitatively coupled electric fields (5,6) by an increased rate of DNA synthesis. When any of these effectors were used, a rapid (5-15min) increase in intracellular cyclic AMP production preceeded the increased DNA synthesis (1-3). Calcitonin (CT), another bone seeking hormone (7) which is an antagonist of PTH action *in vivo*, also increased the intracellular concentration of cAMP in the same cultures; however CT had no effect on DNA synthesis (4). Activation of ornithine decarboxylase (ODC), the rate limiting enzyme in polyamine biosynthesis (8), usually associated with macromolecular synthesis as well as cell growth and division(9), was suggested to be an important prerequisite for the initiation of DNA synthesis (10). It was also suggested that ODC activity can be induced by changes in cyclic nucleotides such as cAMP (11, 12).

We report here that ODC activity in cultured bone cells is closely regulated by changes in cAMP concentration. However increased ODC activity does not necessarily leads to an increase in DNA synthesis.

MATERIALS AND METHODS

Cell cultures.

Bone cell cultures were prepared from rat embryo calvaria as described previously (3). Calvaria from 19-21 day old rat embryos were excised; cells were released by digestion with trypsin-EDTA. Cells were plated at a concentration of 10^6 cells/ 54 mm culture dish and incubated in BGJ_b medium supplemented with 10% FCS. Cells were used after they reached confluence.

Stimulation of cells in culture.

a) PTH (10U/ml), PGE_2 (500 ng/ml), CT (100 ng/ml), 8BrcAMP (10 μg/ml) or 8BrcGMP (1 μg/ml) were added to the cells and ODC activity was measured at the time periods indicated in each experiment.
b) Physical forces stimulation was performed as described previously (1) by using an orthodontic screw.
c) Electric field stimulation of bone cells was obtained as described previously (6) by a series of rectangular voltage pulses (25 μ sec width, 3Hz repetition rate) of varying amplitude lasting 5 minutes.

ODC activity.

After treatment for various periods, cells were collected by scraping and were sonicated in cold homogenizing medium (10 mM Tris HCl pH 7.5, 250 mM sucrose, 5 mM dithiothreitol, 1 mM Na_2EDTA and 0.1 mM pyridoxal phosphate) for a few seconds.

The homogenate was centrifuged at 38,000 xg for 30 min. at 4°C and the supernatant was used to determine enzymic activity (13). Enzyme activity was assayed (14) by measuring the $^{14}CO_2$ released from [^{14}C]-DL ornithine. The protein concentration of enzyme preparations were determined by the method of Lowry (15).

RESULTS AND DISCUSSION

Bone cells in culture incubated with either PTH, CT or PGE_2 showed an increase in ODC activity. In all cases, the maximal activity was reached at 4 hours after treatment followed by a rapid decrease at 5-6 hours (Fig. 1). ODC activity was increased 8-9 fold by PGE_2, 6-7 fold by PTH and 2-3 fold by CT (Fig. 1). Addition of PTH and CT to the cultures had an additive effect on the activity of ODC (Fig. 2) as has been previously shown for cAMP production (4,16,17). These results suggest that the cultured bone cell system is derived from a mixed population of CT and PTH responsive cells (18). Though it was postulated that an increase in ODC activity is an important prerequisite of DNA synthesis (10) the CT induced increase in cAMP production and in ODC activity was not accompanied by enhanced DNA synthesis. In order to study the mechanism of ODC stimulation, specifically the possibility that CT induces ODC activity via production of cAMP, we have tested the ability of exogenous 8BrcAMP and 8BrcGMP to raise ODC activity. While 10 μg/ml of 8BrcAMP increased ODC activity by 2.5 fold, 8BrcGMP failed to do so (Fig. 2). At the same time, exogenous 8BrcAMP did not increase DNA syntehsis in this system (4). Moreover, indomethacin, a PG synthesis inhibitor, totally inhibited cAMP production in CT or in physically stimulated cells and to a lesser extent in PTH stimulated cells (1,3). Indomethacin was also shown to inhibit to the same degree ODC activity in the same cells (Fig. 3). This correlation between

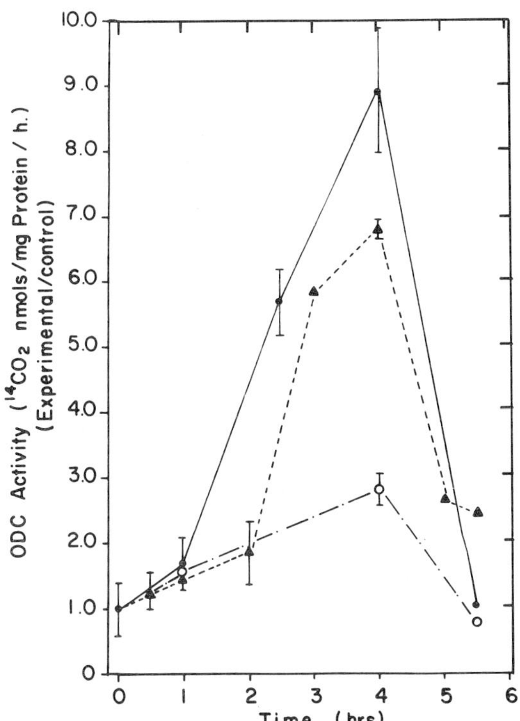

FIG.1. Time course of induction of ODC activity in cultured bone cells by PTH (10 U/ml ▲---▲), CT (100 ng/ml o---o) and PG (500 ng/ml ●——●). Conditions of cell growth, hormonal stimulation as well as ODC preparation and assay are described in Materials and Methods. The activity of the enzyme from untreated cells was 1.06 nmoles$^{14}CO_2$/mg Prot/h. The results are expressed as means ± SEM for n ≧ 4.

cAMP production and ODC activity reinforces our hypothesis of a close relationship between cAMP production and the induction of ODC activity. Stimulation by capacitative electric fields, which was shown to increase cAMP production and to enhance DNA synthesis as a function of field strength (5,6) is able to induce changes in ODC activity as well (Fig. 4). A low voltage electric field causes a 2 fold increase in ODC activity while a high voltage electric field causes almost a 3 fold increase, but with no changes in the middle voltage electric field (Fig. 4). This is well in accord with the changes in cAMP production and DNA synthesis in this system that were previously reported (5,6).

In conclusion, all the effectors tested (physical, electrical and chemical) were able to stimulate increases in both cAMP production and ODC activity but not always increased DNA synthesis.

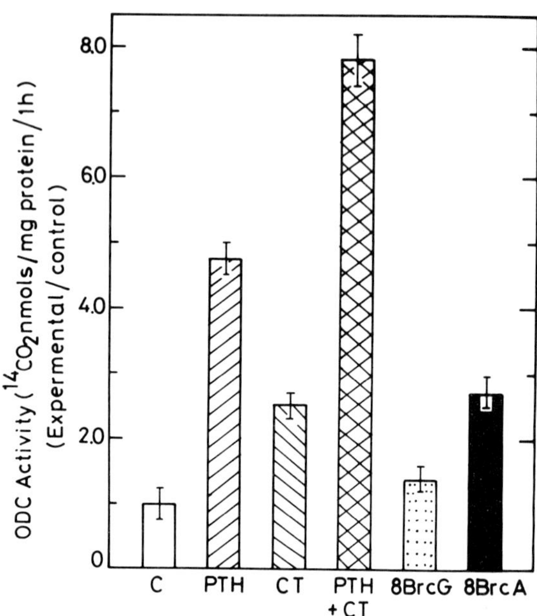

FIG.2. Induction of ODC activity in cultured bone cells by PTH (10 U/ml), CT (100 ng/ml), PTH + CT, 8BrcGMP (1 μg/ml) or 8BrcAMP (10 μg/ml). Cells were grown as described in Materials and Methods and treated with chemicals for 4 hours before harvesting and assay for ODC. The activity of the enzyme from untreated cells was 1.66 nmoles $^{14}CO_2$/mg Prot/h. Results are expressed as means ± SEM for n = 5.

The data presented here support the generalization that while an increase in cAMP concentration, caused by either hormones or physical agents, can trigger increased ODC activity in cultured bone cells, increased ODC activity 4 hours after stimulation is not sufficient to trigger subsequent DNA synthesis in all cases.

ACKNOWLEDGEMENT

We want to thank Mrs. E. Berger for excellent technical help. This research was supported in part by a grant from the United States-Israel Binational Science Foundation (BSF, Jerusalem, Israel). A.M.K. is the incumbent of the Joseph Moss Professorial Chair of Molecular Endocrinology.

R.K. is an incumbent of the Association of Friends of the Weizmann Institute of Science in Israel Career Development Chair.

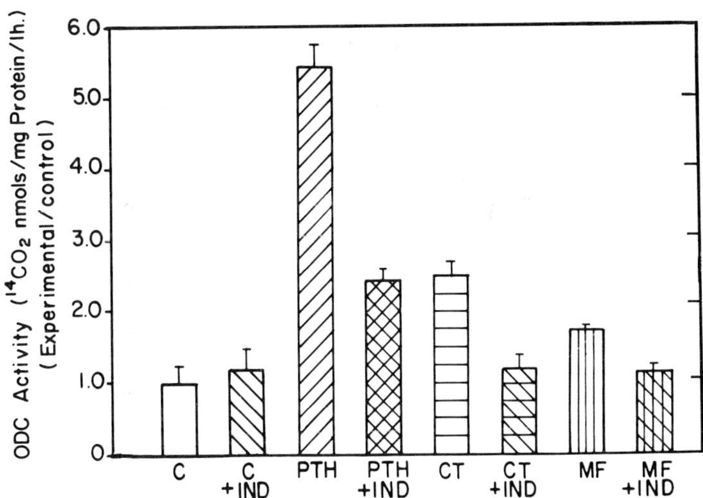

FIG.3. The effect of indomethacin (pretreatment for 20 min with 5 μg/ml) on the induction of ODC activity in cultured bone cells induced by either PTH (10 U/ml), CT (100 ng/ml) or mechanical forces. Cells were grown as described and treated for 4 hours before harvesting and assay for ODC. Activity of ODC from untreated cells was 1.07 nmoles $^{14}CO_2$/mg Prot/h). Results are expressed as mean ± SEM for n = 5.

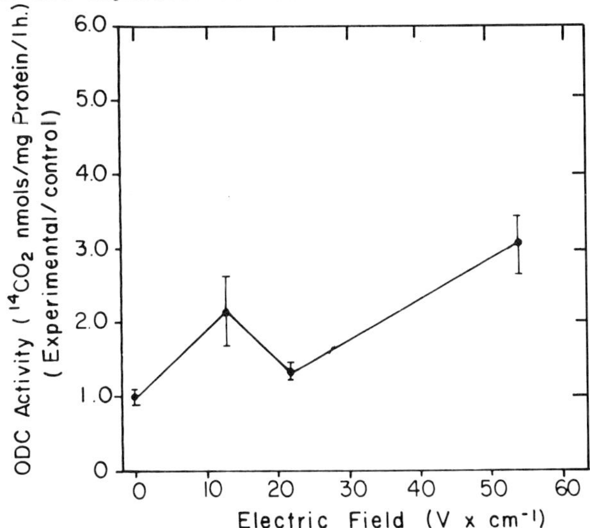

FIG.4. The effect of electric field stimulation on ODC activity in cultured bone cells. Cells were grown as described, and subjected to electric field stimulation for 5 min; ODC activity was measured 4 hours later. ODC activity from untreated cells was 0.077 nmoles $^{14}CO_2$/mg Prot/h. Results are expressed as mean ± SEM for n = 6.

REFERENCES

1. Sömjen, D., Binderman, I., Berger, E. and Harell, A. (1980): Biochim. Biophys. Acta, 627:91.

2. Binderman, I., Sömjen, D., Shimshoni, Z. and Harell, A.(1982): Proc.Inter.Symp. on Osteoporosis, John Willey & Sons Ltd., in press.

3. Sömjen, D., Fine, N., Harell, A. and Binderman, I.(1982):Proc. Inter.Symp. on Osteoporosis, John Willey & Sons Ltd., in press.

4. Sömjen, D., Fine, N., Harell, A., Schlessinger, J. and Binderman, I.: Biochem.Biophys.Res.Commun. Submitted for publication.

5. Sömjen, D., Korenstein, R., Fischler, H. and Binderman, I. (1982); Current Advances in Skeletogenesis: Development,Biomineralization, Mediators and Metabolic Bone Diseases. Excerpta Medica, edited by M. Silberman and H.C. Slavkin, pp.412-416. Amsterdam.

6. Korenstein, R., Sömjen, D., Fischler, H. and Binderman, I. (1982): Science, submitted for publication.

7. Binderman, I., Duksin, D., Harell, A., Sachs, L. and Katchalsky, E. (1974): J. Cell Biol. 61: 425.

8. Raina, A. and Janne, J. (1968): Acta Chem. Scand., 22:2375.

9. Fillingame, R.H., Jorstad, C.M. and Morris, D.R. (1975): Proc. Natl. Acad. Sci. U.S.A., 72:4042.

10. Stastny, N. and Cohen, S.S. (1970): Biochim. Biophys. Acta, 204:578.

11. Byus, C. and Russell, D.H. (1975): Science, 187:670.

12. Byus, C. and Russell, D.H. (1975): Life Sciences, 15: 1991.

13. Icekson, I. and Kaye, A.M. (1976): FEBS Lett, 61:54.

14. Kaye, A.M., Icekson, I., Lamprecht, S.A., Gruss, R., Tsafriri, A. and Linder, H.R. (1973); Biochemistry, 12:3072.

15. Lowry, D.H., Rosenbrough, J.H., Farr, A.L. and Randall, R.J. (1951): J. Biol. Chem. 193:165.

16. Peck, W.A., Burks, J.K., Wilkin, J., Rodan, S.B. and Rodan, G.A. (1977): Endocrinology, 100: 1357.

17. Heersche, J.N.M., Heyboer, M.P.M. and Ng, B. (1978): Endocrinology, 103:330.

18. Luben, R.A., Wong, G.L. and Cohn, D.V. (1975): Endocrinology, 99: 526.

Advances in Polyamine Research, Vol. 4, edited by U. Bachrach, A. Kaye, and R. Chayen. Raven Press, New York © 1983.

Whole Body- and Microautoradiographic Localization of Ornithine Decarboxylase in the Kidneys of Nandrolone-Treated Mice Using Tritium Labelled α-Difluoromethyl Ornithine

A.-Ch. Henningsson, S. Henningsson, *H. Tjälve, **L. Hammar, and G. Löwendahl

*Department of Physiology and Biophysics, University of Lund, S-223 62 Lund, Sweden; *Department of Toxicology, University of Uppsala, S-751 23 Uppsala, Sweden; **Institute of Biochemistry, University of Uppsala, S-751 23, Uppsala Sweden*

The aliphatic diamine putrescine and the polyamines spermidine and spermine appear to be ubiquitously present in mammalian tissues. Increased intracellular concentrations of these amines and elevated activities of their biosynthetic enzymes are among the earliest biochemical responses to hormones, drugs and other growth promoting stimuli (For ref. see 5,9).

Ornithine decarboxylase catalyzes the formation of putrescine, a reaction which is the first step in the biosynthesis of the polyamines. In the kidneys of gonadectomized male mice treated with the anabolic steroid nandrolone the ornithine decarboxylase activity was increased thousand-fold coinciding in time with an increase in the weight of the kidneys (1). Parallel with these events elevated concentrations of diamines and polyamines were detected in the kidneys and the urine.

Methods used for enzyme localization are mainly based either on the visualization of products formed from the reaction catalyzed by the enzyme or on the visualization of an antibody raised against the enzyme. The use of a specific irreversible inhibitor for cytological localization of an enzyme represents another approach, which has not been much developed so far.

Alpha halomethyl amino acids act as enzyme-activated irreversible inhibitors of the corresponding pyridoxal phosphate dependent amino acid decarboxylase (6,8). These inactivators are relatively unreactive molecules with sufficient structural similarity to the substrate so that they react with the active site of the enzyme. They are then converted, at the active site, through the catalytic action of the enzyme to a reactive species which can form a covalent bond with a functional group at the active site and inactivate the enzyme. Thus, the specificity of these inhibitors is not only based on their similarity of the substrates, but also on the mechanism of action of the target enzyme which implies a high degree of enzyme selectivity.

The availability of radioactively labelled enzyme activated irreversible inhibitors is apt to provide a specific and accurate tool for localization of enzymes in vitro as well as in vivo.

In the present study a selective radioactive labelling of structures in the kidneys of the nandrolone-treated mouse after intravenous injection of tritiated α-difluoromethyl ornithine is shown.

METHODS

Animals

Adult male mice of the NMRI strain (weight about 30 g), were used. They were fed a diet of commercial mouse chow and water ad libitum. Gonadectomy was performed 4 weeks before the experiments. Nandrolone phenpropionate (Durabolin®, Organon, Oss, the Netherlands) was suspended in arachis oil (100 µg/50 µl) and injected s.c. for 4 days (100 µg/mouse x day). Control mice received equivalent volumes of arachis oil.

Labelled Compound

α-(5-^{3}H)-Difluoromethyl ornithine (^{3}H-α-DFMO) was manufactured by New England Nuclear, Boston, Mass., USA at a specific activity of 22.5 Ci/mmol and a radiochemical purity of more than 98%.

Determination of Protein-Bound Radioactivity in the Kidneys

Nandrolone-treated and control mice were given 8 µCi ^{3}H-α-DFMO and the kidneys were removed 4 h later and homogenized in 8 volumes of 5% trichloroacetic acid (TCA). Aliquots of the homogenates were sequentially washed and extracted respectively with 95% aq. ethanol, absolute ethanol, chloroform:methanol (2:1) and ether to remove unbound radioactivity (2). The pellet was wetted by addition of 50 µl H_2O and solubilization was accomplished by addition of 0.5 ml Soluene100® (Packard Instrument Company, Inc.). After addition of 10 ml scintillation solution (5.5 g Permablend III®, Packard/l toluene) the radioactivity was determined in a liquid scintillation spectrometer (Kontron MR 300).

Whole Body Autoradiography

Nandrolone-treated and control mice were injected through a tail vein with 100 µCi ^{3}H-α-DFMO in 0.2 ml 0.9% NaCl. The mice were sacrificed 4 h after the injections. Killing of the animals was accomplished by CO_2-asphyxiation, followed by embedding in carboxymethyl cellulose gel and immersion in carbon dioxide-cooled n-hexane (-78 ^{0}C). Twenty µm thick sections through the frozen animals were cut sagitally in a microtome at -15 ^{0}C, attached onto tape and dried, all according to Ullberg (10). Twenty duplicate sections were taken from each mouse and freeze-dried. To localize tissue-bound radioactivity every other section was washed successively with 5% TCA, water, methanol and heptane for 0.5 min, respectively. The sections were dried and apposed to X-ray film together with the adjacent non-extracted freeze-dried sections. The time of exposure was 10 weeks.

Microautoradiography

Nandrolone-treated and control mice were injected in a tail vein with 100 μCi ^{3}H-α-DFMO. The mice were killed after 4 h and pieces of the kidneys were removed and fixed in 4% formaldehyde in a phosphate buffer (pH 7.0). After fixation, the tissues were dehydrated in an ethanol series and embedded in paraffin. Five μm thick sections were cut and mounted on glass slides. After deparaffinization of the sections in ethanol, the slides were dipped in NTB-2 (Eastman-Kodak) liquid film emulsion and stored for 10 weeks at 4 °C. After exposure, the slides were developed and stained with haematoxylin-eosin. In microautoradiograms made under these conditions, it is assumed that only tissue-bound radioactivity is left in the sections.

RESULTS

Protein-Bound Radioactivity in the Kidneys

The total radioactivity in kidney homogenates from nandrolone-treated mice following administration of ^{3}H-α-DFMO was similar to that found in the control kidneys exceeding the control value by only 20%. However, the amount of radioactivity recovered in the protein fraction after washing of the tissue homogenates showed a clear difference between the two groups; a 13-fold higher labelling was found in the kidney homogenate from the nandrolone-treated mice.

Whole Body Autoradiography

In FIG. 1 whole body autoradiograms from mice injected with ^{3}H-α-DFMO are shown. In the autoradiograms obtained from unwashed sections similar distribution patterns were seen in the nandrolone-treated and the control mice, with labelling of the kidneys, the liver and the intestinal contents (FIG. 1a and 1c). After the tissue-washing the radioactivity was lost from all tissues of the control mouse (FIG. 1d). However, in the nandrolone-treated mouse radioactivity was retained in the kidneys, but it was lost from all other tissues (FIG. 1b). The retained radioactivity in the kidneys of nandrolone-treated mice was strongest in a zone in the cortex adjacent to the medulla (FIG. 1b and FIG. 2). Most other areas of the cortex were also labelled, although to a lower extent, whereas the medulla was non-labelled.

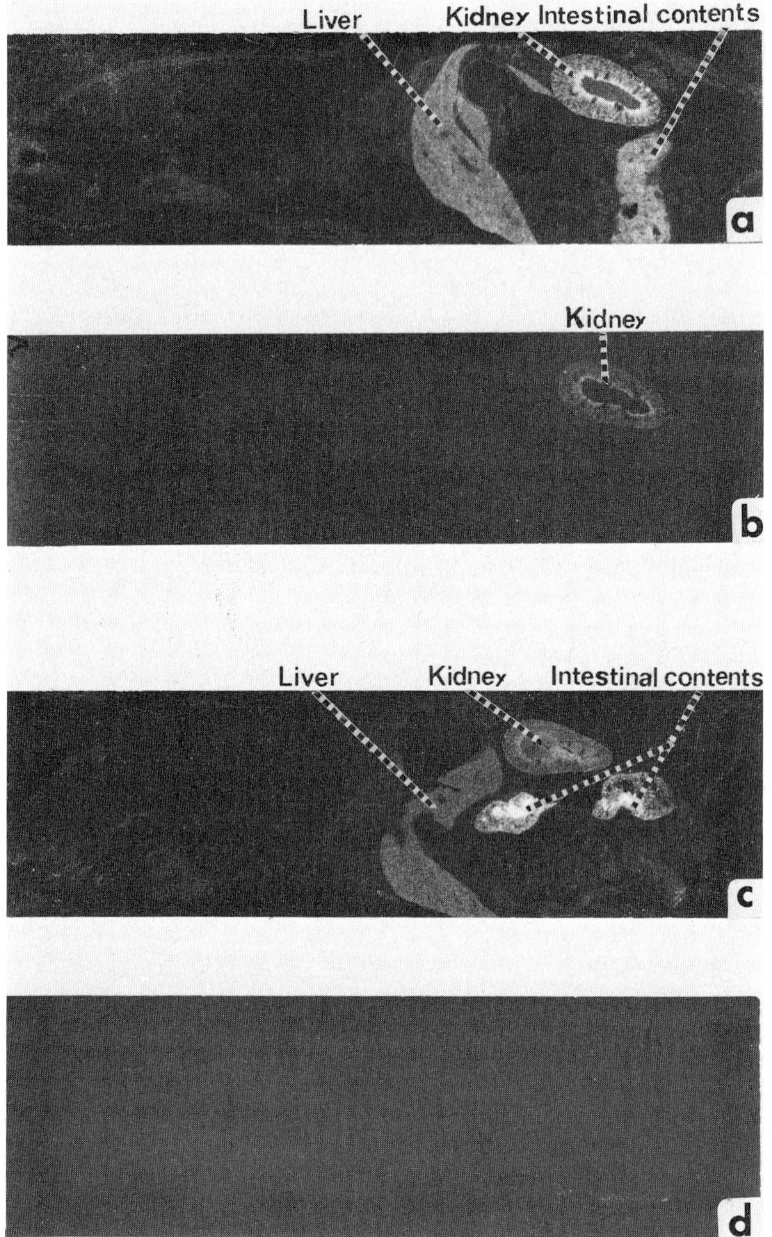

FIG. 1. Whole body autoradiograms of male mice given 100 μCi ^{3}H-α-DFMO 4 h before killing.
a) and b): Autoradiograms from a nandrolone-treated mouse. a) unwashed section; b) washed section.
c) and d): Autoradiograms from a control mouse. c) unwashed section; d) washed section.

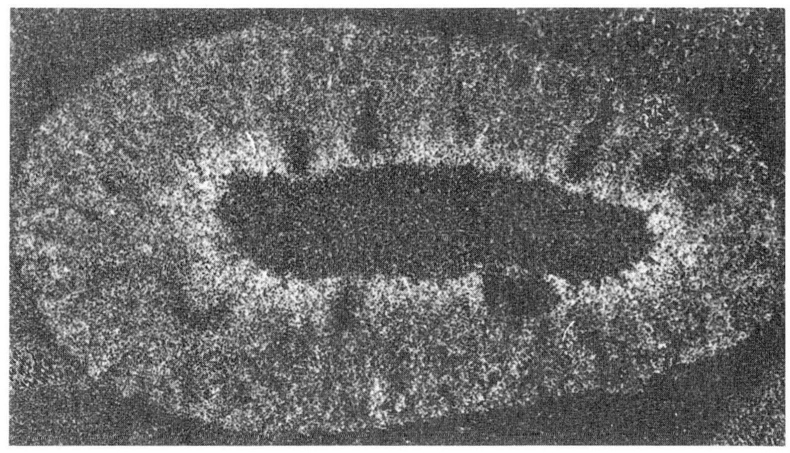

FIG. 2. Enlargement of the kidney in a whole body autoradiogram, obtained by exposure of a washed section of a nandrolone-treated mouse killed 4 h after the injection of 100 μCi ^{3}H-α-DFMO. (Magnification x 10).

Detailed Localization of ^{3}H-α-DFMO-Radioactivity in the Kidney

The microautoradiograms obtained after the administration of ^{3}H-α-DFMO to nandrolone-treated mice showed labelling of specific areas of the kidney cortex (FIG. 3). The strongest labelled portions were identified as the thick segments of the descending loops of Henle. The proximal convoluted tubuli were also labelled, whereas other parts of the nephrons, including the glomeruli, the thin and the thick ascending segments of the loops of Henle, the distal convoluted tubuli and the collecting tubuli, were non-labelled. Microautoradiograms from control mice showed no labelling.

DISCUSSION

In the present study it was for the first time demonstrated that ^{3}H-α-DFMO, a radioactively labelled enzyme activated irreversible inhibitor of ornithine decarboxylase appears to be useful for autoradiographic localization of the enzyme.

In the kidney of gonadectomized male mice the ornithine decar-

FIG. 3. Microautoradiograms from the kidneys of a nandrolone-treated mouse given 100 μCi ^{3}H-α-DFMO 4 h before sacrifice. Hl=thick segments of the descending loops of Henle; ct=collecting tubuli; pct=proximal convoluted tubuli; dct=distal convoluted tubuli; g=glomerulus. Asterisks show the junction between the thick and thin segments of the descending loops of Henle. (Magnification, a): x 780; b) x 725).

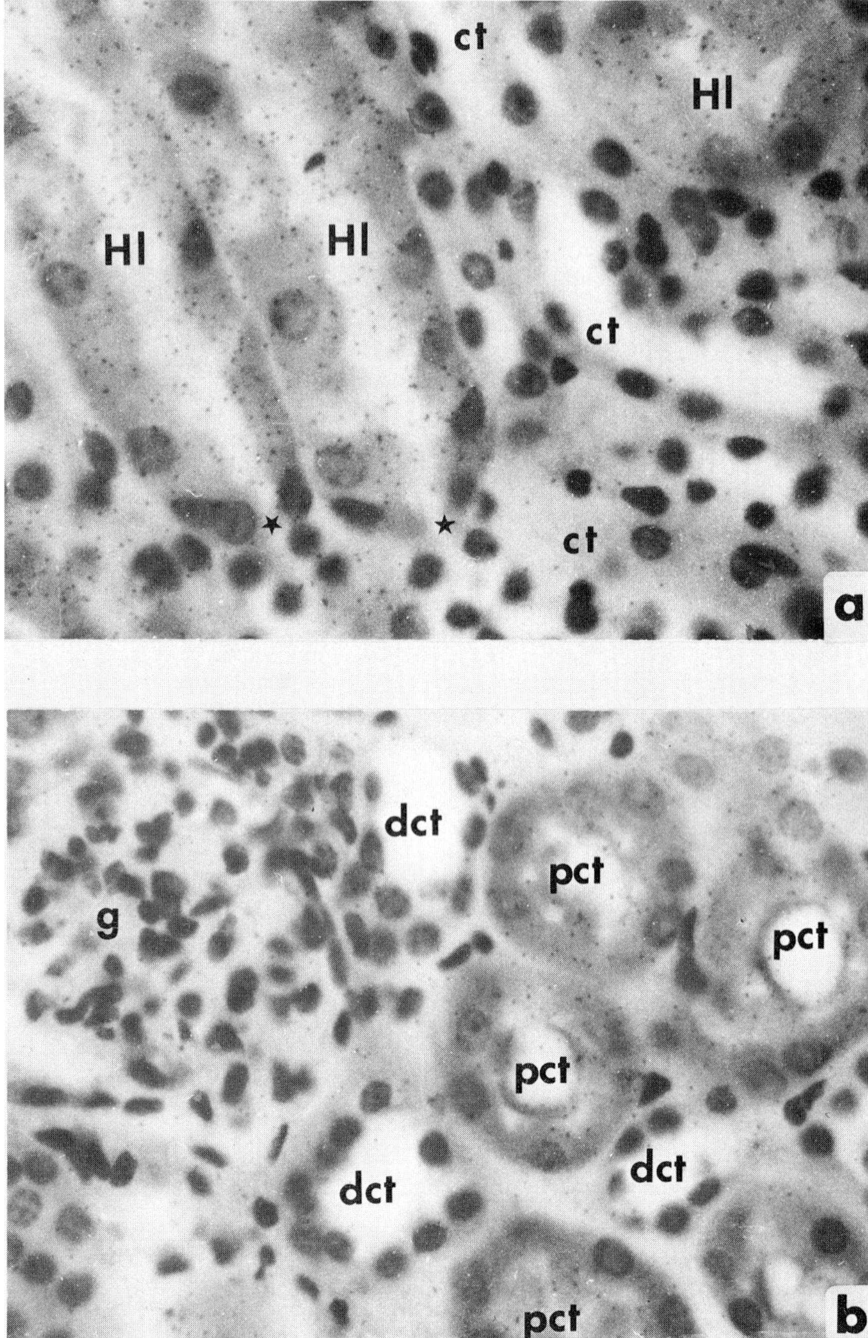

FIG. 3. For legend see preceding page.

boxylase activity is reduced to very low levels while in the kidneys of nandrolone-treated mice the ornithine decarboxylase activity is highly elevated (1). These differences in enzyme activity is reflected in differences in protein-incorporated labelling in tissue homogenates and tissue sections after in vivo administration of ^{3}H-α-DFMO. The kidney microautoradiograms showed localization of ^{3}H-α-DFMO-radioactivity to the thick segments of the descending loops of Henle. The proximal convoluted tubuli were also labelled. Thus, a very detailed localization of the kidney ornithine decarboxylase was achieved. Using antibodies against kidney ornithine decarboxylase it was not established if the immunoreactive cells were located in proximal and/or distal tubuli; but they seemed to be confined to the proximal ones (7). With a tritiated irreversible inhibitor of histidine decarboxylase (^{3}H-α-fluoromethyl histidine) for localization of histidine decarboxylase in the kidney of the pregnant mouse a labelling pattern obviously different from that obtained with ^{3}H-α-DFMO was seen in whole body as well as microautoradiograms (4).

A rhodamine conjugate of α-DFMO has recently been used to label ornithine decarboxylase for localization of the enzyme in histological sections of rat cerebellum (3). A disadvantage of this method as well as of immunocytochemical methods is that these techniques can only be used for in vitro localization of the enzyme. The specificity of the antibodies used for immunocytochemistry raises another problem.

In isoelectrofocusing experiments performed after administration of ^{3}H-α-DFMO to nandrolone-treated mice the labelled kidney enzyme focused as the active enzyme (to be published), indicating the specificity of the binding of ^{3}H-α-DFMO to ornithine decarboxylase.

The presented technique appears most valuable, representing a new approach to show whole body and micro localization of an enzyme in vivo.

ACKNOWLEDGEMENTS

This study was supported by the Swedish Medical Research Council (Grants Nos. 02212 and 05676) and the Faculty of Medicine, University of Lund.

REFERENCES

1. Andersson, A.-Ch., Hammar, L., Henningsson, S., and Löwendahl, G.O. (1982): Acta Physiol. Scand., 114:225-233.
2. Brittebo, E., Löfberg, B., and Tjälve, H. (1981): Xenobiotica, 11:619-625.
3. Gilad, G.M., and Gilad, V.H. (1981): In: Advances in Polyamine Research, Vol. 3, edited by C.M. Caldarera, V. Zappia, and U. Bachrach, pp. 111-121, Raven Press, New York.
4. Hammar, L., Henningsson, A.-Ch., Henningsson, S., Tjälve, H., Appelgren, L.-E., Hökfelt, T., Fuxe, K., Karobath, M., and Kollonitsch, J. Manuscript.

5. Jänne, J., Pösö, H., and Raina, A. (1978): Biochim. Biophys. Acta, 473: 241-293.
6. Metcalf, B.W., Bey, P., Danzin, C., Jung, M.J., Casara, P., and Vevert, J.P. (1978): J. Am. Chem. Soc., 100:2551-2553.
7. Persson, L., Rosengren, E., and Sundler, F. (1982): Biochem. Biophys. Res. Commun., 104:1196-1201.
8. Rando. R.R. (1974): Science, 185:320-324.
9. Tabor, C.W., and Tabor, H. (1976): A. Rev. Biochem. 45:285-306.
10. Ullberg, S. (1977). Science Tools, The LKB Instrument Journal Special Issue, pp. 2-29.

Advances in Polyamine Research, Vol. 4, edited by U. Bachrach, A. Kaye, and R. Chayen. Raven Press, New York © 1983.

A Role for the Polyamines in Mouse Embryonal Carcinoma (F9 and PCC3) Cell Differentiation but Not in Human Promyelocytic Leukemia (HL-60) Cell Differentiation

O. Heby, S. M. Oredsson, *I. Olsson, and L. J. Marton

*Department of Zoophysiology, University of Lund, S-223 62 Lund, Sweden; *Department of Internal Medicine, University of Lund, S-221 85 Lund, Sweden*

An increased rate of polyamine synthesis characterizes cells that are stimulated to grow and divide (for references, see 17) and cells that are induced to differentiate (14, 26, 34, 50). The importance of polyamine synthesis for cell growth and proliferation has been clearly demonstrated in studies using DL-α-difluoromethylornithine (DFMO), an enzyme-activated, irreversible inhibitor of ornithine decarboxylase (29). Inhibition of polyamine synthesis slows the growth rate of cells in culture (for references, see 16). DNA synthesis and cytokinesis seem to be the events most sensitive to polyamine limitation in the cell cycle (16, 42, 47, 48), a conclusion supported by experiments with a polyamine auxotrophic mammalian cell line (38, 39).

Whether an increased rate of polyamine synthesis is also an obligatory step in cell differentiation has not been as thoroughly investigated. The results of experiments with early invertebrate (3, 12, 18) and vertebrate (1, 18, 24) embryos, mouse erythroleukemia cells (14) and rabbit costal chondrocytes (49) are indicative of a role for polyamines in differentiation. However, difficulties in evaluating the processes of proliferation and differentiation separately, have partly prevented a definitive conclusion.

L.J.M. is on sabbatical leave from the Department of Laboratory Medicine, and the Brain Tumor Research Center of the Department of Neurological Surgery, School of Medicine, University of California, San Francisco, California 94143.

The present studies were undertaken to determine the possible involvement of polyamines in the control of cell growth and differentiation in human promyelocytic leukemia (HL-60) cells and in mouse embryonal carcinoma (F9 and PCC3) cells. These cell lines can all be induced to differentiate with retinoic acid.

HUMAN PROMYELOCYTIC LEUKEMIA CELL DIFFERENTIATION

Cells and Culture Conditions

The human promyelocytic leukemia (HL-60) cell line was kindly provided by Dr. Robert C. Gallo (National Cancer Institute, Bethesda, Md). This cell line was derived from the peripheral blood leukocytes of an adult female with acute promyelocytic leukemia (7, 9, 13). In culture HL-60 cells grow in single-cell suspension and display distinct myeloid characteristics, morphologically, histochemically as well as functionally (Fig. 1) (7, 9, 13). The majority of the cultured cells are myeloblasts and promyelocytes, but up to 10 % of these cells spontaneously differentiate into more mature granulocytes, including myelocytes, metamyelocytes and banded and segmented neutrophils (7, 9, 13).

HL-60 cells also can be induced to differentiate into mature granulocytes by incubation with a variety of compounds such as butyric acid, hypoxanthine, actinomycin D, dimethyl sulfoxide, hexamethylene bisacetamide (8, 9) and retinoic acid (5). Induced HL-60 cells express morphological and functional characteristics of normal polymorphonuclear granulocytes. Moreover, induction into monocyte/macrophage-like cells can be obtained with phorbol esters (23, 40, 41). Differentiation inducing factors from human mononuclear blood cells induce HL-60 cells to differentiate into myelomonocytic cells (35).

HL-60 cells maintain their myeloid characteristics in long-term culture and form colonies in semisolid medium as characteristic of transformed cells (13). The fact that these cells produce subcutaneous myeloid tumors in athymic nude mice (13), establishes their oncogenicity.

Our experiments used cells between passages 20 and 40. They were subcultured weekly at an initial density of 2.0×10^5 cells/ml in RPMI-1640 medium (Flow Labs) supplemented with 10 % fetal calf serum, 50 µg/ml penicillin (GIBCO) and 100 µg/ml streptomycin (GIBCO). These cultures reached a saturation density of about 2.5×10^6 cells/ml in 5 to 6 days.

For the experiments, cells in exponential growth were transferred to serum-free RPMI-1640 medium supplemented with 5 µg/ml bovine insulin (Sigma), 5 µg/ml human transferrin (Sigma), 0.6 mg/ml fatty acid-free bovine albumin (Sigma) and antibiotics (4) and seeded at a concentration of 2.0×10^5 cells/ml in 35 mm plastic petri dishes (Falcon). The cells were incubated at 37 °C in a humidified 5 % CO_2 atmosphere. In this serum-free medium the doubling time of untreated exponentially growing HL-60 cells was about 44 hr, i.e. only slightly longer than in serum-

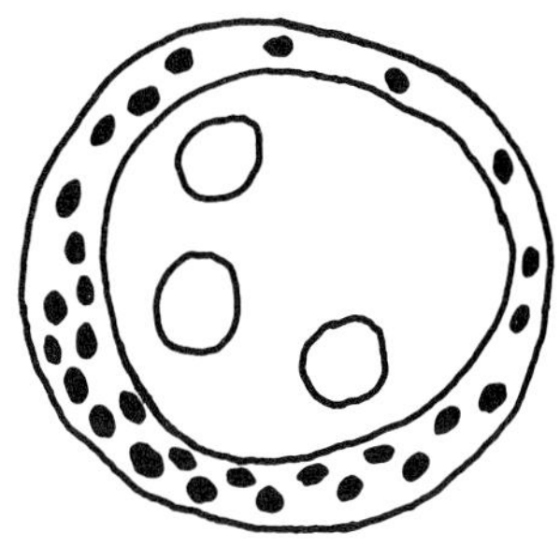

UNINDUCED HL-60	INDUCED HL-60
Morphological Differentiation	
Predominantly promyelocytes	Predominantly metamyelocytes and banded neutrophils
Large cells	Small cells
High nuclear:cytoplasmic ratio	Low nuclear:cytoplasmic ratio
Large round nucleus with dispersed chromatin	Pyknotic changes in the chromatin and marked indentation, convolution and segmentation of the nucleus
Two to four nucleoli per nucleus	Marked reduction or complete disappearance of nucleoli
Basophilic cytoplasm with prominent azurophilic granules	Decreased cytoplasmic basophilia, less prominent cytoplasmic granules
Functional Differentiation	
Active DNA synthesis	DNA synthesis ceases
High mitotic activity	Mitotic activity ceases
Low phagocytic activity	High phagocytic activity

FIG. 1. Characteristics of uninduced and induced human promyelocytic leukemia cells (HL-60) cells (7, 9, 13). Inducers such as retinoic acid, dimethyl sulfoxide, hexamethylene bisacetamide, hypoxanthine, actinomycin D and butyric acid produce the mature granulocytes described (5, 8, 9), while 12-0-tetradecanoylphorbol-13-acetate produces monocyte/macrophage-like cells (23, 40, 41) and differentiation inducing factors produce myelomonocytic cells (35).

supplemented medium. Cells in serum-free medium retain the distinct myeloid characteristics of HL-60, and display spontaneous and induced morphological and functional granulocytic differentiation similar to that found in serum-supplemented medium (4).

Induction of Terminal Differentiation with Retinoic Acid

A remarkable feature of the cultured HL-60 cells is that they are capable of terminally differentiating to functionally mature granulocytes, despite their aggressive malignant growth potential. Addition of appropriate concentrations of retinoic acid (all-*trans*-β-retinoic acid; vitamin A acid) to the cell culture induces, in the majority of HL-60 cells, a striking change characteristic of terminal differentiation of myeloid cells (Fig. 1) (5).

Induction of differentiation was carried out as described by Breitman et al. (5), using 0.01 and 1.0 μM retinoic acid (Sigma). Functional differentiation was determined 4 days after treatment with retinoic acid using nitro blue tetrazolium (NBT) dye reduction as a differentiation marker (Table 1) (10, 31).

Assessment of Functional Differentiation

The ability of mature HL-60 cells to reduce NBT dye is a physiological marker for myeloid differentiation (10, 31). A burst of respiratory activity induced in mature cells either by phagocytosis or by a phorbol ester, results in the formation of superoxide anion (O_2^-), which in turn reduces the water-soluble yellow NBT dye to insoluble blue-black formazan (10, 31). Mature myeloid cells, but not immature myeloblasts and promyelocytes, reduce NBT when stimulated by 12-O-tetradecanoyl-phorbol-13-acetate (4β-phorbol-12-myristate-13-acetate). A close correlation has been found between the percentage of cells induced to differentiate morphologically beyond promyelocytes and the percentage of cells capable of reducing NBT (8). NBT reduction is a sensitive as well as an easily quantitated marker which eliminates observer subjectivity associated with morphological assessment alone.

A total of 2 x 10^6 cells was washed twice with serum-containing RPMI-1640 medium and resuspended in 2 ml of serum-containing RPMI-1640 medium supplemented with 0.1 % freshly diluted NBT (Sigma) and 100 ng/ml 12-O-tetradecanoyl-phorbol-13-acetate (Sigma). This cell suspension was incubated for 25 min at 37 °C and cytospin slides were prepared by centrifuging 0.5 ml aliquots of the cell suspension at 500 r.p.m. for 5 min in a Shandon-Elliott SCA-0030 Cytospin centrifuge. Cells were stained with Wright-Giemsa. For each experimental point a total of 300 cells were examined using light microscopy and the percentage of cells containing intracellular blue-black formazan deposits, the result of NBT dye reduction, was calculated.

TABLE 1. Induction of differentiation of human promyelocytic leukemia cells (HL-60) with retinoic acid

Treatment[a]	Cell density (cells/ml)		Growth inhibition (% of control)	Functional differentiation (%)[b]
	Day 0	Day 4		
-	2.0×10^5	7.8×10^5	0	6
0.01 μM RA	2.0×10^5	6.2×10^5	28	18
1.0 μM RA	2.0×10^5	3.0×10^5	83	98

[a]Retinoic acid (RA) was added at the time of seeding (Day 0).
[b]Percentage of NBT-positive cells at Day 4.

Functional Changes Induced by Retinoic Acid

The higher dose of retinoic acid was more effective inducer of HL-60 cell differentiation than the lower dose (Table 1). Thus, after 4 days of incubation in the presence of 1 μM retinoic acid, almost all the cells were NBT-positive (Table 1). This induction of myeloid maturation was associated with inhibition of cell growth. Such retinoic acid-treated cells no longer proliferate even when resuspended in retinoic acid-free growth medium (5). This ability of HL-60 cells to undergo maturation indicates that despite karyotypic abnormalities (aneuploidy), the genetic information coding for terminal differentiation in these cells is present and capable of being phenotypically expressed.

Effects of Polyamine Synthesis Inhibition on Cell Proliferation and Differentiation

DFMO treatment resulted in suppression of HL-60 cell proliferation, but did not change the degree of functional differentiation (Table 2). That is, in DFMO-treated cultures, where the cells have ceased to proliferate, the percentage of mature cells is about the same as in untreated control cultures. These facts demonstrate that cessation of proliferation is not necessarily associated with differentiation.

When DFMO was added to HL-60 cultures in combination with retinoic acid a pronounced antiproliferative effect was observed (Table 2). Growth inhibition was accompanied by functional differentiation (Table 2), to an extent comparable to that seen in cultures treated with retinoic acid alone (Table 1). Despite the fact that DFMO inhibits retinoic acid-induced ODC activity (25) it does not prevent retinoic acid from inducing terminal differentiation (Table 2).

The inhibitory effect of DFMO on HL-60 cell proliferation,

TABLE 2. Effects of DFMO on retinoic acid-induced differentiation of human promyelocytic leukemia cells (HL-60)

Treatment[a]	Cell density (cells/ml) Day 0	Cell density (cells/ml) Day 4	Growth inhibition (% of control)	Functional differentiation (%)[b]
-	2.0×10^5	7.8×10^5	0	6
5 mM DFMO	2.0×10^5	2.8×10^5	86	10
5 mM DFMO + 0.01 μM RA	2.0×10^5	1.7×10^5	100	26
5 mM DFMO + 1.0 μM RA	2.0×10^5	2.7×10^5	88	90

[a]DFMO and retinoic acid (RA) were added at the time of seeding (Day 0).

[b]Percentage of NBT-positive cells at Day 4.

TABLE 3. Prevention of polyamine depletion in DFMO-treated human promyelocytic leukemia cells (HL-60) by putrescine. Effects on retinoic acid-induced differentiation

Treatment[a]	Cell density (cells/ml) Day 0	Cell density (cells/ml) Day 4	Growth inhibition (% of control)	Functional differentiation (%)[b]
-	2.0×10^5	7.8×10^5	0	6
10 μM Pu	2.0×10^5	7.5×10^5	5	7
10 μM Pu + 5 mM DFMO + 0.01 μM RA	2.0×10^5	5.6×10^5	38	25
10 μM Pu + 5 mM DFMO + 1.0 μM RA	2.0×10^5	3.7×10^5	71	96

[a]Putrescine (Pu), DFMO and retinoic acid (RA) were added at the time of seeding (Day 0).

[b]Percentage of NBT-positive cells at Day 4.

observed in retinoic acid-treated cultures (cf. Table 1 and 2), was prevented by the concomitant addition of putrescine (Table 3). However, the inhibitory effect on proliferation that retinoic acid exerted by itself (Table 1), was not prevented by putrescine addition (Table 3). These findings demonstrate that the antiproliferative effects of DFMO are due to polyamine limitation.

Putrescine alone did not affect proliferation or the state of differentiation of the control cells (Table 3). In combination with DFMO, putrescine did not affect the extent of retinoic acid-induced HL-60 cell differentiation (cf. Tables 1 to 3).

These results are consistent with recent data from other laboratories (20, 25), and suggest that polyamine synthesis is required for HL-60 cell proliferation but not for differentiation. Not only retinoic acid-induced differentiation into granulocytes, but also the phorbol ester-induced differentiation of HL-60 cells into monocytes proceed normally during polyamine synthesis inhibition (20, 25).

MOUSE EMBRYONAL CARCINOMA CELL DIFFERENTIATION

Teratocarcinomas as a Model System for Studying Embryogenesis and Neoplasia

The majority of cell lines that have been isolated retain their initial characteristics during subsequent growth, but some are capable of terminally differentiating. As a rule, only one terminal cell type is formed. Teratocarcinoma-derived cell lines (27, 28, 33), however, constitute exceptions to this rule. Thus, teratocarcinoma stem cells, so-called embryonal carcinoma (EC) cells, are generally pluripotent. Like the cells of an early embryo, EC cells are capable of differentiating into a wide variety of cell types. Therefore, teratocarcinoma cells in culture offer a system for studying early biochemical events involved in the induction and regulation of embryonic differentiation and development.

Although EC cells usually have the potential of differentiating into cell types representing all three germ layers, they are initially highly malignant. As they differentiate, however, they lose their malignancy. In fact, following injection into early mouse embryos (blastocysts) EC cells from *in vitro* cell lines (11, 36, 37) and from *in vivo* transplanted tumors (6, 21, 30) contribute to the formation of normal tissues.

Cells and Culture Conditions

Two clonal lines (F9 and PCC3) of mouse teratocarcinoma-derived EC cells were kindly provided by Dr. Nils R. Ringertz (Karolinska Institutet, Stockholm, Sweden). They were originally obtained from the Pasteur Institute (Paris, France). F9 is a

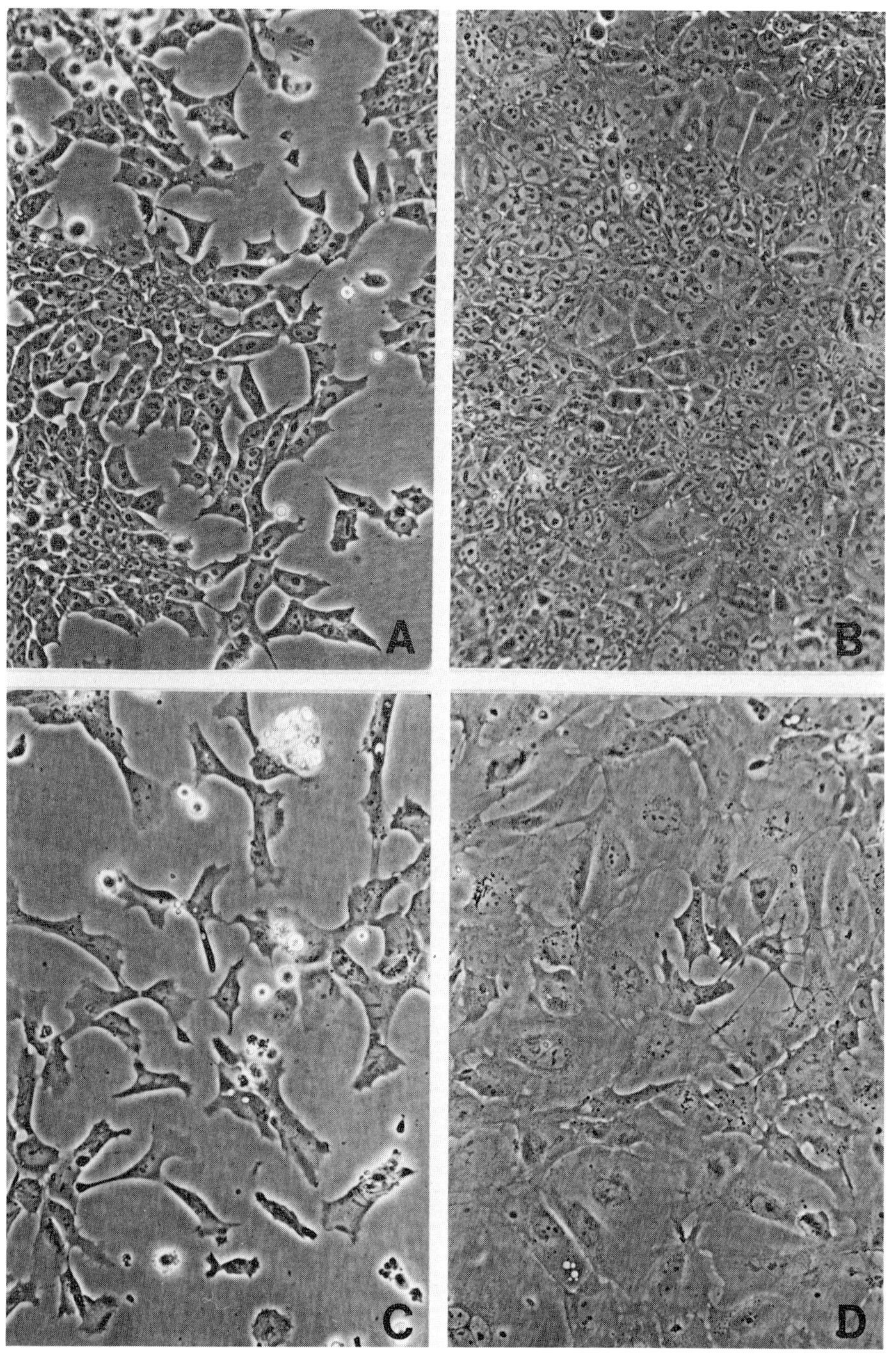

FIG. 2A-D.

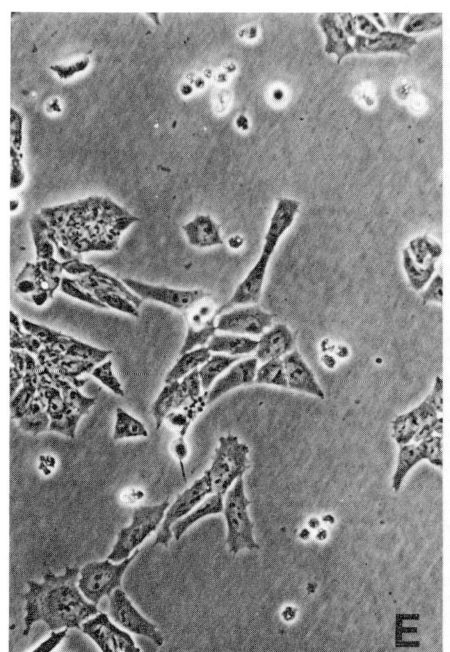

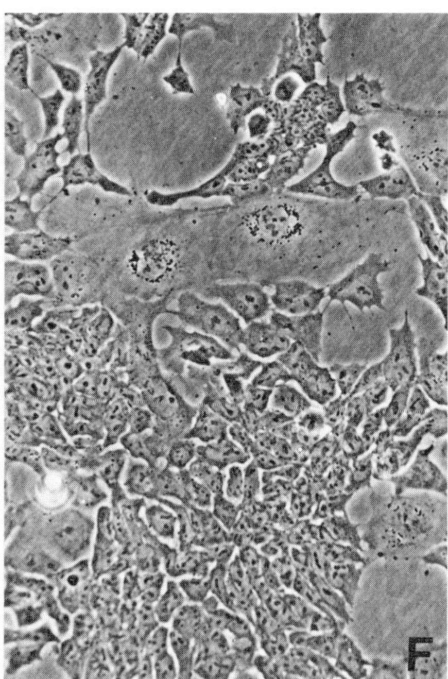

FIG. 2. Morphology of F9 embryonal carcinoma cells treated with retinoic acid (1 µM) or DFMO (5 mM). Both retinoic acid and DFMO were added at the time of seeding. (A-B) Untreated control cells at 48 hr (A) and at 96 hr (B) after seeding. (C-D) Retinoic acid-treated cells at 48 hr (C) and at 96 hr (D) after seeding. (E-F) DFMO-treated cells at 48 hr (E) and at 96 hr (F) after seeding. Magnification, 165X.

"nullipotent" and PCC3 is a pluripotent cell line. The F9 cell line (2) as well as the PCC3 cell line (22) were isolated from embryoid bodies of the transplantable tumor OTT 6050, a pluripotent teratoma, which originated from the grafting of a 6-day male mouse embryo to the testis of a strain 129 mouse (45).

In order to retain the "undifferentiated", malignant state characteristic for EC cells, the F9 and PCC3 cells are frequently subcultured - in the exponential phase of growth they undergo only very limited differentiation. When cultured under conditions favoring formation of aggregates (growth in bacteriological culture dishes), EC cells are capable of differentiating; F9 aggregates give rise to a single differentiated cell type with morphological and histochemical properties similar to those of parietal endoderm cells (43, 44), whereas PCC3 aggregates give rise to derivatives of all three germ layers (32). Differentiation can be induced in these cell lines even in monolayer cultures by exposing the cells to retinoic acid (19, 46).

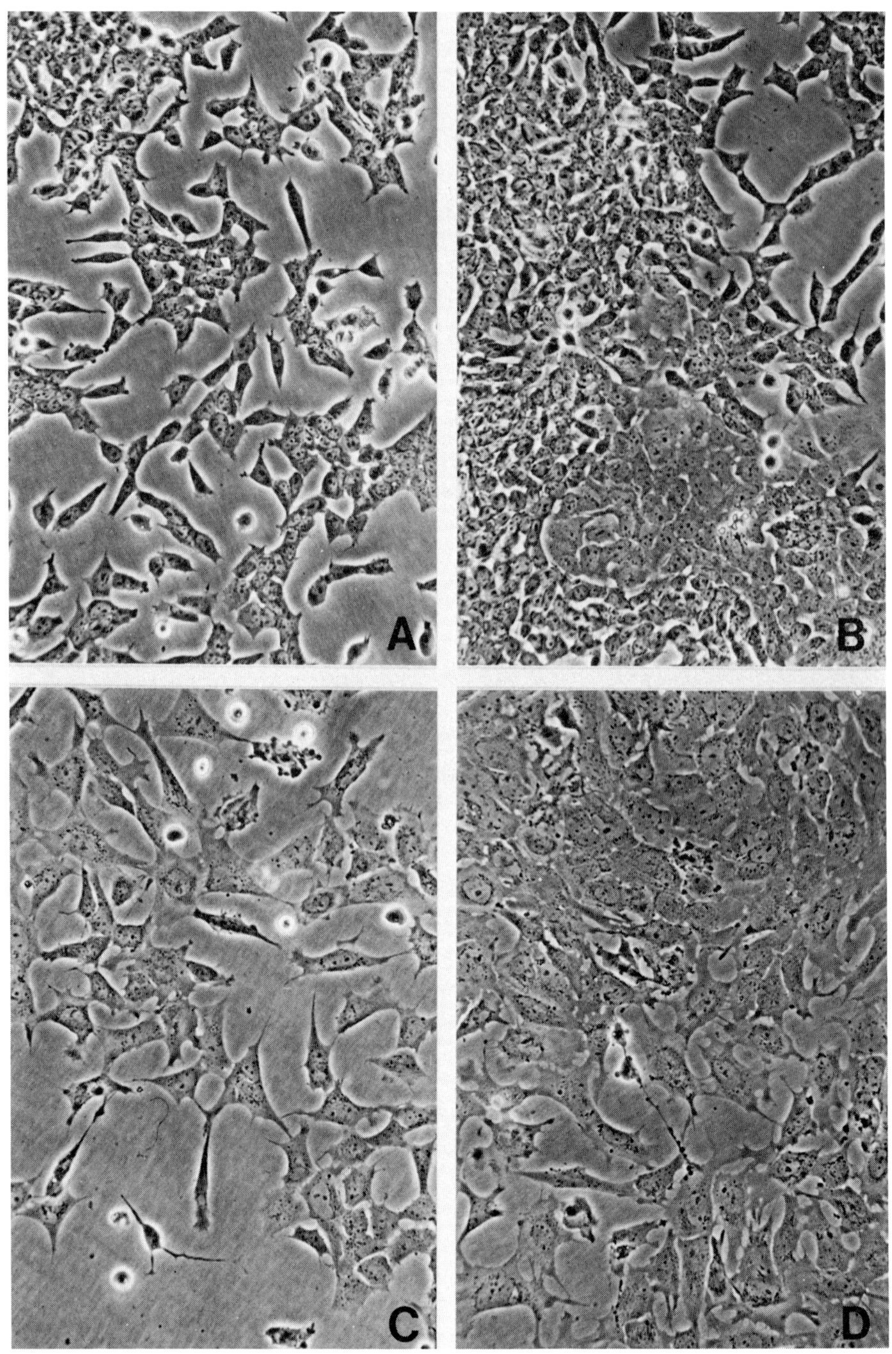

FIG. 3A-D.

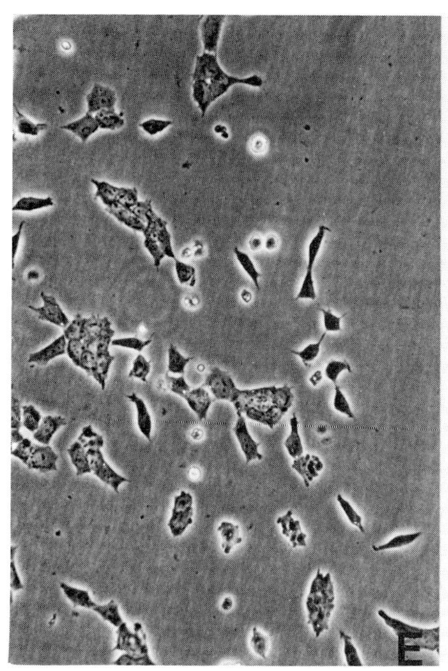

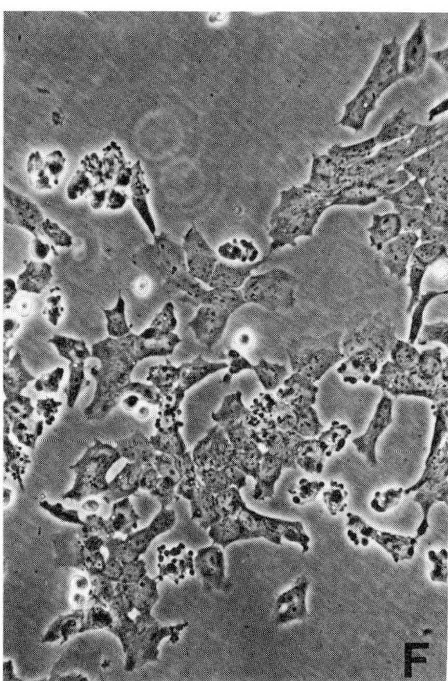

FIG. 3. Morphology of PCC3 embryonal carcinoma cells treated with retinoic acid (1 μM) or DFMO (5 mM). Retinoic acid was added at the time of seeding, while DFMO was added 24 hr after seeding. (A-B) Untreated control cells at 72 hr (A) and at 120 hr (B) after seeding. (C-D) Retinoic acid-treated cells at 72 hr (C) and at 120 hr (D) after seeding. (E-F) DFMO-treated cells at 72 hr (E) and at 120 hr (F) after seeding. Magnification, 165X.

For maintenance passage, F9 and PCC3 monolayers were thoroughly dissociated with trypsin, and the cells plated in gelatinized 50 mm plastic petri dishes at a density of 2 x 10^5 cells per 5 ml of culture medium. The cells were subcultured every second day. All incubations were carried out at 37 °C in a humidified atmosphere of 5 % CO_2 in air.

Induction of F9 Cell Differentiation

F9 cells were plated at a density of 5 x 10^4 cells per 50 mm dish in the presence of 1 μM retinoic acid or 5 mM DFMO. Cells were treated with retinoic acid for 2 days, i.e. retinoic acid-free growth medium was used for subculture. For DFMO treatment the fresh medium contained DFMO at each subculture.

Under normal culture conditions, the F9 cell line gives rise primarily to a single cell type, characteristic of embryonal carcinoma. F9 cells grow in culture as tightly packed colonies of EC cells. The cell population appears predominantly homoge-

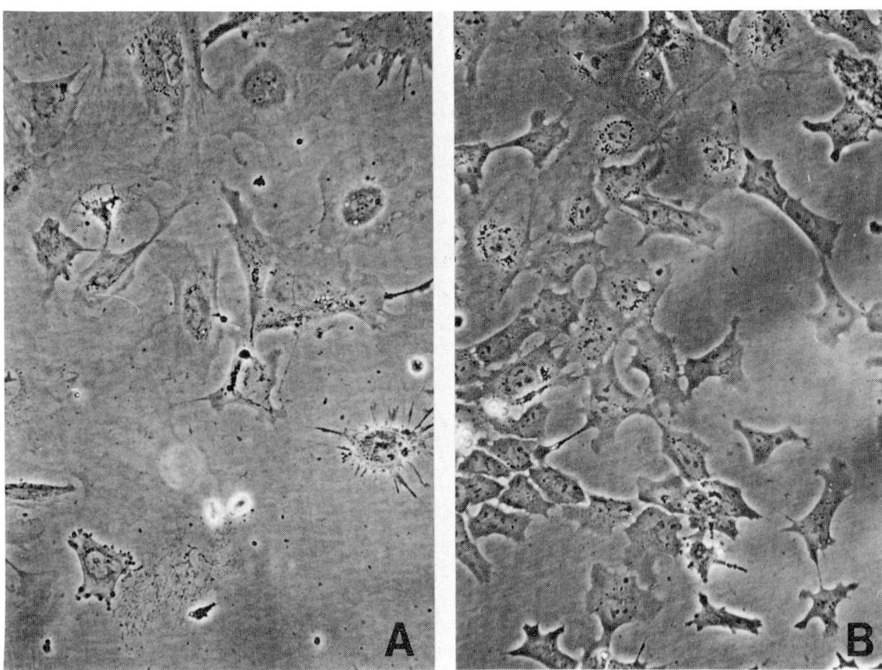

FIG. 4. Comparison between morphologically differentiated F9 embryonal carcinoma cells induced by retinoic acid (1 μM) or DFMO (5 mM). Both retinoic acid and DFMO were added at the time of seeding. (A) Retinoic acid-treated cells at 168 hr after seeding. (B) DFMO-treated cells at 168 hr after seeding. Magnification, 165X.

nous initially, but patches of cells with different morphology develop during the course of growth (Fig. 2A, B). Apparently, a certain degree of differentiation occurs spontaneously (44).

Retinoic acid, at concentrations as low as 1 nM, has been shown to induce multiple phenotypic changes in F9 cultures (46). These changes include morphological alteration, elevated levels of plasminogen activator production, sensitivity to dibutyryl cAMP and increased synthesis of collagen-like proteins (46). The nature of these changes indicates that retinoic acid induces differentiation of F9 cells into endoderm (19, 46).

In the present study the addition of 1 μM retinoic acid to the F9 cultures, reduced the growth rate and caused a pronounced morphological change (Fig. 2C, D). Within 48 hr (Fig. 2C), the cells move apart from one another and the colonies become less compact. The cellular morphology alters gradually to triangular, flat cells with cytoplasmic granules (Fig. 2C). The differentiating cells greatly increase in size (Fig. 2D) so that the nucleus of some of these giant cells is larger than an entire EC stem cell (cf. Fig. 2A and D).

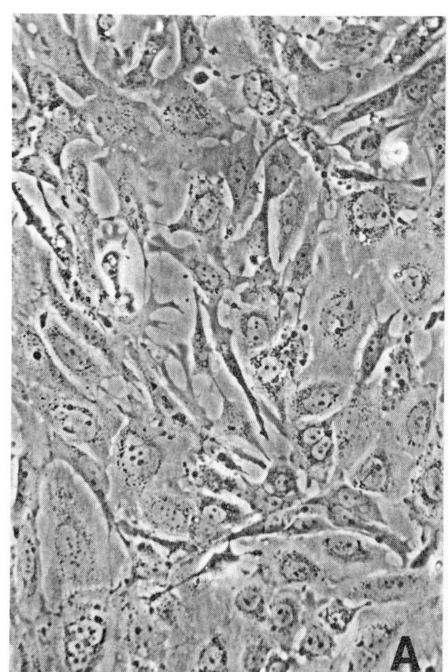

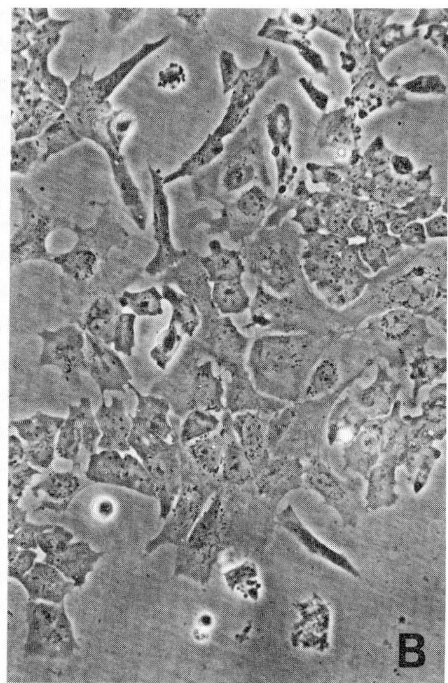

FIG. 5. Comparison between morphologically differentiated PCC3 embryonal carcinoma cells induced by retinoic acid (1 μM) or DFMO (5 mM). Retinoic acid was added at the time of seeding, while DFMO was added 24 hr after seeding. (A) Retinoic acid-treated cells at 168 hr after seeding. (B) DFMO-treated cells at 168 hr after seeding. Magnification, 165X.

The addition of 5 mM DFMO to the F9 cultures (Fig. 2E, F) caused an effect that was about the same as that caused by retinoic acid (Fig. 2C, D). However, DFMO was somewhat more inhibitory to cell proliferation and somewhat less stimulatory to differentiation. Further differentiation into flat, giant cells with a low nuclear:cytoplasmic ratio proceeded not only in retinoic acid-induced cultures, but also in DFMO-induced cultures (Fig. 4A, B), although at a somewhat lower rate in the latter. In the giant cells most of the organelles are sequestered around the perinuclear region as seen in Fig. 2D, 2F and 4B.

Induction of PCC3 Cell Differentiation

To test the generality of the response of EC cells to retinoic acid and DFMO, the PCC3 cell line was tested in a manner identical to that described for F9 cells, with one exception - DFMO was added 24 hr after seeding. If added at the time of seeding, the concentration of DFMO used (5 mM) almost completely prevented the cells from growing. Addition of DFMO at 24 hr, however, moderated the antiproliferative effect of the inhibitor.

The PCC3 cell line was also capable of spontaneously differentiating to some extent when grown in its normal medium - notably at higher cell densities (Fig. 3A, B). In agreement with the results obtained for the F9 cells, retinoic acid and DFMO induced differentiation in the PCC3 cell line as well (Fig. 3C-F and 5A, B). In this cell line, differentiation followed a more complex pattern, yielding cell types corresponding not only to derivatives of one germ layer (as in the case of F9 cells), but to all three (32).

Role of the Polyamines in Expression of the Differentiated Phenotype of Embryonal Carcinoma Cells

Our experiments with DFMO strongly suggest that the polyamines play an important role in EC cell proliferation and differentiation. To ascertain that the effects observed during DFMO treatment are due to polyamine limitation, attempts were made to prevent the effects exerted by DFMO by adding putrescine to the cultures, either at a 10 μM or at a 100 μM final concentration. The 10 μM putrescine concentration prevented only to some extent the effects of DFMO, but the 100 μM concentration completely eradicated the effects of DFMO on both cell proliferation and differentiation.

The present data indicate that inhibition of ODC activity and resulting polyamine deficiency facilitates the expression of differentiated cell functions both in a "nullipotent" (F9) and in a pluripotent (PCC3) EC cell line. The fact that retinoids are capable of inhibiting ODC induction (15) may indicate that retinoic acid, at least partly, exerts its effects on cell proliferation and differentiation by interfering with polyamine synthesis.

ACKNOWLEDGEMENTS

This investigation was supported by grants from the Swedish Natural Science Research Council and the Swedish Cancer Society. DL-α-Difluoromethylornithine was generously donated by Centre de Recherche Merrell International, Strasbourg, France.

REFERENCES

1. Alexandre, H. (1979): J. Embryol. Exp. Morph., 53:145-162.
2. Bernstine, E.G., Hooper, M.L., Grandchamp, S., and Ephrussi, B. (1973): Proc. Natl. Acad. Sci. USA, 70:3899-3903.
3. Brachet, J., Mamont, P., Boloukhère, M., Baltus, E., and Hanocq-Quertier, J. (1978): C.R. Acad. Sci. Ser. D, 287: 1289-1292.

4. Breitman, T.R., Collins, S.J., and Keene, B.R. (1980): Exp. Cell Res., 126:494-498.
5. Breitman, T.R., Selonick, S.E., and Collins, S.J. (1980): Proc. Natl. Acad. Sci. USA, 77:2936-2940.
6. Brinster, R.L. (1974): J. Exp. Med., 140:1049-1056.
7. Collins, S.J., Gallo, R.C., and Gallagher, R.E. (1977): Nature, 270:347-349.
8. Collins, S.J., Bodner, A., Ting, R., and Gallo, R.C. (1980): Int. J. Cancer, 25:213-218.
9. Collins, S.J., Ruscetti, F.W., Gallagher, R.E., and Gallo, R.C. (1978): Proc. Natl. Acad. Sci. USA, 75:2458-2462.
10. Collins, S.J., Ruscetti, F.W., Gallagher, R.E., and Gallo, R.C. (1979): J. Exp. Med., 149:969-974.
11. Dewey, M.J., Martin, Jr., D.W., Martin, G.R., and Mintz, B. (1977): Proc. Natl. Acad. Sci. USA, 74:5564-5568.
12. Emanuelsson, H., and Heby, O. (1978): Proc. Natl. Acad. Sci. USA, 75:1039-1042.
13. Gallagher, R., Collins, S., Trujillo, J., McCredie, K., Ahearn, M., Tsai, S., Metzgar, R., Aulakh, G., Ting, R., Ruscetti, F., and Gallo, R. (1979): Blood, 54:713-733.
14. Gazitt, Y., and Friend, C. (1980): Cancer Res., 40: 1727-1732.
15. Haddox, M.K., Frasier Scott, K.F., and Russell, D.H. (1979): Cancer Res., 39:4930-4938.
16. Heby, O. (1981): Differentiation, 19:1-20.
17. Heby, O., and Andersson, G. (1980): In: Polyamines in Biomedical Research, edited by J.M. Gaugas, pp. 17-34. John Wiley & Sons, New York.
18. Heby, O., and Emanuelsson, H. (1981): Med. Biol., 59: 417-422.
19. Hogan, B.L.M., Taylor, A., and Adamson, E. (1981): Nature, 291:235-237.
20. Huberman, E., Weeks, C., Herrmann, A., Callaham, M., and Slaga, T. (1981): Proc. Natl. Acad. Sci. USA, 78:1062-1066.
21. Illmensee, K., and Mintz, B. (1976): Proc. Natl. Acad. Sci. USA, 73:549-553.
22. Jakob, H., Boon, T., Gaillard, J., Nicolas, J.-F., and Jacob, F. (1973): Ann. Microbiol. (Inst. Pasteur), 124B: 269-282.
23. Lotem, J., and Sachs, L. (1979): Proc. Natl. Acad. Sci. USA, 76:5158-5162.
24. Löwkvist, B., Emanuelsson, H., and Heby, O. (1982): Cell Differentiation (in press)
25. Luk, G.D., Civin, C.I., Weissman, R.M., and Baylin, S.B. (1982): Science, 216:75-77.
26. MacDonnell, P.C., Nagaiah, K., Lakshmanan, J., and Guroff, G. (1977): Proc. Natl. Acad. Sci. USA, 74:4681-4684.
27. Martin, G.R. (1975): Cell, 5:229-243.
28. Martin, G.R. (1980): Science, 209:768-776.
29. Metcalf, B.W., Bey, P., Danzin, C., Jung, M.J., Casara, P., and Vevert, J.P. (1978): J. Am. Chem. Soc., 100:2551-2553.

30. Mintz, B., and Illmensee, K. (1975): Proc. Natl. Acad. Sci. USA, 72:3585-3589.
31. Newburger, P.E., Chovaniec, M.E., Greenberger, J.S., and Cohen, H.J. (1979): J. Cell Biol., 82:315-322.
32. Nicolas, J.F., Dubois, P., Jakob, H., Gaillard, J., and Jacob, F. (1975): Ann. Microbiol. (Inst. Pasteur), 126A: 3-22.
33. Nicolas, J.F., Avner, P., Gaillard, J., Guenet, J.L., Jakob, H., and Jacob, F. (1976): Cancer Res., 36:4224-4231.
34. Oka, T., and Perry, J.W. (1974): J. Biol. Chem., 249: 7647-7652.
35. Olsson, I., Olofsson, T., and Mauritzon, N. (1981): J. Natl. Cancer Inst., 67:1225-1230.
36. Papaioannou, V.E., McBurney, M.W., Gardner, R.L., Evans, M.J. (1975): Nature, 258:70-73.
37. Papaioannou, V.E., Gardner, R.L., McBurney, M.W., Babinet, C., and Evans, M.J. (1978): J. Embryol. Exp. Morph., 44: 93-104.
38. Pohjanpelto, P., Virtanen, I., and Hölttä, E. (1981): Nature, 293:475-477.
39. Pohjanpelto, P., Långström, E., Linden, M., Anehus, S., Knuutila, S., Hölttä, E., and Heby, O. (1982): Proc. Am. Assoc. Cancer Res., 23:33.
40. Rovera, G., O'Brien, T.G., and Diamond, L. (1979): Science, 204:868-870.
41. Rovera, G., Santoli, D., and Damsky, C. (1979): Proc. Natl. Acad. Sci. USA, 76:2779-2783.
42. Seyfried, C.E., and Morris, D.R. (1979): Cancer Res., 39: 4861-4867.
43. Sherman, M.I., and Miller, R.A. (1978): Dev. Biol., 63: 27-34.
44. Sherman, M.I., Strickland, S., and Reich, E. (1976): Cancer Res., 36:4208-4216.
45. Stevens, L.C. (1970): Dev. Biol., 21:364-382.
46. Strickland, S., and Mahdavi, V. (1978): Cell, 15:393-403.
47. Sunkara, P.S., Pargac, M.B., Nishioka, K., and Rao, P.N. (1979): J. Cell. Physiol., 98:451-457.
48. Sunkara, P.S., Rao, P.N., Nishioka, K., and Brinkley, B.R. (1979): Exp. Cell Res., 119:63-68.
49. Takano, T., Takigawa, M., and Suzuki, F. (1981): Med. Biol., 59:423-427.
50. Takigawa, M., Ishida, H., Takano, T., and Suzuki, F. (1980): Proc. Natl. Acad. Sci. USA, 77:1481-1485.

Advances in Polyamine Research, Vol. 4, edited by U. Bachrach, A. Kaye, and R. Chayen. Raven Press, New York © 1983.

The Role of Polyamines in the Differentiation of Mouse Neuroblastoma Cells

Kuang Yu Chen and *Alice Y.-C. Liu

*Department of Chemistry, Rutgers University, Wright Chemistry Laboratory, Piscataway, New Jersey 08854; *Department of Pharmacology, Harvard Medical School, Boston, Massachusetts 02115*

Mouse neuroblastoma cells in tissue culture was first established in 1940 (for a review, see ref. 38). Although these murine neuroblastoma cells possess some of the biochemical properties of neurons (e.g. low level of neurotransmitter synthesizing enzymes), they generally remain in a relatively immature state of differentiation (1). Nevertheless, in the presence of cAMP analogs or agents which increase cellular cAMP content, these cells can undergo differentiation, a process defined by the morphological appearance of neurites and the biochemical expression of neurotransmitter metabolizing enzymes (e.g., catechol-O-methyl transferase, acetylcholinesterase, choline acetyltransferase, tyrosine hydroxylase, etc.). Tumorigenicity of neuroblastoma cells has been shown to be abolished after differentiation (37). The ability of cAMP to produce large changes in morphology as well as enzymatic and electrical activities in mouse neuroblastoma cells renders these cells an excellent system for studying the action of cAMP in the regulation of neuronal differentiation. An in depth analysis of the action of cAMP in neuroblastoma cells differentiation may also provide insights into the mechanism of cAMP-induced differentiation of other cell types such as melanoma (34), PC-12 pheochromocytoma (6) and embryonic carcinoma (42).

In studying the cAMP-induced differentiation of mouse neuroblastoma cells, it appears likely that regulation of polyamine metabolism may be a potential target of cAMP action based on the following considerations: (a). The effect of cAMP in the regulation of ornithine decarboxylase (ODC, EC. 4.1.1.17), the rate-controlling enzyme for the biosynthesis of polyamines, has been well documented (for reviews see 18,24,39). (b). Tissue polyamines (putrescine, spermidine and spermine) generally increase under conditions of rapid growth and are considered to be functionally important since their synthesis is sequentially regulated during hypertrophy and hyperplasia (39). (c). The induction of ODC is one of the early events in growth stimulation (2,8,44). Although there are many studies which suggest a causal relationship of elevated ODC activity/polyamine contents and proliferative/neoplastic growth, there have been until recently relatively few studies on the possible involvement of polyamines and ODC in cell differentiation. In this chapter, we wish to summarize our studies on the role of polyamines in neuroblastoma cell differentiation.

DIFFERENTIATION OF MOUSE NEUROBLASTOMA CELLS

Mouse neuroblastoma cells (NB-15, N-18 or N2a clone) were grown as monolayer cultures in Dulbecco's modified Eagle medium (with 4,500 mg glucose per liter) supplemented with 10% fetal calf serum (FCS). To induce the differentiation of neuroblastoma cells, 1mM dibutyryl cAMP (Bt_2cAMP) was added to sparse cultures of neuroblastoma cells at 15 hours after subculture (seeding density $2x10^4$ cells/cm^2). In some experiments 0.5mM 3-isobutyl-1-methylxanthine (IBMX) was added together with Bt_2cAMP. In both cases, greater than 90% of the cells were differentiated.

Cells grown in normal growth medium (i.e. Dulbecco's medium puls 10% FCS) will be designated as NB cells, denoting control neuroblastoma culture, whereas cells grown in the "differentiation" medium (i.e. normal growth medium plus 1mM Bt_2cAMP with or without 0.5mM IBMX) will be designated as ND cells, denoting differentiating neuroblastoma culture.

When seeded at $2x10^4$ cells/cm^2, NB cells reached confluency (saturation density $3\text{~}4x10^5$ cells/cm^2) at 90-100 hours after seeding. In contrast, ND cells reached stationary phase of growth ~70 hours after the addition of Bt_2cAMP (saturation density $0.8\text{~}1.0x10^5$ cells/cm^2). Although neurite outgrowth of the ND cells was detectable 2~6 hours after the addition of Bt_2cAMP, other differentiation phenotypes, such as increases of acetylcholinesterase activity and cAMP-binding activity, were not fully expressed until the ND cells have reached a stationary phase of growth (28). In this article, the term "differentiated neuroblastoma cells" is used to refer to ND cells at the stationary phase of growth.

CHANGES OF ODC ACTIVITY AND POLYAMINES DURING NEUROBLASTOMA CELL DIFFERENTIATION

As an initial effort to understand the role of polyamines in neuroblastoma differentiation we carried out time-course studies of the changes of ODC activity and polyamine contents in both the NB and ND cells during a 5 to 6 days culture period (17). ODC activity was determined according to procedure described previously (8). One unit of ODC activity is defined as 1 nmole CO_2 evolved per hour. Fig. 1 shows that the differentiation was accompanied by an attenuation, both in magnitude and in duration, of cellular ODC activity. At t=70-80 hours the ODC activity of

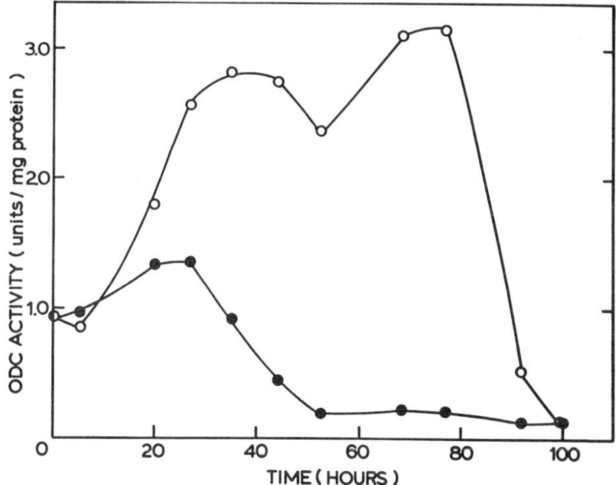

FIG. 1., Changes of ODC activity in NB(-o-) and ND(-•-) cells as a function of time in tissue culture. Time zero refers to the time when Bt_2cAMP (1mM) and IBMX (0.5mM) were added to the ND culture 15 hours after plating. (Data from ref. 17)

ND cells was less than 5% of that of the NB cells. It should be noted that our finding of an early attenuation of ODC activity after the addition of Bt_2cAMP, either with or without IBMX, to sparse cultures of mouse neuroblastoma cells is in apparent contradiction to previous observations that cAMP induces ODC activity in confluent neuroblastoma cultures (3, 4). This apparent dichotomy of cAMP action appears to be related to cell growth and cell density. Thus, in rapidly dividing sparse culture (cell density ~$2x10^4$ cells/cm^2), the addition of Bt_2cAMP results in, in chronological order, neurite outgrowth, attenuation of ODC activity, slowing of growth, and maximal expression of various differentiated phenotypes (17). In contrast, the addition of Bt_2cAMP to confluent culture of neuroblastoma cells (cell density ~$6x10^5$ cells/cm^2) causes a greater than 100-fold stimulation of ODC activity (13) but was ineffective in inducing differentiation of these cells.

For the quantitation of polyamines, we adopted the procedure of Seiler and Wiechman (40) to prepare dansyl derivatives of polyamines which were then quantitatively separated by reverse-phase liquid chromatography (17). Fig. 2 is a representative chromatogram. Under the experimental condition used, complete separation of putrescine, spermidine and spermine was achieved within 8 min. Using this quantitation method, the individual polyamine contents of NB and ND cells were determined at various time points over a 5-day culture period. Fig. 3 illustrates the results obtained using the NB-15 cells. Similar data were also obtained with N-18 and N2a mouse neuroblastoma clonal cells (D. Nau and K.Y. Chen, unpublished). Among three polyamines, the difference in spermidine content between the NB and ND cells is most notable. The spermidine content of the NB cells increased

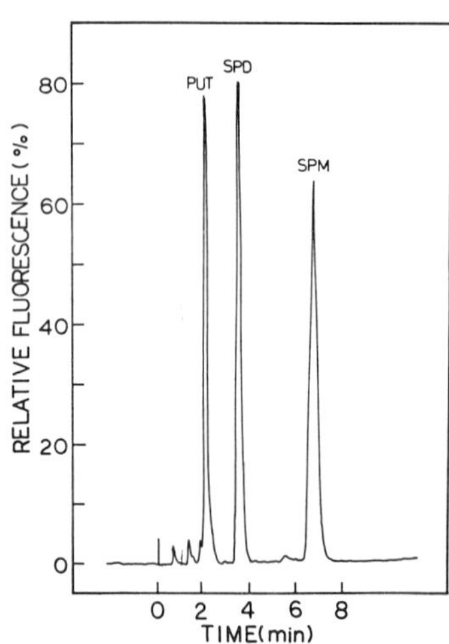

FIG. 2., Elution pattern of dansylated putrescine (PUT), spermidine (SPD), and spermine (SPM). Samples were applied to a reverse-phase column (RP-18, 7μm ODS) connected to a Beckman model 110A pump. Each peak corresponds to 50 pmoles of sample. Solvent system: CH_3CN: H O (85:15 v/v), flow rate 2.3 ml/min.

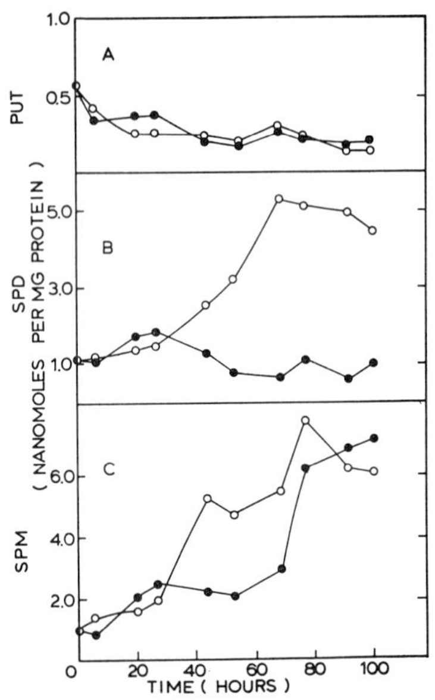

FIG. 3., Changes of polyamine contents in NB(-o-) and ND(-●-) cells as a function of time in tissue culture. Time zero refers to the time when Bt_2cAMP and IBMX were added to the ND culture 15 hours after plating. (Data from ref. 17)

steadily from t=0 reaching a maximal value of 5.3 nmoles/mg protein at t=~70 hours. The elevated level of spermidine in the NB cells was sustained for the remainder of the 5-day culture period. The pattern of change in spermidine content of the ND cells was distinct from that of the NB cells; except a small peak of spermidine (1.8 nmoles/mg protein) at t=~20 hours the spermidine level of ND cells hovered at the basal level (1.0 nmole/mg protein) throughtout the 5-day culture period.

In both NB and ND cells, the putrescine level decreased steadily as a function of time in culture. There was no significant difference in the putrescine content of the NB and ND cells Quantitation of the spermine content in the NB and ND cells. showed a progressive time-dependent increase in both the NB and ND cells. However, the increase in spermine content of the NB cells was more pronounced than that of the ND cells. The significance of the late increase of spermine content in the ND cells as illustrated in Fig. 3 is not clear. It should be noted that the magnitude of such increase varied from experiment to experiment (D. Nau and K.Y. Chen, unpublished).

PUTRESCINE TRANSPORT SYSTEM

Studies have shown that extracellular polyamines can be transported into cells and can regulate the activities of polyamine biosynthetic enzymes (7, 23). While the physiological significance of polyamine transport system is not clear, several studies have suggested that it may play some role in the regulation of growth. For example, Pohjanpelto (35) has found that putrescine transport is greatly increased in human fibroblasts which are stimulated to grow. Kano and Oka (25) have reported that the active transport system for polyamines in mouse mammary explants can be stimulated by insulin and prolactin.

Our studies with the mouse neuroblastoma cells have demonstrated a common transport system for spermidine, spermine and putrescine in mouse neuroblastoma cells (14). The time course of the rate of putrescine uptake in both NB and ND cells (NB-15 clone) over a 5-day culture period is shown in Fig. 4A. The results clearly indicate that the cAMP-induced neuroblastoma differentiation was associated with a significant decrease of

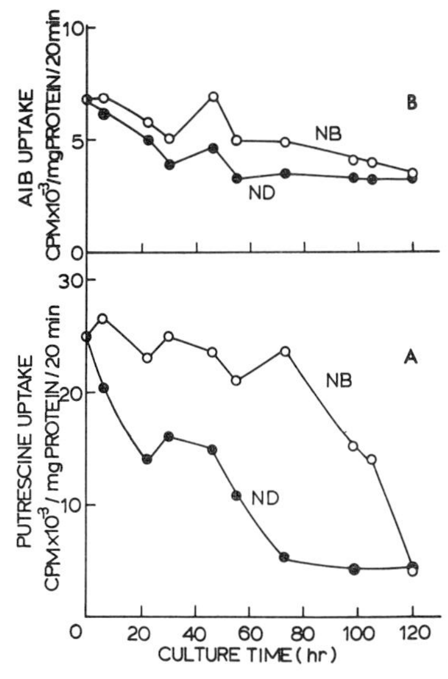

FIG. 4., Initial rate of transport of putrescine (A) and aminoisobutyrate (B) into NB and ND cells as a function of time in tissue culture. (Data from ref. 14)

putrescine transport. Six hours after the addition of Bt_2cAMP and IBMX, the putrescine transport rate of the ND cells was reduced by 27% when compared to that of the NB cells. At the same time, the transport of aminoisobutyric acid (AIB) in the ND cells was only slightly inhibited (Fig. 4B). Kinetic analysis indicated that the decrease in the rate of putrescine transport in the ND cells may be attributable to a change of apparent Km value of the putrescine transport system. We found the the Vmax values of the NB and ND cells at day-4 were identical (5.3 mnoles/mg protein/hr) but the Km values were different; the Km's were 2.75 μM for the NB and 28.5 μM for the ND cells. Differentiation-associated changes of plasma membrane structure of mouse neuroblastoma cells have been previously reported (16, 27, 42), it appears that the change of Km value of putrescine transport system in the ND cells may represent one of earlier detectable membrane changes associated with the differentiation of mouse neuroblastoma cells.

THE EFFECTS OF ASPARAGINE ON ODC INDUCTION IN UNDIFFERENTIATED AND DIFFERENTIATED MOUSE NEUROBLASTOMA CELLS

It has been previously suggested that plasma membrane/cytoskeletal structure may be involved in the regulation of ODC activity in mammalian cells (12, 21, 32). Work from many laboratories has also indicated changes of plasma membrane/cytoskeletal structure upon differentiation of the mouse neuroblastoma cells (9, 16, 27, 42); our work on the putrescine transport also supports differences in membrane structure of the undifferentiated and differentiated cells. In view of these considerations, we studied and compared the induction of ODC of the undifferentiated and differentiated mouse neuroblastoma cells.

Our previous studies have shown that FCS, Bt_2cAMP, and asparagine can elicit maximal increases of ODC activity of confluent neuroblastoma cells maintained in fresh Dulbecco's medium. We have also shown that when neuroblastoma cells are maintained in a salts/glucose medium (e.g. Earle's balanced slats solution), only asparagine can elicit a maximal increase of ODC activity; serum factors, hormones or agents which increase intracellular cAMP content are completely ineffective (13). These results suggest a possible primary function of asparagine in the induction of ODC activity in cultured cells.

Using the salts/glucose solution as an induction medium, we have compared the effects of asparagine on the induction of ODC activity in the undifferentiated and differentiated mouse neuroblastoma cells (10). The results shown in Fig. 5 demonstrated that 10mM asparagine alone produced a marked increase in ODC activity of the undifferentiated neuroblastoma cells; an effect that was not dependent on or potentiated by the addition of FCS or Bt_2cAMP. This is to be contrasted with results obtained with the differentiated neuroblastoma cells. In the differentiated neuroblastoma cells, 10 mM asparagine only produced a slight increase in ODC activity, which was, however, potentiated by the simultaneous addition of either 10% FCS or 1 mM Bt_2cAMP. The molecular basis for the differences in regulation of ODC activity in the undifferentiated and differentiated neuroblastoma cells, and particularly the response to asparagine as an inducer of ODC activity; is not clear. It is possible that the sensitivity towards asparagine may be related to the dynamic state of the plasma membrane/cytoskeletal structure.

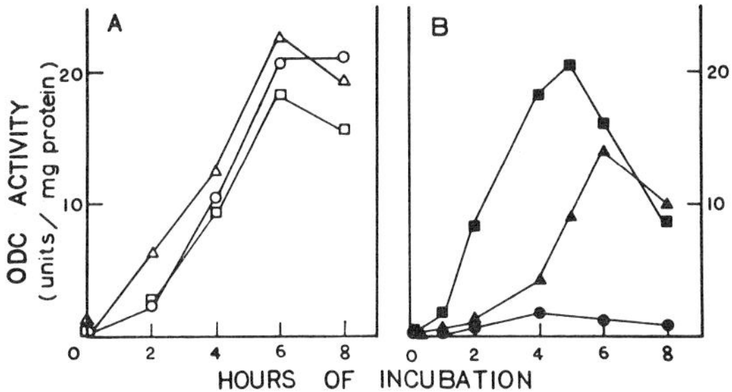

FIG. 5., (A) ODC induction in NB cells (stationary phase) by 10mM asparagine (o), 10mM asparagine plus 10% FCS (□) or 10mM asparagine plus 1mM Bt_2cAMP (Δ). (B) ODC induction in ND cells (stationary phase) by 10mM asparagine (●), 10mM asparagine plus 10% FCS (■) or 10mM asparagine plus 1mM Bt_2cAMP (▲). (Data from ref. 10)

METABOLIC LABELING OF AN 18,000 Dalton PROTEIN BY [^{14}C]PUTRESCINE AND THE CONVERSION OF PUTRESCINE TO GABA

In the previous sections of this chapter we have discussed differences in the intracellular concentration and transport of polyamines as well as the regulation of ODC activity of the NB and ND cells. To further define the precise function(s) of polyamines in growth regulation, it is desirable to identify and characterize the specific biochemical reactions involving polyamines. In this connection, we have investigated the incorporation of radioactive polyamines into cellular proteins. Two specific biochemical events associated with polyamines have thus been identified in mouse neuroblastoma cells, namely: (a) Specific labeling of an 18,000 dalton protein, and (b) Conversion of putrescine to amino acids via GABA (15). Preliminary studies of these two biochemical events in both undifferentiated and differentiated mouse neuroblastoma cells have been made. The result of a representative experiment is shown in the form of an autoradiogram in Fig. 6. The addition of [^{14}C]putrescine to cultures of neuroblastoma cells resulted in the labeling of an 18,000-dalton protein band and other cellular proteins. The labeling of the 18,000-dalton protein appears to be a specific metabolic incorporation reaction of putrescine and spermidine (11, 15). Folk and co-workers (20, 33) have reported a [^{3}H]putrescine-labeled 18,000 -dalton protein in human lymphocytes.

The labeling of various cellular proteins other than the 18,000-dalton protein may be attributable to general protein synthesis using radioactive amino acids derived from [^{14}C]putrescine through its conversion to GABA, the GABA shunt and Krebs cycle. This notion is supported by the findings that (a) aminoguanidine, a diamine oxidase inhibitor which inhibits the conversion of

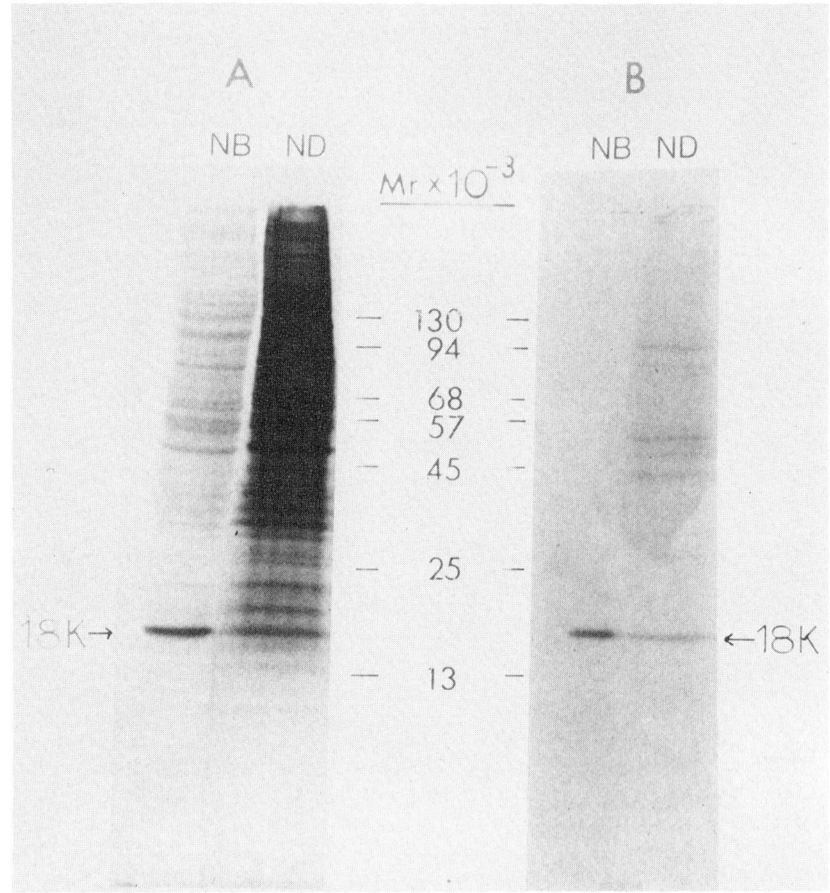

FIG. 6., Fluorogram of the radioactively labeled proteins in the undifferentiated and differentiated neuroblastoma cells exposed to 1μCi/ml [^{14}C]putrescine in fresh Dulbecco's medium with (A) or without (B) 10% FCS for 20 hours. Each lane contained 50μg protein. (Data from ref. 15)

putrescine to GABA, completely abolishes the labeling of all cellular proteins except the 18,000-dalton protein, and (b) with the exception of the 18,000-dalton protein, the labeling pattern of the undifferentiated or differentiated cells can be reproduced by using [^{14}C]GABA as precursor (15). The data shown in Fig. 6 indicate that the labeling of the 18,000-dalton protein is more prominent in the undifferentiated cells than in the differentiated cells whereas the conversion of putrescine to amino acids via GABA is more prominent in the differentiated cells than in the undifferentiated cells. The physiological significance of the changes of these two polyamine-associated biochemical events upon cell differentiation is currently under investigation.

THE EFFECT OF α-FLUOROMETHYLORNITHINE ON THE DIFFERENTIATION OF MOUSE NEUROBLASTOMA CELLS

While the data presented above may suggest that polyamines play some role in the differentiation of mouse neuroblastoma cells, they do not provide direct evidence of a cause-effect relationship. Recently, highly specific irreversible inhibitors of ODC such as D,L-α difluoromethylornithine (α-DFMO) and (R,S)-α-fluoromethylornithine (α-FMO) have been developed (26, 31). Studies have shown these compounds are potent and effective inhibitors of ODC in many cultured cell systems and animal tissues (19, 29, 30, 36). The availability of these inhibitors enabled us to examine directly the effect of inhibition of polyamine biosynthesis on the differentiation of mouse neuroblastoma cells. We have recently found that the presence of either α-DFMO or α-FMO at 0.5 ~ 1.5 mM can reduce the amount of Bt_2cAMP needed for maximal induction of neuroblastoma differentiation by 5 to 10 fold. Thus, 1 mM Bt_2cAMP was needed to induce maximal differentiation of the neuroblastoma cells in the absence of α-DFMO or α-FMO, while only 0.1 ~ 0.2 mM Bt_2cAMP was required to elicit a similar if not greater degree of differentiation of neuroblastoma cells in the presence of 0.5 ~ 1.5 mM α-DFMO or α-FMO. It should be noted that Bt_2cAMP at 0.1 ~ 0.2 mM neither affect the cell growth nor induce cell differentiation.

Fig. 7 shows the photomicrographs of N2a mouse neuroblastoma cells grown in the presence of 0.2 mM Bt_2cAMP alone (Fig. 7A) and cells grown in the presence of a combination of α-FMO and Bt_2cAMP at various concentrations (Fig. 7 B-D). It can be seen that cells grown in the presence of 0.5 - 1.5 mM α-FMO plus 0.1 - 0.2 mM Bt_2cAMP exhibited thick and long processes, sometimes greater than 600 μm in length, and forming extensive neurite webs. Exogenous putrescine or spermidine at 0.1 mM completely blocked the action of 1 mM α-FMO, either alone or with 0.1 - 0.2 mM Bt_2cAMP, on neuroblastoma cell differentiation (unpublished observation). These preliminary studies suggest that modulation of intracellular ODC activity, and thus polyamine levels may bear a causal relationship to neuroblastoma differentiation. However, the fact that in the presence of 0.5 ~ 1.5 mM α-FMO, a suboptimal concentration of Bt_2cAMP (0.1 ~ 0.2 mM) is needed to bring about maximal neuroblastoma differentiation, also indicates that changes of polyamine metabolism alone is not sufficient for neuroblastoma differentiation.

The mechanism through which suboptimal concentrations of Bt_2cAMP potentiate the effect of α-FMO on neuroblastoma cell differentiation is not clear. Recently Gunning et al. (22) have reported a synergistic effect of Bt_2cAMP and β-nerve growth factor on the neurite outgrowth of PC-12 pheochromocytoma cells and suggested that Bt_2cAMP is responsible for the initiation of neurite outgrowth. In light of their study, the potentiating role of Bt_2cAMP on the α-FMO-induced neuroblastoma differentiation may be related to microtubule assembly. The possible involvement

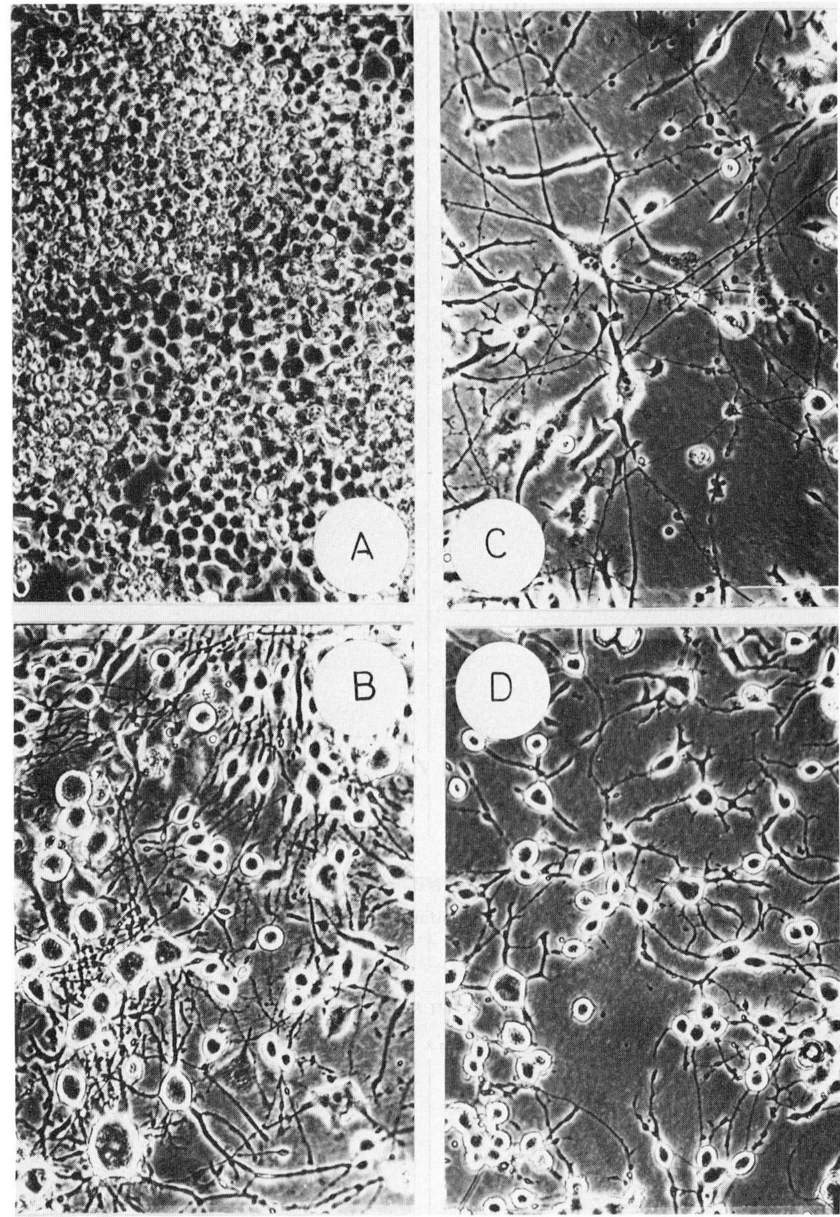

FIG. 7. N2a mouse neuroblastoma cells treated with (A) 0.2mM Bt_2cAMP, (B) 1 mM -FMO plus 0.2 mM Bt_2cAMP, (C) 1.5 mM -FMO plus 0.1 mM Bt_2cAMP, and (D) 0.5 mM -FMO plus 0.2 mM Bt_2cAMP. Phase contrast photomicrographs of cells were taken 8 days after initiation of treatment. (Chen, K.Y., unpublished data.)

of cAMP and protein phosphorylation in microtubule polymerization has been proposed (41). Alternatively, it is possible that low concentrations of Bt_2cAMP may modify certain biochemical event(s) which by itself has no effect on tumor growth or differentiation but can "prime" or prepare cells for maximal differentiation when cellular ODC activity and polyamine contents are reduced by α-FMO treatment.

CONCLUSION

The data presented in this chapter have demonstrated that the differentiation of mouse neuroblastoma cells is accompanied by significant changes of polyamine metabolism. These changes include (a) a decrease of cellular ODC activity and spermidine level (17), (b) a decrease in putrescine uptake (14), (c) a decrease in the asparagine-dependent ODC induction (10), (d) a decrease in the metabolic labeling of the 18,000-dalton protein by $[^{14}C]$putrescine (15), and (e) an increase of the conversion of putrescine to amino acids via GABA (15). The essential role of polyamines and ODC in the differentiation of mouse neuroblastoma cells is suggested by the findings that α-FMO, a suicidal enzyme inhibitor of ODC, plus suboptimal concentration of Bt_2cAMP (0.1 - 0.2 mM) can cause maximal differentiation of neuroblastoma cells and the differentiation can be blocked by exogenously added putrescine or spermidine.

Tumor differentiation is undoubtedly a complex phenomenon. Our data emphasize that the modulation of ODC/polyamines may be essential for the differentiation of mouse neuroblastoma cells. Further studies are needed to elucidate the precise mechanisms involved. Recently, Bethell and Pegg (5) showed that α-DFMO inhibits the differentiation of 3T3-L1 fibroblasts into adipose cells. Their findings suggest the involvement of ODC/polyamines in normal cell differentiation. In view of these considerations, it is plausible that modulation of cellular ODC activity and polyamine centents is of general significance in the differentiation of both normal and tumor cells.

ACKNOWLEDGMENT

The research reported here was supported in part by USPHS grants CA-24479 and PHSRR-7058, by the Rutgers Research Council and by a grant from American Cancer Society, Massachusetts Chapter.

REFERENCES

1. Amano, T., Richelson, E., and Nirenberg, M. (1972): Proc. Natl. Acad. Sci. USA., 69: 258-263.
2. Bachrach, U. (1973): Function of Naturally Occurring Polyamines. Academic Press, New York.

3. Bachrach, U. (1975): Proc. Natl. Acad. Sci. USA., 72: 3087 -3091.
4. Bachrach, U. (1976): FEBS Lett., 68: 63-37.
5. Bethell, D.R. and Pegg, A.E. (1981) Biochem. Biophys. Res. Commun., 102: 272-278.
6. Bothwell, M.A., Schechter, A.L., and Vaughn, K.M. (1980): Cell, 21: 857-866.
7. Canellakis, E.S., Heller, J.S., Kyriakidis, D., and Chen, K.Y. (1978): In: Advances in Polyamine Research, Vol. 1, edited by R.A. Campbell, et. al., pp. 17-31, Raven Press, New York.
8. Cannellakis, E.S., Viceps - Madore, D., Kyriakidis, D.A., and Heller, J.S. (1979): Current Topics in Cellular Regulation, 15: 155-202.
9. Chan, K.Y. and Baxter, C.F. (1979): Brain Res., 174: 135-152.
10. Chen, K.Y. (1980): FEBS Lett., 119: 307-311.
11. Chen, K.Y. (1981): The Third Pan American Biochemistry Congress, Mexico City, Abstracts p. 242.
12. Chen, K.Y., Heller, J.S., and Canellakis, E.S. (1976): Biochem. Biophys. Res. Commun., 68: 401-408.
13. Chen, K.Y. and Canellakis, E.S. (1977): Proc. Natl. Acad. Sci. USA., 74: 3791-3795.
14. Chen, K.Y. and Rinehart, C.A. Jr. (1981): Biochem. Biophys. Res. Commun., 101: 243-249.
15. Chen, K.Y. and Liu, A.Y-C. (1981): FEBS Lett., 134: 71-74.
16. Chen, K.Y. and Rinehart, C.A. Jr. (1982):Biochim. Biophys. Acta., 685: 61-70.
17. Chen, K.Y., Presepe, V., Parken, N. and Liu, A.Y-C. (1982) J. Cell. Physiol., 110: 285-290.
18. Cohen, S.S. (1971): Introduction to the Polyamines. Prentice Hall, New Jersey.
19. Danzin, C., Jung, J.J., Grove, J., and Bey, P. (1979): Life Sci., 24: 519-524.
20. Folk, J.E., Park, M.H., Chung, S.I., Schrode, J., Lester, E.P., and Cooper, H.L. (1980): J. Biol. Chem., 255: 3695-3700.
21. Gibbs, J.B., Hsu, C.Y., Terasaki, W.L., and Brooker, G. (1980): Proc. Natl. Acad. Sci. USA., 77: 995-999.
22. Gunning, P.W., Landreth, G.E., Bothwell, M.A., and Shotter, E.M. (1981): J. Cell Biol., 89: 240-245.
23. Heller, J.S., Chen, K.Y., Kyriakidis, D.A., Fong, W.F., and Canellakis, E.S. (1978): J. Cell. Physiol., 96: 225-234.
24. Jänne, J. Pösö, H. and Raina, A. (1977): Biochim. Biophys. Acta., 473: 241-293.
25. Kano, K. and Oka, T. (1976): J. Biol. Chem., 251: 2795-2800.
26. Kollonitsch, J., Patchett, A.A., Marburg, S., Maycock, A.L. Perkins, L.M., Doldouras, G.A., Duggan, D.E., and Aster, S.D. (1978): Nature, 274: 906-908.
27. Littauer, U.Z., Giovanni, M.Y., and Glick, M.C. (1980): J. Biol. Chem., 255: 5448-5453.

28. Liu, A.Y-C., Fiske, W. and Chen, K.Y. (1981): Cancer Res., 40:4100-4108
29. Luk, G.D., Marton, L.J., and Baylin, S.B. (1980): Science, 210: 195-198.
30. Mamont, P.S., Duchesne, M.C., Grove, J., and Bey, P. (1978) Biochem. Biophys. Res. Commun., 81: 58-66.
31. Metcalf, B.W., Bey, P., Danzin, C., Jung, J.J., Casara, P., and Vevert, J.P. (1978): J. Am. Chem. Soc., 100: 2551-2553.
32. O'Brien, T.G., Simsiman, R.C., and Boutwell, R.K. (1976): Cancer Res., 36: 3766-3770.
33. Park, M.H., Cooper, H.L., and Folk, J.E. (1981): Proc. Natl Acad. Sci. USA., 78: 2869-2873.
34. Pawelek, J.M. (1976): J. Invest. Dermatol., 66: 201-209.
35. Pohjanpelto, P. (1976) J. Cell Biol., 68: 512-520.
36. Prakash, J.J., Schechter, P.J., Mamont, P.S., Grove, J., Koch-Weser, J., and Szoerdsma, A. (1980): Life Sci., 26: 181-194.
37. Prasad, K.N. (1972): Cytobiol., 6: 103-116.
38. Prasad, K.N. (1980): Regulation of Differentiation in Mammalian Nerve Cells. Plenum Press, New York
39. Russell, D.H. and Durie, B.G.M. (1978): Polyamines as Biochemical Markers of Normal and Malignant Growth. Raven Press, New York.
40. Seiler, N. and Wiechman, M. (1970): In: Progress in Thin-Layer Chromatography and Related Methods, edited by A. Niederwieser and G. Pataki, Vol. 1, pp. 94-145. Humphrey Science Publishers, Michigan.
41. Sloboda, R.D., Rudolph, S.A., Rosenbaum, J.L., and Greengard, P. (1975): Proc. Natl. Acad. Sci. USA., 72: 177-181.
42. Strickland, S., Smith, K.K. and Marotti, K.R. (1980): Cell, 347-355.
43. Trudding, R., Shelanski, M.L., Daniel, M.P., and Morell, P. (1974): J. Biol. Chem., 249: 3973-3982.
44. Williams-Ashman, H.G. and Canellakis, Z.N. (1979): Prespect. Biol. Med., 22: 421-453.

Advances in Polyamine Research, Vol. 4, edited by U. Bachrach, A. Kaye; and R. Chayen. Raven Press, New York © 1983.

Turnover and Analysis of Polyamines and Their Acetyl Derivatives in Differentiating Friend Erythroleukemia Cells

Yair Gazitt and *Uriel Bachrach

*The Lautenberg Center for General and Tumor Immunology and the *Department of Molecular Biology, The Hebrew University-Hadassah Medical School, Jerusalem, Israel*

In previous publications we presented evidence that erythrodifferentiation induced by dimethylsulfoxide (Me_2SO) and hexamethylene bisacetamide (HMBA) is probably triggered by early stimulation of ornithine decarboxylase (ODC) (5,6). It was also shown that the most commonly studied inhibitors of induced differentiation such as dexamethasone (DEX). (19), 12-O-tetradecanoyl-phorbol-13-acetate (TPA) (17) and high doses of interferon (16), inhibit ODC stimulation induced by HMBA or Me_2SO with no apparent effect on cell growth. Inhibition of erythrodifferentiation by DEX and TPA could be prevented by the addition of polyamines such as putrescine, spermidine and spermine to the culture medium (5,6). Erythrodifferentiation was also inhibited by inhibitors of spermidine-biosynthesis such as methylglyoxal bis (guanylhydrazone) and by analogues of ornithine such as methylornithine and α- hydrazinoornithine. This inhibition was also prevented by addition of polyamines (5,6). These findings led to the conclusion that the polyamines might play an important role in triggering erythrodifferentiation of Friend erythroleukemia cells (FL).

In this paper evidence is presented that Me_2SO and HMBA induce an early and rapid turnover of spermidine and spermine which results in a net decrease in the pool of these polyamines. As a consequence, N-acetylated derivatives of putrescine and spermidine are formed in the presence of both inducers. These results give further support to our previous conclusion on the possible role of ODC and polyamines in triggering differentiation of FL cells.

MATERIALS AND METHODS

Cells: Friend erythroleukemia cells strain 745A (4) were employed throughout this study. Stationary phase cells were cultued at 1×10^5 cells/ml in Eagle's Minimal Essential Medium

(MEM, GIBCO) supplemented with 15% fetal calf serum (GIBCO). Cells were grown at 37°C in a humidified incubator in an atmosphere of 5% CO_2-95% air, and treated with the test compounds on the followinging day. Cells were counted daily and the degree of differentiation was measured on day 3 by the method of Orkin et al. (15).

Uptake of [C^{14}] putrescine: To cells in the log phase ($2x10^5$ cells/ml), [1,4-C^{14}] putrescine (Radiochemical Centre, Amersham, 100 mCi/mmol) was added to a final concentration of 0.25mμCi/ml $2x10^5$ cells. Aliqouts of $4x10^6$ cells were sampled at the time points indicated, washed three times with phosphate buffered-saline, and treated with 300μl of perchloric acid (3%V/V). The supernatant fluid obtained after centrifugation at 6000 rpm for 10 min was analyzed for polyamines, and the protein content of the precipitate was determined according to Lowry et al. (11).

Polyamine analysis: To 0.2 ml samples of the perchloric acid extracts of labeled or unlabeled polyamines, 20 mg of sodium carbonate and 0.4 ml of dansyl chloride solution (30 mg/ml in acetone) were added. After keeping in thedark for 15-20 hours, dansyl derivatives were extracted and analyzed as previously described by Seiler (20). They were separated on silica-gel plates (Kodak 13179, Eastman chromatogram sheets, 100μm thick) using ethyl acetate-cyclohexane (2:3 V/V) as a solvent (13). For separation of acetyl polyamines the plates were developed in benzene-methanol (19:1), or in ethylacetate (22). When labeled polyamines were separated, the plates were apposed to X-Ray films (Kodak,No screen) for 10-20 days. The spots of the various polyamines were identified by autoradiography and by authentic dansyl-derivatives. The spots corresponding to dansyl polyamines were scraped off the plate and radioactivity of the material was counted. When unlabeled polyamines were separated in the same way, the spots were scraped off and the dansylated polyamines extracted in ethyl acetate and the fluorescence intensity was measured (excit. 360nm; emission 500nm).

Chemicals: Dimethylsulfoxide (Me_2SO) was from Fisher Scientific Co., Springfield, N.J. Hexamethylene bisacetamide (HMBA) was prepared by acetylation of 1,6-diaminohexane. Dexamethasone, dansylchloride and polyamines were from Sigma Chemical Co., St. Louis, Mo.

RESULTS

Effect of inducers on the pool-size of polyamines and their interconversion

The effect of HMBA and Me_2SO on the level of free polyamines was studied. The results in Fig. 1 indicate an early and rapid decrease in the pool size of putrescine in the absence and presence of both inducers. However, in the presence of inducers a slower decrease in putrescine level was observed. Moreover, both inducers caused a rapid decrease

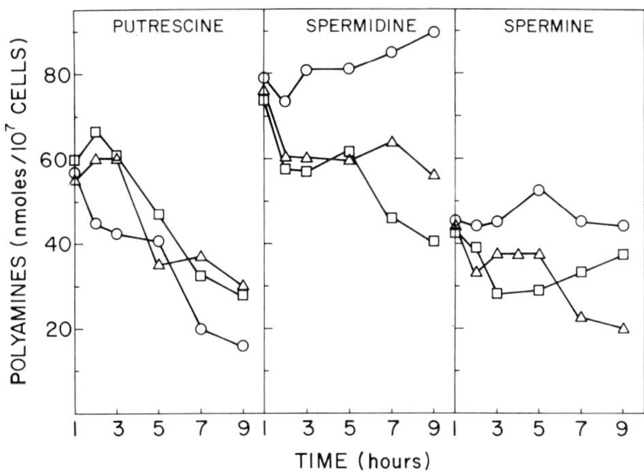

FIG. 1. Effect of inducers on polyamines levels. Samples of 4×10^6 cells were treated with perchloric acid to extract polyamines. Polyamines were dansylated and separated in the form of their dansyl derivatives by thin layer chromatography. The results are the average of two experiments done in duplicates.
No addition (-o-o-); 4mM HMBA (-□-□-); 280mM Me_2SO (-△-△-).

in the level of spermidine and spermine. These effects lead to a drastic change in the ratio of putrescine to both spermine and spermidine mainly after incubation of 9 hours (Fig. 1).

The effect of inducers on the metabolism of polyamines was next followed by incubating the cells with $[C^{14}]$ putrescine. In Fig. 2a, the effect of HMBA and Me_2SO on the metabolism of tracer putrescine is demonstrated. The results indicated a rapid metabolism of putrescine to spermidine and spermine. By 9 hours practically all cellular label is converted from putrescine to spermidine and spermine. In the presence of inducers, however, the rate of conversion is slowed down by 50% or more. This is illustrated in Fig. 2b. The results imply that the total pool of free polyamines might be affected in the same way. Indeed, when the total pool of free polyamines was measured, a drop in the level of spermidine and spermine was observed in the presence of HMBA and Me_2SO. This is reflected as an increase in the specific radioactivity of spermidine and spermine as shown in Fig. 3. The changes observed in the metabolism of

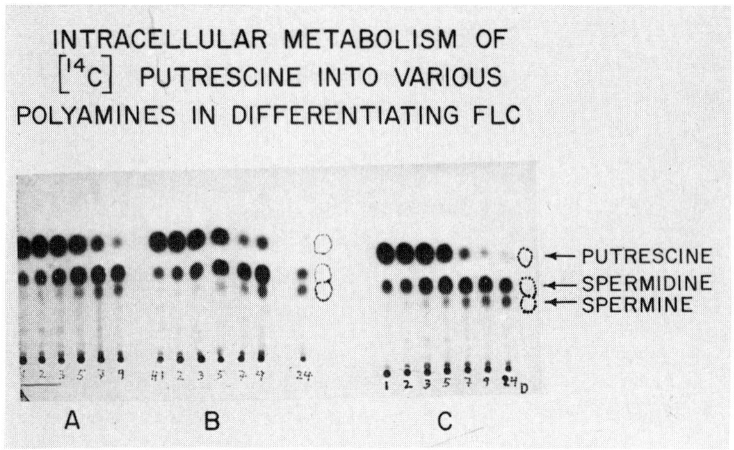

FIG. 2a. Effect of inducers on the rate of [C^{14}] putrescine metabolism in differentiating FL cells. Log phase cells were cultured with test compounds and [C^{14}] putrescine was added as described under Materials and Methods. Polyamines were extracted, dansylated and separated by thin layer chromatography. Autoradiography was performed as described under Materials and Methods. (Exposure of the X-Ray film was for 10 days).
Control cells (A); 4mM HMBA (B); Me_2SO (C).

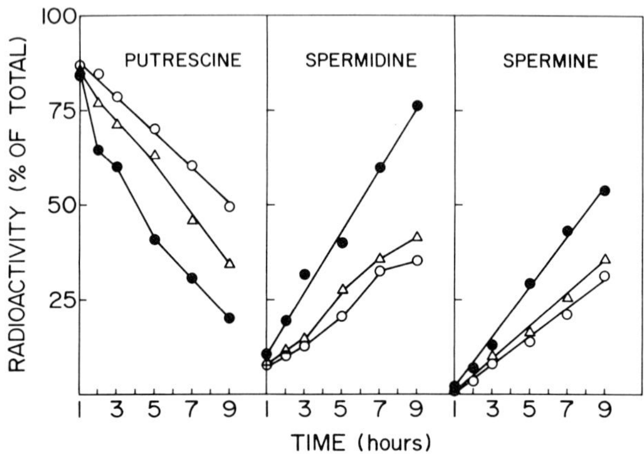

FIG. 2b. Quantitation of the results shown in Fig. 2a. Spots were sampled and radioactivity measured as described under Materials and Methods. The results are presented as radioactivity of each of the polyamines relative to the combined radioactivity found in the spots corresponding to putrescine+ spermidine+spermine.
No addition (●-●); 4mM HMBA (Δ-Δ); 280mM Me_2SO (o-o).

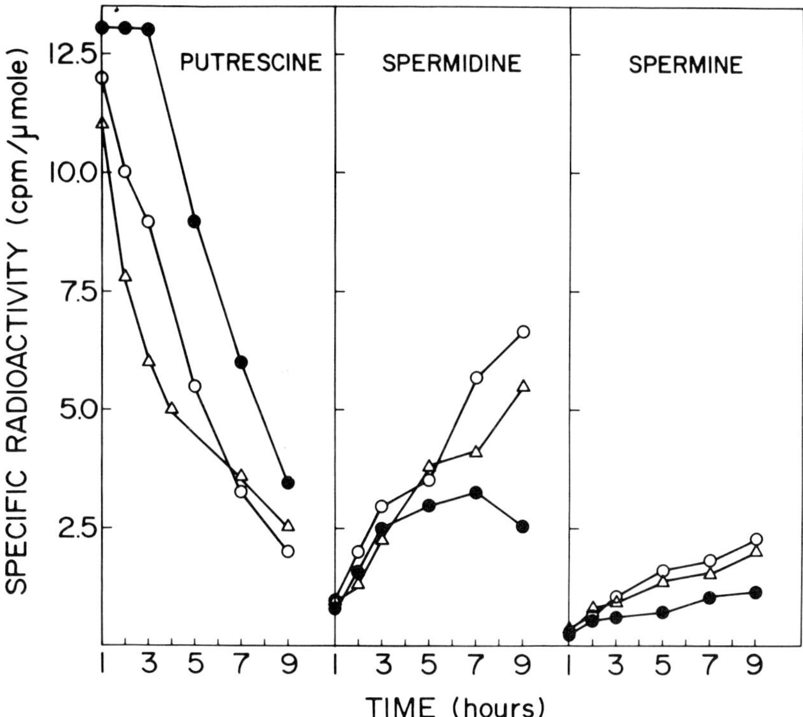

FIG. 3. Effect of HMBA and Me_2SO on the specific radioactivity of free polyamines in differentiating FL cells. Log phase cells were cultured with or without inducers. Aliquots containing $4x10^6$ cells were sampled every hour for the first 9 hours and polyamines were extracted and analyzed as described under Materials and Methods. Data were calculated from the results illustrated in Figs. 1 and 2a. Control cells with no addition (●-●); 4mM HMBA (Δ-Δ); 280mM Me_2SO (o-o). The experiment was done twice in duplicate samples.

labeled putrescine in the presence of inducers is not due to differences in uptake of [C^{14}] putrescine which is not affected by the inducers (Fig. 4). The growth medium contained only putrescine. Its concentration declined steadily and by 9 hours reached less than 5% of the amount found after the first hour of incubation (results not shown).

Conversion of [C^{14}] putrescine to acetylated polyamines

Since acetyl derivatives of polyamines are believed to play a role in regulating many cellular functions, we examined the conversion of radioactive putrescine to acetylated derivatives. The results shown in Fig. 5a indicate the

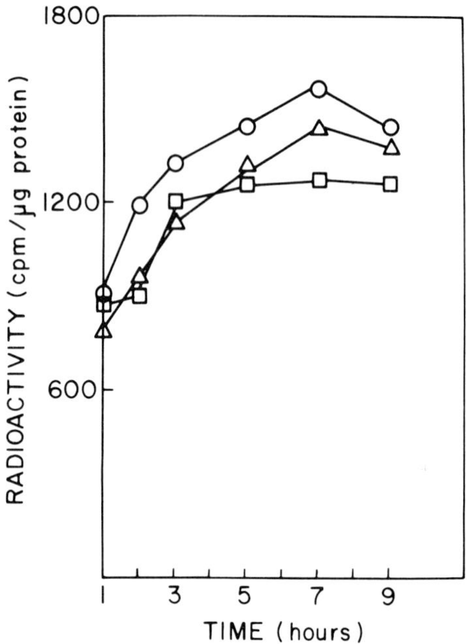

FIG. 4. Uptake of [C^{14}] putrescine by FL cells induced to differentiate by Me_2SO and HMBA. Samples were taken at various times, solubilized with 1% (V/V) SDS and aliqouts were taken for measurement of radioactivity and for protein determination. 4mM HMBA (Δ-Δ); 280mM Me_2SO (□-□); No inducer (o-o).

appearance of N-acetylputrescine and N^1-acetylspermidine in the cells in the presence of inducers. Quantitative analysis shows a 3-4 fold increase in the level of N-acetyl putrescine in the presence of HMBA but no increase was observed in the presenc of Me_2SO (Fig. 5b). On the other hand, a 2-3 fold increase in N^1-acetylspermidine in the presence of both inducers was observed between 5-9 hours after their addition. In contrast, no significant change in the cellular level of N^8-acetylspermidine was observed in the presence of inducers (Fig. 5b).

Acetyl derivatives of polyamines were also detected in the growth medium. Thus, N-acetylputrescine (0.1-0.2 nmole/ml) were detected in the medium during the first 7 hours of induction in the presence of HMBA and, to a smaller extent, in the presence of Me_2SO (Fig. 6).

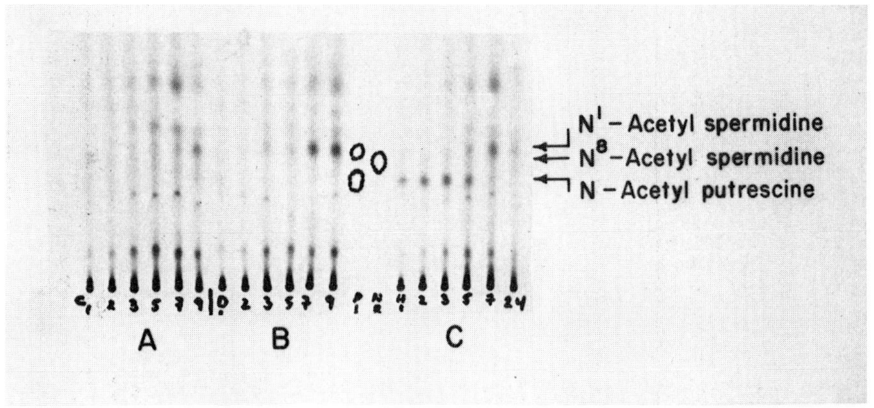

FIG. 5a. Intracellular conversion of putrescine into N-acetylated polyamines in differentiating FL cells. Cell extracts were prepared as described in Fig. 1. Plates were developed in a solvent suitable for separation of acetyl compounds, as specified under Materials and Methods. No addition (A); 280mM Me_2SO (B); 4mM HMBA (C). Plates were apposed to X-Ray films for 20 days.

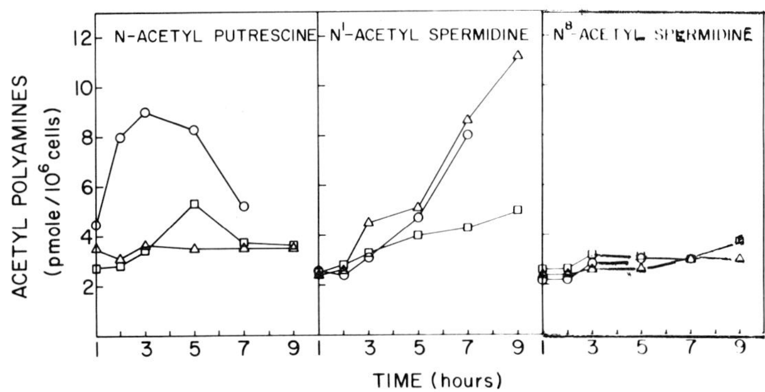

Fig. 5b. Quantitation of the results from Fig. 5a. The methods for quantitation were those detailed in the legend to Fig. 2b.
No addition (□-□); 4mM HMBA (o-o); 280mM Me_2SO (-Δ-Δ-).

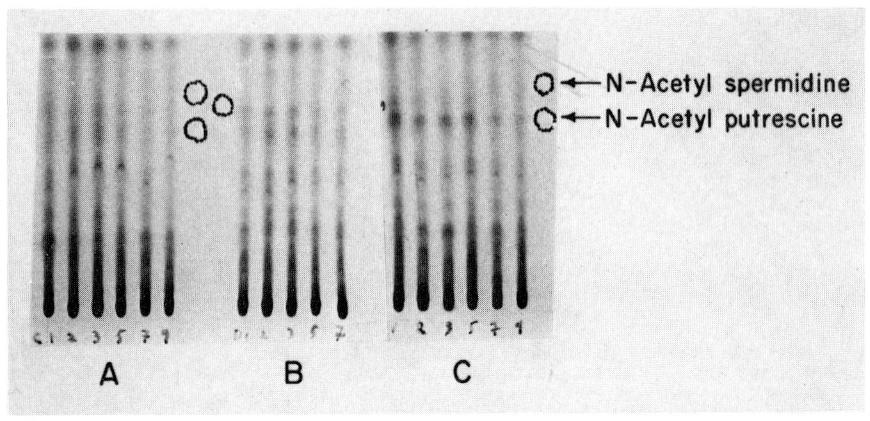

FIG. 6. Accumulation of radioactive N-acetylated polyamines in the extracellular medium during differentiation of FL cells. Cells were cultured with [C^{14}] putrescine as described in the legend to Fig. 2. The extracellular medium was analyzed for the presence of metabolites of labeled putrescine. Ten milliliter of culture medium was freed of cells and the resulting supernatant was freeze-dried.

Perchloric acid (1 ml) was then added to extract polyamines. Dansylation and separation of dansylated derivatives were done as for cellular polyamines. TLC plates were apposed to X-Ray films for 20 days.
No addition (A); 280 mM Me_2SO (B); 4 mM HMBA (C).

Since N^1 and N^8 acetyl spermidine spots were very close, the sum of the two compounds is given as "N-acetyl spermidine". N^1-acetylspermidine was also detected in the medium derived from cells treated with Me_2SO and, to a lesser extent, in the medium of cells treated with HMBA. Thus, between 50-70% of the labeled putrescine added is found in the medium in the form of N-acetylputrescine. The amount of N-acetylspermidine in the medium is much smaller and accounts for less than 5% of [C^{14}] putrescine added (quantitation not shown).

Dexamethasone (2µM) prevented the effect of inducers on all parameters measured. In its presence,the levels of N-acetyl derivatives did not increase and the levels of spermidine and spermine were not bellow those found in control cells without inducers (results are not shown).

DISCUSSION

In previous publications we showed that compounds such as Me_2SO and HMBA induce erythrodifferentiation of FL cells via stimulation of ODC (5,6). Stimulation of ODC was observed 3 hours after onset of induction and thus coincided with the increase in cellular cAMP content (7). All other reported effects of these inducers on FL cells, namely, the effect on transport of small molecules (8,12) and the effect on cell cycle (9) occurred much later (10-24 hours). Evidence is presented in this paper that inducers affect both the pool size of polyamines (notably that of spermidine and spermine), the rate of putrescine conversion into spermidine, and spermidine conversion into spermine.

It is evident from Fig. 1 that the level of putrescine decreased during the first hours of incubation both in the presence and in the absence of inducers. However, in the presence of inducers the rate of degradation of putrescine is somewhat slower than in the control cells without inducers added. On the other hand, the inducers stimulated the rate of degradation of spermidine and spermine.

The net effect of inducers is to change the ratio of putrescine to spermidine and spermine. The ratio of putrescine to spermidine and spermine might be important with regard to the mechanism by which growth and differentiation is regulated. In this case a higher ratio of spermidine and spermine to putrescine might be a signal for growth while the reverse situation might be a signal for differentiation. In agreement with this hypothesis is the finding of Russell et al. (18) showing a 3-4 fold increase in the ratio of spermidine to putrescine in regenerating rat liver as compared to resting normal liver.

Similar results were obtained when the metabolism of $[C^{14}]$ putrescine was followed. A significant reduction in the conversion of $[C^{14}]$ putrescine into $[C^{14}]$ spermidine or $[C^{14}]$ spermine is noticed after induction (Figs. 2-4). As polyamine uptake is not affected by the inducers, and as inducer enhance the disappearence of spermidine from FL cells (Fig.1) the rate of polyamine secretion out of the cells may increase during the differentiation process. Studies from our laboratory demonstrated that acetyl putrescine and N^1-acetyl spermidine are formed in cultured lymphocytes after stimulation with lectins. A significant proportion of cellular polyamines is excreted from the stimulated lymphocytes as N^1-acetylspermidine (1). Acetyl putrescine and N^1-acetylspermidine are also excreted from chick embryo fibroblasts following transformation by Rous sarcoma virus (2).

Here we present evidence that acetyl putrescine and N^1-acetylspermidine are formed in FL cells induced by either HMBA or Me_2SO. A substantial amount of these compounds is secreted into the medium.

The appearance of these derivatives in the medium is the result of degradation of labeled spermidine and spermine. This shows that one effect of inducers on polyamine metabolism is to stimulate degradation of spermidine and of spermine. Acetylation of polyamines may be involved in regulating cellular functions (3,10). Thus, polyamine acetylation may affect the conformation of DNA (14) by removal of polyamines from binding sites. This, in turn, may lead to changes in the overall pattern of gene expression. Such mechanisms of regulation of cell growth and differentiation have been proposed (21).

ACKNOWLEDGEMENT

We would like to acknowledge the excellent technical assistance of Ms. H. Wiener. This work supported by the Lautenberg Center Endowment and the Concern Foundation, Inc. of Los Angeles.

REFERENCES

1. Bachrach, U., Menashe, M., Faber, J., Desser, H., and Seiler, N. (1981): In: Advances in Polyamine Research, edited by C.M Caldarera, V. Zappia, and U. Bachrach, Vol. 3, pp. 259-274. Raven Press, New York.

2. Bachrach, U., and Seiler, N. (1981): Cancer Res., 41: 1205-1208.

3. Blankenship, J., and Walle, T. (1978): In: Advances in Polyamine Research, edited by R.A. Campbell, D.R. Morris, D. Bartos G.D. Daves and F.Bartos, Vol. 2, pp. 97-120, Raven Press, New York.

4. Friend, C., Scher, W., Holland, J.G., and Sato,T. (1971): Proc. Natl. Acad. Sci. USA, 68:378-382.

5. Gazitt Y., and Friend, C. (1980): Cancer Res., 40:1727-1737.

6. Gazitt, Y., and Friend, C. (1980): In: In vivo and in vitro Erythroposesis: The Friend System, edited by G.B. Rossi, pp. 209-217. Elsevier, North-Holland Biomedical Press.

7. Gazitt, Y., Reuben, R.C., Deitch, A.D., Marks, P.A., and Rifkind, R.A. (1978): Cancer Res., 38:3779-3783.

8. Gazitt, Y. (1979): J. Cell Physiol., 99:407-416.

9. Gazitt, Y., Deitch, A.D., Marks, P.A., and Rifkind, R.A. (1978): Expt. Cell Res., 117:413-420.

10. Libby, P.R. (1978): J. Biol. Chem., 253:233-237.

11. Lowry, O.H., Rosebrough, N.J., Farr, A.L., and Randall, J.R. (1951): J. Biol. Chem., 193:266-275.

12. Mager, D., and Bernstein, A. (1978): J. Cell. Physiol., 94:275-286.

13. Menashe, M., Faber, J., and Bachrach, U. (1980): Biochem J., 188: 263-267.

14. Nordheim, A., Pardue, M.L., Lafer, E.M. Möller, A., Stollar, B.D., and Rich, A. (1981): Nature, 294:417-422.

15. Orkin, J.H., Harosi, F.I., and Leder, P. (1975): Proc. Natl. Acad. Sci. USA, 72:98-102.

16. Rossi, G.B., dolei, A., Cioe, L., Benedetto, A., Matarese, G.P., and Belardelli, F. (1977): Proc. Natl. Acad. Sci. USA, 74:2036-2040.

17. Rovera, G., O'Brien, T., and Diamond, L (1977): Proc. Natl. Acad. Sci. USA, 74:2894-2898.

18. Russell, D.H., Medina, D.J., and Snyder, S.H. (1970): J. Biol. Chem. 245:6728-6732.

19. Scher, W., Tsuei, D., Sassa, S., Price, P., Gabalman, N., and Friend, C. (1978): Proc. Natl. Acad. Sci. USA, 75:3851-3855.

20. Seiler, N., and Wiechmann, M. (1967): Hoppe Seylers' Physiol. Chem., 348:1285-1290.

21. Seiler, N. (1979): In: Polyamines in Biology and Medicine, edited by L.J. Marton, and D.R. Morris, pp. 169-182. Marcel Dekker Inc., New York.

22. Seiler, N., and Knodgen, B. (1979): J. Chromat., 164: 155-168.

Advances in Polyamine Research, Vol. 4, edited by U. Bachrach, A. Kaye, and R. Chayen. Raven Press, New York © 1983.

Diacetylputrescine Induces Differentiation and is Metabolized in Friend Erythroleukemia Cells

*Zoe Nakos-Canellakis and Philip K. Bondy

*West Haven Veterans Administration Medical Center, West Haven Connecticut and the Departments of Internal Medicine and *Pharmacology, Yale University School of Medicine, New Haven, Connecticut 06510*

ABSTRACT

Friend murine erythroleukemia cells (MELC) can be induced to differentiate and produce hemoglobin by exposure to N,N'-diacetylputrescine (tetramethylene*bis*acetamide; TMBA). We have synthesized ^{14}C-methylene labeled TMBA and studied its metabolism by MELC during induction of differentiation. We have also investigated the effect of dexamethasone in TMBA-induced differentiation.

INTRODUCTION

The murine erythroleukemia cell line (MELC) originally described by Friend (3) characteristically grows as a poorly differentiated proerythrocytic cell. Many reagents can induce it to differentiate, as reflected by termination of cell division, combined with initiation of hemoglobin synthesis (7). Few of the reagents producing this effect are likely to occur as natural metabolic products. An exception may be the N,N'-diacetylated derivative of putrescine, tetramethylene*bis*acetamide (TMBA). Because the two primary amino groups of this compound are blocked by acetyl groups, it cannot be measured by the usual methods of polyamine analysis which depend on the presence of a free primary amino group. This may explain the fact that, to date, no information has been published concerning the presence or metabolism of this substance in cells. In view of the suggestive evidence that a polyamine derivative is capable of inducing cell differentiation, and of the importance of understanding the mechanism(s) inducing undifferentiated cells to mature, it seemed justifiable to study the relationship between the metabolism of TMBA and induction of differentiation in MELC. In order to carry out this research it was first necessary to synthesize labelled TMBA and then to correlate its metabolic conversion to the polyamines and their metabolites by MELC.

It has been demonstrated that TMBA and its analogue, hexamethylenebisacetamide (HMBA) can induce differentiation in MELC (11) and that the differentiating effect of HMBA can be blocked by dexamethasone (5). It therefore seemed relevant to determine whether dexamethasone also blocks the differentiating effects of TMBA and to study the relationship between this inhibiting effect and the metabolism of TMBA.

METHODS

Cells of clone DS19 of MELC obtained from Dr. Scher in Dr. Charlotte Friend's laboratory were cultured in a 5% CO_2/air atmosphere in BME enriched with 15% fetal bovine serum in the presence of streptomycin and penicillin (12). Cells were diluted so that 16 to 18 hours later (which marked the onset of the experiment) they were in log growth phase and had reached a concentration of 5 x 10^5 cells/ml. Thereafter, they were studied during the first 24 hours of growth by harvesting at intervals as indicated for analysis. The cell culture was continued for two additional days, counted with a hemocytometer and then evaluated by benzidine staining for the presence of hemoglobin (8). Unstimulated cells doubled in approximately 24 hours and less than 1% were benzidine positive. Viability, which was estimated by trypan blue exclusion, always exceeded 99%. TMBA, synthesized as described later, was added in 5 mM concentration, since this was found to cause the optimal percent of benzidine positive conversion. The concentration of dexamethasone used for inhibition was 2 μM. Addition of TMBA or dexamethasone did not alter the growth rate during the first 24 hours.

At appropriate intervals, cells were harvested, washed exhaustively and fractionated into cytosol, membrane and nuclear fractions (2). These were precipitated with 10% TCA and the protein-free supernatant fractions were analysed by high performance liquid chromatography (HPLC) as previously described (1). Total polyamines were determined by orthophthalaldehyde fluorescence. Fractions which were collected simultaneously for liquid scintillation counting provided a measurement of the radioactive polyamines. The content of non-radioactive polyamines was then determined by difference calculation.

^{14}C-TMBA was prepared by reacting 1,4-^{14}C-putrescine with excess acetic anhydride. The final specific activity of the ^{14}C-TMBA was 1 µCi/µMol with a 30% yield. On HPLC analysis, less than 0.5% of the counts were present as putrescine; the rest eluted in a sharp peak midway between putrescine and the void volume. After hydrolysis of the ^{14}C-TMBA in a sealed tube in 8.3 M HCl at 110° for 16 hours, the radioactivity was identified quantitatively as putrescine. Nonradioactive TMBA was also prepared by the same procedure. TMBA which was recrystalized from hot water had a melting point of 135-136° (uncorr). It contained less than 0.7% putrescine on HPLC analysis, and was converted quantitatively to putrescine by acid hydrolysis.

RESULTS

MELC cells took up isotope from ^{14}C-TMBA present in the medium in a linear manner during the early hours of incubation. By nine hours the intracellular ^{14}C reached a plateau level of approximately 173,000 cpm/50 x 10^6 cells. The distribution of the ^{14}C derived from ^{14}C-TMBA in the cytosol is shown in Table 1.

TABLE 1. Cytosolic ^{14}C-TMBA Derivatives

	HPLC analysis; % total cpm in peak				Total cpm/
Hrs	Front	Put	Spd	Spm	50 x 10^6 cells
3	83	4.2	1.7	0.5	62,500
6	84	4.7	4.9	0.3	66,600
9	84	5.6	4.5	0.3	99,500
24	65	13.2	13.5	3.3	119,900

Mean of 3 separate experiments; 80% of the cells were benzidine (+) at 72 hrs.
TMBA: 5 mM (s.a. = 1,000 cpm/nmol)

The "front," which is present in all cell fractions, consisted of at least two components: a minor one, approximately 2% of the total counts, which ran close to the void volume and a large fraction, which had a mobility identical to TMBA. Acid hydrolysis converted the radioactivity of the large fraction quantitatively to ^{14}C-putrescine (Figure 1). The "front" appeared in each cell fraction and, after 24 hours of incubation, represented about 65% of the total cytosolic counts, 34% of the total counts in the membrane fraction and 13% of the total counts in the nuclei.

Routine analysis of the medium after incubation showed the presence of ^{14}C-TMBA (5 mM) and ^{14}C-monoacetylputrescine (30 µM). Since no other radioactive compounds were present, we felt justified in re-using the medium for incubation, fortifying it at each subsequent use by supplementation with 15% fetal bovine serum.

The distribution of the major polyamines in the cytosol, membrane fraction and nuclei of MELC after 16-18 hours growth in log phase is presented in Table 2.

TABLE 2. Intracellular Distribution of Polyamines

	nMol/50 x 10^6 cells			
FRACTION:	Cytosol	Membrane	Nuclei	Total
Monoacetylputrescine	0.0	0.7	0.0	0.7
Putrescine	1.2	2.2	0.7	4.1
Spermidine	8.8	20.4	12.4	41.6
Spermine	7.3	22.1	14.0	43.4

Putrescine (1.6 μM) was the only polyamine present in the medium. The results represent the means of determinations performed on two sets of 50 ml cell cultures.

The intracellular distribution of isotope from ^{14}C-TMBA in MELC induced to differentiate by TMBA was compared with that seen when differentiation was blocked by dexamethasone (Table 3).

TABLE 3. Effect of Dexamethasone on Uptake of ^{14}C-TMBA

	(% of total cellular counts)			cell-associated cpm as % of
FRACTION:	Cytosol	Membr.	Nuclei	cpm in medium
Experiment No. 1				
TMBA	19.3	67.3	13.4	0.062
TMBA + Dex.	22.9	61.7	15.4	0.062
Experiment No. 2				
TMBA	33.0	48.7	18.3	0.059
TMBA + Dex	30.7	49.8	19.5	0.059

Cells (40 ml cultures) were incubated for 24 hours with 5 mM ^{14}C-TMBA (1 μCi/μmol) +/- dexamethasone, 2 μM. The number of cells doubled in each case.

The patterns of intracellular polyamines in the various cell fractions after MELC cells had been exposed to ^{14}C-TMBA for 24 hours in the absence and presence of 2 μM dexamethasone are described in Tables 4 and 5 respectively.

TABLE 4. Intracellular Distribution of Polyamines in the Presence of TMBA#

FRACTION:	Cytosol		Membrane		Nuclei	
	nMol/50 x 10^6 cells					
	total	mol-frac ^{14}C*	total	mol-frac ^{14}C*	total	mol-frac ^{14}C*
Experiment No. 1:						
Monoacetyl-putrescine	0.0	--	2.0	0.05	0.0	--
Putrescine	2.3	0.23	7.5	0.37	1.3	0.29
Spermidine	3.6	0.25	15.8	0.22	8.4	0.16
Spermine	5.2	0.03	27.2	0.02	13.4	0.01
Experiment No. 2						
Monoacetyl-putrescine	0.0	--	1.5	0.04	0.0	--
Putrescine	2.3	0.30	6.2	0.32	2.0	0.31
Spermidine	6.8	0.18	16.1	0.21	11.1	0.17
Spermine	8.7	0.02	23.9	0.02	16.3	0.02

\# The medium contained ^{14}C-TMBA, 5 mM and ^{14}C-monoacetylputrescine, 30 μM. The total polyamine level was determined as described in the Methods section.

* In every case, "mol-frac ^{14}C" was calculated as the ratio of the moles of each ^{14}C polyamine to the total for the particular polyamine.

TABLE 5. Intracellular Distribution of Polyamines in the Presence of TMBA plus Dexamethasone#

FRACTION:	Cytosol		Membrane		Nuclei	
	nMol/50 x 10^6 cells					
	total	mol-frac ^{14}C*	total	mol-frac ^{14}C*	total	mol-frac ^{14}C*
Experiment No. 1:						
Monoacetyl-putrescine	0.0	--	1.3	0.14	0.0	--
Putrescine	5.1	0.19	18.0	0.28	6.7	0.17
Spermidine	4.6	0.09	15.2	0.13	11.4	0.13
Spermine	4.7	0.03	22.4	0.03	15.9	0.03
Experiment No. 2						
Monoacetyl-putrescine	0.0	--	1.5	0.06	0.0	--
Putrescine	2.4	0.39	9.1	0.40	3.2	0.41
Spermidine	6.0	0.19	18.1	0.25	13.4	0.21
Spermine	6.5	0.02	28.2	0.02	21.7	0.02

Footnotes as for Table 4.

A typical HPLC profile of the distribution of the radioactivity before and after acid hydrolysis is presented in Figure 1.

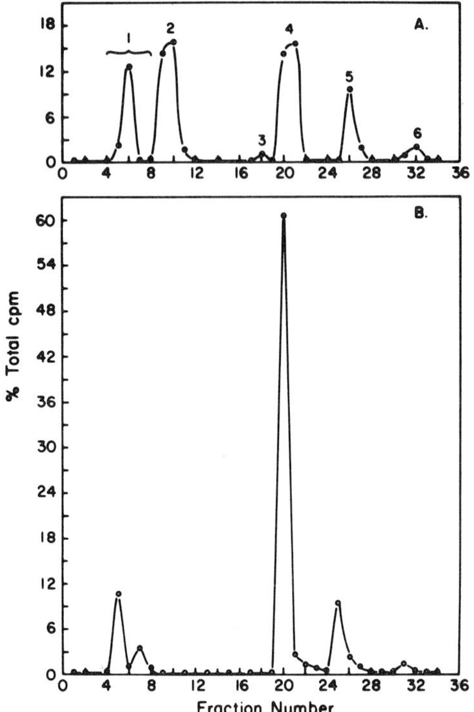

Figure 1. HPLC analysis of the nuclear fraction of cells induced to differentiate with ^{14}C-TMBA (5 mM TMBA; 24 hours 25 x 10^6 cells;). The upper panel (A) represents the profile before acid hydrolysis, and the lower panel (B) represents the profile after hydrolysis. Note that the sum of the TMBA (2), monoacetyl putrescine (3) and putrescine (4) areas in the unhydrolysed sample is equal to the putrescine peak after hydrolysis. Peaks marked 1,5 and 6 represent "front," spermidine and spermine respectively.

DISCUSSION

During induction of differentiation in MELC, the inducing agent, TMBA, remains stable in the extracellular medium. After 24 hours exposure to cells the radioactive TMBA peak remains a single component which is quantitatively hydrolyzed by hot acid to putrescine. The trace amount of ^{14}C-putrescine (less than 1%) which contaminates the preparation of ^{14}C-TMBA disappears during the initial incubation with MELC. Thereafter the only ^{14}C-containing component of the medium, aside from TMBA itself, is approximately 2% of monoacetylputrescine. TMBA seems to pass through the cell membrane and appears intracellularly as TMBA. The entire intracellular ^{14}C-TMBA peak is converted quantitatively to putrescine by hot acid hydrolysis.

In addition, TMBA is converted intracellularly to a range of metabolites including putrescine, monoacetylputrescine spermidine, spermine and trace amounts of monoacetylspermidine (Tables 1, 2, 4 and 5).

MELC cells in log phase culture have an endogenous level of spermidine and spermine of approximately 40 nMol/50x10^6 cells. These levels are significantly higher than putrescine, which is approximately 4 nMol/50 x 10^6 cells, and monoacetyl putrescine, which is an order of magnitude less than putrescine (Table 2). These values yield an intracellular concentration of 0.16 mM for putrescine, 1.6 mM for spermidine and spermine and 0.03 mM for monoacetylputrescine (4).

In MELC growing in log phase, the highest concentrations of acid soluble polyamines are found in the membrane fraction. Indeed, monoacetylputrescine is not demonstrable in any other fraction. The intracellular distribution of the polyamines is interesting, since the cytosol, representing about 60% of the cell volume, contains only approximately 20% of the polyamines. This implies that most of the acid soluble intracellular polyamines are adsorbed to cell organelles, rather than existing freely in solution. In contrast, most of the "front" running material is found in the cytosol.

During differentiation approximately 0.06% of the ^{14}C-TMBA counts in the medium are taken up by cells after 24 hours incubation (Table 3). More than half of the intracellular radioactivity is present in the membrane fraction, whereas approximately 20-30% is present in the cytosol and less than 20% is in the nuclear fraction. Both the total uptake of radioactivity and the intracellular distribution of label remain unaffected by dexamethasone.

The effectiveness of extracellular TMBA as a precursor to intracellular polyamines is seen in the radioactive labeling of cellular putrescine (Tables 4 and 5). The membrane putrescine pool is the most highly labeled, followed by the appearance of label in spermidine. Monoacetylputrescine and spermine are much less highly labeled. It seems reasonable to anticipate that ^{14}C-TMBA could be deacetylated to form monoacetylputrescine which, in turn, would be converted to putrescine. However, the fact that the specific activity of monoacetylputrescine is consistently much lower than that of putrescine provides evidence that monoacetylputrescine is not a metabolic precursor of putrescine.

Dexamethasone blocks the TMBA-induced differentiation of MELC, an effect similar to that observed with HMBA induction (5). Its major effect on polyamines is a striking elevation in the level of intracellular putrescine (Table 6).

TABLE 6. Ratios of Distribution of Polyamines in Cell Fractions

	cytosol	:membrane	:nuclei	Total nMol/ 50 x 10^6 cells
CONTROL:				
Putrescine	1	1.8	0.6	4.1
Spermidine	1	2.3	1.4	41.6
Spermine	1	3.0	1.9	43.4
TMBA:				
Putrescine	1	3.0	0.8	10.8
Spermidine	1	3.4	2.0	30.9
Spermine	1	4.0	2.3	47.3
TMBA + DEXAMETHASONE:				
Putrescine	1	3.7	1.3	22.3
Spermidine	1	3.2	2.4	34.3
Spermine	1	4.6	3.4	49.7

All data are the mean of two experiments and are derived by calculation from tables 3, 4 and 5.

We would call attention to the enrichment of spermidine and spermine in the membrane and nuclear fractions relative to cytosol when cells are induced to differentiate by the addition of TMBA. This redistribution occurs even in the presence of markedly elevated levels of putrescine which are observed with either TMBA or the combined addition of TMBA and dexamethasone. Total intracellular spermidine levels drop during the first 24 hours of induction, an effect that seems to be partially reversed by the presence of dexamethasone.

Induction of hemoglobin production during differentiation seems to occur independently of ornithine decarboxylase (ODC) activity. When MELC cells in culture are exposed to 100 μM putrescine, there is a time-dependent decrease of ODC activity during the first three hours of exposure of the cells to putrescine (data not shown). The endogenous level of ODC activity drops to about 20% of normal after three hours of exposure of MELC cells in log phase. Yet, after exposure to this level of putrescine, and in spite of the early effect of putrescine on ODC, the cells differentiate as measured by hemoglobin synthesis. The presence of up to 10^{-4}M putrescine has no effect on the ODC assay itself. Thus we conclude that although the polyamine derivative, diacetylputrescine (TMBA), can bring about differentiation of MELC, there is no apparent association with the level of ODC activity. Our findings are consistent with the observation that HL-60 cells can be induced to differentiate even when ODC activity has been blocked by the presence of D,L-α-difluoromethylornithine (6).

The data reported here are similar in some respects to those previously reported by Reuben, who studied the metabolism of HMBA with radioactive label on either the acetyl or the methylene positions (10). Since hexamethylenediamine is not metabolized by mammalian cells, however, our results differed from hers in indicating that the tetramethylenediamine moiety of TMBA enters actively into the normal pathways of polyamine metabolism.

TMBA in effecting differentiation is taken up and metabolized. It causes an intracellular redistribution of the major polyamines which is accompanied by increases in total putrescine and monoacetylputrescine and a decrease in spermidine. Dexamethasone blocks the effect of TMBA by inhibiting differentiation which is accompanied by a further increase in the level of putrescine and a partial reversal of the drop in spermidine. It seems likely that the parent polyamines themselves are not the immediate effectors in this system, but rather that one of their metabolic products might be involved in the process of differentiation. Our preliminary evidence indicating that TMBA is a naturally occurring metabolite of polyamines in MELC cells suggests that TMBA itself may be the metabolic product associated with differentiation.

Very small amounts of label from ^{14}C-TMBA also appear regularly as acid-precipitable material (data not shown). These observations led us to look at labelled putrescine as a potential precursor of the acid-precipitable material. In preliminary experiments we find a single radioactively labeled band on SDS polyacrylamide gel electrophoresis which stains with Coomassie blue, has a low molecular weight and is present after 5 hours' exposure of MELC cells to isotopically labeled putrescine. We are currently investigating whether there is any similarity between this protein band and the protein containing hypusine observed in mitogen-induced lymphocytes (9).

This project was initiated and the early experiments were performed in collaboration with Dr. Yair Gazitt when he was a guest in our laboratory. We are grateful to Leslie Lande Marsh and Paul Young for their expert technical assistance. We acknowledge the VA merit review program and the National Cancer Institute (CA 08341) for support of this research.

BIBLIOGRAPHY

1. Bondy, P.K. and Canellakis, Z.N. (1980) J. Chromatography 244:371-379.

2. Busch, H., Ballal, N.R., Olson, M.O.J. and Yeoman, L.C. (1975) in Methods in Cancer Research, H. Busch, editor, 11: 44-111.

3. Friend, C., Scher, W., Holland, J.G. and Sato, T. (1971) Proc. Natl. Acad. Sci. 68:378-382.

4. Gazitt, Y., Deitch, A.D., Marks, P.A. and Rifkind, R.A. (1978) Exper. Cell Res. 117: 413-420.

5. Gazitt, Y. and Friend, C. (1980) Cancer Res. 40:1727-1732.

6. Luk, G.D., Civin, C.I., Weissman, R.M. and Baylin, S.B. (1982) Science, 216:75-77.

7. Marks, P.A. and Rifkind, R.A. (1978) Ann. Rev. Biochem. 47:419-448.

8. Orkin, S.H., Harosi, F.I. and Leder, P. (1975) Proc. Natl. Acad. Sci., 72:98-102.

9. Park, M.H., Cooper, H.L. and Folk, J.E. (1981) Proc. Natl. Acad. Sci., 78:2869-2873.

10. Reuben, R.C. (1979) Biochim Biophys Acta 588: 310-321.

11. Reuben, R.C., Wife, R.L., Breslow, R. Rifkind, R.A. and Marks, P.A. (1976) Proc. Natl. Acad. Sci. 73: 862-866.

12. Scher, W., Tsuei, D., Sassa, S., Price, P., Gabelman, N., and Friend, C. (1978) Proc. Natl. Acad. Sci., 75: 3851-3855.

Advances in Polyamine Research, Vol. 4, edited by U. Bachrach, A. Kaye, and R. Chayen. Raven Press, New York © 1983.

Effect of 5′-Methylthioadenosine and Its Analogs on Friend Erythroleukemic Cell Proliferation and Differentiation

M. Cartenì-Farina, M. Porcelli, G. Cacciapuoti, V. Zappia, *M. Grieco, and *P. P. DiFiore

*Department of Biochemistry, First Medical School, University of Naples, 80138 Naples, Italy; and *Department of Viral Oncology, Institute of General Pathology, Medical School, University of Naples, 80131 Naples, Italy*

5'-Methylthioadenosine (MTA) is a natural sulfur-containing nucleoside ubiquitously distributed in mammalian tissues (1,2).

In the eukaryotes polyamine biosynthesis apparently represents the most important pathway for MTA formation (3,4): two mol of MTA are released/mol of spermine and 1 mol/mol of spermidine. A second pathway involves the direct cleavage of adenosylmethionine (AdoMet) into MTA and α-amino-γ-butyrolactone through the action of a specific cyclotransferase (5,6).

A number of evidences from several laboratories suggest that this sulfur nucleoside, mistakenly regarded in the past as a simple byproduct of polyamine biosynthesis, could play in vivo a number of physiological functions related to growth control. A significant antiproliferative activity is exerted by MTA in several systems: the molecule inhibits RNA synthesis in salivary glands of Drosophila melanogaster (7), DNA synthesis and cellular proliferation of stimulated human lymphocytes (1), the growth rate of virally transformed mouse fibroblasts (8) and murine lymphoid cells (9). Likewise several structural analogs of MTA i.e. 5'-methylthiotubercidin (MTT) (10), 5'-isobutylthioadenosine (SIBA) (11-14) and 5'-iodoformicin (15) are endowed with antimitotic activity. The mechanism(s) of these antiproliferative effects have not yet been elucidated and could be related to the inhibition of a number of enzyme systems: it is interesting to note in this respect that spermine and spermidine synthases (16), AdoMet-dependent methyltransferases (2,17,18) and adenosylhomocysteine hydrolase (19) are inhibited in vitro by MTA. Indirect in vivo evidence of a physiological growth inhibiting role of MTA stems also from the fact that MTA phosphorylase, the main enzyme responsible for the thioether degradation in eukaryotes (20-22), has been found to increase several fold after a number of growth stimuli such as androgenic stimulation in rat prostate (23), estrogenic stimulation in rat uterus (23) and lectin-induced lymphocyte transformation (1).

To further explore the biological role and the possible mechanism of action of MTA and SIBA, we have studied the effects of these two substances upon Friend erythroleukemic cells (FLC) proliferation. These cells are murine erythroid elements which replicate in vitro with a very low doubling time, thus representing a valuable model system for studying eukaryotic cell growth regulation. Furthermore, these cells are arrested at an early stage of the erythroid differentiation and can be induced to fully differentiate in vitro by a variety of chemical compounds, thus providing also a valuable model for the investigation of the biochemical events involved in cell differentiation (24). The results presented here demonstrate that both MTA and SIBA have a profound inhibitory effect on the growth of Friend cells. This inhibitory growth effect is not mediated by adenine or by polyamines depletion. MTA and SIBA are not capable of inducing the FLC to differentiate in vitro to haemoglobin containing erythroblasts.

The transport of the molecule as well as the main metabolic events occurring in this system have been studied by a newly developed high performance liquid chromatographic (HPLC) procedure

Materials and Methods

Chemicals - AdoMet was prepared by biosynthesis with yeast (25) and isolated by a procedure developed in this laboratory (26); S-adenosyl-L-(methyl-^{14}C)methionine (specific activity 50mCi/mmol) was supplied by the Radiochemical Centre (Amersham, Bucks, U.K.); MTA and 5'-(methyl-^{14}C)MTA were prepared by acid hydrolysis of AdoMet at pH 4.5 for 30 min at 100°C (27); SIBA was obtained from SIGMA Chemical (St. Louis, Mo., USA). 5-(Methyl-^{14}C)methylthioribose-1-phosphate was obtained from the phosphorolytic cleavage of 5'-(methyl-^{14}C)MTA with MTA phosphorylase purified from human placenta (22). 5-Methylthioribose (MTR) and 5-(methyl-^{14}C)MTR were prepared by acid hydrolysis of labeled and unlabeled MTA in HCl 1 N at 100°C for 1 h. Adenine , spermine, spermidine and dimethylsulfoxide (Me_2SO) were purchased from SIGMA Chemical Co. (St.Louis Mo., USA). The chemical and radiochemical purity of the compounds obtained by synthesis was checked by paper and thin layer chromatography (17), high voltage electrophoresis and HPLC (28).

Cell culture procedures - The F4-6 clone of FLC was a generous gift of Dr. Wolfram Ostertag (Hamburg, West Germany). It was cultured in suspension in 75 cm^2 flasks in minimum essential medium containing 10% foetal calf serum (FCS) in 5% CO_2 in air (29). In some experiments, where indicated, cells were cultured in 10%

horse serum (HS) instead of FCS.

Treatment of cells with drugs - All the drugs used in these experiments (i.e. MTA, SIBA, adenine, spermine, spermidine) were dissolved in bidistilled water and added to the culture medium at the time of cell seeding. Initial cell concentrations were 0.5 to 2.0 x 10^5/ml. No medium changes were performed during the experiments, except where indicated. Me_2SO was added at a final concentration of 1.5% v/v (0.21 M); the other drugs were added at the concentrations indicated in the figure legends.

MTA phosphorylase assay - Red cell membranes were prepared according to Galletti et al. (30). Cell-free extract was obtained by haemolysis of 250 μl of resuspended cells in 2 ml of H_2O. After addition of 250 μl of 1 M potassium phosphate buffer, pH 7.4 and 200 μl of 40 mM dithiothreitol, the mixture was centrifuged at 15000 g for 30 min. The supernatant was employed to investigate MTA phosphorylase activity (31).

5'-Methylthioadenosine uptake by erythroleukemic Friend cells - In order to obtain phosphate-depleted cells, the FLC were washed 10 times in 50 vol buffered saline (5 mM glucose, 150 mM NaCl and 5 mM Tris-HCl, pH 7.4). The packed cells were then resuspended in 5 vol of the same buffer: 100 μl of this cell suspension corresponded to 1.1 x 10^7 cells. The uptake measurements were performed according to Zappia et al. (31).

Determination of MTA and its metabolites by high performance liquid chromatography - Intracellular and extracellular ^{14}C-labeled metabolites were identified by high performance liquid chromatographic-isotopic methodology. A Beckman liquid chromatograph (model 324) and a Perkin-Elmer liquid chromatograph (model LC657, series 2), both equipped with an ultraviolet detector operating at 254 nm, were used. Aliquots of 20 μl were injected into the columns via a model 70-10 sample injector valve (Rheodyne). Two different methods were selected to separate 5'-(methyl-^{14}C)MTA and its radioactive metabolites i.e. 5-(methyl-^{14}C)MTR and 5-(methyl-^{14}C)MTR-1-P. The first employed a reversed-phase column (Ultrasphere ODS, 5 μm particle size) connected to a short (5 cm) precolumn packed with the same phase, equilibrated with 42% anhydrous methanol in bidistilled water. The elution was carried out at 0.8 ml/min with the same solution. Eluent from the column was directed through a low dead-volume connection into a LKB-Redirac fractions collector (model 2112) and timed fractions of 0.5 min were collected into miniscintillation vials over 20 min. This method allows a quantitative separation of MTR-1-P, MTR and MTA, the elution time being 180 sec, 270 sec and 684 sec respectively.

Since MTR-1-P is not retained by the resin in this condition, a second method has been developed to confirm the identity of non-retained labeled compounds. The second procedure employs a column pre-packed with anion-exchange resin (Partisil 10-SAX) equilibrated with 0.25 M ammonium formate buffer, pH 4. The elution was carried out at a flow rate of 1 ml/min with the same buffer. Under this condition MTA and MTR (200 sec) were not resolved while MTR-1-P was retained for a longer time (890 sec). The samples of intracellular and extracellular compartments for HPLC analysis were prepared according to Zappia et al. (31). For a better resolution of the chromatographic analyses unlabeled MTA (final concentration 0.5 mM) was added to the samples of intracellular compartment. Routinely 20 μl aliquots of the samples were analyzed by the two chromatographic methods. The stability of the compounds during these procedures was checked employing standard labeled compounds.

Assay [illegible] lobinproducing cells - The number of differentiating cells was determined by scoring benzidine-positive cells as described by Singer et al. (32). Five to ten minutes after addition of the benzidine reagent, an average of 300 cells were scored for benzidine positivity.

Other experimental procedures - Protein concentration was estimated by the method of Lowry et al. (33) using cristalline serum albumin as standard.

Radioactivity was measured in a Tri-Carb liquid scintillation spectrometer (Packard, model 4600) equipped with an absolute radioactivity analyzer. The scintillation liquid was InstaGel (Packard). Quenching was corrected by external standardization.

Results

Transport and metabolism of MTA in Friend cells

Transport and metabolism of MTA in intact erythroleukemic Friend cells have been preliminarly investigated. The sulfur nucleoside is rapidly transported across the plasma membrane by a facilitated-diffusion mechanism (data not shown), similar to that reported for human erythrocytes (31) and it is intracellularly cleaved by a specific phosphorylase In order to investigate at molecular level the metabolic events following transmembrane transport, a chromatographic analysis employing high pressure liquid chromatography has been developed. In Fig. 1 are depicted the HPLC analyses of intracellular and extracellular compartments at different time intervals of exposure of the cells to $^{14}CH_3$-labeled MTA. Fig. 1 (upper) illustrates the high pressure liquid chromatographic elution profiles of labeled intracellular species at 5 sec, 1 min and 10 min after cells exposure to 200 μM 5'-(methyl-^{14}C)MTA. The lower portion of the figure refers to the labeled compounds in the incubation medium at the same time intervals. At 5 sec the major intracellular constituent is MTA, but small amounts of MTR-1-P and MTR are also detectable, while in the outer compartment only MTA is present. After 1 min a significant increase of the cellular concentration of MTR-1-P is observable, while MTR does not accumulate in intracellular space, presumably because of its ready permeation; this explanation was confirmed by the presence of large amounts of MTR in the medium at 10 min. MTR-1-P increases continously over the intervals of observation and accounts for the major portion of intracellular ^{14}C.

The data reported indicate that the phosphorolytic cleavage of MTA begins very soon after its entry in the cell although at a lower rate than the rate of transport.

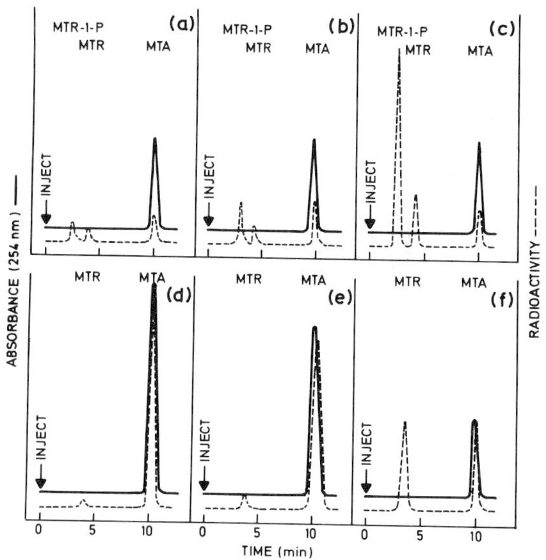

Fig. 1 - HPLC analysis of MTA and its labeled metabolites detectable in the intracellular (upper part) and extracellular (lower part) compartment after 5 sec (a,d), 1 min (b,e) and 10 min (c,f) exposure of cells to 200 μM 5'-(methyl-^{14}C)MTA. Retention times are indicated on the abscissa. The amounts of radioactivity applied to the column was different at various times. The column employed was reversed-phase (Ultrasphere ODS). Other details of the experiment are reported under Materials and Methods.

Table 1 summarizes the results of experiments performed in order to determine the Km and Vmax for the two events: MTA transport and intracellular phosphorolytic breakdown.

A mathematical approach has been employed to calculate $V^{zt}_{1,2}$ values of the transport of MTA in phosphate-depleted FLC. An apparent Km of 2,209 ± 289 μM and a Vmax of 456 ± 36 pmol/10^6cells /min have been calculated by the least squares method. The obtained kinetic parameters are of the same order of magnitude of those reported in human erythrocytes, for MTA (31) and for some purines and other nucleosides (34,35).

The kinetic parameters of MTA phosphorylase reported in Table 1, have been calculated graphically.

The data reported permit us to conclude that MTA uptake is a result of a tandem action of a transport step of high capacity and low affinity followed by a metabolic step of low capacity and high affinity.

TABLE 1. Kinetic parameters of transport and phosphorolytic cleavage of MTA in Friend erythroleukemic cells.

	Km	Vmax
	(μM)	(pmol/10^6 cells/min)
Transport in intact cells	2,209 ± 289	456 ± 36
	(μM)	(nmol/h/mg of protein)
Phosphorolytic cleavage in cell-free extract (MTA phosphorylase)	26 ± 3	87 ± 7

Results are ± S.E.

Effect of different MTA concentrations on FLC growth and viability

After having assessed that MTA is transported across the plasma membranes of Friend cells, the effect of different doses of this compound upon the growth of FLC has been investigated. The results of this experiment are presented in Fig. 2. The experiment was performed in the presence of HS in order to avoid the degradation of MTA since, in contrast with FCS, HS lacks MTA phosphorylase. In these conditions MTA-induced reduction of cell growth reaches values of 50-60% for drug concentrations varying from 180 to 300 μM.

To eliminate the possibility that MTA action on cell growth is mediated by an aspecific toxic mechanism, the drug was washed out from the medium either during the logarithmic (2 days) or stationary (5 days) phase on cell growth (Fig. 3). The concentration of MTA used was 400 μM, the maximal drug concentration used in the above experiment. The treated cells were harvested, washed free of the inhibitory compound and resuspended in fresh medium without MTA. Recovery was assayed by cell counting up to the seventh day of culture. After MTA removal, FLC reacquire a similar growth rate and reach the same saturation density $3x10^6$ cells/ml as those of untreated cells.

MTA effect on cell proliferation is not mediated via its conversion into adenine.

It has been proposed (8) that MTA action is mediated via its conversion into adenine by MTA phosphorylase (20-22). Based on such an hypothesis, we studied the proliferation pattern of F4-6 cells in the presence of either MTA or adenine added at the same molar concentration to the culture medium. No modification of the growth pattern of FLC is observable upon adenine addition, even at a concentration as high as 1 mM. At such a concentration MTA exerts a dramatic reduction of growth in F4-6 cells, as previously described (see Fig. 4).

Polyamines do not revert the effect of MTA on cell growth.

Since MTA is a powerful inhibitor of polyamine synthases (16), we checked whether spermidine and spermine, at concentration of 10 and 100 μM, revert the inhibitory effect of MTA on FLC growth. As shown in Fig. 5A both polyamines do not exert significant effect on FLC growth and, as shown in Fig. 5B, they are not capable of reverting the effect of 400 μM MTA.

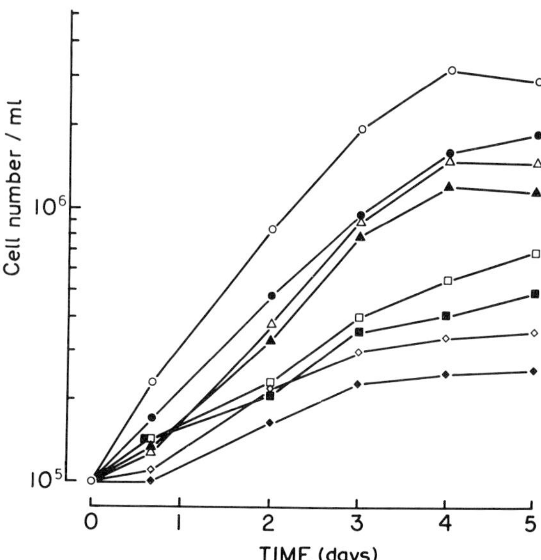

Fig. 2 - Effect of MTA on growth of erythroleukemic cell line (FLC). Cells were seeded at a concentration of $1x10^5$/ml in the absence and presence of MTA; where indicated horse serum (HS) was substituted for foetal calf serum (FCS). (o), control FLC in FCS; (●) control FLC in HS; (△) MTA 1 μM in HS; (▲) MTA 10 μM in HS; (□) MTA 75 μM in HS; (■) MTA 180 μM in HS; (◇) MTA 300 μM in HS; (◆) MTA 400 μM in HS.

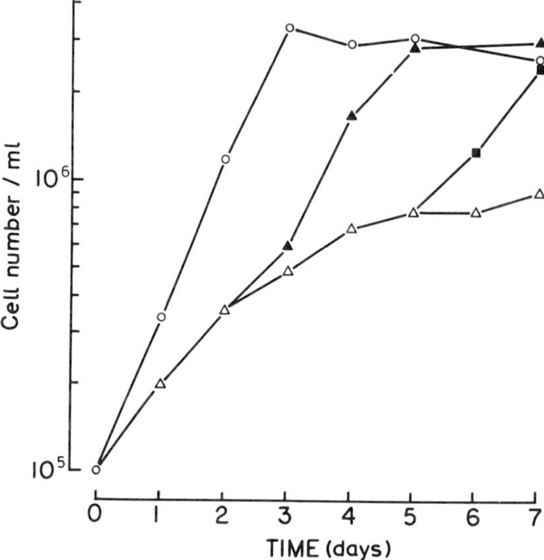

Fig. 3 - Recovery of FLC from MTA-mediated inhibition. Cells were seeded at a concentration of $1x10^5$/ml in the absence (o) and presence of MTA 400 μM (△). On days 2 (▲) and 5 (■) of treatment, cells were washed free of the compound and resuspended in fresh medium lacking MTA.

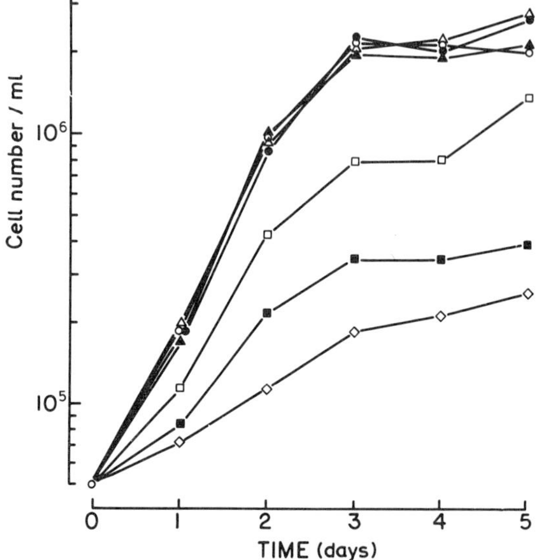

Fig. 4 - Effect of adenine on cell growth compared with that of MTA. Cells were seeded at a concentration of $5x10^4$/ml and treated with adenine or MTA. (o) FLC control; (●) adenine 100 μM; (△) adenine 400 μM; (▲) adenine 1 mM; (□) MTA 100 μM; (■) MTA 400 μM; (◇) MTA 1 mM.

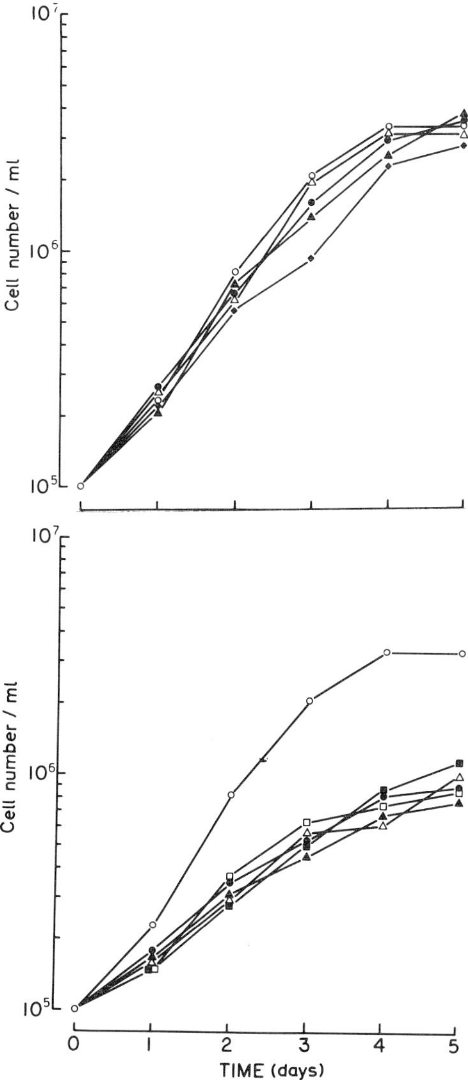

Fig. 5 - Effect of polyamines and MTA on cell growth of FLC. Cells were seeded at a concentration of $1x10^5$/ml and treated with polyamines alone, MTA alone or MTA plus polyamines. In this experiments HS was substituted for FCS.
A) (o) FLC control; (●) spermine 10 μM; (▲) spermine 100 μM; (△) spermidine 10 μM; (◆) spermidine 100 μM;
B) (o) FLC control; (●) MTA 400 μM; (△) MTA 400 μM plus spermine 10 μM; (△) MTA 400 μM plus spermine 100 μM; (□) MTA 400 μM plus spermidine 10 μM; (■) MTA 400 μM plus spermidine 100 μM.

Effect of different concentrations of SIBA on FLC growth.

Since it is well documented that SIBA, a synthetic structural analog of MTA, is also capable of interfering with the growth of several cell systems. (11), we investigated the effect of this compound in our model. As shown in Fig. 6, SIBA exerts growth inhibitory action on FLC similar to that of MTA at the same molar concentration.

Effect of MTA and SIBA upon erythroid differentiation.

MTA and SIBA (at concentration ranging from 1 to 1000 μM) do not induced erythroid differentiation of FLC as evidenced by benzidine staining of cultured cell up to 5 days of growth in presence of either compounds (results not shown). In contrast, preliminary results indicate that both compounds inhibit Me_2SO induced-differentiation of FLC.

A detailed analysis of these effects of MTA and SIBA will be published elsewhere

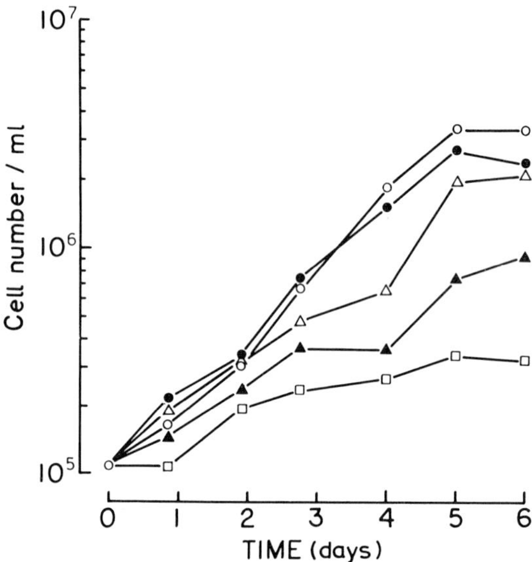

Fig. 6 - Effect of SIBA on growth of FLC. Cells were seeded at a concentration of $1x10^5$/ml in the absence and presence of SIBA.
(o) Control FLC; (●) SIBA 1 μM; (△) SIBA 10 μM; (▲) SIBA 100 μM; (□) SIBA 400 μM.

Discussion

The reported results indicate that MTA as well as its structural analog, SIBA, profoundly inhibit the growth of F4-6 erythroleukemic Friend cells. The inhibitory effect is dose-dependent and is not caused by cytotoxicity, since the inhibition is completely reversible after the removal of the thioether.

The results obtained on the transport and metabolism of MTA in the cells exposed to unphysiologically high concentrations of the nucleoside, indicate that MTA is rapidly incorporated into FLC cells. The HPLC analysis of the intracellular and extracellular compounds demonstrate that the transport of this sulfur-nucleoside is followed by its phosphorolytic cleavage: the kinetic values of the transport, compared with kinetic parameters of MTA phosphorylase, allow to predict a steady-state levels of the molecule after its cell entry in quantitative terms. In fact the rate of phosphorolytic cleavage may or not approach the rate of transport, depending on the concentration of exogenous MTA. At physiological level of the thioether (below 5 μM), the initial velocity values of transport represent the limiting step of MTA influx and virtually no amounts of the thioether are detectable intracellularly. Conversely, when the cells were exposed to cytostatic concentrations of the thioether (∿ 200 μM), the initial velocity values of the transport exceed that of MTA phosphorylase and the cellular pool of free nucleoside is significantly increased. Since at these concentrations a number of enzyme systems are inhibited *in vitro* by the nucleoside, these effects are conceivably operative also *in vivo* and could be related to the antiproliferative action of the molecule.

It has been postulated by Pegg (8) that the growth inhibition exerted by MTA can be ascribed to adenine, released through the action of MTA phosphorylase. The results here reported indicating the lack of any effect of adenine on cell growth, together with the observation that the phosphorolytic degradation of MTA decreases its inhibitory action, permit to exclude this possibility.

Based upon the inhibition of spermine and spermidine synthases exerted *in vitro* by MTA (16), is the hypothesis that the growth inhibitory effect is mediated *via* the depletion of polyamine pool. Since spermine and spermidine (Fig. 5B) do not reverse this inhibition, the MTA-dependent decrease of polyamine pool cannot be taken into account as the sole cause of its cytostatic effect. A similar result has been obtained by Pegg in mouse fibroblasts (8).

A third hypothesis relates the antiproliferative effect of MTA with an inhibition of transmethylation reactions either direct or mediated by an increase of AdoHcy. In fact, although less effectively than AdoHcy, MTA inhibits *in vitro* several methyltransferases and it has been recently demonstrated that the molecule irreversibly inactivates AdoHcy hydrolase purified from human erythrocytes (19). Therefore it is tempting to postulate that the effects on living cells, exposed to high levels of MTA and SIBA, are in part due to suicide-inactivation of AdoHcy hydrolase, with subsequent accumulation of AdoHcy, given the well documented antiviral and antimitotic properties of analogs of AdoHcy (11).

An other possibility that cannot be rouled out is that MTA and SIBA could act at cellular level by modulating cAMP metabolism, also in this case the effect could be either direct or mediated via AdoHcy. In fact it has been demonstrated that MTA is a competitive antagonist of the adenosine response of cAMP systems in brain slices and in fibroblasts (36). Furthermore it has been recently reported that SIBA and AdoHcy are potent competitive inhibitors of the high affinity cAMP phosphodiesterase and enhance the cAMP response of intact cells to several activators of adenylate cyclase (37). Such stimuli-caused elevations of cAMP, prolonged and enhanced in presence of inhibitors of cAMP phosphodiesterase, can then result in the inhibition of cellular functions sensitive to elevated cAMP.

In preliminary experiments we have observed that MTA and SIBA also inhibit Me_2SO-induced erythroid differentiation of FLC; this effect could be related to the hypothesized role of polyamines in the differentiation process (38). A detailed analysis of such effects will be presented elsewhere.

ACKNOWLEDGEMENTS

This work was supported by grants from the Consiglio Nazionale delle Ricerche, Rome, Italy - progetto finalizzato "Controllo della Crescita Neoplastica".

REFERENCES

1. Ferro, A.J. (1979): In: Transmethylation, edited by E. Usdin, R.T. Borchardt and C.R. Creveling, pp. 117-126, Elsevier/ North Holland, New York.
2. Zappia, V., Cartenì-Farina, M., Cacciapuoti, G., Oliva, A. and Gambacorta, A. (1980): In: Natural Sulfur Compounds: Novel Biochemical and Structural Aspects, edited by D. Cavallini, G.E. Gaull and V. Zappia, pp. 133-148, Plenum Press, New York.
3. Tabor, H. and Tabor, C.W. (1976): In: Ann. Rev. Biochem., edited by E.E. Snell, P.D. Boyer, A.Meister and C.C. Richardson, vol. 45, pp. 285-306, Annual Review Inc., Palo Alto, California.
4. Zappia, V., Cacciapuoti, G., Pontoni, G., Della Ragione, F. and Cartenì-Farina, M. (1981): In: Advances in Polyamine Research, edited by C.M. Caldarera, V. Zappia and U. Bachrach, vol. 3, pp. 39-53, Raven Press, New York.
5. Mudd, S.H. (1959): J. Biol. Chem., 234:87-92.
6. Wilson, J., Corti, A., Hawking, M., Williams-Ashman, H.G. and Pegg, A.E. (1979): Biochem. J., 180: 515-522.
7. Law, R.E., Sinibaldi, R.M., Cumming , M.R. and Ferro, A.J. (1976): Biochem. Biophys. Res. Commun., 73: 600-606

8. Pegg, A.E., Borchardt, R.T. and Coward, J.K. (1981): Biochem. J., 194: 79-89.
9. Wolford, R.W., MacDonald, M.R., Zehfus, B., Rogers, T.J. and Ferro, A.J. (1981): Cancer Res., 41: 3035-3039.
10. Coward, J.K., Motola, N.C. and Moyer, J.D. (1977): J. Med. Chem., 20: 500-505.
11. Robert-Géro, M., Blanchard, P., Lawrence, F., Pierré, A., Vidal, M., Vuilhorgne, M. and Lederer, E. (1979) In: Transmethylation, edited by E. Usdin, R.T. Borchardt and C.R. Creveling, pp. 207-214, Elsevier/North Holland, New York.
12. Cartenì-Farina, M., Della Ragione, F., Ragosta, G., Oliva, A. and Zappia, V. (1979): FEBS Lett., 104: 266-270.
13. Bachrach, U., Schnur, L.F., El-On, J., Greenblatt, C.L., Pearlman, E., Robert-Géro, M. and Lederer, E. (1980): FEBS Lett., 121: 287-291.
14. Zimmerman, T.P., Schmitges, C.J., Wolberg, G., Deeprose, R.D. and Duncan, G.S. (1980): Life Sci., 28: 647-652.
15. Savarese, T.M., Dexter, D.L., Spremulli, E.N., Crabtree, G.W. Chu, S.H., Montgomery, J.A., Parks, R.E. Jr. and Calabresi, P. (1981): 72th Meeting of the American Association for Cancer Research, Abstr. 827, Washington D.C.
16 Pajula, R.L. and Raina, A. (1979): FEBS Lett., 99: 343-345.
17. Zappia, V., Zydek-Cwick, C.R. and Schlenk, F. (1969): J. Biol. Chem., 244: 4499-4509.
18. Williams-Ashman, H.G., Seidenfeld, J. and Galletti, P. (1982): Biochem. Pharmacol., 31: 277- 287.
19. Ferro, A.J., Vandenbark, A.A. and MacDonald, M.R. (1981): Biochem. Biophys. Res. Commun., 100: 523-531.
20. Garbers, D.L. (1978): Biochim. Biophys. Acta, 523: 82-93.
21. Zappia, V., Oliva, A., Cacciapuoti, G., Galletti, P., Mignucci, G. and Cartenì-Farina, M. (1978): Biochem. J., 175: 1043-1050.
22. Cacciapuoti, G., Oliva, A. and Zappia, V. (1978): Int. J. Biochem., 9: 35-41.
23. Nicolette, J.A., Wrobel, M.C. and Ferro, A.J. (1980): Biochim. Biophys. Acta, 627: 190-198.
24. Marks, P.A. and Rifkind, R.A. (1978): Ann. Rev. Biochem., 47: 419-448.
25. Schlenk, F. and De Palma, R.E. (1957): J. Biol. Chem., 229: 1037-1050.
26. Salvatore, F., Utili, R., Zappia, V. and Shapiro, S.K. (1971): Annal. Biochem., 41: 16-28.
27. Schlenk, F. and Ehninger, D.J. (1964): Arch. Biochem. Biophys. 106: 95-100.
28. Della Ragione, F., Cartenì-Farina, M., Porcelli, M., Cacciapuoti, G. and Zappia, V. (1981): J. Chromatogr., 226: 243-249.
29. Colletta, G., Fragomele, F., Sandomenico, L.M. and Vecchio, G. (1979): Exp. Cell. Res., 119: 253-264.
30. Galletti, P., Paik, W.K. and Kim, S. (1980): Biochemistry, 17: 4272-4276.

31. Zappia, V., Della Ragione, F., Porcelli, M., Cacciapuoti, G. and Cartenì-Farina, M. (1981): Med. Biol., 59: 354-358.
32. Singer, D., Cooper, M., Maniatis, G.M., Marks, P.A. and Rifkind, R.A. (1974): Proc. Natl. Acad. Sci. USA, 71: 2668-2670.
33. Lowry, O.K., Rosebrough, N.J., Farr, A.L., Randall, R.J. (1951): J. Biol. Chem., 193: 265-275.
34. Plagemann, P.G.W., Richey, D.P. (1974): Biochim. Biophys. Acta, 344: 263-305.
35. Marz, R., Wohlhueter, R.M., Plagemann, P.G.W. (1979): J. Biol. Chem., 254: 2329-2338.
36. Bruns, R.F. (1980): Canad. J. Physiol. Pharmacol., 58: 673-691.
37. Zimmerman, T.P., Schmitges, C.J., Wolberg, G., Deeprose, R.D., Duncan, G.S., Cuatrecasas, P. and Elian, G.B. (1980): Proc. Natl. Acad. Sci. USA, 77: 5639-5643.
38. Bethell, D.R. and Pegg, A.E. (1981): Biochem. Biophys. Res. Commun., 102: 272-278.

Subject Index

Abortion
  plasma DAO in, 193–196, 197
  polyamine excretion in, 196–197
Acetylation
  amine oxidase and, 138
  carbon tetrachloride and, 138
  DNA and, 766
  FL differentiation and, 761–766
  of spermidine, 138, 321–329
  of spermine, 138
  of urinary polyamines, 114
Acetylputrescine
  diacetylated form, 769, 774
  FL differentiation and, 761–764, 769–777
  in polyamine catabolism, 138
  putrescine and, 775
Acetylspermidine
  FL differentiation and, 761–764
  nuclear protein kinase and, 632, 636, 638
  putrescine from, 321
S-adenosylmethionine
  E. coli deficient, 521–523
S-adenosylmethionine decarboxylase, (SAMDC)
  of E. coli, 456
  gene for, 456
  half-life of, 129
  induction of, 129, 131
  MGBG and, 19–20
  mutants of, 471
  phospholipids and, 339–341
  preparation of, 245
  properties of, 456
  pyruvate and, 456
  of *S. cerevisiae,* 471
  spermidine acetyltransferase and, 327
  in tumor induction, 129
S-adenosylmethionine synthetase
  E. coli mutant, 521–523
  deficiency of, 521–523
β_2-adrenoceptor
  catecholamines and, 549–550
  ODC and, 551, 557–559
  subtypes of, 550, 552–553
African sleeping sickness
  polyamine metabolism and, 221–222, 228
Agmatine
  arginine decarboxylase and, 381–392
  in plant hordatines, 358–359
  in plants, 385–386, 397
  putrescine and, 397
Aldehyde metabolizing enzymes
  of brain tumors, 189–190
  DAO activity and, 176, 199–201
  inhibitors of, 176
  in pregnancy, 199–201
Alicyclic amines
  spermidine synthase inhibition and, 213–214
Amine oxidase, *see also* specific amine oxidases
  assay of, 136–137, 185
    in calf serum, 146–147, 149, 169–174
    function of, 161–164
    inhibition of, 161–162, 188–189
    MGBG and, 162–163
    parasitic infection and, 169–174
    of placenta, 155–157
  in pregnancy, 155–164
    putreanine and, 148–153
    SAMDC and, 187
    substrates of, 135, 148–153, 163–164
Aminoacetonitrile
  brain and, 147–148, 151
  diamine oxidase and, 143–146
  polyamine levels and, 143–145
  putreanine and, 140–143, 147–148, 151
Aminoacetylation of tRNA
  in E. coli, 307–317
  Mg^{++} and, 313–314, 317
  in plants, 307–317
  polyamines and, 307–317
  tRNA concentration and, 314–315
  spermine binding and, 308–310, 315–318
γ-aminobutyric acid, (GABA)
  in kidney, 200–201
  in liver, 200
  in neuroblastoma cells, 749–750
  in ovary, 200
  in placenta, 200–201
  in pregnancy, 200–202
  putrescine metabolism and, 199–201, 749–750
  Δ^1-pyrroline and, 199–201

Aminoguanidine
in brain, 147–148, 151
cadaverine and, 197
calf serum oxidase and, 146–147
as DAO inhibitor, 135
placental amine oxidase and, 156–157, 197
polyamine metabolism and, 138–139
in pregnancy, 195–197
putreanine and, 135, 138–143, 150–151
spermine oxidase and, 160
tumor amine oxidase and, 188–189
Antibiotics
aminoglycosides, 11–12, 119–120
as antitumor agents, 2–3, 12–13
biosynthesis of, 1–2,
cadaverine in, 355–356
DNA and, 4, 5, 7
polyamines in, 1–13, 116, 119
structures of, 4–5
Antibodies
to ODC, 571–572, 581
Antifungal agents
structure-activity, 356–363
Anti-viral effect
of DFMO, 509–511, 515–516
of diaminopropane, 510
of MGBG, 509–512, 515–516
of polyamine inhibitors, 507–516
Antizyme, *see* ODC antizyme
Arginine
in plants, 397–406, 412
putrescine and, 397
Arginine decarboxylase, (ADC)
assay of, 382, 410–411
bacterial, 495, 498–504
control of, 381, 399–402
cycloheximide and, 383
DFMA and, 409, 411–416, 498–504
DFMO and, 388–389, 411–416, 493, 497–498
gene for, 456
half-life of, 498–499
hormone regulation of, 386–387
induction of, 383–386, 391
light control of, 383–386
osmotic stress and, 387–390
pH effects, 390–391, 449
in plants, 381–392, 397–400, 409–416, 443–444, 449–450
polyamines and, 387–390
properties of, 390–391, 449
putrescine and, 387–389, 409
stimulation of, 381
Astrocytoma
amine oxidase in, 185–186, 188
ODC in, 187
SAMDC in, 187
Autoradiographic localization
in kidney, 719–721, 723–725
methods, 720–721
of ODC, 719–725
of putrescine, 749–750
in whole body, 719–722

Babesia rodhaini
oxidized polyamines and, 169–174
Bacillus Calmette-Guerin, (BCG)
oxidized polyamines and, 173
Bacteria, *see also* specific type
ADC inhibition, 498–499
DFMA in, 498–504
DFMO in, 497–498, 500–503
inhibition of polyamine synthesis in, 495, 498
polyamine synthesis in, 495
putrescine synthesis in, 495–496
Bacteriophage
multiplication of, 462
mutants of, 460
polyamine depletion in, 460–462
Barley seed
ODC activator in, 432–435
ODC antizyme of, 430–432
ODC purification from, 428–430
Binding
(1,25-$(OH)_2D_3$) effect, 267, 270–273, 283–284
of polyamines, 267
proteins, 267–276, 279–284
of spermine, 115, 267–276, 279
1,3-bis (2-chloroethyl)-1-nitrosourea, (BCNU)
DFMO and, 34, 41
MGBG and, 35–36
polyamines and, 34
Bleomycin
biosynthesis of, 4
cancer and, 8, 37
copper complex of, 4, 6
DNA binding, 4, 5, 7
metabolism of, 7–8
polyamines in, 3–4, 6, 37
structure of, 3, 4–5
toxicity of, 7–8
trypanosomes, 37, 223
tumor spheroids and, 97
Bone cells
cyclic AMP and, 713–715
electric field, 715, 717
hormonal influences, 713–717
ODC in, 713–717
parathyroid hormone and, 713–716
Brain
aminoguanidine and, 147–148, 151
polyamines in, 209, 214–215

putreanine in, 147–148, 151
spermidine acetyltransferase in, 323
spermidine synthase of, 211–214
Brain tumor
amine oxidase in, 184–190
BCNU and, 33–35
DFMO and, 41–42
9L, 9L cells, 33–34, 41–42
polyamines and, 33, 34
Breast carcinoma
bone metastasis in, 60–63
cadaverine in, 65–71
mastectomy in, 66–71
metastases in, 59–63
polyamine excretion in, 59–63, 65–71
radiation and, 60–63

Cadaverine
aminoguanidine and, 197
antibiotic of, 355–356
in cystinuria, 195
DFMO and, 23–24
in growth, 236–237
GCG and, 198–199
lysine decarboxylase and, 456–457
in ovary, 73, 197
polyamine oxidase and, 161
in pregnancy, 195–197
propylamine transferase and, 249
protein synthesis and, 237–239
toxicity of, 108
urinary, 65–71
Caffeoylputrescine
in plants, 351
in tobacco plant, 348–350
Calcium, (Ca^{++})
heart ODC and, 541–542, 545–546
ODC and, 622–627
polyamine-dependent kinase and, 622–627
polyamine synthesis and, 622–623
Caldopentamine
derivatives of, 481–483
occurrence, 484–486
structure, 483
synthesis of, 484
as unusual polyamine, 480–481
Calmodulin
ODC and, 622–627
polyamine-dependent kinase and,
622–627, 647–653
Canaline, plant ODC and, 414–415
Canavanine
plant effects, 403–405
plant ODC and, 414–415
polyamine levels and, 403–405
Cancer, *see also* specific type
BCNU in, 34, 35, 41
of breast, 59–63, 65–71
CENU and, 36
chemical induction, 82, 91–92
DFMO and, 33–38, 41–42
histamine in, 49
hyperthermia in, 97
liver and, 56
metastases, 59–63
nitrosoureas in, 41–42
ODC in, 87–92
polyamine depletion in, 33–38
radiation and, 60–63, 73
serum polyamines in, 49, 51–58, 73–77
urinary polyamines in, 45–46, 59–63,
65–71, 73–77
Carbon tetrachloride, acetylation and,
138–139, 151
Carboxymethyl cellulose, in spermine
binding, 279–282, 284
Carnitine metabolism, polyamine effects,
300–303
Carnitine palmitoyl transferase, polyamine
effects, 300–302
Catecholamines
fetal heart and, 553–557
hypertrophy and, 549–550
ODC activity and, 537, 539–546, 553,
556–559
proliferation and, 550
Cell cycle
DFMO and, 35, 294
in Ehrlich ascites, 707
hyperthermia and, 98–99
ODC and, 591, 592, 707–711
ODC antizyme and, 707–711
phospholipids and, 337
in plants, 419–420, 449–450
polyamines and, 286, 294
putrescine and, 93
Cerebral spinal fluid, (CSF)
amine oxidase in, 189–190
in brain tumors, 189–190
Chinese hamster ovary cells, (CHO)
culture media and, 598–600
ODC mutants in, 592–601
Chlorella
ADC of, 449
cell cycle of, 449–450
ODC of, 449
Chloroethylnitrosoureas, (CENU),
polyamines and, 34
Choriocarcinoma
methotrexate and, 73–77
serum polyamines in, 73–77
urinary polyamines in, 73–77
Chromatid exchange, DFMO and, 37
Cinnamic acid
formation of, 347–350
in plants, 347–350

polyamine amides, 347–366
in tobacco plant, 348–350
Cis-platinum
DFMO and, 36–37
DNA and, 36–37
Cleomycin, structure of, 5–6
Cloning
bacterial, 456
of ODC mutants, 592–601
Clostridium thermohydrosulfuricum
ODC activity in, 489–494
ODC properties, 490
Coumaroylputrescine
hydroxy derivatives of, 351
methyl derivatives of, 351
in tobacco plant, 348–350
Cyclic AMP
in bone cells, 713–715
catecholamines and, 553
cell proliferation and, 667–668
cyclase assay and, 669
differentiation and, 743
DNA synthesis and, 713–715
ODC and, 553–554, 713, 743
phospholipids and, 335–338
polyamines and, 672–679, 743
neuroblastoma cells and, 743–744
serum effects, 670–671
Cyclic GMP
cell proliferation and, 667–668
cyclase and, 673–674
cyclase assay for, 669–670
phosphodiesterase and, 672–673
phospholipids and, 335–338
polyamine regulation of, 673–679
serum effects, 670–671
Cyclic nucleotides
cell proliferation and, 667–668
DFMO and, 338–339
heart cell and, 670
MTA and, 790
phospholipid effects, 335–338
polyamines and, 338, 667–679, 743
serum effects, 670–672
Cycloheximide
arginine decarboxylase and, 383
DAO and, 178
in half-life determinations, 129, 179
ODC and, 129, 541
spermine-binding protein and, 271
Cyclopentylamine
spermidine synthase inhibition, 213
Cystinuria
polyamine metabolism in, 114
Cytomegalovirus, (CMV)
DFMO in, 516
growth curve for, 513
MGBG and, 510–512
polyamines in, 511, 513–514
Decarboxylated adenosylmethionine, *see* Methyl thioadenosine
Dexamethasone
diacetylputrescine and, 772–773
FL differentiation and, 757, 775–777
FL polyamines and, 772–773, 775–776
HMBA and, 770
Diacetylputrescine
dexamethasone and, 770
FL differentiation and, 769
ODC and, 776
metabolism of, 771–777
polyamine synthesis from, 771–774
synthesis of, 770–771
uptake of, 771
Diamine oxidase, (DAO)
abortion and, 193
aminoacetonitrile and, 143–146, 149–150
assay of, 137, 175–176, 193–194, 199–201
in brain, 151
cadaverine and, 187–198
carbon dioxide and, 201–202
in cell cultures, 183
GABA and, 199–201
growth processes and, 175–179, 183
half-life of, 179
induction of, 175–179
inhibition of, 156, 158–159, 176
in liver, 177–179, 183
MGBG and, 20
ODC activity and, 179
in plasma, 157–159, 193–194
in pregnancy, 155, 157–159, 193–194
putreanine and, 151
putrescine and, 150, 175
Δ^1-pyrroline and, 200
in small intestine, 143–146, 150
substrates of, 135, 149, 159, 177
in thyroid, 184
tissue sources of, 175, 177
in transformed cells, 183
in tumors, 184
of uterus, 177
1,8-Diaminooctane
tRNA aminoacylation and, 310–312
4,4′ Diaminodiphenylurea-bis(guanylhydrazone), (DDUG)
for trypanosomes, 224–225
1,3-Diaminopropane
amine oxidase and, 164
as anti-viral agent, 510
growth activity of, 237, 411–413
ODC activity and, 581
1,3-Diaminopropanal
insulin and, 564
liver polyamines and, 565–566
lipogenesis and, 566–568
ODC activity and, 565–566

ODC inhibition and, 20
in plants, 411–412
Dibutyryl cAMP
in bone cells, 713–715
DFMO and, 751–753
heart ODC and, 529–530
neurite outgrowth and, 751–753
neuroblastoma differentiation and, 743–746
putrescine transport and, 747–748
2,4-Dichlorophenoxyacetic acid, (2,4-D)
plant effects, 419
Dicyclohexylamine
brain polyamines and, 214–215
convulsions and, 215–216
spermidine synthase inhibition and, 213–214
Dicyclohexylammonium sulfate
as spermidine synthase inhibitor, 501–502
Differentiation
acetylpolyamines and, 761–766
criteria for, 729–730, 735
cyclic AMP and, 743
cytoskeletal system and, 748
dexamethasone and, 757, 770
DFMO and, 731–732, 737–740, 751–753
diacetylputrescine and, 769–777
in embryonal carcinoma cells, 733–740
in FL, 757–766
HMBA and, 757–766
in leukemia cells, 729–734
Me_2SO and, 757–766
in neuroblastoma cells, 743
ODC and, 731–732, 737–740, 748–749
polyamine synthesis and, 727, 731–733, 738–740, 753, 757, 758–759
putrescine transport and, 747–748, 764–766
retinoic acid and, 730–734, 737
TPA and, 730
α-Difluoromethylarginine, (DFMA)
in bacteria, 498–504
growth inhibition by, 411–413
MFMP and, 501
in plant culture, 409, 411–413
DL-2-Difluoromethylornithine, (DFMO)
anti-tumor activity, 19, 21–22
in autoradiographic localization, 608, 719–725
in bacteria, 497–498
BCNU and, 34–36, 41
cadaverine and, 23–24
cancer and, 33–38
cell cycle and, 35, 294
CENU and, 36
cyclic nucleotides and, 338–339
cytotoxicity and, 41–42
derivatives of, 497
differentiation and, 731–732, 737–740, 751–753
DNA synthesis and, 290–291
effects of, 19
embryogenesis and, 203–204
growth rate and, 21
in heart cells, 338–342
incorporation into protein, 604–606
kinetics of, 604
in leukemia, 19, 23–24, 28–29, 731–732
mechanism of action, 604, 611–612
MGBG and, 19–25, 290–291
in neuroblasts, 585–589
nitrosoureas and, 41–42
nuclei and, 295
ODC binding and, 474, 582, 595–596, 603
ODC induction and, 132, 592, 595, 606
ODC turnover and, 609–610
phospholipids and, 338–339, 343
in plants, 350, 388–389, 404, 411–416
polyamine levels and, 289–291, 294
polyamine uptake and, 23–24, 291–292
in psoriasis, 25, 29
radiolabeled, 604–605
retinoids and, 731–732, 738–740
rhodamine labeled, 585–589, 725
species variation to, 497, 607–608
specificity of, 603, 719–720, 725
structure of, 405
subcellular polyamines and, 289–290, 292
in tobacco plant, 350
topical use of, 25–28
trypanosomes and, 37, 222–224
X-rays and, 33–34
1,25-dihydroxycholecalciferol, (1,25-$(OH)_2D_3$)
binding of spermine and, 267, 270–273, 275, 283–284
Ca^{++} and, 270
function of, 273–276
Dimethylsulfoxide, (Me_2SO)
FL differentiation and, 757–766
MTA and, 790
ODC and, 757, 764–765
polyamine acetylation and, 761–764
polyamine levels and, 758–759, 765
polyamine metabolism and, 759–761, 765
DNA
bleomycin binding, 4, 5, 7
in chlorella, 450
histograms, 709–710
nitrosoureas and, 41–42
polyamines and, 33, 34, 285
repair enzymes, 34
tumor induction and, 128
DNA synthesis
agmatine and, 391
DFMO and, 285, 290–291, 292–293
MGBG and, 285, 290–293

ODC and, 713, 716
phospholipids and, 331, 334–335
polyamine depletion and, 285–286, 290–293, 727

Ehrlich ascites carcinoma
cellular classification of, 708–709
DFMO and, 21
DNA of, 709–710
MGBG and, 21
ODC activity in, 707–711
ODC antizyme and, 707–711
polyamine-dependent kinase in, 618, 620–622
Embryogenesis
DFMO and, 203–204
heart ODC and, 555–557
MGBG and, 203
polyamine synthesis and, 203–204
Embryonal carcinoma
cell differentiation in, 733–740
culture of, 733–737
DFMO in, 738–740
morphology of, 735–737
ODC in, 738–740
putrescine in, 740
retinoids in, 738–740
Endocytosis
in kidney, 110–112, 115–120
spermine and, 117–119
Epinephrine
in aortic constriction, 550, 552
heart ODC and, 537, 540, 543, 558–559
hypertrophy and, 554–555
plasma levels of, 552
Erythrocytes, polyamines in, 49
Escherichia coli
S-adenosylmethione deficient, 521–523
aminoacylation in, 307–317
DFMA and, 500, 501
DFMO and, 500–502
genes of, 455, 457, 519–520
growth of, 235–240, 458–459
methionine and, 519–520
MFMP and, 502
mutants of, 457–459, 521–523
polyamine deficiency and, 457–462
polyamine genes of, 456
polyamines and, 235–240, 308, 519–520
protein biosynthesis in, 457–462
putrescine depletion in, 235–240
putrescine derivatives in, 236–240
ribosomes in, 240–242
Ethidium bromide, tRNA and, 311–312
Ethylene
biosynthesis of, 438–441
in plants, 437–441
polyamines and, 438–441
Exocytosis, in kidney, 120
Extraction, of polyamines, 255, 256–257, 262–263

Fatty acid metabolism
in platelets, 299–303
polyamines and, 299–303
Feruloylputrescine
hydroxy derivatives of, 351
in tobacco plant, 348–350
Friend erythroleukemia cells, (FL)
diacetylputrescine and, 769–777
differentiation of, 757–766, 769, 779, 788–790
growth of, 784–787, 789–790
HMBA and, 758–766
inducers and, 758–766
intracellular distribution of polyamines in, 772
Me_2SO and, 758
MTA metabolism in, 781–783
MTA transport in, 782–784
polyamine acetylation and, 761–764, 765–766
polyamine levels in, 758–759
polyamine metabolism in, 759–761, 764–766, 771–777

Genes
ADC, 456
cloning of, 456
of E. coli, 519
for enzymes of polyamines synthesis, 456–457
expression of, 631
mutants for, 457
ODC, 456, 591–592, 597–598, 601
SAMDC, 456
of *S. cerevisiae*, 468–474
Gibberellin
arginine decarboxylase activity and, 386–387
ODC activity and, 428
Glucagon, spermidine acetyltransferase and, 324–325
Glysperins, spermidine in, 9
Growth regulation
ADC and, 411–415, 498–504
amine oxidase and, 164
in bacteria, 495
cadaverine and, 236–237
cyclic nucleotides and, 668–679
DAO and, 175–179
DFMO and, 411–413, 497–498
in E. coli, 458–459
histamine and, 194
ODC and, 411–416, 495–505, 599–600
phospholipids and, 331, 343
of plants, 395, 399–401, 406

by polyamines, 395, 458–459, 496, 667, 727–728
putrescine and, 236–242, 747
in *S. cerevisiae* mutants, 470–473, 477
Guanine triphosphate, (GTP)
analogues of, 491–492
ODC activity and, 491–492
Guanylate cyclase
of heart cells, 673–677
kinetics of, 675–676
manganese and, 677
polyamines and, 673–677
serum effects, 674

Heart
Ca^{++} and, 540–542
catecholamines and, 537–546
cells in culture, 331–333, 667–679
cyclic nucleotides and, 335–336, 537–538, 667–669
DFMO and, 338–342
dibutyryl cAMP and, 529–530
fetal, 553–557
guanylate cyclase, 673–677
hypertrophy of, 550–551
hyperoxia and, 526–528, 537
isoproterenol and, 527–528, 529, 537, 540–546
ODC, 525–534, 537–546, 549, 553, 555–556
oxygen tension and, 528
perfusion of, 537–546
phospholipids in, 332–335
polyamines in, 335–336, 342, 527–528, 667–679
superoxide radicals and, 530–531
Helianthus tuber
ADC activity in, 411–415
DFMO, 411–415
explants of, 420–423
growth inhibition of, 411–415
ODC activity and, 411–415
tissue culture of, 410
Herpes simplex virus, (HSV)
DFMO and, 511, 515–516
MGBG and, 510–512
polyamine synthesis and, 510
Hexamethylene bisacetamide, (HMBA)
dexmethylene and, 757, 770
FL differentiation and, 757
ODC and, 757, 764–765
polyamine acetylation and, 761–764, 765–766
polyamine levels and, 758–759, 765
polyamine metabolism and, 759–761, 765
Histamine
in fetal liver, 194
in growth, 194
in leukemia, 49, 51–57
methylation of, 196
placental amine oxidase and, 157
plasma amine oxidase and, 159
in serum, 49, 51–57
species variability, 193–194
in urine, 194
Hordatines
agmatine in, 358–359
biosynthesis of, 363
coumaroylagmatine, 359–361
fungitoxicity of, 356
gauzatine, 358–359
metabolism of, 357–358
plant distribution of, 363–366
structure activity of, 356–363
Human chronic gonadotropin, (HCG)
cadaverine and, 198–199
ovary ODC and, 576–577
putrescine and, 73, 198–199
Hyperproliferation
polyamines and, 25–28
of skin, 25–28
Hyperthermia
in cancer, 97
spermine and, 97–103
in tumor spheroids, 97–103

Immune response, amine oxidase and, 164
Immunocytochemical localization
of kidney ODC, 571, 574
methods, 573–574
of nerve ODC, 580–581, 582
of ovary ODC, 574–575
of placental ODC, 578–579
Indole-3-acetic acid (IAA)
plant explants and, 423–424
polyamines and, 425
Indomethacin, ODC induction and, 130, 715, 717
Insulin
mammary gland and, 658, 660
polyamine depletion and, 564
putrescine formation and, 321
spermidine and, 568
spermidine acetyltransferase and, 325–326, 327–328
Interferon, polyamine-dependent kinase and, 618
Intracellular parasites
amine oxidase and, 169–174
polyamine oxidation and, 169–174
5-Isobutylthioadenosine
FL differentiation and, 780, 788–790
FL growth and, 788–790
MTA and, 779
Isoproterenol
Ca^{++} and, 542
cardiac ODC and, 527–528, 540–546, 556–557
β_2-adrenoceptor and, 552

Kidney
aminoacetonitrile and, 143–145
brush border of, 115
cadaverine in, 200
diamine oxidase in, 150, 177–179
endocytoxic vesicles in, 110–112, 115–116, 118–120
green monkey, 285, 286–287
hypertrophy of, 178–179
myeloid bodies in, 112–114
nandrolone and, 719–725
ODC, 571, 574, 576, 581–582, 605, 608, 609, 719–725
ODC localization in, 719–725
polyamine metabolism in, 114, 150
proximal tubule of, 110–112
spermidine acetyltransferase in, 328
spermine binding of, 115
spermine toxicity, 108–121
testosterone and, 120–121, 574–575, 581, 605, 609
Kukoamine, as spermine derivative, 352

Leukemia
acute form, 51–53
blast crisis in, 51–53, 56
cell culture of, 728–729
DFMO and, 19, 23–24, 28–29, 731–732
differentiation in, 729–733
MGBG in, 18, 23–24, 28–29
myeloid and, 730
polyamines in, 49, 51–57, 731–733
promyelocytic, 727–733
retinoic acid and, 730–734
Leupeptin
biosynthesis of, 1–2
Lipid metabolism
polyamine effects, 297
Lipogenesis
diabetic liver and, 564
diaminopropanol and, 566–568
in starvation, 566–568
Liver
aminoacetonitrile and, 143–145
DAO of, 177–179
DFMO and, 604–605
in diabetic, 564
histamine in, 194
lipogenesis in, 564
ODC activity in, 80–93, 563, 683–691
ODC antizyme in, 80, 83, 87–93, 683
protein kinases of, 650–653
putreanine in, 140–143, 150
putrescine and, 87–90, 92, 565
pyruvate kinase of, 563
regeneration in, 84–87, 91–93, 177–179
SAMDC in, 565
in starvation, 563
tumor of, 80
Luteinizing hormone, (LH)
ovary ODC and, 574, 576–577
putrescine and, 73
Lymphoma
histamine in, 51–54
serum polyamines in, 51–54
Lysosomes
spermine and, 111–112, 115–119

Magnesium, (Mg^{++})
in cell free systems, 255
nucleic acids and, 255
polyamines and, 255–264
protein synthesis and, 257–259
ribosomes and, 259–262
tRNA and, 313–314, 317
Mammary gland
autoradiographic protein analysis in, 657–658, 662–665
hormonal effects, 658–661
insulin and, 658–661
lactogenesis in, 656
milk protein synthesis in, 661, 664
organ culture of, 656
in pregnancy, 658–660
progesterone and, 661–665
protein kinases of, 655–665
Maytenine
as putrescine derivative, 351–352
Medulloblastoma
amine oxidase in, 185–186, 188
Membranes
phospholipids of, 117–118, 335
stabilization of, 117–118
Meningiomas
amine oxidase in, 185–186
Methionine adenosyltransferase
gene for, 456
purification of, 456
Methionine sulfoximine
brain polyamines in, 214–217
MGBG and, 216
seizures and, 209, 215–216
3′-methyl-4-(dimethylamino) azobenzine, (3′-Me-DAB)
liver ODC and, 82, 87–90
ODC antizyme and, 89–90, 91–92
1,1′-(methylethanediylidene)-dinitrilo bis(3-aminoguanidine)
actions of, 20
Methylglyoxal bis(guanylhydrazone), (MGBG)
adsorption of, 26
amine oxidase and, 162–163
as antimetabolite, 18
antitumor activity, 19–25

anti-viral activity of, 509–512, 515–516
BCNU and, 35–36
cell cycle and, 294
clinical use of, 17, 28–29
DFMO and, 19–25, 224, 290–291
diamine oxidase and, 20
DNA synthesis and, 290–292
effects of, 18–19
embryogenesis and, 203
formula of, 405
growth rate and, 19
in leukemia, 18, 23–24, 28–29
in nerve injury, 705
in plants, 404
polyamine transport and, 20, 23
polyamine uptake and, 23–24, 291–292
psoriasis and, 25, 29
subcellular polyamines and, 289–290, 292
therapeutic index of, 18
topical use, 25–28
for trypanosomes, 224–225
uptake of, 18–19, 291–292
X-rays and, 33–34
α-Methylornithine
ODC mutants and, 592, 595, 598–599
Methyl putrescine
growth and, 239–240
protein synthesis and, 238–240
Methyl thioadenosine, (MTA)
biosynthesis of, 779
determination of, 781–782
DFMO and, 34–35
FL differentiation and, 779–780
FL proliferation and, 780, 784–790
functions of, 779–780
growth and, 784–787, 789–790
metabolites of, 781–783, 789
in plants, 371
polyamines and, 785, 787
preparation of, 246, 248
propylamine transferase and, 248
putrescine and, 34
tissue levels, 248
transport of, 782–784, 789
Mezerein
as ODC inducer, 128
Mitochondria
fatty acid metabolism in, 299–303
polyamine content of, 303
polyamine effects, 300–303
Monoamine oxidase
inhibitors of, 156, 158
substrates of, 156, 158
Monofluoromethylornithine
as ODC inhibitor, 497, 502–505
Monofluoromethyl putrescine
E. coli growth and, 501
polyamine levels and, 501
Morphogenesis
plant polyamines and, 419–425
Mutants
of bacteriophage, 460
cloning of, 592
of E. coli, 457–459
DFMO and, 592, 599
growth of, 599–600
ODC defect, 469–471, 597, 601
in ODC expression, 591–601
ODC overproduction in, 592–595, 601
of Saccharomyces, 467, 469–474
selection methods, 595–597
spermidine deficient, 469
spermine deficient, 469
Myeloid bodies
in leukemic cells, 730
spermine toxicity and, 112–114, 116–118, 120
testosterone and, 120–121

α-Naphthylisothiocynate, (α-NIT)
liver ODC and, 82, 87–90
ODC antizyme and, 89–90, 91–92
Nerve
injury of, 705–706
ODC localization in, 580–581, 582
Neuroblastoma cells
asparagine and, 748–749
cyclic nucleotides and, 743–746
differentiation in, 743–753
DFMO in, 751–753
GABA and, 749–750
ODC activity in, 744–746
polyamines and, 744–746
putrescine incorporation into protein in, 749–750, 753
putrescine transport in, 747–748
Neuroblasts
ODC activity in, 587–588
ODC localization in, 585–589
Nicotine
putrescine incorporation in, 348
Nitrosoureas
cytotoxicity of, 41–42
Nocardamine
as cadaverine antibiotic, 355–356
Norepinephrine
aortic constriction and, 550, 552
heart ODC and, 537, 540–543
Nuclear protein kinase
acetyl spermidine and, 632, 636, 638
kinetics of, 632, 634–635
phosphorylation and, 638–639
polyamine effect, 631–644
RNA polymerase and, 631
specificity of, 633–634
spermidine and, 631–644

Nucleic acids
  Mg^{++} and, 255
  putrescine effects, 240–242
  spermine and, 285, 288
Nuclei
  DFMO and, 292
  DNA synthesis in, 285–286
  MGBG and, 292
  in plants, 421
  polyamines in, 288
  spermine in, 285, 288
Nucleoli
  in plants, 419, 421

Ornithine
  analogues of, 599
  DFMO and, 604
  metabolism of, 596, 598
Ornithine decarboxylase, (ODC)
  activator of, 432–435
  active site of, 611–612
  antibodies to, 571–572
  anti-viral effect, 509–516
  assay of, 82–83, 526–527, 538–539
  asparagine and, 748–749
  autoradiographic localization, 608
  bacterial, 497
  in bone cells, 713–717
  Ca^{++} and, 541–542, 545–546
  cell cycle and, 707–711
  control mechanisms for, 129–130
  cyclooxygenase and, 130
  DAO and, 179
  dephosphorylation of, 625–627
  differentiation and, 743–745
  DFMO and, 19–20, 132, 474, 493, 595, 603–612
  in fetal heart, 553–557
  genes for, 456
  growth and, 599–600
  GTP and, 491–492
  half-life of, 129, 609–610, 693
  in heart, 529–538, 537–546, 549, 553, 555–556
  immunoassay of, 609, 611
  immunoprecipitation of, 474–476, 609
  induction of, 91–93, 127–128, 132, 592, 594, 606, 748–749
  inhibitors of, 398–399
  kidney and, 571, 574, 576, 605, 608
  K_m of, 398
  lack of, 595–598
  in liver, 80–93, 604–605
  3′ Me DAB and, 82
  molecular forms of, 81, 605
  molecules, cell, 605
  mutants, 469–471, 591–601
  negative control of, 617
  in neuroblasts, 585–589
  ovary localization, 574, 576–577
  overproduction of, 592–595
  ovulation and, 574–577
  phospholipids of, 339–341
  phosphorylation of, 615–617, 622–627
  in plants, 397–400, 411–416, 428–430
  polyamine influences, 427, 430
  post-translational modification of, 476
  proliferation and, 79–80
  protease inhibitors of, 399–400
  purification of, 428–430, 474–476, 526, 606–607, 697
  putrescine and, 80, 81, 84, 87–90, 91, 132, 604, 617
  pyridoxal and, 434
  regulation of, 80, 91–93, 476
  rhodamine-labeled, 585
  SAMDC and, 20
  in skin, 80, 127, 129–133
  species differences, 497, 607–608
  specific activity of, 606
  stability of, 694–695
  in starvation, 563, 565–566
  subunits of, 605
  testosterone and, 574–575
  of thermophiles, 489–496
  thiol groups in, 525, 531–532
  titration of, 604–607
  TPA and, 127–133
  tumor promotion and, 127–133
  turnover of, 609–610
  of yeast, 474–476
ODC antizyme
  assay of, 82–84, 685, 688–689
  in barley seeds, 427, 430–435
  carcinogenesis and, 79, 82, 87–88, 91–92
  cell cycle and, 707–711
  characteriziation of, 683, 686
  diet and, 87–89
  induction of, 430–431, 707–708
  in liver, 79–93, 683–691
  in liver regeneration, 84–87, 91
  mode of action, 686–688
  ODC complex, 688–689
  polyamine regulation of, 80–93
  purification of, 431–432, 683–685
  putrescine and, 690, 707
  synthesis, 90
  yeast and, 476
ODC inactivating factor
  antizyme and, 699–700
  assay of, 695–697
  ODC half-life and, 693
  properties of, 698–699
  in prostate, 695–699
  protease activity of, 700–702
  purification of, 698–699

Osteomyelofibrosis
 serum polyamines in, 51–54
Ovary
 cadaverine in, 73, 197–199
 GABA in, 200
 HCG in, 571, 576–577
 ODC activity in, 693–695
 ODC localization in, 574, 576–577
 polyamines in, 73
 putrescine metabolism in, 198–201
 Δ^1-pyrroline in, 200–201
Oxidized polyamines
 BCG and, 173
 cell toxicity of, 169, 184
 parasitic infection and, 169–174
Oxygen tension
 cardiac ODC and, 528–529, 532–533

Palmitoyal Co A
 polyamine effects, 301–302
Pancreas
 spermidine acetyltransferase in, 323
 spermidine metabolism in, 321
Papillomas, ODC half-life in, 129
Phleomycin, structure of, 5–6
Phospholipids
 cell cycle and, 337
 cyclic nucleotides and, 335–336, 337–338
 DNA and, 334–335, 341–342
 heart cells in culture and, 333–346
 ODC and, 339–341
 membrane effects, 335
 protein synthesis and, 334–335
 putrescine and, 338
 RNA and, 334–335
 SAMDC and, 339–341
 spermidine and, 335–336, 338
 spermine and, 335–336, 338
Phosphorylation
 dibutyl cyclic AMP and, 662–665
 of mammary proteins, 657–658, 662–665
 of nuclear protein kinase and, 631–644
 ODC and, 615–617
 polyamine requirement for, 615, 617, 631–644
 of serine, 638, 640
 of threonine, 638, 640
Physarum polycephalum
 ODC of, 616, 623–624
 ODC phosphorylation in, 615–617, 623–625
Placenta
 GABA in, 200
 ODC localization in, 578–579
 in pregnancy, 578
 putrescine metabolism in, 200
Placental amine oxidase
 characterization of, 155–157
 function of, 163–164
 immune response and, 164
 inhibition of, 155–157
 in pregnancy, 177
 purification of, 155–156
 regulation of, 164
 substrates of, 155–157, 158
Plants
 aminoacylation of tRNA in, 307–317
 arginine, 397–406
 arginine decarboxylase in, 381–392, 396–400, 411–416, 443–444, 449–450
 cell division in, 443, 444, 445–446, 447, 449
 cinnamic acid-polyamine conjugates in, 347–366
 development, 395
 DFMO effects, 350, 388–389, 398, 404, 411–413, 445–446
 DNA synthesis in, 391–392, 447
 ethylene and, 437, 438
 exogenous polyamines and, 437–441
 fruit development in, 397–398, 437, 441, 443–444, 445–446, 447
 gibberellin effects, 386–387, 400, 428
 leaves of, 371
 morphogenesis, 419–425
 ODC antizyme of, 430–432
 ODC in, 397–400, 410–416, 427–435, 443–444, 449–450
 osmotic stress in, 387–390, 402
 phytochrome of, 383–385
 polyamine levels in, 385–386, 388–390, 402, 405–406, 444–445, 451–452
 polyamine metabolism in, 396–405
 protein synthesis in, 439–440
 putrescine incorporation in, 348–350, 388, 392
 putrescine synthesis in, 396, 451–453
 RNA synthesis in, 420–421, 439
 RNase activity in, 437
 root initiation in, 402, 403–404, 451–453
 seeds of, 351
 seed germination, 427–435
 senescence of, 395, 403, 437, 441
 spermidine synthesis in, 371
 tissue culture of, 409–416, 447
Plasmodium, oxidized polyamines and, 171–174
Platelets
 acid hydrolysis of, 297
 aggregation of, 299
 fatty acid metabolism in, 297, 299–303
 mitochondria of, 299
 polyamine content of, 297–298
 polyamine synthesis in, 297–298
 SAMDC in, 298

Polyamines
alkaloids of, 353–356
analytical procedures, 136, 210–211, 257, 382, 745–746
in antibiotics, 1–13
as antitumor agents, 8–9
bacteriophage and, 460–462
cancer and, 33–38, 45–46, 49–58, 59–63, 65–71
cell cycle and, 286
cinnamic acid derivatives of, 347–350
compartmentalization of, 285, 286, 288
conjugates of, 348–349
cyclic nucleotides and, 667–669, 743
depletion of, 285–286, 292, 460
derivatives of, 233–242
diacetylputrescine and, 771–777
differentiation and, 727–740, 743–746, 757–759, 769–777
DNA and, 33
exogenous applied, 437
extraction of, 255, 262–263
and FL differentiation, 757–759
gas chromatography of, 74
growth and, 236–242, 395, 458–459
guanylate cyclase and, 673–677
homeostasis of, 107
incorporation into protein, 749–750, 777
intracellular distribution, 772–773
in kidney, 143–145
in liver, 143–145
Mg^{++} and, 255–264
MTA and, 785, 787
nucleic acids and, 233, 240–242
parenteral administration of, 108
phospholipids and, 335–336, 338
plant levels of, 385–386, 388–390, 402–406, 444, 451–455
in platelets, 297–303
protein synthesis and, 238–240, 334–335
in serum, 49–58
in skin, 25–28
in small intestine, 143–145
structure-activity of, 233–240, 242
subcellular distribution of, 262–263, 274, 285–286, 288
toxicity of, 108
transport of, 20, 23
in trypanosomes, 221–222
tumor induction and, 129
unusual, 479–486
uptake of, 18–19, 20, 23–24, 102–103, 107
urinary, 45–46, 59–63
in viruses, 513–514
in yeast, 467–468
Polyamine alkaloids
function of, 355–356
in plants, 353–356
structures of, 353–356
Polyamine-dependent protein kinase
Ca^{++} and, 622–627
calmodulin and, 622–627
ODC and, 623–624
polyamine effect on, 617, 619–620
properties of, 615–617
purification of, 618–621, 623
RNA requirements, 618–621
Polyamine oxidase, of pea seedling, 376–377
Polyamine synthesis
Ca^{++} and, 622–623
in bacteria, 495
in E. coli, 235–240, 457–462
FL differentiation and, 758–759
inhibition of, 17–29
in plants, 395, 402–404
plant growth and, 395, 402–404
Polycythemia vera, serum polyamines in, 53–54, 56
Pregnancy
abortion of, 193
amine oxidase in, 155–164
aminoguanidine in, 195–197, 201–202
cadaverine in, 195–198
DAO in, 193–194, 197–198, 578
GABA in, 200–202
histamine in, 193–197
mammary gland in, 658–660
ODC in, 578–579
origin of polyamines in, 196–197
placenta in, 196, 197, 578
plasma diamine oxidase, 157–159
plasma histamine in, 194
putrescine metabolism in, 195–197, 200–202
Δ^1-pyrroline in, 199–201
serum polyamine oxidase, 159–164
urinary histamine in, 194–197
urinary polyamines in, 194–197
Prokaryotic tRNA aminoacylation, vs eukaryotic, 307–317
Propylamine transferase, *see also* Spermidine synthase
assay of, 245–246
cadaverine and, 249
inhibition of, 250
mutants, 469
properties of, 247–250
purification of, 246–247
of *S. cerevisiae*, 468
Prostaglandins, ODC induction and, 130
Protein kinase
activators of, 651–652
assay, 648–649, 656–657
Ca^{++} and, 622–627
calmodulin and, 622–627, 647–653
cyclic AMP dependent, 647–653

cyclic nucleotide independent, 655–665
cytoplasmic, 647–653
of hepatoma, 650–653
hormonal regulation of, 655–665
induction of, 661–662
inhibitors of, 650–651
interferon and, 618–621
of mammary gland, 655–665
nuclear, 631–644
ODC inactivating factor and, 699–700
polyamines and, 615–617, 627, 652–653, 655
polylysine and, 652, 658, 660
properties of, 616, 621
purification of, 649
specificity of, 616, 621
Protein synthesis
cadaverine effects, 237–239
in cell free systems, 255, 257–259
in E. coli, 457–462,
Mg^{++} and, 255, 257–259
phospholipids and, 334–335
polyamines and, 255
in polyamine deficiency, 457–463
putrescine effects, 238–240
spermidine and, 257–259
Psoriasis
DFMO for, 25, 29
MGBG for, 25, 29
platelets and, 299
P. aeruginosa
DFMA and, 502–504
DFMO and, 500–504
growth inhibition, 500–504
ODC inhibition in, 500–504
Putreanine
aminoacetonitrile and, 140–143, 149–150
aminoguanidine and, 135, 140–143
in brain, 147–148, 151
DAO and, 135, 150
determination of, 136
in liver, 140–143, 150
ODC and, 493
spermidine and, 140–143, 148, 151
Putrescine
arginine and, 396–400
auxotroph selection, 595, 597
cell division and, 445–446
derivatives, 234–236, 312–313
diamine oxidase and, 150, 175–178
growth activity of, 235–240
half-life of, 132
HCG and, 73
LH and, 73
liver, 80, 91–93
metabolites of, 199–201
ODC and, 80, 81, 84–87, 91–93
ODC antizyme and, 81, 83–87, 91–93, 690
ovulation and, 199
in plants, 348–350, 388, 392, 396, 438–439, 443–444, 445–446
in pregnancy, 200–202
protein synthesis and, 238
Δ^1-pyrroline, 199–201
transport of, 747–748
toxicity of, 108
tumor promotion and, 132–133
in yeast, 467–468
Δ^1-Pyrroline
polyamine oxidase and, 377
from putrescine, 199–201
Pyruvate kinase
assay of, 565
in liver, 564–568
spermidine and, 568

Quinacrin, amine oxidase inhibition and, 160, 162

Reticulocyte lysate
preparation of, 255–256
protein synthesis in, 256, 257–259
translation in, 258–259
Retinoids
cell differentiation and, 730–734
DFMO and, 731–732, 738–740
in embryonal carcinoma, 738–740
ODC induction and, 130–132, 731–732, 738–740
polyamines and, 731–733
tumor promotion and, 131–132
Rhodamine
labeled DFMO, 585–589, 725
Ribosomes
of E. coli, 457, 463
Mg^{++} and, 259
polyamines and, 459, 463
putrescine derivatives and, 240–242
putrescine effects, 240–242
spermidine and, 259–262, 264
spermine and, 259–262, 264
stoichiometry of binding, 260–262
of wheat germ, 256
RNA polymerase
ATP and, 641–642
nuclear protein kinase and, 631
polyamine stimulation of, 639–643
RNase
in plants, 437, 441
polyamine inhibition of, 437, 441
RNA synthesis
phospholipids and, 334–335
in plants, 420–421

Saccharomyces cerevisiae
DFMO and, 605, 607
genetic studies in, 467–474

Saccharomyces cervisiae (contd.)
mutants of, 469–474
ODC of, 469–471, 474–476, 605, 607
polyamine content of, 467
polyamine synthesis in, 468–477
propylamine transferase of, 468
spermidine synthase of, 472
Seizures
methionine sulfoximine induced, 209, 215–216
MGBG and, 216
polyamine induced, 209
Senescence
in plants, 395, 403, 437, 441
Serum effects
on cyclic nucleotides, 670–679
on guanylate cyclase, 674–675
on heart cell cultures, 670–679
on polyamines, 672–679
Serum polyamines
in diver's disease, 56
in leukemia, 49–57
Skin
ODC of, 25, 127–133
polyamines of, 25–28
tumor promotion in, 127
UV light and, 25, 129
Small intestine
amine oxidase in, 143–146, 150
polyamine levels in, 143–145
Spergualin
as anti-tumor agent, 8–9
Spermidine
acetylation of, 138–139, 150
in antibiotics, 4, 8–9
guanylate cyclase and, 673–677
insulin and, 568
in ODC regulation, 617
in oxidase, 160–162
phosphorylation and, 631
in pregnancy, 195–197
pregnancy polyamine oxidase and, 159
protein kinase and, 631–633, 652, 658–659
protein synthesis and, 257–259
putreanine and, 135, 140–143, 148, 151–152
pyruvate kinase and, 568
translation and, 258–259
tumor induction and, 129
in viruses, 513–514
Spermidine acetyltransferase
assay of, 322
in brain, 323
fasting and, 323–324
glucagon and, 324–325
half-life of, 321, 326, 328
induction of, 321
insulin and, 325–326, 327–328
K_m of, 323
organ distribution, 323
putrescine and, 321
SAMDC activity and, 327
spermine and, 327
subcellular distribution of, 323
substrate specificity, 323–324, 328
Spermidine synthase
assay of, 211–212, 245, 372, 376–377, 378
of brain, 211–214
of chinese cabbage, 371–378
decarboxylated adenosylmethionine and, 248
genes for, 468
inhibition of, 213–214, 377–378, 501–502
kinetics of, 248, 251, 374
MTA inhibition of, 779
mutants of, 469, 472
in plants, 371
properties of, 247, 375–376
purification of, 246–247, 372, 375, 378–379
sources of, 246–247
Spermine
acetylation of, 138
analysis of, 65–66
in antibiotics, 4, 8–9
binding of, 115, 267–276, 279–284, 308–310
in breast cancer, 65–71
convulsions and, 216–217
guanylate cyclase and, 673–677
hyperthermia and, 97–103
kidney and, 108–121
in ODC regulation, 617
oxidized, 169–174
parenteral administration, 108
phosphorylation and, 631
plant effects, 439–441
in pregnancy, 195–197
protein kinase and, 652, 658–659
protein synthesis and, 439–441
specificity of, 307, 317–318
in thermophiles, 479
toxicity of, 108–121
tRNA and, 308–310, 315–318
tumor spheroids and, 97–103
uptake of, 102–103
in viruses, 513–514
Spermine-binding protein
assay, 267–270, 279–282, 284
cycloheximide and, 271–272
1,25-(OH) D effects, 267, 270–273, 275, 283–284
function of, 273–276

isolation of, 268–269
kinetics of, 281–283
in nucleoplasm, 274–275
ornithine and, 282–283
serum albumin and, 281
source of, 271
specificity of, 275
spermidine and, 282–283
in vitamin D deficiency, 283–284
Spermine oxidase
acetylation and, 138
aminoacetonitrile and, 140–143
determination of, 136–137, 159–160
function of, 163–164
half-life of, 142
inhibitors of, 160–162
in pregnancy serum, 159–160
parasitic infection and, 169–174
purification of, 159–160
putreanine and, 148–153
substrates of, 160
Spermine synthase
assay, 145, 212
of brain, 246, 247, 249
genes for, 468
inhibitors of, 250
kinetics of, 248, 251
K_m of, 248
MTA inhibition of, 779
mutants, 472
in plants, 373
properties of, 247
purification of, 246–247
sources of, 246–247
specific activity of, 246
Starvation
ODC activity in, 563, 565
polyamine levels in, 565
SAMDC activity in, 563, 565
Subcellular distribution
DMFO and, 292
MGBG and, 292
of polyamines, 262–263, 274, 285, 286, 288, 292
Superior cervical ganglion
inujury of, 705–706
ODC in, 705–706
Superoxide
heart ODC and, 531–532

Testosterone
myeloid bodies and, 120–121
polyamines and, 120–121
12-01 Tetradecanoylphorbol-13-acetate, (TPA)
DFMO and, 132
differentiation and, 730
DNA and, 128
ODC and, 128–133
SAMDC and, 129
as tumor promotor, 127–128
Thallysomycin, structure of, 5–6
Theophylline, spermidine acetyltransferase and, 325
Thermine
occurence, 481–484, 486, 494
structure, 483
synthesis of, 484
Thermus thermophilus
spermine in, 479, 481
temperature effects, 481
unusual polyamines of, 479–486
Thermophiles
DFMO in, 493
ODC in, 489
polyamines of, 479–486
unusual amines of, 479–486
Thermospermine
occurrence, 481–484
structure, 483
Thyroid, DAO in, 184
Tobacco mosaic visus, (TMV), putrescine in, 349–350
Tobacco plant
arginine decarboxylase in, 402
cinnamic acid-polyamine conjugates, 347
DFMO effects, 350
ODC in, 398
polyamines conjugates, 348–349
polyamine levels in, 349
putrescine incorporation in, 348, 402
virus of, 349–350
Tomato fruit
ADC inhibition and, 445–446
ODC inhibition and, 445–446, 447
polyamines and, 443–444, 447
Transcription, polyamine effect, 639–642
Translation
polyamine deficiency and, 457–463
spermidine and, 258–259
Tricoumaroylspermidine, from plants, 352
$tRNA^{val}$
aminoacylation of, 307–317
diamines and, 312
of E. coli, 307–317
eukaryotic, 307–317
Mg^{++} and, 313–314
polyamines and, 307, 309
in plants, 308
prokaryotic, 307–317
spermine and, 308, 310
valine concentration and, 313
Trypanosoma brucei
chemotherapy of, 221–228
Trypanosoma gambiense
chemotherapy of, 221–224

Trypanosomes
as animal parasites, 221
Antrycide and, 226–227
berenil and, 226–227
bleomycin and, 37, 223
control of, 222
DFMO and, 37, 223–224, 228
diamidines and, 225–228
MGBG and, 224–225
nucleic acid synthesis in, 223
ODC in, 222, 223
oxidized polyamines and, 169
pentamidine and, 225–228
polyamine content, 222
polyamine synthesis in, 221, 222, 225–226
trypanocides and, 226–228
Tumors
amine oxidase in, 184–190
MGBG use, 18, 28–29
Tumor promotion
cyclic AMP and, 127
invitation, 127–129
ODC and, 127–133
promotion of, 127–133
putrescine and, 132–133
Tumor spheroids
bleomycin and, 97
hyperthermia in, 97–103
spermine and, 97
Turnip yellow mosaic virus, (TYMV)
spermidine in, 371

Uremia, polyamine metabolism in, 108, 114
Uridine incorporation, spermine and, 439
Urinary polyamines
acetylation of, 114
in cancer, 45–46, 59–63, 65–71
in choriocarcinoma, 73–77
in cystinuria, 114
kidney and, 114
in pregnancy, 195–198
Urine
histamine in, 194
in pregnancy, 195–198
UV-irradiation, skin polyamines and, 25–28, 129

Vaccinia virus
DFMO and, 509–510, 515
MGBG and, 509–510, 515
ODC and, 515–516
Valyl-tRNA synthetase
of E. coli, 308–317
of plants, 308–309
polyamine binding to, 310–311
polyamine stimulation, 307, 309–310
properties of, 309
specificity of, 309
spermine and, 308, 309–310
Verapamil
Ca^{++} and, 541–542, 545
heart ODC, 541–542, 544–545
Viruses
cytomegalovirus, 508, 510–516
herpes simplex, 508–511, 515–516
inhibition of polyamine synthesis in, 507–516
polyamine content, 513–514, 516
replication of, 507–516
spermidine synthesis in, 371
TMV, 349–350
TYMV, 371
vaccinia, 508–510
Vitamin D, spermine-binding and, 283–284

X-ray irradiation
DFMO and, 33–34
MGBG and, 33–34
polyamine and, 33–34
urinary polyamines in, 60–63

Zea mays, root polyamines of, 451–453